Autodata®

TECH SERIES

Diagnostic Trouble Codes

Engine Management Systems

Domestic Vehicles 1992-2002

2003 EDITION

Caution

The information used to compile this manual has been sourced from the vehicle manufacturers and was believed to be the latest available in October 2002. Manufacturers' information is liable to change at any time and there may have been updates or revisions which applied to the content of this manual but were not available at the time of compilation. Later revisions published by the manufacturers' may contain revisions which would be applied retrospectively to the information contained in this manual.

The information contained in this manual applies only to standard models and does not apply to vehicles fitted with equipment other than the standard production options.

Autodata Publications Inc.

Published in the U.S.A. by
Autodata Publications Inc. under license.
Manufactured in the U.S.A.

Autodata Publications Inc.
19 Bonazzoli Avenue,
Hudson, MA 01749,

Phone: 978 562-9511
Fax: 978 562-9533
email: autodata@autodatapubs.com
www.autodatapubs.com

Product No.: 03-340	ISBN: 1-893026-17-5	November 2002

Contents

This manual is a comprehensive single source of information on diagnostic trouble codes for engine management systems for US domestic vehicles introduced or revised during the period 1992 to 2002.

The manual is part of a series from Autodata and deals specifically with trouble codes related to the engine management system. It has been written and presented in a way which enables any professional automotive technician, with appropriate skills and competence, to make accurate diagnoses of fuel and ignition system related components and circuits.

Detailed knowledge of self-diagnosis or fuel injection and ignition systems is not required in order to make full use of this manual. With a basic understanding of fuel and electrical systems, tests and diagnoses can be made using the minimum of specialised test equipment.

Where possible, procedures for accessing and erasing trouble codes without special diagnostic equipment are given, but increasingly some form of 'scan tool' is required to successfully read the fault memories of control modules. This particularly applies to OBD-II systems where 'freeze frame' data and manufacturer 'PID' (Parameter Identification) information is accessed.

Each chapter covers a range of models sharing the same trouble code table and lists the codes in numerical or alphabetical order along with their fault locations and probable causes.

The probable causes column identifies the various different areas of the system that should be investigated in addition to the primary component listed in the fault location column.

If a trouble code has been logged, the fault location can be looked up in the trouble code table. This will give suggested probable causes, which should be the primary areas for checking. Most of these probable causes will suggest checking the wiring as well as the primary component in the circuit. The multi-plugs, wiring continuity and insulation should all be verified before replacing components.

Index

Index

Index

Index

DODGE

Index

Index

Index

Index

Index

Index

PLYMOUTH

Index

PONTIAC

About this manual

Manufacturer name

Model identification

Manufacturer's system name

Engine identification

Year range

Reference to further information and troubleshooters

Accessing procedure

Accessing method

MIL flash detail

Data link connector location

Data link connector detail

Erasing method

PLYMOUTH

Model:	Engine identification:	Year:
Voyager 2.4L	EDZ	1996-97
Voyager 3.0L	EFA	1996-97
Voyager 3.3L	EGA/EGM	1996-97
Voyager 3.8L	EGH	1996-97

System: **SBEC III**

Self-diagnosis

General information

- Refer to the front of this manual for general test conditions, terminology, detailed descriptions of wiring faults and a general trouble shooter for electrical and mechanical faults.
- Malfunction indicator lamp (MIL) will illuminate or flash if certain faults are recorded.
- ECM operates in backup mode if sensors fail, to enable vehicle to be driven to repair shop.
- Trouble codes can be displayed by the MIL or by using a Scan Tool connected to the data link connector (DLC).

Accessing

Flash codes

- Within 5 seconds switch the ignition key ON-OFF-ON-OFF-ON.
- Count MIL flashes. Note trouble codes.
- Each trouble code consists of two groups **1** [A].
- The first group of flashes indicate the 'tens' of the trouble code.
- The second group of flashes indicate the 'units' of the trouble code.
- A short pause separates 'tens' and 'units' **1** [B].
- A long pause separates each trouble code **1** [C].
- For example: Trouble code 21 displayed **1**.
- End of test sequence indicated by trouble code 55.

P codes & Hex codes

- Trouble codes can be displayed by using a Scan Tool connected to the data link connector (DLC) **2**.

1

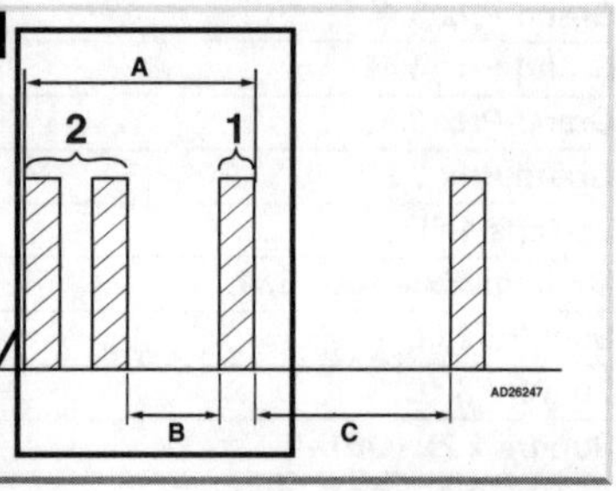

2

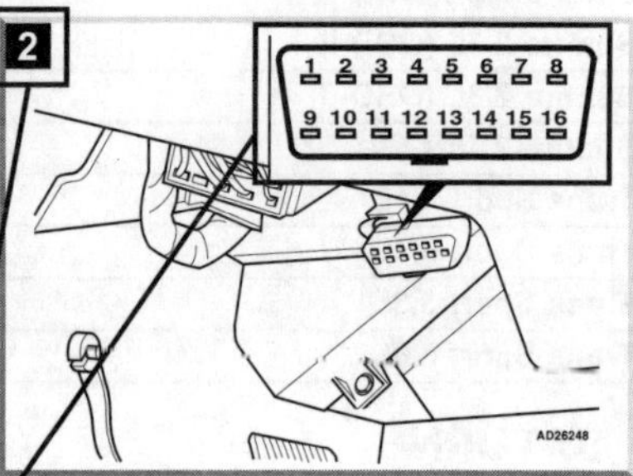

Erasing

- Trouble codes can only be erased by using a Scan Tool connected to the data link connector (DLC).

686 Autodata

Voyager 2.4L • Voyager 3.0L • Voyager 3.3L • Voyager 3.8L

PLYMOUTH

Trouble code identification

Reference to standard OBD-II trouble code table

OBD-II trouble code

Primary fault location

Probable cause

Scan code P type	Fault location	Probable cause
P0000-P0999	Refer to OBD-II trouble code table at the front of this manual	–
P1290	Compressed natural gas pressure sensor – pressure high	Compressed natural gas system, wiring, compressed natural gas pressure sensor
P1294	Idle speed control (ISC) system – rpm lower/higher than expected	Wiring, air leak, throttle valve tight/sticking, IAC valve
P1297	Manifold absolute pressure (MAP) sensor	Wiring, hose connection(s), MAP sensor
P1391	Crankshaft position (CKP) sensor/camshaft position (CMP) sensor – intermittent signal	Wiring, poor connection, air gap, CKP sensor, CMP sensor
P1398	Engine control module (ECM) – misfire detection limit reached	Wiring, poor connection, CKP sensor
P1486	Evaporative emission (EVAP) system – blockage detected	Hose connection(s), hose blocked/trapped, EVAP canister
P1491	Engine coolant blower motor relay – low/high speed	Wiring open circuit/short circuit, engine coolant blower motor relay – low/high speed
P1492	Ambient/battery temperature sensor – voltage high	Wiring, ambient/battery temperature sensor
P1493	Ambient/battery temperature sensor – voltage low	Wiring, ambient/battery temperature sensor
P1494	Evaporative emission (EVAP) leak detection pump – pressure switch	Wiring, hose leak/blockage, EVAP leak detection pump
P1495	Evaporative emission (EVAP) leak detection pump solenoid	Wiring open circuit/short circuit, hose leak/blockage, EVAP leak detection pump solenoid
P1496	Engine control module (ECM) – 5 volt output low	ECM
P1696	Engine control module (ECM) – EEPROM error	ECM
P1697	Engine control module (ECM) – EEPROM error	ECM
P1698	CAN data bus, TCM – no signal	Wiring, TCM
P1899	Park/neutral position (PNP) switch – input circuit malfunction	Wiring, PNP switch, ECM/TCM

Trouble code type

OBD or MIL flash code

Scanner code 'HEX type'

Flash code	Scan code Hex type	Fault location	Probable cause
11		Crankshaft position (CKP) sensor/camshaft position (CMP) sensor	Wiring, poor connection, air gap, CKP sensor, CMP sensor
	9D	Crankshaft position (CKP) sensor/camshaft position (CMP) sensor – intermittent signal	Wiring, poor connection, air gap, CKP sensor, CMP sensor
	28	Crankshaft position (CKP) sensor – no signal	Wiring, air gap, CKP sensor

Autodata 687

Safety precautions

Airbags (Supplementary Restraint System – SRS)

Many of the models covered by this manual are fitted with airbags as standard equipment. When working on a vehicle fitted with such a system, extreme caution must be taken to avoid accidental firing of the airbag, which could result in personal injury. Unauthorised repairs to the system could render it inoperative, or cause it to inflate accidentally.

NOTE: ***All related wiring is encased in a yellow outer covering.***

When the engine is started the AIRBAG warning lamp should go out after approximately 5-10 seconds, if not this indicates a fault in the system. The system should be checked and the fault corrected by a competent technician before any other work is undertaken.

- NEVER attempt to test the system using a multi-meter.
- NEVER tamper with or disconnect the airbag wiring harness.
- NEVER make extra connections to any part of the system wiring harness or terminals.
- ALWAYS ensure that the airbag wiring harness has not been trapped or damaged in any way when working on adjacent components or systems.

Electrical

> **CAUTION:** ***To prevent the engine starting and to avoid damaging the catalytic converter(s), disconnect the fuel injector valve multi-plug(s) before cranking tests.***

- ALWAYS ensure that the battery is properly connected before attempting to start the engine.
- DO NOT attempt to start the engine using a source in excess of 12 volts, such as a fast charger (16 volts) or by connecting two batteries in series (24 volts). ALWAYS disconnect the battery before charging it.
- DO NOT disconnect the battery while the engine is running.
- DO NOT connect the battery with reverse polarity.
- DO NOT disconnect or touch the HT leads when the engine is being cranked or when it is running.
- DO NOT connect or disconnect the electronic control module (ECM), or any other component of the fuel injection system while the ignition is switched ON.
- DO NOT disconnect ECM Multi-plug within 30 seconds of switching ignition OFF.
- DO NOT connect or disconnect multi-meters, voltmeters, ammeters or ohmmeters with the ignition switched ON.
- DO NOT reverse the polarity of the fuel pump.
- ALWAYS ensure that all electrical connections are in good condition and making good contact, PARTICULARLY the ECM connector.
- ALWAYS disconnect the ignition coil, ECM, fuel pump relay/fuse before carrying out a compression test.
- DO NOT flash a wire or circuit to ground to check that continuity exists.
- Modern ignition systems operate at very high voltages and these high voltages can severely damage transistorised components such as a wrist-watch if electrical contact is made. Wearers of heart pacemaker devices, therefore, should not at any time carry out work involving ignition systems. In addition to the danger from electric shock, further hazards can arise through sudden uncontrolled body movement causing involuntary contact with moving parts of the engine, i.e. fan blades, pulleys and drive belts.
- ALWAYS ensure that any replacement fuel or ignition system parts are correct for the application in question. Many units share common external features, but differ internally.

Mechanical

> **CAUTION:** ***To minimise fire risk, fuel system must be depressurised before disconnecting any fuel lines or fuel system components.***

- ALWAYS disconnect the distributor before carrying out a fuel pump pressure or delivery check.
- AVOID the risk of fire – ALWAYS disconnect the ignition coil supply and ground the coil HT lead, so that NO HT spark can be emitted, before checking the fuel injector valves, or any other component of the fuel injection system likely to result in the presence of fuel in or around the engine bay.
- AVOID the risk of fire – NEVER work on the fuel injection system when SMOKING or close to a NAKED FLAME.
- ALWAYS keep a fire extinguisher close at hand when working on the fuel injection system.
- ALWAYS ensure that test equipment, leads, tools and especially items of clothing, are clear of moving parts and are not liable to fall into the engine bay, due to vibration, when the engine is running.

A/C	Air conditioning
AIR	Secondary air
APP	Accelerator pedal position
ASR	Acceleration skid control
AT	Automatic transmission
ATF	Automatic transmission fluid
AWD	All wheel drive
BARO	Barometric pressure
Batt	Battery
BBDC	Before bottom dead centre
BCM	Body control module
BPP	Brake pedal position
Cat	Catalytic converter
CID	Cylinder identification
CKP	Crankshaft position
CMP	Camshaft position
CPP	Clutch pedal position
CPU	Central processing unit
CTP	Closed throttle position
DERM	Diagnostic energy reserve module
DI	Direct ignition
DLC	Data link connector
DOHC	Double overhead camshaft
DPI	Dual plug inhibit
DRB	Data read-out box (Chrysler)
DVM	Digital volt meter
ECCS	Electronic concentrated control system
ECI	Electronically controlled injection
ECM	Engine control module
ECT	Engine coolant temperature (sensor)
EDIS	Electronic distributorless ignition system
EEC	Electronic engine control
EFI	Electronic fuel injection
EFP	Electronic accelerator pedal
EGI	Electronic gasoline injection
EGR	Exhaust gas recirculation
EGRT	Exhaust gas recirculation temperature
EGRTVV	Exhaust gas recirculation thermal vacuum valve
Eng.	Engine
EPI	Electronic petrol injection
EPROM	Electronically programmable read only memory
EOT	Engine oil temperature
EPT	Exhaust pressure transducer
EST	Electronic spark timing
ETC	Electronic traction control
ETS	Electronic throttle system
ETV	Electronic throttle valve
EVAP	Evaporative emission
FICM	Fuel injector control module
FL	Fuse link
FP	Fuel pump
FPCM	Fuel pump control module
FWD	Front wheel drive
GVW	Gross vehicle weight
HEI	High energy ignition
HEX	Hexadecimal
HO	High output
HO2S	Heated oxygen sensor
HT	High tension
I/O	Input-output unit
IAC	Idle air control
IAT	Intake air temperature
IC	Ignition control
IC	Integrated circuit
ICM	Ignition control module
IDM	Ignition diagnostic module
IFZ	Infra-red remote control for central locking
Ign	Ignition
IMA	Idling mixture adjustment sensor
IMRC	Intake manifold runner control
ISC	Idle speed control
JTEC	Jeep/Truck engine controller
KAM	Keep alive memory
KOEO	Key on engine off
KOER	Key on engine running
LED	Light emitting diode
LH	Left hand
MAF	Mass air flow
MAP	Manifold absolute pressure
MECS	Mazda engine control system

Abbreviations

MFI	Multi-port fuel injection
MIL	Malfunction indicator light
MPFI	Multi-port fuel injection
MPI	Multi-port injection
MT	Manual transmission
N	Neutral position, automatic transmission
NP	Neutral position
NTC	Negative temperature coefficient
O2S	Oxygen sensor
OHC	Overhead camshaft
OSS	Output shaft speed
P	Park position, automatic transmission
PAIR	Pulsed secondary air injection
PCV	Positive crankcase ventilation
PGM-FI	Programmed fuel injection
PID	Parameter identification
PIP	Profile ignition pickup
PNP	Park/neutral position
PROM	Programmable read only memory
PS	Power steering
PSI	Pounds per square inch
PSP	Power steering pressure
PTC	Positive temperature coefficient
PWM	Pulse width modulated
RAM	Random access memory
RH	Right hand
RHD	Right-hand drive
ROM	Read only memory
RPM	Revolutions per minute
SAE	Society of Automotive Engineers
SBEC	Single board engine controller
SDM	Sensing & diagnostic module
SEFI	Sequential electronic fuel injection
SFI	Sequential fuel injection
SHO	Super high output
SOHC	Single overhead camshaft
SPFI	Single-point fuel injection
SPI	Single port injection
SPOUT	Spark output
SS	Shift solenoid
ST	Start
STI	Self test input
STO	Self test output
TAC	Throttle actuator control
TBI	Throttle body injection
TC	Turbocharger
TCC	Torque converter clutch
TCM	Transmission control module
TCS	Traction control system
TFP	Transmission fluid pressure
TFT	Transmission fluid temperature
TP	Throttle position
TPM	Tempomat cruise control
TPS	Throttle position switch/sensor
TR	Transmission range
TSS	Turbine shaft speed
VAF	Volume air flow sensor
VICS	Variable intake control system
VIN	Vehicle identification number
VIS	Variable intake system
VRIS	Variable resonance intake system
VSS	Vehicle speed sensor
VSV	Vacuum switching valve
WOT	Wide open throttle
8V	Eight valve
10V	Ten valve
12V	Twelve valve
16V	Sixteen valve
20V	Twenty valve
24V	Twenty four valve
1X	1 pulse per crankshaft revolution
2X	2 pulses per crankshaft revolution
3X	3 pulses per crankshaft revolution
4X	4 pulses per crankshaft revolution
6X	6 pulses per crankshaft revolution
7X	7 pulses per crankshaft revolution
18X	18 pulses per crankshaft revolution
24X	24 pulses per crankshaft revolution
58X	58 pulses per crankshaft revolution

General trouble shooter

Despite the sophistication of modern engine management systems, basic mechanical faults can still cause unsatisfactory starting, running and driveability problems as well as the possibility of recording numerous trouble codes.

Before assuming that a problem is electronic, it should be established that the engine is in good mechanical condition and that the basic fuel and electrical circuits are satisfactory.

Check the following:

Mechanical	• Compression pressure • Valve clearances • Oil filler cap sealing • No excessive crankcase fumes	• Manifold vacuum • Valve timing • Not burning excess oil
Electrical	• Battery fully charged • Ignition coil(s) • HT leads (where applicable) • Engine ground connections	• Plug leads connected correctly • Ignition timing • Adequate spark at the spark plugs • Engine control module (ECM) ground connections
Fuel system	• Air filter for blockage • Fuel delivery rate • Fuel regulated pressure • Vacuum hoses not split or collapsed	• Fuel filter for blockage • Fuel system pressure • Fuel injector spray pattern • Catalytic converter and exhaust system for blockage

Terminology and locations

Throughout this manual the following standard descriptions and terminology have been used for identification purposes, together with SAE J1930 component descriptions.

- ECM – the acronym used to describe either an engine control module (ECM) or a powertrain control module (PCM).
- Bank 1 – Cylinder bank or group including No.1 cylinder (e.g. cylinders 1, 2 & 3 of a six cylinder engine)
- Bank 2 – Cylinder bank or group not including No.1 cylinder (e.g. cylinders 4, 5 & 6 of a six cylinder engine)
- HO2S 1 – Heated oxygen sensor (HO2S) single or nearest to engine (in front of catalytic converter)
- HO2S 2 – Heated oxygen sensor (HO2S) after catalytic converter
- KS 1 – Knock sensor (KS) single or nearest No.1 cylinder
- Left-hand (LH) and right-hand (RH) – As seen from the driver's seat facing forward

Typical sensor locations and descriptions:

1

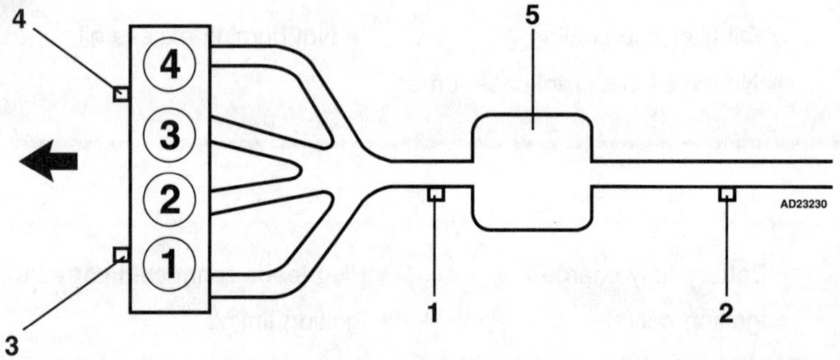

4 cylinder in-line

1. *Heated oxygen sensor (HO2S) 1*
2. *Heated oxygen sensor (HO2S) 2*
3. *Knock sensor (KS) 1*
4. *Knock sensor (KS) 2*
5. *Catalytic converter*

2

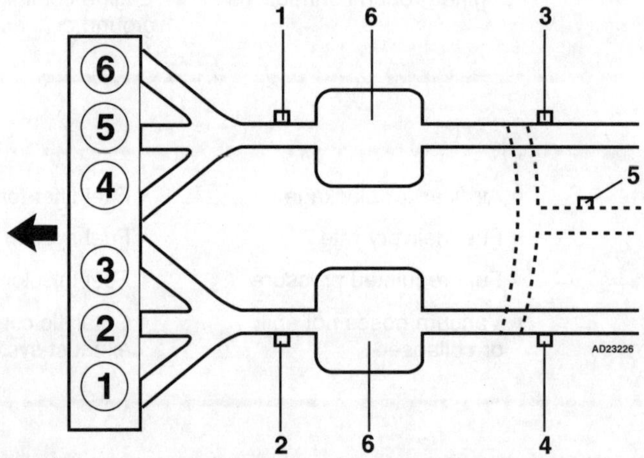

6 cylinder in-line

1. *Heated oxygen sensor (HO2S) 1, bank 2*
2. *Heated oxygen sensor (HO2S) 1, bank 1*
3. *Heated oxygen sensor (HO2S) 2, bank 2*
4. *Heated oxygen sensor (HO2S) 2, bank 1*
5. *Heated oxygen sensor (HO2S) 2 (if only single heated oxygen sensor (HO2S) after cat)*
6. *Catalytic converter*

3

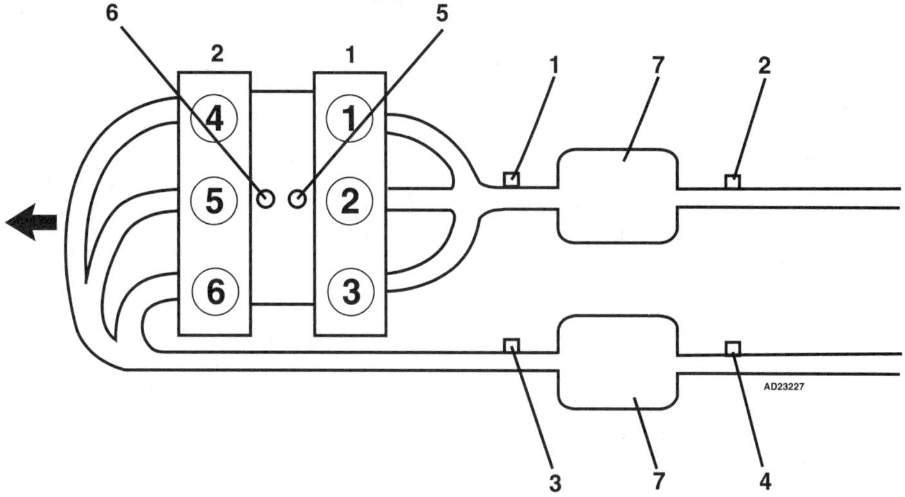

V6

1. *Heated oxygen sensor (HO2S) 1, bank 1*
2. *Heated oxygen sensor (HO2S) 2, bank 1*
3. *Heated oxygen sensor (HO2S) 1, bank 2*
4. *Heated oxygen sensor (HO2S) 2, bank 2*
5. *Knock sensor (KS) 1*
6. *Knock sensor (KS) 2*
7. *Catalytic converter*

4

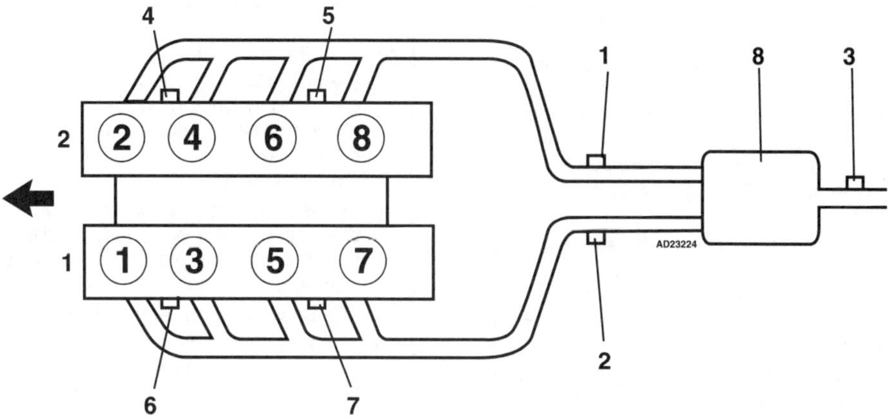

V8

1. *Heated oxygen sensor (HO2S) 1, bank 2*
2. *Heated oxygen sensor (HO2S) 1, bank 1*
3. *Heated oxygen sensor (HO2S) 2*
4. *Knock sensor (KS) 1, bank 2*
5. *Knock sensor (KS) 2, bank 2*
6. *Knock sensor (KS) 1, bank 1*
7. *Knock sensor (KS) 2, bank 1*
8. *Catalytic converter*

How to use this manual

General information

All the fuel and ignition systems covered by this manual are controlled by an engine control module (ECM) which incorporates a self-diagnosis function. These ECMs are capable of detecting certain faulty signals and storing them for access later.

In many cases the detection of a faulty signal will cause the ECM to invoke a backup mode, which adopts a mean signal value (programmed into the ECM memory), to enable the vehicle to be driven until the fault can be diagnosed and rectified.

In many early models faults are signalled by numerical trouble codes comprising two, three or four digits, which can be interpreted by reference to the code tables in this manual. Later models have OBD-II 'P-codes'.

Trouble codes on early models may be displayed by an LED on the ECM, or on the fascia mounted malfunction indicator lamp (MIL). If this is not possible then an LED tester can be used at the data link connector (DLC). If access to trouble codes is possible in this manner it is described along with the relevant code table.

Most later models that comply with OBD-II regulations require a scan tool to enable the trouble code information to be displayed.

OBD-I systems

- Prior to 1988 there was no uniformity in the display, format or method of retrieval of diagnostic trouble codes, each vehicle manufacturer developing a dedicated system with special diagnostic equipment and varying amounts of data logging. Some systems having just a few one or two digit codes while others had more comprehensive three and four digit coding systems.
- In many cases the codes for these early systems could be displayed by connecting an LED tester between the appropriate terminals of the diagnostic socket and counting the flashes or groups of flashes.
- For example **1** represents three short flashes, with a short pause separating each flash, displaying trouble code 3. If more than one code has been logged, a long pause would be used to separate each code.
- If the trouble code is greater than 9, long flashes are used for the tens and short flashes for the units. For example **2** one long flash, and two short flashes separated by a short pause, indicating trouble code 12.

1

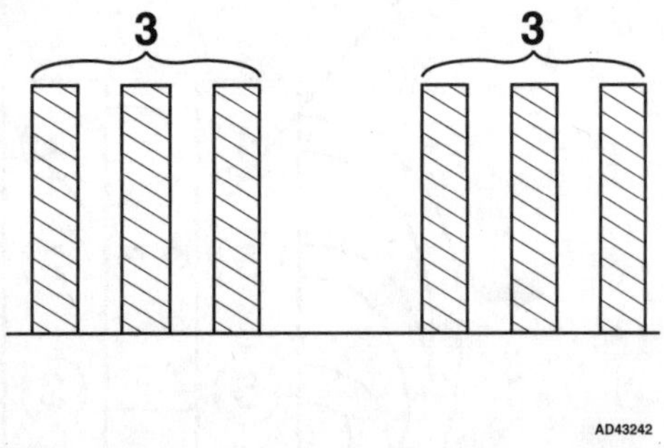

2

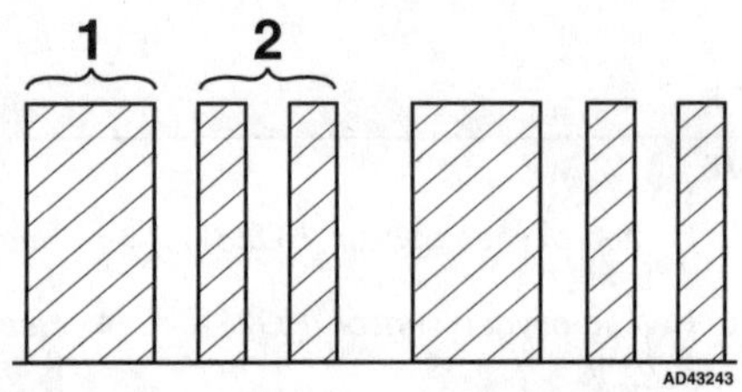

- In 1988 the California Air Resources Board (CARB) introduced legislation covering all new cars sold in California. These new laws governed on-board diagnostic requirements for emissions related faults and began the process of standardization.
- These regulations were known as onboard diagnostic systems - phase I (OBD-I) and were subsequently adopted by the vehicle manufacturers, for States other than California.
- The basic requirements of OBD-I were as follows:
- A malfunction indicator lamp (MIL), to provide a visible warning that an emissions related fault had been logged by the engine control module (ECM).
- Provision for the logging and storage, for subsequent retrieval, of diagnostic trouble codes (DTC)/flash codes for emissions related faults.
- The location of the diagnostic socket (data-link connector), the method of access to the trouble codes and the configuration and scope of the self-diagnosis was left to individual vehicle manufacturers.

- This resulted in an almost complete lack of standardization, requiring the technician to adopt different procedures and to use different equipment for each vehicle.

OBD-II systems

- In 1995 the California Air Resources Board (CARB), after consultation with the Society of Automotive Engineers (SAE), introduced further legislation covering all new cars sold in California.
- These regulations are known as OBD-II and provide standard diagnostic procedures and protocols which have subsequently been adopted for all States. They closely define trouble code logging and retrieval. More significantly they require monitoring of the operation of the system and components, in addition to detecting component failures.
- OBD-II also covers systems other than engine management.

Basic requirements of OBD-II

Configuration of 16 pin data link Connector (DLC)

- The 16 pins of the DLC are arranged in two parallel rows, with certain pin connections defined by the OBD-II regulations and the rest allocated by the vehicle manufacturer – 3
- The pins are designated as follows:

TERMINAL	DESIGNATION
Pin 1	Manufacturer specific
Pin 2	SAE J1850 bus +ve
Pin 3	Manufacturer specific
Pin 4	Chassis ground
Pin 5	Signal ground
Pin 6	SAE J2284 CAN data-bus (high)
Pin 7	ISO 9141-2 - K-Line
Pin 8	Manufacturer specific
Pin 9	Manufacturer specific
Pin 10	SAE J1850 bus -ve
Pin 11	Manufacturer specific
Pin 12	Manufacturer specific
Pin 13	Manufacturer specific
Pin 14	SAE J2284 CAN data-bus (low)
Pin 15	ISO 9141-2 - L-Line
Pin 16	Battery positive

3

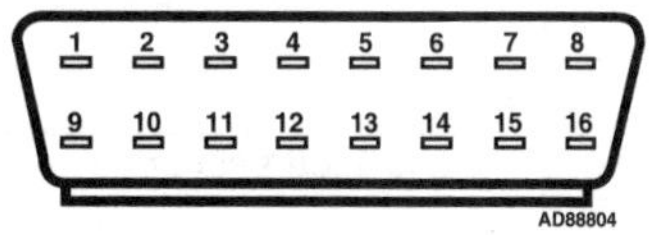

Location for the data link connector (DLC)

- The standard location for the DLC is defined as being between the driver's end of the instrument panel and 300 mm beyond the vehicle centreline, accessible from the driver's seat.
- The recommended location is between the steering column and the vehicle centreline.
- Access to the connector should not require special tools and should permit one-handed connection and disconnection.

NOTE: ***Some exemptions were granted for existing vehicle designs.***

OBD-II trouble code format

- Standard, 5 element alpha-numeric format, trouble codes, made up as follows:

Letter prefix defines system designation

B	Body
C	Chassis
P	Powertrain
U	Network communication

First digit defines code type

0	Powertrain - SAE defined (OBD-II)
1	Powertrain - Manufacturer defined
2	Powertrain - SAE defined
3	Powertrain - P3000-P3399 - Manufacturer defined
3	Powertrain - P3400-P3999 - SAE defined

Second digit defines system identification

0	Fuel and air metering and auxiliary emission control
1	Fuel and air metering
2	Fuel and air metering
3	Ignition
4	Emission control
5	Vehicle or idle speed control
6	ECM or output
7	Transmission
8	Transmission
9	Transmission

Third and fourth digits define specific fault

- These digits define general malfunction, range or performance of component or system, low or high input.

Example code P0108

- This code is logged when a high input is detected in the circuit of the manifold absolute pressure (MAP) sensor or barometric pressure (BARO) sensor.

P	System – Powertrain
0	Code type – OBD-II
1	System identification (fuel and air)
08	MAP/BARO sensor - circuit, high input

Standard areas monitored:

The following areas are monitored continuously:

- Misfire detection
- Fuel system performance
- Component performance

The following areas are monitored once per 'trip'

NOTE: ***The definition of a 'trip' varies, but is basically – key ON, vehicle driven, key OFF.***

- Catalytic converter
- Evaporative emission system
- Secondary air (AIR) system
- Air conditioning system
- Heated oxygen sensor (HO2S)
- Oxygen sensor (O2S) heater
- Exhaust gas recirculation (EGR) system

- In cases where flash codes and OBD-II P codes are listed for the same model range, only one flash code will be generated for a particular component, but there may be a list of many P codes related to this component. Alternatively there may be some P codes with no equivalent flash code(s).
- These OBD-II P codes will provide more specific details about the fault location, wiring, voltage signals and other information relative to the circuit or system.

Fault logging and drive cycles with OBD-II

- Fault logging is based on inputs received during each 'trip' which may comprise a number of 'drive cycles'.
- Each 'drive cycle' is initiated when the engine is started and is terminated when the engine is switched off.
- For a complete 'trip' to be completed the following, complex, conditions need to be met (and may take several 'drive cycles' to complete):
  - □ Before starting the engine, coolant temperature should be below 50°C and +/-6°C of ambient temperature.
    - ■ This ensures that the ECM logs a cold start.
  - □ Allow engine to idle for 2-3 minutes with an electrical load switched on (such as headlamps or heated rear window).
    - ■ This will ensure that the misfire monitoring and fuel trim monitoring programmes run.
  - □ Drive vehicle up to at least 50 mph and maintain this speed for at least 3 minutes.
    - ■ Misfire and fuel trim monitoring will run.
  - □ Allow the speed to reduce to around 20 mph without changing gear, using the brakes or the clutch.
    - ■ Fuel trim monitoring will run.
  - □ Accelerate to at least 50 mph and maintain for at least 5 minutes.
    - ■ Misfire and fuel trim monitoring will run.
  - □ Allow the speed to reduce to around 20 mph without changing gear, using the brakes or the clutch.
    - ■ Fuel trim monitoring will run.
- Each 'trip' starts when the engine is started and continues through a number of 'drive cycles' until all the OBD-II monitors have completed a self-test.

- The OBD-II monitoring programme provides operating tolerance checks on all emissions related sensors and actuators. Some components/

circuits are continuously monitored and some are only activated under predetermined operating conditions.

- Misfire monitoring detects irregularities in the crankshaft position (CKP) sensor signal pattern and identifies which cylinder misfired and if the misfire is frequent enough to cause catalytic converter damage due to excess internal temperatures. If this is the case the MIL lamp will flash.
- If the misfire is likely to increase emissions above the OBD-II limits and occurs on each of two consecutive 'trips' the MIL will flash. If the misfire is absent during the next three 'trips' the MIL will be extinguished.
- When a trouble code is logged a number of data parameters are stored with it, to assist in accurate fault diagnosis. These are:
  - □ Vehicle speed
  - □ Engine coolant temperature
  - □ Engine rpm
  - □ Engine load
  - □ Oxygen sensor system status – open/closed loop
  - □ Distance since fault first logged
  - □ Long term fuel trim (LTFT) level
- OBD-II scan tools are capable of capturing this 'snapshot' of data and displaying (or printing) it for analysis to aid the fault diagnosis process.

Emissions problems and fuel trim

- Short and long term fuel trim refers to the strategy used to reduce exhaust emissions after the basic computation of injection period, using engine load as the major parameter.
- Both front and rear oxygen sensor signals are used to fine tune the fuel/air mixture by increasing or decreasing the injection period +/-25% above or below the basic level.
- Any fault requiring a correction beyond this level will result in a trouble code being logged.
- When the engine is new and running satisfactorily a level of fuel trim will be established – represented by 100%.
- The fuel trim will oscillate the injection period +/-5% above and below the mean level [**A**].
- Manufacturing and in-service tolerances of load sensors (MAP, MAF or VAF) and injectors in particular and faults such as intake air leaks will affect the fuel/air mixture and cause the fuel trim to quickly compensate.
- An intake air leak for instance would result in the injection period increasing, for example to 115-125%. This level will also oscillate +/-5% as before [**B**].
- This new short term fuel trim (STFT) level will be stored in the ECM if it is established as a new basic mixture level. It will then become a long term fuel trim (LTFT) correction and results in the correct mixture level immediately after starting, even when the HO2S has had insufficient time to heat up.
- Long term fuel trim (LTFT) values can be erased by disconnecting the ECM power supply for a suitable length of time.
- If the memory is not erased after repairs the ECM will eventually learn the new LTFT values, but this will take some time and probably cause high emissions and some driveability problems.
- For example if an intake air leak causes the engine to run lean this will be compensated by a change in the LTFT value, which will be stored in the ECM memory.
- After the leak has been repaired this memorised LTFT value will still be used to compute the injection period, resulting in excessively rich running, until new LTFT values have been learned.

4

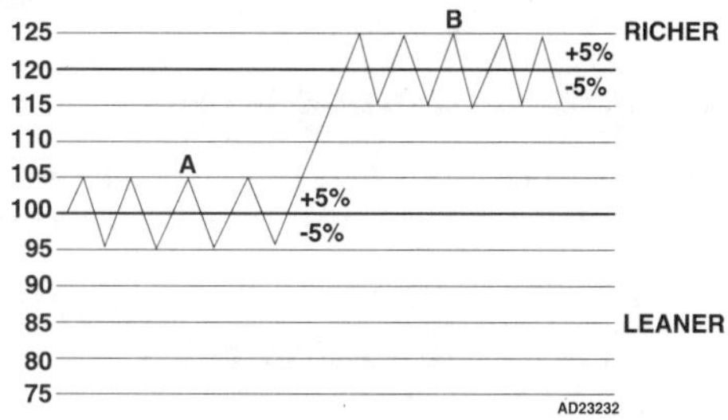

Probable causes

Wiring

- Wiring – refers not only to the wiring harness, but also to any associated component multi-plugs, relay plates, welded, crimped or soldered joints and bulkhead connectors.
- Short to ground – refers to wiring that has a 'leak' to ground somewhere along its length e.g. if the insulation has chaffed through and the wiring is in contact with the body or engine.

How to use this manual

- Short to positive – refers to wiring that would normally not be carrying 12 volts, but is shorted to the 12 volt supply e.g. a sensor wire that would normally be fed with approx. 5 volts from the ECM, shorted to a 12 volt battery positive supply.
- Open circuit – refers to wiring that is either disconnected or broken, but is not shorted to either ground or a positive supply.
- Poor connection – refers to a loose, intermittent or high resistance connection.

Mechanical

- Mechanical fault - refers to items such as: fuel leaks, air leaks, crankcase pressure and cylinder compression. Refer to the general trouble shooting section at the front of this manual for further information.

After rectification

- OBD-II monitoring is very sensitive and replacement components may still be identified as faulty if their specifications do not exactly match the originals.
- The trouble code memory must be completely erased and a 'trip' completed.
- The trouble code memory should be accessed to establish that no further codes have been logged.

General test procedures

- Find the model related chapter from vehicle manufacturer, model name, engine code, year of manufacture and system.

Observe the following relevant test conditions:

- All auxiliary equipment, including air conditioning, switched OFF.
- Ensure battery voltage fully charged.
- Ensure battery ground wires in good condition.
- If vehicle fitted with engine malfunction indicator lamp (MIL):
- Check MIL operation – if MIL does not flash or illuminate, replace bulb or repair circuit.

NOTE: ***If MIL remains illuminated or flashes, this indicates a fault, but if it goes out there could still be trouble codes logged in the ECM. Scan tool may not respond if MIL is not working.***

- Engine at normal operating temperature.
- Carry out road test.
- Operate accelerator pedal several times over complete travel range.
- Allow engine to idle.
- If engine does not start: Briefly crank engine and leave ignition switched ON.
- Observe any model specific information.
- Find the location of the data link connector (DLC).
- Bridge DLC terminals, connect an LED tester or scan tool, as appropriate and observe any special conditions. If a scan tool is used, follow the manufacturer's operating instructions.
- Follow procedure for accessing trouble codes and note any codes displayed.
- Refer to the appropriate trouble code table. If the vehicle has an OBD-II system the standard SAE defined trouble codes (P0000-P9999) are listed in a table at the front of this manual - refer to relevant section.

NOTE: ***The trouble code may not identify the exact cause of any problem, but serves as a guide to the component or system to be investigated.***

- Refer to probable causes and carry out systematic check of components and circuits until cause of problem is identified and rectified. Where the probable cause is a non-specific engine fault, fuel system or ignition system fault, refer to the General Trouble Shooter section at the front of this manual.
- Check for further trouble codes.
- Follow procedure for erasing trouble codes.

NOTE: ***If the battery is disconnected to erase trouble codes, data stored, such as radio security codes and ECM adaptive memory may be lost.***

- Road test vehicle and recheck for any logged trouble codes.

NOTE: ***If a trouble code is output, but not listed in the appropriate code table, suspect a control module fault. In the case of signal or component malfunction, the ECM may substitute a fixed value from its internal limited operating strategy (LOS) software. This enables the vehicle to be driven in limp-home mode until the fault can be rectified. Depending on the component or circuit in question, it may not be obvious to the driver that the engine management system is operating in limp-home mode.***

OBD-II trouble code table

- The following list covers all SAE defined standard P zero codes allocated at the time of publication.
- All OBD-II codes starting with P zero have standard meanings irrespective of vehicle make or model.
- For OBD-II codes, other than those starting P zero, refer to model specific chapters.

Trouble code	Fault location	Probable cause
P0000	No fault found	–
P0001	Fuel volume regulator control – circuit open	**Wiring, regulator control solenoid**
P0002	Fuel volume regulator control – circuit range/ performance	**Wiring, regulator control solenoid**
P0003	Fuel volume regulator control – circuit low	**Wiring short to ground, regulator control solenoid**
P0004	Fuel volume regulator control – circuit high	**Wiring open circuit/short to positive, regulator control solenoid**
P0005	Fuel shut-off valve control – circuit open	**Wiring open circuit, fuel shut-off valve**
P0006	Fuel shut-off valve control – circuit low	**Wiring short to ground, fuel shut-off valve**
P0007	Fuel shut-off valve control – circuit high	**Wiring short to positive, fuel shut-off valve**
P0008	Engine position system, bank 1 – performance	**Mechanical fault**
P0009	Engine position system, bank 2 – performance	**Mechanical fault**
P0010	Camshaft position (CMP) actuator, intake/left/front, bank 1 – circuit malfunction	**Wiring, CMP actuator, ECM**
P0011	Camshaft position (CMP), intake/left/front, bank 1 – timing over-advanced/system performance	**Valve timing, engine mechanical fault, CMP actuator**
P0012	Camshaft position (CMP), intake/left/front, bank 1 – timing over-retarded	**Valve timing, engine mechanical fault, CMP actuator**
P0013	Camshaft position (CMP) actuator, intake/left/front, bank 1 – circuit malfunction	**Wiring, CMP actuator, ECM**
P0014	Camshaft position (CMP) actuator, exhaust/right/ rear, bank 1 – timing over-advanced/system performance	**Valve timing, engine mechanical fault, CMP actuator**
P0015	Camshaft position (CMP) actuator, exhaust/right/ rear, bank 1 – timing over-retarded	**Valve timing, engine mechanical fault, CMP actuator**
P0016	Crankshaft position/camshaft position, bank 1 sensor A – correlation	**Wiring, CKP sensor, CMP sensor, mechanical fault**
P0017	Crankshaft position/camshaft position, bank 1 sensor B – correlation	**Wiring, CKP sensor, CMP sensor, mechanical fault**
P0018	Crankshaft position/camshaft position, bank 2 sensor A – correlation	**Wiring, CKP sensor, CMP sensor, mechanical fault**
P0019	Crankshaft position/camshaft position, bank 2 sensor B – correlation	**Wiring, CKP sensor, CMP sensor, mechanical fault**
P0020	Camshaft position (CMP) actuator, intake/left/front, bank 2 – circuit malfunction	**Wiring, CMP actuator, ECM**
P0021	Camshaft position (CMP), intake/left/front, bank 2 – timing over-advanced/system performance	**Valve timing, engine mechanical fault, CMP actuator**
P0022	Camshaft position (CMP), intake/left/front, bank 2 – timing over-retarded	**Valve timing, engine mechanical fault, CMP actuator**
P0023	Camshaft position (CMP) actuator, exhaust/right/ rear, bank 2 – circuit malfunction	**Wiring, CMP actuator, ECM**
P0024	Camshaft position (CMP), exhaust/right/rear, bank 2 – timing over-advanced/system performance	**Valve timing, engine mechanical fault, CMP actuator**

OBD-II trouble code table

Trouble code	Fault location	Probable cause
P0025	Camshaft position (CMP), exhaust/right/rear, bank 2 – timing over-retarded	**Valve timing, engine mechanical fault, CMP actuator**
P0026	Intake valve control solenoid circuit, bank 1 – range/performance	**Wiring, intake valve control solenoid**
P0027	Exhaust valve control solenoid circuit, bank 1 – range/performance	**Wiring, exhaust valve control solenoid**
P0028	Intake valve control solenoid circuit, bank 2 – range/performance	**Wiring, intake valve control solenoid**
P0029	Exhaust valve control solenoid circuit, bank 2 – range/performance	**Wiring, exhaust valve control solenoid**
P0030	Heated oxygen sensor (HO2S) 1, bank 1, heater control – circuit malfunction	**Wiring, HO2S, ECM**
P0031	Heated oxygen sensor (HO2S) 1, bank 1, heater control – circuit low	**Wiring short to ground, HO2S, ECM**
P0032	Heated oxygen sensor (HO2S) 1, bank 1, heater control – circuit high	**Wiring short to positive, HO2S, ECM**
P0033	Turbocharger (TC) wastegate regulating valve – circuit malfunction	**Wiring, TC wastegate regulating valve, ECM**
P0034	Turbocharger (TC) wastegate regulating valve – circuit low	**Wiring short to ground, TC wastegate regulating valve, ECM**
P0035	Turbocharger (TC) wastegate regulating valve – circuit high	**Wiring short to positive, TC wastegate regulating valve, ECM**
P0036	Heated oxygen sensor (HO2S) 2, bank 1, heater control – circuit malfunction	**Wiring, HO2S, ECM**
P0037	Heated oxygen sensor (HO2S) 2, bank 1, heater control – circuit low	**Wiring short to ground, HO2S, ECM**
P0038	Heated oxygen sensor (HO2S) 2, bank 1, heater control – circuit high	**Wiring short to positive, HO2S, ECM**
P0039	Turbo/super charger bypass valve, control circuit – range/performance	**Wiring, bypass valve**
P0040	Oxygen sensor signals swapped, bank 1 sensor 1/ bank 2 sensor 1	**Wiring**
P0041	Oxygen sensor signals swapped, bank 1 sensor 2/ bank 2 sensor 2	**Wiring**
P0042	Heated oxygen sensor (HO2S) 3, bank 1, heater control – circuit malfunction	**Wiring, HO2S, ECM**
P0043	Heated oxygen sensor (HO2S) 3, bank 1, heater control – circuit low	**Wiring short to ground, HO2S, ECM**
P0044	Heated oxygen sensor (HO2S) 3, bank 1, heater control – circuit high	**Wiring short to positive, HO2S, ECM**
P0045	Turbo/super charger boost control solenoid – circuit open	**Wiring, boost control solenoid**
P0046	Turbo/super charger boost control solenoid – circuit range/performance	**Wiring, boost control solenoid, mechanical fault**
P0047	Turbo/super charger boost control solenoid – circuit low	**Wiring short to ground, boost control solenoid**
P0048	Turbo/super charger boost control solenoid – circuit high	**Wiring short to positive, boost control solenoid**
P0049	Turbo/super charger turbine – over-speed	**Mechanical fault**
P0050	Heated oxygen sensor (HO2S) 1, bank 2, heater control – circuit malfunction	**Wiring, HO2S, ECM**

OBD-II trouble code table

Trouble code	Fault location	Probable cause
P0051	Heated oxygen sensor (HO2S) 1, bank 2, heater control – circuit low	**Wiring short to ground, HO2S, ECM**
P0052	Heated oxygen sensor (HO2S) 1, bank 2, heater control – circuit high	**Wiring short to positive, HO2S, ECM**
P0053	Heated oxygen sensor (HO2S), bank 1, sensor 1 – heater resistance	**Wiring, HO2S**
P0054	Heated oxygen sensor (HO2S), bank 1, sensor 2 – heater resistance	**Wiring, HO2S**
P0055	Heated oxygen sensor (HO2S), bank 1, sensor 3 – heater resistance	**Wiring, HO2S**
P0056	Heated oxygen sensor (HO2S) 2, bank 2, heater control – circuit malfunction	**Wiring, HO2S, ECM**
P0057	Heated oxygen sensor (HO2S) 2, bank 2, heater control – heater circuit low	**Wiring short to ground, HO2S, ECM**
P0058	Heated oxygen sensor (HO2S) 2, bank 2, heater control – circuit high	**Wiring short to positive, HO2S, ECM**
P0059	Heated oxygen sensor (HO2S), bank 2, sensor 1 – heater resistance	**Wiring, HO2S**
P0060	Heated oxygen sensor (HO2S), bank 2, sensor 2 – heater resistance	**Wiring, HO2S**
P0061	Heated oxygen sensor (HO2S), bank 2, sensor 3 – heater resistance	**Wiring, HO2S**
P0062	Heated oxygen sensor (HO2S) 3, bank 2, heater control – circuit malfunction	**Wiring, HO2S, ECM**
P0063	Heated oxygen sensor (HO2S) 3, bank 2, heater control – circuit low	**Wiring short to ground, HO2S, ECM**
P0064	Heated oxygen sensor (HO2S) 3, bank 2, heater control – circuit high	**Wiring short to positive, HO2S, ECM**
P0065	Air assisted injector – range/performance problem	**Air assisted injector**
P0066	Air assisted injector – circuit malfunction/circuit low	**Wiring short to ground, air assisted injector, ECM**
P0067	Air assisted injector – circuit high	**Wiring short to positive, air assisted injector, ECM**
P0068	Manifold pressure (MAP) sensor/Mass air flow (MAF) sensor – throttle position correlation	**Wiring, MAP sensor, MAF sensor, mechanical fault**
P0069	Manifold pressure (MAP) sensor – barometric pressure correlation	**MAF sensor, mechanical fault**
P0070	Outside air temperature sensor – circuit malfunction	**Wiring, outside air temperature sensor, ECM**
P0071	Outside air temperature sensor – range/ performance problem	**Outside air temperature sensor**
P0072	Outside air temperature sensor – low input	**Wiring short to ground, outside air temperature sensor, ECM**
P0073	Outside air temperature sensor – high input	**Wiring short to positive, outside air temperature sensor, ECM**
P0074	Outside air temperature sensor – circuit intermittent	**Wiring, poor connection, outside air temperature sensor, ECM**
P0075	Intake valve control solenoid, bank 1 – circuit malfunction	**Wiring, intake valve control solenoid, ECM**
P0076	Intake valve control solenoid, bank 1 – circuit low	**Wiring short to ground, intake valve control solenoid, ECM**

OBD-II trouble code table

Trouble code	Fault location	Probable cause
P0077	Intake valve control solenoid, bank 1 – circuit high	**Wiring short to positive, intake valve control solenoid, ECM**
P0078	Exhaust valve control solenoid, bank 1 – circuit malfunction	**Wiring, exhaust valve control solenoid, ECM**
P0079	Exhaust valve control solenoid, bank 1 – circuit low	**Wiring short to ground, exhaust valve control solenoid, ECM**
P0080	Exhaust valve control solenoid, bank 1 – circuit high	**Wiring short to positive, exhaust valve control solenoid, ECM**
P0081	Intake valve control solenoid, bank 2 – circuit malfunction	**Wiring, intake valve control solenoid, ECM**
P0082	Intake valve control solenoid, bank 2 – circuit low	**Wiring short to ground, intake valve control solenoid, ECM**
P0083	Intake valve control solenoid, bank 2 – circuit high	**Wiring short to positive, intake valve control solenoid, ECM**
P0084	Exhaust valve control solenoid, bank 2 – circuit malfunction	**Wiring, exhaust valve control solenoid, ECM**
P0085	Exhaust valve control solenoid, bank 2 – circuit low	**Wiring short to ground, exhaust valve control solenoid, ECM**
P0086	Exhaust valve control solenoid, bank 2 – circuit high	**Wiring short to positive, exhaust valve control solenoid, ECM**
P0087	Fuel rail/system pressure – too low	**Fuel pump, fuel pressure regulator, fuel supply pipe blockage, mechanical fault**
P0088	Fuel rail/system pressure – too high	**Fuel pump, fuel pressure regulator, fuel return pipe blockage, mechanical fault**
P0089	Fuel pressure regulator – performance problem	**Fuel pressure regulator, mechanical fault**
P0090	Fuel metering solenoid – open circuit	**Wiring open circuit, fuel metering solenoid, ECM**
P0091	Fuel metering solenoid – short to ground	**Wiring short to ground, fuel metering solenoid, ECM**
P0092	Fuel metering solenoid – short to positive	**Wiring short to positive, fuel metering solenoid, ECM**
P0093	Fuel system leak detected – large leak	**Wiring, fuel pressure sensor, mechanical fault**
P0094	Fuel system leak detected – small leak	**Wiring, fuel pressure sensor, mechanical fault**
P0095	Intake air temperature (IAT) sensor 2 – circuit malfunction	**Wiring, poor connection, IAT sensor, ECM**
P0096	Intake air temperature (IAT) sensor 2 – circuit range/performance	**Wiring, poor connection, IAT sensor, ECM**
P0097	Intake air temperature (IAT) sensor 2 – circuit low input	**Wiring short to ground, IAT sensor, ECM**
P0098	Intake air temperature (IAT) sensor 2 – circuit high input	**Wiring short to positive, IAT sensor, ECM**
P0099	Intake air temperature (IAT) sensor 2 – circuit intermittent/erratic	**Wiring, poor connection, IAT sensor, ECM**
P0100	Mass air flow (MAF) sensor/volume air flow (VAF) sensor – circuit malfunction	**Wiring, MAF/VAF sensor, ECM**
P0101	Mass air flow (MAF) sensor/volume air flow (VAF) sensor – range/performance problem	**Intake leak/blockage, MAF/VAF sensor**
P0102	Mass air flow (MAF) sensor/volume air flow (VAF) sensor – low input	**Wiring short to ground, MAF/VAF sensor, ECM**

Trouble code	Fault location	Probable cause
P0103	Mass air flow (MAF) sensor/volume air flow (VAF) sensor – high input	Wiring short to positive, MAF/VAF sensor, ECM
P0104	Mass air flow (MAF) sensor/volume air flow (VAF) sensor – circuit intermittent	Wiring, poor connection, MAF/VAF sensor, ECM
P0105	Manifold absolute pressure (MAP) sensor/ barometric pressure (BARO) sensor – circuit malfunction	Wiring, MAP sensor, BARO sensor, ECM
P0106	Manifold absolute pressure (MAP) sensor/ barometric pressure (BARO) sensor – range/ performance problem	Intake/exhaust leak, wiring, MAP sensor, BARO sensor
P0107	Manifold absolute pressure (MAP) sensor/ barometric pressure (BARO) sensor – low input	Wiring short to ground, MAP sensor, BARO sensor, ECM
P0108	Manifold absolute pressure (MAP) sensor/ barometric pressure (BARO) sensor – high input	Wiring short to positive, MAP sensor, BARO sensor, ECM
P0109	Manifold absolute pressure (MAP) sensor/ barometric pressure (BARO) sensor – circuit intermittent	Wiring, poor connection, MAP sensor, BARO sensor, ECM
P0110	Intake air temperature (IAT) sensor – circuit malfunction	Wiring, IAT sensor, ECM
P0111	Intake air temperature (IAT) sensor – range/ performance problem	IAT sensor
P0112	Intake air temperature (IAT) sensor – low input	Wiring short to ground, IAT sensor, ECM
P0113	Intake air temperature (IAT) sensor – high input	Wiring open circuit/short to positive, ground wire defective, IAT sensor, ECM
P0114	Intake air temperature (IAT) sensor – circuit intermittent	Wiring, poor connection, IAT sensor, ECM
P0115	Engine coolant temperature (ECT) sensor – circuit malfunction	Wiring, ECT sensor, ECM
P0116	Engine coolant temperature (ECT) sensor – range/ performance problem	Coolant thermostat, poor connection, wiring, ECT sensor
P0117	Engine coolant temperature (ECT) sensor – low input	Coolant thermostat, wiring short to ground, ECT sensor
P0118	Engine coolant temperature (ECT) sensor – high input	Coolant thermostat, wiring open circuit/short to positive, ground wire defective, ECT sensor
P0119	Engine coolant temperature (ECT) sensor – circuit intermittent	Wiring, poor connection, ECT sensor, ECM
P0120	Throttle position (TP) sensor A/accelerator pedal position (APP) sensor A – circuit malfunction	Wiring, TP/APP sensor, ECM
	Throttle position (TP) switch A/accelerator pedal position (APP) switch A – circuit malfunction	Wiring, TP/APP switch, ECM
P0121	Throttle position (TP) sensor A/accelerator pedal position (APP) sensor A – range/performance problem	Accelerator cable adjustment, TP/APP sensor
	Throttle position (TP) switch A/accelerator pedal position (APP) switch A – range/performance problem	Accelerator cable adjustment, TP/APP switch
P0122	Throttle position (TP) sensor A/accelerator pedal position (APP) sensor A – low input	Wiring short to ground, TP/APP sensor, ECM
	Throttle position (TP) switch A/accelerator pedal position (APP) switch A – low input	Wiring short to ground, TP/APP switch, ECM

OBD-II trouble code table

Trouble code	Fault location	Probable cause
P0123	Throttle position (TP) sensor A/accelerator pedal position (APP) sensor A – high input	**Wiring short to positive, TP/APP sensor, ECM**
	Throttle position (TP) switch A/accelerator pedal position (APP) switch A – high input	**Wiring short to positive, TP/APP switch, ECM**
P0124	Throttle position (TP) sensor A/accelerator pedal position (APP) sensor A – circuit intermittent	**Wiring, poor connection, TP/APP sensor, ECM**
	Throttle position (TP) switch A/accelerator pedal position (APP) switch A – circuit intermittent	**Wiring, poor connection, TP/APP switch, ECM**
P0125	Insufficient coolant temperature for closed loop fuel control	**Wiring, cooling system, coolant thermostat, ECT sensor**
P0126	Insufficient coolant temperature for stable operation	**Wiring, cooling system, coolant thermostat, ECT sensor**
P0127	Intake air temperature – too high	**Wiring short to ground, IAT sensor 2, mechanical fault, ECM**
P0128	Coolant thermostat – coolant temp below thermostat regulating temperature	**Mechanical fault** REPLACE THERMOSTAT
P0129	Barometric pressure – too low	**Wiring, BARO sensor, mechanical fault**
P0130	Heated oxygen sensor (HO2S) 1, bank 1 – circuit malfunction	**Heating inoperative, poor connection, wiring, HO2S**
	Oxygen sensor (O2S) 1, bank 1 – circuit malfunction	**Wiring, O2S, ECM**
P0131	Heated oxygen sensor (HO2S) 1, bank 1 – low voltage	**Exhaust leak, wiring short to ground, HO2S, ECM**
	Oxygen sensor (O2S) 1, bank 1 – low voltage	**Exhaust leak, wiring short to ground, O2S, ECM**
P0132	Heated oxygen sensor (HO2S) 1, bank 1 – high voltage	**Wiring short to positive, HO2S, ECM**
	Oxygen sensor (O2S) 1, bank 1 – high voltage	**Wiring short to positive, O2S, ECM**
P0133	Heated oxygen sensor (HO2S) 1, bank 1 – slow response	**Heating inoperative, wiring, HO2S**
	Oxygen sensor (O2S) 1, bank 1 – slow response	**Wiring, O2S**
P0134	Heated oxygen sensor (HO2S) 1, bank 1 – no activity detected	**Wiring open circuit, heating inoperative, HO2S**
	Oxygen sensor (O2S) 1, bank 1 – no activity detected	**Wiring, O2S**
P0135	Heated oxygen sensor (HO2S) 1, bank 1, heater control – circuit malfunction	**Fuse, wiring, HO2S, ECM**
P0136	Heated oxygen sensor (HO2S) 2, bank 1 – circuit malfunction	**Heating inoperative, wiring, HO2S, ECM**
	Oxygen sensor (O2S) 2, bank 1 – circuit malfunction	**Wiring, O2S, ECM**
P0137	Heated oxygen sensor (HO2S) 2, bank 1 – low voltage	**Exhaust leak, wiring short to ground, HO2S, ECM**
	Oxygen sensor (O2S) 2, bank 1 – low voltage	**Exhaust leak, wiring short to ground, O2S, ECM**
P0138	Heated oxygen sensor (HO2S) 2, bank 1 – high voltage	**Wiring short to positive, HO2S, ECM**
	Oxygen sensor (O2S) 2, bank 1 – high voltage	**Wiring short to positive, O2S, ECM**
P0139	Heated oxygen sensor (HO2S) 2, bank 1 – slow response	**Heating inoperative, wiring, HO2S**
	Oxygen sensor (O2S) 2, bank 1 – slow response	**Wiring, O2S**

Trouble code	Fault location	Probable cause
P0140	Heated oxygen sensor (HO2S) 2, bank 1 – no activity detected	Wiring, heating inoperative, HO2S, ECM
	Oxygen sensor (O2S) 2, bank 1 – no activity detected	Wiring, O2S, ECM
P0141	Heated oxygen sensor (HO2S) 2, bank 1, heater control – circuit malfunction	Wiring, HO2S, ECM
P0142	Heated oxygen sensor (HO2S) 3, bank 1 – circuit malfunction	Wiring, HO2S, ECM
P0143	Heated oxygen sensor (HO2S) 3, bank 1 – low voltage	Exhaust leak, wiring short to ground, HO2S, ECM
	Oxygen sensor (O2S) 3, bank 1 – low voltage	Exhaust leak, wiring short to ground, O2S, ECM
P0144	Heated oxygen sensor (HO2S) 3, bank 1 – high voltage	Wiring short to positive, HO2S, ECM
	Oxygen sensor (O2S) 3, bank 1 – high voltage	Wiring short to positive, O2S, ECM
P0145	Heated oxygen sensor (HO2S) 3, bank 1 – slow response	Heating inoperative, wiring, HO2S
	Oxygen sensor (O2S) 3, bank 1 – slow response	Wiring, O2S
P0146	Heated oxygen sensor (HO2S) 3, bank 1 – no activity detected	Wiring, HO2S, ECM
	Oxygen sensor (O2S) 3, bank 1 – no activity detected	Wiring, O2S, ECM
P0147	Heated oxygen sensor (HO2S) 3, bank 1, heater control – circuit malfunction	Wiring, HO2S, ECM
P0148	Fuel delivery error	Fuel pump/fuel injection pump
P0149	Fuel timing error	Fuel pump/fuel injection pump
P0150	Heated oxygen sensor (HO2S) 1, bank 2 – circuit malfunction	Wiring, HO2S, ECM
	Oxygen sensor (O2S) 1, bank 2 – circuit malfunction	Wiring, O2S, ECM
P0151	Heated oxygen sensor (HO2S) 1, bank 2 – low voltage	Exhaust leak, wiring short to ground, HO2S, ECM
	Oxygen sensor (O2S) 1, bank 2 – low voltage	Exhaust leak, wiring short to ground, O2S, ECM
P0152	Heated oxygen sensor (HO2S) 1, bank 2 – high voltage	Wiring short to positive, HO2S, ECM
	Oxygen sensor (O2S) 1, bank 2 – high voltage	Wiring short to positive, O2S, ECM
P0153	Heated oxygen sensor (HO2S) 1, bank 2 – slow response	Heating inoperative, wiring, HO2S
	Oxygen sensor (O2S) 1, bank 2 – slow response	Wiring, O2S
P0154	Heated oxygen sensor (HO2S) 1, bank 2 – no activity detected	Wiring, HO2S, ECM
	Oxygen sensor (O2S) 1, bank 2 – no activity detected	Wiring, O2S, ECM
P0155	Heated oxygen sensor (HO2S) 1, bank 2, heater control – circuit malfunction	Wiring, HO2S, ECM
P0156	Heated oxygen sensor (HO2S) 2, bank 2 – circuit malfunction	Heating inoperative, wiring, HO2S, ECM
	Oxygen sensor (O2S) 2, bank 2 – circuit malfunction	Wiring, O2S, ECM

OBD-II trouble code table

Trouble code	Fault location	Probable cause
P0157	Heated oxygen sensor (HO2S) 2, bank 2 – low voltage	**Exhaust leak, wiring short to ground, HO2S, ECM**
	Oxygen sensor (O2S) 2, bank 2 – low voltage	**Exhaust leak, wiring short to ground, O2S, ECM**
P0158	Heated oxygen sensor (HO2S) 2, bank 2 – high voltage	**Wiring short to positive, HO2S, ECM**
	Oxygen sensor (O2S) 2, bank 2 – high voltage	**Wiring short to positive, O2S, ECM**
P0159	Heated oxygen sensor (HO2S) 2, bank 2 – slow response	**Heating inoperative, wiring, HO2S**
	Oxygen sensor (O2S) 2, bank 2 – slow response	**Wiring, O2S**
P0160	Heated oxygen sensor (HO2S) 2, bank 2 – no activity detected	**Wiring, HO2S, ECM**
	Oxygen sensor (O2S) 2, bank 2 – no activity detected	**Wiring, O2S, ECM**
P0161	Heated oxygen sensor (HO2S) 2, bank 2, heater control – circuit malfunction	**Wiring, HO2S, ECM**
P0162	Heated oxygen sensor (HO2S) 3, bank 2 – circuit malfunction	**Wiring, HO2S, ECM**
	Oxygen sensor (O2S) 3, bank 2 – circuit malfunction	**Wiring, O2S, ECM**
P0163	Heated oxygen sensor (HO2S) 3, bank 2 – low voltage	**Exhaust leak, wiring short to ground, HO2S, ECM**
	Oxygen sensor (O2S) 3, bank 2 – low voltage	**Exhaust leak, wiring short to ground, O2S, ECM**
P0164	Heated oxygen sensor (HO2S) 3, bank 2 – high voltage	**Wiring short to positive, HO2S, ECM**
	Oxygen sensor (O2S) 3, bank 2 – high voltage	**Wiring short to positive, O2S, ECM**
P0165	Heated oxygen sensor (HO2S) 3, bank 2 – slow response	**Heating inoperative, wiring, HO2S**
	Oxygen sensor (O2S) 3, bank 2 – slow response	**Wiring, O2S**
P0166	Heated oxygen sensor (HO2S) 3, bank 2 – no activity detected	**Wiring, HO2S, ECM**
	Oxygen sensor (O2S) 3, bank 2 – no activity detected	**Wiring, O2S, ECM**
P0167	Heated oxygen sensor (HO2S) 3, bank 2, heater control – circuit malfunction	**Wiring, HO2S, ECM**
P0168	Fuel temperature – too high	**Wiring, fuel temperature sensor, mechanical fault**
P0169	Incorrect fuel composition	**Wiring, fuel composition sensor, mechanical fault**
P0170	Fuel trim (FT), bank 1 – malfunction	**Intake leak, AIR system, fuel pressure/pump, injector(s), EVAP canister purge valve, HO2S**
P0171	System too lean, bank 1	**Intake/exhaust leak, AIR system, MAF/VAF sensor, fuel pressure/pump, injector(s), HO2S**
P0172	System too rich, bank 1	**Intake blocked, EVAP canister purge valve, fuel pressure, EGR system, injector(s), HO2S**
P0173	Fuel trim (FT), bank 2 – malfunction	**Intake leak, AIR system, fuel pressure/pump, injector(s), EVAP canister purge valve, HO2S**
P0174	System too lean, bank 2	**Intake/exhaust leak, fuel pressure/pump, injector(s), AIR system, hose connection(s)**

Trouble code	Fault location	Probable cause
P0175	System too rich, bank 2	Intake blocked, EVAP canister purge valve, fuel pressure, EGR system, injector(s), HO2S
P0176	Fuel composition sensor – circuit malfunction	Wiring, fuel composition sensor, ECM
P0177	Fuel composition sensor – range/performance problem	Fuel composition sensor
P0178	Fuel composition sensor – low input	Wiring short to ground, fuel composition sensor, ECM
P0179	Fuel composition sensor – high input	Wiring short to positive, fuel composition sensor, ECM
P0180	Fuel temperature sensor A – circuit malfunction	Wiring, fuel temperature sensor, ECM
P0181	Fuel temperature sensor A – range/performance problem	Fuel temperature sensor
P0182	Fuel temperature sensor A – low input	Wiring short to ground, fuel temperature sensor, ECM
P0183	Fuel temperature sensor A – high input	Wiring short to positive, fuel temperature sensor, ECM
P0184	Fuel temperature sensor A – circuit intermittent	Wiring, poor connection, fuel temperature sensor, ECM
P0185	Fuel temperature sensor B – circuit malfunction	Wiring, fuel temperature sensor, ECM
P0186	Fuel temperature sensor B – range/performance problem	Fuel temperature sensor
P0187	Fuel temperature sensor B – low input	Wiring short to ground, fuel temperature sensor, ECM
P0188	Fuel temperature sensor B – high input	Wiring short to positive, fuel temperature sensor, ECM
P0189	Fuel temperature sensor B – circuit intermittent	Wiring, poor connection, fuel temperature sensor, ECM
P0190	Fuel rail pressure (FRP) sensor – circuit malfunction	Wiring, fuel rail pressure sensor, ECM
P0191	Fuel rail pressure (FRP) sensor – range/ performance problem	Wiring, FRP sensor
P0192	Fuel rail pressure (FRP) sensor – low input	Wiring short to ground, FRP sensor
P0193	Fuel rail pressure (FRP) sensor – high input	Wiring short to positive, FRP sensor
P0194	Fuel rail pressure (FRP) sensor – circuit intermittent	Wiring, poor connection, FRP sensor
P0195	Engine oil temperature (EOT) sensor – circuit malfunction	Wiring, EOT sensor, ECM
P0196	Engine oil temperature (EOT) sensor – range/ performance problem	EOT sensor
P0197	Engine oil temperature (EOT) sensor – low input	Wiring short to ground, EOT sensor
P0198	Engine oil temperature (EOT) sensor – high input	Wiring short to positive, EOT sensor
P0199	Engine oil temperature (EOT) sensor – circuit intermittent	Wiring, poor connection, EOT sensor, ECM
P0200	Injector – circuit malfunction	Wiring, injector, ECM
P0201	Injector 1 – circuit malfunction	Wiring, injector, ECM
P0202	Injector 2 – circuit malfunction	Wiring, injector, ECM
P0203	Injector 3 – circuit malfunction	Wiring, injector, ECM

OBD-II trouble code table

Trouble code	Fault location	Probable cause
P0204	Injector 4 – circuit malfunction	**Wiring, injector, ECM**
P0205	Injector 5 – circuit malfunction	**Wiring, injector, ECM**
P0206	Injector 6 – circuit malfunction	**Wiring, injector, ECM**
P0207	Injector 7 – circuit malfunction	**Wiring, injector, ECM**
P0208	Injector 8 – circuit malfunction	**Wiring, injector, ECM**
P0209	Injector 9 – circuit malfunction	**Wiring, injector, ECM**
P0210	Injector 10 – circuit malfunction	**Wiring, injector, ECM**
P0211	Injector 11 – circuit malfunction	**Wiring, injector, ECM**
P0212	Injector 12 – circuit malfunction	**Wiring, injector, ECM**
P0213	Cold start injector 1 – circuit malfunction	**Wiring, cold start injector, ECM**
P0214	Cold start injector 2 – circuit malfunction	**Wiring, cold start injector, ECM**
P0215	Fuel shut-off solenoid – circuit malfunction	**Wiring, fuel shut-off solenoid, ECM**
P0216	Fuel injection timing control – circuit malfunction	**Wiring, fuel injection timing control solenoid, ECM**
P0217	Engine over temperature condition	**Wiring, cooling system, coolant thermostat, ECT sensor**
P0218	Transmission over temperature condition	**Wiring, TFT sensor, ECM**
P0219	Engine over speed condition	**Incorrect gear change**
P0220	Throttle position (TP) sensor B/accelerator pedal position (APP) sensor B – circuit malfunction	**Wiring, TP/APP sensor, ECM**
	Throttle position (TP) switch B/accelerator pedal position (APP) switch B – circuit malfunction	**Wiring, TP/APP switch, ECM**
P0221	Throttle position (TP) sensor B/accelerator pedal position (APP) sensor B – range/performance problem	**Accelerator cable adjustment, TP/APP sensor**
	Throttle position (TP) switch B/accelerator pedal position (APP) switch B – range/performance problem	**Accelerator cable adjustment, TP/APP switch**
P0222	Throttle position (TP) sensor B/accelerator pedal position (APP) sensor B – low input	**Wiring short to ground, TP/APP sensor, ECM**
	Throttle position (TP) switch B/accelerator pedal position (APP) switch B – low input	**Wiring short to ground, TP/APP switch, ECM**
P0223	Throttle position (TP) sensor B/accelerator pedal position (APP) sensor B – high input	**Wiring short to positive, TP/APP sensor, ECM**
	Throttle position (TP) switch B/accelerator pedal position (APP) switch B – high input	**Wiring short to positive, TP/APP switch, ECM**
P0224	Throttle position (TP) sensor B/accelerator pedal position (APP) sensor B – circuit intermittent	**Wiring, poor connection, TP/APP sensor, ECM**
	Throttle position (TP) switch B/accelerator pedal position (APP) switch B – circuit intermittent	**Wiring, poor connection, TP/APP switch, ECM**
P0225	Throttle position (TP) sensor C/accelerator pedal position (APP) sensor C – circuit malfunction	**Wiring, TP/APP sensor, ECM**
	Throttle position (TP) switch C/accelerator pedal position (APP) switch C – circuit malfunction	**Wiring, TP/APP switch, ECM**
P0226	Throttle position (TP) sensor C/accelerator pedal position (APP) sensor C – range/performance problem	**Accelerator cable adjustment, TP/APP sensor**

Trouble code	Fault location	Probable cause
	Throttle position (TP) switch C/accelerator pedal position (APP) switch C – range/performance problem	**Accelerator cable adjustment, TP/APP switch**
P0227	Throttle position (TP) sensor C/accelerator pedal position (APP) sensor C – low input	**Wiring short to ground, TP/APP sensor, ECM**
	Throttle position (TP) switch C/accelerator pedal position (APP) switch C – low input	**Wiring short to ground, TP/APP switch, ECM**
P0228	Throttle position (TP) sensor C/accelerator pedal position (APP) sensor C – high input	**Wiring short to positive, TP/APP sensor, ECM**
	Throttle position (TP) switch C/accelerator pedal position (APP) switch C – high input	**Wiring short to positive, TP/APP switch, ECM**
P0229	Throttle position (TP) sensor C/accelerator pedal position (APP) sensor C – circuit intermittent	**Wiring, poor connection, TP/APP sensor, ECM**
	Throttle position (TP) switch C/accelerator pedal position (APP) switch C – circuit intermittent	**Wiring, poor connection, TP/APP switch, ECM**
P0230	Fuel pump relay – circuit malfunction	**Wiring, fuel pump relay, ECM**
P0231	Fuel pump relay – circuit low	**Wiring short to ground, fuel pump relay, ECM**
P0232	Fuel pump relay – circuit high	**Wiring short to positive, fuel pump relay, ECM**
P0233	Fuel pump relay – circuit intermittent	**Wiring, poor connection, fuel pump relay, ECM**
P0234	Engine boost condition – limit exceeded	**Hose connection(s), wiring, TC wastegate regulating valve, TC wastegate**
P0235	Engine boost condition – limit not reached	**Hose connection(s), wiring, TC wastegate regulating valve, TC wastegate, TC**
P0236	Manifold absolute pressure (MAP) sensor A, TC system – range/performance problem	**Intake/exhaust leak, hose connection(s), MAP sensor**
P0237	Manifold absolute pressure (MAP) sensor A, TC system – low input	**Wiring short to ground, MAP sensor, ECM**
P0238	Manifold absolute pressure (MAP) sensor A, TC system – high input	**Wiring short to positive, MAP sensor, ECM**
P0239	Manifold absolute pressure (MAP) sensor B, TC system – circuit malfunction	**Wiring, MAP sensor, ECM**
P0240	Manifold absolute pressure (MAP) sensor B, TC system – range/performance problem	**Intake/exhaust leak, hose connection(s), MAP sensor**
P0241	Manifold absolute pressure (MAP) sensor B, TC system – low input	**Wiring short to ground, MAP sensor, ECM**
P0242	Manifold absolute pressure (MAP) sensor B, TC system – high input	**Wiring short to positive, MAP sensor, ECM**
P0243	Turbocharger (TC) wastegate regulating valve A – circuit malfunction	**Wiring, TC wastegate regulating valve, ECM**
P0244	Turbocharger (TC) wastegate regulating valve A – range/performance problem	**TC wastegate regulating valve**
P0245	Turbocharger (TC) wastegate regulating valve A – circuit low	**Wiring short to ground, TC wastegate regulating valve, ECM**
P0246	Turbocharger (TC) wastegate regulating valve A – circuit high	**Wiring short to positive, TC wastegate regulating valve, ECM**
P0247	Turbocharger (TC) wastegate regulating valve B – circuit malfunction	**Wiring, TC wastegate regulating valve, ECM**
P0248	Turbocharger (TC) wastegate regulating valve B – range/performance problem	**TC wastegate regulating valve**

OBD-II trouble code table

Trouble code	Fault location	Probable cause
P0249	Turbocharger (TC) wastegate regulating valve B – circuit low	**Wiring short to ground, TC wastegate regulating valve, ECM**
P0250	Turbocharger (TC) wastegate regulating valve B – circuit high	**Wiring short to positive, TC wastegate regulating valve, ECM**
P0251	Injection pump A, rotor/cam – circuit malfunction	**Wiring, injection pump, ECM**
P0252	Injection pump A, rotor/cam – range/performance problem	**Injection pump**
P0253	Injection pump A, rotor/cam – circuit low	**Wiring short to ground, injection pump, ECM**
P0254	Injection pump A, rotor/cam – circuit high	**Wiring short to positive, injection pump, ECM**
P0255	Injection pump A, rotor/cam – circuit intermittent	**Wiring, poor connection, injection pump, ECM**
P0256	Injection pump B, rotor/cam – circuit malfunction	**Wiring, injection pump, ECM**
P0257	Injection pump B, rotor/cam – range/performance problem	**Injection pump**
P0258	Injection pump B, rotor/cam – circuit low	**Wiring short to ground, injection pump, ECM**
P0259	Injection pump B, rotor/cam – circuit high	**Wiring short to positive, injection pump, ECM**
P0260	Injection pump B, rotor/cam – circuit intermittent	**Wiring, poor connection, injection pump, ECM**
P0261	Injector 1 – circuit low	**Wiring short to ground, injector, ECM**
P0262	Injector 1 – circuit high	**Wiring short to positive, injector, ECM**
P0263	Cylinder 1 – contribution/balance fault	**Wiring, fuel system, ECM**
P0264	Injector 2 – circuit low	**Wiring short to ground, injector, ECM**
P0265	Injector 2 – circuit high	**Wiring short to positive, injector, ECM**
P0266	Cylinder 2 – contribution/balance fault	**Wiring, fuel system, ECM**
P0267	Injector 3 – circuit low	**Wiring short to ground, injector, ECM**
P0268	Injector 3 – circuit high	**Wiring short to positive, injector, ECM**
P0269	Cylinder 3 – contribution/balance fault	**Wiring, fuel system, ECM**
P0270	Injector 4 – circuit low	**Wiring short to ground, injector, ECM**
P0271	Injector 4 – circuit high	**Wiring short to positive, injector, ECM**
P0272	Cylinder 4 – contribution/balance fault	**Wiring, fuel system, ECM**
P0273	Injector 5 – circuit low	**Wiring short to ground, injector, ECM**
P0274	Injector 5 – circuit high	**Wiring short to positive, injector, ECM**
P0275	Cylinder 5 – contribution/balance fault	**Wiring, fuel system, ECM**
P0276	Injector 6 – circuit low	**Wiring short to ground, injector, ECM**
P0277	Injector 6 – circuit high	**Wiring short to positive, injector, ECM**
P0278	Cylinder 6 – contribution/balance fault	**Wiring, fuel system, ECM**
P0279	Injector 7 – circuit low	**Wiring short to ground, injector, ECM**
P0280	Injector 7 – circuit high	**Wiring short to positive, injector, ECM**
P0281	Cylinder 7 – contribution/balance fault	**Wiring, fuel system, ECM**
P0282	Injector 8 – circuit low	**Wiring short to ground, injector, ECM**
P0283	Injector 8 – circuit high	**Wiring short to positive, injector, ECM**
P0284	Cylinder 8 – contribution/balance fault	**Wiring, fuel system, ECM**

Trouble code	Fault location	Probable cause
P0285	Injector 9 – circuit low	Wiring short to ground, injector, ECM
P0286	Injector 9 – circuit high	Wiring short to positive, injector, ECM
P0287	Cylinder 9 – contribution/balance fault	Wiring, fuel system, ECM
P0288	Injector 10 – circuit low	Wiring short to ground, injector, ECM
P0289	Injector 10 – circuit high	Wiring short to positive, injector, ECM
P0290	Cylinder 10 – contribution/balance fault	Wiring, fuel system, ECM
P0291	Injector 11 – circuit low	Wiring short to ground, injector, ECM
P0292	Injector 11 – circuit high	Wiring short to positive, injector, ECM
P0293	Cylinder 11 – contribution/balance fault	Wiring, fuel system, ECM
P0294	Injector 12 – circuit low	Wiring short to ground, injector, ECM
P0295	Injector 12 – circuit high	Wiring short to positive, injector, ECM
P0296	Cylinder 12 – contribution/balance fault	Wiring, fuel system, ECM
P0297	Vehicle over-speed condition	Wiring, VSS, mechanical fault
P0298	Engine oil temperature – too high	Wiring, EOT sensor, mechanical fault
P0299	Turbo/super charger – low boost	Mechanical fault
P0300	Random/multiple cylinder(s) – misfire detected	Spark plug(s), HT lead(s), injector(s), ignition coil(s), low compression, wiring
P0301	Cylinder 1 – misfire detected	Engine mechanical fault, wiring, ignition/fuel system, injector, ECT/MAF sensor, ECM
P0302	Cylinder 2 – misfire detected	Engine mechanical fault, wiring, ignition/fuel system, injector, ECT/MAF sensor, ECM
P0303	Cylinder 3 – misfire detected	Engine mechanical fault, wiring, ignition/fuel system, injector, ECT/MAF sensor, ECM
P0304	Cylinder 4 – misfire detected	Engine mechanical fault, wiring, ignition/fuel system, injector, ECT/MAF sensor, ECM
P0305	Cylinder 5 – misfire detected	Engine mechanical fault, wiring, ignition/fuel system, injector, ECT/MAF sensor, ECM
P0306	Cylinder 6 – misfire detected	Engine mechanical fault, wiring, ignition/fuel system, injector, ECT/MAF sensor, ECM
P0307	Cylinder 7 – misfire detected	Engine mechanical fault, wiring, ignition/fuel system, injector, ECT/MAF sensor, ECM
P0308	Cylinder 8 – misfire detected	Engine mechanical fault, wiring, ignition/fuel system, injector, ECT/MAF sensor, ECM
P0309	Cylinder 9 – misfire detected	Engine mechanical fault, wiring, ignition/fuel system, injector, ECT/MAF sensor, ECM
P0310	Cylinder 10 – misfire detected	Engine mechanical fault, wiring, ignition/fuel system, injector, ECT/MAF sensor, ECM
P0311	Cylinder 11 – misfire detected	Engine mechanical fault, wiring, ignition/fuel system, injector, ECT/MAF sensor, ECM
P0312	Cylinder 12 – misfire detected	Engine mechanical fault, wiring, ignition/fuel system, injector, ECT/MAF sensor, ECM
P0313	Misfire detected – low fuel level	Fuel system, mechanical fault
P0314	Single cylinder misfire – cylinder not specified	Engine mechanical fault, wiring, ignition/fuel system, injector
P0315	Crankshaft position system – variation not learned	Engine mechanical fault, wiring

OBD-II trouble code table

Trouble code	Fault location	Probable cause
P0316	Misfire detected during startup – first 1000 revolutions	**Engine mechanical fault, wiring, ignition/fuel system, injector**
P0317	Rough road hardware not present	**Wiring, ECM**
P0318	Rough road sensor signal A – circuit malfunction	**Wiring, rough road sensor A, mechanical fault**
P0319	Rough road sensor signal B – circuit malfunction	**Wiring, rough road sensor B, mechanical fault**
P0320	Crankshaft position (CKP) sensor/engine speed (RPM) sensor – circuit malfunction	**Wiring, CKP/RPM sensor, ECM**
P0321	Crankshaft position (CKP) sensor/engine speed (RPM) sensor – range/performance problem	**Air gap, metal particle contamination, insecure sensor/rotor, wiring, CKP/RPM sensor**
P0322	Crankshaft position (CKP) sensor/engine speed (RPM) sensor – no signal	**Wiring, CKP/RPM sensor, ECM**
P0323	Crankshaft position (CKP) sensor/engine speed (RPM) sensor – circuit intermittent	**Wiring, poor connection, CKP/RPM sensor, ECM**
P0324	Knock control system error	**Wiring, poor connection, KS, ECM**
P0325	Knock sensor (KS) 1, bank 1 – circuit malfunction	**Wiring, poor connection, KS**
P0326	Knock sensor (KS) 1, bank 1 – range/performance problem	**Wiring, KS incorrectly tightened, KS**
P0327	Knock sensor (KS) 1, bank 1 – low input	**Insecure KS, poor connection, wiring short to ground, incorrectly tightened, KS, ECM**
P0328	Knock sensor (KS) 1, bank 1 – high input	**Wiring short to positive, KS incorrectly tightened, KS, ECM**
P0329	Knock sensor (KS) 1, bank 1 – circuit intermittent	**Wiring, poor connection, KS, ECM**
P0330	Knock sensor (KS) 2, bank 2 – circuit malfunction	**Wiring, KS, ECM**
P0331	Knock sensor (KS) 2, bank 2 – range/performance problem	**Wiring, KS incorrectly tightened, KS**
P0332	Knock sensor (KS) 2, bank 2 – low input	**Insecure KS, poor connection, wiring short to ground, KS incorrectly tightened, KS, ECM**
P0333	Knock sensor (KS) 2, bank 2 – high input	**Wiring short to positive, KS incorrectly tightened, KS, ECM**
P0334	Knock sensor (KS) 2, bank 2 – circuit intermittent	**Wiring, poor connection, KS, ECM**
P0335	Crankshaft position (CKP) sensor – circuit malfunction	**Wiring, CKP sensor, ECM**
P0336	Crankshaft position (CKP) sensor – range/ performance problem	**Insecure sensor/rotor, air gap, wiring, CKP sensor**
P0337	Crankshaft position (CKP) sensor – low input	**Wiring short to ground, CKP sensor, ECM**
P0338	Crankshaft position (CKP) sensor – high input	**Wiring short to positive, CKP sensor, ECM**
P0339	Crankshaft position (CKP) sensor – circuit intermittent	**Wiring, poor connection, CKP sensor, ECM**
P0340	Camshaft position (CMP) sensor A, bank 1 – circuit malfunction	**Wiring, CMP sensor, ECM**
P0341	Camshaft position (CMP) sensor A, bank 1 – range/performance problem	**Insecure sensor/rotor, air gap, wiring, CMP sensor**
P0342	Camshaft position (CMP) sensor A, bank 1 – low input	**Wiring short to ground, CMP sensor, ECM**
P0343	Camshaft position (CMP) sensor A, bank 1 – high input	**Wiring short to positive, CMP sensor, ECM**
P0344	Camshaft position (CMP) sensor A, bank 1 – circuit intermittent	**Wiring, poor connection, CMP sensor, ECM**

Trouble code	Fault location	Probable cause
P0345	Camshaft position (CMP) sensor A, bank 2 – circuit malfunction	**Wiring, CMP sensor, ECM**
P0346	Camshaft position (CMP) sensor A, bank 2 – range/performance problem	**Insecure sensor/rotor, air gap, wiring, CMP sensor**
P0347	Camshaft position (CMP) sensor A, bank 2 – low input	**Wiring short to ground, CMP sensor, ECM**
P0348	Camshaft position (CMP) sensor A, bank 2 – high input	**Wiring short to positive, CMP sensor, ECM**
P0349	Camshaft position (CMP) sensor A, bank 2 – circuit intermittent	**Wiring, poor connection, CMP sensor, ECM**
P0350	Ignition coil, primary/secondary – circuit malfunction	**Wiring, ignition coil, ECM**
P0351	Ignition coil A, primary/secondary – circuit malfunction	**Wiring, ignition coil, ECM**
P0352	Ignition coil B, primary/secondary – circuit malfunction	**Wiring, ignition coil, ECM**
P0353	Ignition coil C, primary/secondary – circuit malfunction	**Wiring, ignition coil, ECM**
P0354	Ignition coil D, primary/secondary – circuit malfunction	**Wiring, ignition coil, ECM**
P0355	Ignition coil E, primary/secondary – circuit malfunction	**Wiring, ignition coil, ECM**
P0356	Ignition coil F, primary/secondary – circuit malfunction	**Wiring, ignition coil, ECM**
P0357	Ignition coil G, primary/secondary – circuit malfunction	**Wiring, ignition coil, ECM**
P0358	Ignition coil H, primary/secondary – circuit malfunction	**Wiring, ignition coil, ECM**
P0359	Ignition coil I, primary/secondary – circuit malfunction	**Wiring, ignition coil, ECM**
P0360	Ignition coil J, primary/secondary – circuit malfunction	**Wiring, ignition coil, ECM**
P0361	Ignition coil K, primary/secondary – circuit malfunction	**Wiring, ignition coil, ECM**
P0362	Ignition coil L, primary/secondary – circuit malfunction	**Wiring, ignition coil, ECM**
P0363	Misfire detected – fuelling disabled	**Fuel system, mechanical fault**
P0365	Camshaft position (CMP) sensor B, bank 1 – circuit malfunction	**Wiring, poor connection, CMP sensor, ECM**
P0366	Camshaft position (CMP) sensor B, bank 1 – circuit range/performance	**Wiring, poor connection, CMP sensor**
P0367	Camshaft position (CMP) sensor B, bank 1 – circuit low input	**Wiring short to ground, CMP sensor, ECM**
P0368	Camshaft position (CMP) sensor B, bank 1 – circuit high input	**Wiring short to positive, CMP sensor, ECM**
P0369	Camshaft position (CMP) sensor B, bank 1 – circuit intermittent	**Wiring, poor connection, ECM**
P0370	Timing reference, high resolution signal A – malfunction	**Wiring, CKP/RPM/CMP sensor, ECM**
P0371	Timing reference, high resolution signal A – too many pulses	**Wiring, CKP/RPM/CMP sensor, ECM**

OBD-II trouble code table

Trouble code	Fault location	Probable cause
P0372	Timing reference, high resolution signal A – too few pulses	**Wiring, CKP/RPM/CMP sensor, ECM**
P0373	Timing reference, high resolution signal A – intermittent erratic pulses	**Wiring, poor connection, CKP/RPM/CMP sensor, ECM**
P0374	Timing reference, high resolution signal A – no pulses	**Wiring, CKP/RPM/CMP sensor, ECM**
P0375	Timing reference, high resolution signal B – malfunction	**Wiring, CKP/RPM/CMP sensor, ECM**
P0376	Timing reference, high resolution signal B – too many pulses	**Wiring, CKP/RPM/CMP sensor, ECM**
P0377	Timing reference, high resolution signal B – too few pulses	**Wiring, CKP/RPM/CMP sensor, ECM**
P0378	Timing reference, high resolution signal B – intermittent erratic pulses	**Wiring, poor connection, CKP/RPM/CMP sensor, ECM**
P0379	Timing reference, high resolution signal B – no pulses	**Wiring, CKP/RPM/CMP sensor, ECM**
P0380	Glow plugs, circuit A – malfunction	**Wiring, glow plug relay, fuse, glow plugs, ECM**
P0381	Glow plug warning lamp – circuit malfunction	**Wiring, glow plug warning lamp, ECM**
P0382	Glow plugs, circuit B – malfunction	**Wiring, glow plug relay, glow plugs, ECM**
P0385	Crankshaft position (CKP) sensor B – circuit malfunction	**Wiring, CKP sensor, ECM**
P0386	Crankshaft position (CKP) sensor B – range/ performance problem	**Insecure sensor/rotor, air gap, wiring, CKP sensor**
P0387	Crankshaft position (CKP) sensor B – low input	**Wiring short to ground, CKP sensor, ECM**
P0388	Crankshaft position (CKP) sensor B – high input	**Wiring short to positive, CKP sensor, ECM**
P0389	Crankshaft position (CKP) sensor B – circuit intermittent	**Wiring, poor connection, CKP sensor, ECM**
P0390	Camshaft position (CMP) sensor B, bank 2 – circuit malfunction	**Wiring, poor connection, CMP sensor, ECM**
P0391	Camshaft position (CMP) sensor B, bank 2 – circuit range/performance	**Wiring, poor connection, CMP sensor**
P0392	Camshaft position (CMP) sensor B, bank 2 – circuit low input	**Wiring short to ground, CMP sensor, ECM**
P0393	Camshaft position (CMP) sensor B, bank 2 – circuit high input	**Wiring short to positive, CMP sensor, ECM**
P0394	Camshaft position (CMP) sensor B, bank 2 – circuit intermittent	**Wiring, poor connection, ECM**
P0400	Exhaust gas recirculation (EGR) system – flow malfunction	**Hose leak/blockage, basic setting not carried out (if applicable), wiring, EGR valve, EGR solenoid, ECM**
P0401	Exhaust gas recirculation (EGR) system – insufficient flow detected	**Hose leak/blockage, basic setting not carried out (if applicable), wiring, EGR valve, EGR solenoid, ECM**
P0402	Exhaust gas recirculation (EGR) system – excessive flow detected	**Hose leak/blockage, basic setting not carried out (if applicable), wiring, EGR valve, EGR solenoid, ECM**
P0403	Exhaust gas recirculation (EGR) – circuit malfunction	**Wiring, EGR solenoid, ECM**
P0404	Exhaust gas recirculation (EGR) system – range/ performance problem	**Hose leak/blockage, wiring, EGR valve/solenoid**

Trouble code	Fault location	Probable cause
P0405	Exhaust gas recirculation (EGR) valve position sensor A – low input	Wiring short to ground, EGR valve position sensor, ECM
P0406	Exhaust gas recirculation (EGR) valve position sensor A – high input	Wiring short to positive, EGR valve position sensor, ECM
P0407	Exhaust gas recirculation (EGR) valve position sensor B – low input	Wiring short to ground, EGR valve position sensor, ECM
P0408	Exhaust gas recirculation (EGR) valve position sensor B – high input	Wiring short to positive, EGR valve position sensor, ECM
P0409	Exhaust gas recirculation (EGR) sensor A – circuit malfunction	Wiring, poor connection, EGR sensor, ECM
P0410	Secondary air injection (AIR) system – malfunction	Wiring, AIR valve, AIR solenoid, ECM
P0411	Secondary air injection (AIR) system – incorrect flow detected	AIR pump, AIR valve, AIR hose(s)
P0412	Secondary air injection (AIR) solenoid A – circuit malfunction	Wiring, AIR solenoid, ECM
P0413	Secondary air injection (AIR) solenoid A – open circuit	Wiring open circuit, AIR solenoid, ECM
P0414	Secondary air injection (AIR) solenoid A – short circuit	Wiring short circuit, AIR solenoid, ECM
P0415	Secondary air injection (AIR) solenoid B – circuit malfunction	Wiring, AIR solenoid, ECM
P0416	Secondary air injection (AIR) solenoid B – open circuit	Wiring open circuit, AIR solenoid, ECM
P0417	Secondary air injection (AIR) solenoid B – short circuit	Wiring short circuit, AIR solenoid, ECM
P0418	Secondary air injection (AIR) pump relay A – circuit malfunction	Wiring, AIR pump relay, ECM
P0419	Secondary air injection (AIR) pump relay B – circuit malfunction	Wiring, AIR pump relay, ECM
P0420	Catalytic converter system, bank 1 – efficiency below threshold	Catalytic converter, wiring, HO2S 2
P0421	Warm up catalytic converter, bank 1 – efficiency below threshold	Catalytic converter, wiring, HO2S 2
P0422	Main catalytic converter, bank 1 – efficiency below threshold	Catalytic converter, wiring, HO2S 2
P0423	Heated catalytic converter, bank 1 – efficiency below threshold	Catalytic converter, wiring, HO2S 2
P0424	Heated catalytic converter, bank 1 – temperature below threshold	Catalytic converter, wiring, HO2S 2
P0425	Catalytic converter temperature sensor, bank 1	Wiring, poor connection, catalytic converter temperature sensor, ECM
P0426	Catalytic converter temperature sensor, bank 1 – range/performance	Wiring, poor connection, catalytic converter temperature sensor
P0427	Catalytic converter temperature sensor, bank 1 – low input	Wiring short to ground, catalytic converter temperature sensor, ECM
P0428	Catalytic converter temperature sensor, bank 1 – high input	Wiring short to positive, catalytic converter temperature sensor, ECM
P0429	Catalytic converter heater, bank 1 – control circuit malfunction	Wiring, relay, ECM
P0430	Catalytic converter system, bank 2 – efficiency below threshold	Catalytic converter, wiring, HO2S 2

OBD-II trouble code table

Trouble code	Fault location	Probable cause
P0431	Warm up catalytic converter, bank 2 – efficiency below threshold	**Catalytic converter, wiring, HO2S 2**
P0432	Main catalytic converter, bank 2 – efficiency below threshold	**Catalytic converter, wiring, HO2S 2**
P0433	Heated catalytic converter, bank 2 – efficiency below threshold	**Catalytic converter, wiring, HO2S 2**
P0434	Heated catalytic converter, bank 2 – temperature below threshold	**Catalytic converter, wiring, HO2S 2**
P0435	Catalytic converter temperature sensor, bank 2	**Wiring, poor connection, catalytic converter temperature sensor, ECM**
P0436	Catalytic converter temperature sensor, bank 2 – range/performance	**Wiring, poor connection, catalytic converter temperature sensor**
P0437	Catalytic converter temperature sensor, bank 2 – low input	**Wiring short to ground, catalytic converter temperature sensor, ECM**
P0438	Catalytic converter temperature sensor, bank 2 – high input	**Wiring short to positive, catalytic converter temperature sensor, ECM**
P0439	Catalytic converter heater, bank 2 – control circuit malfunction	**Wiring, relay, ECM**
P0440	Evaporative emission (EVAP) system – malfunction	**Hose connection(s), intake leak, EVAP canister purge valve**
P0441	Evaporative emission (EVAP) system – incorrect flow detected	**Hose connection(s), intake leak, EVAP canister purge valve**
P0442	Evaporative emission (EVAP) system – small leak detected	**Hose connection(s), intake leak, EVAP canister, EVAP canister purge valve**
P0443	Evaporative emission (EVAP) canister purge valve – circuit malfunction	**Wiring, EVAP canister purge valve, ECM**
P0444	Evaporative emission (EVAP) canister purge valve – open circuit	**Wiring open circuit, EVAP canister purge valve, ECM**
P0445	Evaporative emission (EVAP) canister purge valve – short circuit	**Wiring short circuit, EVAP canister purge valve, ECM**
P0446	Evaporative emission (EVAP) system, vent control – circuit malfunction	**Wiring, EVAP canister purge valve, ECM**
P0447	Evaporative emission (EVAP) system, vent control – open circuit	**Wiring open circuit, EVAP canister purge valve, ECM**
P0448	Evaporative emission (EVAP) system, vent control – short circuit	**Wiring short circuit, EVAP canister purge valve, ECM**
P0449	Evaporative emission (EVAP) system, vent valve – circuit malfunction	**Wiring, EVAP canister purge valve, ECM**
P0450	Evaporative emission (EVAP) pressure sensor – circuit malfunction	**Wiring, EVAP pressure sensor, ECM**
P0451	Evaporative emission (EVAP) pressure sensor – range/performance problem	**EVAP pressure sensor**
P0452	Evaporative emission (EVAP) pressure sensor – low input	**Wiring short to ground, EVAP pressure sensor, ECM**
P0453	Evaporative emission (EVAP) pressure sensor – high input	**Wiring short to positive, EVAP pressure sensor, ECM**
P0454	Evaporative emission (EVAP) pressure sensor – circuit intermittent	**Wiring, poor connection, EVAP pressure sensor, ECM**
P0455	Evaporative emission (EVAP) system – large leak detected	**Hose connection(s), intake leak, EVAP canister, EVAP canister purge valve**

Trouble code	Fault location	Probable cause
P0456	Evaporative emission system – very small leak detected	**Mechanical fault, hose connection(s), EVAP pressure sensor**
P0457	Evaporative emission system – leak detected (fuel cap loose/off)	**Mechanical fault, hose connection(s), EVAP pressure sensor**
P0458	Evaporative emission system, purge control valve – circuit low	**Wiring short to ground, EVAP valve**
P0459	Evaporative emission system, purge control valve – circuit high	**Wiring short to positive, EVAP valve**
P0460	Fuel tank level sensor – circuit malfunction	**Wiring, fuel tank level sensor, ECM**
P0461	Fuel tank level sensor – range/performance problem	**Wiring, fuel tank level sensor**
P0462	Fuel tank level sensor – low input	**Wiring short to ground, fuel tank level sensor, ECM**
P0463	Fuel tank level sensor – high input	**Wiring short to positive, fuel tank level sensor, ECM**
P0464	Fuel tank level sensor – circuit intermittent	**Wiring, poor connection, fuel tank level sensor, ECM**
P0465	Evaporative emission (EVAP) canister purge flow sensor – circuit malfunction	**Wiring, EVAP canister purge flow sensor, ECM**
P0466	Evaporative emission (EVAP) canister purge flow sensor – range/performance problem	**EVAP canister purge flow sensor**
P0467	Evaporative emission (EVAP) canister purge flow sensor – low input	**Wiring short to ground, EVAP canister purge flow sensor, ECM**
P0468	Evaporative emission (EVAP) canister purge flow sensor – high input	**Wiring short to positive, EVAP canister purge flow sensor, ECM**
P0469	Evaporative emission (EVAP) canister purge flow sensor – circuit intermittent	**Wiring, poor connection, EVAP canister purge flow sensor, ECM**
P0470	Exhaust gas pressure sensor – circuit malfunction	**Wiring, exhaust gas pressure sensor, ECM**
P0471	Exhaust gas pressure sensor – range/performance problem	**Exhaust gas pressure sensor**
P0472	Exhaust gas pressure sensor – low input	**Wiring short to ground, exhaust gas pressure sensor, ECM**
P0473	Exhaust gas pressure sensor – high input	**Wiring short to positive, exhaust gas pressure sensor, ECM**
P0474	Exhaust gas pressure sensor – circuit intermittent	**Wiring, poor connection, exhaust gas pressure sensor, ECM**
P0475	Exhaust gas pressure control valve – circuit malfunction	**Wiring, exhaust gas pressure control valve, ECM**
P0476	Exhaust gas pressure control valve – range/ performance problem	**Exhaust gas pressure control valve**
P0477	Exhaust gas pressure control valve – low input	**Wiring short to ground, exhaust gas pressure control valve, ECM**
P0478	Exhaust gas pressure control valve – high input	**Wiring short to positive, exhaust gas pressure control valve, ECM**
P0479	Exhaust gas pressure control valve – circuit intermittent	**Wiring, poor connection, exhaust gas pressure control valve, ECM**
P0480	Engine coolant blower motor 1 – circuit malfunction	**Wiring, engine coolant blower motor, ECM**
P0481	Engine coolant blower motor 2 – circuit malfunction	**Wiring, engine coolant blower motor, ECM**
P0482	Engine coolant blower motor 3 – circuit malfunction	**Wiring, engine coolant blower motor, ECM**

OBD-II trouble code table

Trouble code	Fault location	Probable cause
P0483	Engine coolant blower motor, rationality check – malfunction	**Wiring, engine coolant blower motor, ECM**
P0484	Engine coolant blower motor – circuit over current	**Wiring, engine coolant blower motor, ECM**
P0485	Engine coolant blower motor, power/earth – circuit malfunction	**Wiring, engine coolant blower motor, ECM**
P0486	Exhaust gas recirculation (EGR) valve position sensor B – circuit malfunction	**Wiring, poor connection, EGR valve position sensor, ECM**
P0487	Exhaust gas recirculation (EGR) system, throttle position control – circuit malfunction	**Wiring, poor connection, ECM**
P0488	Exhaust gas recirculation (EGR) system, throttle position control – range/performance	**Wiring, poor connection, ECM**
P0489	Exhaust gas recirculation (EGR) system – circuit low	**Wiring short to ground, EGR valve**
P0490	Exhaust gas recirculation (EGR) system – circuit high	**Wiring short to positive, EGR valve**
P0491	Secondary air injection system, bank 1 – malfunction	**Wiring, AIR solenoid, hose connections, mechanical fault**
P0492	Secondary air injection system, bank 2 – malfunction	**Wiring, AIR solenoid, hose connections, mechanical fault**
P0493	Fan over-speed (clutch locked)	**Fan clutch, mechanical fault**
P0494	Fan speed – low	**Wiring, relay, fan motor, mechanical fault**
P0495	Fan speed – high	**Wiring, relay, fan motor, mechanical fault**
P0496	Evaporative emission system – high purge flow	**Wiring, EVAP valve, mechanical fault**
P0497	Evaporative emission system – low purge flow	**Wiring, EVAP valve, hoses blocked, mechanical fault**
P0498	Evaporative emission system, vent control – circuit low	**Wiring short to ground, EVAP valve,**
P0499	Evaporative emission system, vent control – circuit high	**Wiring short to positive, EVAP valve**
P0500	Vehicle speed sensor (VSS) – circuit malfunction	**Wiring, VSS, ECM**
P0501	Vehicle speed sensor (VSS) – range/performance problem	**Wiring, speedometer, VSS, CAN data bus**
P0502	Vehicle speed sensor (VSS) – low input	**Wiring short to ground, VSS, ECM**
P0503	Vehicle speed sensor (VSS) – intermittent/erratic/ high input	**Wiring, poor connection, other connected system, instrument panel, VSS**
P0504	Brake switch – A/B correlation	**Wiring, mechanical fault**
P0505	Idle speed control (ISC) system – malfunction	**Wiring, ISC actuator/IAC valve, throttle motor, throttle valve tight/sticking, ECM**
P0506	Idle speed control (ISC) system – rpm lower than expected	**Wiring, ISC actuator/IAC valve, throttle motor, throttle valve tight/sticking, ECM**
P0507	Idle speed control (ISC) system – rpm higher than expected	**Wiring, ISC actuator/IAC valve, throttle motor, throttle valve tight/sticking, ECM**
P0508	Idle air control (IAC) system – circuit low	**Wiring short to ground, IAC valve, ECM**
P0509	Idle air control (IAC) system – circuit high	**Wiring short to positive, IAC valve, ECM**
P0510	Closed throttle position (CTP) switch – circuit malfunction	**Wiring, CTP switch, ECM**
P0511	Idle air control (IAC) system – circuit malfunction	**Wiring, poor connection, IAC valve, ECM**

Trouble code	Fault location	Probable cause
P0512	Starter request circuit – malfunction	Wiring, immobilizer system, relay
P0513	Incorrect immobilizer key	Immobilizer system
P0514	Battery temperature sensor – circuit range/ performance	Wiring, poor connection, battery temperature sensor
P0515	Battery temperature sensor – circuit malfunction	Wiring, poor connection, battery temperature sensor
P0516	Battery temperature sensor – circuit low	Wiring short to ground, battery temperature sensor, ECM
P0517	Battery temperature sensor – circuit high	Wiring short to positive, battery temperature sensor, ECM
P0518	Idle air control (IAC) system – circuit intermittent	Wiring, poor connection, IAC valve, ECM
P0519	Idle air control (IAC) system – circuit performance	Wiring, poor connection, IAC valve, ECM
P0520	Engine oil pressure sensor/switch – circuit malfunction	Wiring, engine oil pressure sensor/switch, ECM
P0521	Engine oil pressure sensor/switch – range/ performance problem	Engine oil pressure sensor/switch
P0522	Engine oil pressure sensor/switch – low voltage	Wiring short to ground, engine oil pressure sensor/switch, ECM
P0523	Engine oil pressure sensor/switch – high voltage	Wiring short to positive, engine oil pressure sensor/switch, ECM
P0524	Engine oil pressure too low	Mechanical fault
P0525	Cruise control, servo control – circuit range/ performance	Wiring, poor connection, cruise control servo
P0526	Fan speed sensor – circuit malfunction	Wiring, poor connection, fan speed sensor, ECM
P0527	Fan speed sensor – circuit range/performance	Wiring, poor connection, fan speed sensor
P0528	Fan speed sensor – no signal	Wiring, poor connection, fan speed sensor, ECM
P0529	Fan speed sensor – circuit intermittent	Wiring, poor connection, ECM
P0530	AC refrigerant pressure sensor – circuit malfunction	Wiring, AC refrigerant pressure sensor, ECM
P0531	AC refrigerant pressure sensor – range/ performance problem	AC refrigerant pressure sensor
P0532	AC refrigerant pressure sensor – low input	AC refrigerant pressure too low (incorrectly charged), wiring, AC refrigerant pressure sensor, ECM
P0533	AC refrigerant pressure sensor – high input	AC refrigerant pressure too high (cooling fault/ incorrectly charged), wiring, AC refrigerant pressure sensor, ECM
P0534	AC refrigerant charge loss	AC leak, wiring, AC refrigerant pressure sensor
P0535	A/C evaporator temperature sensor – circuit malfunction	Wiring, poor connection, A/C evaporator temperature sensor, ECM
P0536	A/C evaporator temperature sensor – circuit range/ performance	Wiring, poor connection, A/C evaporator temperature sensor, ECM
P0537	A/C evaporator temperature sensor – circuit low	Wiring short to ground, A/C evaporator temperature sensor, ECM
P0538	A/C evaporator temperature sensor – circuit high	Wiring short to positive, A/C evaporator temperature sensor, ECM

OBD-II trouble code table

Trouble code	Fault location	Probable cause
P0539	A/C evaporator temperature sensor – circuit intermittent	**Wiring, poor connection, A/C evaporator temperature sensor, ECM**
P0540	Intake air heater A – circuit malfunction	**Wiring, relay, intake air heater**
P0541	Intake air heater A – circuit low	**Wiring short to ground, intake air heater**
P0542	Intake air heater A – circuit high	**Wiring short to positive, intake air heater**
P0543	Intake air heater A – circuit open	**Wiring, intake air heater**
P0544	Exhaust gas recirculation temperature (EGRT) sensor, bank 1 – circuit malfunction	**Wiring, EGRT sensor, ECM**
P0545	Exhaust gas recirculation temperature (EGRT) sensor, bank 1 – low input	**Wiring short to ground, EGRT sensor, ECM**
P0546	Exhaust gas recirculation temperature (EGRT) sensor, bank 1 – high input	**Wiring short to positive, EGRT sensor, ECM**
P0547	Exhaust gas temperature sensor, bank 2 sensor 1 – circuit malfunction	**Wiring, poor connection, exhaust gas temperature sensor, ECM**
P0548	Exhaust gas temperature sensor, bank 2 sensor 1 – circuit low	**Wiring short to ground, exhaust gas temperature sensor, ECM**
P0549	Exhaust gas temperature sensor, bank 2 sensor 1 – circuit high	**Wiring short to positive, exhaust gas temperature sensor, ECM**
P0550	Power steering pressure (PSP) sensor/switch – circuit malfunction	**Wiring, PSP sensor/switch, ECM**
P0551	Power steering pressure (PSP) sensor/switch – range/performance problem	**PAS system, PSP sensor/switch**
P0552	Power steering pressure (PSP) sensor/switch – low input	**Wiring short to ground, PSP sensor/switch, ECM**
P0553	Power steering pressure (PSP) sensor/switch – high input	**Wiring short to positive, PSP sensor/switch, ECM**
P0554	Power steering pressure (PSP) sensor/switch – circuit intermittent	**Wiring, poor connection, PSP sensor/switch, ECM**
P0555	Brake booster pressure sensor – circuit malfunction	**Wiring, poor connection, brake booster pressure sensor, ECM**
P0556	Brake booster pressure sensor – circuit range/ performance	**Wiring, poor connection, brake booster pressure sensor, ECM**
P0557	Brake booster pressure sensor – circuit low input	**Wiring short to ground, brake booster pressure sensor, ECM**
P0558	Brake booster pressure sensor – circuit high input	**Wiring short to positive, brake booster pressure sensor, ECM**
P0559	Brake booster pressure sensor – circuit intermittent	**Wiring, poor connection, brake booster pressure sensor, ECM**
P0560	System voltage – malfunction	**Wiring, poor connection, battery, alternator**
P0561	System voltage – unstable	**Wiring, poor connection, battery, alternator**
P0562	System voltage – low	**Wiring, poor connection, battery, alternator**
P0563	System voltage – high	**Alternator**
P0564	Cruise control system, multi-function input A – circuit malfunction	**Wiring, poor connection, multi-function switch, mechanical fault**
P0565	Cruise control master switch, ON signal – malfunction	**Wiring, cruise control master switch, ECM**
P0566	Cruise control master switch, OFF signal – malfunction	**Wiring, cruise control master switch, ECM**

Trouble code	Fault location	Probable cause
P0567	Cruise control selector switch, RESUME signal – malfunction	**Wiring, cruise control selector switch, ECM**
P0568	Cruise control master switch, SET signal – malfunction	**Wiring, cruise control master switch, ECM**
P0569	Cruise control selector switch, COAST signal – malfunction	**Wiring, cruise control selector switch, ECM**
P0570	Cruise control system, APP sensor signal – malfunction	**Wiring, APP sensor, ECM**
P0571	Cruise/brake switch A – circuit malfunction	**Wiring, cruise/brake switch, ECM**
P0572	Cruise/brake switch A – circuit low	**Wiring short to ground, cruise/brake switch, ECM**
P0573	Cruise/brake switch A – circuit high	**Wiring short to positive, cruise/brake switch, ECM**
P0574	Cruise control system – vehicle speed too high	**Mechanical fault**
P0575	Cruise control system – input circuit malfunction	**Wiring, poor connection, mechanical fault, ECM**
P0576	Cruise control system – input circuit low	**Wiring short to ground**
P0577	Cruise control system – input circuit high	**Wiring short to positive**
P0578	Cruise control system, multi-function input A – circuit stuck	**Wiring, poor connection, multi-function switch, mechanical fault**
P0579	Cruise control system, multi-function input A – circuit range/performance	**Wiring, poor connection, multi-function switch, mechanical fault**
P0580	Cruise control system, multi-function input A – circuit low	**Wiring short to ground, multi-function switch, mechanical fault**
P0581	Cruise control system, multi-function input A – circuit high	**Wiring short to positive, multi-function switch, mechanical fault**
P0582	Cruise control system, vacuum control – circuit open	**Wiring, vacuum control solenoid**
P0583	Cruise control system, vacuum control – circuit low	**Wiring short to ground, vacuum control solenoid**
P0584	Cruise control system, vacuum control – circuit high	**Wiring short to positive, vacuum control solenoid**
P0585	Cruise control system, multi-function input A/B – correlation	**Mechanical fault**
P0586	Cruise control system, vent control – circuit open	**Wiring, vent control solenoid**
P0587	Cruise control system, vent control – circuit low	**Wiring short to ground, vent control solenoid**
P0588	Cruise control system, vent control – circuit high	**Wiring short to positive, vent control solenoid**
P0589	Cruise control system, multi-function input B – circuit malfunction	**Wiring, poor connection, multi-function switch, mechanical fault**
P0590	Cruise control system, multi-function input B – circuit stuck	**Wiring, poor connection, multi-function switch, mechanical fault**
P0591	Cruise control system, multi-function input B – circuit range/performance	**Wiring, poor connection, multi-function switch, mechanical fault**
P0592	Cruise control system, multi-function input B – circuit low	**Wiring short to ground, multi-function switch, mechanical fault**
P0593	Cruise control system, multi-function input B – circuit high	**Wiring short to positive, multi-function switch, mechanical fault**
P0594	Cruise control system, servo control – circuit open	**Wiring, servo control solenoid**
P0595	Cruise control system, servo control – circuit low	**Wiring short to ground, servo control solenoid**

OBD-II trouble code table

Trouble code	Fault location	Probable cause
P0596	Cruise control system, servo control – circuit high	**Wiring short to positive, servo control solenoid**
P0597	Thermostat heater control system – circuit open	**Wiring, relay, thermostat heater**
P0598	Thermostat heater control system – circuit low	**Wiring short to ground, relay, thermostat heater**
P0599	Thermostat heater control system – circuit high	**Wiring short to positive, relay, thermostat heater**
P0600	CAN data bus – malfunction	**Wiring, connected system, ECM**
P0601	Engine control module (ECM) – memory check sum error	**ECM**
P0602	Engine control module (ECM) – programming error	**ECM**
P0603	Engine control module (ECM) – KAM error	**ECM**
P0604	Engine control module (ECM) – RAM error	**ECM**
P0605	Engine control module (ECM) – ROM error	**ECM**
P0606	Engine control module (ECM) – processor fault	**ECM**
P0607	Control module – performance	**Control module**
P0608	Engine control module (ECM), VSS output A – malfunction	**ECM**
P0609	Engine control module (ECM), VSS output B – malfunction	**ECM**
P0610	Control module – vehicle options error	**Control module**
P0611	Fuel injector control module – performance	**Fuel injector control module**
P0612	Fuel injector control module – relay control circuit	**Wiring, relay, fuel injector control module**
P0613	Transmission control module (TCM) processor error	**TCM**
P0614	Engine control module (ECM)/Transmission control module (TCM) – mismatch	**ECM/TCM**
P0615	Starter relay – circuit malfunction	**Wiring, poor connection, starter relay, ECM**
P0616	Starter relay – circuit low	**Wiring short to ground, starter relay, ECM**
P0617	Starter relay – circuit high	**Wiring short to positive, starter relay, ECM**
P0618	Alternative fuel control module – KAM error	**Alternative fuel control module**
P0619	Alternative fuel control module – RAM/ROM error	**Alternative fuel control module**
P0620	Alternator, control – circuit malfunction	**Wiring, alternator, battery, ECM**
P0621	Alternator warning lamp – circuit malfunction	**Wiring, alternator warning lamp, ECM**
P0622	Alternator, field control – circuit malfunction	**Wiring, alternator, battery, ECM**
P0623	Generator lamp control – circuit malfunction	**Wiring, poor connection, bulb, ECM**
P0624	Fuel cap lamp control – circuit malfunction	**Wiring, poor connection, bulb, ECM**
P0625	Generator field terminal – circuit low	**Wiring short to ground, generator**
P0626	Generator field terminal – circuit high	**Wiring short to positive, generator**
P0627	Fuel pump control – circuit open	**Wiring, relay, fuel pump**
P0628	Fuel pump control – circuit low	**Wiring short to ground, relay, fuel pump**
P0629	Fuel pump control – circuit high	**Wiring short to positive, relay, fuel pump**
P0630	VIN not programmed or mismatch – ECM	**ECM**

Trouble code	Fault location	Probable cause
P0631	VIN not programmed or mismatch – TCM	TCM
P0632	Odometer not programmed – ECM	ECM
P0633	Immobilizer key not programmed – ECM	ECM
P0634	ECM/TCM – internal temperature too high	Mechanical fault, ECM/TCM
P0635	Power steering (PAS) control – circuit malfunction	Wiring, poor connection, PAS pressure switch, ECM
P0636	Power steering (PAS) control – circuit low	Wiring short to ground, PAS pressure switch, ECM
P0637	Power Steering (PAS) control – circuit high	Wiring short to positive, PAS pressure switch, ECM
P0638	Throttle actuator control, bank 1 – range/ performance problem	Basic setting not carried out (if applicable), ISC actuator/throttle motor, APP sensor
P0639	Throttle actuator control, bank 2 – range/ performance	Wiring, throttle control unit
P0640	Intake air heater control – circuit malfunction	Wiring, relay, intake air heater
P0641	Sensor reference voltage A – circuit open	Wiring short to positive
P0642	Engine control module (ECM), knock control – defective	ECM
P0643	Sensor reference voltage A – circuit high	Wiring short to positive
P0644	Driver display, serial communication – circuit malfunction	Wiring, CAN data bus, ECM
P0645	Air conditioning (AC)	Wiring, AC system
P0646	A/C clutch relay control – circuit low	Wiring short to ground, A/C clutch relay
P0647	A/C clutch relay control – circuit high	Wiring short to positive, A/C clutch relay
P0648	Immobilizer lamp control – circuit malfunction	Wiring, poor connection, bulb, ECM
P0649	Cruise control lamp control – circuit	Wiring, poor connection, bulb, ECM
P0650	Malfunction indicator lamp (MIL) – circuit malfunction	Wiring, MIL, ECM
P0651	Sensor reference voltage B – circuit open	Wiring short to positive
P0652	Sensor reference voltage B – circuit low	Wiring short to ground
P0653	Sensor reference voltage B – circuit high	Wiring short to positive
P0654	Engine rpm, output – circuit malfunction	Wiring, ECM
P0655	Engine hot lamp output – circuit malfunction	Wiring, engine hot lamp, ECM
P0656	Fuel level output – circuit malfunction	Wiring, ECM
P0657	Actuator supply voltage – circuit open	Wiring
P0658	Actuator supply voltage – circuit low	Wiring short to ground, actuator
P0659	Actuator supply voltage – circuit high	Wiring short to positive, actuator
P0660	Intake manifold tuning valve, bank 1 – circuit open	Wiring, intake manifold tuning valve
P0661	Intake manifold tuning valve, bank 1 – circuit low	Wiring short to ground, intake manifold tuning valve
P0662	Intake manifold tuning valve, bank 1 – circuit high	Wiring short to positive, intake manifold tuning valve
P0663	Intake manifold tuning valve, bank 2 – circuit open	Wiring, intake manifold tuning valve

OBD-II trouble code table

Trouble code	Fault location	Probable cause
P0664	Intake manifold tuning valve, bank 2 – circuit low	**Wiring short to ground, intake manifold tuning valve**
P0665	Intake manifold tuning valve, bank 2 – circuit high	**Wiring short to positive, intake manifold**
P0666	ECM/TCM internal temperature sensor – circuit malfunction	**Wiring, poor connection, internal temperature sensor, ECM/TCM**
P0667	ECM/TCM internal temperature sensor – range/ performance	**Wiring, poor connection, internal temperature sensor, ECM/TCM**
P0668	ECM/TCM internal temperature sensor – circuit low	**Wiring short to ground, internal temperature sensor, ECM/TCM**
P0669	ECM/TCM internal temperature sensor – circuit high	**Wiring short to positive, internal temperature sensor, ECM/TCM**
P0670	Glow plug module control – circuit malfunction	**Wiring, poor connection, control module, glow plug, ECM**
P0671	Glow plug, cylinder 1 – circuit malfunction	**Wiring, poor connection, relay, control module, glow plug, ECM**
P0672	Glow plug, cylinder 2 – circuit malfunction	**Wiring, poor connection, relay, control module, glow plug, ECM**
P0673	Glow plug, cylinder 3 – circuit malfunction	**Wiring, poor connection, relay, control module, glow plug, ECM**
P0674	Glow plug, cylinder 4 – circuit malfunction	**Wiring, poor connection, relay, control module, glow plug, ECM**
P0675	Glow plug, cylinder 5 – circuit malfunction	**Wiring, poor connection, relay, control module, glow plug, ECM**
P0676	Glow plug, cylinder 6 – circuit malfunction	**Wiring, poor connection, relay, control module, glow plug, ECM**
P0677	Glow plug, cylinder 7 – circuit malfunction	**Wiring, poor connection, relay, control module, glow plug, ECM**
P0678	Glow plug, cylinder 8 – circuit malfunction	**Wiring, poor connection, relay, control module, glow plug, ECM**
P0679	Glow plug, cylinder 9 – circuit malfunction	**Wiring, poor connection, relay, control module, glow plug, ECM**
P0680	Glow plug, cylinder 10 – circuit malfunction	**Wiring, poor connection, relay, control module, glow plug, ECM**
P0681	Glow plug, cylinder 11 – circuit malfunction	**Wiring, poor connection, relay, control module, glow plug, ECM**
P0682	Glow plug, cylinder 12 – circuit malfunction	**Wiring, poor connection, relay, control module, glow plug, ECM**
P0683	Glow plug control module communication to ECM – malfunction	**Wiring, poor connection, control module, ECM**
P0684	Glow plug control module communication to ECM – range/performance	**Wiring, poor connection, control module, ECM**
P0685	ECM power relay control – circuit open	**Wiring, ECM relay**
P0686	ECM power relay control – circuit low	**Wiring short to ground, ECM relay, ECM**
P0687	Engine control relay – short to ground	**Wiring short to ground, engine control relay, ECM**
P0688	Engine control relay – short to positive	**Wiring short to positive, engine control relay, ECM**
P0689	ECM power relay sense – circuit low	**Wiring short to ground, ECM relay, ECM**
P0690	ECM power relay sense – circuit high	**Wiring short to positive, ECM relay, ECM**

Trouble code	Fault location	Probable cause
P0691	Engine coolant blower motor 1 – short to ground	Wiring short to ground, engine coolant blower motor, ECM
P0692	Engine coolant blower motor 1 – short to positive	Wiring short to positive, engine coolant blower motor, ECM
P0693	Engine coolant blower motor 2 – short to ground	Wiring short to ground, engine coolant blower motor, ECM
P0694	Engine coolant blower motor 2 – short to positive	Wiring short to positive, engine coolant blower motor, ECM
P0695	Fan 3 control – circuit low	Wiring short to ground, fan motor
P0696	Fan 3 control – circuit high	Wiring short to positive, fan motor
P0697	Sensor reference voltage C – circuit open	Wiring short to positive
P0698	Sensor reference voltage C – circuit low	Wiring short to ground
P0699	Sensor reference voltage C – circuit high	Wiring short to positive
P0700	Transmission control system – malfunction	Wiring, TCM
P0701	Transmission control system – range/performance problem	Wiring, TCM
P0702	Transmission control system – electrical	Wiring, TCM
P0703	Torque converter/brake switch B – circuit malfunction	Wiring, torque converter/brake switch, ECM/TCM
P0704	Clutch pedal position (CPP) switch – circuit malfunction	Wiring, CPP switch, ECM/TCM
P0705	Transmission range (TR) sensor/switch, PRNDL input – circuit malfunction	Wiring, TR sensor/switch, ECM/TCM
P0706	Transmission range (TR) sensor/switch – range/performance problem	Wiring, TR sensor/switch
P0707	Transmission range (TR) sensor/switch – low input	Wiring short to ground, TR sensor/switch, ECM/TCM
P0708	Transmission range (TR) sensor/switch – high input	Wiring short to positive, TR sensor/switch, ECM/TCM
P0709	Transmission range (TR) sensor/switch – circuit intermittent	Wiring, poor connection, TR sensor/switch, ECM/TCM
P0710	Transmission fluid temperature (TFT) sensor – circuit malfunction	Wiring, TFT sensor, ECM, ECM/TCM
P0711	Transmission fluid temperature (TFT) sensor – range/performance problem	Wiring, TFT sensor
P0712	Transmission fluid temperature (TFT) sensor – low input	Wiring short to ground, TFT sensor, ECM/TCM
P0713	Transmission fluid temperature (TFT) sensor – high input	Wiring short to positive, TFT sensor, ECM/TCM
P0714	Transmission fluid temperature (TFT) sensor – circuit intermittent	Wiring, poor connection, TFT sensor, ECM/TCM
P0715	Turbine shaft speed (TSS) sensor – circuit malfunction	Wiring, TSS sensor, ECM/TCM
P0716	Turbine shaft speed (TSS) sensor – range/performance problem	Wiring, TSS sensor
P0717	Turbine shaft speed (TSS) sensor – no signal	Wiring, TSS sensor, ECM/TCM
P0718	Turbine shaft speed (TSS) sensor – circuit intermittent	Wiring, poor connection, TSS sensor, ECM/TCM

OBD-II trouble code table

Trouble code	Fault location	Probable cause
P0719	Torque converter/brake switch B – circuit low	**Wiring short to ground, torque converter/brake switch, ECM/TCM**
P0720	Output shaft speed (OSS) sensor – circuit malfunction	**Wiring, OSS sensor, ECM/TCM**
P0721	Output shaft speed (OSS) sensor – range/ performance problem	**Wiring, OSS sensor**
P0722	Output shaft speed (OSS) sensor – no signal	**Wiring, OSS sensor, ECM/TCM**
P0723	Output shaft speed (OSS) sensor – circuit intermittent	**Wiring, poor connection, OSS sensor, ECM/ TCM**
P0724	Torque converter/brake switch B – circuit high	**Wiring short to positive, torque converter/brake switch, ECM/TCM**
P0725	Engine RPM input – circuit malfunction	**Wiring, CKP/RPM sensor, ECM/TCM**
P0726	Engine RPM input – range/performance problem	**Wiring, CKP/RPM sensor**
P0727	Engine RPM input – no signal	**Wiring, CKP/RPM sensor, ECM/TCM**
P0728	Engine RPM input – circuit intermittent	**Wiring, poor connection, CKP/RPM sensor, ECM/TCM**
P0730	Incorrect gear ratio	**Wiring, TR sensor/switch, shift solenoids, transmission mechanical fault**
P0731	Gear 1 – incorrect ratio	**Wiring, TR sensor/switch, shift solenoids, transmission mechanical fault**
P0732	Gear 2 – incorrect ratio	**Wiring, TR sensor/switch, shift solenoids, transmission mechanical fault**
P0733	Gear 3 – incorrect ratio	**Wiring, TR sensor/switch, shift solenoids, transmission mechanical fault**
P0734	Gear 4 – incorrect ratio	**Wiring, TR sensor/switch, shift solenoids, transmission mechanical fault**
P0735	Gear 5 – incorrect ratio	**Wiring, TR sensor/switch, shift solenoids, transmission mechanical fault**
P0736	Reverse – incorrect ratio	**Wiring, TR sensor/switch, shift solenoids, transmission mechanical fault**
P0737	TCM engine speed output – circuit	**Wiring, TCM**
P0738	TCM engine speed output – circuit low	**Wiring, TCM**
P0739	TCM engine speed output – circuit high	**Wiring, TCM**
P0740	Torque converter clutch (TCC) solenoid – circuit malfunction	**Wiring, TCC solenoid, ECM/TCM**
P0741	Torque converter clutch (TCC) solenoid – performance or stuck off	**Wiring, TCC solenoid**
P0742	Torque converter clutch (TCC) solenoid – stuck on	**Wiring, TCC solenoid**
P0743	Torque converter clutch (TCC) solenoid – electrical	**Wiring, TCC solenoid, ECM/TCM**
P0744	Torque converter clutch (TCC) solenoid – circuit intermittent	**Wiring, poor connection, TCC solenoid, ECM/ TCM**
P0745	Transmission fluid pressure (TFP) solenoid – circuit malfunction	**Wiring, TFP solenoid, ECM/TCM**
P0746	Transmission fluid pressure (TFP) solenoid – performance or stuck off	**Wiring, TFP solenoid**
P0747	Transmission fluid pressure (TFP) solenoid – stuck on	**Wiring, TFP solenoid**

Trouble code	Fault location	Probable cause
P0748	Transmission fluid pressure (TFP) solenoid – electrical	**Wiring, TFP solenoid, ECM/TCM**
P0749	Transmission fluid pressure (TFP) solenoid – circuit intermittent	**Wiring, poor connection, TFP solenoid, ECM/TCM**
P0750	Shift solenoid (SS) A – circuit malfunction	**Wiring, shift solenoid, ECM/TCM**
P0751	Shift solenoid (SS) A – performance or stuck off	**Wiring, shift solenoid**
P0752	Shift solenoid (SS) A – stuck on	**Wiring, shift solenoid**
P0753	Shift solenoid (SS) A – electrical	**Wiring, shift solenoid, ECM/TCM**
P0754	Shift solenoid (SS) A – circuit intermittent	**Wiring, poor connection, shift solenoid, ECM/TCM**
P0755	Shift solenoid (SS) B – circuit malfunction	**Wiring, shift solenoid, ECM/TCM**
P0756	Shift solenoid (SS) B – performance or stuck off	**Wiring, shift solenoid**
P0757	Shift solenoid (SS) B – stuck on	**Wiring, shift solenoid**
P0758	Shift solenoid (SS) B – electrical	**Wiring, shift solenoid, ECM/TCM**
P0759	Shift solenoid (SS) B – circuit intermittent	**Wiring, poor connection, shift solenoid, ECM/TCM**
P0760	Shift solenoid (SS) C – circuit malfunction	**Wiring, shift solenoid, ECM/TCM**
P0761	Shift solenoid (SS) C – performance or stuck off	**Wiring, shift solenoid**
P0762	Shift solenoid (SS) C – stuck on	**Wiring, shift solenoid**
P0763	Shift solenoid (SS) C – electrical	**Wiring, shift solenoid, ECM/TCM**
P0764	Shift solenoid (SS) C – circuit intermittent	**Wiring, poor connection, shift solenoid, ECM/TCM**
P0765	Shift solenoid (SS) D – circuit malfunction	**Wiring, shift solenoid, ECM/TCM**
P0766	Shift solenoid (SS) D – performance or stuck off	**Wiring, shift solenoid**
P0767	Shift solenoid (SS) D – stuck on	**Wiring, shift solenoid**
P0768	Shift solenoid (SS) D – electrical	**Wiring, shift solenoid, ECM/TCM**
P0769	Shift solenoid (SS) D – circuit intermittent	**Wiring, poor connection, shift solenoid, ECM/TCM**
P0770	Shift solenoid (SS) E – circuit malfunction	**Wiring, shift solenoid, ECM/TCM**
P0771	Shift solenoid (SS) E – performance or stuck off	**Wiring, shift solenoid**
P0772	Shift solenoid (SS) E – stuck on	**Wiring, shift solenoid**
P0773	Shift solenoid (SS) E – electrical	**Wiring, shift solenoid, ECM/TCM**
P0774	Shift solenoid (SS) E – circuit intermittent	**Wiring, poor connection, shift solenoid, ECM/TCM**
P0775	Pressure control solenoid B – malfunction	**Pressure control solenoid**
P0776	Pressure control solenoid B – performance or stuck off	**Wiring, pressure control solenoid**
P0777	Pressure control solenoid B – stuck on	**Wiring, pressure control solenoid**
P0778	Pressure control solenoid B – electrical malfunction	**Wiring, pressure control solenoid**
P0779	Pressure control solenoid B – intermittent	**Wiring, poor connection, pressure control solenoid**
P0780	Gear selection – shift malfunction	**Wiring, TR sensor, shift solenoids, transmission mechanical fault**

OBD-II trouble code table

Trouble code	Fault location	Probable cause
P0781	Gear selection, 1-2 – shift malfunction	**Wiring, TR sensor, shift solenoids, transmission mechanical fault**
P0782	Gear selection, 2-3 – shift malfunction	**Wiring, TR sensor, shift solenoids, transmission mechanical fault**
P0783	Gear selection, 3-4 – shift malfunction	**Wiring, TR sensor, shift solenoids, transmission mechanical fault**
P0784	Gear selection, 4-5 – shift malfunction	**Wiring, TR sensor, shift solenoids, transmission mechanical fault**
P0785	Shift/timing solenoid – circuit malfunction	**Wiring, shift/timing solenoid, ECM/TCM**
P0786	Shift/timing solenoid – range/performance problem	**Wiring, shift/timing solenoid**
P0787	Shift/timing solenoid – low	**Wiring short to ground, shift/timing solenoid, ECM/TCM**
P0788	Shift/timing solenoid – high	**Wiring short to positive, shift/timing solenoid, ECM/TCM**
P0789	Shift/timing solenoid – intermittent	**Wiring, poor connection, shift/timing solenoid, ECM/TCM**
P0790	Transmission mode selection switch – circuit malfunction	**Wiring, transmission mode selection switch, ECM/TCM**
P0791	Intermediate shaft speed sensor – circuit malfunction	**Wiring, poor connection, intermediate shaft speed sensor, ECM/TCM**
P0792	Intermediate shaft speed sensor – range/ performance problem	**Wiring, poor connection, intermediate shaft speed sensor, ECM/TCM**
P0793	Intermediate shaft speed sensor – no signal	**Wiring, poor connection, short to ground, intermediate shaft speed sensor, ECM/TCM**
P0794	Intermediate shaft speed sensor – intermittent circuit malfunction	**Wiring, poor connection, intermediate shaft speed sensor, ECM/TCM**
P0795	Transmission fluid pressure (TFP) solenoid C – circuit malfunction	**Wiring, poor connection, TFP solenoid, ECM/ TCM**
P0796	Transmission fluid pressure (TFP) solenoid C – performance or stuck off	**Wiring, poor connection, TFP solenoid, ECM/ TCM**
P0797	Transmission fluid pressure (TFP) solenoid C – stuck on	**Wiring, poor connection, TFP solenoid, ECM/ TCM**
P0798	Transmission fluid pressure (TFP) solenoid C – electrical malfunction	**Wiring, poor connection, TFP solenoid, ECM/ TCM**
P0799	Transmission fluid pressure (TFP) solenoid C – intermittent circuit malfunction	**Wiring, poor connection, ECM/TCM**
P0800	Transfer case control system, MIL request – malfunction	**Wiring, mechanical fault**
P0801	Reverse inhibit circuit – malfunction	**Wiring, poor connection**
P0802	Transmission control system, MIL request – circuit open	**Wiring, mechanical fault**
P0803	1-4 Upshift (Skip shift) solenoid – circuit malfunction	**Wiring, poor connection, upshift solenoid**
P0804	1-4 Upshift (Skip shift) warning lamp – circuit malfunction	**Wiring, poor connection**
P0805	Clutch position sensor – circuit malfunction	**Wiring, poor connection, clutch position sensor, ECM/TCM**
P0806	Clutch position sensor – range/performance problem	**Wiring, poor connection, clutch position sensor, ECM/TCM**

OBD-II trouble code table

Trouble code	Fault location	Probable cause
P0807	Clutch position sensor – low input	**Wiring, short to ground, clutch position sensor, ECM/TCM**
P0808	Clutch position sensor – high input	**Wiring, short to positive, clutch position sensor, ECM/TCM**
P0809	Clutch position sensor – intermittent circuit malfunction	**Wiring, poor connection, clutch position sensor, ECM/TCM**
P0810	Clutch position control error	**Wiring, poor connection, ECM/TCM**
P0811	Excessive clutch slip	**Wiring, poor connection, mechanical fault, ECM/TCM**
P0812	Reverse gear – input circuit malfunction	**Wiring, poor connection, ECM/TCM**
P0813	Reverse gear – output circuit malfunction	**Wiring, poor connection, ECM/TCM**
P0814	Transmission range (TR) display – circuit malfunction	**Wiring, poor connection, TR sensor, ECM/TCM**
P0815	Upshift switch – circuit malfunction	**Wiring, poor connection, upshift switch, ECM/TCM**
P0816	Downshift switch – circuit malfunction	**Wiring, poor connection, downshift switch, ECM/TCM**
P0817	Starter disable circuit – malfunction	**Wiring, poor connection, ECM/TCM**
P0818	Driveline disconnect switch – circuit malfunction	**Wiring, poor connection, upshift switch, ECM/TCM**
P0819	Up and down shift switch to transmission range – correlation	**Wiring, poor connection, TR sensor, ECM/TCM**
P0820	Gear lever X-Y position sensor – circuit malfunction	**Wiring, poor connection, gear lever position sensor, ECM/TCM**
P0821	Gear lever X position sensor – circuit malfunction	**Wiring, poor connection, gear lever position sensor, ECM/TCM**
P0822	Gear lever Y position sensor – circuit malfunction	**Wiring, poor connection, gear lever position sensor, ECM/TCM**
P0823	Gear lever X position sensor – circuit intermittent	**Wiring, poor connection, gear lever position sensor, ECM/TCM**
P0824	Gear lever Y position sensor – circuit intermittent	**Wiring, poor connection, gear lever position sensor, ECM/TCM**
P0825	Gear lever push-pull switch – circuit malfunction	**Wiring, poor connection, gear lever push-pull switch, ECM/TCM**
P0826	Up and down switch – input circuit	**Wiring, Up/down switch**
P0827	Up and down switch – input circuit low	**Wiring short to ground, Up/down switch**
P0828	Up and down switch – input circuit high	**Wiring short to positive, Up/down switch**
P0829	5-6 Shift	**Mechanical fault**
P0830	Clutch pedal position (CPP) switch A – circuit malfunction	**Wiring, poor connection, CPP switch, ECM/TCM**
P0831	Clutch pedal position (CPP) switch A – low input	**Wiring, short to ground, CPP switch, ECM/TCM**
P0832	Clutch pedal position (CPP) switch A – high input	**Wiring, short to positive, CPP switch, ECM/TCM**
P0833	Clutch pedal position (CPP) switch B – circuit malfunction	**Wiring, poor connection, CPP switch, ECM/TCM**
P0834	Clutch pedal position (CPP) switch B – low input	**Wiring, short to ground, CPP switch, ECM/TCM**
P0835	Clutch pedal position (CPP) switch B – high input	**Wiring, short to positive, CPP switch, ECM/TCM**

OBD-II trouble code table

Trouble code	Fault location	Probable cause
P0836	Four wheel drive switch – circuit malfunction	**Wiring, poor connection, four wheel drive switch, ECM/TCM**
P0837	Four wheel drive switch – range/performance problem	**Wiring, poor connection, four wheel drive switch, ECM/TCM**
P0838	Four wheel drive switch – low input	**Wiring, short to ground, four wheel drive switch, ECM/TCM**
P0839	Four wheel drive switch – high input	**Wiring, short to positive, four wheel drive switch, ECM/TCM**
P0840	Transmission fluid pressure (TFP) sensor A – circuit malfunction	**Wiring, poor connection, TFP sensor, ECM/TCM**
P0840	Transmission fluid pressure (TFP) switch A – circuit malfunction	**Wiring, poor connection, TFP switch, ECM/TCM**
P0841	Transmission fluid pressure (TFP) sensor A – range/performance problem	**Wiring, poor connection, TFP sensor, ECM/TCM**
	Transmission fluid pressure (TFP) switch A – range/performance problem	**Wiring, poor connection, TFP switch, ECM/TCM**
P0842	Transmission fluid pressure (TFP) sensor A – low input	**Wiring, short to ground, TFP sensor, ECM/TCM**
	Transmission fluid pressure (TFP) switch A – low input	**Wiring, short to ground, TFP switch, ECM/TCM**
P0843	Transmission fluid pressure (TFP) sensor A – high input	**Wiring, short to positive, TFP sensor, ECM/TCM**
	Transmission fluid pressure (TFP) switch A – high input	**Wiring, short to positive, TFP switch, ECM/TCM**
P0844	Transmission fluid pressure (TFP) sensor A – intermittent circuit malfunction	**Wiring, poor connection, TFP sensor, ECM/TCM**
	Transmission fluid pressure (TFP) switch A – intermittent circuit malfunction	**Wiring, poor connection, TFP switch, ECM/TCM**
P0845	Transmission fluid pressure (TFP) sensor B – circuit malfunction	**Wiring, poor connection, TFP sensor, ECM/TCM**
	Transmission fluid pressure (TFP) switch B – circuit malfunction	**Wiring, poor connection, TFP switch, ECM/TCM**
P0846	Transmission fluid pressure (TFP) sensor B – range/performance problem	**Wiring, poor connection, TFP sensor, ECM/TCM**
	Transmission fluid pressure (TFP) switch B – range/performance problem	**Wiring, poor connection, TFP switch, ECM/TCM**
P0847	Transmission fluid pressure (TFP) sensor B – low input	**Wiring, short to ground, TFP sensor ,ECM/TCM**
	Transmission fluid pressure (TFP) switch B – low input	**Wiring, short to ground, TFP switch, ECM/TCM**
P0848	Transmission fluid pressure (TFP) sensor B – high input	**Wiring, short to positive, TFP sensor, ECM/TCM**
	Transmission fluid pressure (TFP) switch B – high input	**Wiring, short to positive, TFP switch, ECM/TCM**
P0849	Transmission fluid pressure (TFP) sensor B – intermittent circuit malfunction	**Wiring, poor connection, TFP sensor, ECM/TCM**
	Transmission fluid pressure (TFP) switch B – intermittent circuit malfunction	**Wiring, poor connection, TFP switch, ECM/TCM**
P0850	Park/neutral position (PNP) switch – input circuit malfunction	**Wiring, PNP switch, ECM/TCM**
P0851	Park/neutral position (PNP) switch – input circuit low	**Wiring, short to ground, PNP switch, ECM/TCM**

Trouble code	Fault location	Probable cause
P0852	Park/neutral position (PNP) switch – input circuit high	Wiring, short to positive, PNP switch, ECM/TCM
P0853	Drive switch – input circuit malfunction	Wiring, drive switch, ECM/TCM
P0854	Drive switch – input circuit low	Wiring, short to ground, drive switch, ECM/TCM
P0855	Drive switch – input circuit high	Wiring, short to positive, drive switch, ECM/TCM
P0856	Traction control input signal – malfunction	Wiring, poor connection, ECM/TCM
P0857	Traction control input signal – range/performance problem	Wiring, poor connection, ECM/TCM
P0858	Traction control input signal – low	Wiring, short to ground, ECM/TCM
P0859	Traction control input signal – high	Wiring, short to positive, ECM/TCM
P0860	Gear shift module communication circuit – malfunction	Wiring, poor connection, gear shift module, ECM/TCM
P0861	Gear shift module communication circuit – low input	Wiring, short to ground, gear shift module, ECM/TCM
P0862	Gear shift module communication circuit – high input	Wiring, short to positive, gear shift module, ECM/TCM
P0863	Transmission control module (TCM) communication circuit – malfunction	Wiring, poor connection, TCM
P0864	Transmission control module (TCM) communication circuit – range/performance problem	Wiring, poor connection, TCM
P0865	Transmission control module (TCM) communication circuit – low input	Wiring, short to ground, TCM
P0866	Transmission control module (TCM) communication circuit – high input	Wiring, short to positive, TCM
P0867	Transmission fluid pressure (TFP) sensor	Wiring, poor connection, TFP sensor, ECM/TCM
P0868	Transmission fluid pressure (TFP) sensor – low	Wiring, short to ground, TFP sensor, ECM/TCM
P0869	Transmission fluid pressure (TFP) sensor – high	Wiring, short to positive, TFP sensor, ECM/TCM
P0870	Transmission fluid pressure (TFP) sensor C – circuit malfunction	Wiring, poor connection, TFP sensor, ECM/TCM
	Transmission fluid pressure (TFP) switch C – circuit malfunction	Wiring, poor connection, TFP switch, ECM/TCM
P0871	Transmission fluid pressure (TFP) sensor C – range/performance	Wiring, poor connection, TFP sensor, ECM/TCM
	Transmission fluid pressure (TFP) switch C – range/performance	Wiring, poor connection, TFP switch, ECM/TCM
P0872	Transmission fluid pressure (TFP) sensor C – circuit low	Wiring, short to ground, TFP sensor, ECM/TCM
	Transmission fluid pressure (TFP) switch C – circuit low	Wiring, short to ground, TFP switch, ECM/TCM
P0873	Transmission fluid pressure (TFP) sensor C – circuit high	Wiring, short to positive, TFP sensor, ECM/TCM
	Transmission fluid pressure (TFP) switch C – circuit high	Wiring, short to positive, TFP switch, ECM/TCM
P0874	Transmission fluid pressure (TFP) sensor C – intermittent circuit malfunction	Wiring, poor connection, TFP sensor, ECM/TCM
	Transmission fluid pressure (TFP) switch C – intermittent circuit malfunction	Wiring, poor connection, TFP switch, ECM/TCM

OBD-II trouble code table

Trouble code	Fault location	Probable cause
P0875	Transmission fluid pressure (TFP) sensor D – circuit malfunction	**Wiring, poor connection, TFP sensor, ECM/TCM**
	Transmission fluid pressure (TFP) switch D – circuit malfunction	**Wiring, poor connection, TFP switch, ECM/TCM**
P0876	Transmission fluid pressure (TFP) sensor D – range/performance	**Wiring, poor connection, TFP sensor, ECM/TCM**
	Transmission fluid pressure (TFP) switch D – range/performance	**Wiring, poor connection, TFP switch, ECM/TCM**
P0877	Transmission fluid pressure (TFP) sensor D – circuit low	**Wiring, short to ground, TFP sensor, ECM/TCM**
	Transmission fluid pressure (TFP) switch D – circuit low	**Wiring, short to ground, TFP switch, ECM/TCM**
P0878	Transmission fluid pressure (TFP) sensor D – circuit high	**Wiring, short to positive, TFP sensor, ECM/TCM**
	Transmission fluid pressure (TFP) switch D – circuit high	**Wiring, short to positive, TFP switch, ECM/TCM**
P0879	Transmission fluid pressure (TFP) sensor D – intermittent circuit malfunction	**Wiring, poor connection, TFP sensor, ECM/TCM**
	Transmission fluid pressure (TFP) switch D – intermittent circuit malfunction	**Wiring, poor connection, TFP switch, ECM/TCM**
P0880	Transmission control module (TCM) – power input signal malfunction	**Wiring, poor connection, TCM**
P0881	Transmission control module (TCM) – power input signal range/performance	**Wiring, poor connection, TCM**
P0882	Transmission control module (TCM) – power input signal low	**Wiring, short to ground, TCM**
P0883	Transmission control module (TCM) – power input signal high	**Wiring, short to positive, TCM**
P0884	Transmission control module (TCM) – power input signal intermittent malfunction	**Wiring, poor connection, TCM**
P0885	Transmission control module (TCM) power relay – control circuit open	**Wiring, poor connection, TCM power relay, TCM**
P0886	Transmission control module (TCM) power relay – control circuit low	**Wiring, short to ground, TCM power relay, TCM**
P0887	Transmission control module (TCM) power relay – control circuit high	**Wiring, short to positive, TCM power relay, TCM**
P0888	Transmission control module (TCM) power relay – sense circuit malfunction	**Wiring, poor connection, TCM power relay, TCM**
P0889	Transmission control module (TCM) power relay – sense circuit range/performance	**Wiring, poor connection, TCM power relay, TCM**
P0890	Transmission control module (TCM) power relay – sense circuit low	**Wiring, short to ground, TCM power relay, TCM**
P0891	Transmission control module (TCM) power relay – sense circuit high	**Wiring, short to positive, TCM power relay, TCM**
P0892	Transmission control module (TCM) power relay – sense circuit intermittent malfunction	**Wiring, poor connection, TCM power relay, TCM**
P0893	Multiple gears engaged	**Mechanical fault**
P0894	Transmission component slipping	**Mechanical fault**
P0895	Shift time too short	**Mechanical fault**
P0896	Shift time too long	**Mechanical fault**

Trouble code	Fault location	Probable cause
P0897	Transmission fluid deteriorated	Mechanical fault
P0898	Transmission control system – MIL request – circuit low	Wiring, poor connection, short to ground
P0899	Transmission control system – MIL request – circuit high	Wiring, poor connection, short to positive
P0900	Clutch actuator – circuit open	Wiring, clutch actuator, ECM/TCM
P0901	Clutch actuator – circuit range/performance	Wiring, poor connection, clutch actuator, ECM/TCM
P0902	Clutch actuator – circuit low	Wiring, short to ground, clutch actuator, ECM/TCM
P0903	Clutch actuator – circuit high	Wiring, short to positive, clutch actuator, ECM/TCM
P0904	Transmission gate select position circuit – malfunction	Wiring, poor connection, ECM/TCM
P0905	Transmission gate select position circuit – range/performance	Wiring, poor connection, ECM/TCM
P0906	Transmission gate select position circuit – low	Wiring, short to ground, ECM/TCM
P0907	Transmission gate select position circuit – high	Wiring, short to positive, ECM/TCM
P0908	Transmission gate select position circuit – intermittent circuit malfunction	Wiring, poor connection, ECM/TCM
P0909	Transmission gate select control error	Mechanical fault
P0910	Transmission gate select actuator – circuit open	Wiring, transmission gate select actuator, ECM/TCM
P0911	Transmission gate select actuator – circuit range/performance	Wiring, poor connection, transmission gate select actuator, ECM/TCM
P0912	Transmission gate select actuator – circuit low	Wiring, short to ground, transmission gate select actuator, ECM/TCM
P0913	Transmission gate select actuator – circuit high	Wiring, short to positive, transmission gate select actuator, ECM/TCM
P0914	Gear shift position circuit – malfunction	Wiring, poor connection, ECM/TCM
P0915	Gear shift position circuit – range/performance	Wiring, poor connection, ECM/TCM
P0916	Gear shift position circuit – low	Wiring, short to ground, ECM/TCM
P0917	Gear shift position circuit – high	Wiring, short to positive, ECM/TCM
P0918	Gear shift position circuit – intermittent malfunction	Wiring, poor connection, ECM/TCM
P0919	Gear shift position control – error	Wiring, poor connection, ECM/TCM
P0920	Gear shift forward actuator – circuit open	Wiring, gear shift forward actuator, ECM/TCM
P0921	Gear shift forward actuator – circuit range/performance	Wiring, poor connection, gear shift forward actuator, ECM/TCM
P0922	Gear shift forward actuator – circuit low	Wiring, short to ground, gear shift forward actuator, ECM/TCM
P0923	Gear shift forward actuator – circuit high	Wiring, short to positive, gear shift forward actuator, ECM/TCM
P0924	Gear shift reverse actuator – circuit open	Wiring, gear shift reverse actuator, ECM/TCM
P0925	Gear shift reverse actuator – circuit range/performance	Wiring, poor connection, gear shift reverse actuator, ECM/TCM
P0926	Gear shift reverse actuator – circuit low	Wiring, short to ground, gear shift reverse actuator, ECM/TCM

OBD-II trouble code table

Trouble code	Fault location	Probable cause
P0927	Gear shift reverse actuator – circuit high	**Wiring, short to positive, gear shift reverse actuator, ECM/TCM**
P0928	Gear shift lock solenoid – circuit open	**Wiring, gear shift lock solenoid, ECM/TCM**
P0929	Gear shift lock solenoid – circuit range/ performance	**Wiring, gear shift lock solenoid, ECM/TCM**
P0930	Gear shift lock solenoid – circuit low	**Wiring, short to ground, gear shift lock solenoid, ECM/TCM**
P0931	Gear shift lock solenoid – circuit high	**Wiring, short to positive, gear shift lock solenoid, ECM/TCM**
P0932	Hydraulic pressure sensor – circuit malfunction	**Wiring, poor connection, hydraulic pressure sensor, ECM/TCM**
P0933	Hydraulic pressure sensor – range/performance	**Wiring, hydraulic pressure sensor, ECM/TCM**
P0934	Hydraulic pressure sensor – circuit low input	**Wiring, short to ground, hydraulic pressure sensor, ECM/TCM**
P0935	Hydraulic pressure sensor – circuit high input	**Wiring, short to positive, hydraulic pressure sensor, ECM/TCM**
P0936	Hydraulic pressure sensor – circuit intermittent	**Wiring, poor connection, hydraulic pressure sensor, ECM/TCM**
P0937	Hydraulic oil temperature sensor – circuit malfunction	**Wiring, poor connection, hydraulic oil temperature sensor, ECM/TCM**
P0938	Hydraulic oil temperature sensor – range/ performance	**Wiring, hydraulic oil temperature sensor, ECM/ TCM**
P0939	Hydraulic oil temperature sensor – circuit low input	**Wiring, short to ground, hydraulic oil temperature sensor, ECM/TCM**
P0940	Hydraulic oil temperature sensor – circuit high input	**Wiring, short to positive, hydraulic oil temperature sensor, ECM/TCM**
P0941	Hydraulic oil temperature sensor – circuit intermittent	**Wiring, poor connection, hydraulic oil temperature sensor, ECM/TCM**
P0942	Hydraulic pressure unit	**Mechanical fault**
P0943	Hydraulic pressure unit – cycling period too short	**Mechanical fault**
P0944	Hydraulic pressure unit – loss of pressure	**Mechanical fault**
P0945	Hydraulic pump relay – circuit open	**Wiring, hydraulic pump relay, ECM/TCM**
P0946	Hydraulic pump relay – circuit range/performance	**Wiring, hydraulic pump relay, ECM/TCM**
P0947	Hydraulic pump relay – circuit low	**Wiring, short to ground, hydraulic pump relay, ECM/TCM**
P0948	Hydraulic pump relay – circuit high	**Wiring, short to positive, hydraulic pump relay, ECM/TCM**
P0949	ASM – adaptive learning not done	**ECM/TCM**
P0950	ASM control circuit	**Wiring, poor connection, ECM/TCM**
P0951	ASM control circuit – range/performance	**Wiring, poor connection, ECM/TCM**
P0952	ASM control circuit – low	**Wiring, poor connection, short to ground, ECM/ TCM**
P0953	ASM control circuit – high	**Wiring, poor connection, short to positive, ECM/TCM**
P0954	ASM – intermittent circuit malfunction	**Wiring, poor connection, ECM/TCM**
P0955	ASM mode circuit – malfunction	**Wiring, poor connection, ECM/TCM**
P0956	ASM mode circuit – range/performance	**Wiring, poor connection, ECM/TCM**

Trouble code	Fault location	Probable cause
P0957	ASM mode circuit – low	**Wiring, poor connection, short to ground, ECM/TCM**
P0958	ASM mode circuit – high	**Wiring, poor connection, short to positive, ECM/TCM**
P0959	ASM mode circuit – intermittent circuit malfunction	**Wiring, poor connection, ECM/TCM**
P0960	Pressure control (PC) solenoid A – control circuit open	**Wiring, poor connection, pressure control solenoid, ECM/TCM**
P0961	Pressure control (PC) solenoid A – control circuit range/performance	**Wiring, poor connection, pressure control solenoid, ECM/TCM**
P0962	Pressure control (PC) solenoid A – control circuit low	**Wiring, short to ground, pressure control solenoid, ECM/TCM**
P0963	Pressure control (PC) solenoid A – control circuit high	**Wiring, short to positive, pressure control solenoid, ECM/TCM**
P0964	Pressure control (PC) solenoid B – control circuit open	**Wiring, poor connection, pressure control solenoid, ECM/TCM**
P0965	Pressure control (PC) solenoid B – control circuit range/performance	**Wiring, poor connection, pressure control solenoid, ECM/TCM**
P0966	Pressure control (PC) solenoid B – control circuit low	**Wiring, short to ground, pressure control solenoid, ECM/TCM**
P0967	Pressure control (PC) solenoid B – control circuit high	**Wiring, short to positive, pressure control solenoid, ECM/TCM**
P0968	Pressure control (PC) solenoid C – control circuit open	**Wiring, poor connection, pressure control solenoid, ECM/TCM**
P0969	Pressure control (PC) solenoid C – control circuit range/performance	**Wiring, poor connection, pressure control solenoid, ECM/TCM**
P0970	Pressure control (PC) solenoid C – control circuit low	**Wiring, short to ground, pressure control solenoid, ECM/TCM**
P0971	Pressure control (PC) solenoid C – control circuit high	**Wiring, short to positive, pressure control solenoid, ECM/TCM**
P0972	Shift solenoid (SS) A – control circuit range/performance	**Wiring, poor connection, shift solenoid, ECM/TCM**
P0973	Shift solenoid (SS) A – control circuit low	**Wiring, short to ground, shift solenoid, ECM/TCM**
P0974	Shift solenoid (SS) A – control circuit high	**Wiring, short to positive, shift solenoid, ECM/TCM**
P0975	Shift solenoid (SS) B – control circuit range/performance	**Wiring, poor connection, shift solenoid, ECM/TCM**
P0976	Shift solenoid (SS) B – control circuit low	**Wiring, short to ground, shift solenoid, ECM/TCM**
P0977	Shift solenoid (SS) B – control circuit high	**Wiring, short to positive, shift solenoid, ECM/TCM**
P0978	Shift solenoid (SS) C – control circuit range/performance	**Wiring, poor connection, shift solenoid, ECM/TCM**
P0979	Shift solenoid (SS) C – control circuit low	**Wiring, short to ground, shift solenoid, ECM/TCM**
P0980	Shift solenoid (SS) C – control circuit high	**Wiring, short to positive, shift solenoid, ECM/TCM**
P0981	Shift solenoid (SS) D – control circuit range/performance	**Wiring, poor connection, shift solenoid, ECM/TCM**
P0982	Shift solenoid (SS) D – control circuit low	**Wiring, short to ground, shift solenoid, ECM/TCM**
P0983	Shift solenoid (SS) D – control circuit high	**Wiring, short to positive, shift solenoid, ECM/TCM**
P0984	Shift solenoid (SS) E – control circuit range/performance	**Wiring, poor connection, shift solenoid, ECM/TCM**

OBD-II trouble code table

Trouble code	Fault location	Probable cause
P0985	Shift solenoid (SS) E – control circuit low	**‘Wiring, short to ground, shift solenoid, ECM/TCM**
P0986	Shift solenoid (SS) E – control circuit high	**Wiring, short to positive, shift solenoid, ECM/TCM**
P0987	Transmission fluid pressure (TFP) sensor E – circuit malfunction	**Wiring, poor connection, TFP sensor, ECM/TCM**
	Transmission fluid pressure (TFP) switch E – circuit malfunction	**Wiring, poor connection, TFP switch, ECM/TCM**
P0988	Transmission fluid pressure (TFP) sensor E – circuit range/performance	**Wiring, poor connection, TFP sensor, ECM/TCM**
	Transmission fluid pressure (TFP) switch E – circuit range/performance	**Wiring, poor connection, TFP switch, ECM/TCM**
P0989	Transmission fluid pressure (TFP) sensor E – circuit low	**Wiring, short to ground, TFP sensor, ECM/TCM**
	Transmission fluid pressure (TFP) switch E – circuit low	**Wiring, short to ground, TFP switch, ECM/TCM**
P0990	Transmission fluid pressure (TFP) sensor E – circuit high	**Wiring, short to positive, TFP sensor, ECM/TCM**
	Transmission fluid pressure (TFP) switch E – circuit high	**Wiring, short to positive, TFP switch, ECM/TCM**
P0991	Transmission fluid pressure (TFP) sensor E – circuit intermittent	**Wiring, poor connection, TFP sensor, ECM/TCM**
	Transmission fluid pressure (TFP) switch E – circuit intermittent	**Wiring, poor connection, TFP switch, ECM/TCM**
P0992	Transmission fluid pressure (TFP) sensor F – circuit malfunction	**Wiring, poor connection, TFP sensor, ECM/TCM**
	Transmission fluid pressure (TFP) switch F – circuit malfunction	**Wiring, poor connection, TFP switch, ECM/TCM**
P0993	Transmission fluid pressure (TFP) sensor F – circuit range/performance	**Wiring, poor connection, TFP sensor, ECM/TCM**
	Transmission fluid pressure (TFP) switch F – circuit range/performance	**Wiring, poor connection, TFP switch, ECM/TCM**
P0994	Transmission fluid pressure (TFP) sensor F – circuit low	**Wiring, short to ground, TFP sensor , ECM/TCM**
	Transmission fluid pressure (TFP) switch F – circuit low	**Wiring, short to ground, TFP switch, ECM/TCM**
P0995	Transmission fluid pressure (TFP) sensor F – circuit high	**Wiring, short to positive, TFP sensor, ECM/TCM**
	Transmission fluid pressure (TFP) switch F – circuit high	**Wiring, short to positive, TFP switch, ECM/TCM**
P0996	Transmission fluid pressure (TFP) sensor F – circuit intermittent	**Wiring, poor connection, TFP sensor, ECM/TCM**
	Transmission fluid pressure (TFP) switch F – circuit intermittent	**Wiring, poor connection, TFP switch, ECM/TCM**
P0997	Shift solenoid (SS) F – control circuit range/ performance	**Wiring, poor connection, shift solenoid, ECM/TCM**
P0998	Shift solenoid (SS) F – control circuit low	**Wiring short to ground, shift solenoid, ECM/ TCM**
P0999	Shift solenoid (SS) F – control circuit high	**Wiring short to positive, shift solenoid, ECM/TCM**

BUICK

Model:	Engine identification:	Year:
Century 2.2L	**VIN position 8 = 4**	**1993-95**
Century 2.5L	**VIN position 8 = R**	**1992**
LeSabre 3.8L	**VIN position 8 = L**	**1992-93**
Park Avenue 3.8L	**VIN position 8 = L/1**	**1993**
Regal 3.1L	**VIN position 8 = T/M**	**1992-95**
Regal 3.8L	**VIN position 8 = L**	**1992-93**

System: **GM**

Self-diagnosis

General information

- Refer to the front of this manual for general test conditions, terminology, detailed descriptions of wiring faults and a general trouble shooter for electrical and mechanical faults.
- Malfunction indicator lamp (MIL) will illuminate continuously if permanent faults are recorded.
- Malfunction indicator lamp (MIL) may illuminate briefly and then switch off if intermittent faults are recorded.
- ECM operates in backup mode if sensors fail, to enable vehicle to be driven to repair shop.
- Trouble codes can be displayed by the MIL or by using a Scan Tool connected to the data link connector (DLC).

Accessing

Models with data link connector (DLC) terminal B connected

- Access data link connector (DLC).
- Jump data link connector (DLC) terminals A and B **1**.
- Switch ignition ON. Do NOT start engine.
- Count MIL flashes. Note trouble codes. Compare with trouble code table.
- Firstly trouble code 12 will be displayed 3 times.
- Trouble codes consist of 2 groups, displayed as follows:
  - First group indicates the 'tens' of the trouble code **2** [**A**].
  - Second group indicates the 'units' of the trouble code **2** [**C**].
  - A short pause separates each group **2** [**B**].
  - A long pause separates each trouble code **2** [**D**].
  - For example: Trouble code 12 displayed **2**. Start of diagnosis.
- Each additional trouble code will now be displayed 3 times, in ascending order.

NOTE: ***If no additional trouble codes are stored, code 12 will continue to be displayed.***

- Trouble codes can also be displayed by using a Scan Tool connected to the data link connector (DLC) **1**.

1

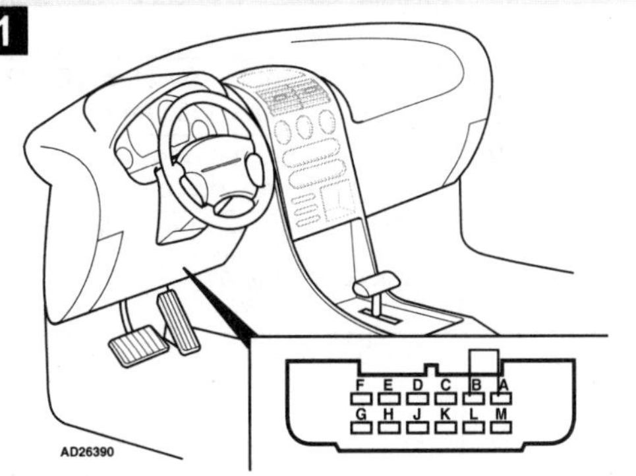

2

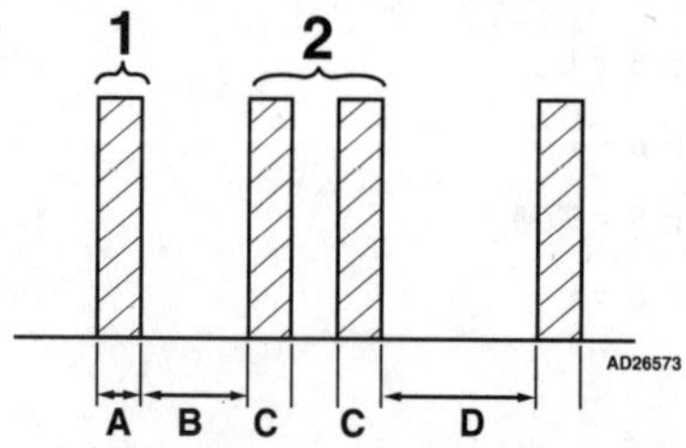

Erasing

NOTE: ***Disconnecting battery ground lead may cause driveability problems until the engine management system relearns basic values.***

- Ensure ignition switched OFF.
- Disconnect battery ground cable for at least 30 seconds.
- Reconnect battery ground cable.
- Repeat accessing procedure to ensure no trouble codes remain in ECM fault memory.
- Trouble codes can also be erased by using a Scan Tool connected to the data link connector (DLC).

NOTE: ***Disconnecting battery ground lead may also erase memory from electronic units, such as radio.***

Models without data link connector (DLC) terminal B connected

- Trouble codes can only be displayed by using a Scan Tool connected to the data link connector (DLC) [1].

Trouble code identification

Flash code	Fault location	Probable cause
12	Start of diagnosis	**No engine speed (RPM) signal received**
13	Oxygen sensor (O2S)/heated oxygen sensor (HO2S)	**Wiring open circuit, O2S/HO2S, ECM**
14	Engine coolant temperature (ECT) sensor – high temperature indicated	**Wiring short circuit to ground, ECT sensor, ECM**
15	Engine coolant temperature (ECT) sensor – low temperature indicated	**Wiring open circuit, ECT sensor, ECM**
16	Battery voltage – high or low	**Wiring, alternator, ECM**
17 [9]	Spark reference circuit	**Wiring open circuit/short circuit to ground, ICM, ECM**
17 [7]	Camshaft position (CMP) sensor	**Wiring open circuit/short circuit to ground/ short circuit to positive, ignition feed circuit, CMP sensor, mechanical fault, ECM**
18	Camshaft position (CMP) signal/crankshaft position (CKP) signal error	**Wiring open circuit/short circuit to ground, CKP sensor, CMP sensor, crankshaft pulley/ balancer**
19	Crankshaft position (CKP) sensor intermittent 7X signal	**Wiring, CKP sensor, PROM, ECM**
21	Throttle position (TP) sensor – voltage high	**Wiring open circuit/short circuit to positive, TP sensor, ECM**
22	Throttle position (TP) sensor – voltage low	**Wiring open circuit/short circuit to ground, TP sensor, ECM**
23	Intake air temperature (IAT) sensor – low temperature	**Wiring open circuit, IAT sensor, ECM**

Flash code	Fault location	Probable cause
24	Vehicle speed sensor (VSS)	**Wiring open circuit/short circuit to ground, PROM, VSS, ECM**
25	Intake air temperature (IAT) sensor – high temperature	**Wiring short circuit to ground, IAT sensor, ECM**
26	Engine control module quad driver/output driver	**Wiring open circuit/short circuit to ground/ short circuit to positive, ECM controlled components, ECM [16]**
27 [10]	Engine control module quad driver/output driver	**Wiring open circuit/short circuit to ground/ short circuit to positive, ECM controlled components, ECM [16]**
27 [8]	Transmission gear switch signal circuit 2nd gear	**Wiring open circuit/short circuit to ground, 2nd gear switch, ECM**
28 [11]	Engine control module quad driver/output driver	**Wiring open circuit/short circuit to ground/ short circuit to positive, ECM controlled components, ECM [16]**
28 [7]	Transmission range (TR) switch	**Wiring open circuit/short circuit to ground/ short circuit to positive, TR switch, ECM**
28 [8]	Transmission gear switch signal circuit 3rd gear	**Wiring open circuit/short circuit to ground, 3rd gear switch, ECM**
29	Transmission gear switch signal circuit 4th gear	**Wiring open circuit/short circuit to ground, 4th gear switch, ECM**
31	Park/neutral position (PNP) switch	**Wiring open circuit/short circuit to ground, incorrectly adjusted, PNP switch, ECM**
32	Exhaust gas recirculation (EGR) solenoid	**Wiring open circuit/short circuit to ground, EGR valve, ECM**
33	Manifold absolute pressure (MAP) sensor – circuit malfunction	**Wiring open circuit/short circuit to positive, hose blocked/leaking, MAP sensor, ECM**
34 [13]	Manifold absolute pressure (MAP) sensor – circuit malfunction	**Wiring open circuit/short circuit to ground, MAP sensor, ECM**
34 [2] [8] [12]	Mass air flow (MAF) sensor – circuit malfunction	**Wiring open circuit/short circuit to ground/ short circuit to positive, MAF sensor, ECM**
35	Idle speed control (ISC) system – malfunction	**Wiring open circuit/short circuit to ground/ short circuit to positive, IAC valve, ECM**
36 [9]	Shift solenoids (SS)	**Wiring open circuit/short circuit to ground, SS**
36 [7]	24X crankshaft position sensor (CKP) – circuit malfunction	**Wiring open circuit/short circuit to ground/ short circuit to positive, 24X CKP sensor, ignition feed circuit, ECM**
37	Brake pedal position (BPP) switch for torque converter clutch (TCC)	**Wiring open circuit/short circuit to ground, incorrectly adjusted BPP switch, BPP switch, ECM**
38 [10]	Brake pedal position (BPP) switch/ torque converter clutch (TCC)	**Wiring open circuit/short circuit to ground, supply wire, BPP switch, TCC switch, ECM**
38 [8]	Brake pedal position (BPP) switch	**Wiring open circuit/short circuit to ground/ short circuit to positive, BPP switch incorrectly adjusted, BPP switch, ECM**

Flash code	Fault location	Probable cause
39	Torque converter clutch (TCC)	**Wiring open circuit/short circuit to ground, TCC solenoid, BPP switch, cooling fan fuse, transmission mechanical fault, ECM**
41 [14]	Camshaft position (CMP) sensor	**Wiring open circuit/short circuit to ground, CMP sensor, ICM, ECM**
41 [6]	Engine control module (ECM) – PROM	**Insecure PROM, incorrect PROM, ECM**
41 [7]	Ignition control module (ICM) – electronic spark timing (EST)/ignition control (IC) circuit malfunction	**Wiring open circuit, ICM, ECM**
42	Ignition control module (ICM) – electronic spark timing (EST)/ignition control (IC)/bypass circuit malfunction	**Wiring open circuit/short circuit to ground, ICM, ECM**
43	Knock sensor (KS)	**Wiring open circuit/short circuit to ground/ short to positive, KS, PROM, ECM**
44	Oxygen sensor (O2S)/heated oxygen sensor (HO2S) – mixture lean	**Wiring short circuit to ground, fuel pressure low, MAP sensor, misfire, fuel contamination, exhaust system leaking, O2S/HO2S, ECM**
45	Oxygen sensor (O2S)/heated oxygen sensor (HO2S) – mixture rich	**Fuel pressure high, injector(s) leaking, O2S contamination, EVAP canister purge valve, TP sensor, MAP sensor, EGR system, O2S/ HO2S, ECM**
46	Immobilizer control module – fuel enable signal	**Incorrect ignition key, wiring open circuit/ short circuit to ground/short circuit to positive, immobilizer control module, ECM**
51	Engine control module (ECM) – PROM	**Insecure PROM, incorrect PROM, ECM**
53 [3] [7]	Battery voltage – high/low	**Wiring open circuit, alternator, ECM**
53 [4] [8]	Exhaust gas recirculation (EGR) valve solenoid 1	**Wiring open circuit/short circuit to ground, exhaust system blocked, EGR valve, ECM**
54 [14]	Exhaust gas recirculation (EGR) valve solenoid 2	**Wiring open circuit/short circuit to ground, exhaust system blocked, EGR valve, ECM**
54 [6] [7]	Fuel pump circuit – voltage low	**Wiring open circuit/short circuit to ground, fuel pump relay, fuel pump, engine oil pressure warning lamp switch, ECM**
55 [5]	Fuel lean monitor – lean during acceleration	**Fuel contamination, injector(s) blocked, fuel system, misfire**
55 [4] [8]	Exhaust gas recirculation (EGR) valve solenoid 3	**Wiring open circuit/short circuit to ground, exhaust system blocked, EGR valve, ECM**
55 [6]	Engine control module (ECM)	**Ground wiring open circuit, insecure PROM, ECM**
56	Engine control module quad driver/output driver	**Wiring, ECM controlled components, ECM [16]**
57	Boost control problem	**Wiring short circuit to ground, hoses blocked/leaking, boost bypass actuator, boost control solenoid, ECM**
58 [9]	Immobilizer contol module – fuel enable signal	**Incorrect ignition key, wiring open circuit/ short circuit to ground/short circuit to positive, immobilizer control module, ECM**

Flash code	Fault location	Probable cause
58 [7]	Transmission fluid temperature (TFT) sensor – voltage low	Wiring short circuit to ground, TFT sensor, ECM
59	Transmission fluid temperature (TFT) sensor – voltage high	Wiring open circuit, TFT sensor, ECM
61 [15]	Cruise control vent solenoid	Wiring open circuit/short circuit to ground/ short circuit to positive, cruise control servo, ECM
61 [6]	Oxygen sensor (O2S)/heated oxygen sensor (HO2S)	O2S/HO2S
62 [15]	Cruise control vacuum solenoid	Wiring open circuit/short circuit to ground/ short circuit to positive, cruise control servo, ECM
62 [6]	Transmission gear switch signal circuit	Wiring open circuit/short circuit to ground, 2nd gear switch, 3rd gear switch, ECM
63	Cruise control servo position sensor – circuit malfunction	Wiring open circuit/short circuit to ground/ short circuit to positive, hoses leaking, cruise control servo, ECM
65	Cruise control servo position sensor – circuit malfunction	Wiring open circuit/short circuit to ground/ short circuit to positive, cruise control servo, ECM
66 [1]	A/C refrigerant pressure sensor – circuit malfunction	Wiring open circuit/short circuit to ground/ short circuit to positive, A/C refrigerant pressure sensor, ECM
66 [2] [8]	Low A/C refrigerant charge	Wiring open circuit/short circuit to ground, low A/C refrigerant charge, blocked A/C orifice tube, A/C refrigerant low pressure switch, A/C refrigerant high pressure switch, A/C amplifier, ECM
67	Cruise control switch – circuit malfunction	Wiring open circuit/short circuit to positive, cruise control switch, brake pedal vacuum valve, ECM
68	Cruise control servo – servo position sensor high	Hoses blocked, brake pedal vacuum valve, cruise control servo, cruise control cable binding
69	A/C refrigerant pressure switch – engine coolant blower motors	Wiring short circuit to ground, A/C refrigerant pressure switch, ECM
70	A/C refrigerant pressure sensor – voltage high	Wiring open circuit/short circuit to positive, A/C refrigerant pressure sensor, ECM
72	Vehicle speed sensor (VSS)	Wiring open circuit/short circuit to ground, VSS, ECM
75	Exhaust gas recirculation (EGR) valve solenoid 1	Wiring open circuit/short circuit to ground, supply wire, EGR valve, ECM
76	Exhaust gas recirculation (EGR) valve solenoid 2	Wiring open circuit/short circuit to ground, supply wire, EGR valve, ECM
77	Exhaust gas recirculation (EGR) valve solenoid 3	Wiring open circuit/short circuit to ground, supply wire, EGR valve, ECM
79	Transmission fluid temperature (TFT) – high	Transmission mechanical fault, towing

Flash code	Fault location	Probable cause
80	Torque converter clutch (TCC) – slipping	**Wiring open circuit/short circuit to ground, supply wire, transmission internal wiring, TCC solenoid, transmission mechanical fault, ECM**
82	Ignition control module (ICM) – 3X reference circuit malfunction	**Wiring open circuit/short circuit to ground, CKP signal sensor, ICM, ECM**
85	Engine control module (ECM) – PROM	**Insecure PROM, incorrectly coded**
86	Engine control module (ECM) – analogue/ digital converter	**A/C refrigerant pressure sensor/KS/ECT sensor/fuel pump monitor circuit short circuit to battery voltage**
87	Engine control module (ECM) – PROM error	**Insecure PROM, incorrectly coded**
90	Torque converter clutch (TCC) – stuck ON	**Wiring open circuit/short circuit to ground, transmission internal wiring, TCC solenoid, transmission mechanical fault, ECM**
96	Engine control module (ECM) – ignition supply voltage low	**Wiring open circuit, alternator, ECM**
98	Engine control module (ECM) – incorrectly coded	**ECM incorrectly coded, ECM**
99 [9] [10]	Cruise control – power management mode	**Hoses blocked, brake pedal vacuum valve, cruise control servo, cruise control cable binding**
99 [7]	Engine control module (ECM) – incorrectly coded	**ECM incorrectly coded, ECM**

[1] Century, Regal 3.1L
[2] Le Sabre 3.8L
[3] Century 2.2L VIN 4, Century 2.5L VIN R
[4] 1993 Le Sabre 3.8L/1993 Park Avenue 3.8L, VIN L/1
[5] Century 2.2L
[6] Regal 3.1L VIN T
[7] Regal 3.1L VIN M
[8] Regal 3.8L VIN L
[9] Except Regal 3.1L VIN M
[10] Except Regal 3.8L VIN L
[11] Except Regal 3.1L VIN M & Regal 3.8L VIN L
[12] Park Avenue 3.8L VIN L, Park Avenue 3.8L VIN 1
[13] Except Le Sabre 3.8L VIN L, Regal 3.8L VIN L, Park Avenue 3.8L VIN L & Park Avenue 3.8L VIN 1
[14] Except Regal 3.1L VIN T & Regal 3.1L VIN M
[15] Except Regal 3.1L VIN T
[16] Depending on model, year and equipment level, the following components may be connected to this circuit. Use of the vehicle wiring diagram is recommended to confirm which components are fitted.

- □ A/C compressor clutch relay
- □ A/C condenser blower motor relay
- □ Engine coolant blower motor relay
- □ Engine coolant 'hot' warning lamp
- □ Evaporative emission (EVAP) canister purge valve
- □ Exhaust gas recirculation (EGR) solenoid
- □ Malfunction indicator lamp (MIL)
- □ Transmission shift warning light (MT)
- □ Torque converter clutch (TCC) solenoid
- □ Transmission shift solenoids (SS)

BUICK

Model:	Engine identification:	Year:
Century 2.2L	**VIN position 8 = 4**	**1996**
Century 3.1L	**VIN position 8 = M**	**1994-02**
LeSabre 3.8L	**VIN position 8 = L/K/1**	**1994-02**
Park Avenue 3.8L	**VIN position 8 = L/K/1**	**1994-02**
Regal 3.1L	**VIN position 8 = M**	**1996-97**
Regal 3.8L	**VIN position 8 = L/K/1**	**1994-02**

System: **GM**

Self-diagnosis

General information

- Refer to the front of this manual for general test conditions, terminology, detailed descriptions of wiring faults and a general trouble shooter for electrical and mechanical faults.
- Malfunction indicator lamp (MIL) will illuminate or flash if certain faults are recorded.
- ECM operates in backup mode if sensors fail, to enable vehicle to be driven to repair shop.

Accessing

- Trouble codes can only be displayed by using a Scan Tool connected to the data link connector (DLC).
- 12 pin connector **1** :
  - Century →95.
  - Regal →95.
- 16 pin connector **2** :
  - Century 96→.
  - LeSabre.
  - Park Avenue.
  - Regal 96→.

1

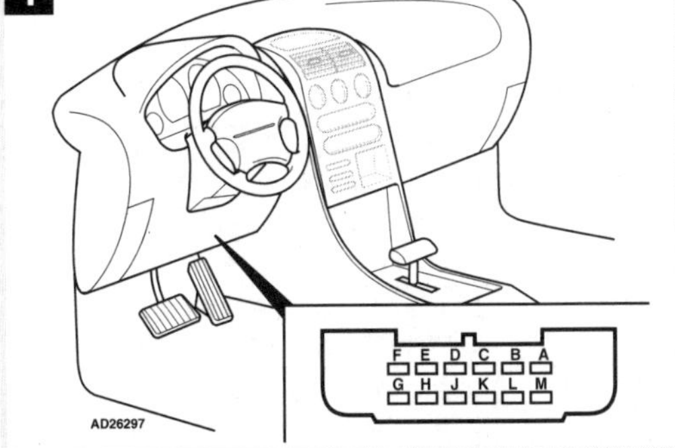

2

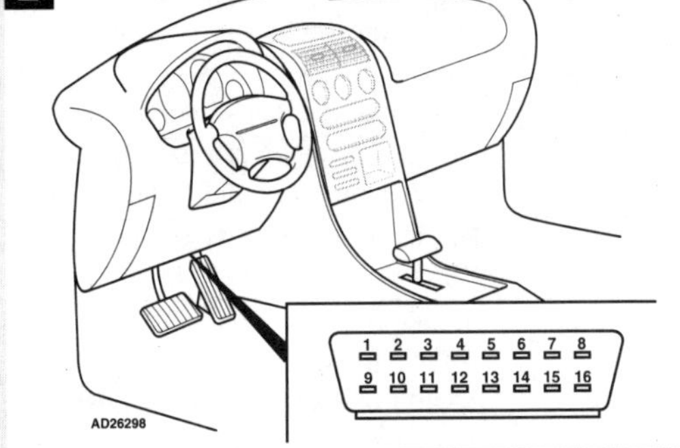

Erasing

→2000 – method 1

- After the faults have been rectified, the manufacturer recommends that trouble codes are erased by using a Scan Tool connected to the data link connector (DLC).

→2000 – method 2

NOTE: ***Disconnecting battery ground lead may cause driveability problems until the engine management system relearns basic values.***

- Ensure ignition switched OFF for at least 10 seconds.
- Disconnect battery ground cable for at least 30 seconds.
- Reconnect battery ground cable.
- Repeat accessing procedure to ensure no trouble codes remain in ECM fault memory.

NOTE: ***Disconnecting battery ground lead may also erase memory from electronic units, such as radio.***

2001→

- After the faults have been rectified, the manufacturer recommends that trouble codes are erased by using a Scan Tool connected to the data link connector (DLC).

Trouble code identification

Scan code	Fault location	Probable cause
P0000-P0999	Refer to OBD-II trouble code table at the front of this manual	–
P1106	Manifold absolute pressure (MAP) sensor – circuit intermittent – voltage high	**Wiring open circuit/short circuit to positive**
P1107	Manifold absolute pressure (MAP) sensor – circuit intermittent – voltage low	**Wiring open circuit/short circuit to ground**
P1111	Intake air temperature (IAT) sensor – circuit intermittent – voltage high	**Wiring open circuit/short circuit to positive, ground lead defective**
P1112	Intake air temperature (IAT) sensor – circuit intermittent – voltage low	**Wiring short circuit to ground**
P1114	Engine coolant temperature (ECT) sensor – circuit malfunction – voltage low	**Wiring short circuit to ground**
P1115	Engine coolant temperature (ECT) sensor – circuit malfunction – voltage high	**Wiring open circuit/short circuit to positive, ground lead defective**
P1121	Throttle position (TP) sensor – circuit malfunction – voltage high	**Wiring open circuit/short circuit to positive, ground lead defective**
P1122	Throttle position (TP) sensor – circuit malfunction – voltage low	**Wiring open circuit/short circuit to ground**
P1133	Heated oxygen sensor (HO2S) 1 – insufficient switching	**Wiring open circuit/short circuit to ground, exhaust system leaking, HO2S**
P1134	Heated oxygen sensor (HO2S) 1 – slow response	**Wiring open circuit/short circuit to ground, exhaust system leaking, HO2S**
P1171	Fuel system – lean during acceleration	**Fuel filter, fuel pump, HO2S, injectors**
P1189	Engine oil pressure warning lamp switch – circuit malfunction	**Wiring, engine oil pressure warning lamp switch, ECM**
P1200	Injector(s) – circuit malfunction	**Wiring open circuit/short circuit to ground/ short circuit to positive, ECM**

Scan code	Fault location	Probable cause
P1257	Boost control problem	**Wiring short circuit to ground, hoses blocked/leaking, boost bypass actuator, boost control solenoid, ECM**
P1260	Fuel pump speed relay – circuit malfunction	**Wiring open circuit/short circuit to ground/ short circuit to positive, fuel pump relay, fuel pump control module, ECM**
P1336	Crankshaft position sensor (CKP) – variation not learned	**Learning procedure not performed**
P1350	Ignition control module (ICM) – electronic spark timing (EST)/ignition control (IC)/bypass circuit malfunction	**Wiring open circuit/short circuit to ground/ short circuit to positive, ICM, ECM**
P1351	Ignition control module, ignition control (IC) circuit – voltage high	**Wiring open circuit, ICM, ECM**
P1352	Ignition control circuit – bypass line – voltage high	**Wiring open circuit, ICM, ECM**
P1361	Ignition control module (ICM) – electronic spark timing (EST)/ignition control (IC)/bypass circuit malfunction	**Wiring open circuit/short circuit to ground/ short circuit to positive, ICM, ECM**
P1362	Ignition control circuit, bypass line – voltage low	**Wiring short circuit to ground/short circuit to positive, ICM, ECM**
P1374	Crankshaft position (CKP)	**Wiring open circuit/short circuit to ground/ short circuit to positive, wiring routed near spark plug wires, ignition coils arcing/ tracking, ICM, ECM**
P1380	Rough road signal – signal error	**ABS control module trouble code(s) stored**
P1381	Rough road signal – no communication from antilock brake system (ABS) CAN data bus	**Wiring open circuit, ABS system**
P1403	Exhaust gas recirculation (EGR) valve solenoid 1	**Wiring open circuit/short circuit to ground, exhaust system blocked, EGR valve, ECM**
P1404 [1]	Exhaust gas recirculation (EGR) valve solenoid 2	**Wiring open circuit/short circuit to ground, exhaust system blocked, EGR valve, ECM**
P1404 [3] [9] [14]	Exhaust gas recirculation (EGR) valve – stuck open	**Wiring open circuit/short circuit to positive, EGR valve, exhaust system blocked, ECM**
P1405	Exhaust gas recirculation (EGR) valve solenoid 3	**Wiring open circuit/short circuit to ground, exhaust system blocked, EGR valve, ECM**
P1406	Exhaust gas recirculation (EGR) valve position sensor – circuit malfunction	**Wiring open circuit/short circuit to ground/ short circuit to positive, supply wire, EGR valve, ECM**
P1441	Evaporative emission (EVAP) system – leak detected	**Wiring short circuit to ground, hose connection(s), EVAP system vacuum switch, EVAP canister purge valve, ECM**
P1442	Evaporative emission (EVAP) system vacuum switch – circuit malfunction	**Wiring open circuit/short circuit to ground, EVAP system vacuum switch, ECM**
P1483	Engine cooling system – performance	**Cooling system, engine coolant temperature (ECT) sensor, intake air temperature (IAT) sensor**

Scan code	Fault location	Probable cause
P1520	Park/neutral position (PNP) switch	**Wiring short circuit to ground, incorrectly adjusted, ECM**
P1530	A/C refrigerant pressure sensor– circuit malfunction	**A/C refrigerant pressure, wiring open circuit/short circuit to ground/short circuit to positive, A/C refrigerant pressure sensor, ECM**
P1531	Low A/C refrigerant charge	**Low A/C refrigerant charge, wiring open circuit/short circuit to ground/short circuit to positive, A/C refrigerant low pressure switch, ECM**
P1546	A/C compressor clutch relay – circuit malfunction	**Wiring open circuit/short circuit to ground/ short circuit to positive, A/C compressor clutch relay, ECM**
P1550	Cruise control stepper motor, circuit malfunction	**Wiring open circuit/ short circuit to positive, cruise control module, ECM**
P1554	Cruise control status circuit	**Wiring short circuit to ground, cruise control module, ECM**
P1558	Cruise control servo position sensor – circuit malfunction	**Wiring open circuit/short circuit to ground/ short circuit to positive, hoses leaking, cruise control servo, ECM**
P1561	Cruise control vent solenoid	**Wiring open circuit/short circuit to ground/ short circuit to positive, cruise control servo, ECM**
P1562	Cruise control vacuum solenoid	**Wiring open circuit/short circuit to ground/ short circuit to positive, cruise control servo, ECM**
P1565	Cruise control servo position sensor – circuit malfunction	**Wiring open circuit/short circuit to ground/ short circuit to positive, cruise control servo, ECM**
P1567	Cruise control switch – circuit malfunction	**Wiring open circuit/short circuit to positive, cruise control switch, brake pedal vacuum valve, ECM**
P1568	Cruise control servo – servo position sensor high	**Hoses blocked, brake pedal vacuum valve, cruise control servo, cruise control cable binding**
P1571	Traction control system (TCS), torque request circuit	**Wiring open circuit/short circuit to ground/ short circuit to positive, ABS control module trouble code(s) stored, ECM**
P1573	Data bus – antilock brake system (ABS)/ traction control system (TCS)	**Wiring, ABS control module trouble code(s) stored**
P1575	Brake pedal position (BPP) switch – extended travel	**Wiring short circuit to positive, BPP switch, ECM**
P1585	Cruise control system – inhibit output circuit	**Wiring open circuit/short circuit to ground/ short circuit to positive, supply wire, cruise control motor/actuator, ECM**
P1599	Cruise control – power management mode	**Hoses blocked, brake pedal vacuum valve, cruise control servo, cruise control cable binding**

Scan code	Fault location	Probable cause
P1602	CAN data bus – antilock brake system (ABS)/ traction control system (TCS)	Wiring, ABS control module trouble code(s) stored
P1604	CAN data bus	Wiring, BCM trouble code(s) stored
P1605	CAN data bus – A/C	Wiring, A/C system trouble code(s) stored
P1610	CAN data bus – body control module (BCM)	Wiring, BCM trouble code(s) stored
P1619	Engine oil life monitor reset switch – circuit malfunction	Wiring short circuit to ground, engine oil life monitor reset switch, ECM
P1623	Engine control module (ECM) – PROM	Insecure PROM, ECM
P1626 11 12	Data bus, immobilizer	Wiring, BCM trouble code(s) stored
P1626 7 10 13 14	Immobilizer control module – fuel enable signal	Incorrect ignition key, wiring open circuit/ short circuit to ground/short circuit to positive, immobilizer control module, ECM
P1629	Immobilizer control module – fuel enable signal	Incorrect ignition key, wiring open circuit/ short circuit to ground/short circuit to positive, immobilizer control module, ECM
P1630 1 5 12	Battery voltage – high/low	Wiring, alternator, ECM
P1630 2 7 10 15	Engine control module (ECM), immobilizer code learning mode	Immobilizer code not programmed
P1631	Engine control module (ECM) – incorrect immobilizer code	Incorrect key, immobilizer code not programmed
P1635	Sensor supply voltage – circuit A malfunction	Wiring short circuit to positive, ground wiring open circuit, TP sensor, MAP sensor, EGR valve, ECM
P1639	Sensor supply voltage – circuit B malfunction	Wiring short circuit to ground/positive, ground wiring open circuit, A/C refrigerant pressure sensor, ECM
P1640	Engine control module quad driver/output driver	Wiring, ECM controlled components, ECM 16
P1641 4 8 12 13	Malfunction indicator lamp (MIL) – circuit malfunction	Wiring open circuit/short circuit to ground/ short circuit to positive, MIL, supply wire, ECM
P1641 3 6 14	A/C compressor clutch relay – circuit malfunction	Wiring open circuit/short circuit to positive, supply wire, A/C compressor relay, ignition main relay, ECM
P1643	Fuel pump control module – circuit malfunction	Wiring open circuit/short circuit to ground/ short circuit to positive, fuel pump control module, ECM
P1645	Supercharger (SC) control valve solenoid – circuit malfunction	Wiring open circuit/short circuit to ground/ short circuit to positive, SC control valve solenoid, ECM
P1646	Supercharger (SC) control valve solenoid – circuit malfunction	Wiring open circuit/short circuit to ground/ short circuit to positive, SC control valve solenoid, ECM

Scan code	Fault location	Probable cause
P1650	Engine control module, quad driver/output driver	**Wiring, ECM controlled components, ECM 16**
P1651	Engine coolant blower motor relay 1 – circuit malfunction	**Wiring open circuit/short circuit to ground/ short circuit to positive, supply wire, engine coolant blower motor relay 1, ECM**
P1652	Engine coolant blower motor relay 2 – circuit malfunction	**Wiring open circuit/short circuit to ground/ short circuit to positive, supply wire, engine coolant blower motor relay 2, ECM**
P1653	Traction control system (TCS) – delivered torque circuit	**Wiring open circuit/short circuit to ground/ short circuit to positive, ABS control module trouble code(s) stored, ECM**
P1654	A/C compressor clutch relay – circuit malfunction	**Wiring open circuit/short circuit to ground/ short circuit to positive, supply wire, A/C compressor clutch relay, ECM**
P1655	Evaporative emission (EVAP) canister purge valve – circuit malfunction	**Wiring open circuit/short circuit to ground/ short circuit to positive, supply wire, EVAP canister purge valve, ECM**
P1660	Engine control module quad driver/output driver	**Wiring, ECM controlled components, ECM 16**
P1662	Cruise control inhibit circuit – circuit malfunction	**Wiring open circuit/short circuit to ground, cruise control motor/actuator, cruise control module, ECM**
P1663	Alternator warning lamp – circuit malfunction	**Wiring open circuit/short circuit to ground/ short circuit to positive, supply wire, instrument panel, ECM**
P1665	Evaporative emission (EVAP) canister vent valve – circuit malfunction	**Wiring open circuit/short circuit to ground/ short circuit to positive, supply wire, EVAP canister vent valve, ECM**
P1667	Fuel pump control module – circuit malfunction	**Wiring open circuit/short circuit to ground/ short circuit to positive, supply wire, fuel pump control module, ECM, fuel pump relay**
P1670	Engine control module quad driver/output driver	**Wiring, ECM controlled components, ECM 16**
P1671	Malfunction indicator lamp (MIL) – circuit malfunction	**Wiring open circuit/short circuit to ground/ short circuit to positive, supply wire, ECM**
P1672	Engine oil level warning lamp – circuit malfunction	**Wiring open circuit/short circuit to ground/ short circuit to positive, supply wire, ECM**
P1673	Engine coolant 'hot' warning lamp – circuit malfunction	**Wiring open circuit/short circuit to ground, cooling system, ECM**
P1674	Tachometer – circuit malfunction	**Wiring open circuit/short circuit to ground/ short circuit to positive, instrument panel, ECM**
P1675	Evaporative emission (EVAP) canister vent valve – circuit malfunction	**Wiring open circuit/short circuit to ground, EVAP canister vent valve, ECM**
P1676	Evaporative emission (EVAP) canister purge valve – circuit malfunction	**Wiring open circuit/short circuit to ground/ short circuit to positive, supply wire, EVAP canister purge valve, ECM**

Scan code	Fault location	Probable cause
P1689	Traction control system (TCS) – delivered torque output circuit	**Wiring open circuit/short circuit to ground/ short circuit to positive, supply wire, ABS trouble codes stored, ECM**

1 Century 3.1L →1995
2 Century 3.1L 1997→
3 Century 3.1L 1998→
4 Century 3.1L 1996, 1997
5 Le Sabre 3.8L/Park Avenue 3.8L 1994, 1995
6 Le Sabre 3.8L/Park Avenue 3.8L 1998→
7 Le Sabre/Park Avenue
8 Le Sabre 3.8L 1996, Park Avenue 3.8L 1996, Regal 3.1L 1996-97
9 Le Sabre 3.8L/ Park Avenue 3.8L 1998→
10 Regal 3.1L
11 Le Sabre/Park Avenue 3.8L VIN 1/L/M 1994-95
12 Regal 3.8L 1994-95
13 Regal 3.8L 1996-97
14 Regal 3.8L 1998→
15 Regal 3.8L 1997-00
16 Depending on model, year and equipment level, the following components may be connected to this circuit. Use of the vehicle wiring diagram is recommended to confirm which components are fitted.

- ☐ A/C compressor clutch relay
- ☐ A/C condenser blower motor relay
- ☐ Engine coolant blower motor relay
- ☐ Engine coolant 'hot' warning lamp
- ☐ Evaporative emission (EVAP) canister purge valve
- ☐ Exhaust gas recirculation (EGR) solenoid
- ☐ Malfunction indicator lamp (MIL)
- ☐ Transmission shift warning light (MT)
- ☐ Torque converter clutch (TCC) solenoid
- ☐ Transmission shift solenoids (SS)

BUICK

Model:	Engine identification:	Year:
Riviera 3.8L	VIN position 8 = L	1992-94
Roadmaster 5.7L	VIN position 8 = 7	1992-93
Roadmaster 5.7L	VIN position 8 = P	1994-95
Skylark 2.3L	VIN position 8 = A/D/3	1992-95
Skylark 3.3L	VIN position 8 = N	1992-93
Skylark 3.1L	VIN position 8 = M	1994-95

System: **GM**

Self-diagnosis

General information

- Refer to the front of this manual for general test conditions, terminology, detailed descriptions of wiring faults and a general trouble shooter for electrical and mechanical faults.
- Malfunction indicator lamp (MIL) will illuminate continuously if permanent faults are recorded.
- Malfunction indicator lamp (MIL) may illuminate briefly and then switch off if intermittent faults are recorded.
- ECM operates in backup mode if sensors fail, to enable vehicle to be driven to repair shop.
- Trouble codes can be displayed by the MIL or by using a Scan Tool connected to the data link connector (DLC).

Accessing

Models with data link connector (DLC) terminal B connected

- Access data link connector (DLC).
- Jump data link connector (DLC) terminals A and B **1**.
- Switch ignition ON. Do NOT start engine.
- Count MIL flashes. Note trouble codes. Compare with trouble code table.
- Firstly trouble code 12 will be displayed 3 times.
- Trouble codes consist of 2 groups, displayed as follows:
  - First group indicates the 'tens' of the trouble code **2** [**A**].
  - Second group indicates the 'units' of the trouble code **2** [**C**].
  - A short pause separates each group **2** [**B**].
  - A long pause separates each trouble code **2** [**D**].
  - For example: Trouble code 12 displayed **2**. Start of diagnosis.
- Each additional trouble code will now be displayed 3 times, in ascending order.

NOTE: ***If no additional trouble codes are stored, code 12 will continue to be displayed.***

- Trouble codes can also be displayed by using a Scan Tool connected to the data link connector (DLC) **1**.

1

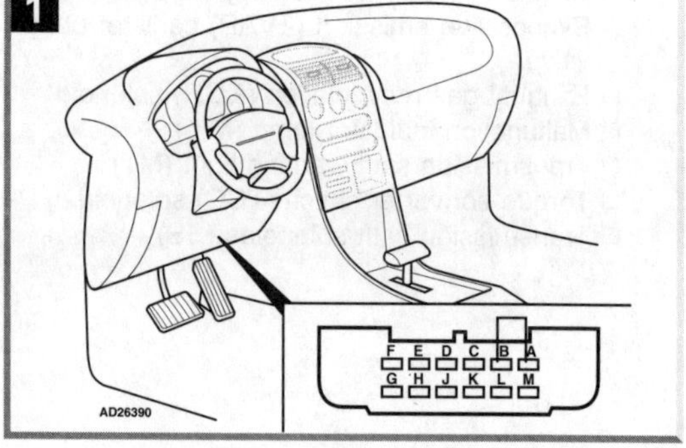

2

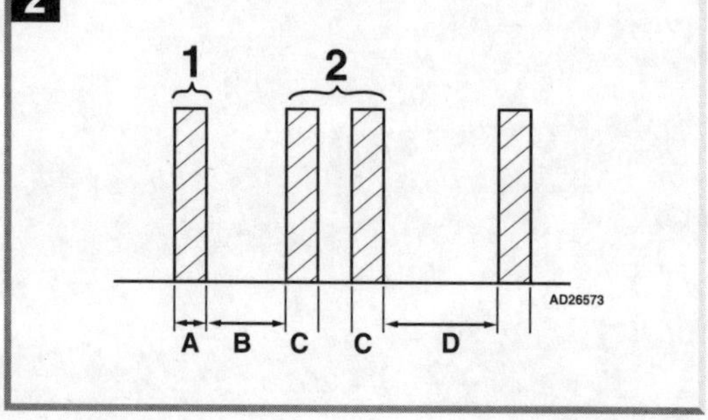

Models without data link connector (DLC) terminal B connected

- Trouble codes can only be displayed by using a Scan Tool connected to the data link connector (DLC) [1].

Erasing

NOTE: ***Disconnecting battery ground lead may cause driveability problems until the engine management system relearns basic values.***

- Ensure ignition switched OFF.
- Disconnect battery ground cable for at least 30 seconds.
- Reconnect battery ground cable.
- Repeat accessing procedure to ensure no trouble codes remain in ECM fault memory.
- Trouble codes can also be erased by using a Scan Tool connected to the data link connector (DLC).

NOTE: ***Disconnecting battery ground lead may also erase memory from electronic units, such as radio.***

Trouble code identification

Flash code	Fault location	Probable cause
11	Malfunction indicator lamp (MIL) – circuit malfunction	**Wiring open circuit/short circuit to ground, MIL, supply wire, ECM**
12	Start of diagnosis	**No engine speed (RPM) signal received**
13 [8]	Oxygen sensor (O2S)/heated oxygen sensor (HO2S)	**Wiring open circuit, O2S/HO2S, ECM**
13 [4]	Heated oxygen sensor (HO2S) bank 1 – open circuit	**Wiring open circuit/short circuit to ground/ short circuit to positive, HO2S, ECM**
14	Engine coolant temperature (ECT) sensor – high temperature indicated	**Wiring short circuit to ground, ECT sensor, ECM**
15	Engine coolant temperature (ECT) sensor – low temperature indicated	**Wiring open circuit, ECT sensor, ECM**
16 [9]	Battery voltage – high or low	**Wiring, alternator, ECM**
16 [4]	Ignition system – low resolution pulse	**Wiring open circuit/short circuit to ground/ short circuit to positive, distributor, ECM**
16 [5]	2X reference cicuit – circuit malfunction	**Wiring open circuit/short circuit to ground/ short circuit to positive, ICM, ECM**
17 [10]	Spark reference circuit	**Wiring open circuit/short circuit to ground, ICM, ECM**
17 [7]	Camshaft position (CMP) sensor – circuit malfunction	**Wiring open circuit/short circuit to ground/ short circuit to positive, mechanical fault, CMP sensor, ECM**
18	Injector(s) – circuit malfunction	**Wiring open circuit/short circuit to ground, supply wire, injectors, ECM**
19	7X reference – circuit malfunction	**Wiring, mechanical fault, CKP sensor, PROM, ECM**
21	Throttle position (TP) sensor – voltage high	**Wiring open circuit/short circuit to positive, TP sensor, ECM**
22	Throttle position (TP) sensor – voltage low	**Wiring open circuit/short circuit to ground, TP sensor, ECM**
23	Intake air temperature (IAT) sensor – low temperature	**Wiring open circuit, IAT sensor, ECM**

Flash code	Fault location	Probable cause
24	Vehicle speed sensor (VSS)	**Wiring open circuit/short circuit, PROM, VSS, ECM**
25	Intake air temperature (IAT) sensor – high temperature	**Wiring short circuit to ground, IAT sensor, ECM**
26 8	Engine control module quad driver/output driver	**Wiring open circuit/short circuit to ground/ short circuit to positive, ECM controlled components, ECM 16**
26 4	Evaporative emission (EVAP) canister purge valve – circuit malfunction	**Wiring open circuit/short circuit to ground/ short circuit to positive, EVAP canister purge valve, ECM**
27 11 12	Exhaust gas recirculation (EGR) vacuum control – circuit malfunction	**Wiring open circuit/short circuit to ground/ short circuit to positive, EGR valve, ECM**
27 5	Engine control module quad driver/output driver	**Wiring open circuit/short circuit to ground/ short circuit to positive, ECM controlled components, ECM 16**
27 6	Transmission gear switch signal circuit 2nd gear	**Wiring open circuit/short circuit to ground, 2nd gear switch, ECM**
28 10 11	Transmission range (TR) pressure switch	**Wiring open circuit/short circuit to ground, ECM**
28 5	Engine control module quad driver/output driver	**Wiring open circuit/short circuit to ground/ short circuit to positive, ECM controlled components, ECM 16**
28 6	Transmission gear switch signal circuit 3rd gear	**Wiring open circuit/short circuit to ground, 3rd gear switch, ECM**
28 7	Transmission range switch – circuit malfunction	**Wiring open circuit/short circuit to ground, transmission range switch, ECM**
29 11	Secondary air injection (AIR) system	**Wiring open circuit/short circuit to ground/ short circuit to positive, AIR valve relay, ECM**
29 5	Engine control module quad driver/output driver	**Wiring open circuit/short circuit to ground/ short circuit to positive, ECM controlled components, ECM 16**
31	Park/neutral position (PNP) switch	**Wiring open circuit/short circuit to ground, incorrectly adjusted, PNP switch, ECM**
32	Exhaust gas recirculation (EGR) solenoid	**Wiring open circuit/short circuit to ground, exhaust system blocked, EGR valve, ECM**
33	Manifold absolute pressure (MAP) sensor – circuit malfunction	**Wiring open circuit/short circuit to positive, hose blocked/leaking, MAP sensor, ECM**
34 10 11 13	Mass air flow (MAF) sensor – circuit malfunction	**Wiring open circuit/short circuit to ground/ short circuit to positive, MAF sensor, ECM**
34 1 5 7	Manifold absolute pressure (MAP) sensor – circuit malfunction	**Wiring open circuit/short circuit to ground, MAP sensor, ECM**
35	Idle air control (IAC) valve	**Wiring open circuit/short circuit to ground/ short circuit to positive, intake system leaking, mechanical fault, IAC valve, ECM**
36 8 10	Shift solenoids (SS)	**Wiring open circuit/short circuit to ground, SS**
36 4	Ignition system high resolution or extra low resolution pulse – circuit malfunction	**Wiring open circuit/short circuit to ground/ short circuit to positive, distributor, ECM**

Flash code	Fault location	Probable cause
36 [7]	Ignition control module (ICM) – 24X signal circuit malfunction	**Wiring open circuit/short circuit to ground, CKP sensor, ECM**
37 [10]	Brake switch stuck ON	**Wiring open circuit/short circuit to ground/ short circuit to positive, brake switch, ECM**
37 [7]	Brake pedal position (BPP) switch for torque converter clutch (TCC) – circuit malfunction	**Wiring open circuit/short circuit to ground/ short circuit to positive, TCC brake switch, ECM**
38 [8]	Brake pedal position (BPP) switch	**Wiring open circuit/short circuit to ground/ short circuit to positive, BPP switch incorrectly adjusted, BPP switch, ECM**
38 [4]	Brake switch stuck OFF	**Wiring open circuit/short circuit to ground/ short circuit to positive, brake switch, ECM**
39	Torque converter clutch (TCC) – circuit malfunction	**Wiring open circuit, BPP switch, TCC solenoid, TCC switch, 3rd gear switch, ECM, mechanical fault**
41 [9]	Camshaft position (CMP) sensor	**Wiring open circuit/short circuit to ground, CMP sensor, ICM, ECM**
41 [4]	Ignition control module, ignition control (IC) circuit – open circuit	**Wiring open circuit, ICM, ECM**
41 [5]	1X reference cicuit – circuit malfunction	**Wiring open circuit/short circuit to ground/ short circuit to positive, ICM, ECM**
42 [14]	Ignition control module (ICM) – electronic spark timing (EST)/ignition control (IC)/bypass circuit malfunction	**Wiring open circuit/short circuit to ground, ICM, ECM**
42 [2]	Ignition control module (ICM) – electronic spark timing (EST)/ignition control (IC)/bypass circuit malfunction	**Wiring open circuit/short circuit to ground, DI control module, ECM**
43 [15]	Knock sensor (KS)	**Wiring open circuit/short circuit to ground/ short to positive, KS, PROM, ECM**
43 [3]	Knock sensor (KS)	**Wiring open circuit/short circuit to ground/ short to positive, KS module, KS, ECM**
44 [8]	Oxygen sensor (O2S)/heated oxygen sensor (HO2S) – mixture lean	**Wiring short circuit to ground, fuel pressure low, fuel contamination, exhaust system leaking, O2S/HO2S, ECM**
44 [4]	Heated oxygen sensor (HO2S) bank 1 – mixture lean	**Wiring short circuit to ground, injector(s) blocked, fuel contamination, fuel pressure low, exhaust system leaking, HO2S, ECM**
45 [8]	Oxygen sensor (O2S)/heated oxygen sensor (HO2S) – mixture rich	**Fuel pressure high, injector(s) leaking, shield wiring open circuit, O2S contamination, EVAP canister purge valve, TP sensor, MAP sensor, EGR system, excessive fuel in engine oil, ECM**
45 [4]	Heated oxygen sensor (HO2S) bank 1 – mixture rich	**Wiring open circuit, fuel pressure high, injector(s) leaking, HO2S contamination, EVAP canister purge valve, TP sensor, EGR system, HO2S, MAF sensor, ECM**
46	Immobilizer control module – fuel enable signal	**Incorrect ignition key, wiring open circuit/ short circuit to ground, immobilizer control module, ECM**

Flash code	Fault location	Probable cause
47 8	Body control module (BCM), CAN data bus – communication malfunction	Wiring, PROM, BCM
47 4	Knock sensor (KS) – circuit malfunction	insecure PROM, ECM
48 15	Mass air flow (MAF) sensor – circuit malfunction	Wiring open circuit/short circuit to ground/ short circuit to positive, MAF sensor, ECM
48 3	Misfire	ignition system, fuel system, O2S
50	System voltage low	Wiring open circuit/short circuit to ground, alternator, ECM
51	Engine control module (ECM) – PROM	Insecure PROM, incorrect PROM, ECM
53 11	System voltage high	Alternator, ECM
53	Battery voltage error	Wiring open circuit, alternator, ECM
54	Fuel pump circuit – voltage low	Wiring open circuit/short circuit to ground, fuel pump relay, fuel pump, engine oil pressure warning lamp switch, ECM
55 8	Engine control module (ECM)	Ground wiring open circuit, insecure PROM, EC
55 4 5	Engine control module (ECM), fuel lean monitor	Fuel filter/lines blocked, fuel contamination
58 8	Immobilizer contol module – fuel enable signal	Incorrect ignition key, wiring open circuit/ short circuit to ground/short circuit to positive, immobilizer control module, ECM
58 4 7	Transmission fluid temperature (TFT) sensor – high temperature indicated	Wiring short circuit to ground, TFT sensor, ECM
59	Transmission fluid temperature (TFT) sensor – low temperature indicated	Wiring open circuit, TFT sensor, ECM
61	Cruise control vent solenoid	Wiring open circuit/short circuit to ground/ short circuit to positive, cruise control servo, ECM
62	Cruise control vacuum solenoid	Wiring open circuit/short circuit to ground/ short circuit to positive, cruise control servo, ECM
63	Heated oxygen sensor (HO2S) bank 2 – open circuit	Wiring open circuit/short circuit to ground/ short circuit to positive, HO2S contamination, HO2S, ECM
64	Heated oxygen sensor (HO2S) bank 2 – mixture lean	Wiring short circuit to ground, injector(s) blocked, fuel contamination, fuel pressure low, exhaust system leaking, ECM
65 9	Cruise control servo position sensor – circuit malfunction	Wiring open circuit/short circuit to ground/ short circuit to positive, cruise control servo, ECM
65 4	Heated oxygen sensor (HO2S) bank 2 – mixture rich	Fuel pressure high, injector(s) leaking, EVAP canister purge valve, TP sensor, EGR system, HO2S contamination, MAF sensor, ECM
65 5	Fuel injector – circuit malfunction	Wiring open circuit/short circuit to ground/ short circuit to positive, injector, ECM
66 10	A/C refrigerant pressure sensor circuit – open or short circuit	Wiring open circuit/short circuit to ground/ short circuit to positive, A/C refrigerant pressure sensor, ECM

Flash code	Fault location	Probable cause
66 7	A/C refrigerant pressure sensor – low pressure	**Wiring open circuit, A/C refrigerant pressure sensor, ECM**
67 8	Cruise control switch – circuit malfunction	**Wiring open circuit/short circuit to positive, cruise control switch, brake pedal vacuum valve, ECM**
67 4	A/C refrigerant pressure sensor circuit – pressure sensor or A/C clutch circuit malfunction	**Wiring open circuit/short circuit to ground/ short circuit to positive, A/C refrigerant pressure sensor, A/C compressor clutch**
68 8	Cruise control servo – servo position sensor high	**Hoses blocked, brake pedal vacuum valve, cruise control servo, cruise control cable binding**
68 4	A/C compressor clutch relay – circuit malfunction	**Wiring short circuit to ground/short circuit to positive, A/C compressor clutch relay, ECM**
69	A/C compressor clutch relay – circuit malfunction	**Wiring open circuit, A/C compressor clutch relay, ECM**
70 10	A/C compressor clutch relay – circuit malfunction	**Wiring open circuit/short circuit to ground/ short to positive, A/C compressor clutch relay, ECM**
70 7	A/C refrigerant pressure sensor – high pressure	**Wiring open circuit/short circuit to positive, A/C refrigerant pressure sensor, ECM**
72	Vehicle speed sensor (VSS) – circuit malfunction	**Wiring open circuit/short circuit to ground, VSS, ECM**
73	Pressure control solenoid – circuit malfunction	**Wiring open circuit/short circuit to ground/ short to positive, ECM**
75 10	Transmission system – voltage low	**supply wire, alternator, ECM**
75 7	Exhaust gas recirculation (EGR) valve solenoid 1	**Wiring open circuit/short circuit to ground, supply wire, EGR valve, ECM**
76	Exhaust gas recirculation (EGR) valve solenoid 2	**Wiring open circuit/short circuit to ground, supply wire, EGR valve, ECM**
77 10	Engine coolant blower motor relay 1 – circuit malfunction	**Wiring open circuit/short circuit to ground/ short circuit to positive, engine coolant blower motor relay, ECM**
77 7	Exhaust gas recirculation (EGR) valve solenoid 3	**Wiring open circuit/short circuit to ground, supply wire, EGR valve, ECM**
78	Engine coolant blower motor relay 2 – circuit malfunction	**Wiring open circuit/short circuit to ground/ short circuit to positive, engine coolant blower motor relay, ECM**
79	Transmission fluid – overtemp	**Wiring open circuit/short circuit to ground, TFT sensor, ECM**
80	Transmission component – slipping	**Wiring open circuit/short circuit to ground/ short circuit to positive, TR pressure switch, transmission mechanical fault, ECM**
81	Transmission shift solenoid (SS), 2nd-3rd – circuit malfunction	**Wiring open circuit/short circuit to ground, shift solenoid, ECM**
82 10	Transmission shift solenoid (SS), 1st-2nd – circuit malfunction	**Wiring open circuit/short circuit to ground, shift solenoid, ECM**
82 7	Ignition control module (ICM) – 3X reference circuit malfunction	**Wiring open circuit/short circuit to ground, CKP sensor, ICM, ECM**

Flash code	Fault location	Probable cause
83	Torque converter clutch (TCC) solenoid, pulse width modulated – circuit malfunction	**Wiring open circuit/short circuit to ground/ short circuit to positive, TCC solenoid, ECM**
84	Transmission shift solenoid (SS), 3rd-2nd – circuit malfunction	**Wiring open circuit/short circuit to ground, shift solenoid, ECM**
85 [10]	Torque converter clutch (TCC) – stuck ON	**Wiring open circuit/short circuit to ground, transmission mechanical fault, TCC solenoid**
85 [7]	Engine control module (ECM) – PROM	**Insecure PROM, incorrect PROM, ECM**
86	ECM analog/digital – signal error	**Wiring short to positive, A/C refrigerant pressure sensor, KS, ECT sensor, other DTC codes**
87	Engine control module (ECM) – EEPROM	**Insecure EEPROM, incorrect EEPROM, ECM**
90	Torque converter clutch (TCC) solenoid – circuit malfunction	**Wiring open circuit/short circuit to ground, TCC solenoid, ECM**
95	Change oil lamp – circuit malfunction	**Wiring open circuit/short circuit to ground/ short circuit to positive, instrument panel, engine oil life lamp bulb, ECM**
96 [10]	Engine oil level warning lamp – circuit malfunction	**Wiring open circuit/short circuit to ground/ short circuit to positive, instrument panel, engine oil level lamp bulb, oil level sensor, ECM**
96 [7]	Transmission system – voltage low	**Wiring open circuit, alternator, ECM**
97	Vehicle speed sensor (VSS) – circuit malfunction	**Wiring open circuit/short circuit to ground/ short circuit to positive, VSS, ECM**
98	Invalid ECM program	**ECM**
99	Invalid ECM program	**ECM**

[1] Riviera 3.8L
[2] Roadmaster 5.7L VIN 7 1993
[3] Roadmaster 5.7L VIN 7/P 1993-95
[4] Roadmaster 5.7L VIN P 1994-95
[5] Skylark 2.3L
[6] Skylark 3.3L
[7] Skylark 3.1L
[8] Except Roadmaster 5.7L VIN P 1994-95
[9] Except Roadmaster 5.7L VIN P 1994-95 & Skylark 2.3L
[10] Except Skylark 3.1L
[11] Except Skylark 2.3L
[12] Except Skylark 3.3L
[13] Except Riviera 3.8L
[14] Except Roadmaster 5.7L VIN 7 1993
[15] Except Roadmaster 5.7L VIN 7 1993-95
[16] Depending on model, year and equipment level, the following components may be connected to this circuit. Use of the vehicle wiring diagram is recommended to confirm which components are fitted.

- ☐ A/C compressor clutch relay
- ☐ A/C condenser blower motor relay
- ☐ Engine coolant blower motor relay
- ☐ Engine coolant 'hot' warning lamp
- ☐ Evaporative emission (EVAP) canister purge valve
- ☐ Exhaust gas recirculation (EGR) solenoid
- ☐ Malfunction indicator lamp (MIL)
- ☐ Transmission shift warning light (MT)
- ☐ Torque converter clutch (TCC) solenoid
- ☐ Transmission shift solenoids (SS)

BUICK

Model:	Engine identification:	Year:
Riviera 3.8L	**VIN position 8 = K/1**	**1995-99**
Roadmaster 5.7L	**VIN position 8 = P**	**1996**
Skylark 2.4L	**VIN position 8 = T**	**1996-98**
Skylark 3.1L	**VIN position 8 = M**	**1996-98**

System: **GM**

Self-diagnosis

General information

- Refer to the front of this manual for general test conditions, terminology, detailed descriptions of wiring faults and a general trouble shooter for electrical and mechanical faults.
- Malfunction indicator lamp (MIL) will illuminate or flash if certain faults are recorded.
- ECM operates in backup mode if sensors fail, to enable vehicle to be driven to repair shop.

Accessing

- Trouble codes can only be displayed by using a Scan Tool connected to the data link connector (DLC) **1**.

1

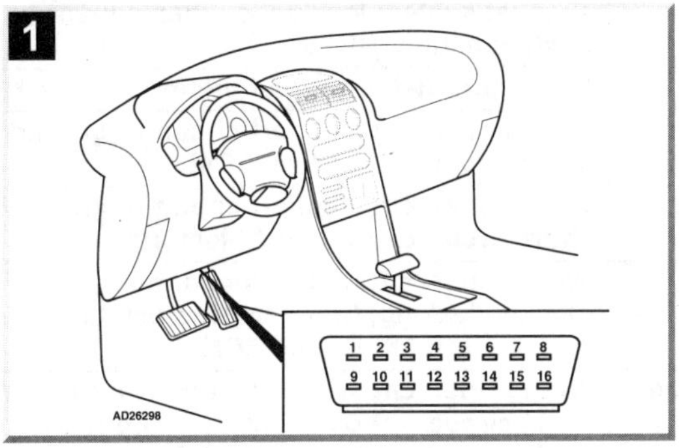

Erasing

Method 1

- After the faults have been rectified, the manufacturer recommends that trouble codes are erased by using a Scan Tool connected to the data link connector (DLC).

Method 2

NOTE: ***Disconnecting battery ground lead may cause driveability problems until the engine management system relearns basic values.***

- Ensure ignition switched OFF for at least 10 seconds.
- Disconnect battery ground cable for at least 30 seconds.
- Reconnect battery ground cable.
- Repeat accessing procedure to ensure no trouble codes remain in ECM fault memory.

NOTE: ***Disconnecting battery ground lead may also erase memory from electronic units, such as radio.***

Trouble code identification

Scan code	Fault location	Probable cause
P0000-P0999	Refer to OBD-II trouble code table at the front of this manual	–
P1106	Manifold absolute pressure (MAP) sensor – circuit intermittent – voltage high	**Wiring open circuit/short circuit to positive**
P1107	Manifold absolute pressure (MAP) sensor – circuit intermittent – voltage low	**Wiring open circuit/short circuit to ground**

Scan code	Fault location	Probable cause
P1111	Intake air temperature (IAT) sensor – circuit intermittent – voltage high	**Wiring open circuit/short circuit to positive, ground lead defective**
P1112	Intake air temperature (IAT) sensor – circuit intermittent – voltage low	**Wiring short circuit to ground**
P1114	Engine coolant temperature (ECT) sensor – circuit malfunction – voltage low	**Wiring short circuit to ground**
P1115	Engine coolant temperature (ECT) sensor – circuit malfunction – voltage high	**Wiring open circuit/short circuit to positive, ground lead defective**
P1121	Throttle position (TP) sensor – circuit malfunction – voltage high	**Wiring open circuit/short circuit to positive, ground lead defective**
P1122	Throttle position (TP) sensor – circuit malfunction – voltage low	**Wiring open circuit/short circuit to ground**
P1133 5	Heated oxygen sensor (HO2S) 1 – insufficient switching	**Wiring open circuit/short circuit to ground, exhaust system leaking, HO2S**
P1133 3	Heated oxygen sensor (HO2S) 1, bank 1 – insufficient switching	**Wiring open circuit/short circuit to ground, exhaust system leaking, oxygen sensor heater, oxygen sensor contamination, HO2S**
P1134 5	Heated oxygen sensor (HO2S) 1 – slow response	**Wiring open circuit/short circuit to ground, exhaust system leaking, HO2S**
P1134 3	Heated oxygen sensor (HO2S) 1, bank 1 – slow response	**Wiring open circuit/short circuit to ground, exhaust system leaking, oxygen sensor heater, oxygen sensor contamination, HO2S**
P1153	Heated oxygen sensor (HO2S) 1, bank 2 – insufficient switching	**Wiring open circuit/short circuit to ground, exhaust system leaking, oxygen sensor heater, oxygen sensor contamination, HO2S**
P1154	Heated oxygen sensor (HO2S) 1, bank 2 – slow response	**Wiring open circuit/short circuit to ground, exhaust system leaking, oxygen sensor contamination, HO2S**
P1171	Fuel system – lean during acceleration	**Fuel filter, fuel lines, fuel pump**
P1200	Injector(s) – circuit malfunction	**Wiring open circuit/short circuit to ground/ short circuit to positive, ECM**
P1222	Injector control – circuit malfunction	**Wiring open circuit/short circuit to ground/ short circuit to positive, injector(s)**
P1257	Boost control problem	**Wiring short circuit to ground, hoses blocked/leaking, boost bypass actuator, boost control solenoid, ECM**
P1260	Fuel pump speed relay – circuit malfunction	**Wiring open circuit/short circuit to ground/ short circuit to positive, fuel pump relay, Fuel pump control module, ECM**
P1336	Crankshaft position sensor (CKP) – variation not learned	**Learning procedure not performed**
P1350	Ignition control module (ICM) – electronic spark timing (EST)/ignition control (IC)/bypass circuit malfunction	**Wiring open circuit/short circuit to ground/ short circuit to positive, ICM, ECM**
P1351	Ignition control module, ignition control (IC) circuit – voltage high	**Wiring open circuit, ICM, ECM**
P1352	Ignition control (IC) circuit – bypass line – voltage high	**Wiring open circuit, ICM, ECM**

Scan code	Fault location	Probable cause
P1361	Ignition control module (ICM) – electronic spark timing (EST)/ignition control (IC)/bypass circuit malfunction	**Wiring open circuit/short circuit to ground/ short circuit to positive, ICM, ECM**
P1362	Ignition control circuit, bypass line – voltage low	**Wiring short circuit to ground/short circuit to positive, ICM, ECM**
P1371	Direct ignition (DI) low resolution circuit malfunction	**Wiring open circuit/short circuit to ground/ short circuit to positive, distributor, ICM, ECM**
P1374	Crankshaft position (CKP)	**Wiring open circuit/short circuit to ground/ short circuit to positive, wiring routed near spark plug wires, ignition coils arcing/ tracking, ICM, ECM**
P1380	Rough road signal – signal error	**ABS control module trouble code(s) stored**
P1381	Rough road signal – no communication from antilock brake system (ABS) CAN data bus	**Wiring open circuit, ABS system**
P1404	Exhaust gas recirculation (EGR) valve – stuck open	**Wiring open circuit/short circuit to positive, EGR valve, exhaust system blocked, ECM**
P1406	Exhaust gas recirculation (EGR) valve position sensor – circuit malfunction	**Wiring open circuit/short circuit to ground/ short circuit to positive, EGR valve position sensor, EGR valve, ECM**
P1415	Secondary air injection (AIR) system – bank 1	**Wiring, hoses blocked/leaking, AIR valve**
P1416	Secondary air injection (AIR) system – bank 2	**Wiring, hoses blocked/leaking, AIR valve**
P1441	Evaporative emission (EVAP) system – leak detected	**Wiring short circuit to ground, hose connection(s), EVAP system vacuum switch, EVAP canister purge valve, ECM**
P1442	Evaporative emission (EVAP) system vacuum switch – circuit malfunction	**Wiring open circuit/short circuit to ground, EVAP system vacuum switch, ECM**
P1508	Idle air control (IAC) valve – rpm low	**Wiring open circuit/short circuit to ground/ short circuit to positive, IAC passages blocked, IAC valve**
P1509	Idle air control (IAC) valve – rpm high	**Wiring open circuit/short circuit to ground/ short circuit to positive, IAC passages blocked, IAC valve**
P1530	A/C refrigerant pressure sensor– circuit malfunction	**A/C refrigerant pressure, wiring open circuit/ short circuit to ground/short circuit to positive, A/C refrigerant pressure sensor, ECM**
P1539	A/C compressor clutch relay – voltage high	**Wiring open circuit/short circuit to ground/ short circuit to positive, A/C compressor clutch relay, ECM**
P1545	A/C compressor clutch relay – circuit malfunction	**Wiring open circuit/short circuit to ground/ short circuit to positive, A/C compressor clutch relay, ECM**
P1546	A/C compressor clutch relay – circuit malfunction	**Wiring open circuit/short circuit to ground/ short circuit to positive, A/C compressor clutch relay, ECM**
P1554	Cruise control status circuit	**Wiring short circuit to ground, cruise control module, ECM**

Scan code	Fault location	Probable cause
P1558	Cruise control servo position sensor – circuit malfunction	**Wiring open circuit/short circuit to ground/ short circuit to positive, hoses leaking, cruise control servo, ECM**
P1561	Cruise control vent solenoid	**Wiring open circuit/short circuit to ground/ short circuit to positive, cruise control servo, ECM**
P1562	Cruise control vacuum solenoid	**Wiring open circuit/short circuit to ground/ short circuit to positive, cruise control servo, ECM**
P1565	Cruise control servo position sensor – circuit malfunction	**Wiring open circuit/short circuit to ground/ short circuit to positive, TCC solenoid, ECM**
P1567	Cruise control switch – circuit malfunction	**Wiring open circuit/short circuit to positive, cruise control switch, brake pedal vacuum valve, ECM**
P1568	Cruise control servo – servo position sensor high	**Hoses blocked, brake pedal vacuum valve, cruise control servo, cruise control cable binding**
P1571	Traction control system (TCS), torque request circuit	**Wiring open circuit/short circuit to ground/ short circuit to positive, ABS control module trouble code(s) stored, ECM**
P1573	Data bus – antilock brake system (ABS)/traction control system (TCS)	**Wiring, ABS control module trouble code(s) stored**
P1585	Cruise control system – inhibit output circuit	**Wiring open circuit/short circuit to ground/ short circuit to positive, supply wire, cruise control motor/actuator, ECM**
P1599	Cruise control – power management mode	**Hoses blocked, brake pedal vacuum valve, cruise control servo, cruise control cable binding**
P1601	CAN data bus – circuit malfunction	**Wiring open circuit/short circuit to ground/ short circuit to positive, instrument panel, ECM**
P1619	Engine oil life monitor reset switch – circuit malfunction	**Wiring short circuit to ground, engine oil life monitor reset switch, ECM**
P1621	Engine control module (ECM) – memory performance	**ECM**
P1623	Engine control module (ECM) – PROM	**Insecure PROM, ECM**
P1626	Immobilizer control module – fuel enable signal	**Incorrect ignition key, wiring open circuit/ short circuit to ground/short circuit to positive, immobilizer control module, ECM**
P1629	Immobilizer control module – fuel enable signal	**Incorrect ignition key, wiring open circuit/ short circuit to ground/short circuit to positive, immobilizer control module, ECM**
P1630	Battery voltage – high/low	**Wiring, alternator, ECM**
P1635	Sensor supply voltage – circuit A malfunction	**Wiring short circuit to positive, ground wiring open circuit, TP sensor, MAP sensor, EGR valve, ECM**
P1639	Sensor supply voltage – circuit B malfunction	**Wiring short circuit to ground/positive, ground wiring open circuit, A/C refrigerant pressure sensor, ECM**

Scan code	Fault location	Probable cause
P1640	Engine control module quad driver/output driver	Wiring, ECM controlled components, ECM 8
P1641 6	Malfunction indicator lamp (MIL) – circuit malfunction	Wiring open circuit/short circuit to ground/ short circuit to positive, MIL, supply wire, ECM
P1641 1 4	A/C compressor clutch relay – circuit malfunction	Wiring open circuit/short circuit to positive, supply wire, A/C compressor relay, ignition Main relay, ECM
P1641 3	Engine coolant blower motor relay 1 – circuit malfunction	Wiring open circuit/short circuit to ground/ short circuit to positive, supply wire, engine coolant blower motor relay 1, ECM
P1642	Engine coolant blower motor relay 2 – circuit malfunction	Wiring open circuit/short circuit to ground/ short circuit to positive, supply wire, engine coolant blower motor relay 2, ECM
P1643 5	Fuel pump control module – circuit malfunction	Wiring open circuit/short circuit to ground/ short circuit to positive, fuel pump control module, ECM
P1643 3	Engine speed output circuit – circuit malfunction	Wiring open circuit/short circuit to ground/ short circuit to positive, ECM
P1645	Supercharger (SC) control valve solenoid – circuit malfunction	Wiring open circuit/short circuit to ground/ short circuit to positive, SC control valve solenoid, ECM
P1646	Supercharger (SC) control valve solenoid – circuit malfunction	Wiring open circuit/short circuit to ground/ short circuit to positive, SC control valve solenoid, ECM
P1650	Engine control module, quad driver/output driver	Wiring, ECM controlled components, ECM 8
P1651	Engine coolant blower motor relay 1 – circuit malfunction	Wiring open circuit/short circuit to ground/ short circuit to positive, supply wire, engine coolant blower motor relay 1, ECM
P1652 5	Engine coolant blower motor relay 2 – circuit malfunction	Wiring open circuit/short circuit to ground/ short circuit to positive, supply wire, engine coolant blower motor relay 2, ECM
P1652 3	Vehicle speed output signal – circuit malfunction	Wiring open circuit/short circuit to ground/ short circuit to positive, instrument panel, ECM
P1653 7	Traction control system (TCS) – delivered torque circuit	Wiring open circuit/short circuit to ground/ short circuit to positive, ABS control module trouble code(s) stored, ECM
P1653 2 3	Engine oil level warning lamp – circuit malfunction	Wiring open circuit/short circuit to ground/ short circuit to positive, instrument panel, ECM
P1654	A/C compressor clutch relay – circuit malfunction	Wiring open circuit/short circuit to ground/ short circuit to positive, supply wire, A/C compressor clutch relay, ECM
P1655	Evaporative emission (EVAP) canister purge valve – circuit malfunction	Wiring open circuit/short circuit to ground/ short circuit to positive, supply wire, EVAP canister purge valve, ECM
P1661	Malfunction indicator lamp (MIL) – circuit malfunction	Wiring open circuit/short circuit to ground/ short circuit to positive, MIL lamp bulb, ECM
P1662	Cruise control inhibit circuit – circuit malfunction	Wiring open circuit/short circuit to ground, cruise control motor/actuator, cruise control module, ECM

Scan code	Fault location	Probable cause
P1663 5	Alternator warning lamp – circuit malfunction	**Wiring open circuit/short circuit to ground/ short circuit to positive, supply wire, instrument panel, ECM**
P1663 3	Engine oil level warning lamp – circuit malfunction	**Wiring open circuit/short circuit to ground/ short circuit to positive, engine oil level lamp bulb, ECM**
P1665	Evaporative emission (EVAP) canister vent valve – circuit malfunction	**Wiring open circuit/short circuit to ground/ short circuit to positive, supply wire, EVAP canister vent valve, ECM**
P1667	Fuel pump control module – circuit malfunction	**Wiring open circuit/short circuit to ground/ short circuit to positive, supply wire, fuel pump control module, ECM, fuel pump relay**
P1670	Engine control module quad driver/output driver	**Wiring, ECM controlled components, ECM 8**
P1671	Malfunction indicator lamp (MIL) – circuit malfunction	**Wiring open circuit/short circuit to ground/ short circuit to positive, supply wire, ECM**
P1673	Engine coolant 'hot' warning lamp – circuit malfunction	**Wiring open circuit/short circuit to ground, cooling system, ECM**
P1672	Engine oil level warning lamp – circuit malfunction	**Wiring open circuit/short circuit to ground/ short circuit to positive, supply wire, ECM**
P1675	Evaporative emission (EVAP) canister vent valve – circuit malfunction	**Wiring open circuit/short circuit to ground/ short to positive, EVAP canister vent valve, ECM**
P1676	Evaporative emission (EVAP) canister purge valve – circuit malfunction	**Wiring open circuit/short circuit to ground/ short circuit to positive, supply wire, EVAP canister purge valve, ECM**
P1860	Torque converter clutch (TCC) solenoid – circuit malfunction	**Wiring open circuit/short circuit to ground, TCC solenoid, ECM**
P1689	Traction control system (TCS) – delivered torque output circuit	**Wiring open circuit/short circuit to ground/ short circuit to positive, supply wire, ABS trouble codes stored, ECM**
U1016	Instrument panel, CAN data bus – communication malfunction	**Wiring open circuit/short circuit to ground/ short circuit to positive, instrument panel, ECM**

1 Riviera 3.8L 1998 VIN 1
2 Riviera 3.8L 1999 VIN 1
3 Roadmaster 5.7L 1996 VIN P
4 Skylark 3.1L 1998 VIN M
5 Except Roadmaster 5.7L 1996 VIN P
6 Except Riviera 3.8L 1998 VIN 1, Skylark 3.1L 1998 VIN M & Roadmaster 5.7L 1996 VIN P
7 Except Riviera 3.8L 1999 VIN 1 & Roadmaster 5.7L 1996 VIN P
8 Depending on model, year and equipment level, the following components may be connected to this circuit. Use of the vehicle wiring diagram is recommended to confirm which components are fitted.

- ☐ A/C compressor clutch relay
- ☐ A/C condenser blower motor relay
- ☐ Engine coolant blower motor relay
- ☐ Engine coolant 'hot' warning lamp
- ☐ Evaporative emission (EVAP) canister purge valve
- ☐ Exhaust gas recirculation (EGR) solenoid
- ☐ Malfunction indicator lamp (MIL)
- ☐ Transmission shift warning light (MT)
- ☐ Torque converter clutch (TCC) solenoid
- ☐ Transmission shift solenoids (SS)

CADILLAC

Model:	Engine identification:	Year:
Allanté 4.5L	**VIN position 8 = 8**	**1991-92**

System: **GM**

Self-diagnosis

General information

- Refer to the front of this manual for general test conditions, terminology, detailed descriptions of wiring faults and a general trouble shooter for electrical and mechanical faults.
- Malfunction indicator lamp (MIL) will illuminate if certain faults are recorded.
- ECM operates in backup mode if sensors fail, to enable vehicle to be driven to repair shop.
- Trouble codes can be displayed by the driver information center.

1

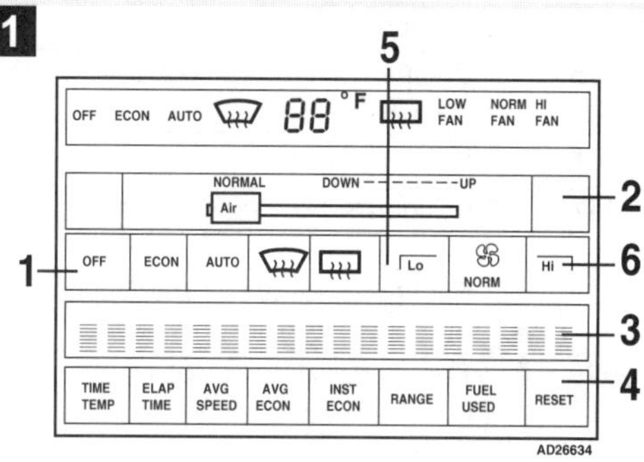

Accessing

- Switch ignition ON.
- Simultaneously press and hold the A/C control panel 'OFF' and 'WARMER' buttons **1** [**1**] & [**2**].
- When all segments of the display **1** [**3**] illuminate: Release the A/C control panel 'OFF' and 'WARMER' buttons **1** [**1**] & [**2**].
- Trouble codes will be displayed in the driver information center **1** [**3**].
- Trouble codes will be displayed in ascending order.

NOTE: ***All ECM codes start with the letter 'E'. Trouble codes for other systems may also be displayed.***

- All current trouble codes end with the letter 'C'.
- All history trouble codes end with the letter 'H'.
- If no ECM trouble codes are stored, 'NO ECM CODE' will be displayed.
- When all trouble codes have been displayed:
- Press the 'RESET' button **1** [**4**].
- Switch ignition OFF.

Erasing

- Switch ignition ON.
- Simultaneously press and hold the A/C control panel 'OFF' and 'WARMER' buttons **1** [**1**] & [**2**].
- When all segments of the display **1** [**3**] illuminate: Release the A/C control panel 'OFF' and 'WARMER' buttons **1** [**1**] & [**2**].
- Trouble codes will be displayed in the driver information center **1** [**3**].
- Trouble codes will be displayed in ascending order.

NOTE: ***Pressing and releasing the 'LOW' button* 1 [5] *will bypass trouble code display.***

- When all trouble codes have been displayed:
- Press and release the 'LOW' button **1** [**5**] until 'ECM' is displayed.
- Press and release the 'HI' button **1** [**6**].
- Press and release the 'LOW' button **1** [**5**] until 'ECM CLEAR CODES' is displayed.
- Press and release the 'HI' button **1** [**6**].
- 'ECM CODES CLEAR' will be displayed for 3 seconds.
- Press the 'RESET' button **1** [**4**].
- Switch ignition OFF.

Trouble code identification

Trouble code	Fault location	Probable cause
E012	Distributor – no reference signal	**Wiring open circuit/ short circuit to ground/ positive, ICM, CKP, ECM**
E013	Oxygen sensor (O2S) 1, bank 1	**Wiring open circuit/short circuit to positive, O2S, ECM**
E014	Engine coolant temperature (ECT) sensor – voltage low	**Wiring short circuit to ground, ECT sensor, ECM**
E015	Engine coolant temperature (ECT) sensor – voltage high	**Wiring open circuit, ECT sensor, ECM**
E016	Battery voltage – high/low	**Wiring, alternator, ECM**
E017	Oxygen sensor (O2S) 1, bank 2	**Wiring open circuit/short circuit to positive, O2S, ECM**
E019	Fuel pump circuit – short circuit	**Wiring short circuit to positive, fuel pump relay, engine oil pressure switch, ECM**
E020	Fuel pump circuit – open circuit	**Wiring open circuit/short circuit to ground/ positive, fuel pump relay, engine oil pressure switch, ECM**
E021	Throttle position (TP) sensor – voltage high	**Wiring open circuit/short circuit to positive, TP sensor, ECM**
E022	Throttle position (TP) sensor – voltage low	**Wiring open circuit/short circuit to ground, TP sensor, ECM**
E023	Ignition control module (ICM) – circuit malfunction	**Wiring open circuit/short circuit to ground, CKP sensor, ICM**
E024	Vehicle speed sensor (VSS)	**Wiring open circuit/short circuit to ground, VSS, ECM**
E026	Closed position throttle (CPT) switch – short circuit	**Wiring, CPT switch, ECM**
E027	Closed position throttle (CPT) switch – open circuit	**Wiring, CPT switch, ECM**
E030	Idle speed control (ISC) actuator	**Wiring, misfire, throttle valve tight/sticking, intake system leaking, TPS, ISC actuator, ECM**
E031	Manifold absolute pressure (MAP) sensor – short circuit	**Wiring short circuit to positive, MAP sensor**
E032	Manifold absolute pressure (MAP) sensor – open circuit	**Wiring open circuit/short circuit to ground, MAP sensor, ECM**
E034	Manifold absolute pressure (MAP) sensor – voltage high	**Wiring open circuit/short circuit to positive, MAP sensor**
E037	Intake air temperature (IAT) sensor – short circuit	**Wiring short circuit, IAT sensor, ECM**
E038	Intake air temperature (IAT) sensor – open circuit	**Wiring open circuit, IAT sensor, ECM**
E039	Torque converter clutch (TCC) – slipping	**Wiring open circuit/short circuit to ground, supply wire, transmission internal wiring, TCC solenoid, transmission mechanical fault, ECM**

Trouble code	Fault location	Probable cause
E040	Power steering pressure (PSP) switch – open circuit	Wiring open circuit/short circuit to ground/ positive, PSP switch, ECM
E041	Camshaft position (CMP) sensor – circuit malfunction	Wiring open circuit/short circuit to ground, CMP sensor, ECM
E042	Oxygen sensor (O2S) 1, bank 2 – mixture lean	Wiring short circuit to ground, injector(s) blocked, fuel contamination, fuel pressure low, exhaust system leaking, O2S, ECM
E043	Oxygen sensor (O2S) 1, bank 2 – mixture rich	Wiring open circuit, fuel pressure high, injector(s) leaking, O2S contamination, EVAP canister purge valve, MAP sensor hoses, O2S, ECM
E044	Oxygen sensor (O2S) 1, bank 1 – mixture lean	Wiring short circuit to ground, injector(s) blocked, fuel contamination, fuel pressure low, exhaust system leaking, O2S, ECM
E045	Oxygen sensor (O2S) 1, bank 1 – mixture rich	Wiring open circuit, fuel pressure high, injector(s) leaking, O2S contamination, EVAP canister purge valve, MAP sensor hoses, O2S, ECM
E046	Long term fuel trim – imbalance bank 1/bank 2	Injector(s) leaking, HO2S, intake/exhaust system leaking, spark plug(s), ignition system
E047	Body control module (BCM), CAN data bus – communication malfunction	Wiring, BCM
E050	Transmission 2nd gear pressure switch	Wiring open circuit/short circuit to ground, vacuum modulator/hoses, transmission mechanical fault
E051	Engine control module (ECM) – PROM	Insecure PROM, incorrect PROM, ECM
E052	Engine control module (ECM) – KAM reset	Battery disconnected, ECM disconnected
E053	Distributor – intermittent reference signal	Wiring open circuit/ short circuit to ground/ positive, CKP
E055	Throttle position (TP) sensor – incorrectly adjusted	Throttle valve tight/sticking, TP sensor incorrectly adjusted
E058	Immobilizer control module – fuel enable signal	Incorrect ignition key, wiring open circuit/ short circuit to ground/short circuit to positive, immobilizer control module, ECM
E060	Park/neutral input with cruise control engaged	Driver error, wiring open circuit/short circuit to ground, PNP switch
E061	Cruise control vent solenoid	Wiring open circuit/short circuit to ground/ short circuit to positive, cruise control vent solenoid, cruise control servo, ECM
E062	Cruise control vacuum solenoid	Wiring open circuit/short circuit to ground/ short circuit to positive, cruise control vacuum solenoid, cruise control servo, ECM
E063	Cruise control – set speed/vehicle speed mismatch	Wiring open circuit/short circuit to ground/ short circuit to positive, hoses leaking, cruise control servo, ECM

Trouble code	Fault location	Probable cause
E064	Cruise control – acceleration too high	**Slippery road conditions**
E065	Cruise control servo position sensor – circuit malfunction	**Wiring open circuit/short circuit to ground/ short circuit to positive, cruise control servo, ECM**
E066	Cruise control – engine speed too high	**Slippery road conditions, transmission mechanical fault**
E067	Cruise control switch – circuit malfunction	**Wiring open circuit/short circuit to ground/ short circuit to positive, cruise control switch, ECM**
E068	Cruise control servo position sensor – range	**Wiring, hoses blocked/leaking, brake pedal vacuum valve, cruise control servo, ECM**
E069	Cruise control disabled – traction control system (TCS) active	**Slippery road conditions**
E070	Throttle position (TP) sensor – intermittent signal	**Wiring open circuit/short circuit to ground/ short circuit to positive, TP sensor**
E071	Manifold absolute pressure (MAP) sensor – intermittent signal	**Wiring open circuit/short circuit to ground/ short circuit to positive, MAP sensor**
E073	Engine coolant temperature (ECT) sensor – intermittent signal	**Wiring open circuit/short circuit to ground/ short circuit to positive, ECT sensor**
E074	Intake air temperature (IAT) sensor – intermittent signal	**Wiring open circuit/short circuit to ground/ short circuit to positive, IAT sensor**
E075	Vehicle speed sensor (VSS) – intermittent signal	**Wiring open circuit/short circuit, VSS**
E080	Fuel system rich	**Fuel pressure high, injector(s) leaking, EVAP system**
E085	Throttle body service required	**throttle valve tight/sticking, TP sensor, ISC actuator**
E090	Brake pedal position (BPP) switch for torque converter clutch (TCC)	**Wiring open circuit/short circuit to ground, incorrectly adjusted, BPP switch, ECM**
E091	Park/neutral position (PNP) switch – circuit malfunction	**Wiring open circuit/short circuit to ground, incorrectly adjusted, PNP switch, ECM**
E096	Torque converter clutch (TCC) – over stressed	**ECM trouble codes stored, driver error**
E097	Transmission shifted from P/N to D/R at high throttle angle	**Driver error, wiring, TR switch**
E098	Transmission shifted from P/N to D/R in ISC range	**Driver error, wiring, TR switch**
E099	Cruise control servo – position high after disengagement	**Wiring, throttle valve tight/sticking, hoses blocked/leaking, cruise control servo**

CADILLAC

Model:	Engine identification:	Year:
Allanté 4.6L	**VIN position 8 = 9**	**1993**

System: **GM**

Self-diagnosis

General information

- Refer to the front of this manual for general test conditions, terminology, detailed descriptions of wiring faults and a general trouble shooter for electrical and mechanical faults.
- Malfunction indicator lamp (MIL) will illuminate if certain faults are recorded.
- ECM operates in backup mode if sensors fail, to enable vehicle to be driven to repair shop.
- Trouble codes can be displayed by the driver information center.

1

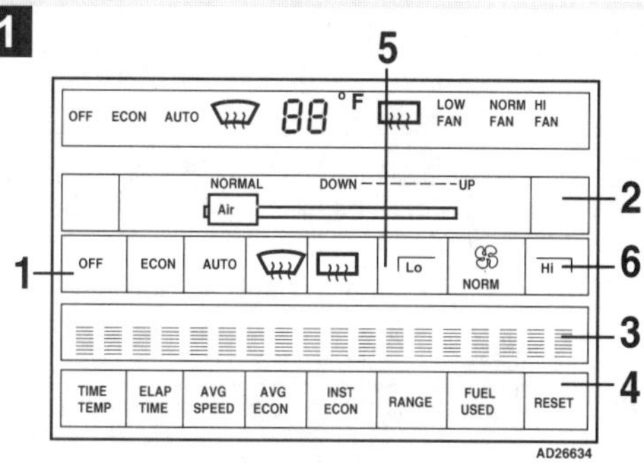

Accessing

- Switch ignition ON.
- Simultaneously press and hold the A/C control panel 'OFF' and 'WARMER' buttons **1** [**1**] & [**2**].
- When all segments of the display **1** [**3**] illuminate: Release the A/C control panel 'OFF' and 'WARMER' buttons **1** [**1**] & [**2**].
- Trouble codes will be displayed in the driver information center **1** [**3**].
- Trouble codes will be displayed in ascending order.

NOTE: ***All ECM codes start with the letter 'P'. Trouble codes for other system's may also be displayed.***

- All current trouble codes end with the letter 'C'.
- All history trouble codes end with the letter 'H'.
- If no ECM trouble codes are stored, 'NO PCM CODE' will be displayed.

NOTE: ***If 'NO PCM DATA' is be displayed there is a communication problem between the ECM and BCM.***

- When all trouble codes have been displayed:
- Press the 'RESET' button **1** [**4**].
- Switch ignition OFF.

Erasing

- Switch ignition ON.
- Simultaneously press and hold the A/C control panel 'OFF' and 'WARMER' buttons **1** [**1**] & [**2**].
- When all segments of the display **1** [**3**] illuminate: Release the A/C control panel 'OFF' and 'WARMER' buttons **1** [**1**] & [**2**].
- Trouble codes will be displayed in the driver information center **1** [**3**].
- Trouble codes will be displayed in ascending order.

NOTE: ***Pressing and releasing the 'LOW' button 1 [5] will bypass trouble code display.***

- When all trouble codes have been displayed:
- Press and release the 'LOW' button **1** [**5**] until 'ECM' is displayed.
- Press and release the 'HI' button **1** [**6**].
- Press and release the 'LOW' button **1** [**5**] until 'ECM CLEAR CODES' is displayed.
- Press and release the 'HI' button **1** [**6**].
- 'ECM CODES CLEAR' will be displayed for 3 seconds.
- Press the 'RESET' button **1** [**4**].
- Switch ignition OFF.

Trouble code identification

Trouble code	Fault location	Probable cause
P012	Ignition control module (ICM) – 4X reference circuit malfunction	**Wiring open circuit/short circuit to ground/ short circuit to positive, ICM, ECM**
P013	Heated oxygen sensor (HO2S) 2, bank 1 – open circuit	**Wiring open circuit/short circuit to ground/ short circuit to positive, HO2S, ECM**
P014	Engine coolant temperature (ECT) sensor – short circuit	**Wiring short circuit to ground, ECT sensor, ECM**
P015	Engine coolant temperature (ECT) sensor – open circuit	**Wiring open circuit, ECT sensor, ECM**
P016	Battery voltage – high/low	**Wiring, alternator, ECM**
P017	Heated oxygen sensor (HO2S) 1, bank 2 – open circuit	**Wiring open circuit/short circuit to ground/ short circuit to positive, HO2S, ECM**
P019	Fuel pump circuit – short circuit	**Wiring short circuit to positive, fuel pump relay, engine oil pressure warning lamp switch**
P020	Fuel pump circuit – open circuit	**Wiring open circuit/short circuit to ground, fuel pump relay, fuel pump, engine oil pressure warning lamp switch, engine oil pressure too low**
P021	Throttle position (TP) sensor – voltage high	**Wiring open circuit/short circuit to positive, TP sensor, ECM**
P022	Throttle position (TP) sensor – voltage low	**Wiring open circuit/short circuit to ground, TP sensor, ECM**
P023	Ignition control module (ICM) – circuit malfunction	**Wiring open circuit/short circuit to positive, ICM**
P024	Vehicle speed sensor (VSS)	**Wiring open circuit/short circuit to ground, VSS, ECM**
P025	Ignition control module (ICM) – 24X signal circuit malfunction	**Wiring open circuit/short circuit to ground, ICM, ECM**
P026	Closed throttle position (CTP) switch – short circuit	**Wiring short circuit to ground, ISC actuator, ECM**
P027	Closed throttle position (CTP) switch – open circuit	**Wiring open circuit, ISC actuator, ECM**
P028	Transmission pressure switch – circuit malfunction	**Wiring open circuit/short circuit to ground, transmission pressure switch, ECM**
P029	Shift solenoid (SS) B	**Wiring open circuit/short circuit to ground, transmission mechanical fault, SS, ECM**
P030	Idle speed (ISC) actuator	**Wiring open circuit/short circuit to ground, throttle valve tight/sticking, ISC actuator, ECM**
P031	Manifold absolute pressure (MAP) sensor – short circuit	**Wiring short circuit to positive, hose blocked/leaking, MAP sensor, ECM**
P032	Manifold absolute pressure (MAP) sensor – open circuit	**Wiring open circuit/short circuit to ground, MAP sensor, ECM**
P033	Brake pedal position switch (BPP) – extended travel contacts – circuit malfunction	**Wiring open circuit/short circuit to ground/ positive**

Trouble code	Fault location	Probable cause
P034	Manifold absolute pressure (MAP) sensor – voltage high	Wiring open circuit/short circuit to positive, vacuum leak, MAP sensor, ECM
P035	Ignition control module (ICM) – reference low circuit malfunction	Wiring open circuit/short circuit to positive, ICM, ECM
P036	Exhaust gas recirculation (EGR) solenoid position sensor	Wiring open circuit/short circuit to ground/ short circuit to positive, EGR valve, ECM
P037	Intake air temperature (IAT) sensor – short circuit	Wiring short circuit to ground, IAT sensor, ECM
P038	Intake air temperature (IAT) sensor – open circuit	Wiring open circuit, IAT sensor, ECM
P039	Torque converter clutch (TCC) – not engaging	Wiring open circuit/short circuit to ground, BPP switch for TCC, supply wire, TCC solenoid, transmission mechanical fault, ECM
P040	Power steering pressure (PSP) switch – open circuit	Wiring open circuit/short circuit to ground, PSP switch, ECM
P041	Ignition control module (ICM) – camshaft position (CMP) signal	Wiring open circuit/short circuit to ground, CMP sensor, ICM, ECM
P042	Heated oxygen sensor (HO2S) 1, bank 2 – mixture lean	Wiring short circuit to ground, injector(s) blocked, fuel contamination, fuel pressure low, exhaust system leaking, HO2S
P043	Heated oxygen sensor (HO2S) 1, bank 2 – mixture rich	Wiring open circuit, fuel pressure high, injector(s) leaking, HO2S contamination, EVAP canister purge valve, TP sensor, EGR system, HO2S
P044	Heated oxygen sensor (HO2S) 2, bank 1 – mixture lean	Wiring short circuit to ground, injector(s) blocked, fuel contamination, fuel pressure low, exhaust system leaking, HO2S
P045	Heated oxygen sensor (HO2S) 2, bank 1 – mixture rich	Wiring open circuit, fuel pressure high, injector(s) leaking, HO2S contamination, EVAP canister purge valve, TP sensor, EGR system, HO2S
P046	Long term fuel trim – imbalance bank 1/bank 2	Injector(s) leaking, HO2S, intake/exhaust system leaking
P047	Body control module (BCM), CAN data bus – communication malfunction	Wiring, BCM
P048	Exhaust gas recirculation (EGR) system	Wiring open circuit/short circuit to ground, EGR system blocked, EGR valve, ECM
P051	Engine control module (ECM) – PROM	Insecure PROM, incorrect PROM, ECM
P052	Engine control module (ECM) – memory reset	Battery disconnected, wiring
P053	Ignition control module (ICM) – 4X reference circuit malfunction	Wiring open circuit/short circuit to ground, ICM, ECM
P055	Closed throttle position (CPT) – range	Throttle valve tight/sticking, ISC actuator, TP sensor, cruise control cable
P056	Turbine shaft speed (TSS) sensor – circuit malfunction	Wiring, TSS sensor, ECM

Trouble code	Fault location	Probable cause
P057	Transmission fluid temperature (TFT) sensor – short circuit	**Wiring short circuit to ground, TFT sensor, ECM**
P058	Immobilizer control module – fuel enable signal	**Wiring open circuit/short circuit to ground, immobilizer control module, ECM**
P059	Transmission fluid temperature (TFT) sensor – open circuit	**Wiring open circuit, TFT sensor, ECM**
P060	Park/neutral input with cruise control engaged	**Driver error, wiring open circuit/short circuit to ground, TFP switch**
P061	Cruise control vent solenoid	**Wiring open circuit/short circuit to ground/ short circuit to positive, cruise control vent solenoid, cruise control servo, ECM**
P062	Cruise control vacuum solenoid	**Wiring open circuit/short circuit to ground/ short circuit to positive, cruise control vacuum solenoid, cruise control servo, ECM**
P063	Cruise control – set speed/vehicle speed mismatch	**Wiring open circuit/short circuit to ground/ short circuit to positive, hoses leaking, cruise control servo, ECM**
P064	Cruise control – acceleration too high	**Slippery road conditions**
P065	Cruise control servo position sensor – circuit malfunction	**Wiring open circuit/short circuit to ground/ short circuit to positive, cruise control servo, ECM**
P066	Cruise control – engine speed too high	**Slippery road conditions, transmission mechanical fault**
P067	Cruise control switch – circuit malfunction	**Wiring open circuit/short circuit to ground/ short circuit to positive, cruise control switch, ECM**
P068	Cruise control servo position sensor – range	**Wiring, hoses blocked/leaking, brake pedal vacuum valve, cruise control servo, ECM**
P069	Cruise control disabled – traction control system active (TCS) active	**Slippery road conditions**
P070	Throttle position (TP) sensor – intermittent signal	**Wiring open circuit/short circuit to ground/ short circuit to positive, TP sensor**
P071	Manifold absolute pressure (MAP) sensor – intermittent signal	**Wiring open circuit/short circuit to ground/ short circuit to positive, MAP sensor**
P073	Engine coolant temperature (ECT) sensor – intermittent signal	**Wiring open circuit/short circuit to ground/ short circuit to positive, ECT sensor**
P074	Intake air temperature (IAT) sensor – intermittent signal	**Wiring open circuit/short circuit to ground/ short circuit to positive, IAT sensor**
P075	Vehicle speed sensor (VSS) – circuit malfunction	**Wiring open circuit/short circuit, VSS**
P076	Transmission pressure control (PC) solenoid – circuit malfunction	**Wiring, PC solenoid, ECM**
P080	Throttle position (TP) sensor – idle learned incomplete	**TP sensor/idle learn procedure not completed**
P081	Camshaft position (CMP) sensor/4X reference circuit – signal variation	**Electromagnetic interference, wiring, CMP sensor, mechanical fault**

Trouble code	Fault location	Probable cause
P083	Ignition control module (ICM) – 24X reference circuit malfunction	**Wiring open circuit/short circuit/short circuit to positive, CKP sensor(s), ICM, ECM**
P085	Idle throttle angle – too high	**throttle valve tight/sticking, TP sensor, ISC actuator**
P086	Transmission gear ratio	**TSS sensor, VSS, transmission mechanical fault, ECM**
P088	Torque converter clutch (TCC) – stuck ON	**Wiring short circuit to ground, transmission internal wiring, TCC solenoid, ECM**
P089	Extended shift – adaptation limit reached	**Transmission mechanical fault**
P090	Brake pedal position (BPP) switch for torque converter clutch (TCC)	**Wiring open circuit/short circuit to ground, incorrectly adjusted, BPP switch, ECM**
P091	Transmission range (TR) switch – circuit malfunction	**Wiring open circuit/short circuit to ground, incorrectly adjusted, TR switch, ECM**
P093	Traction control system (TCS) – torque request circuit	**Wiring open circuit/short circuit to ground/short circuit to positive, ABS control module trouble code(s) stored, ECM**
P094	Shift solenoid (SS) A – ratio problem	**Wiring open circuit/short circuit to ground, SS, transmission mechanical fault, ECM**
P095	Engine stall or near stall detected	**ECM trouble code(s) stored, wiring, mechanical fault, transmission mechanical fault**
P096	Torque converter clutch (TCC) – over-stressed	**ECM trouble codes stored, driver error**
P097	Transmission shifted from P/N to D/R at high throttle angle	**Driver error, wiring, TR switch**
P099	Cruise control servo – position high after disengagement	**Wiring, throttle valve tight/sticking, cruise control servo**
P102	Brake servo vacuum sensor – short circuit	**Wiring open circuit/short circuit to positive, brake servo vacuum sensor, ECM**
P103	Brake servo vacuum sensor – open circuit	**Wiring open circuit/short circuit to ground, brake servo vacuum sensor, ECM**
P105	Brake servo vacuum sensor – voltage low	**Wiring short circuit to ground, brake servo vacuum sensor, ECM**
P106	Brake pedal position (BPP) switch, brake lamps – circuit malfunction	**Wiring, BPP switch**
P110	Alternator 'L' terminal – circuit malfunction	**Wiring open circuit/short circuit to ground, alternator, battery temperature sensor, ECM**
P112	Engine control module (ECM) – EEPROM	**ECM**
P117	Shift solenoid (SS) A/B – circuit malfunction	**Wiring open circuit/short circuit to ground, transmission mechanical fault, ECM**
P0131	Knock sensor (KS) – test failure	**Wiring, KS incorrectly fitted, KS, ECM**
P0132	Knock sensor (KS) – circuit malfunction	**Wiring open circuit/short circuit to ground/short circuit to positive, KS incorrectly fitted, KS, insecure PROM, ECM**
P0137	CAN data bus – antilock brake system (ABS)/traction control system (TCS)	**Wiring, ABS control module trouble code(s) stored**

CADILLAC

Model:	Engine identification:	Year:
Brougham 5.0L	**VIN position 8 = E**	**1991-92**
Brougham 5.7L	**VIN position 8 = 7**	**1991-92**

System: **GM**

Self-diagnosis

General information

- Refer to the front of this manual for general test conditions, terminology, detailed descriptions of wiring faults and a general trouble shooter for electrical and mechanical faults.
- Malfunction indicator lamp (MIL) will illuminate if certain faults are recorded.
- ECM operates in backup mode if sensors fail, to enable vehicle to be driven to repair shop.
- Trouble codes can be displayed by the MIL, the A/C control panel, or by using a Scan Tool connected to the data link connector (DLC).

Accessing

Without using A/C control panel

- Access data link connector (DLC).
- Jump data link connector (DLC) terminals A and B **1**.
- Switch ignition ON. Do NOT start engine.
- Count MIL flashes. Note trouble codes. Compare with trouble code table.
- Firstly trouble code 12 will be displayed 3 times.
- Trouble codes consist of 2 groups, displayed as follows:
  - First group indicates the 'tens' of the trouble code **2** [**A**].
  - Second group indicates the 'units' of the trouble code **2** [**C**].
  - A short pause separates each group **2** [**B**].
  - A long pause separates each trouble code **2** [**D**].
  - For example: Trouble code 12 displayed **2**. Start of diagnosis.
- Each additional trouble code will now be displayed 3 times, in ascending order.

NOTE: ***If no additional trouble codes are stored, code 12 will continue to be displayed.***

- Trouble codes can also be displayed by using a Scan Tool connected to the data link connector (DLC) **1** , located below center of fascia near SRS control module.

1

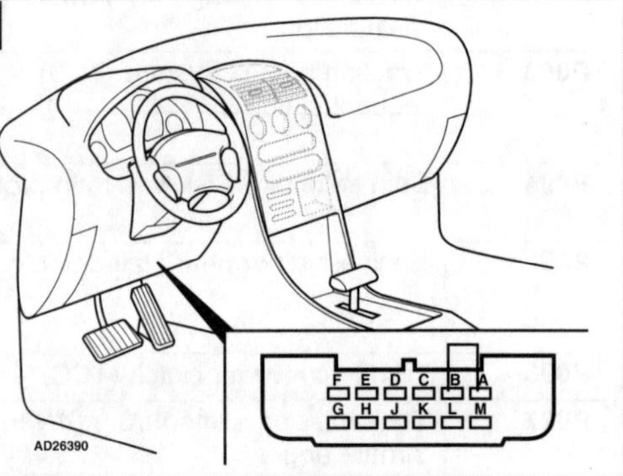

2

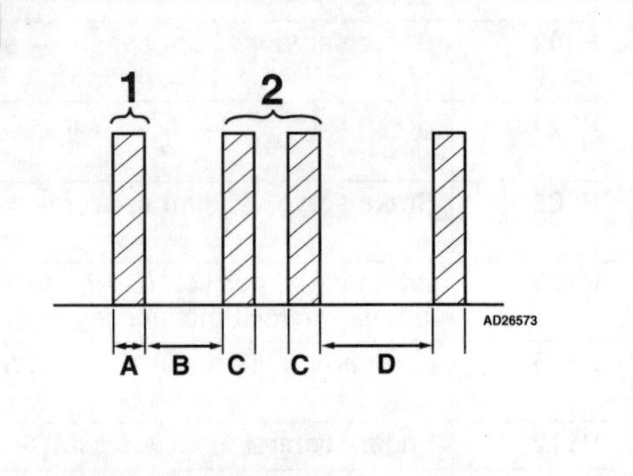

Using A/C control panel

- Switch ignition ON.
- Simultaneously press and hold 'OFF' and 'WARMER' buttons **3** [**1**] & [**2**].
- When all segments of the display **3** [**3**] illuminate:
  - Release the 'OFF' and 'WARMER' buttons **3** [**1**] & [**2**].

- Parameter number '-00' (A/C codes) will be displayed **3** [**3**].
- Press and release the 'HI' button **3** [**4**].
- Parameter number '-01' (ECM codes) will be displayed **3** [**3**].
- Press and release the 'HI' button **3** [**4**].
- Display will alternate between ECM parameter number '-01' and the trouble code at 1.6 second intervals **3** [**3**].
- Trouble codes will be displayed in ascending order.
- Cycle will be repeated after all ECM trouble codes have been displayed.
- If no ECM trouble codes are stored, trouble code '00' will be displayed.

NOTE: ***All ECM codes start with the parameter number '-01'.***

- To end diagnosis:
  - Press the 'AUTO' button **3** [**5**], or switch ignition OFF for 10 seconds.

3

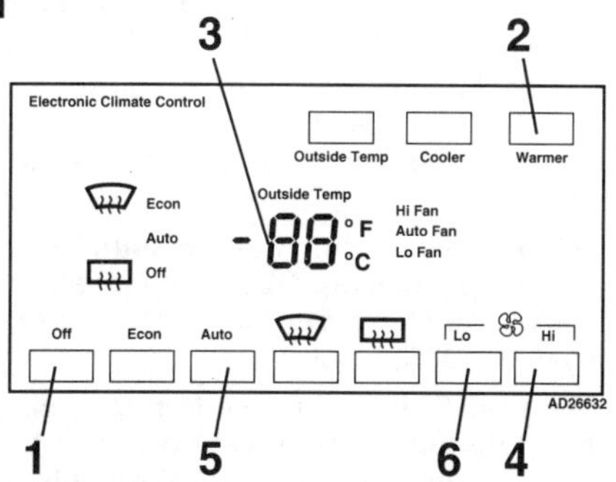

Erasing

Without using A/C control panel

NOTE: ***Disconnecting battery ground lead may cause driveability problems until the engine management system relearns basic values.***

- Ensure ignition switched OFF.
- Disconnect battery ground cable for 30 seconds.
- Reconnect battery ground cable.
- Repeat checking procedure to ensure no trouble codes remain in ECM fault memory.

NOTE: ***Disconnecting battery ground lead may also erase memory from electronic units, such as radio.***

- Trouble codes can also be erased by using a Scan Tool connected to the data link connector (DLC).

Using A/C control panel

- Access ECM trouble codes.
- During trouble code display:
- Simultaneously press and hold 'OFF' and 'LO' buttons **3** [**1**] & [**6**].
- When '00' is displayed **3** [**3**]:
  - Release the 'OFF' and 'LO' buttons **3** [**1**] & [**6**].
- To end diagnosis:
  - Press the 'AUTO' button **3** [**5**], or switch ignition OFF for 10 seconds.

Trouble code identification

Trouble code	Fault location	Probable cause
00 **1**	No fault found	–
12 **2**	Start of diagnosis	**No engine speed (RPM) signal received**
13	Oxygen sensor (O2S)	**Wiring open circuit, O2S, ECM**
14	Engine coolant temperature (ECT) sensor – high temperature indicated	**Wiring short circuit to ground, ECT sensor, ECM**
15	Engine coolant temperature (ECT) sensor – low temperature indicated	**Wiring open circuit, ECT sensor, ECM**
21	Throttle position (TP) sensor – voltage high	**Wiring open circuit/short circuit to positive, TP sensor, ECM**

Trouble code	Fault location	Probable cause
22	Throttle position (TP) sensor – voltage low	**Wiring open circuit/short circuit to ground, TP sensor, ECM**
23	Intake air temperature (IAT) sensor – low temperature indicated	**Wiring open circuit, IAT sensor, ECM**
24	Vehicle speed sensor (VSS)	**Wiring open circuit/short circuit to ground, VSS, ECM**
25	Intake air temperature (IAT) sensor – high temperature indicated	**Wiring short circuit to ground, IAT sensor, ECM**
32	Exhaust gas recirculation (EGR) solenoid	**Wiring open circuit/short circuit to ground, supply wire, EGR valve, ECM**
33	Manifold absolute pressure (MAP) sensor – voltage high, low vacuum indicated	**Wiring short circuit to positive, hose blocked/leaking, MAP sensor, ECM**
34	Manifold absolute pressure (MAP) sensor – voltage low, high vacuum indicated	**Wiring open circuit/short circuit to ground, MAP sensor, ECM**
42	Ignition control module (ICM) – circuit malfunction	**Wiring open circuit/short circuit to ground, ICM, ECM**
43	Knock sensor (KS)	**Wiring open circuit/short circuit to ground/ short circuit to positive, KS incorrectly fitted, KS, insecure PROM, ECM**
44	Oxygen sensor (O2S) – mixture lean	**Wiring open circuit/short circuit to ground, injector(s) blocked, fuel contamination, fuel pressure low, exhaust system leaking, O2S, ECM**
45	Oxygen sensor (O2S)– mixture rich	**Wiring open circuit, fuel pressure high, injector(s) leaking, O2S contamination, EVAP canister purge valve, TP sensor, EGR system, O2S, ECM**
46	Immobilizer control module – fuel enable signal	**Incorrect ignition key, wiring open circuit/ short circuit to ground/short circuit to positive, immobilizer control module, ECM**
51	Engine control module (ECM) – PROM	**Insecure PROM, incorrect PROM, ECM**
54	Fuel pump circuit – voltage low	**Wiring open circuit/short circuit to ground, fuel pump relay, engine oil pressure warning lamp switch**
55	Engine control module (ECM)	**ECM**

1 Code 00 is displayed when accessing trouble codes with the A/C control panel.
2 Code 12 is displayed when accessing trouble codes with the MIL.

CADILLAC

Model:	Engine identification:	Year:
Catera 3.0L	VIN position 8 = R	1997-01
DeVille 4.6L	VIN position 8 = Y/9	1996-02
Eldorado 4.6L	VIN position 8 = Y/9	1996-02
Escalade 5.3L	VIN position 8 = T	2001-02
Escalade 5.7L	VIN position 8 = R	1999-00
Escalade 6.0L	VIN position 8 = U	2001-02
Seville 4.6L	VIN position 8 = Y/9	1996-02

System: **GM**

Self-diagnosis

General information

- Refer to the front of this manual for general test conditions, terminology, detailed descriptions of wiring faults and a general trouble shooter for electrical and mechanical faults.
- Malfunction indicator lamp (MIL) will illuminate or flash if certain faults are recorded.
- ECM operates in backup mode if sensors fail, to enable vehicle to be driven to repair shop.

Accessing

- Trouble codes can only be displayed by using a Scan Tool connected to the data link connector (DLC) **1**.

1

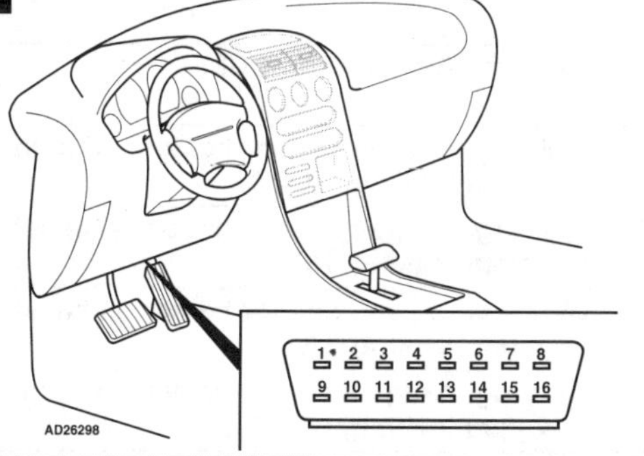

Erasing

→1999 – method 1

- After the faults have been rectified, the manufacturer recommends that trouble codes are erased by using a Scan Tool connected to the data link connector (DLC).

→1999 – method 2

NOTE: ***Disconnecting battery ground lead may cause driveability problems until the engine management system relearns basic values.***

- Ensure ignition switched OFF.
- Disconnect battery ground cable for at least 30 seconds.
- Reconnect battery ground cable.
- Repeat accessing procedure to ensure no trouble codes remain in ECM fault memory.

NOTE: ***Disconnecting battery ground lead may also erase memory from electronic units, such as radio.***

2000→

- After the faults have been rectified, the manufacturer recommends that trouble codes are erased by using a Scan Tool connected to the data link connector (DLC).

CADILLAC

Catera 3.0L • DeVille 4.6L • Eldorado 4.6L • Escalade 5.3L
Escalade 5.7L • Escalade 6.0L • Seville 4.6L

Trouble code identification

Scan code	Fault location	Probable cause
P0000-0999	Refer to OBD-II trouble code table in the front of this manual	–
P1031	Heated oxygen sensor (HO2S) 1, banks 1 & 2 – heater current monitor control circuit	**Wiring short circuit to positive, HO2S, ECM**
P1032	Heated oxygen sensor (HO2S) 1, banks 1 & 2 – heater warm up control circuit	**Wiring short circuit to positive, HO2S, ECM**
P1106	Manifold absolute pressure (MAP) sensor – circuit intermittent – voltage high	**Wiring open circuit/short circuit to ground/ short circuit to positive**
P1107	Manifold absolute pressure (MAP) sensor – circuit intermittent – voltage low	**Wiring short circuit to ground**
P1108	Barometric pressure (BARO) sensor/manifold absolute pressure (MAP) sensor – comparison too high	**Wiring open circuit/short circuit to positive, MAP sensor, ECM**
P1111	Intake air temperature (IAT) sensor – circuit intermittent – voltage high	**Wiring open circuit, ground lead defective**
P1112 [1]	Intake air valve control valve – circuit malfunction	**Wiring open circuit/short circuit to ground/ positive, intake air valve control valve, ECM**
P1112 [2]	Intake air temperature (IAT) sensor – circuit intermittent – voltage low	**Wiring short circuit to ground**
P1113	Intake air flap control actuator – circuit malfunction	**Wiring open circuit/short circuit to ground/ positive, intake air valve control valve, ECM**
P1114	Engine coolant temperature (ECT) sensor – circuit malfunction – voltage low	**Wiring short circuit to ground**
P1115	Engine coolant temperature (ECT) sensor – circuit malfunction – voltage high	**Wiring open circuit, ground lead defective**
P1120	Throttle motor position sensor 1 – circuit malfunction	**Wiring open circuit/short circuit to ground/ short circuit to positive, throttle control unit, ECM**
P1121	Throttle position (TP) sensor – circuit malfunction – voltage high	**Wiring short circuit to positive, ground lead defective**
P1122	Throttle position (TP) sensor – circuit malfunction – voltage low	**Wiring open circuit/short circuit to ground**
P1125	Accelerator pedal position (APP) sensor 1/2/3 – signal malfunction	**Wiring, APP sensor, ECM**
P1133	Heated oxygen sensor (HO2S) 1, bank 1 – insufficient switching	**Wiring, oxygen sensor heater, exhaust system leaking, HO2S**
P1134	Heated oxygen sensor (HO2S) 1, bank 1 – slow response	**Wiring, oxygen sensor heater, exhaust system leaking, HO2S**
P1139	Heated oxygen sensor (HO2S) 2, bank 1 – insufficient switching	**Wiring, oxygen sensor heater, exhaust system leaking, HO2S**
P1140	Heated oxygen sensor (HO2S) 2, bank 1 – slow response	**Wiring, oxygen sensor heater, exhaust system leaking, HO2S**
P1141	Heated oxygen sensor (HO2S) 2, bank 1 – heater control – circuit malfunction	**Fuse, wiring, HO2S, ECM**

Scan code	Fault location	Probable cause
P1153	Heated oxygen sensor (HO2S) 1, bank 2 – insufficient switching	**Wiring, oxygen sensor heater, exhaust system leaking, HO2S**
P1154	Heated oxygen sensor (HO2S) 1, bank 2 – slow response	**Wiring, oxygen sensor heater, exhaust system leaking, HO2S**
P1161	Heated oxygen sensor (HO2S) 2, bank 2 – heater control – circuit malfunction	**Fuse, wiring, HO2S, ECM**
P1172	Fuel level sensors, primary and secondary – no fuel level change during fuel transfer from secondary to primary tank	**Wiring, secondary fuel pump/relay, fuel level sensors**
P1189	Engine oil pressure warning switch – circuit malfunction	**Wiring open circuit, engine oil pressure warning switch, ECM**
P1220	Throttle motor position sensor 2 – circuit malfunction	**Wiring open circuit/short circuit to ground/ short circuit to positive, throttle control unit, ECM**
P1221	Throttle motor position sensor 1/2 – signal variation	**Wiring, throttle control unit, ECM**
P1258	Engine overheating	**Cooling system, ECT sensor**
P1271	Accelerator pedal position (APP) sensor 1/2 – signal variation	**Wiring, APP sensor, ECM**
P1275	Accelerator pedal position (APP) sensor 1 – circuit malfunction	**Wiring open circuit/short circuit to ground/ short circuit to positive, APP sensor, ECM**
P1276	Accelerator pedal position (APP) sensor 1 – range performance	**Wiring, APP sensor, ECM**
P1280	Accelerator pedal position (APP) sensor 2 – circuit malfunction	**Wiring open circuit/short circuit to ground/ short circuit to positive, APP sensor, ECM**
P1281	Accelerator pedal position (APP) sensor 2 – range performance	**Wiring, APP sensor, ECM**
P1285	Accelerator pedal position (APP) sensor 3 – circuit malfunction	**Wiring, APP sensor, ECM**
P1286	Accelerator pedal position (APP) sensor 3 – range performance	**Wiring, APP sensor, ECM**
P1320	Ignition control module (ICM) – 4X reference circuit, intermittent	**Wiring open circuit/short circuit to ground, ICM, ECM**
P1323	Ignition control module (ICM) – 24X reference circuit, low frequency	**Wiring open circuit/short circuit to ground, ICM, ECM**
P1336	Crankshaft position sensor (CKP) – variation not learned	**Learning procedure not performed**
P1345	Crankshaft position sensor (CKP)/camshaft position (CMP) sensor – signal variation	**CKP sensor, distributor/camshaft alignment, mechanical fault**
P1350	Ignition control module (ICM) – bypass line monitor circuit malfunction	**Wiring open circuit/short circuit to ground/ short circuit to positive, ICM, ECM**
P1351	Ignition control module (ICM) – ignition control (IC) circuit, voltage high	**Wiring open circuit, ICM, ECM**
P1359	Ignition control module (ICM) bank 1 – circuit malfunction	**Wiring open circuit/short circuit to positive, ICM, ECM**

Scan code	Fault location	Probable cause
P1360	Ignition control module (ICM) bank 2 – circuit malfunction	Wiring open circuit/short circuit to positive, ICM, ECM
P1361	Ignition control module (ICM) – ignition control (IC) circuit, voltage low	Wiring short circuit to ground, ICM, ECM
P1370	Ignition control module (ICM) – 4X reference circuit, too many pulses	Wiring, electromagnetic interference, ICM
P1371	Ignition control module (ICM) – 4X reference circuit, too few pulses	Wiring, electromagnetic interference, ICM
P1372	Crankshaft position sensor (CKP)	Wiring open circuit/short circuit to ground/ short circuit to positive, CKP, ECM
P1375	Ignition control module (ICM) – 24X reference circuit, voltage high	Wiring open circuit/short circuit to ground/ short circuit to positive, CKP sensor(s), ICM
P1376	Ignition control module (ICM) – ground circuit	Wiring open circuit/short circuit to positive, ground lead defective, ICM
P1377	Ignition control module (ICM) – cam pulse/ 4X reference pulse	Wiring open circuit/short circuit to ground, electromagnetic interference, valve timing
P1380	Rough road signal – signal error	ABS control module trouble code(s) stored
P1381	Rough road signal – no communication from antilock brake system (ABS) CAN data bus	Wiring open circuit, ABS system
P1404	Exhaust gas recirculation (EGR) valve – stuck open	Wiring open circuit/short circuit to positive, EGR valve position sensor, EGR solenoid
P1406	Exhaust gas recirculation (EGR) valve position sensor – circuit malfunction	Wiring open circuit/short circuit to ground/ short circuit to positive, EGR valve position sensor, EGR valve
P1415	Secondary air injection (AIR) system, bank 1	Wiring, hoses blocked/leaking, AIR valve
P1416	Secondary air injection (AIR) system, bank 2	Wiring, hoses blocked/leaking, AIR valve, AIR pump relay
P1441	Evaporative emission (EVAP) system – leak detected	Hose connection(s), EVAP canister, EVAP canister purge valve
P1460	Misfire – low fuel level	Fuel level low, misfire
P1483	Engine cooling system – performance	Cooling system, ECT sensor, IAT sensor
P1501	Immobilizer control module – circuit malfunction	Wiring open circuit/short circuit to ground/ short circuit to positive, MIL circuit
P1502	Immobilizer control module – fuel enable signal	Wiring open circuit/short circuit to ground/ short circuit to positive, immobilizer control module, ECM
P1503	Immobilizer control module – fuel enable signal	Incorrect ignition key, wiring open circuit/ short circuit to ground/short circuit to positive, immobilizer control module, ECM
P1508	Idle air control (IAC) valve – low rpm	Wiring open circuit/short circuit to ground/ short circuit to positive, IAC valve, throttle valve tight/sticking
P1509	Idle air control (IAC) valve – high rpm	Wiring open circuit/short circuit to ground/ short circuit to positive, IAC valve, throttle valve tight/sticking

Scan code	Fault location	Probable cause
P1510	Throttle position motor – throttle limitation active	**Wiring open circuit/short circuit to ground/ short circuit to positive, throttle control unit, ECM**
P1511	Throttle position motor – limp home system performance	**Throttle valve tight/sticking, TP sensor, wiring, throttle control unit, ECM**
P1514	Intake air flow – excessive flow	**Throttle control unit, MAP sensor, CMP sensor signal, ECM**
P1515	Throttle command/actual throttle position – signal variation	**Wiring, throttle valve sticking, throttle control unit, ECM**
P1516	Throttle position motor – position performance	**Throttle valve tight/sticking, TP sensor, wiring, throttle control unit, ECM**
P1517	Throttle control module – malfunction	**Throttle control module**
P1518	Throttle control module/engine control module (ECM) – communication malfunction	**Wiring, throttle control module, ECM**
P1519	Throttle position motor – circuit malfunction	**Wiring open circuit/short circuit to ground, throttle control unit, ECM**
P1520	Park/neutral position (PNP) switch/ transmission range (TR) switch	**Wiring short circuit to ground, incorrectly adjusted, PNP switch/TR switch, ECM**
P1523	Throttle position motor – closed position performance	**Throttle valve tight/sticking, TP sensor, wiring open circuit/short circuit to ground, throttle control unit, ECM**
P1524	Throttle position (TP) sensor – learned closed throttle angle	**Accelerator cable adjustment, throttle valve tight/sticking, throttle valve incorrectly adjusted**
P1526	Minimum throttle position not learned	**Initial learn procedure not carried out**
P1527	Transmission range (TR) switch/transmission fluid pressure (TFP) switch – signal variation	**Incorrectly adjusted TR switch, wiring open circuit/short circuit to ground, TR switch, TFP switch, ECM**
P1530	Throttle position motor – circuit malfunction	**Wiring open circuit/short circuit to ground, TP sensor, throttle control unit, ECM**
P1545	A/C compressor clutch relay – circuit malfunction	**Wiring open circuit/short circuit to ground, A/C compressor clutch relay, ECM**
P1554	Cruise control status circuit – high voltage	**Wiring open circuit/short circuit to ground, cruise control module, ECM**
P1560	Cruise control – transmission not in drive	**Driver error, TFP switch circuit**
P1564 1	Battery voltage	**Wiring**
P1564 2	Cruise control – acceleration too high	**Slippery road conditions**
P1566	Cruise control – engine speed too high	**Slippery road conditions, transmission mechanical fault**
P1567	Cruise control – active brake control system active, cruise control disabled	**Active brake control system engaged with cruise control on**
P1570	Cruise control – traction control system (TCS) active, cruise control disabled	**Slippery road surface**

Scan code	Fault location	Probable cause
P1571	Traction control system (TCS), torque request circuit	**Wiring open circuit/short circuit to ground/ short circuit to positive, ABS control module trouble code(s) stored, ECM**
P1574	Brake pedal position switch (BPP) – voltage high	**Wiring, ABS**
P1575	Brake pedal position switch (BPP) – extended travel contacts – voltage high	**Wiring, ABS**
P1576	Brake boost vacuum sensor – voltage high	**Wiring open circuit/short circuit to positive, ECM**
P1577	Brake boost vacuum sensor – voltage low	**Wiring open circuit/short circuit to ground, ECM**
P1578	Brake boost vacuum sensor – vacuum low	**Wiring open circuit/short circuit to ground, vacuum leak, ECM**
P1579	Transmission shifted from P/N to D/R at high throttle angle	**Driver error, wiring, TR switch**
P1585	Cruise control system – inhibit output circuit	**Wiring open circuit/short circuit to ground/ short circuit to positive, supply wire, cruise control motor/actuator, ECM**
P1599	Engine stall or near stall detected	**ECM trouble code(s) stored, mechanical fault, transmission mechanical fault**
P1600	Engine control module (ECM) – memory reset	**ECM**
P1601	Engine control module (ECM) – temperature high	**Module cooling fan, ECM**
P1602 1	Engine control module (ECM) – knock sensor (KS) module	**ECM**
P1602 2	CAN data bus – antilock brake system (ABS)/ traction control system (TCS)	**Wiring, ABS control module trouble code(s) stored**
P1603	CAN data bus – sensing/diagnostic module (SDM)	**Wiring**
P1604	CAN data bus – instrument panel	**Wiring, BCM trouble code(s) stored**
P1605	CAN data bus – A/C	**Wiring, A/C system trouble code(s) stored**
P1610	CAN data bus – body control module (BCM)	**Wiring, BCM trouble code(s) stored**
P1611	Suspension control module	**Wiring open circuit/short circuit to ground/ short circuit to positive, suspension control module, ECM**
P1612	Instrumentation control module	**Wiring open circuit/short circuit to ground/ short circuit to positive, instrumentation control module, ECM**
P1613	Dash integration module, data bus	**Wiring open circuit/short circuit to ground/ short circuit to positive, dash integration module, ECM**
P1614	Rear integration module, data bus	**Wiring open circuit/short circuit to ground/ short circuit to positive, rear integration module, ECM**

Scan code	Fault location	Probable cause
P1615	Immobilizer control module, data bus	**Wiring open circuit/short circuit to ground/ short circuit to positive, immobilizer control module, ECM**
P1617	Engine oil level switch – circuit malfunction	**Wiring short circuit to ground, engine oil level, ECM**
P1621	Engine control module (ECM) – memory performance	**ECM**
P1624	Customer snapshot function operated	**ECM trouble codes stored**
P1626	Immobilizer control module	**Wiring, BCM trouble code(s) stored**
P1627	Engine control module (ECM) – memory performance	**ECM**
P1630	Engine control module (ECM),immobilizer code learning mode	**Immobilizer code not programmed**
P1631	Engine control module (ECM) – incorrect immobilizer code	**Incorrect key, Immobilizer code not programmed**
P1632	Engine control module (ECM), fuel disable signal	**Fuses, BCM, immobilizer code**
P1633	Engine control module (ECM), ignition supply voltage	**Fuse, wiring open circuit, ECM**
P1634	Engine control module (ECM) – ignition supply voltage – voltage low	**Wiring open circuit, ECM**
P1635 3	Sensor supply voltage – circuit A malfunction	**Wiring short circuit to ground/positive, ground wiring open circuit, EGR valve position sensor, MAP sensor, TP sensor, ECM**
P1635 4	Sensor supply voltage – circuit A malfunction	**Wiring short circuit to ground/positive, ground wiring open circuit, EGR valve position sensor, MAP sensor, ECM**
P1637	Alternator 'L' terminal – circuit malfunction	**Wiring open circuit/short circuit to ground, alternator, battery temperature sensor, ECM**
P1638	Alternator 'F' terminal – circuit malfunction	**Wiring open circuit/short circuit to ground, alternator, ECM**
P1639 3	Sensor supply voltage – circuit B malfunction	**Wiring short circuit to ground/positive, ground wiring open circuit, A/C pressure sensor (DeVille/Seville), FTP sensor, ECM**
P1639 4	Sensor supply voltage – circuit B malfunction	**Wiring short circuit to ground/positive, ground wiring open circuit, FTP sensor, TP sensor, ECM**
P1640	Engine control module, output driver 1 – supply voltage high	**Wiring, loose alternator connection, alternator**
P1641	Malfunction indicator lamp (MIL) – circuit malfunction	**Wiring open circuit/short circuit to ground, MIL, supply wire, ECM**
P1642	Vehicle speed sensor (VSS) – circuit malfunction	**Wiring open circuit/short circuit to ground/ short circuit to positive, VSS, ECM**

Scan code	Fault location	Probable cause
P1644	Delivered torque – circuit malfunction	**Wiring open circuit/short circuit to ground/ short circuit to positive, ABS trouble codes stored, ECM**
P1645	Evaporative emission (EVAP) canister purge valve – circuit malfunction	**Wiring open circuit/short circuit to ground, EVAP canister purge valve, ECM**
P1646	Evaporative emission (EVAP) canister purge valve – circuit malfunction	**Wiring open circuit/short circuit to ground, EVAP canister purge valve, ECM**
P1650	Engine control module, output driver 2 – supply voltage high	**Wiring, loose alternator connection, alternator**
P1652	Suspension control module – engine control module (ECM) acceleration/braking signal malfunction	**Wiring open circuit/short circuit to ground/ short circuit to positive, suspension control module, ECM**
P1654	Cruise control disable circuit – circuit malfunction	**Wiring open circuit/short circuit to ground/ short circuit to positive, cruise control module, ECM**
P1660	Engine coolant blower motor relay(s) – circuit malfunction	**Wiring open circuit/short circuit to ground, engine coolant blower motor relay(s), ECM**
P1662	Cruise control indicator lamp – circuit malfunction	**Fuse, wiring, cruise control indicator lamp, ECM**
P1680	Engine control module (ECM) – memory performance	**ECM**
P1681	Engine control module (ECM) – memory performance	**ECM**
P1683	Engine control module (ECM) – memory performance	**ECM**
P1700	Transmission control module (TCM) – MIL request circuit malfunction	**OBD II TCM trouble codes stored**
P1701	Transmission control module (TCM) – MIL request circuit	**Wiring open circuit/short circuit to ground/ positive, ECM**
P1740	Transmission control module (TCM) – torque control circuit	**Wiring open circuit/short circuit to ground/ positive, TCM, ECM**
P1845	Torque limit management	**Other trouble codes stored, ECM**

1 Catera
2 Except Catera
3 DeVille, Eldorado, Seville
4 Escalade

CADILLAC

Model:	Engine identification:	Year:
DeVille 4.6L	VIN position 8 = Y	1994-95
DeVille 4.9L	VIN position 8 = B	1994-95
Eldorado 4.6L	VIN position 8 = Y/9	1993-95
Seville 4.6L	VIN position 8 = Y/9	1993-95

System: **GM**

Self-diagnosis

General information

- Refer to the front of this manual for general test conditions, terminology, detailed descriptions of wiring faults and a general trouble shooter for electrical and mechanical faults.
- Malfunction indicator lamp (MIL) will illuminate if certain faults are recorded.
- ECM operates in backup mode if sensors fail, to enable vehicle to be driven to repair shop.
- Trouble codes can be displayed by the driver information display.

1

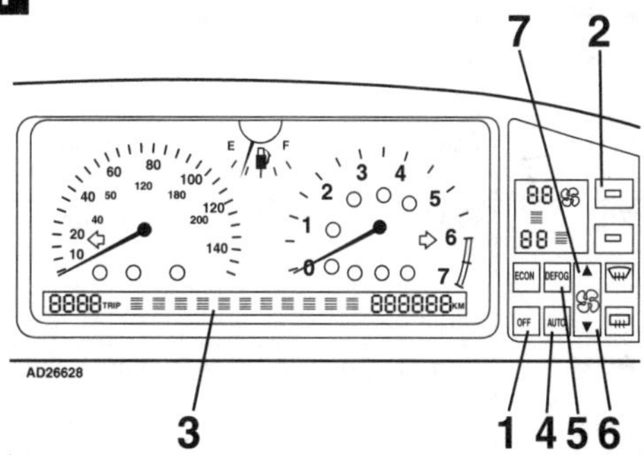

Accessing

- Switch ignition ON.
- Simultaneously press and hold the A/C control panel 'OFF' and 'WARMER' buttons **1** [**1**] & [**2**].
- When all segments of the display illuminate:
  - Release the A/C control panel 'OFF' and 'WARMER' buttons **1** [**1**] & [**2**].
- Trouble codes will be displayed on the driver information display **1** [**3**].
- Trouble codes will be displayed in ascending order.

NOTE: ***All ECM codes start with the letter 'P'. Trouble codes for other systems may also be displayed.***

- All current trouble codes end with the letter 'C'.
- All history trouble codes end with the letter 'H'.
- If no ECM trouble codes are stored, 'NO E CODE' or 'NO P CODE' will be displayed.

NOTE: ***If 'NO E DATA' or 'NO P DATA' is displayed there is a communication problem between the ECM and the instrument panel.***

- When all trouble codes have been displayed:
  - Press the 'AUTO' or 'DEFOG' button **1** [**4**] or [**5**] or switch ignition OFF.

Erasing

- Switch ignition ON.
- Simultaneously press and hold the A/C control panel 'OFF' and 'WARMER' buttons **1** [**1**] & [**2**].
- When all segments of the display illuminate:
  - Release the A/C control panel 'OFF' and 'WARMER' buttons **1** [**1**] & [**2**].
- Trouble codes will be displayed on the driver information display **1** [**3**].

NOTE: ***Pressing and releasing the 'LOW' button* 1 [6] *will bypass trouble code display.***

- When all trouble codes have been displayed:
  - Press and release the 'LOW' button **1** [**6**] until 'PCM' is displayed.
- Press and release the 'HI' button **1** [**7**].
- Press and release the 'LOW' button **1** [**6**] until 'PCM CLEAR CODES' is displayed.
- Press and release the 'HI' button **1** [**7**].
- 'PCM CODES CLEAR' will be displayed for 3 seconds.
- Press the 'AUTO' or 'DEFOG' button **1** [**4**] or [**5**] or switch ignition OFF.

Trouble code identification

Trouble code	Fault location	Probable cause
P012 1	Ignition control module (ICM) – 4X reference circuit malfunction	**Wiring open circuit/short circuit to ground/ short circuit to positive, ICM, ECM**
P012 2	Ignition control module (ICM) – no distributor reference signal	**Wiring open circuit/short circuit to ground/ short circuit to positive, CKP sensor, ICM, ECM**
P013	Heated oxygen sensor (HO2S) 2, bank 1 – open circuit	**Wiring open circuit/short circuit to ground/ short circuit to positive, HO2S, ECM**
P014	Engine coolant temperature (ECT) sensor – short circuit	**Wiring short circuit to ground, ECT sensor, ECM**
P015	Engine coolant temperature (ECT) sensor – open circuit	**Wiring open circuit, ECT sensor, ECM**
P016	Battery voltage – high/low	**Wiring, alternator, ECM**
P017	Heated oxygen sensor (HO2S) 1, bank 2 – open circuit	**Wiring open circuit/short circuit to ground/ short circuit to positive, HO2S, ECM**
P019	Fuel pump circuit – short circuit	**Wiring short circuit to positive, fuel pump relay, engine oil pressure warning lamp switch**
P020	Fuel pump circuit – open circuit	**Wiring open circuit/short circuit to ground, fuel pump relay, fuel pump, engine oil pressure warning lamp switch, engine oil pressure too low**
P021	Throttle position (TP) sensor – voltage high	**Wiring open circuit/short circuit to positive, TP sensor, ECM**
P022	Throttle position (TP) sensor – voltage low	**Wiring open circuit/short circuit to ground, TP sensor, ECM**
P023	Ignition control module (ICM) – circuit malfunction	**Wiring open circuit/short circuit to positive, ICM**
P024	Vehicle speed sensor (VSS)	**Wiring open circuit/short circuit to ground, VSS, ECM**
P025	Ignition control module (ICM) – 24X signal circuit malfunction	**Wiring open circuit/short circuit to ground, ICM, ECM**
P026	Closed throttle position (CTP) switch – short circuit	**Wiring short circuit to ground, ISC actuator, ECM**
P027	Closed throttle position (CTP) switch – open circuit	**Wiring open circuit, ISC actuator, ECM**
P028	Transmission pressure switch – circuit malfunction	**Wiring open circuit/short circuit to ground, transmission pressure switch, ECM**
P029	Shift solenoid (SS) B	**Wiring open circuit/short circuit to ground, transmission mechanical fault, SS, ECM**
P030	Idle speed (ISC) actuator	**Wiring open circuit/short circuit to ground, throttle valve tight/sticking, ISC actuator, ECM**
P031	Manifold absolute pressure (MAP) sensor – short circuit	**Wiring short circuit to positive, hose blocked/ leaking, MAP sensor, ECM**
P032	Manifold absolute pressure (MAP) sensor – open circuit	**Wiring open circuit/short circuit to ground, MAP sensor, ECM**
P033	Brake pedal position switch (BPP) – extended travel contacts – circuit malfunction	**Wiring open circuit/short circuit to ground/ positive**
P034	Manifold absolute pressure (MAP) sensor – voltage high	**Wiring open circuit/short circuit to positive, vacuum leak, MAP sensor, ECM**

Trouble code	Fault location	Probable cause
P035	Ignition control module (ICM) – reference low circuit malfunction	**Wiring open circuit/short circuit to positive, ICM, ECM**
P036	Exhaust gas recirculation (EGR) solenoid position sensor	**Wiring open circuit/short circuit to ground/ short circuit to positive, EGR valve, ECM**
P037	Intake air temperature (IAT) sensor – short circuit	**Wiring short circuit to ground, IAT sensor, ECM**
P038	Intake air temperature (IAT) sensor – open circuit	**Wiring open circuit, IAT sensor, ECM**
P039	Torque converter clutch (TCC) – not engaging	**Wiring open circuit/short circuit to ground, BPP switch for TCC, supply wire, TCC solenoid, transmission mechanical fault, ECM**
P040	Power steering pressure (PSP) switch – open circuit	**Wiring open circuit/short circuit to ground, PSP switch, ECM**
P041	Ignition control module (ICM) – camshaft position (CMP) signal	**Wiring open circuit/short circuit to ground, CMP sensor, ICM, ECM**
P042	Heated oxygen sensor (HO2S) 1, bank 2 – mixture lean	**Wiring short circuit to ground, injector(s) blocked, fuel contamination, fuel pressure low, exhaust system leaking, HO2S**
P043	Heated oxygen sensor (HO2S) 1, bank 2 – mixture rich	**Wiring open circuit, fuel pressure high, injector(s) leaking, HO2S contamination, EVAP canister purge valve, TP sensor, EGR system, HO2S**
P044	Heated oxygen sensor (HO2S) 2, bank 1 – mixture lean	**Wiring short circuit to ground, injector(s) blocked, fuel contamination, fuel pressure low, exhaust system leaking, HO2S**
P045	Heated oxygen sensor (HO2S) 2, bank 1 – mixture rich	**Wiring open circuit, fuel pressure high, injector(s) leaking, HO2S contamination, EVAP canister purge valve, TP sensor, EGR system, HO2S**
P046	Long term fuel trim – imbalance bank 1/bank 2	**Injector(s) leaking, HO2S, intake/exhaust system leaking**
P047	Body control module (BCM), CAN data bus – communication malfunction	**Wiring, BCM**
P048	Exhaust gas recirculation (EGR) system	**Wiring open circuit/short circuit to ground, EGR system blocked, EGR valve, ECM**
P051	Engine control module (ECM) – PROM	**Insecure PROM, incorrect PROM, ECM**
P052	Engine control module (ECM) – memory reset	**Battery disconnected, wiring**
P053 1	Ignition control module (ICM) – 4X reference circuit malfunction	**Wiring open circuit/short circuit to ground, ICM, ECM**
P053 2	Ignition control module (ICM) – intermittent distributor reference signal	**Wiring open circuit/short circuit to ground/ short circuit to positive, CKP sensor**
P055	Closed throttle position (CPT) – range	**Throttle valve tight/sticking, ISC actuator, TP sensor, cruise control cable**
P056	Turbine shaft speed (TSS) sensor – circuit malfunction	**Wiring, TSS sensor, ECM**
P057	Transmission fluid temperature (TFT) sensor – short circuit	**Wiring short circuit to ground, TFT sensor, ECM**
P058	Immobilizer control module – fuel enable signal	**Wiring open circuit/short circuit to ground, immobilizer control module, ECM**

Trouble code	Fault location	Probable cause
P059	Transmission fluid temperature (TFT) sensor – open circuit	**Wiring open circuit, TFT sensor, ECM**
P060	Park/neutral input with cruise control engaged	**Driver error, wiring open circuit/short circuit to ground, TFP switch**
P061	Cruise control vent solenoid	**Wiring open circuit/short circuit to ground/ short circuit to positive, cruise control vent solenoid, cruise control servo, ECM**
P062	Cruise control vacuum solenoid	**Wiring open circuit/short circuit to ground/ short circuit to positive, cruise control vacuum solenoid, cruise control servo, ECM**
P063	Cruise control – set speed/vehicle speed mismatch	**Wiring open circuit/short circuit to ground/ short circuit to positive, hoses leaking, cruise control servo, ECM**
P064	Cruise control – acceleration too high	**Slippery road conditions**
P065	Cruise control servo position sensor – circuit malfunction	**Wiring open circuit/short circuit to ground/ short circuit to positive, cruise control servo, ECM**
P066	Cruise control – engine speed too high	**Slippery road conditions, transmission mechanical fault**
P067	Cruise control switch – circuit malfunction	**Wiring open circuit/short circuit to ground/ short circuit to positive, cruise control switch, ECM**
P068	Cruise control servo position sensor – range	**Wiring, hoses blocked/leaking, brake pedal vacuum valve, cruise control servo, ECM**
P069	Cruise control disabled – traction control system (TCS) active	**Slippery road conditions**
P070	Throttle position (TP) sensor – intermittent signal	**Wiring open circuit/short circuit to ground/ short circuit to positive, TP sensor**
P071	Manifold absolute pressure (MAP) sensor – intermittent signal	**Wiring open circuit/short circuit to ground/ short circuit to positive, MAP sensor**
P073	Engine coolant temperature (ECT) sensor – intermittent signal	**Wiring open circuit/short circuit to ground/ short circuit to positive, ECT sensor**
P074	Intake air temperature (IAT) sensor – intermittent signal	**Wiring open circuit/short circuit to ground/ short circuit to positive, IAT sensor**
P075	Vehicle speed sensor (VSS) – circuit malfunction	**Wiring open circuit/short circuit, VSS**
P076	Transmission pressure control (PC) solenoid – circuit malfunction	**Wiring, PC solenoid, ECM**
P080 1	Throttle position (TP) sensor – idle learned incomplete	**TPS/idle learn procedure not completed**
P080 2	Fuel system rich	**Fuel pressure high, injector(s) leaking, EVAP system**
P081	Camshaft position (CMP) sensor/4X reference circuit – signal variation	**Electromagnetic interference, wiring, CMP sensor, mechanical fault**
P083	Ignition control module (ICM) – 24X reference circuit malfunction	**Wiring open circuit/short circuit/short circuit to positive, CKP sensor(s), ICM, ECM**
P085	Idle throttle angle – too high	**throttle valve tight/sticking, TP sensor, ISC actuator**
P086	Transmission gear ratio	**TSS sensor, VSS, transmission mechanical fault, ECM**

Trouble code	Fault location	Probable cause
P088	Torque converter clutch (TCC) – stuck ON	**Wiring short circuit to ground, transmission internal wiring, TCC solenoid, ECM**
P089	Extended shift – adaptation limit reached	**Transmission mechanical fault**
P090	Brake pedal position (BPP) switch TCC	**Wiring open circuit/short circuit to ground, incorrectly adjusted, BPP switch, ECM**
P091	Transmission range (TR) switch – circuit malfunction	**Wiring open circuit/short circuit to ground, incorrectly adjusted, TR switch, ECM**
P092	Heated windshield request – circuit malfunction	**Wiring short circuit to ground, heated windshield module, ECM**
P093	Traction control system (TCS) – torque request circuit	**Wiring open circuit/short circuit to ground/ short circuit to positive, ABS control module trouble code(s) stored, ECM**
P094	Shift solenoid (SS) A – ratio problem	**Wiring open circuit/short circuit to ground, SS, transmission mechanical fault, ECM**
P095	Engine stall or near stall detected	**ECM trouble code(s) stored, wiring, mechanical fault, transmission mechanical fault**
P096	Torque converter clutch (TCC) – over-stressed	**ECM trouble codes stored, driver error**
P097	Transmission shifted from P/N to D/R at high throttle angle	**Driver error, wiring, TR switch**
P098	Transmission shifted from P/N to D/R in idle speed control (ISC) range	**Driver error, wiring, TR switch**
P099	Cruise control servo – position high after disengagement	**Wiring, throttle valve tight/sticking, cruise control servo**
P102	Brake servo vacuum sensor – short circuit	**Wiring open circuit/short circuit to positive, brake servo vacuum sensor, ECM**
P103	Brake servo vacuum sensor – open circuit	**Wiring open circuit/short circuit to ground, brake servo vacuum sensor, ECM**
P105	Brake servo vacuum sensor – voltage low	**Wiring short circuit to ground, brake servo vacuum sensor, ECM**
P106	Brake pedal position (BPP) switch, brake lamps – circuit malfunction	**Wiring, BPP switch**
P110	Alternator 'L' terminal – circuit malfunction	**Wiring open circuit/short circuit to ground, alternator, battery temperature sensor, ECM**
P112	Engine control module (ECM) – EEPROM	**ECM**
P117	Shift solenoid (SS) A/B – circuit malfunction	**Wiring open circuit/short circuit to ground, transmission mechanical fault, ECM**
P0131	Knock sensor (KS) – test failure	**Wiring, KS incorrectly fitted, KS, ECM**
P0132	Knock sensor (KS) – circuit malfunction	**Wiring open circuit/short circuit to ground/ short circuit to positive, KS incorrectly fitted, KS, insecure PROM, ECM**
P0137	CAN data bus – antilock brake system (ABS)/ traction control system (TCS)	**Wiring, ABS control module trouble code(s) stored**

1 4.6L
2 4.9L

CADILLAC

Model:	Engine identification:	Year:
DeVille 4.9L	**VIN position 8 = B**	**1992-93**
Fleetwood 4.9L	**VIN position 8 = B**	**1992-93**

System: **GM**

Self-diagnosis

General information

- Refer to the front of this manual for general test conditions, terminology, detailed descriptions of wiring faults and a general trouble shooter for electrical and mechanical faults.
- Malfunction indicator lamp (MIL) will illuminate if certain faults are recorded.
- ECM operates in backup mode if sensors fail, to enable vehicle to be driven to repair shop.
- Trouble codes can be displayed by the fuel data center.

Accessing

- Switch ignition ON.
- Simultaneously press and hold the A/C control panel 'OFF' and 'WARMER' buttons **1** [**1**] & [**2**].
- When all segments of the display **1** [**3**] illuminate:
  - ○ Release the 'OFF' and 'WARMER' buttons **1** [**1**] & [**2**].
- Trouble codes are displayed by the fuel data center **2**.
- '8.8.8' will be displayed for 1 second.
- History trouble codes are displayed first followed by current codes.
- If ECM history trouble codes are stored:
  - ○ '..E' will be displayed for 1 second followed by the trouble code for 2 seconds.
- If ECM current trouble codes are stored:
  - ○ '.E.E' will be displayed for 1 second followed by the trouble code for 2 seconds.

NOTE: ***The same sequence will follow for BCM trouble codes with the prefix '..F' and '.F.F'***

- When all trouble codes have been displayed:
  - ○ Fuel data center will display '.7.0'.
- Press the A/C control panel 'Auto' button **1** [**4**] or:
- Switch ignition OFF for a minimum of 10 seconds.

1

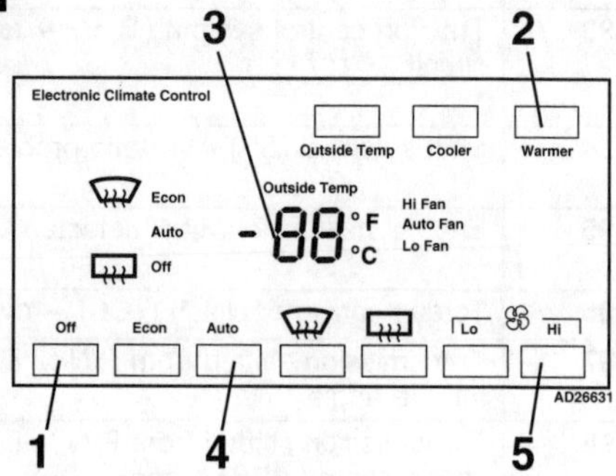

2

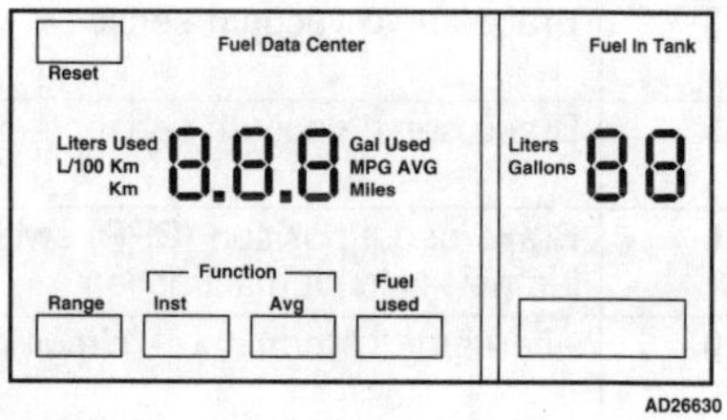

Erasing

- Access ECM trouble codes.
- When all trouble codes have been displayed:
  - ○ Fuel data center **2** will display '.7.0'.
- Simultaneously press and hold the A/C control panel 'OFF' and 'HI' buttons **1** [**1**] & [**5**] until:
  - ○ Fuel data center displays 'E.O.O' followed by '.7.0'.
- Switch ignition OFF for a minimum of 15 seconds.

Trouble code identification

Trouble code	Fault location	Probable cause
E12	Distributor – no reference signal	**Wiring open circuit/ short circuit to ground/ positive, ICM, CKP, ECM**
E13	Oxygen sensor (O2S) 1	**Wiring open circuit/short circuit to positive, O2S, ECM**
E14	Engine coolant temperature (ECT) sensor – voltage low	**Wiring short circuit to ground, ECT sensor, ECM**
E15	Engine coolant temperature (ECT) sensor – voltage high	**Wiring open circuit, ECT sensor, ECM**
E16	Battery voltage – high/low	**Wiring, alternator, ECM**
E19	Fuel pump circuit – short circuit	**Wiring short circuit to positive, fuel pump relay, engine oil pressure switch, ECM**
E20	Fuel pump circuit – open circuit	**Wiring open circuit/short circuit to ground/ positive, fuel pump relay, engine oil pressure switch, ECM**
E21	Throttle position (TP) sensor – voltage high	**Wiring open circuit/short circuit to positive, TP sensor, ECM**
E22	Throttle position (TP) sensor – voltage low	**Wiring open circuit/short circuit to ground, TP sensor, ECM**
E23	Ignition control module (ICM) – circuit malfunction	**Wiring open circuit/short circuit to ground, CKP sensor, ICM**
E24	Vehicle speed sensor (VSS)	**Wiring open circuit/short circuit to ground, VSS, ECM**
E26	Closed position throttle (CPT) switch – short circuit	**Wiring, CPT switch, ECM**
E27	Closed position throttle (CPT) switch – open circuit	**Wiring, CPT switch, ECM**
E30	Idle speed control (ISC) actuator	**Wiring, misfire, throttle valve tight/sticking, intake system leaking, TPS, ISC actuator, ECM**
E31	Manifold absolute pressure (MAP) sensor – short circuit	**Wiring short circuit to positive, MAP sensor**
E32	Manifold absolute pressure (MAP) sensor – open circuit	**Wiring open circuit/short circuit to ground, MAP sensor, ECM**
E34	Manifold absolute pressure (MAP) sensor – voltage high	**Wiring open circuit/short circuit to positive, MAP sensor**
E37	Intake air temperature (IAT) sensor – short circuit	**Wiring short circuit, IAT sensor, ECM**
E38	Intake air temperature (IAT) sensor – open circuit	**Wiring open circuit, IAT sensor, ECM**
E39	Torque converter clutch (TCC) – slipping	**Wiring open circuit/short circuit to ground, supply wire, transmission internal wiring, TCC solenoid, transmission mechanical fault, ECM**
E40	Power steering pressure (PSP) switch – open circuit	**Wiring open circuit/short circuit to ground/ positive, PSP switch, ECM**

Trouble code	Fault location	Probable cause
E41	Camshaft position (CMP) sensor – circuit malfunction	**Wiring open circuit/short circuit to ground, CMP sensor, ECM**
E44	Oxygen sensor (O2S) 1 – mixture lean	**Wiring short circuit to ground, injector(s) blocked, fuel contamination, fuel pressure low, exhaust system leaking, O2S, ECM**
E45	Oxygen sensor (O2S) 1 – mixture rich	**Wiring open circuit, fuel pressure high, injector(s) leaking, O2S contamination, EVAP canister purge valve, MAP sensor hoses, O2S, ECM**
E47	Body control module (BCM), CAN data bus – communication malfunction	**Wiring, BCM**
E48	Exhaust gas recirculation (EGR) system	**Wiring open circuit/short circuit to ground, EGR system blocked, hoses blocked/ leaking, EGR valve, EGR solenoid, ECM**
E52	Engine control module (ECM) – KAM reset	**Battery disconnected, ECM disconnected**
E53	Distributor – intermittent reference signal	**Wiring open circuit/ short circuit to ground/ positive, CKP**
E55	Throttle position (TP) sensor – incorrectly adjusted	**Throttle valve tight/sticking, TP sensor incorrectly adjusted**
E58	Immobilizer control module – fuel enable signal	**Incorrect ignition key, wiring open circuit/ short circuit to ground/short circuit to positive, immobilizer control module, ECM**
E60	Park/neutral input with cruise control engaged	**Driver error, wiring open circuit/short circuit to ground, PNP switch**
E61	Cruise control vent solenoid	**Wiring open circuit/short circuit to ground/ short circuit to positive, cruise control vent solenoid, cruise control servo, ECM**
E62	Cruise control vacuum solenoid	**Wiring open circuit/short circuit to ground/ short circuit to positive, cruise control vacuum solenoid, cruise control servo, ECM**
E63	Cruise control – set speed/vehicle speed mismatch	**Wiring open circuit/short circuit to ground/ short circuit to positive, hoses leaking, cruise control servo, ECM**
E64	Cruise control – acceleration too high	**Slippery road conditions**
E65	Cruise control servo position sensor – circuit malfunction	**Wiring open circuit/short circuit to ground/ short circuit to positive, cruise control servo, ECM**
E66	Cruise control – engine speed too high	**Slippery road conditions, transmission mechanical fault**
E67	Cruise control switch – circuit malfunction	**Wiring open circuit/short circuit to ground/ short circuit to positive, cruise control switch, ECM**
E68	Cruise control servo position sensor – range	**Wiring, hoses blocked/leaking, brake pedal vacuum valve, cruise control servo, ECM**
E70	Throttle position (TP) sensor – intermittent signal	**Wiring open circuit/short circuit to ground/ short circuit to positive, TP sensor**

Trouble code	Fault location	Probable cause
E71	Manifold absolute pressure (MAP) sensor – intermittent signal	**Wiring open circuit/short circuit to ground/ short circuit to positive, MAP sensor**
E73	Engine coolant temperature (ECT) sensor – intermittent signal	**Wiring open circuit/short circuit to ground/ short circuit to positive, ECT sensor**
E74	Intake air temperature (IAT) sensor – intermittent signal	**Wiring open circuit/short circuit to ground/ short circuit to positive, IAT sensor**
E75	Vehicle speed sensor (VSS) – intermittent signal	**Wiring open circuit/short circuit, VSS**
E80	Fuel system rich	**Fuel pressure high, injector(s) leaking, EVAP system**
E85	Idle throttle angle – too high	**throttle valve tight/sticking, TP sensor, ISC actuator**
E90	Brake pedal position (BPP) switch for torque converter clutch (TCC)	**Wiring open circuit/short circuit to ground, incorrectly adjusted, BPP switch, ECM**
E91	Park/neutral position (PNP) switch – circuit malfunction	**Wiring open circuit/short circuit to ground, incorrectly adjusted, PNP switch, ECM**
E92	Heated windshield request – circuit malfunction	**Wiring short circuit to ground, heated windshield module, heated windshield power module, ECM**
E96	Torque converter clutch (TCC) – over stressed	**ECM trouble codes stored, driver error**
E97	Transmission shifted from P/N to D/R at high throttle angle	**Driver error, wiring, TR switch**
E98	Transmission shifted from P/N to D/R in ISC range	**Driver error, wiring, TR switch**
E99	Cruise control servo – position high after disengagement	**Wiring, throttle valve tight/sticking, hoses blocked/leaking, cruise control servo**

CADILLAC

Model:	Engine identification:	Year:
Eldorado 4.9L	**VIN position 8 = B**	**1992**
Seville 4.9L	**VIN position 8 = B**	**1992**

System: **GM**

Self-diagnosis

General information

- Refer to the front of this manual for general test conditions, terminology, detailed descriptions of wiring faults and a general trouble shooter for electrical and mechanical faults.
- Malfunction indicator lamp (MIL) will illuminate if certain faults are recorded.
- ECM operates in backup mode if sensors fail, to enable vehicle to be driven to repair shop.
- Trouble codes can be displayed by the driver information display.

Accessing

- Switch ignition ON.
- Simultaneously press and hold the A/C control panel 'OFF' and 'WARMER' buttons **1** [**1**] & [**2**].
- When all segments of the display illuminate:
  - Release the A/C control panel 'OFF' and 'WARMER' buttons **1** [**1**] & [**2**].
- Trouble codes will be displayed on the driver information display **1** [**3**].
- Trouble codes will be displayed in ascending order.

NOTE: ***All ECM codes start with the letter 'E'. Trouble codes for other systems may also be displayed.***

- All current trouble codes end with the letter 'C'.
- All history trouble codes end with the letter 'H'.
- If no ECM trouble codes are stored, 'NO E CODE' will be displayed.

NOTE: ***If 'NO E DATA' is displayed there is a communication problem between the ECM and the instrument panel.***

- When all trouble codes have been displayed:
  - Press the 'AUTO' or 'DEFOG' button **1** [**4**] or [**5**] or switch ignition OFF.

1

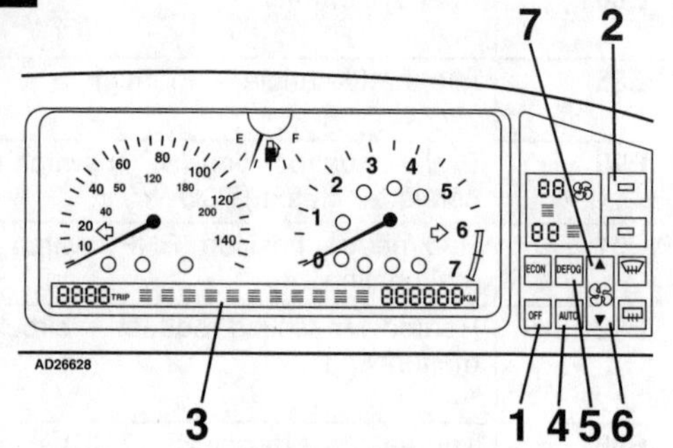

Erasing

- Switch ignition ON.
- Simultaneously press and hold the A/C control panel 'OFF' and 'WARMER' buttons **1** [**1**] & [**2**].
- When all segments of the display illuminate:
  - Release the A/C control panel 'OFF' and 'WARMER' buttons **1** [**1**] & [**2**].
- Trouble codes will be displayed on the driver information display **1** [**3**].

NOTE: ***Pressing and releasing the 'LOW' button* 1 [*6*] *will bypass trouble code display.***

- When all trouble codes have been displayed:
  - Press and release the 'LOW' button **1** [**6**] until 'PCM' is displayed.
- Press and release the 'HI' button **1** [**7**].
- Press and release the 'LOW' button **1** [**6**] until 'PCM CLEAR CODES' is displayed.
- Press and release the 'HI' button **1** [**7**].
- 'PCM CODES CLEAR' will be displayed for 3 seconds.
- Press the 'AUTO' or 'DEFOG' button **1** [**4**] or [**5**] or switch ignition OFF.

Trouble code identification

Trouble code	Fault location	Probable cause
E012	Distributor – no reference signal	**Wiring open circuit/ short circuit to ground/ short circuit to positive, HEI module, HEI pick up coil, ECM**
E013	Oxygen sensor (O2S) 1, bank 1	**Wiring open circuit/short circuit to ground/ short circuit to positive, O2S, ECM**
E014	Engine coolant temperature (ECT) sensor – circuit malfunction	**Wiring short circuit to ground, ECT sensor, ECM**
E015	Engine coolant temperature (ECT) sensor – circuit malfunction	**Wiring open circuit, ECT sensor, ECM**
E016	Battery voltage – high/low	**Wiring, alternator, ECM**
E017	Oxygen sensor (O2S) 1, bank 2	**Wiring open circuit/short circuit to ground/ short circuit to positive, O2S, ECM**
E019	Fuel pump circuit – short circuit	**Wiring short circuit to positive, fuel pump relay, engine oil pressure switch, ECM**
E020	Fuel pump circuit – open circuit	**Wiring open circuit/short circuit to ground/ short circuit to positive, fuel pump relay, fuel pump, engine oil pressure switch, ECM**
E021	Throttle position (TP) sensor – voltage high	**Wiring open circuit/short circuit to positive, TP sensor, ECM**
E022	Throttle position (TP) sensor – voltage low	**Wiring open circuit/short circuit to ground, TP sensor, ECM**
E023	Ignition control module (ICM) – circuit malfunction	**Wiring open circuit/short circuit to ground, HEI pick up coil, ICM, ECM**
E024	Vehicle speed sensor (VSS)	**Wiring open circuit/short circuit to ground, VSS, ECM**
E026	Closed position throttle (CPT) switch – short circuit	**Wiring open circuit/short circuit to ground, TPS, ECM**
E027	Closed position throttle (CPT) switch – open circuit	**Wiring open circuit, ISC actuator, ECM**
E030	Idle speed control (ISC) actuator	**Wiring short to ground, misfire, throttle valve tight/sticking, intake system leaking, TPS, ISC actuator, ECM**
E031	Manifold absolute pressure (MAP) sensor – short circuit	**Wiring open circuit/short circuit to positive, ECM**
E032	Manifold absolute pressure (MAP) sensor – open circuit	**Wiring open circuit/short circuit to ground, MAP sensor, ECM**
E034	Manifold absolute pressure (MAP) sensor – voltage high	**Wiring open circuit/short circuit to positive, MAP sensor, ECM**
E037	Intake air temperature (IAT) sensor – short circuit	**Wiring short circuit to ground/short circuit to positive, IAT sensor, ECM**
E038	Intake air temperature (IAT) sensor – open circuit	**Wiring open circuit, IAT sensor, ECM**

Trouble code	Fault location	Probable cause
E039	Torque converter clutch (TCC) – slipping	**Wiring open circuit/short circuit to ground, supply wire, transmission internal wiring, TCC solenoid, transmission mechanical fault, ECM**
E040	Power steering pressure (PSP) switch – open circuit	**Wiring open circuit/short circuit to ground/ short circuit to positive, PSP switch, ECM**
E041	Camshaft position (CMP) sensor – circuit malfunction	**Wiring open circuit/short circuit to ground, CMP sensor, ECM**
E042	Oxygen sensor (O2S) 1, bank 2 – mixture lean	**Wiring short circuit to ground, injector(s) blocked, fuel contamination, fuel pressure low, exhaust system leaking, O2S, ECM**
E043	Oxygen sensor (O2S) 1, bank 2 – mixture rich	**Wiring open circuit/ short circuit to positive, fuel pressure high, injector(s) leaking, O2S contamination, EVAP canister purge valve, MAP sensor hoses, O2S, ECM**
E044	Oxygen sensor (O2S) 1, bank 1 – mixture lean	**Wiring open circuit/short circuit to ground, injector(s) blocked, fuel contamination, fuel pressure low, exhaust system leaking, O2S, ECM**
E045	Oxygen sensor (O2S) 1, bank 1 – mixture rich	**Wiring open circuit/short circuit to positive, fuel pressure high, injector(s) leaking, O2S contamination, EVAP canister purge valve, MAP sensor hoses, O2S, ECM**
E046	Long term fuel trim – imbalance bank 1/ bank 2	**Injector(s) leaking, HO2S, intake/exhaust system leaking, spark plug(s), ignition system**
E047	Body control module (BCM), CAN data bus – communication malfunction	**Wiring, BCM**
E048	EGR system problem	**Wiring open circuit/short circuit to ground, supply wire, vacuum hoses, EGR solenoid, ECM**
E051	Engine control module (ECM) – PROM	**Insecure PROM, incorrect PROM, ECM**
E052	Engine control module (ECM) – KAM reset	**Battery disconnected, ECM disconnected**
E053	Distributor – intermittent reference signal	**Wiring open circuit/short circuit to positive, HEI module pick up coil**
E055	Throttle position (TP) sensor – closed throttle out of range	**Throttle valve tight/sticking, ISC actuator, TP sensor, TP sensor incorrectly adjusted**
E058	Immobilizer control module – fuel enable signal	**Incorrect ignition key, wiring open circuit/ short circuit to ground/short circuit to positive, immobilizer control module, ECM**
E060	Park/neutral input with cruise control engaged	**Driver error, wiring open circuit/short circuit to ground, PNP switch**
E061	Cruise control vent solenoid	**Wiring open circuit/short circuit to ground/ short circuit to positive, cruise control servo, ECM**
E062	Cruise control vacuum solenoid	**Wiring open circuit/short circuit to positive, cruise control servo, ECM**

Trouble code	Fault location	Probable cause
E063	Cruise control – set speed/vehicle speed mismatch	**Wiring short circuit to positive, hoses leaking, cruise control servo, throttle binding**
E064	Cruise control – acceleration too high	**Slippery road conditions**
E065	Cruise control servo position sensor – circuit malfunction	**Wiring open circuit/short circuit to ground , cruise control servo**
E066	Cruise control – engine speed too high	**Slippery road conditions, transmission mechanical fault**
E067	Cruise control switch – circuit malfunction	**Wiring short circuit to ground/short circuit to positive, supply wire, cruise control switch, ECM**
E068	Cruise control servo position sensor – range	**Wiring short to positive, hoses blocked/ leaking, brake pedal vacuum valve, cruise control servo, throttle binding**
E070	Throttle position (TP) sensor – intermittent signal	**Wiring open circuit/short circuit to ground/ short circuit to positive, TP sensor, throttle binding**
E071	Manifold absolute pressure (MAP) sensor – intermittent signal	**Wiring open circuit/short circuit to ground/ short circuit to positive, MAP sensor**
E073	Engine coolant temperature (ECT) sensor – intermittent signal	**Wiring open circuit/short circuit to ground/ short circuit to positive, ECT sensor**
E074	Intake air temperature (IAT) sensor – intermittent signal	**Wiring open circuit/short circuit to ground/ short circuit to positive, IAT sensor**
E075	Vehicle speed sensor (VSS) – intermittent signal	**Wiring open circuit/short circuit to ground/ short circuit to positive, VSS**
E080	Fuel system rich	**Fuel pressure high, injector(s) leaking, EVAP system**
E085	Idle throttle angle too high	**throttle valve tight/sticking, TP sensor, ISC actuator**
E090	Brake pedal position (BPP) switch for torque converter clutch (TCC)	**Wiring open circuit/short circuit to positive, incorrectly adjusted BPP switch, BPP switch, ECM**
E091	Park/neutral position (PNP) switch – circuit malfunction	**Wiring open circuit/short circuit to ground, incorrectly adjusted PNP switch, PNP switch, ECM**
E092	Heated windshield request problem	**Wiring open circuit/short circuit to ground/ short circuit to positive, console switch, windshield control module**
E096	Torque converter clutch (TCC) – over stressed	**ECM trouble codes stored, driver error**
E097	Transmission shifted from P/N to D/R at high throttle angle	**Driver error, wiring, transmission range (TR) switch**
E098	Transmission shifted from P/N to D/R in idle speed control (ISC) range	**Driver error, wiring, transmission range (TR) switch**
E099	Cruise control servo – position high after disengagement	**Wiring short to positive, throttle valve tight/ sticking, hoses blocked/leaking, cruise control servo**

CADILLAC

Model:	Engine identification:	Year:
Fleetwood 5.7L	**VIN position 8 = 7**	**1993**
Fleetwood 5.7L	**VIN position 8 = P**	**1994-95**

System: **GM**

Self-diagnosis

General information

- Refer to the front of this manual for general test conditions, terminology, detailed descriptions of wiring faults and a general trouble shooter for electrical and mechanical faults.
- Malfunction indicator lamp (MIL) will illuminate if certain faults are recorded.
- ECM operates in backup mode if sensors fail, to enable vehicle to be driven to repair shop.
- Trouble codes can be displayed by the MIL, the A/C control panel, or by using a Scan Tool connected to the data link connector (DLC).

Accessing

Without using A/C control panel

- Access data link connector (DLC).
- Jump data link connector (DLC) terminals A and B **1**.
- Switch ignition ON. Do NOT start engine.
- MIL should flash.
- Long flashes indicate the 'tens' of the trouble code **2** [**A**].
- Short flashes indicate the 'units' of the trouble code **2** [**C**].
- A short pause separates each flash **2** [**B**].
- A long pause separates each trouble code **2** [**D**].
- For example: Trouble code 12 displayed **2**.
- Trouble code 12 will be displayed 3 times.
- Each trouble code will be displayed three times.
- If no trouble codes are stored, code 12 will continue to be displayed.
- Trouble codes can also be displayed by using a Scan Tool connected to the data link connector (DLC) **1** , located below center of fascia near SRS control module.

1

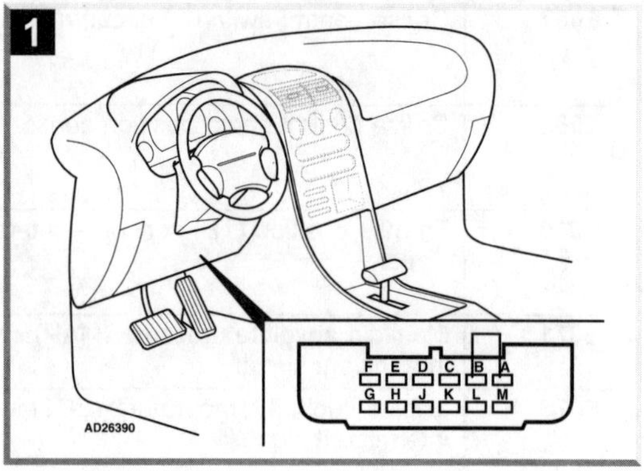

2

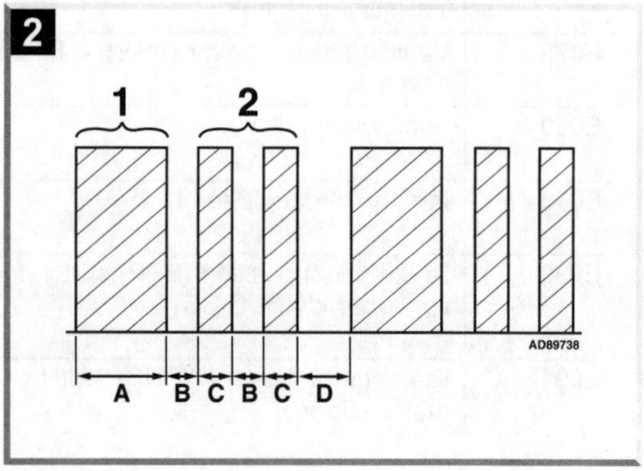

Using A/C control panel

- Switch ignition ON.
- Simultaneously press and hold 'OFF' and 'TEMP UP' buttons **3** [**1**] & [**2**].
- When all segments of the display **3** [**3**] illuminate: Release the 'OFF' and 'TEMP UP' buttons **3** [**1**] & [**2**].
- Parameter number '-00' (ECM codes) will be displayed **3** [**3**].
- Press and release the 'OUT TEMP' button **3** [**4**].

- Buttons 'FAN SPEED UP' and 'FAN SPEED DOWN' **3** [**5**] increase or decrease the parameter numbers.
- Display will alternate between ECM parameter number '-00' and the trouble code at 1.6 second intervals **3** [**3**].
- Trouble codes will be displayed in ascending order.
- Cycle will be repeated after all ECM trouble codes have been displayed.
- If no ECM trouble codes are stored, trouble code '00' will be displayed.
- To end diagnosis:
- Press the 'AUTO' button **3** [**6**].

3

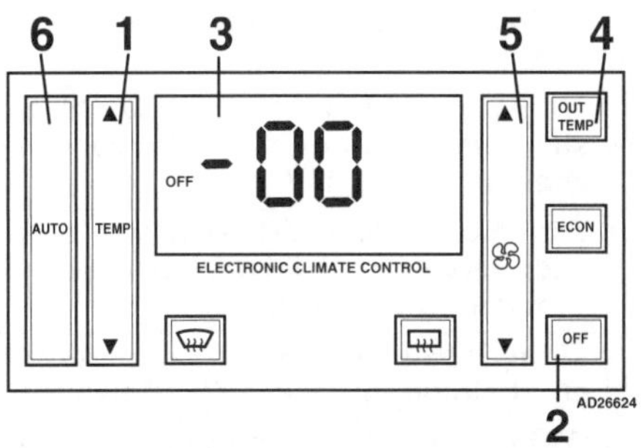

Erasing

Without using A/C control panel

- Ensure ignition switched OFF.
- Disconnect battery ground cable for 30 seconds.
- Reconnect battery ground cable.
- Repeat checking procedure to ensure no trouble codes remain in ECM fault memory.

NOTE: ***Disconnecting battery ground lead may also erase memory from electronic units, such as radio.***

- Trouble codes can also be erased by using a Scan Tool connected to the data link connector (DLC) **1**.

Using A/C control panel

- Access ECM trouble codes.
- During trouble code display:
  - Press and hold 'OFF' **3** [**2**].
- When '00' is displayed **3** [**3**]:
  - Release the 'OFF' button **3** [**2**].
- To end diagnosis:
  - Press the 'AUTO' button **3** [**5**].

Trouble code identification

Trouble code	Fault location	Probable cause
00 **1**	No fault found	–
11	Malfunction indicator lamp (MIL) – circuit malfunction	**Wiring open circuit/short circuit to ground/ short circuit to positive, MIL bulb, supply wire, ECM**
12	Start of diagnosis	**No RPM signal received**
13 **2**	Oxygen sensor (O2S)	**Wiring open circuit, O2S, ECM**
13 **3**	Heated oxygen sensor (HO2S) Bank 1 – circuit malfunction	**Wiring open circuit, HO2S, ECM**
14	Engine coolant temperature (ECT) sensor – high temperature indicated	**Wiring short circuit to ground, ECT sensor, ECM**
15	Engine coolant temperature (ECT) sensor – low temperature indicated	**Wiring open circuit, ECT sensor, ECM**
16	Ignition system – low resolution pulse	**Wiring open circuit/short circuit to ground/ short circuit to positive, distributor, ECM**

Trouble code	Fault location	Probable cause
18	Injector(s) – circuit malfunction	**Wiring open circuit/short circuit to ground/ short circuit to ground, supply wire, injectors, ECM**
21	Throttle position (TP) sensor – voltage high	**Wiring open circuit/short circuit to positive, TP sensor, TP sensor interface module, ECM**
22	Throttle position (TP) sensor – voltage low	**Wiring open circuit/short circuit to ground, TP sensor, TP sensor interface module, ECM**
23	Intake air temperature (IAT) sensor – low temperature indicated	**Wiring open circuit, IAT sensor, ECM**
24	Vehicle speed sensor (VSS)	**Wiring open circuit/short circuit to ground, VSS, VSS buffer module, park neutral position circuit, ECM**
25	Intake air temperature (IAT) sensor – high temperature indicated	**Wiring short circuit to ground, IAT sensor, ECM**
26 [2]	Engine control module quad driver/output driver	**Wiring open circuit/short circuit to ground/ short circuit to positive, ECM controlled components, ECM [4]**
26 [3]	Evaporative emission (EVAP) canister purge valve – circuit malfunction	**Wiring open circuit/short circuit to ground/ short circuit to positive, supply wire, EVAP canister purge valve, ECM**
27	Exhaust gas recirculation (EGR) vacuum control – circuit malfunction	**Wiring open circuit/short circuit to ground/ short circuit to positive, EGR valve, ECM**
28	Transmission range (TR) pressure switch	**Wiring open circuit/short circuit to ground, ECM**
29	Secondary air injection (AIR) system	**Wiring open circuit/short circuit to ground/ short circuit to positive, AIR valve relay, ECM**
32	Exhaust gas recirculation (EGR) solenoid	**Wiring open circuit/short circuit to ground, supply wire, EGR valve, ECM, hoses blocked/leaking, exhaust system blocked**
33	Manifold absolute pressure (MAP) sensor – voltage high, low vacuum indicated	**Wiring short circuit to positive, hose blocked/leaking, MAP sensor, ECM**
34	Manifold absolute pressure (MAP) sensor – voltage low, high vacuum indicated	**Wiring open circuit/short circuit to ground, MAP sensor, ECM**
36	Ignition system high resolution or extra low resolution pulse – circuit malfunction	**Wiring open circuit/short circuit to ground/ short circuit to positive, distributor, ECM**
37	Brake switch stuck ON	**Wiring open circuit/short circuit to ground/ short circuit to positive, brake switch, ECM**
38	Brake switch stuck OFF	**Wiring open circuit/short circuit to ground/ short circuit to positive, brake switch, ECM**
41	Ignition control module (ICM) – electronic spark timing (EST)/ignition control (IC) circuit malfunction	**Wiring open circuit, DI control module, ECM**

Trouble code	Fault location	Probable cause
42	Ignition control module (ICM) – electronic spark timing (EST)/ignition control (IC)/bypass circuit malfunction	**Wiring open circuit/short circuit to ground, DI control module, ECM**
43	Knock sensor (KS)	**Wiring open circuit/short circuit to ground/ short circuit to positive, KS, insecure PROM, ECM**
44 [2]	Oxygen sensor (O2S) – mixture lean	**Wiring short circuit to ground, injector(s) blocked, fuel contamination, fuel pressure low, exhaust system leaking, MAP sensor, O2S, ECM**
44 [3]	Heated oxygen sensor (HO2S) bank 1 – mixture lean	**Wiring short circuit to ground, injector(s) blocked, fuel contamination, fuel pressure low, exhaust system leaking, secondary air system, HO2S, ECM**
45 [2]	Oxygen sensor (O2S) – mixture rich	**Wiring open circuit, fuel pressure high, injector(s) leaking, EVAP canister purge valve, TP sensor, MAP sensor, ECM**
45 [3]	Heated oxygen sensor (HO2S) bank 1 – mixture rich	**Wiring open circuit, fuel pressure high, fuel pressure regulator, injector(s) leaking, EVAP canister purge valve, TP sensor, EGR system, HO2S, MAF sensor, ECM**
46	Immobilizer control module – fuel enable signal	**Incorrect ignition key, wiring open circuit/ short circuit to ground/short circuit to positive, immobilizer control module, ECM**
47	Knock sensor (KS) – circuit malfunction	**insecure PROM, ECM**
48	Mass air flow (MAF) sensor – circuit malfunction	**Wiring open circuit/short circuit to ground/ short circuit to positive, MAF sensor, ECM**
50	System voltage low	**Wiring open circuit/short circuit to ground, alternator, ECM**
51 [2]	Engine control module (ECM) – PROM	**Insecure PROM, incorrect PROM, ECM**
51 [3]	Engine control module (ECM) – EEPROM	**Insecure EEPROM, incorrect EEPROM, ECM**
53	System voltage high	**alternator, ECM**
54	Fuel pump circuit – voltage low	**Wiring open circuit/short circuit to ground, fuel pump relay, engine oil pressure warning lamp switch, fuel pump, ECM**
55	Engine control module (ECM), fuel lean monitor	**Fuel filter/lines blocked, fuel pump**
58	Transmission fluid temperature (TFT) sensor – high temperature indicated	**Wiring short circuit to ground, TFT sensor, ECM**
59	Transmission fluid temperature (TFT) sensor – low temperature indicated	**Wiring open circuit, TFT sensor, ECM**
63	Heated oxygen sensor (HO2S) bank 2 – open circuit	**Wiring open circuit/short circuit to ground/ short circuit to positive, HO2S, ECM**
64	Heated oxygen sensor (HO2S) bank 2 – mixture lean	**Wiring short circuit to ground, injector(s) blocked, fuel contamination, fuel pressure low, exhaust system leaking, secondary air system, ECM**

Trouble code	Fault location	Probable cause
65	Heated oxygen sensor (HO2S) bank 2 – mixture rich	**Fuel pressure high, injector(s) leaking, EVAP canister purge valve, TP sensor, EGR system, HO2S contamination, MAF sensor, ECM**
66	A/C refrigerant pressure sensor circuit – open or short circuit	**Wiring open circuit/short circuit to ground/ short circuit to positive, A/C refrigerant pressure sensor, ECM**
67	A/C refrigerant pressure sensor circuit – circuit malfunction	**Wiring open circuit/short circuit to ground/ short circuit to positive, A/C refrigerant pressure sensor**
68	A/C compressor clutch relay – circuit malfunction	**Wiring short circuit to ground/short circuit to positive, A/C compressor clutch relay, ECM**
69	A/C compressor clutch – circuit malfunction	**Wiring open circuit, A/C compressor clutch relay, ECM**
70	A/C compressor clutch relay – circuit malfunction	**Wiring open circuit/short circuit to ground/ short to positive, A/C compressor clutch relay, ECM**
72	Vehicle speed sensor (VSS) – circuit malfunction	**Wiring open circuit, VSS, ECM**
73	Pressure control solenoid – circuit malfunction	**Wiring open circuit/short circuit to ground/ short to positive, ECM**
74	Traction control system (TCS), circuit malfunction	**Wiring short circuit to ground, ABS control module trouble code(s) stored, ECM**
75	Transmission system – voltage low	**Supply wire, alternator, ECM**
77	Engine coolant blower motor relay 1 – circuit malfunction	**Wiring open circuit/short circuit to ground/ short circuit to positive, engine coolant blower motor relay, ECM**
78	Engine coolant blower motor relay 2 – circuit malfunction	**Wiring open circuit/short circuit to ground/ short circuit to positive, engine coolant blower motor relay, ECM**
79	Transmission fluid – overtemp	**Wiring open circuit/short circuit to ground, TFT sensor, ECM**
81	Transmission shift solenoid (SS), 2nd-3rd – circuit malfunction	**Wiring open circuit/short circuit to ground, SS, ECM**
82	Transmission shift solenoid (SS), 1st-2nd – circuit malfunction	**Wiring open circuit/short circuit to ground, SS, ECM**
83	Torque converter clutch (TCC) solenoid, pulse width modulated – circuit malfunction	**Wiring open circuit/short circuit to ground/ short circuit to positive, TCC solenoid, ECM**
84	Transmission shift solenoid (SS), 3rd-2nd – circuit malfunction	**Wiring open circuit/short circuit to ground, SS, ECM**
85	Torque converter clutch (TCC) – stuck ON	**Wiring open circuit/short circuit to ground, transmission mechanical fault, TCC solenoid**
90	Transmission TCC solenoid – circuit malfunction	**Wiring open circuit/short circuit to ground, TCC solenoid, ECM**

Trouble code	Fault location	Probable cause
95	Change oil lamp – circuit malfunction	**Wiring open circuit/short circuit to ground/ short circuit to positive, instrument panel, engine oil life lamp bulb, ECM**
96	Engine oil level warning lamp – circuit malfunction	**Wiring open circuit/short circuit to ground/ short circuit to positive, instrument panel, engine oil level lamp bulb, oil level sensor, ECM**
97	Vehicle speed sensor (VSS) – circuit malfunction	**Wiring open circuit/short circuit to ground/ short circuit to positive, VSS, ECM**

1 Code 00 is displayed when accessing trouble codes with the A/C control panel.

2 Except 5.7L 1994-95

3 5.7L 1994-95

4 Depending on model, year and equipment level, the following components may be connected to this circuit. Use of the vehicle wiring diagram is recommended to confirm which components are fitted.

- □ A/C compressor clutch relay
- □ A/C condenser blower motor relay
- □ Air bypass valve
- □ Engine coolant blower motor relay
- □ Engine coolant 'hot' warning lamp
- □ Evaporative emission (EVAP) canister purge valve
- □ Exhaust gas recirculation (EGR) solenoid
- □ Malfunction indicator lamp (MIL)
- □ Secondary air pump relay
- □ Transmission shift warning light (MT)
- □ Torque converter clutch (TCC) solenoid
- □ Transmission shift solenoids (SS)

CHEVROLET

Model:	Engine identification:	Year:
Astro/Blazer 4.3L	**VIN position 8 = W**	**1996-02**
Camaro/Caprice 3.8/5.7L	**VIN position 8 = G/K/P**	**1995-02**
C/K Pickup 4.3/5.0/5.7/7.4L	**VIN position 8 = J/M/R/W**	**1996-00**
Chevy Van/Express 4.3/5.0/5.7/7.4/8.1L	**VIN position 8 = G/J/M/R/W**	**1996-02**
Corvette 5.7L (OBD-II)	**VIN position 8 = G/P/S**	**1994-02**
Metro 1.0/1.3L	**VIN position 8 = 2/6**	**1998-01**
Silverado/Suburban 4.3/4.8/5.0/5.3/5.7/6.0/7.4/8.1L	**VIN position 8 = G/J/M/R/T/U/V/W**	**1996-02**
S10 Pickup 2.2/4.3L	**VIN position 8 = W/X/4/5**	**1996-02**
Tahoe 4.8/5.3/5.7L	**VIN position 8 = R/T/V**	**1996-02**
Tracker 1.6/2.0/2.5L	**VIN position 8 = 1/6/C**	**1998-02**

System: **GM**

Self-diagnosis

General information

- Refer to the front of this manual for general test conditions, terminology, detailed descriptions of wiring faults and a general trouble shooter for electrical and mechanical faults.
- Malfunction indicator lamp (MIL) will illuminate or flash if certain faults are recorded.
- ECM operates in backup mode if sensors fail, to enable vehicle to be driven to repair shop.

Accessing

- Trouble codes can only be displayed by using a Scan Tool connected to the data link connector (DLC) **1**.

1

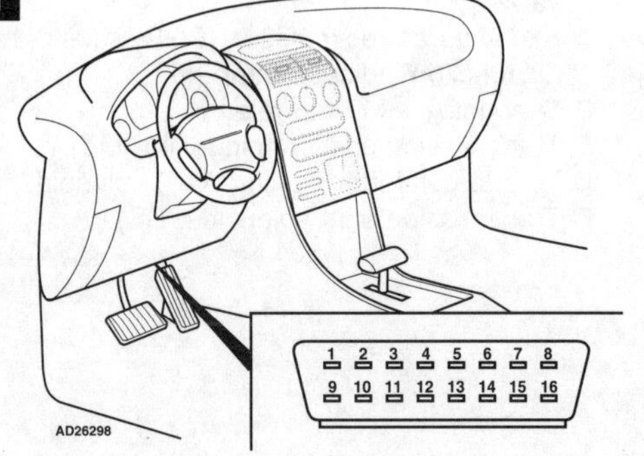

Erasing

- After the faults have been rectified, the manufacturer recommends that trouble codes are erased by using a Scan Tool connected to the data link connector (DLC).

Trouble code identification

Scan code	Fault location	Probable cause
P0000-P0999	Refer to OBD-II trouble code table at the front of this manual	–
P1106	Manifold absolute pressure (MAP) sensor – voltage high, intermittent	**Wiring open circuit/short circuit to positive, poor connections**
P1107	Manifold absolute pressure (MAP) sensor – voltage low, intermittent	**Wiring short circuit to ground, poor connections**
P1111	Intake air temperature (IAT) sensor – voltage high, intermittent	**Wiring open circuit/short to positive, poor connections**
P1112	Intake air temperature (IAT) sensor – voltage low, intermittent	**Wiring short circuit to ground**
P1114	Engine coolant temperature (ECT) sensor – voltage low, intermittent	**Wiring short circuit to ground, poor connections**
P1115	Engine coolant temperature (ECT) sensor – voltage high, intermittent	**Wiring open circuit/short to positive, poor connections**
P1120	Throttle motor position sensor 1 – circuit malfunction	**Wiring, throttle control module**
P1121	Throttle position (TP) sensor – voltage high, intermittent	**Wiring open circuit/short to positive, poor connections**
P1122	Throttle position (TP) sensor – voltage low, intermittent	**Wiring open circuit/short to ground, poor connections**
P1125	Accelerator pedal position (APP) sensor 1/2/3 – signal malfunction	**Wiring, APP sensor, ECM**
P1133	Oxygen sensor (O2S)/heated oxygen sensor (HO2S) 1, bank 1 – insufficient switching	**Wiring, exhaust system leaking, O2S/HO2S**
P1134	Heated oxygen sensor (HO2S) 1, bank 1 – slow response	**Wiring, oxygen sensor heater, exhaust system leaking, HO2S**
P1137	Heated oxygen sensor (HO2S) 2, bank 1 – voltage low	**Wiring, exhaust system leaking, HO2S**
P1138	Heated oxygen sensor (HO2S) 2, bank 1 – voltage high	**Wiring, HO2S**
P1153	Heated oxygen sensor (HO2S) 1, bank 2 – insufficient switching	**Wiring, oxygen sensor heater, exhaust system leaking, HO2S**
P1154	Heated oxygen sensor (HO2S) 1, bank 2 – slow response	**Wiring, oxygen sensor heater, exhaust system leaking, HO2S**
P1158	Heated oxygen sensor (HO2S) 2, bank 2 – not switching, fuel trim (FT) rich mixture	**Wiring, exhaust system leaking, HO2S**
P1171	Fuel system – lean during acceleration	**Fuel filter, fuel pump, HO2S, injectors**
P1172	Fuel level sensors, primary and secondary – no fuel level change during fuel transfer from secondary to primary tank	**Wiring, fuel pump relays, secondary fuel pump, fuel level sensors, ECM**
P1187	Engine oil temperature (EOT) sensor – voltage low	**Wiring open circuit/short circuit to ground, EOT sensor, ECM**
P1188	Engine oil temperature (EOT) sensor – voltage high	**Wiring open circuit/short circuit to positive, EOT sensor, ECM**

Scan code	Fault location	Probable cause
P1220 [3]	Injector control, circuit monitor – malfunction	**Wiring, injector, ECM**
P1220 [4]	Throttle motor position sensor 2 – circuit malfunction	**Wiring, throttle control module**
P1221	Throttle motor position sensor 1/2 – signal variation	**Wiring, throttle control module**
P1222	Injector control, circuit monitor – malfunction	**Injector, wiring**
P1250	Intake manifold heater – circuit malfunction	**Wiring, intake manifold heater, intake manifold heater relay**
P1258	Engine overheating	**Cooling system, ECT sensor**
P1271	Accelerator pedal position (APP) sensor 1/2 – signal variation	**Wiring, APP sensor, ECM**
P1272	Accelerator pedal position (APP) sensor 2/3 – signal variation	**Wiring, APP sensor, ECM**
P1273	Accelerator pedal position (APP) sensor 1/3 – signal variation	**Wiring, APP sensor, ECM**
P1275	Accelerator pedal position (APP) sensor 1 – circuit malfunction	**Wiring, APP sensor, ECM**
P1276	Accelerator pedal position (APP) sensor 1 – range performance	**Wiring, APP sensor, ECM**
P1280	Accelerator pedal position (APP) sensor 2 – circuit malfunction	**Wiring, APP sensor, ECM**
P1281	Accelerator pedal position (APP) sensor 2 – range performance	**Wiring, APP sensor, ECM**
P1285	Accelerator pedal position (APP) sensor 3 – circuit malfunction	**Wiring, APP sensor, ECM**
P1286	Accelerator pedal position (APP) sensor 3 – range performance	**Wiring, APP sensor, ECM**
P1336	Crankshaft position sensor (CKP) – variation not learned	**Learning procedure not performed**
P1345	Crankshaft position sensor (CKP)/camshaft position sensor (CMP) – signal variation	**CKP sensor, distributor/camshaft alignment, mechanical fault**
P1350	Ignition control system, bypass circuit – circuit malfunction	**Wiring, poor connections, ICM, ECM**
P1351	Ignition control system, coil – voltage high	**Wiring open circuit, ICM, ECM**
P1352	Ignition control system, bypass circuit – voltage high	**Wiring open circuit, ICM, ECM**
P1361	Ignition control system, ignition control (IC) circuit – voltage low	**Wiring, circuit short circuit to ground, poor connections, ICM, ECM**
P1362	Ignition control, bypass circuit – voltage low	**Wiring short circuit to ground, ICM, ECM**
P1371	Ignition control reference signal, low/high resolution – low signal lost	**Wiring, ICM**
P1374	Crankshaft position sensor (CKP), high to low resolution frequency – signal variation	**Wiring, wiring routed near spark plug wires, ignition coils arcing/tracking, ICM, ECM**

Scan code	Fault location	Probable cause
P1380	Antilock brake system (ABS)/traction control system (TCS), rough road signal to engine control module (ECM) – malfunction	ABS trouble code(s) stored
P1381	Antilock brake system (ABS)/traction control system (TCS), rough road signal to engine control module (ECM) – no communication	Wiring, poor connections, ABS
P1404	Exhaust gas recirculation (EGR) valve – not fully closing	Wiring, poor connections, EGR valve position sensor, EGR solenoid
P1406	Exhaust gas recirculation (EGR) valve position sensor – range performance	Wiring, EGR valve tight/sticking, EGR valve position sensor
P1408	Manifold absolute pressure (MAP) sensor – voltage out range	Wiring, poor connections, MAP sensor, ECM
P1410	Evaporative emission (EVAP) canister purge valve – circuit malfunction	Wiring, poor connectios, EVAP canister vent valve, ECM
P1415	Secondary air injection (AIR) system, bank 1	Wiring, hoses blocked/leaking, AIR valve
P1416	Secondary air injection (AIR) system, bank 2	Wiring, hoses blocked/leaking, AIR valve
P1431	Fuel level sensor – circuit malfunction	Wiring, fuel level sensor
P1432	Fuel level sensor – low voltage	Wiring open circuit, fuel level sensor
P1433	Fuel level sensor – voltage high	Wiring short circuit to positive, fuel level sensor
P1441	Evaporative emission (EVAP) system – leak detected	Hose connection(s), EVAP canister, EVAP canister purge valve
P1442	Evaporative emission (EVAP) system vacuum switch – circuit malfunction	Hoses blocked/leaking, wiring, EVAP canister, ECM
P1450	Barometric pressure (BARO) sensor – voltage out of range	Wiring, BARO sensor, ECM
P1451	Barometric pressure (BARO) sensor – performance range	Wiring, poor connections, BARO sensor, ECM
P1500	Ignition switch, cranking signal/ system voltage – circuit malfunction	Wiring, poor connections, ECM
P1504	Engine control module (ECM), vehicle speed sensor (VSS) output signal – circuit malfunction	Wiring, poor connections, ABS, ECM
P1508	Idle air control (IAC) valve – low rpm	Wiring open circuit/short circuit to ground/ short circuit to positive, IAC valve, throttle valve tight/sticking
P1509	Idle air control (IAC) valve – high rpm	Wiring open circuit/short circuit to ground/ short circuit to positive, IAC valve, throttle valve tight/sticking
P1510	Engine control module (ECM), KAM supply voltage – circuit malfunction	Wiring, poor connections, ECM
P1514	Throttle control unit, intake air flow – excessive	Throttle control unit, MAP sensor, CMP sensor signal, ECM
P1515	Throttle command/actual throttle position – signal variation	Wiring, throttle valve sticking, throttle control module

Scan code	Fault location	Probable cause
P1516	Throttle command/actual throttle position – signal variation	Wiring, throttle valve sticking, throttle control module
P1517	Throttle control module – malfunction	Throttle control module
P1518	Throttle control module/engine control module (ECM) – communication malfunction	Wiring, throttle control module, ECM
P1519	Throttle motor position sensor 1 – signal malfunction	Throttle control module
P1520	Park/neutral position (PNP) switch – circuit malfunction	Wiring short circuit to ground, PNP switch incorrectly adjusted
P1523	Throttle control unit, closed throttle position – performance	Throttle valve movement restricted, throttle control unit
P1530	Ignition timing adjustment connector	Wiring short circuit to ground, connector short to ground
P1532	A/C refrigerant temperature sensor – voltage low	Wiring open circuit/short circuit to ground, A/C refrigerant temperature sensor, ECM
P1533	A/C refrigerant temperature sensor – voltage high	Wiring open circuit/short circuit to ground, A/C refrigerant temperature sensor, ECM
P1539	A/C compressor clutch relay – voltage high	Wiring, A/C compressor clutch relay, ECM
P1543	A/C system – performance	Refrigerant low, wiring, A/C refrigerant pressure sensor, A/C evaporator temperature sensor, ECM
P1545	A/C compressor clutch relay – voltage low	Wiring, A/C compressor clutch relay, ECM
P1546	A/C compressor clutch relay – circuit malfunction	Wiring, A/C compressor clutch relay, ECM
P1554	Cruise control status circuit – high voltage	Wiring open circuit/short circuit to ground, cruise control module, ECM
P1571	Traction control system (TCS), torque request signal – malfunction	Wiring, poor connections, ABS/TCS
P1572	Traction control system (TCS), active signal – malfunction	Wiring short to ground, ABS/TCS, ECM
P1574	Brake pedal position switch (BPP), voltage high	Wiring, ABS
P1575	Brake pedal position switch (BPP), extended travel contacts – voltage high	Wiring, ABS
P1586	Cruise control system, brake switch 2 – circuit malfunction	Wiring, poor connections, BPP switch
P1600	Engine control module (ECM) – malfunction	ECM
P1601	CAN data bus, engine control module (ECM)/ antilock brake system (ABS)/traction control system (TCS)/instrumentation control module/ diagnostic control module – some communication lost	Wiring, poor connections, ECM, ABS/TCS, diagnostic control module, instrumentation control module
P1602	Engine control module (ECM) to antilock brake system (ABS)/traction control system (TCS), CAN data bus – communication failure	Wiring, poor connections, ABS control module

Scan code	Fault location	Probable cause
P1621	Engine control module (ECM) – malfunction	ECM
P1624	Engine control module (ECM) – memory performance	ECM
P1626	Engine control module (ECM) – fuel enable signal lost	Wiring, poor connections, immobilizer control module, ECM
P1627	Engine control module (ECM) – malfunction	ECM
P1629	Engine control module (ECM) – fuel enable signal not received	Incorrect ignition key, wiring, poor connections, immobilizer control module, ECM
P1630	Engine control module (ECM), immobilizer code – learning mode	ECM immobilizer code programming not completed
P1631	Engine control module (ECM) – incorrect immobilizer code	Incorrect key, immobilizer code not programmed
P1632	Engine control module (ECM) – fuel disable signal received	Fuses, BCM, immobilizer code
P1635	Engine control module (ECM), sensor supply voltage – circuit malfunction	Wiring, poor connections, TP sensor, MAP sensor, EGR valve position sensor, FTP sensor, ECM
P1637	Alternator 'L' terminal – circuit malfunction	Wiring open circuit/short circuit to ground, alternator, battery temperature sensor, ECM
P1638	Alternator 'F' terminal – circuit malfunction	Wiring open circuit/short circuit to ground, alternator, ECM
P1639 [3]	Engine control module (ECM), sensor supply voltage – circuit malfunction	Wiring, poor connections, A/C refrigerant pressure sensor, ECM
P1639 [4]	Engine control module (ECM), sensor supply voltage – circuit malfunction	Wiring, poor connections, fuel tank level sensor, ECM
P1639 [8]	Engine control module (ECM), sensor supply voltage – circuit malfunction	Wiring, poor connections, FTP sensor, ECM
P1641 [3]	Malfunction indicator lamp (MIL) – circuit malfunction	Wiring, poor connections, MIL, ECM
P1641 [4]	Engine coolant blower motor relay 1 – circuit malfunction	Wiring, engine coolant blower motor relay, ECM
P1642	Engine coolant blower motor relay 2 & 3 – circuit malfunction	Wiring, engine coolant blower motor relay(s), ECM
P1643	Engine speed (RPM), engine control module (ECM) output signal – circuit malfunction	Wiring, tachometer, ECM
P1644	Traction control system (TCS), delivered torque signal – malfunction	Wiring, poor connections, ECM
P1651	Engine coolant blower motor relay 1 – circuit malfunction	Wiring, engine coolant blower motor relay, ECM
P1652 [1] [5]	Engine control module(ECM), vehicle speed sensor (VSS) output signal – circuit malfunction	Wiring, poor connections, ECM
P1652 [2]	Engine control module (ECM), firm suspension request signal – circuit malfunction	Wiring, poor connections, suspension control module, ECM

Scan code	Fault location	Probable cause
P1652 3	Engine coolant blower motor relay 2 & 3 – circuit malfunction	**Wiring, engine coolant blower motor relay(s), ECM**
P1653 6	Fuel gauge, control signal – circuit malfunction	**Wiring, ECM**
P1653 7	Engine oil level 'low' warning lamp, control signal – circuit malfunction	**Wiring short circuit to ground, engine oil level, ECM**
P1654	A/C compressor clutch relay – circuit malfunction	**Wiring open circuit/short circuit to ground, A/C compressor clutch relay, ECM**
P1655	Evaporative emission (EVAP) canister purge valve – circuit malfunction	**Wiring open circuit/short circuit to ground, EVAP canister vent valve, ECM**
P1657	Skip shift solenoid – circuit malfunction	**Wiring, skip shift solenoid, ECM**
P1661	Malfunction indicator lamp (MIL) – circuit malfunction	**Wiring, MIL, ECM**
P1662	Cruise control, inhibit output circuit – malfunction	**Wiring open circuit/short circuit to ground, cruise control motor/actuator, cruise control module, ECM**
P1663 6	Alternator warning lamp, control signal – circuit malfunction	**Wiring, instrument panel, ECM**
P1663 7	Engine oil life monitor reset switch – circuit malfunction	**Wiring short circuit to ground, engine oil life monitor reset switch, ECM**
P1664	Skip shift warning lamp – circuit malfunction	**Wiring, skip shift warning lamp, ECM**
P1665	Evaporative emission (EVAP) canister vent valve – circuit malfunction	**Wiring open circuit/short circuit to ground, EVAP canister vent valve, ECM**
P1667	Reverse inhibit solenoid – circuit malfunction	**Wiring, reverse inhibit solenoid, ECM**
P1671	Malfunction indicator lamp (MIL) – circuit malfunction	**Wiring open circuit/short circuit to ground/ short circuit to positive, MIL, ECM**
P1672	Engine oil level 'low' warning lamp, control signal – circuit malfunction	**Wiring short circuit to ground, engine oil level, ECM**
P1676	Evaporative emission (EVAP) canister purge valve – circuit malfunction	**Wiring open circuit/short circuit to ground, EVAP canister purge valve, ECM**
P1680	Engine control module (ECM) – malfunction	**ECM**
P1681	Engine control module (ECM) – malfunction	**ECM**
P1683	Engine control module (ECM) – malfunction	**ECM**
P1689	Traction control system (TCS), delivered torque signal – malfunction	**Wiring, poor connections, ECM**
P1740	Cruise control system, overdrive disable signal – voltage low	**Wiring short to ground, ECM**
P1781	Transmission control module (TCM), malfunction indicator lamp (MIL) request signal – voltage high/low	**Wiring TCM, ECM**
P1810	Transmission gear pressure switch, signal – circuit malfunction	**Wiring open circuit/short circuit to ground, reverse/D1/D2/D3/D4 gear pressure switch(es), ECM**
P1860	Torque converter clutch (TCC) solenoid, pulse width modulation (PWM) – circuit malfunction	**Wiring open circuit/short circuit to ground, TCC solenoid, ECM**

Scan code	Fault location	Probable cause
P1870	Torque converter clutch (TCC) – slip speed excessive	**Transmission fluid level, wiring, TCC solenoid, ECM**
P1875	Transmission, 4WD low selection signal – circuit malfunction	**Wiring, transmission, four wheel drive control module, ECM**
U1000	CAN data bus – communication malfunction	**Wiring, BCM, ECM, instrumentation control module, SDM**
U1026	Transmission control module (TCM)/four wheel drive control module, CAN data bus – communication malfunction	**Wiring, BCM**
U1041	Antilock brake system (ABS)/traction control system (TCS), CAN data bus – communication malfunction	**Wiring, ABS/TCS trouble codes stored**
U1064	Body control module (BCM), CAN data bus – communication malfunction	**Wiring, BCM**
U1192	Engine control module (ECM)/immobilizer control module, CAN data bus – signal malfunction	**Wiring, immobilizer control module, ECM**
U1193	Engine control module (ECM)/immobilizer control module, CAN data bus – signal malfunction	**Wiring, immobilizer control module, ECM**
U1300	CAN data bus – signal low	**Wiring**
U1301	CAN data bus – signal high	**Wiring**

1 1996
2 1997→
3 Camaro 3.8L
4 Except Camaro 3.8L
5 Camaro 5.7L
6 Except Caprice
7 Caprice
8 S10 Pick-up/Express

CHEVROLET

Model:	Engine identification:	Year:
Beretta/Corsica 2.2L	**VIN position 8 = 4**	**1992-95**
Beretta 2.3L	**VIN position 8 = A**	**1992-94**
Cavalier 2.2/2.3L	**VIN position 8 = D/4**	**1992-95**
Cavalier 3.1L	**VIN position 8 = T**	**1992-94**
Lumina 2.5L	**VIN position 8 = R**	**1992**
Lumina/APV 3.1L	**VIN position 8 = D/M/T**	**1992-95**
Lumina 3.4L	**VIN position 8 = X**	**1992-93**
Lumina APV 3.8L	**VIN position 8 = L**	**1992-93**
Monte Carlo 3.1L	**VIN position 8 = M**	**1995**

System: **GM**

Self-diagnosis

General information

- Refer to the front of this manual for general test conditions, terminology, detailed descriptions of wiring faults and a general trouble shooter for electrical and mechanical faults.
- Malfunction indicator lamp (MIL) will illuminate if certain faults are recorded.
- ECM operates in backup mode if sensors fail, to enable vehicle to be driven to repair shop.
- Trouble codes can be displayed by the MIL or by using a Scan Tool connected to the data link connector (DLC).

Accessing

- Access data link connector (DLC).
- Jump data link connector (DLC) terminals A and B **1**.
- Switch ignition ON. Do NOT start engine.
- Count MIL flashes. Note trouble codes. Compare with trouble code table.
- Firstly trouble code 12 will be displayed 3 times.
- Trouble codes consist of 2 groups, displayed as follows:
  - First group indicates the 'tens' of the trouble code **2** [**A**].
  - Second group indicates the 'units' of the trouble code **2** [**C**].
  - A short pause separates each group **2** [**B**].
  - A long pause separates each trouble code **2** [**D**].
  - For example: Trouble code 12 displayed **2**. Start of diagnosis.
- Each additional trouble code will now be displayed 3 times, in ascending order.

NOTE: ***If no additional trouble codes are stored, code 12 will continue to be displayed.***

- Trouble codes can also be displayed by using a Scan Tool connected to the data link connector (DLC) **1**.

1

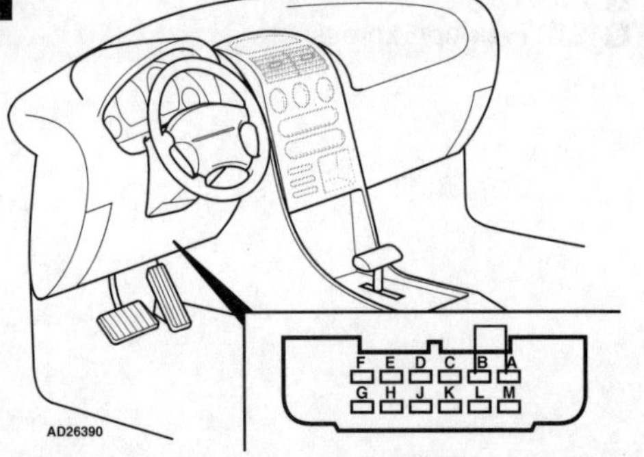

2

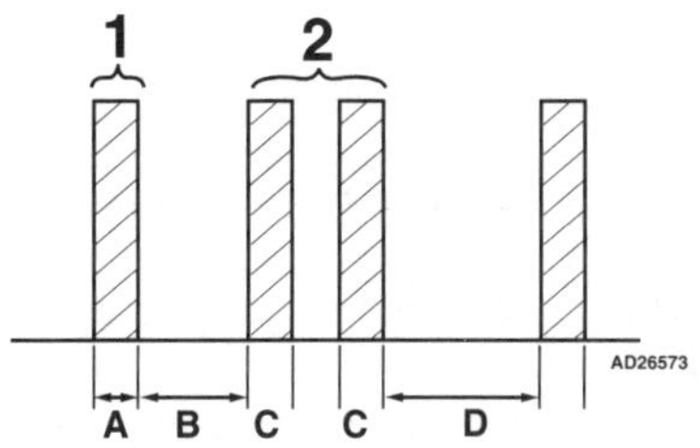

Erasing

NOTE: ***Erasing trouble codes may cause driveability problems until the engine management system relearns basic values.***

- Ensure ignition switched OFF.
- Disconnect battery ground cable for at least 30 seconds.
- Reconnect battery earth cable.
- Repeat accessing procedure to ensure no trouble codes remain in ECM fault memory.

NOTE: ***Disconnecting battery ground lead may also erase memory from electronic units, such as radio.***

- Trouble codes can only be erased by using a Scan Tool connected to the data link connector (DLC) [1].

Trouble code identification

Flash code	Fault location	Probable cause
12 [1]	Start of diagnosis	**No engine speed (RPM) signal received**
13	Oxygen sensor (O2S)/heated oxygen sensor (HO2S)	**Wiring open circuit, O2S/HO2S, ECM**
14	Engine coolant temperature (ECT) sensor – voltage high/low	**Wiring, ECT sensor, ECM**
15	Engine coolant temperature (ECT) sensor – voltage high	**Wiring open circuit, ECT sensor, ECM**
16	System voltage – high/low	**Wiring, battery, alternator**
16 [9]	Ignition control module (ICM) – 2X reference circuit malfunction	**Wiring open circuit/short circuit to ground/ short circuit to positive, ICM, ECM**
17 [2]	Camshaft position (CMP) sensor – circuit malfunction	**Wiring, CMP sensor, mechanical fault, ECM**
17 [3]	Ignition control system, reference signal – circuit malfunction	**Wiring, poor connections, ICM, ECM**
18	Camshaft position (CMP) signal/fuel control signal	**Wiring open circuit/short circuit to ground, CKP sensor, CMP sensor, ICM, ECM**
19	Crankshaft position (CKP) sensor, 7X signal – circuit malfunction	**Wiring, CKP sensor, ICM, ECM**
21	Throttle position (TP) sensor – voltage high/ low	**Wiring, TP sensor, ECM**
22	Throttle position (TP) sensor – voltage low	**Wiring open circuit/short circuit to ground, TP sensor, ECM**
23	Intake air temperature (IAT) sensor – voltage high/low	**Wiring, IAT sensor, ECM**

Flash code	Fault location	Probable cause
24	Vehicle speed sensor (VSS) – no signal	**Wiring, VSS, ECM**
25	Intake air temperature (IAT) sensor – voltage low	**Wiring short circuit to ground, IAT sensor, ECM**
26	Engine control module (ECM), quad driver/ output driver – circuit malfunction	**Wiring, ECM controlled components, ECM 12**
27	Engine control module (ECM), quad driver/ output driver – circuit malfunction	**Wiring, ECM controlled components, ECM 12**
28 7 10 11	Engine control module (ECM), quad driver/ output driver – circuit malfunction	**Wiring, ECM controlled components, ECM 12**
28 8	Transmission range (TR) switch – malfunction	**Wiring, TR switch, ECM**
31	Transmission range (TR) switch – malfunction	**Wiring, TR switch, ECM**
32	Exhaust gas recirculation (EGR) solenoid	**Wiring, poor connections, EGR valve, ECM**
33	Manifold absolute pressure (MAP) sensor – voltage high/low	**Wiring, hose blocked/leaking, MAP sensor, ECM**
34	Manifold absolute pressure (MAP) sensor – voltage low	**Wiring open circuit/short circuit to ground, MAP sensor, ECM**
35	Idle air control (IAC) valve	**Wiring, IAC valve, intake system leaking, fuel system, PCV valve, A/C system, mechanical fault, ECM**
36 2	Crankshaft position (CKP) sensor, at crankshaft pulley – 24X signal circuit malfunction	**Wiring, CKP sensor, ECM**
36 3	Shift solenoids (SS)	**Wiring open circuit/short circuit to ground, SS**
37	Brake pedal position (BPP) switch for torque converter clutch (TCC)	**Wiring, incorrectly adjusted, BPP switch, ECM**
38	Brake pedal position (BPP) switch for torque converter clutch (TCC)	**Wiring open circuit/short circuit to ground/ short circuit to positive, BPP switch incorrectly adjusted, BPP switch, ECM**
39	Clutch pedal position (CPP) switch	**Wiring, incorrectly adjusted, CPP switch, ECM**
41 2	Engine control module (ECM), programming – malfunction	**ECM incorrect programming, ECM failure**
41 3	Camshaft position (CMP) sensor	**Wiring, CMP sensor, mechanical fault, ECM**
41 5	Ignition control, timing signal – circuit malfunction	**Wiring, poor connections, ICM, ECM**
41 9	Ignition control module (ICM) – 1X circuit malfunction	**Wiring open circuit/short circuit to ground/ short circuit to positive, ICM, ECM**
42	Ignition control, bypass/timing signal – circuit malfunction	**Wiring, poor connections, ICM, ECM**
43	Knock sensor (KS)/knock control module – circuit malfunction	**Wiring, KS, knock control module, ECM**

Flash code	Fault location	Probable cause
44	Oxygen sensor (O2S)/heated oxygen sensor (HO2S) – mixture lean	**Wiring open circuit/short circuit to ground, injector(s) blocked, fuel contamination, fuel pressure low, exhaust system leaking, O2S/HO2S, ECM**
45	Oxygen sensor (O2S)/heated oxygen sensor (HO2S) – mixture rich	**Wiring open circuit, fuel pressure high, injector(s) leaking, O2S/HO2S, EVAP purge valve, TP sensor, EGR system, ECM**
46	Immobilizer system, code – malfunction	**Incorrect key, ECM**
51	Engine control module (ECM), PROM – malfunction	**Poor connections, incorrect programming, ECM**
52	Engine control module (ECM) – malfunction	**ECM**
53 [2]	System voltage – high	**Alternator**
53 [3]	Exhaust gas recirculation (EGR) valve solenoid 1	**Wiring open circuit/short circuit to ground, supply wire, EGR valve, ECM**
54 [2]	Fuel pump circuit – voltage low	**Wiring open circuit/short circuit to ground, fuel pump relay, engine oil pressure warning lamp/fuel pump switch**
54 [3]	Exhaust gas recirculation (EGR) valve solenoid 2	**Wiring open circuit/short circuit to ground, supply wire, EGR valve, ECM**
55 [2]	Engine control module (ECM) – malfunction	**ECM**
55 [3]	Exhaust gas recirculation (EGR) valve solenoid 3	**Wiring open circuit/short circuit to ground, supply wire, EGR valve, ECM**
55 [7] [10] [11]	Engine control module (ECM), fuel lean monitor	**Fuel filter/lines blocked, fuel contamination**
56	Engine control module (ECM), quad driver/ output driver – circuit malfunction	**Wiring, ECM controlled components, ECM [12]**
58	Transmission fluid temperature (TFT) sensor – voltage low	**Wiring short circuit to ground, TFT sensor, ECM**
59	Transmission fluid temperature (TFT) sensor – voltage high	**Wiring open circuit, TFT sensor, ECM**
61 [2]	Oxygen sensor (O2S) – malfunction	**O2S**
61 [3]	Cruise control vent solenoid	**Wiring open circuit/short circuit to ground/ short circuit to positive, cruise control vent solenoid, cruise control servo, ECM**
62 [2]	Transmission gear switch signal circuit	**Wiring open circuit/short circuit to ground, 2nd gear switch, 3rd gear switch, ECM**
62 [3]	Cruise control vacuum solenoid	**Wiring open circuit/short circuit to ground/ short circuit to positive, cruise control vacuum solenoid, cruise control servo, ECM**
63	Cruise control servo – servo position sensor low	**Hoses leaking, cruise control servo**
65 [5]	Cruise control servo position sensor – circuit malfunction	**Wiring open circuit/short circuit to ground/ short circuit to positive, cruise control servo, ECM**

Flash code	Fault location	Probable cause
65 [6]	Injectors – circuit malfunction	**Wiring open circuit/short circuit to ground/ short circuit to positive, injectors, ECM**
66	A/C refrigerant pressure sensor – voltage low	**Wiring open circuit/short circuit to ground, A/C refrigerant pressure sensor, ECM**
67	Cruise control switch – circuit malfunction	**Wiring open circuit/short circuit to ground/ short circuit to positive, cruise control switch, ECM**
68	Cruise control servo – servo position sensor high	**Hoses blocked/leaking, brake pedal vacuum valve, cruise control servo**
69	A/C refrigerant pressure switch, engine coolant blower motors	**Wiring open circuit, A/C refrigerant pressure switch, ECM**
70	A/C refrigerant pressure sensor – voltage high	**Wiring open circuit/short circuit to positive, A/C refrigerant pressure sensor, ECM**
72 [5]	Vehicle speed sensor (VSS) – circuit malfunction	**Wiring, VSS, ECM**
72 [11]	CAN data bus, engine control module (ECM) to instrumentation control module – lost communication	**Wiring, poor connections, instrumentation control module, ECM**
75	Exhaust gas recirculation (EGR) solenoid 1 – circuit malfunction	**Wiring, EGR valve, ECM**
76	Exhaust gas recirculation (EGR) solenoid 2 – circuit malfunction	**Wiring, EGR valve, ECM**
77	Exhaust gas recirculation (EGR) solenoid 3 – circuit malfunction	**Wiring, EGR valve, ECM**
79 [4]	Vehicle speed sensor (VSS) – voltage high	**Wiring, VSS, ECM**
79 [5]	Transmission fluid temperature (TFT) – high	**Transmission mechanical fault, towing weight limit exceeded**
80 [4]	Vehicle speed sensor (VSS) – voltage low	**Wiring, VSS, ECM**
80 [5]	Torque converter clutch (TCC) – slipping	**Wiring, TCC solenoid, transmission mechanical fault, ECM**
81	Brake pedal position (BPP) switch for torque converter clutch (TCC) – circuit malfunction	**Wiring, BPP switch, ECM**
82	Crankshaft position (CKP) sensor, in cylinder block – 3X signal circuit malfunction	**Wiring open circuit/short circuit to ground, CKP sensor, ICM**
85	Engine control module (ECM), PROM – malfunction	**Poor connections, incorrect programming, ECM**
86	Engine control module (ECM) – analogue/ digital converter	**A/C refrigerant pressure sensor/KS/ECT sensor/fuel pump monitor circuit short circuit to battery voltage**
87	Engine control module (ECM), EEPROM – malfunction	**Poor connections, incorrect programming, ECM**
90	Torque converter clutch (TCC) – stuck ON	**Wiring open circuit/short circuit to ground, transmission internal wiring, TCC solenoid, transmission mechanical fault, ECM**
96	Engine control module (ECM) – ignition supply voltage low	**Wiring open circuit, alternator, ECM**

Flash code	Fault location	Probable cause
98	Engine control module (ECM) – incorrectly coded	**ECM incorrectly coded, ECM**
99	Cruise control system – engine management has operated in limited power mode	**Throttle cable sticking, brake vacuum release valve, cruise control actuator**

1 Trouble code 12 is recorded when the ECM has not received an engine speed signal.
2 Except Lumina 3.8L
3 Lumina 3.8L
4 Lumina →94
5 Lumina 3.1L/Monte Carlo 3.1L 95
6 Except Lumina 3.1L/Monte Carlo 3.1L 95
7 Beretta/Corsica 2.2L/2.3L 94→
8 Except Beretta/Corsica 2.2L/2.3L 94→
9 Beretta 2.3L
10 Cavalier 2.2L 94→
11 Cavalier 2.3L 95
12 Depending on model, year and equipment level, the following components may be connected to this circuit. Use of the vehicle wiring diagram is recommended to confirm which components are fitted.

- □ A/C compressor clutch relay
- □ A/C condenser blower motor relay
- □ Engine coolant blower motor relay
- □ Engine coolant 'hot' warning lamp
- □ Evaporative emission (EVAP) canister purge valve
- □ Exhaust gas recirculation (EGR) solenoid
- □ Malfunction indicator lamp (MIL)
- □ Torque converter clutch (TCC) solenoid
- □ Transmission shift solenoids (SS)

CHEVROLET

Model:	Engine identification:	Year:
Beretta/Corsica 2.2/3.1L	VIN position 8 = M/4	1996
Cavalier 2.2L	VIN position 8 = F/4	1996-02
Cavalier 2.3/2.4L	VIN position 8 = D/T	1992-02
Lumina 3.1L	VIN position 8 = T/M/J	1996-01
Lumina 3.4L	VIN position 8 = X	1994-96
Lumina APV/Minivan 3.4L	VIN position 8 = E	1996
Lumina APV/Minivan 3.8L	VIN position 8 = L	1994-95
Monte Carlo 3.1L	VIN position 8 = M	1996-99
Monte Carlo/Z34 3.4L	VIN position 8 = X	1995-97
Monte Carlo 3.8L	VIN position 8 = K	1998-02

System: **GM**

Self-diagnosis

General information

- Refer to the front of this manual for general test conditions, terminology, detailed descriptions of wiring faults and a general trouble shooter for electrical and mechanical faults.
- Malfunction indicator lamp (MIL) will illuminate or flash if certain faults are recorded.
- ECM operates in backup mode if sensors fail, to enable vehicle to be driven to repair shop.

Accessing

- Trouble codes can only be displayed by using a Scan Tool connected to the data link connector (DLC) 1 or 2.

1

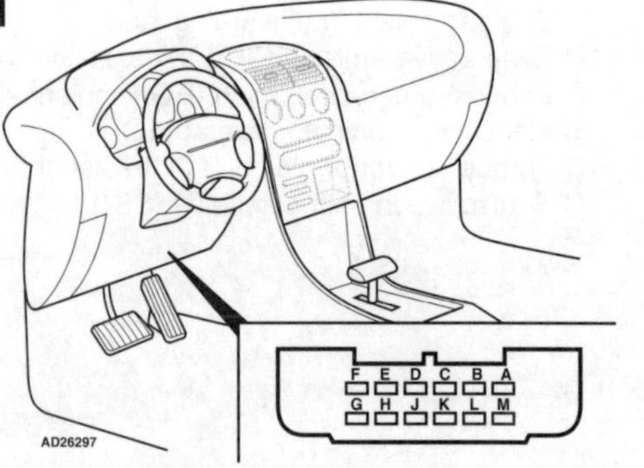

2

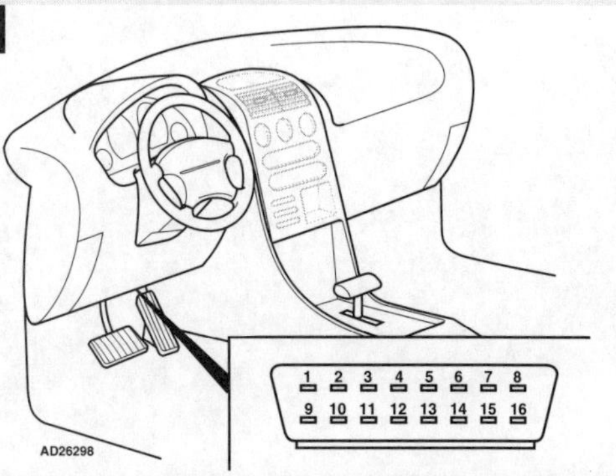

Erasing

- After the faults have been rectified, the manufacturer recommends that trouble codes are erased by using a Scan Tool connected to the data link connector (DLC) **1** or **2**.

Trouble code identification

Scan code	Fault location	Probable cause
P0000-P0999	Refer to OBD-II trouble code table at the front of this manual	–
P1106	Manifold absolute pressure (MAP) sensor – voltage high, intermittent	**Wiring open circuit/short circuit to positive, poor connections**
P1107	Manifold absolute pressure (MAP) sensor – voltage low, intermittent	**Wiring short circuit to ground, poor connections**
P1111	Intake air temperature (IAT) sensor – voltage high, intermittent	**Wiring open circuit/short to positive, poor connections**
P1112	Intake air temperature (IAT) sensor – voltage low, intermittent	**Wiring short circuit to ground**
P1114	Engine coolant temperature (ECT) sensor – voltage low, intermittent	**Wiring short circuit to ground, poor connections**
P1115	Engine coolant temperature (ECT) sensor, voltage high, intermittent	**Wiring open circuit/short to positive, poor connections**
P1121	Throttle position (TP) sensor – voltage high, intermittent	**Wiring open circuit/short to positive, poor connections**
P1122	Throttle position (TP) sensor – voltage low, intermittent	**Wiring open circuit/short to ground, poor connections**
P1133	Oxygen sensor (O2S)/heated oxygen sensor (HO2S) 1, bank 1 – insufficient switching	**Wiring, exhaust system leaking, O2S/HO2S**
P1134	Heated oxygen sensor (HO2S) 1, bank 1 – slow response	**Wiring, oxygen sensor heater, exhaust system leaking, HO2S**
P1137	Heated oxygen sensor (HO2S) 2, bank 1 – voltage low	**Wiring, exhaust system leaking, HO2S**
P1138	Heated oxygen sensor (HO2S) 2, bank 1 – voltage high	**Wiring, HO2S**
P1171	Fuel system – lean during acceleration	**Fuel filter, fuel pump, HO2S, injectors**
P1189	Engine oil pressure warning lamp switch – circuit malfunction	**Wiring, engine oil pressure warning lamp switch**
P1200	Injectors – circuit malfunction	**Wiring open circuit/short circuit to ground/ short circuit to positive, injectors, ECM**
P1336	Crankshaft position sensor (CKP) – variation not learned	**Learning procedure not performed**
P1350	Ignition control system, bypass monitor circuit – malfunction	**Wiring, poor connections, ICM, ECM**
P1351	Ignition control system, coil – voltage high	**Wiring open circuit, ICM, ECM**
P1352	Ignition control system, bypass circuit – voltage high	**Wiring open circuit, ICM, ECM**
P1361	Ignition control system, ignition control (IC) circuit – voltage low	**Wiring, circuit short circuit to ground, poor connections, ICM, ECM**

CHEVROLET

Beretta/Corsica 2.2/3.1L • Cavalier 2.2L • Cavalier 2.3/2.4L • Lumina 3.1L
Lumina 3.4L • Lumina APV/Minivan 3.4L • Lumina APV/Minivan 3.8L
Monte Carlo 3.1L • Monte Carlo/Z34 3.4L • Monte Carlo 3.8L

Scan code	Fault location	Probable cause
P1362	Ignition control, bypass circuit – voltage low	**Wiring short circuit to ground, ICM, ECM**
P1374	Crankshaft position sensor (CKP), high to low resolution frequency – signal variation	**Wiring, wiring routed near spark plug wires, ignition coils arcing/tracking, ICM, ECM**
P1380	Antilock brake system (ABS)/traction control system (TCS), rough road signal to engine control module (ECM) – malfunction	**ABS trouble code(s) stored**
P1381	Antilock brake system (ABS)/traction control system (TCS), rough road signal to engine control module (ECM) – no communication	**Wiring, poor connections, ABS**
P1403	Exhaust gas recirculation (EGR) solenoid 1 – circuit malfunction	**Wiring, EGR valve, ECM**
P1404 1 9	Exhaust gas recirculation (EGR) valve – not fully closing	**Wiring, poor connections, EGR valve position sensor, EGR solenoid**
P1404 2 10	Exhaust gas recirculation (EGR) solenoid 2 – circuit malfunction	**Wiring, EGR valve, ECM**
P1405	Exhaust gas recirculation (EGR) solenoid 3 – circuit malfunction	**Wiring, EGR valve, ECM**
P1406	Exhaust gas recirculation (EGR) valve position sensor – range performance	**Wiring, EGR valve tight/sticking, EGR valve position sensor**
P1441	Evaporative emission (EVAP) system – leak detected	**Hose connection(s), EVAP canister, EVAP canister purge valve**
P1442	Evaporative emission (EVAP) system vacuum switch – circuit malfunction	**Hoses blocked/leaking, wiring, EVAP canister, ECM**
P1483	Engine cooling system – performance	**Cooling system , ECT sensor, IAT sensor**
P1520	Park/neutral position (PNP) switch – circuit malfunction	**Wiring short circuit to ground, PNP switch incorrectly adjusted**
P1530	A/C refrigerant pressure sensor/switch – circuit malfunction	**Wiring, A/C refrigerant pressure sensor/ switch, ECM**
P1531	A/C refrigerant pressure sensor – voltage low	**Incorrect refrigerant charge, wiring, A/C refrigerant pressure sensor, ECM**
P1546	A/C compressor clutch relay – circuit malfunction	**Wiring, A/C compressor clutch relay, ECM**
P1550	Cruise control system, motor actuator – circuit malfunction	**Wiring, poor connections, cruise control motor/actuator, ECM**
P1554	Cruise control status circuit – high voltage	**Wiring open circuit/short circuit to ground, cruise control module, ECM**
P1558	Cruise control system – performance low	**Wiring, poor connections, throttle cable sticking, electrical interference**
P1561	Cruise control, vent solenoid – circuit malfunction	**Wiring, solenoid valve**
P1562	Cruise control, vacuum solenoid – circuit malfunction	**Wiring, solenoid valve**
P1565	Cruise control servo position sensor – circuit malfunction	**Wiring, poor connections, servo position sensor**
P1567	Cruise control master switch – circuit malfunction	**Wiring, poor connections, cruise control master switch**
P1568	Cruise control system – performance high	**Wiring, poor connections, throttle cable sticking**

Scan code	Fault location	Probable cause
P1571	Traction control system (TCS), torque request signal – malfunction	Wiring, poor connections, ABS/TCS
P1573	Engine control module (ECM), antilock brake system (ABS)/traction control system (TCS) communication – malfunction	Wiring, poor connections, ABS control module, ECM
P1585	Cruise control system, inhibit output circuit	Wiring, poor connections, cruise control motor/actuator, ECM
P1599	Cruise control system – engine management has operated in limited power mode	Throttle cable sticking, brake vacuum release valve, cruise control actuator, electrical interference
P1600	Engine control module (ECM) – malfunction	ECM
P1601	CAN data bus, ECM/ABS/TCS/instrumentation control module/diagnostic control module – some communication lost	Wiring, poor connections, ECM, ABS/TCS, diagnostic control module, instrumentation control module
P1621	Engine control module (ECM) – memory performance	ECM
P1623	Engine control module (ECM), PROM – malfunction	Poor connections, incorrect programming, ECM
P1626	Engine control module (ECM) – fuel enable signal lost	Wiring, poor connections, immobilizer control module, ECM
P1627	Engine control module (ECM) – malfunction	ECM
P1629	Engine control module (ECM) – fuel enable signal not received	Incorrect ignition key, wiring, poor connections, immobilizer control module, ECM
P1630 [10]	System voltage – high/low	Wiring, battery, alternator
P1630 [8] [9]	Engine control module (ECM), immobilizer code – learning mode	ECM immobilizer code programming not completed
P1631	Engine control module (ECM) – incorrect immobilizer code	Incorrect key, immobilizer code not programmed
P1632	Engine control module (ECM) – fuel disable signal received	Fuses, BCM, immobilizer code
P1635	Engine control module (ECM), sensor supply voltage – circuit malfunction	Wiring, poor connections, TP sensor, MAP sensor, EGR valve position sensor, FTP sensor, ECM
P1637	Alternator warning lamp, control signal – circuit malfunction	Wiring, instrument panel, ECM
P1639	Engine control module (ECM), sensor supply voltage – circuit malfunction	Wiring, poor connections, A/C refrigerant pressure sensor, FTP sensor, ECM
P1640	Engine control module (ECM), quad driver/ output driver – circuit malfunction	Wiring, ECM controlled components, ECM [11]
P1641 [1]	A/C compressor clutch relay – circuit malfunction	Wiring open circuit/short circuit to ground, A/C compressor clutch relay, ECM
P1641 [2]	Malfunction indicator lamp (MIL) – circuit malfunction	Wiring, poor connections, MIL, ECM
P1642	Engine oil change warning lamp – circuit malfunction	Wiring, ECM
P1642 [5]	Secondary air injection (AIR) pump relay – circuit malfunction	Wiring, AIR pump relay, ECM

Scan code	Fault location	Probable cause
P1644	Traction control system (TCS), delivered torque signal – malfunction	**Wiring, poor connections, ECM**
P1650	Engine control module (ECM), quad driver/ output driver – circuit malfunction	**Wiring, ECM controlled components, ECM 11**
P1651	Engine coolant blower motor relay 1 – circuit malfunction	**Wiring, engine coolant blower motor relay, ECM**
P1652	Engine coolant blower motor relay 2 – circuit malfunction	**Wiring, engine coolant blower motor relay(s), ECM**
P1653 4	Fuel gauge, control signal – circuit malfunction	**Wiring, ECM**
P1653 3	Engine oil level 'low' warning lamp, control signal – circuit malfunction	**Wiring short circuit to ground, engine oil level, ECM**
P1654	A/C compressor clutch relay – circuit malfunction	**Wiring open circuit/short circuit to ground , A/C compressor clutch relay, ECM**
P1655	Evaporative emission (EVAP) canister purge valve – circuit malfunction	**Wiring open circuit/short circuit to ground, EVAP canister vent valve, ECM**
P1660	Engine control module (ECM), quad driver/ output driver – circuit malfunction	**Wiring, ECM controlled components, ECM 11**
P1662	Cruise control, inhibit output circuit – malfunction	**Wiring open circuit/short circuit to ground, cruise control motor/actuator, cruise control module, ECM**
P1663 6	Engine oil change warning lamp – circuit malfunction	**Wiring, instrument panel, ECM**
P1663 7	Alternator warning lamp, control signal – circuit malfunction	**Wiring, instrument panel, ECM**
P1665	Evaporative emission (EVAP) canister vent valve – circuit malfunction	**Wiring open circuit/short circuit to ground, EVAP canister vent valve, ECM**
P1670	Engine control module (ECM), quad driver/ output driver – circuit malfunction	**Wiring, ECM controlled components, ECM 11**
P1671	Malfunction indicator lamp (MIL) – circuit malfunction	**Wiring open circuit/short circuit to ground/ short circuit to positive, MIL, ECM**
P1672	Engine oil level 'low' warning lamp, control signal – circuit malfunction	**Wiring short circuit to ground, engine oil level, ECM**
P1673	Engine coolant 'hot' warning lamp, control signal – circuit malfunction	**Wiring, instrument panel, cooling system, ECM**
P1674	Tachometer – circuit malfunction	**Wiring open circuit/short circuit to ground/ short circuit to positive, instrument panel, ECM**
P1675	Evaporative emission (EVAP) canister vent valve – circuit malfunction	**Wiring, poor connection, EVAP canister vent valve, ECM**
P1676	Evaporative emission (EVAP) canister purge valve – circuit malfunction	**Wiring open circuit/short circuit to ground, EVAP canister purge valve, ECM**
P1680	Engine control module (ECM) – malfunction	**ECM**
P1681	Engine control module (ECM) – malfunction	**ECM**
P1683	Engine control module (ECM) – malfunction	**ECM**
P1689	Traction control system (TCS), delivered torque signal – malfunction	**Wiring, poor connections, ECM**

Scan code	Fault location	Probable cause
P1810	Transmission fluid pressure (TFP) switch – manual valve position	**Fluid level low, wiring, selector lever linkage incorrectly adjusted, ECM**
P1811	Shift time too long – adaptation limit reached	**Fluid level low, transmission mechanical fault**
P1814	Torque converter clutch (TCC) – over-stressed	**Fluid level low, fluid condition, transmission mechanical fault**
P1819	Transmission range (TR) switch, signals 'A/B/C & P' – signal variation	**Wiring, TR switch**
P1820	Transmission range (TR) switch, signal 'A' – voltage low	**Wiring short to ground, TR switch**
P1822	Transmission range (TR) switch, signal 'B' – voltage high	**Wiring, TR switch**
P1823	Transmission range (TR) switch, signal 'P' – voltage low	**Wiring short to ground, TR switch**
P1825	Transmission range (TR) switch, signals 'A/B/C & P' – incorrect signal	**Wiring, TR switch**
P1826	Transmission range (TR) switch, signal 'C' – voltage high	**Wiring, TR switch**
P1860	Torque converter clutch (TCC) solenoid – circuit malfunction	**Wiring open circuit/short circuit to ground, TCC solenoid, ECM**
P1887	Torque converter clutch (TCC) release switch – circuit malfunction	**Wiring open circuit, TCC solenoid release switch, ECM**
U1016	CAN data bus, engine control module (ECM)/ antilock brake system (ABS)/traction control system (TCS)/instrumentation control module/ diagnostic control module – loss of serial communication	**Wiring, poor connections, ECM, ABS/TCS, diagnostic control module, instrumentation control module**

1 1998→
2 →1997
3 2000→ Lumina, Monte Carlo
4 →1999
5 1996 Lumina/Monte Carlo 3.4L
6 Lumina, 1999→ Monte Carlo
7 →1998 Monte Carlo
8 2000 Monte Carlo
9 Cavalier
10 Except Cavalier
11 Depending on model, year and equipment level, the following components may be connected to this circuit. Use of the vehicle wiring diagram is recommended to confirm which components are fitted.

- ☐ A/C compressor clutch relay
- ☐ A/C condenser blower motor relay
- ☐ Engine coolant blower motor relay
- ☐ Engine coolant 'hot' warning lamp
- ☐ Evaporative emission (EVAP) canister purge valve
- ☐ Exhaust gas recirculation (EGR) solenoid
- ☐ Malfunction indicator lamp (MIL)
- ☐ Torque converter clutch (TCC) solenoid
- ☐ Transmission shift solenoids (SS)
- ☐ Transmission shift warning lamp (MT)

CHEVROLET

Model:	Engine identification:	Year:
Beretta/Corsica 3.1L	VIN position 8 = AM/T/4	1992-95
Blazer 5.7L	VIN position 8 = K	1992-94
C/K Pickup 4.3/5.0L	VIN position 8 = H/Z	1992-95
C/K Pickup 5.7/7.4L	VIN position 8 = K/N	1992-95
Camaro 3.1/5.0/5.7L	VIN position 8 = E/F/T/8	1992
Caprice Wagon 4.3L	VIN position 8 = Z	1992
Caprice 5.0/5.7L	VIN position 8 = E/7	1992
Chevy Van 4.3/5.0L	VIN position 8 = H/Z	1993-95
Chevy Van 5.7/7.4L	VIN position 8 = K/N	1993-95
Tahoe 5.7L	VIN position 8 = R	1995

System: **GM**

Self-diagnosis

General information

- Refer to the front of this manual for general test conditions, terminology, detailed descriptions of wiring faults and a general trouble shooter for electrical and mechanical faults.
- Malfunction indicator lamp (MIL) will illuminate if certain faults are recorded.
- ECM operates in backup mode if sensors fail, to enable vehicle to be driven to repair shop.
- Trouble codes can be displayed by the MIL or by using a Scan Tool connected to the data link connector (DLC).

Preparatory conditions

Accessing

Models with data link connector (DLC) terminal B connected

- Access data link connector (DLC).
- Jump data link connector (DLC) terminals A and B **1**.
- Switch ignition ON. Do NOT start engine.
- Count MIL flashes. Note trouble codes. Compare with trouble code table.
- Firstly trouble code 12 will be displayed 3 times.
- Trouble codes consist of 2 groups, displayed as follows:
  - First group indicates the 'tens' of the trouble code **2** [**A**].
  - Second group indicates the 'units' of the trouble code **2** [**C**].
  - A short pause separates each group **2** [**B**].
  - A long pause separates each trouble code **2** [**D**].
  - For example: Trouble code 12 displayed **2**. Start of diagnosis.
- Each additional trouble code will now be displayed 3 times, in ascending order.

NOTE: ***If no additional trouble codes are stored, code 12 will continue to be displayed.***

- Trouble codes can also be displayed by using a Scan Tool connected to the data link connector (DLC) **1**.

1

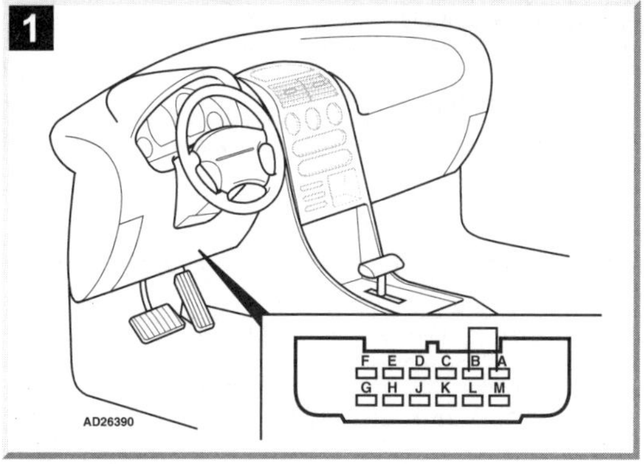

2

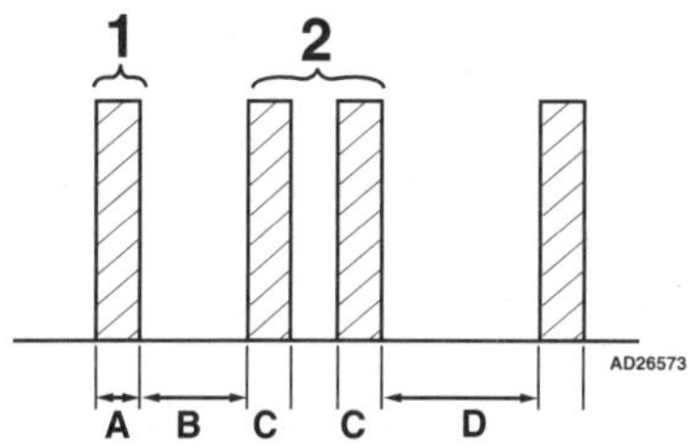

Models without data link connector (DLC) terminal B connected

- Trouble codes can only be displayed by using a Scan Tool connected to the data link connector (DLC) **1**.

Erasing

NOTE: ***Erasing trouble codes may cause driveability problems until the engine management system relearns basic values.***

- Ensure ignition switched OFF.
- Disconnect battery ground cable for at least 30 seconds.
- Reconnect battery ground cable.
- Repeat accessing procedure to ensure no trouble codes remain in ECM fault memory.

NOTE: ***Disconnecting battery ground lead may also erase memory from electronic units, such as radio.***

- Trouble codes can also be erased by using a Scan Tool connected to the data link connector (DLC) **1**.

Trouble code identification

Flash code	Fault location	Probable cause
12 **1**	Start of diagnosis	**No engine speed (RPM) signal received**
13	Oxygen sensor (O2S)/heated oxygen sensor (HO2S)	**Wiring open circuit, O2S/HO2S, ECM**
14	Engine coolant temperature (ECT) sensor – voltage low	**Wiring short circuit to ground, ECT sensor, ECM**
15	Engine coolant temperature (ECT) sensor – voltage high	**Wiring open circuit, ECT sensor, ECM**
16 **3**	System voltage – high/low	**Wiring, battery, alternator**
16	Vehicle speed sensor (VSS) buffer – signal low	**Wiring, VSS buffer, ECM**
17	Camshaft position (CMP) sensor – circuit malfunction	**Wiring, CMP sensor, mechanical fault, ECM**
21	Throttle position (TP) sensor – voltage high	**Wiring open circuit/short circuit to positive, TP sensor, ECM**

Flash code	Fault location	Probable cause
22	Throttle position (TP) sensor – voltage low	**Wiring open circuit/short circuit to ground, TP sensor, ECM**
23	Intake air temperature (IAT) sensor – voltage high	**Wiring open circuit, IAT sensor, ECM**
24	Vehicle speed sensor (VSS) – no signal	**Wiring, VSS, ECM**
25	Intake air temperature (IAT) sensor – voltage low	**Wiring short circuit to ground, IAT sensor, ECM**
28	Transmission range (TR) switch – malfunction	**Wiring, TR switch, ECM**
31	Engine speed control system – malfunction	**Wiring, throttle valve motor, throttle valve control module, ECM**
32	Exhaust gas recirculation (EGR) solenoid	**Wiring, poor connections, EGR valve, ECM**
33	Manifold absolute pressure (MAP) sensor – voltage high	**Wiring short circuit to positive, hose blocked/leaking, MAP sensor, ECM**
34	Manifold absolute pressure (MAP) sensor – voltage low	**Wiring open circuit/short circuit to ground, MAP sensor, ECM**
35	Idle air control (IAC) valve	**Wiring, IAC valve, intake system leaking, fuel system, PCV valve, A/C system, mechanical fault, ECM**
36 [4]	Crankshaft position (CKP) sensor, at crankshaft pulley – 24X signal circuit malfunction	**Wiring, CKP sensor, ECM**
36	Auxiliary idle air control valve	**Wiring, idle air control valve, vacuum pipes, ECM**
37	Brake pedal position (BPP) switch for torque converter clutch (TCC) – stuck ON	**Wiring, BPP switch, ECM**
38	Brake pedal position (BPP) switch, torque converter clutch (TCC) – stuck OFF	**Wiring open circuit/short circuit to ground/ short circuit to positive, BPP switch incorrectly adjusted, BPP switch, ECM**
39	Brake pedal position (BPP) switch – stuck OFF	**Wiring, BPP switch, ECM**
41 [2] [6]	Engine control module (ECM), programming – malfunction	**ECM incorrect programming, ECM failure**
41 [3]	Ignition control, timing signal – circuit malfunction	**Wiring, poor connections, ICM, ECM**
42	Ignition control, bypass signal/timing signal – circuit malfunction	**Wiring, poor connections, ICM, ECM**
43	Knock sensor (KS)/knock control module – circuit malfunction	**Wiring, KS, knock control module, ECM**
44	Oxygen sensor (O2S)/heated oxygen sensor (HO2S) – mixture lean	**Wiring open circuit/short circuit to ground, injector(s) blocked, fuel contamination, fuel pressure low, exhaust system leaking, O2S/ HO2S, ECM**
45	Oxygen sensor (O2S)/heated oxygen sensor (HO2S) – mixture rich	**Wiring open circuit, fuel pressure high, injector(s) leaking, O2S/HO2S, EVAP purge valve, TP sensor, EGR system, ECM**

Flash code	Fault location	Probable cause
46	Immobilizer system, code – malfunction	**Incorrect key, ECM**
51	Engine control module (ECM), PROM – malfunction	**Poor connections, incorrect programming, ECM**
52	Engine control module (ECM) – malfunction	**ECM**
52 [5]	System voltage – high	**Alternator, jump started**
53	System voltage – high	**Alternator**
53 [7]	Immobilizer system, code – malfunction	**Incorrect key, ECM**
54	Fuel pump circuit – voltage low	**Wiring open circuit/short circuit to ground, fuel pump relay, engine oil pressure warning lamp/fuel pump switch**
55	Engine control module (ECM) – malfunction	**ECM**
58	Transmission fluid temperature (TFT) sensor – voltage low	**Wiring short circuit to ground, TFT sensor, ECM**
59	Transmission fluid temperature (TFT) sensor – voltage high	**Wiring open circuit, TFT sensor, ECM**
61	Oxygen sensor (O2S) – malfunction	**O2S**
62	Transmission gear switch signal circuit	**Wiring open circuit/short circuit to ground, 2nd gear switch, 3rd gear switch, ECM**
66	A/C refrigerant pressure sensor – voltage low	**Wiring open circuit/short circuit to ground, A/C refrigerant pressure sensor, ECM**
66 [8]	Transmission shift solenoid (SS), 3rd-2nd – downshift malfunction	**Wiring, SS, ECM**
67 [8]	Torque converter clutch (TCC) – circuit malfunction	**Wiring, TCC solenoid, ECM**
68	Torque converter clutch (TCC) – slipping	**Wiring, TCC solenoid, transmission mechanical fault, ECM**
69	Torque converter clutch (TCC) solenoid – stuck ON	**Wiring open circuit/short circuit to ground, transmission internal wiring, TCC solenoid, transmission mechanical fault, ECM**
70	A/C refrigerant pressure sensor – voltage high	**Wiring open circuit/short circuit to positive, A/C refrigerant pressure sensor, ECM**
72	Vehicle speed sensor (VSS) – circuit malfunction	**Wiring, VSS, ECM**
73	Transmission fluid pressure (TFP) control solenoid – circuit malfunction	**Wiring, poor connection, TFP solenoid**
74	Transmission input shaft speed sensor – circuit malfunction	**Wiring, input shaft speed sensor, ECM**
75 [5]	System voltage, engine control module (ECM) – low	**Wiring, battery, alternator, ECM**
75	Exhaust gas recirculation (EGR) solenoid 1 – circuit malfunction	**Wiring, EGR valve, ECM**
76	Exhaust gas recirculation (EGR) solenoid 2 – circuit malfunction	**Wiring, EGR valve, ECM**

Flash code	Fault location	Probable cause
77	Exhaust gas recirculation (EGR) solenoid 3 – circuit malfunction	**Wiring, EGR valve, ECM**
79	Transmission fluid temperature (TFT) – high	**Transmission mechanical fault, towing weight limit exceeded**
80	Torque converter clutch (TCC) – slipping	**Wiring, TCC solenoid, transmission mechanical fault, ECM**
81	Transmission shift solenoid (SS), 2nd-3rd – circuit malfunction	**Wiring, SS, ECM**
82 [4]	Crankshaft position (CKP) sensor, in cylinder block – 3X signal circuit malfunction	**Wiring open circuit/short circuit to ground, CKP sensor, ICM**
82 [5]	Transmission shift solenoid (SS), 1st-2nd – circuit malfunction	**Wiring, SS, ECM**
83	Torque converter clutch (TCC) solenoid, pulse width modulated – circuit malfunction	**Wiring, TCC solenoid, ECM**
85 [4]	Engine control module (ECM), PROM – malfunction	**Poor connections, incorrect programming, ECM**
85 [5]	Transmission input/output shaft speed, ratio variation – out of range	**Wiring, input shaft speed sensor, output shaft speed sensor, mechanical fault**
86 [5]	Transmission input/output shaft speed, ratio variation – lower than expected	**Wiring, input shaft speed sensor, output shaft speed sensor, mechanical fault**
86 [4]	Engine control module (ECM) – analogue/ digital converter	**A/C refrigerant pressure sensor/KS/ECT sensor/fuel pump monitor circuit short circuit to battery voltage**
87 [4]	Engine control module (ECM), EEPROM – malfunction	**Poor connections, incorrect programming, ECM**
87 [5]	Transmission input/output shaft speed, ratio variation – higher than expected	**Wiring, input shaft speed sensor, output shaft speed sensor, mechanical fault**
90	Torque converter clutch (TCC) – stuck ON	**Wiring open circuit/short circuit to ground, transmission internal wiring, TCC solenoid, transmission mechanical fault, ECM**
96	Engine control module (ECM) – ignition supply voltage low	**Wiring open circuit, alternator, ECM**
98	Engine control module (ECM) – incorrectly coded	**ECM incorrectly coded, ECM**
99	Engine control module (ECM) – incorrectly coded	**ECM incorrectly coded, ECM**

[1] Trouble code 12 is recorded when the ECM has not received an engine speed signal.
[2] Beretta/Corsica →93
[3] Beretta/Corsica 94→
[4] Beretta/Corsica
[5] C/K Pickup 1994→/Chevy Van
[6] Camaro
[7] Camaro 5.0L
[8] Chevy Van

CHEVROLET

Model:	Engine identification:	Year:
Camaro 3.4L	VIN position 8 = S	1993-95
Camaro 5.7L	VIN position 8 = P	1993-94
Caprice Wagon 4.3L	VIN position 8 = W	1993-95
Caprice 5.7L	VIN position 8 = P	1993-95
Corvette 5.7L (OBD-I)	VIN position 8 = P	1994

System: **GM**

Self-diagnosis

General information

- Refer to the front of this manual for general test conditions, terminology, detailed descriptions of wiring faults and a general trouble shooter for electrical and mechanical faults.
- Malfunction indicator lamp (MIL) will illuminate if certain faults are recorded.
- ECM operates in backup mode if sensors fail, to enable vehicle to be driven to repair shop.
- Trouble codes can be displayed by the MIL or by using a Scan Tool connected to the data link connector (DLC).

Accessing

Models with data link connector (DLC) terminal B connected

- Access data link connector (DLC).
- Jump data link connector (DLC) terminals A and B **1**.
- Switch ignition ON. Do NOT start engine.
- Count MIL flashes. Note trouble codes. Compare with trouble code table.
- Firstly trouble code 12 will be displayed 3 times.
- Trouble codes consist of 2 groups, displayed as follows:
  - First group indicates the 'tens' of the trouble code **2** [**A**].
  - Second group indicates the 'units' of the trouble code **2** [**C**].
  - A short pause separates each group **2** [**B**].
  - A long pause separates each trouble code **2** [**D**].
  - For example: Trouble code 12 displayed **2**. Start of diagnosis.
- Each additional trouble code will now be displayed 3 times, in ascending order.

NOTE: ***If no additional trouble codes are stored, code 12 will continue to be displayed.***

- Trouble codes can also be displayed by using a Scan Tool connected to the data link connector (DLC) **1**.

1

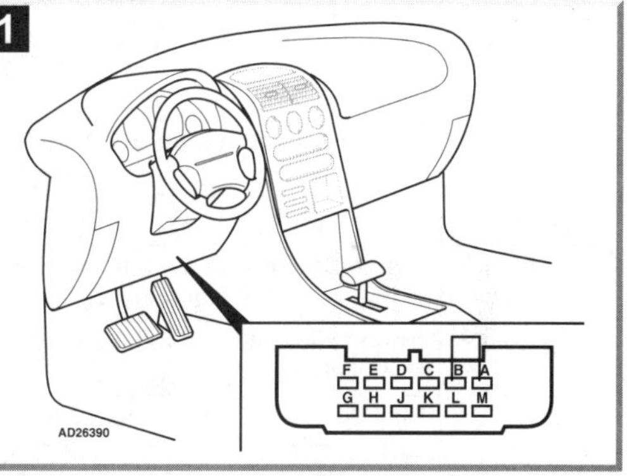

2

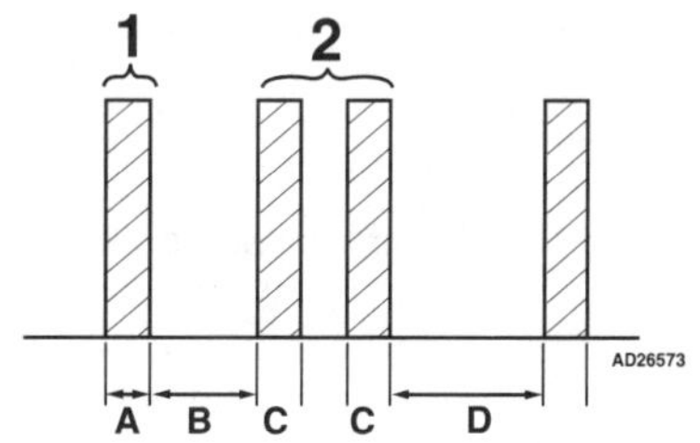

Models without data link connector (DLC) terminal B connected

- Trouble codes can only be displayed by using a Scan Tool connected to the data link connector (DLC) [1].

Erasing

NOTE: ***Disconnecting battery ground lead may cause driveability problems until the engine management system relearns basic values.***

- Ensure ignition switched OFF.
- Disconnect battery ground cable for 30 seconds.
- Reconnect battery ground cable.
- Repeat checking procedure to ensure no trouble codes remain in ECM fault memory.

NOTE: ***Disconnecting battery ground lead may also erase memory from electronic units, such as radio.***

- Trouble codes can also be erased by using a Scan Tool connected to the data link connector (DLC) [1].

Trouble code identification

Flash code	Fault location	Probable cause
11	Malfunction indicator lamp (MIL) – circuit malfunction	**Wiring open circuit/short circuit to ground, MIL, ECM**
12	Start of diagnosis	**No engine speed (RPM) signal received**
13 [1]	Oxygen sensor (O2S), bank 2	**Wiring open circuit, O2S, ECM**
13 [3]	Oxygen sensor (O2S), bank 1	**Wiring open circuit, O2S, ECM**
14	Engine coolant temperature (ECT) sensor – voltage low	**Wiring short circuit to ground, ECT sensor, ECM**
15	Engine coolant temperature (ECT) sensor – voltage high	**Wiring open circuit, ECT sensor, ECM**
16 [1]	System voltage – low	**Wiring, alternator**
16 [3]	Distributor ignition system, low resolution signal – circuit malfunction	**Wiring, distributor, ECM**
17	Camshaft position (CMP) sensor – circuit malfunction	**Wiring, CMP sensor, mechanical fault, ECM**
18	Injector(s) – circuit malfunction	**Wiring, injector(s)**
21	Throttle position (TP) sensor – voltage high	**Wiring open circuit/short circuit to positive, TP sensor, ECM**
22	Throttle position (TP) sensor – voltage low	**Wiring open circuit/short circuit to ground, TP sensor, ECM**
23	Intake air temperature (IAT) sensor – voltage high	**Wiring open circuit, IAT sensor, ECM**
24	Vehicle speed sensor (VSS) – no signal	**Wiring, VSS, ECM**
25	Intake air temperature (IAT) sensor – voltage low	**Wiring short circuit to ground, IAT sensor, ECM**
26 [4]	Evaporative emission (EVAP) canister purge valve – open/short circuit	**Wiring, EVAP canister purge valve, ECM**
26 [2]	Engine control module (ECM), quad driver/ output driver – circuit malfunction	**Wiring, ECM controlled components, ECM [9]**
27 [2]	Engine control module (ECM), quad driver/ output driver – circuit malfunction	**Wiring, ECM controlled components, ECM [9]**

Flash code	Fault location	Probable cause
27 4	Exhaust gas recirculation (EGR) solenoid	Wiring, poor connections, EGR valve, ECM
28 2	Engine control module (ECM), quad driver/ output driver – circuit malfunction	Wiring, ECM controlled components, ECM 9
28 4	Transmission range (TR) switch – malfunction	Wiring, TR switch, ECM
29	Secondary air injection (AIR) pump relay – circuit malfunction	Wiring, AIR pump relay, ECM
32	Exhaust gas recirculation (EGR) solenoid	Wiring, poor connections, EGR valve, ECM
33	Manifold absolute pressure (MAP) sensor – voltage high	Wiring short circuit to positive, hose blocked/leaking, MAP sensor, ECM
34	Manifold absolute pressure (MAP) sensor – voltage low	Wiring open circuit/short circuit to ground, MAP sensor, ECM
35	Idle air control (IAC) valve	Wiring open circuit/short circuit to ground/ short circuit to positive, IAC valve, intake system leaking, fuel system, PCV valve, A/C system, mechanical fault, ECM
36 1	Crankshaft position (CKP) sensor, at crankshaft pulley – 24X signal circuit malfunction	Wiring, CKP sensor, ECM
36 3	Distributor ignition system, high resolution signal – circuit malfunction	Wiring, distributor, ECM
37	Brake pedal position (BPP) switch for torque converter clutch (TCC)	Wiring, incorrectly adjusted, BPP switch, ECM
38	Brake pedal position (BPP) switch, torque converter clutch (TCC) – low voltage	Wiring open circuit, incorrectly adjusted, BPP switch, ECM
39	Clutch pedal position (CPP) switch	Wiring, incorrectly adjusted, CPP switch, ECM
41	Ignition control, timing signal – circuit malfunction	Wiring, poor connections, ICM, ECM
42	Ignition control, bypass signal – circuit malfunction	Wiring, poor connections, ICM, ECM
43	Knock sensor (KS) – circuit malfunction	Wiring, KS, ECM
44 1	Oxygen sensor (O2S), bank 2 – mixture lean	Wiring open circuit/short circuit to ground, injector(s) blocked, fuel contamination, fuel pressure low, exhaust system leaking, O2S, ECM
44 3	Oxygen sensor (O2S)/heated oxygen sensor (HO2S), bank 1 – mixture lean	Wiring open circuit/short circuit to ground, injector(s) blocked, fuel contamination, fuel pressure low, exhaust system leaking, O2S/ HO2S, ECM
45 1	Oxygen sensor (O2S), bank 2 – mixture rich	Wiring open circuit, fuel pressure high, injector(s) leaking, O2S, EVAP purge valve, TP sensor, EGR system, ECM
45 3	Oxygen sensor (O2S)/heated oxygen sensor (HO2S), bank 1 – mixture rich	Wiring open circuit, fuel pressure high, injector(s) leaking, O2S/HO2S, EVAP purge valve, TP sensor, EGR system, ECM
46	Immobilizer system, code – malfunction	Incorrect key, immobilizer control module, ECM

Flash code	Fault location	Probable cause
47	Knock control module, removable part in engine control module (ECM) – circuit malfunction	**Wiring, poor connection, missing knock control module, ECM**
48	Mass airflow (MAF) sensor – no signal	**Wiring, poor connections, MAF sensor, ECM**
50	System voltage – low	**Alternator**
51	Engine control module (ECM), PROM – malfunction	**Poor connections, incorrect programming, ECM**
52	Engine oil temperature sensor – voltage high	**Wiring, engine oil temperature sensor, ECM**
53	System voltage – high	**Alternator**
54	Fuel pump circuit – voltage low	**Wiring open circuit/short circuit to ground, fuel pump relay, engine oil pressure warning lamp/fuel pump switch**
55	Engine control module (ECM), fuel lean monitor	**Fuel filter/lines blocked, fuel contamination**
58	Transmission fluid temperature (TFT) sensor – voltage low	**Wiring short circuit to ground, TFT sensor, ECM**
59	Transmission fluid temperature (TFT) sensor – voltage high	**Wiring open circuit, TFT sensor, ECM**
61	A/C system – performance	**Refrigerant low, wiring, A/C refrigerant pressure sensor, A/C evaporator temperature sensor, ECM**
62	Engine oil temperature sensor – voltage low	**Wiring, engine oil temperature sensor, ECM**
63 [1]	Oxygen sensor (O2S), bank 1	**Wiring open circuit, O2S, ECM**
63 [3]	Oxygen sensor (O2S)/heated oxygen sensor (HO2S), bank 2	**Wiring open circuit, O2S/HO2S, ECM**
64 [1]	Oxygen sensor (O2S), bank 1 – mixture lean	**Wiring open circuit/short circuit to ground, injector(s) blocked, fuel contamination, fuel pressure low, exhaust system leaking, O2S, ECM**
64 [3]	Oxygen sensor (O2S)/heated oxygen sensor (HO2S), bank 2 – mixture lean	**Wiring open circuit/short circuit to ground, injector(s) blocked, fuel contamination, fuel pressure low, exhaust system leaking, O2S/HO2S, ECM**
65 [1]	Oxygen sensor (O2S), bank 1 – mixture rich	**Wiring open circuit, fuel pressure high, injector(s) leaking, O2S, EVAP purge valve, TP sensor, EGR system, ECM**
65 [3]	Oxygen sensor (O2S)/heated oxygen sensor (HO2S), bank 2 – mixture rich	**Wiring open circuit, fuel pressure high, injector(s) leaking, O2S/HO2S, EVAP purge valve, TP sensor, EGR system, ECM**
66	A/C refrigerant pressure sensor – voltage low	**Wiring open circuit/short circuit to ground, A/C refrigerant pressure sensor, ECM**
67	A/C refrigerant pressure sensor – circuit malfunction	**Wiring open circuit, A/C refrigerant pressure sensor, ECM**
68	A/C compressor clutch relay – circuit malfunction	**Wiring short circuit, A/C compressor clutch relay, ECM**
69	A/C compressor clutch relay – circuit malfunction	**Wiring open circuit, A/C compressor clutch relay, ECM**

Flash code	Fault location	Probable cause
70	A/C refrigerant pressure sensor – voltage high	**Wiring open circuit/short circuit to positive, A/C refrigerant pressure sensor, ECM**
71	A/C evaporator temperature sensor – voltage high	**Wiring, poor connections, A/C evaporator temperature sensor**
72	Vehicle speed sensor (VSS) – circuit malfunction	**Wiring, VSS, ECM**
73 [1]	A/C evaporator temperature sensor – voltage low	**Wiring, poor connections, A/C evaporator temperature sensor**
73 [3]	Transmission fluid pressure (TFP) control solenoid – circuit malfunction	**Wiring, poor connection, TFP solenoid**
74	Traction control system (TCS), active signal – malfunction	**Wiring short to ground, ABS/TCS, ECM**
75 [1]	Exhaust gas recirculation (EGR) solenoid 1 – circuit malfunction	**Wiring, EGR valve, ECM**
75 [3]	System voltage, engine control module (ECM) – low	**Wiring, battery, alternator, ECM**
76	Exhaust gas recirculation (EGR) solenoid 2 – circuit malfunction	**Wiring, EGR valve, ECM**
77 [1]	Exhaust gas recirculation (EGR) solenoid 3 – circuit malfunction	**Wiring, EGR valve, ECM**
77 [3]	Engine coolant blower motor relay – circuit malfunction	**Wiring, engine coolant blower motor relay, ECM**
78	Engine coolant blower motor relay 2 – circuit malfunction	**Wiring open circuit/short circuit to ground/ short circuit to positive, engine coolant blower motor relay, ECM**
79	Transmission fluid temperature (TFT) – high	**Transmission mechanical fault, towing weight limit exceeded**
80	Torque converter clutch (TCC) – slipping	**Wiring, TCC solenoid, transmission mechanical fault, ECM**
81 [1]	Brake pedal position (BPP) switch, torque converter clutch (TCC) – circuit malfunction	**Wiring, BPP switch, ECM**
81 [3]	Transmission shift solenoid (SS), 2nd-3rd – circuit malfunction	**Wiring, SS, ECM**
82 [1]	Crankshaft position (CKP) sensor, in cylinder block – 3X signal circuit malfunction	**Wiring open circuit/short circuit to ground, CKP sensor, ICM**
82 [3]	Transmission shift solenoid (SS), 1st-2nd – circuit malfunction	**Wiring, SS, ECM**
83 [6]	Reverse gear inhibit solenoid – circuit malfunction	**Wiring, reverse gear inhibit solenoid, ECM**
83 [5]	Torque converter clutch (TCC) solenoid, pulse width modulated – circuit malfunction	**Wiring open circuit/short circuit to ground/ short circuit to positive, TCC solenoid, ECM**
84 [7]	Transmission 3rd-2nd control solenoid – circuit malfunction	**Wiring, control solenoid, ECM**
84 [8]	2nd-3rd gear inhibit relay – circuit malfunction	**Wiring, inhibit relay, ECM**
85 [1]	Engine control module (ECM), PROM – malfunction	**Poor connections, incorrect programming, ECM**

Flash code	Fault location	Probable cause
85 [3]	Torque converter clutch (TCC) solenoid – stuck ON	**Wiring open circuit/short circuit to ground, transmission internal wiring, TCC solenoid, transmission mechanical fault, ECM**
86	Engine control module (ECM) – analogue/ digital converter	**A/C refrigerant pressure sensor/KS/ECT sensor/fuel pump monitor circuit short circuit to battery voltage**
87	Engine control module (ECM), EEPROM – malfunction	**Poor connections, incorrect programming, ECM**
90	Torque converter clutch (TCC) solenoid – circuit malfunction	**Wiring, TCC solenoid, ECM**
91	Skip shift warning lamp – circuit malfunction	**Wiring, skip shift warning lamp, ECM**
93	Transmission fluid pressure (TFP) control solenoid – circuit malfunction	**Wiring, poor connection, TFP solenoid**
95	Change oil lamp – circuit malfunction	**Wiring open circuit/short circuit to ground/ short circuit to positive, instrument panel, engine oil life lamp bulb, ECM**
96 [6]	Engine control module (ECM) – ignition supply voltage low	**Wiring open circuit, alternator, ECM**
96 [5]	Engine oil level warning lamp – circuit malfunction	**Wiring open circuit/short circuit to ground/ short circuit to positive, instrument panel, engine oil level lamp bulb, oil level sensor, ECM**
97	Engine control module (ECM), vehicle speed sensor (VSS) output signal – circuit malfunction	**Wiring, poor connections, ECM**
98	Engine control module (ECM) – incorrectly coded	**ECM incorrectly coded, ECM**
99 [1]	Engine control module (ECM) – incorrectly coded	**ECM incorrectly coded, ECM**
99 [3]	Engine control module (ECM), tachometer output signal – circuit malfunction	**Wiring, poor connections, ECM**

[1] Camaro 3.4L
[2] Camaro 5.7L 1993
[3] Except Camaro 3.4L
[4] Except Camaro 5.7L 1993
[5] Caprice
[6] Except Caprice
[7] Automatic transmission
[8] Manual transmission
[9] Depending on model, year and equipment level, the following components may be connected to this circuit. Use of the vehicle wiring diagram is recommended to confirm which components are fitted.

- ☐ A/C compressor clutch relay
- ☐ A/C condenser blower motor relay
- ☐ Engine coolant blower motor relay
- ☐ Engine coolant 'hot' warning lamp
- ☐ Evaporative emission (EVAP) canister purge valve
- ☐ Exhaust gas recirculation (EGR) solenoid
- ☐ Malfunction indicator lamp (MIL)
- ☐ Torque converter clutch (TCC) solenoid
- ☐ Transmission shift solenoids (SS)

CHRYSLER

Model:	Engine identification:	Year:
Cirrus 2.4/2.5L	EDZ/EEB	1995
Concorde 3.3/3.5L	EGB/EGE	1993-95
Dynasty 2.5/3.0L	EDM/EDB/EFA	1992-93
Fifth Avenue 3.3L	EGA	1992-93
Imperial 3.3/3.8L	EGA/EGH	1992-93
LeBaron 2.2/2.5L	EDF/EDM/EDS/EDT/EDB/EDN	1992-95
LeBaron 3.0L	EFA	1992-95
LHS 3.5L	EGE	1994-95
New Yorker 3.0/3.3L	EFA/EGA	1992-93
New Yorker 3.5L	EGE	1994-95

System: **SBEC II**

Self-diagnosis

General information

- Refer to the front of this manual for general test conditions, terminology, detailed descriptions of wiring faults and a general trouble shooter for electrical and mechanical faults.
- Malfunction indicator lamp (MIL) will illuminate if certain faults are recorded.
- ECM operates in backup mode if sensors fail, to enable vehicle to be driven to repair shop.
- Trouble codes can be displayed by the MIL or by using a Scan Tool connected to the data link connector (DLC).

Accessing

- Within 5 seconds switch the ignition key ON-OFF-ON-OFF-ON.
- Count MIL flashes. Note trouble codes.
- Each trouble code consists of two groups **1** [**A**].
- The first group of flashes indicate the 'tens' of the trouble code.
- The second group of flashes indicate the 'units' of the trouble code.
- A short pause separates 'tens' and 'units' **1** [**B**].
- A long pause separates each trouble code **1** [**C**].
- For example: Trouble code 21 displayed **1**.
- End of test sequence indicated by trouble code 55.
- Trouble codes can also be displayed by using a Scan Tool connected to the data link connector (DLC):
  - Concorde/LHS/New Yorker **2**.
  - Cirrus **3**.
  - LeBaron Coupe/convertible **4**.
  - LeBaron 4 door **5**.
  - Fifth Avenue/Imperial/New Yorker 1992-93 **6**.

1

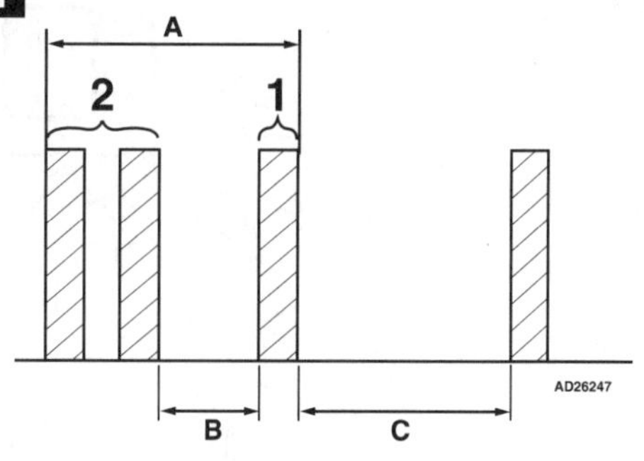

2

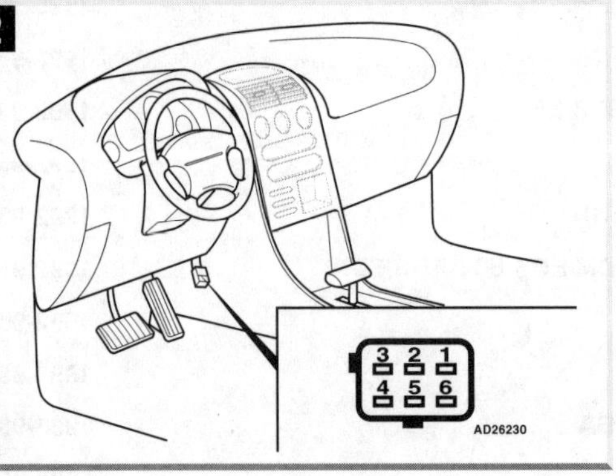

5

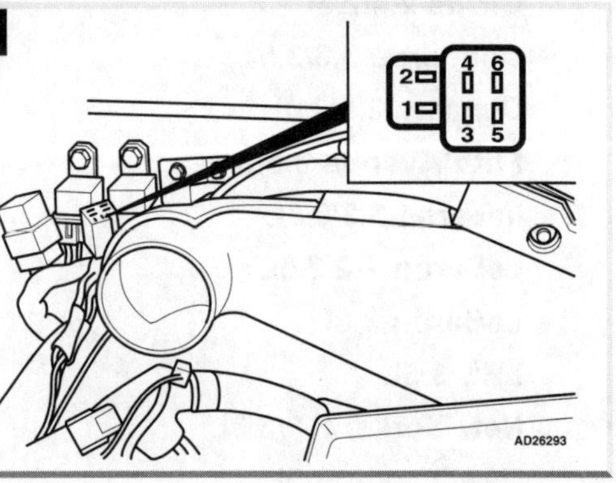

3

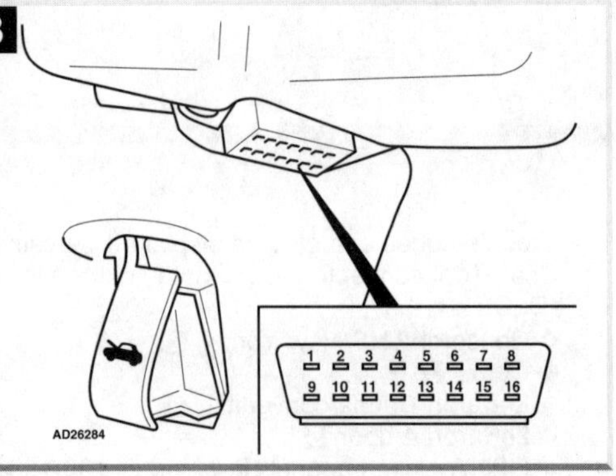

6

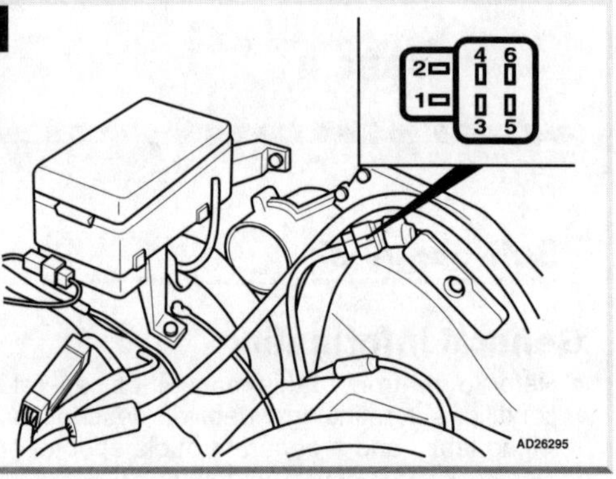

4

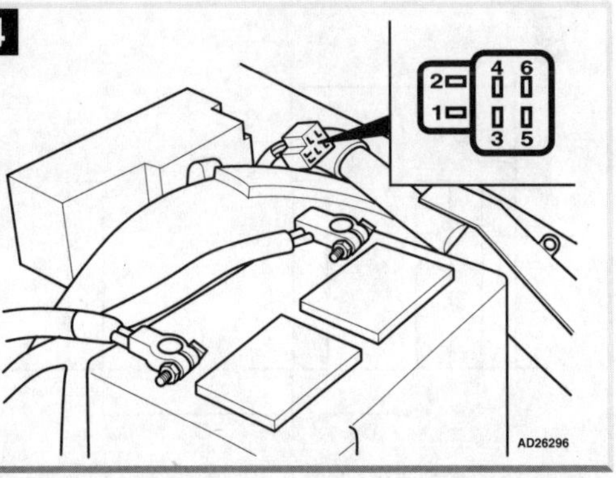

Erasing

NOTE: ***Disconnecting battery ground lead may cause driveability problems until the engine management system relearns basic values.***

- Ensure ignition switched OFF.
- Disconnect battery ground cable for at least 30 seconds.
- Reconnect battery ground cable.
- Repeat accessing procedure to ensure no trouble codes remain in ECM fault memory.

NOTE: ***Disconnecting battery ground lead may also erase memory from electronic units, such as radio.***

- Trouble codes can also be erased by using a Scan Tool connected to the data link connector (DLC).

Trouble code identification

Flash code	Fault location	Probable cause
11	Crankshaft position (CKP) sensor	**Wiring, air gap, CKP sensor**
12	Engine control module (ECM) – supply voltage disconnected within last 50 ignition switch ON-OFF cycles	**Wiring, battery**
13	Manifold absolute pressure (MAP) sensor	**Wiring, MAP sensor**
14	Manifold absolute pressure (MAP) sensor – voltage low/high	**Wiring, MAP sensor**
15	Vehicle speed sensor (VSS) – signal	**Wiring, VSS**
16	Knock sensor (KS) – wiring	**Wiring open/short circuit, KS**
17	Engine temperature – low for long period	**Wiring, coolant thermostat, ECT sensor**
21	Heated oxygen sensor (HO2S) – incorrect signal/voltage high	**Wiring, fuel pressure, injector control, EVAP canister purge valve, compression, air filter blocked, HO2S**
22	Engine coolant temperature (ECT) sensor – voltage low/high	**Wiring, ECT sensor**
23	Intake air temperature (IAT) sensor	**Wiring, IAT sensor**
24	Throttle position (TP) sensor	**Wiring, accelerator cable adjustment, TP sensor**
25	Idle air control (IAC) valve	**Wiring, IAC valve**
27	Injector(s) – circuit malfunction	**Wiring, injector**
31	Evaporative emission (EVAP) canister purge valve – open/short circuit	**Wiring, EVAP canister purge valve**
32	Exhaust gas recirculation (EGR) solenoid/ system – circuit malfunction	**Wiring, hoses, EGR solenoid**
33	A/C compressor clutch relay/A/C refrigerant pressure sensor – voltage low/high	**Wiring, A/C compressor clutch relay, A/C refrigerant pressure sensor**
34	Cruise control	**Wiring open/short circuit, hoses, voltage low/high, cruise control master switch, cruise control vacuum switching valve**
35	Engine coolant blower motor relay – low/high speed	**Wiring, engine coolant blower motor relay – low/high speed**
36	Turbocharger (TC) wastegate regulating valve – circuit malfunction	**Wiring, TC wastegate regulating valve, ECM**
37	Torque converter clutch (TCC) solenoid – circuit malfunction	**Wiring, TCC solenoid**
41	Alternator	**Wiring open/short circuit, alternator**
42	Engine control relay	**Wiring open/short circuit, engine control relay**
43	Ignition coil – primary circuit	**Wiring, ignition coil**
44	Battery temperature sensor/engine coolant temperature (ECT) sensor	**Wiring open circuit/short circuit, battery temperature sensor, ECT sensor, ECM**

Flash code	Fault location	Probable cause
45	Turbo boost limit exceeded	**Wiring, hoses, TC wastegate, TC wastegate regulating valve, MAP sensor, ECM**
46	Alternator – voltage high	**Wiring, battery, alternator**
47	Alternator – voltage low	**Wiring, battery, alternator drive belt, alternator**
51	System too lean	**Wiring, intake/exhaust leak, fuel pressure, injector control, EVAP canister purge valve, HO2S**
52	System too rich	**Wiring, exhaust blockage, fuel pressure, injector control, EVAP canister purge valve, compression, air filter blocked, ignition system, HO2S**
53	Engine control module (ECM) – internal fault	**ECM**
54	Camshaft position (CMP) sensor – no signal	**Wiring, CMP sensor**
55	End of MIL code output	–
61	Barometric pressure (BARO) read solenoid – circuit malfunction	**Wiring, BARO solenoid**
62	Engine control module (ECM) – EEPROM error	**ECM**
63	Engine control module (ECM) – EEPROM error	**ECM**
64	Compressed natural gas pressure sensor – pressure high/low	**Compressed natural gas system, wiring, compressed natural gas pressure sensor**
65	Intake manifold air control solenoid	**Wiring open/short circuit, intake manifold air control solenoid**
66	CAN data bus, transmission control module (TCM)/body control module (BCM) – no signal	**Wiring, TCM, BCM**
71	Engine control module (ECM) – 5 volt output low	**ECM**
77	Cruise control system relay	**Wiring open circuit/short circuit, cruise control system relay**

CHRYSLER

Model:	Engine identification:	Year:
Cirrus 2.4L	EDZ	1996-97
Cirrus 2.5L	EEB	1996-97
Concorde 3.3L	EGB	1996-97
Concorde 3.5L	EGE	1996-97
LHS 3.5L	EGE	1996-97
New Yorker 3.5L	EGE	1996
Sebring Convertible 2.4L	EDZ	1996-97

System: **SBEC III**

Self-diagnosis

General information

- Refer to the front of this manual for general test conditions, terminology, detailed descriptions of wiring faults and a general trouble shooter for electrical and mechanical faults.
- Malfunction indicator lamp (MIL) will illuminate or flash if certain faults are recorded.
- ECM operates in backup mode if sensors fail, to enable vehicle to be driven to repair shop.
- Trouble codes can be displayed by the MIL or by using a Scan Tool connected to the data link connector (DLC).

Accessing

Flash codes

- Within 5 seconds switch the ignition key ON-OFF-ON-OFF-ON.
- Count MIL flashes. Note trouble codes.
- Each trouble code consists of two groups **1** [**A**].
- The first group of flashes indicate the 'tens' of the trouble code.
- The second group of flashes indicate the 'units' of the trouble code.
- A short pause separates 'tens' and 'units' **1** [**B**].
- A long pause separates each trouble code **1** [**C**].
- For example: Trouble code 21 displayed **1**.
- End of test sequence indicated by trouble code 55.

P codes & Hex codes

- Trouble codes can be displayed by using a Scan Tool connected to the data link connector (DLC):
  - Cirrus **2**.
  - Concorde/LHS/New Yorker **3**.
  - Sebring **4**

1

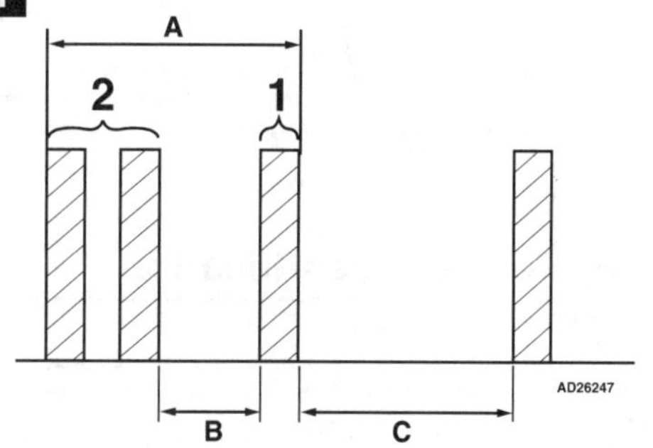

2

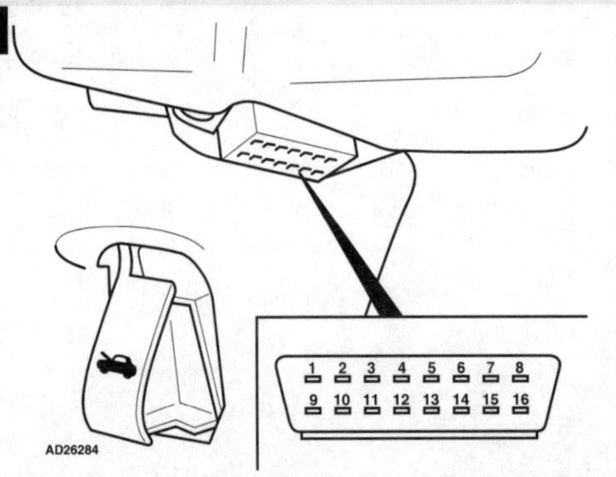

4

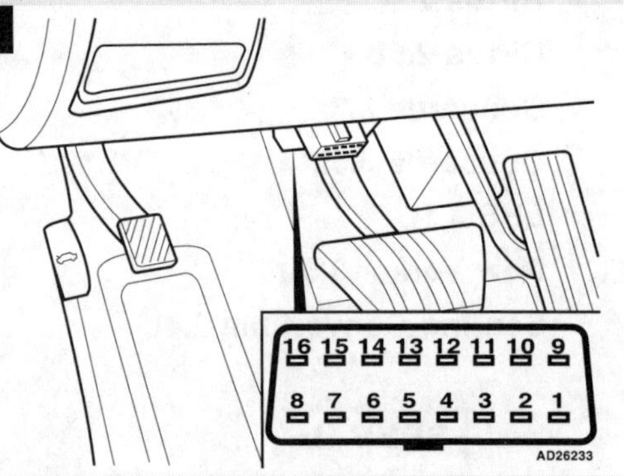

3

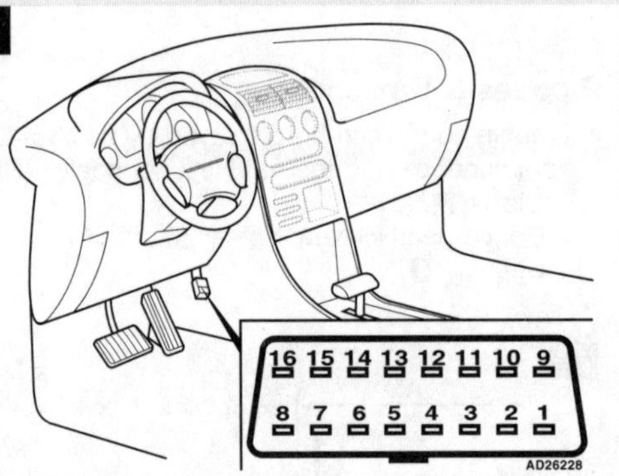

Erasing

- Trouble codes can only be erased by using a Scan Tool connected to the data link connector (DLC).

Trouble code identification

Scan code P type	Fault location	Probable cause
P0000-P0999	Refer to OBD-II trouble code table at the front of this manual	–
P1299	Manifold absolute pressure (MAP) sensor/ throttle position (TP) sensor	**Intake leak, hose connection(s), MAP sensor, TP sensor**
P1289	Intake manifold air control solenoid	**Wiring open/short circuit, intake manifold air control solenoid**
P1294	Idle speed control (ISC) system – rpm lower/ higher than expected	**Wiring, air leak, throttle valve tight/sticking, IAC valve**
P1296	Manifold absolute pressure (MAP) sensor – no 5 volt supply	**Wiring, ECM**
P1297	Manifold absolute pressure (MAP) sensor	**Wiring, hose connection(s), MAP sensor**
P1390	Timing belt jumped 1 tooth or more	**Timing belt loose/jumped position**
P1391	Crankshaft position (CKP) sensor/camshaft position (CMP) sensor – intermittent signal	**Wiring, poor connection, air gap, CKP sensor, CMP sensor**

Scan code P type	Fault location	Probable cause
P1398	Engine control module (ECM) – misfire detection limit reached	Wiring, poor connection, CKP sensor
P1486	Evaporative emission (EVAP) system – blockage detected	Hose connection(s), hose blocked/trapped, EVAP canister
P1489	Engine coolant blower motor relay – high speed	Wiring open/short circuit, engine coolant blower motor relay – high speed
P1490	Engine coolant blower motor relay – low speed	Wiring open/short circuit, engine coolant blower motor relay – low speed
P1492	Ambient/battery temperature sensor – voltage high	Wiring, ambient/battery temperature sensor
P1493	Ambient/battery temperature sensor – voltage low	Wiring, ambient/battery temperature sensor
P1494	Evaporative emission (EVAP) leak detection pump – pressure switch	Wiring, hose leak/blockage, EVAP leak detection pump
P1495	Evaporative emission (EVAP) leak detection pump solenoid	Wiring open/short circuit, hose leak/ blockage, EVAP leak detection pump solenoid
P1496	Engine control module (ECM) – 5 volt output low	ECM
P1496	Manifold absolute pressure (MAP) sensor – 5 volt output low	Wiring, MAP sensor
P1696	Engine control module (ECM) – EEPROM error	ECM
P1697	Engine control module (ECM) – EEPROM error	ECM
P1698	CAN data bus, transmission control module (TCM) – no signal	Wiring, TCM
P1899	Park/neutral position (PNP) switch – input circuit malfunction	Wiring, PNP switch, ECM/TCM

Flash code	Scan code Hex type	Fault location	Probable cause
11		Crankshaft position (CKP) sensor/camshaft position (CMP) sensor	Wiring, poor connection, air gap, CKP sensor, CMP sensor
	85	Timing belt jumped 1 tooth or more	Timing belt loose/jumped position
	9D	Crankshaft position (CKP) sensor/camshaft position (CMP) sensor – intermittent signal	Wiring, poor connection, air gap, CKP sensor, CMP sensor
	28	Crankshaft position (CKP) sensor – no signal	Wiring, air gap, CKP sensor
	BA	Engine control module (ECM) – misfire detection limit reached	Wiring, poor connection, CKP sensor
12		Engine control module (ECM) – supply voltage disconnected within last 50 ignition switch ON-OFF cycles	Wiring, battery
13	27	Manifold absolute pressure (MAP) sensor	Wiring, hose connection(s), MAP sensor

CHRYSLER

Cirrus 2.4L • Cirrus 2.5L • Concorde 3.3L • Concorde 3.5L
LHS 3.5L • New Yorker 3.5L • Sebring Convertible 2.4L

Flash code	Scan code Hex type	Fault location	Probable cause
14		Manifold absolute pressure (MAP) sensor – input low/high	**Wiring, MAP sensor**
	24	Manifold absolute pressure (MAP) sensor – low input	**Wiring short circuit to ground, MAP sensor**
	25	Manifold absolute pressure (MAP) sensor – high input	**Wiring short circuit to positive, MAP sensor**
	87	Manifold absolute pressure (MAP) sensor – no 5 volt supply	**Wiring, ECM**
	92	Manifold absolute pressure (MAP) sensor – 5 volt output low	**Wiring, MAP sensor**
15	23	Vehicle speed sensor (VSS) – signal	**Wiring, VSS**
16		Knock sensor (KS) 1/2 – no signal	**Wiring, KS incorrectly tightened, KS**
	3B	Knock sensor (KS) 1 – no signal	**Wiring, KS incorrectly tightened, KS**
	53	Knock sensor (KS) 2 – no signal	**Wiring, KS incorrectly tightened, KS**
17		Engine temperature – low for long period	**Wiring, cooling system, coolant thermostat, ECT sensor**
	21	Engine temperature – low for long period	**Wiring, cooling system, coolant thermostat, ECT sensor**
	80	Insufficient coolant temperature for closed loop fuel control	**Wiring, cooling system, coolant thermostat, ECT sensor**
21		Heated oxygen sensor (HO2S) 1/2, bank 1/2 – circuit malfunction/incorrect signal/ voltage low/high	**Wiring, fuel pressure, injector control, EVAP canister purge valve, compression, air filter blocked, HO2S**
	9B	Heated oxygen sensor (HO2S) 1, bank 1 – low voltage	**Wiring short circuit to ground, exhaust leak, HO2S**
	3E	Heated oxygen sensor (HO2S) 1, bank 1 – high voltage	**Wiring short circuit to positive, fuel pressure, injector control, EVAP canister purge valve, air filter blocked, HO2S**
	66	Heated oxygen sensor (H02S) 1, bank 1 – slow response	**Wiring, heating inoperative, HO2S**
	20	Heated oxygen sensor (H02S) 1, bank 1 – no activity detected	**Wiring open circuit, heating inoperative, HO2S**
	67	Heated oxygen sensor (HO2S) 1, bank 1 – heater control – circuit malfunction	**Wiring, HO2S**
	9C	Heated oxygen sensor (HO2S) 2, bank 1 – low voltage	**Wiring short circuit to ground, exhaust leak, HO2S**
	7E	Heated oxygen sensor (HO2S) 2, bank 1 – high voltage	**Wiring short circuit to positive, fuel pressure, injector control, EVAP canister purge valve, air filter blocked, HO2S**
	81	Heated oxygen sensor (HO2S) 2, bank 1 – no activity detected	**Wiring, heating inoperative, HO2S**
	69	Heated oxygen sensor (HO2S) 2, bank 1, heater control – circuit malfunction	**Wiring, HO2S**

Flash code	Scan code Hex type	Fault location	Probable cause
	B5	Heated oxygen sensor (HO2S) 1, bank 2 – low voltage	**Wiring short circuit to ground, exhaust leak, HO2S**
	42	Heated oxygen sensor (HO2S) 1, bank 2 – high voltage	**Wiring short circuit to positive, fuel pressure, injector control, EVAP canister purge valve, air filter blocked, HO2S**
	7A	Heated oxygen sensor (HO2S) 1, bank 2 – slow response	**Wiring, heating inoperative, HO2S**
	41	Heated oxygen sensor (HO2S) 1, bank 2 – no activity detected	**Wiring, HO2S**
	7C	Heated oxygen sensor (HO2S) 1, bank 2, heater control – circuit malfunction	**Wiring, HO2S**
	B6	Heated oxygen sensor (HO2S) 2, bank 2 – low voltage	**Wiring short circuit to ground, exhaust leak, HO2S**
	7F	Heated oxygen sensor (HO2S) 2, bank 2 – high voltage	**Wiring short circuit to positive, HO2S**
	82	Heated oxygen sensor (HO2S) 2, bank 2 – no activity detected	**Wiring, HO2S**
	7D	Heated oxygen sensor (HO2S) 2, bank 2, heater control – circuit malfunction	**Wiring, HO2S**
22		Engine coolant temperature (ECT) sensor – voltage low/high	**Wiring, coolant thermostat, ECT sensor**
	1E	Engine coolant temperature (ECT) sensor – low input	**Wiring short circuit to ground, coolant thermostat, ECT sensor**
	1F	Engine coolant temperature (ECT) sensor – high input	**Wiring open circuit/short circuit to positive, coolant thermostat, ground wire defective, ECT sensor**
23		Intake air temperature (IAT) sensor – voltage low/high	**Wiring, IAT sensor**
	39	Intake air temperature (IAT) sensor – low input	**Wiring short circuit to ground, IAT sensor**
	3A	Intake air temperature (IAT) sensor – high input	**Wiring open circuit/short circuit to positive, ground wire defective, IAT sensor**
24		Throttle position (TP) sensor – voltage low/high	**Wiring, accelerator cable adjustment, TP sensor**
	84	Throttle position (TP) sensor – range/performance problem	**Wiring, accelerator cable adjustment, TP sensor**
	1A	Throttle position (TP) sensor – low input	**Wiring short circuit to ground, TP sensor**
	1B	Throttle position (TP) sensor – high input	**Wiring short circuit to positive, TP sensor**
25		Idle air control (IAC) valve	**Wiring open/short circuit, air leak, IAC valve**
	19	Idle speed control (ISC) system – malfunction	**Wiring open/short circuit, IAC valve**

Flash code	Scan code Hex type	Fault location	Probable cause
	8A	Idle speed control (ISC) system – rpm lower/higher than expected	Wiring, air leak, throttle valve tight/sticking, IAC valve
	91	Manifold absolute pressure (MAP) sensor/ throttle position (TP) sensor	Intake leak, hose connection(s), MAP sensor, TP sensor
27		Injector – circuit malfunction	Wiring, injector
	15	Injector 1 – circuit malfunction	Wiring, injector
	14	Injector 2 – circuit malfunction	Wiring, injector
	13	Injector 3 – circuit malfunction	Wiring, injector
	3D	Injector 4 – circuit malfunction	Wiring, injector
	45	Injector 5 – circuit malfunction	Wiring, injector
	46	Injector 6 – circuit malfunction	Wiring, injector
31		Evaporative emission (EVAP) canister purge valve – open/short circuit	Wiring, EVAP canister purge valve
	71	Evaporative emission (EVAP) system – incorrect flow detected	Hose connection(s), intake leak, EVAP canister purge valve
	A0	Evaporative emission (EVAP) system – small leak detected	Hose connection(s), intake leak, EVAP canister, EVAP canister purge valve
	12	Evaporative emission (EVAP) canister purge valve – circuit malfunction	Wiring open/short circuit, EVAP canister purge valve
	A1	Evaporative emission (EVAP) system – large leak detected	Hose connection(s), intake leak, EVAP canister, EVAP canister purge valve
	BB	Evaporative emission (EVAP) system – blockage detected	Hose connection(s), hose blocked/ trapped, EVAP canister
	B8	Evaporative emission (EVAP) leak detection pump – pressure switch	Wiring, hose leak/blockage, EVAP leak detection pump
	B7	Evaporative emission (EVAP) leak detection pump solenoid	Wiring open/short circuit, hose leak/ blockage, EVAP leak detection pump solenoid
32		Exhaust gas recirculation (EGR) solenoid/ system – circuit malfunction	Wiring, hoses, EGR solenoid
	2E	Exhaust gas recirculation (EGR) system – insufficient flow detected	Hose leak/blockage, wiring, EGR valve, EGR solenoid
	11	Exhaust gas recirculation (EGR) – circuit malfunction	Wiring open/short circuit, EGR solenoid
33		A/C compressor clutch relay/A/C refrigerant pressure sensor – voltage low/high	Wiring open/short circuit, A/C refrigerant pressure too low/high (cooling fault/incorrectly charged), A/C compressor clutch relay, A/C refrigerant pressure sensor
	10	A/C compressor clutch relay	Wiring open/short circuit, A/C compressor clutch relay
	5A	A/C refrigerant pressure sensor – high input	Wiring, A/C refrigerant pressure too high (cooling fault/incorrectly charged), A/C refrigerant pressure sensor

Flash code	Scan code Hex type	Fault location	Probable cause
	5B	A/C refrigerant pressure sensor – low input	Wiring, A/C refrigerant pressure too low (incorrectly charged), A/C refrigerant pressure sensor
	86	A/C – no 5 volt supply	Wiring, ECM
34		Cruise control	Wiring open/short circuit, voltage low/ high, cruise control master switch, cruise control vacuum switching valve
	0F	Cruise control	Wiring open/short circuit, voltage low/ high, cruise control vacuum switching valve
	56	Cruise control master switch – high voltage	Wiring, cruise control master switch
	57	Cruise control master switch – low voltage	Wiring, cruise control master switch
35		Engine coolant blower motor relay – low/ high speed	Wiring open/short circuit, engine coolant blower motor relay – low/high speed
	5D	Engine coolant blower motor relay – high speed	Wiring open/short circuit, engine coolant blower motor relay – high speed
	5C	Engine coolant blower motor relay – low speed	Wiring open/short circuit, engine coolant blower motor relay – low speed
37		Torque converter clutch (TCC) solenoid/ park/neutral position (PNP) switch – circuit malfunction	Wiring, TCC solenoid
	94	Torque converter clutch (TCC) solenoid – circuit malfunction	Wiring, TCC solenoid
	0C	Torque converter clutch (TCC) solenoid – electrical	Wiring open/short circuit, TCC solenoid
	72	Park/neutral position (PNP) switch – input circuit malfunction	Wiring, PNP switch, ECM/TCM
41	0B	Alternator, field control – circuit malfunction	Wiring open/short circuit, alternator, battery
42		Fuel pump relay/engine control relay/fuel level sensor – circuit malfunction	Wiring open/short circuit, fuel pump relay, engine control relay, fuel level sensor
	65	Fuel pump relay	Wiring open/short circuit, fuel pump relay
	0A	Engine control relay	Wiring open/short circuit, engine control relay
	2C	Engine control relay	Wiring open circuit, engine control relay
	95	Fuel level sensor	Wiring open circuit, fuel level sensor
	96	Fuel level sensor	Wiring short circuit to positive, fuel level sensor
	97	Fuel level sensor	Wiring, fuel level sensor
43		Random/multiple cylinder(s) – misfire detected	Wiring, spark plug(s), HT lead(s), injector(s), ignition coil(s), low compression

CHRYSLER

Cirrus 2.4L • Cirrus 2.5L • Concorde 3.3L • Concorde 3.5L LHS 3.5L • New Yorker 3.5L • Sebring Convertible 2.4L

Flash code	Scan code Hex type	Fault location	Probable cause
	6A	Random/multiple cylinder(s) – misfire detected	Wiring, spark plug(s), HT lead(s), injector(s), ignition coil(s), low compression
	6B	Cylinder 1 – misfire detected	Wiring, engine mechanical fault, ignition/fuel system, injector
	6C	Cylinder 2 – misfire detected	Wiring, engine mechanical fault, ignition/fuel system, injector
	6D	Cylinder 3 – misfire detected	Wiring, engine mechanical fault, ignition/fuel system, injector
	6E	Cylinder 4 – misfire detected	Wiring, engine mechanical fault, ignition/fuel system, injector
	AE	Cylinder 5 – misfire detected	Wiring, engine mechanical fault, ignition/fuel system, injector
	AF	Cylinder 6 – misfire detected	Wiring, engine mechanical fault, ignition/fuel system, injector
	2B	Ignition coil A, primary/secondary – circuit malfunction	Wiring, ignition coil
	2A	Ignition coil B, primary/secondary – circuit malfunction	Wiring, ignition coil
	29	Ignition coil C, primary/secondary – circuit malfunction	Wiring, ignition coil
44		Ambient/battery temperature sensor – voltage low/high	Wiring, ambient/battery temperature sensor
	9A	Ambient/battery temperature sensor – voltage high	Wiring, ambient/battery temperature sensor
	99	Ambient/battery temperature sensor – voltage low	Wiring, ambient/battery temperature sensor
45	89	Transmission control system – malfunction	Wiring, TCM
46	06	Alternator – voltage high	Wiring, battery, alternator
47	05	Alternator – voltage low	Wiring, battery, alternator drive belt, alternator
51		System too lean	Wiring, intake/exhaust leak, fuel pressure, injector control, EVAP canister purge valve, HO2S
	77	System too lean, bank 1	Intake/exhaust leak, MAP sensor, fuel pressure/pump, injector(s), HO2S
	79	System too lean, bank 2	Intake/exhaust leak, MAP sensor, fuel pressure/pump, injector(s), HO2S
52		System too rich	Wiring, exhaust blockage, fuel pressure, injector control, EVAP canister purge valve, compression, air filter blocked, ignition system, HO2S
	76	System too rich, bank 1	Intake blocked, EVAP canister purge valve, fuel pressure, EGR system, injector(s), HO2S

Flash code	Scan code Hex type	Fault location	Probable cause
	78	System too rich, bank 2	**Intake blocked, EVAP canister purge valve, fuel pressure, EGR system, injector(s), HO2S**
53		Engine control module (ECM) – internal fault	**ECM**
	02	Engine control module (ECM) – internal fault	**ECM**
	44	Engine control module (ECM) – internal fault	**ECM**
54	01	Camshaft position (CMP) sensor – no signal	**Wiring, CMP sensor**
55		End of MIL code output	–
61	3C	Manifold absolute pressure (MAP) sensor – barometric pressure range/ performance problem	**Wiring, intake/exhaust leak, MAP sensor**
62	30	Engine control module (ECM) – EEPROM error	**ECM**
63	31	Engine control module (ECM) – EEPROM error	**ECM**
65		Power steering pressure (PSP) switch/ brake pedal position (BPP) switch/Intake manifold air control solenoid	**Wiring, PAS system, PSP switch, BPP switch, intake manifold air control solenoid**
	73	Power steering pressure (PSP) switch – range/performance problem	**Wiring, PAS system, PSP switch**
	98	Brake pedal position (BPP) switch B – circuit malfunction	**Wiring, BPP switch**
	58	Intake manifold air control solenoid	**Wiring open/short circuit, intake manifold air control solenoid**
66		CAN data bus, transmission control module (TCM)/body control module (BCM) – no signal	**Wiring, TCM, BCM**
	60	CAN data bus, transmission control module (TCM) – no signal	**Wiring, TCM**
	61	CAN data bus, body control module (BCM) – no signal	**Wiring, BCM**
71	92	Engine control module (ECM) – 5 volt output low	**ECM**
72		Catalytic converter system, bank 1/2 – efficiency below threshold	**Wiring, catalytic converter, HO2S 2**
	70	Catalytic converter system, bank 1 – efficiency below threshold	**Wiring, catalytic converter, HO2S 2**
	B4	Catalytic converter system, bank 2 – efficiency below threshold	**Wiring, catalytic converter, HO2S 2**
77	52	Cruise control	**Wiring, cruise control vacuum switching valve**

CHRYSLER

Model:	Engine identification:	Year:
Cirrus 2.4L	**EDZ**	**1998**
Cirrus 2.5L	**EEB**	**1998**
Concorde 2.7L	**EER**	**1998**
Concorde 3.2L	**EGW**	**1998**
Sebring Convertible 2.4L	**EDZ**	**1998**

System: **SBEC III**

Self-diagnosis

General information

- Refer to the front of this manual for general test conditions, terminology, detailed descriptions of wiring faults and a general trouble shooter for electrical and mechanical faults.
- Malfunction indicator lamp (MIL) will illuminate or flash if certain faults are recorded.
- ECM operates in backup mode if sensors fail, to enable vehicle to be driven to repair shop.

Accessing

- Trouble codes can only be displayed by using a Scan Tool connected to the data link connector (DLC):
  - ○ Cirrus **1**.
  - ○ Concorde/Sebring **2**.

1

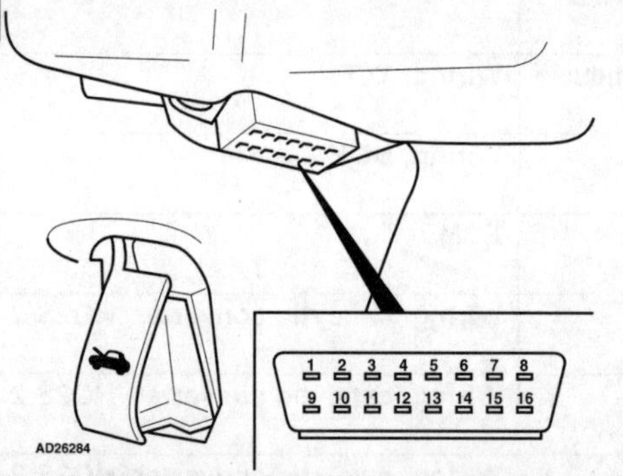

2

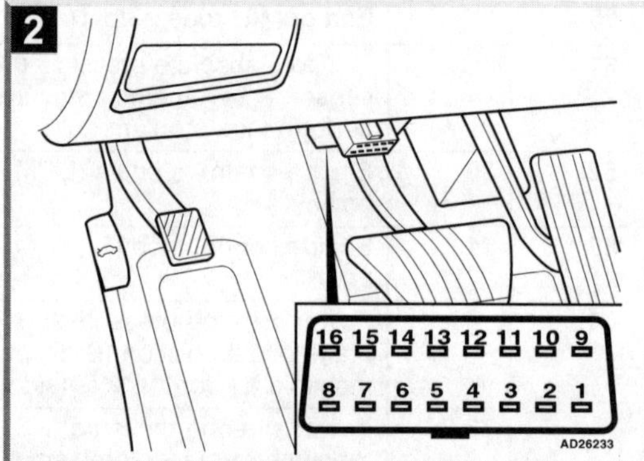

Erasing

- Trouble codes can only be erased by using a Scan Tool connected to the data link connector (DLC).

Trouble code identification

Scan code P type	Fault location	Probable cause
P0000-P0999	Refer to OBD-II trouble code table at the front of this manual	–
P1195	Heated oxygen sensor (HO2S) 1, bank 1 – slow response	**Wiring, heating inoperative, HO2S**
P1196	Heated oxygen sensor (HO2S) 1, bank 2 – slow response	**Wiring, heating inoperative, HO2S**
P1281	Engine temperature – low for long period	**Wiring, cooling system, coolant thermostat, ECT sensor**
P1282	Fuel pump relay	**Wiring open circuit/short circuit, fuel pump relay**
P1289	Intake manifold air control solenoid	**Wiring open circuit/short circuit, intake manifold air control solenoid**
P1294	Idle speed control (ISC) system – rpm lower/higher than expected	**Wiring, air leak, throttle valve tight/sticking, IAC valve**
P1297	Manifold absolute pressure (MAP) sensor	**Wiring, hose connection(s), MAP sensor**
P1299	Manifold absolute pressure (MAP) sensor/throttle position (TP) sensor	**Intake leak, hose connection(s), MAP sensor, TP sensor**
P1388	Engine control relay	**Wiring open circuit/short circuit, engine control relay**
P1389	Engine control relay	**Wiring open circuit, engine control relay**
P1390	Timing belt jumped 1 tooth or more	**Timing belt loose/jumped position**
P1391	Crankshaft position (CKP) sensor/camshaft position (CMP) sensor – intermittent signal	**Wiring, poor connection, air gap, CKP sensor, CMP sensor**
P1398	Engine control module (ECM) – misfire detection limit reached	**Wiring, poor connection, CKP sensor**
P1478	Ambient/battery temperature sensor	**Wiring, ambient/battery temperature sensor**
P1486	Evaporative emission (EVAP) system – blockage detected	**Hose connection(s), hose blocked/trapped, EVAP canister**
P1489	Engine coolant blower motor relay – high speed	**Wiring open circuit/short circuit, engine coolant blower motor relay – high speed**
P1490	Engine coolant blower motor relay – low speed	**Wiring open circuit/short circuit, engine coolant blower motor relay – low speed**
P1492	Ambient/battery temperature sensor – voltage high	**Wiring, ambient/battery temperature sensor**
P1493	Ambient/battery temperature sensor – voltage low	**Wiring, ambient/battery temperature sensor**
P1494	Evaporative emission (EVAP) leak detection pump – pressure switch	**Wiring, hose leak/blockage, EVAP leak detection pump**
P1495	Evaporative emission (EVAP) leak detection pump solenoid	**Wiring open circuit/short circuit, hose leak/blockage, EVAP leak detection pump solenoid**
P1496	Engine control module (ECM) – 5 volt output low	**ECM**

Scan code P type	Fault location	Probable cause
P1594	Alternator – voltage high	Wiring, battery, alternator
P1595	Cruise control	Wiring open circuit/short circuit, hoses, cruise control master switch, cruise control vacuum switching valve
P1596	Cruise control master switch – high voltage	Wiring, cruise control master switch
P1597	Cruise control master switch – low voltage	Wiring, cruise control master switch
P1598	A/C refrigerant pressure sensor – high input	Wiring, A/C refrigerant pressure too high (cooling fault/incorrectly charged), A/C refrigerant pressure sensor
P1599	A/C refrigerant pressure sensor – low input	Wiring, A/C refrigerant pressure too low (incorrectly charged), A/C refrigerant pressure sensor
P1682	Alternator – voltage low	Wiring, battery, alternator drive belt, alternator
P1683	Cruise control	Wiring, cruise control vacuum switching valve
P1685	Key not programmed into immobilizer control module	Wiring, ignition key not matched/damaged, immobilizer control module
P1686	CAN data bus, Immobilizer control module – no signal	Wiring, immobilizer control module
P1687	CAN data bus, instrumentation – no signal	Wiring, instrumentation control module (mechanical)
P1695	CAN data bus, body control module (BCM) – no signal	Wiring, BCM
P1696	Engine control module (ECM) – EEPROM error	ECM
P1697	Engine control module (ECM) – EEPROM error	ECM
P1698	CAN data bus, transmission control module (TCM) – no signal	Wiring, TCM
P1899	Park/neutral position (PNP) switch – input circuit malfunction	Wiring, PNP switch, ECM/TCM

Scan code Hex type	Fault location	Probable cause
01	Camshaft position (CMP) sensor – no signal	Wiring, CMP sensor
02	Engine control module (ECM) – internal fault	ECM
05	Alternator – voltage low	Wiring, battery, alternator drive belt, alternator
06	Alternator – voltage high	Wiring, battery, alternator
0A	Engine control relay	Wiring open circuit/short circuit, engine control relay

Scan code Hex type	Fault location	Probable cause
0B	Alternator, field control – circuit malfunction	**Wiring open circuit/short circuit, alternator, battery**
0F	Cruise control	**Wiring open circuit/short circuit, hoses, cruise control master switch, cruise control vacuum switching valve**
10	A/C compressor clutch relay	**Wiring open circuit/short circuit, A/C compressor clutch relay**
11	Exhaust gas recirculation (EGR) – circuit malfunction	**Wiring open circuit/short circuit, EGR solenoid**
12	Evaporative emission (EVAP) canister purge valve – circuit malfunction	**Wiring open circuit/short circuit, EVAP canister purge valve**
13	Injector 3 – circuit malfunction	**Wiring, injector**
14	Injector 2 – circuit malfunction	**Wiring, injector**
15	Injector 1 – circuit malfunction	**Wiring, injector**
19	Idle speed control (ISC) system – malfunction	**Wiring open circuit/short circuit, IAC valve**
1A	Throttle position (TP) sensor – low input	**Wiring short circuit to ground, TP sensor**
1B	Throttle position (TP) sensor – high input	**Wiring short circuit to positive, TP sensor**
1E	Engine coolant temperature (ECT) sensor – low input	**Wiring short circuit to ground, coolant thermostat, ECT sensor**
1F	Engine coolant temperature (ECT) sensor – high input	**Wiring open circuit/short circuit to positive, coolant thermostat, ground wire defective, ECT sensor**
20	Heated oxygen sensor (H02S) 1, bank 1 – no activity detected	**Wiring open circuit, heating inoperative, HO2S**
21	Engine temperature – low for long period	**Wiring, cooling system, coolant thermostat, ECT sensor**
23	Vehicle speed sensor (VSS) – signal	**Wiring, VSS**
24	Manifold absolute pressure (MAP) sensor – low input	**Wiring short circuit to ground, MAP sensor**
25	Manifold absolute pressure (MAP) sensor – high input	**Wiring short circuit to positive, MAP sensor**
27	Manifold absolute pressure (MAP) sensor	**Wiring, hose connection(s), MAP sensor**
28	Crankshaft position (CKP) sensor – no signal	**Wiring, air gap, CKP sensor**
29	Ignition coil C, primary/secondary – circuit malfunction	**Wiring, ignition coil**
2A	Ignition coil B, primary/secondary – circuit malfunction	**Wiring, ignition coil**
2B	Ignition coil A, primary/secondary – circuit malfunction	**Wiring, ignition coil**
2C	Engine control relay	**Wiring open circuit, engine control relay**
2E	Exhaust gas recirculation (EGR) system – insufficient flow detected	**Wiring, hose leak/blockage, EGR valve, EGR solenoid**

CHRYSLER

Cirrus 2.4L • Cirrus 2.5L • Concorde 2.7L • Concorde 3.2L
Sebring Convertible 2.4L

Scan code Hex type	Fault location	Probable cause
30	Engine control module (ECM) – EEPROM error	**ECM**
31	Engine control module (ECM) – EEPROM error	**ECM**
39	Intake air temperature (IAT) sensor – low input	**Wiring short circuit to ground, IAT sensor**
3A	Intake air temperature (IAT) sensor – high input	**Wiring open circuit/short circuit to positive, ground wire defective, IAT sensor**
3B	Knock sensor (KS) 1 – no signal	**Wiring, KS incorrectly tightened, KS**
3C	Manifold absolute pressure (MAP) sensor – barometric pressure range/performance problem	**Intake/exhaust leak, wiring, MAP sensor**
3D	Injector 4 – circuit malfunction	**Wiring, injector**
3E	Heated oxygen sensor (HO2S) 1, bank 1 – high voltage	**Wiring short circuit to positive, fuel pressure, injector control, EVAP canister purge valve, air filter blocked, HO2S**
41	Heated oxygen sensor (HO2S) 1, bank 2 – no activity detected	**Wiring, heating inoperative, HO2S**
42	Heated oxygen sensor (HO2S) 1, bank 2 – high voltage	**Wiring short circuit to positive, fuel pressure, injector control, EVAP canister purge valve, air filter blocked, HO2S**
44	Engine control module (ECM) – internal fault	**ECM**
45	Injector 5 – circuit malfunction	**Wiring, injector**
46	Injector 6 – circuit malfunction	**Wiring, injector**
47	Ambient/battery temperature sensor	**Wiring, ambient/battery temperature sensor**
4C	Ignition coil D, primary/secondary – circuit malfunction	**Wiring, ignition coil**
4D	Ignition coil E, primary/secondary – circuit malfunction	**Wiring, ignition coil**
52	Cruise control	**Wiring, cruise control vacuum switching valve**
56	Cruise control master switch – high voltage	**Wiring, cruise control master switch**
57	Cruise control master switch – low voltage	**Wiring, cruise control master switch**
58	Intake manifold air control solenoid	**Wiring open circuit/short circuit, intake manifold air control solenoid**
5A	A/C refrigerant pressure sensor – high input	**Wiring, A/C refrigerant pressure too high (cooling fault/incorrectly charged), A/C refrigerant pressure sensor**
5B	A/C refrigerant pressure sensor – low input	**Wiring, A/C refrigerant pressure too low (incorrectly charged), A/C refrigerant pressure sensor**
5C	Engine coolant blower motor relay – low speed	**Wiring open circuit/short circuit, engine coolant blower motor relay – low speed**

Scan code Hex type	Fault location	Probable cause
5D	Engine coolant blower motor relay – high speed	**Wiring open circuit/short circuit, engine coolant blower motor relay – high speed**
60	CAN data bus, transmission control module (TCM) – no signal	**Wiring, TCM**
61	CAN data bus, body control module (BCM) – no signal	**Wiring, BCM**
65	Fuel pump relay	**Wiring open circuit/short circuit, fuel pump relay**
66	Heated oxygen sensor (H02S) 1, bank 1 – slow response	**Heating inoperative, wiring, HO2S**
67	Heated oxygen sensor (HO2S) 1, bank 1 – heater control – circuit malfunction	**Wiring, HO2S**
69	Heated oxygen sensor (HO2S) 2, bank 1, heater control – circuit malfunction	**Wiring, HO2S**
6A	Random/multiple cylinder(s) – misfire detected	**Wiring, spark plug(s), HT lead(s), injector(s), ignition coil(s), low compression**
6B	Cylinder 1 – misfire detected	**Wiring, engine mechanical fault, ignition/ fuel system, injector**
6C	Cylinder 2 – misfire detected	**Wiring, engine mechanical fault, ignition/ fuel system, injector**
6D	Cylinder 3 – misfire detected	**Wiring, engine mechanical fault, ignition/ fuel system, injector**
6E	Cylinder 4 – misfire detected	**Wiring, engine mechanical fault, ignition/ fuel system, injector**
70	Catalytic converter system, bank 1 – efficiency below threshold	**Wiring, HO2S 2, catalytic converter**
71	Evaporative emission (EVAP) system – incorrect flow detected	**Hose connection(s), intake leak, EVAP canister purge valve**
72	Park/neutral position (PNP) switch – input circuit malfunction	**Wiring, PNP switch, ECM/TCM**
73	Power steering pressure (PSP) switch – range/performance problem	**Wiring, PAS system, PSP switch**
76	System too rich	**Wiring, intake blocked, EVAP canister purge valve, fuel pressure, EGR system, injector(s), air filter blocked, ignition system, HO2S**
77	System too lean	**Wiring, intake/exhaust leak, MAP sensor, fuel pressure/pump, EVAP canister purge valve, injector(s), HO2S**
78	System too rich, bank 2	**Intake blocked, EVAP canister purge valve, fuel pressure, EGR system, injector(s), HO2S**
79	System too lean, bank 2	**Intake/exhaust leak, fuel pressure/pump, injector(s), AIR system, hose connection(s)**
7A	Heated oxygen sensor (HO2S) 1, bank 2 – slow response	**Wiring, heating inoperative, HO2S**

Scan code Hex type	Fault location	Probable cause
7C	Heated oxygen sensor (HO2S) 1, bank 2, heater control – circuit malfunction	**Wiring, HO2S**
7D	Heated oxygen sensor (HO2S) 2, bank 2, heater control – circuit malfunction	**Wiring, HO2S**
7E	Heated oxygen sensor (HO2S) 2, bank 1 – high voltage	**Wiring short circuit to positive, catalytic converter, HO2S**
7F	Heated oxygen sensor (HO2S) 2, bank 2 – high voltage	**Wiring short circuit to positive, catalytic converter, HO2S**
80	Insufficient coolant temperature for closed loop fuel control	**Wiring, cooling system, coolant thermostat, ECT sensor**
81	Heated oxygen sensor (HO2S) 2, bank 1 – no activity detected	**Wiring, heating inoperative, HO2S**
82	Heated oxygen sensor (HO2S) 2, bank 2 – no activity detected	**Wiring, heating inoperative, HO2S**
84	Throttle position (TP) sensor – range/ performance problem	**Wiring, accelerator cable adjustment, TP sensor**
85	Timing belt jumped 1 tooth or more	**Timing belt loose/jumped position**
89	Transmission control system – malfunction	**Wiring, ECM/TCM**
8A	Idle speed control (ISC) system – rpm lower/ higher than expected	**Wiring, air leak, throttle valve tight/sticking, IAC valve**
91	Manifold absolute pressure (MAP) sensor/ throttle position (TP) sensor	**Intake leak, hose connection(s), MAP sensor, TP sensor**
92	Engine control module (ECM) – 5 volt output low	**ECM**
95	Fuel level sensor – low input	**Wiring open circuit, fuel level sensor**
96	Fuel level sensor – high input	**Wiring short circuit to positive, fuel level sensor**
97	Fuel level sensor – malfunction	**Wiring, fuel level sensor**
98	Brake pedal position (BPP) switch – circuit malfunction	**Wiring, BPP switch**
99	Ambient/battery temperature sensor – voltage low	**Wiring, ambient/battery temperature sensor**
9A	Ambient/battery temperature sensor – voltage high	**Wiring, ambient/battery temperature sensor**
9B	Heated oxygen sensor (HO2S) 1, bank 1 – low voltage	**Wiring short circuit to ground, exhaust leak, HO2S**
9C	Heated oxygen sensor (HO2S) 2, bank 1 – low voltage	**Wiring short circuit to ground, exhaust leak, HO2S**
9D	Crankshaft position (CKP) sensor/camshaft position (CMP) sensor – intermittent signal	**Wiring, poor connection, air gap, CKP sensor, CMP sensor**
A0	Evaporative emission (EVAP) system – small leak detected	**Hose connection(s), intake leak, EVAP canister, EVAP canister purge valve**
A1	Evaporative emission (EVAP) system – large leak detected	**Hose connection(s), intake leak, EVAP canister, EVAP canister purge valve**

Scan code Hex type	Fault location	Probable cause
AE	Cylinder 5 – misfire detected	**Wiring, engine mechanical fault, ignition/ fuel system, injector**
AF	Cylinder 6 – misfire detected	**Wiring, engine mechanical fault, ignition/ fuel system, injector**
B5	Heated oxygen sensor (HO2S) 1, bank 2 – low voltage	**Wiring short circuit to ground, exhaust leak, HO2S**
B6	Heated oxygen sensor (HO2S) 2, bank 2 – low voltage	**Wiring short circuit to ground, exhaust leak, HO2S**
B7	Evaporative emission (EVAP) leak detection pump solenoid	**Wiring open circuit/short circuit, hose leak/ blockage, EVAP leak detection pump solenoid**
B8	Evaporative emission (EVAP) leak detection pump – pressure switch	**Wiring, hose leak/blockage, EVAP leak detection pump**
BA	Engine control module (ECM) – misfire detection limit reached	**Wiring, poor connection, CKP sensor**
BB	Evaporative emission (EVAP) system – blockage detected	**Hose connection(s), hose blocked/trapped, EVAP canister**
C0	Heated oxygen sensor (HO2S) 1, bank 1 – slow response	**Wiring, heating inoperative, HO2S**
C1	Heated oxygen sensor (HO2S) 1, bank 2 – slow response	**Wiring, heating inoperative, HO2S**
D4	Exhaust gas recirculation (EGR) valve position sensor A – low input	**Wiring short circuit to ground, EGR valve position sensor**
D5	Exhaust gas recirculation (EGR) valve position sensor A – high input	**Wiring short circuit to positive, EGR valve position sensor**
D7	Exhaust gas recirculation (EGR) system – range/performance problem	**Wiring, hose leak/blockage, EGR valve/ solenoid**
D8	Ignition coil F, primary/secondary – circuit malfunction	**Wiring, ignition coil**
E1	CAN data bus, instrumentation – no signal	**Wiring, instrumentation control module (mechanical)**
E2	CAN data bus, Immobilizer control module – no signal	**Wiring, immobilizer control module**
E8	Key not programmed into immobilizer control module	**Wiring, ignition key not matched/damaged, immobilizer control module**

CHRYSLER

Model:	Engine identification:	Year:
Cirrus 2.5L	EEB	1999-00
Concorde 2.7/3.2/3.5L	EER/EES/EGW/EGG/ECG/EGK/EGJ	1999-02
LHS 3.5L	EGG/ECG/EGK/EGJ	1999-01
Prowler 3.5L	VIN position 8 = G	2001-02
PT Cruiser 2.4L	EDZ	2000-02
Sebring Convertible 2.4L	EDZ	1999-00
Sebring Convertible 2.4/2.7L	EDZ/EER/EES	2001-02
Sebring Sedan 2.4L	EDZ	2001-02
Sebring Sedan 2.7L	EER	2001-02
300M 3.5L	EGG/ECG/EGK/EGJ	1999-02

System: **SBEC III**

Self-diagnosis

General information

- Refer to the front of this manual for general test conditions, terminology, detailed descriptions of wiring faults and a general trouble shooter for electrical and mechanical faults.
- Malfunction indicator lamp (MIL) will illuminate or flash if certain faults are recorded.
- ECM operates in backup mode if sensors fail, to enable vehicle to be driven to repair shop.

Accessing

1999-2001

- Trouble codes can only be displayed by using a Scan Tool connected to the data link connector (DLC):
  - Cirrus **1**.
  - Concorde/LHS/300M **2**.

1

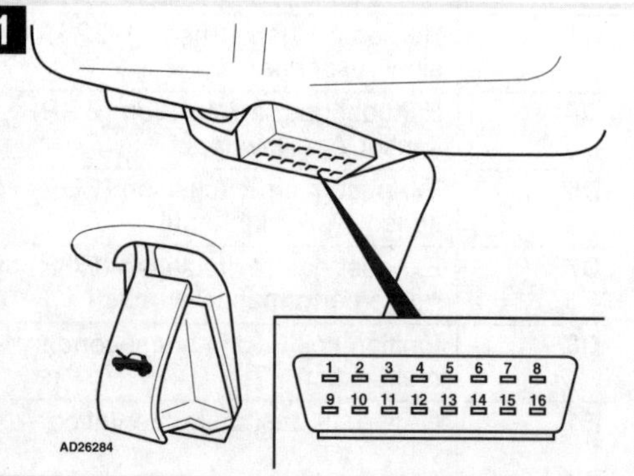

2

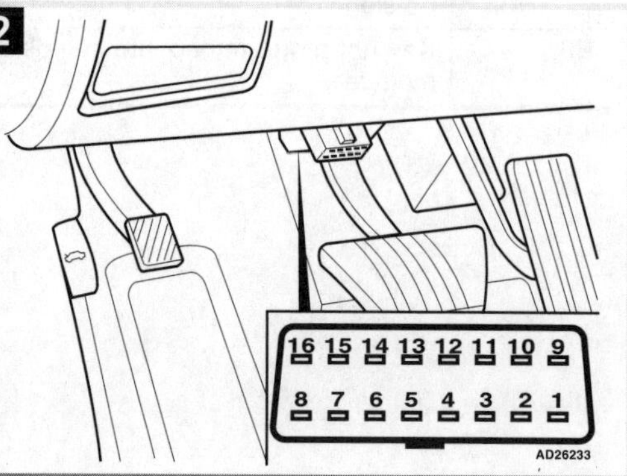

Accessing

2001→

- Within 5 seconds switch the ignition key ON-OFF-ON-OFF-ON.
- Odometer will display trouble codes.
- If no trouble codes are stored in ECM, odometer will display 'P1684' and 'done' or only 'done'.
- Trouble codes can also be displayed by using a Scan Tool connected to the data link connector (DLC):
  - Cirrus **1**.
  - Concorde/LHS/Sebring/300M **2**.
  - Prowler **3**.
  - PT Cruiser **4**.

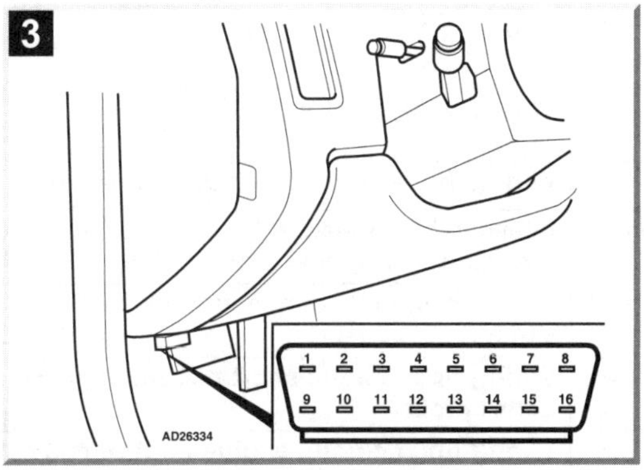

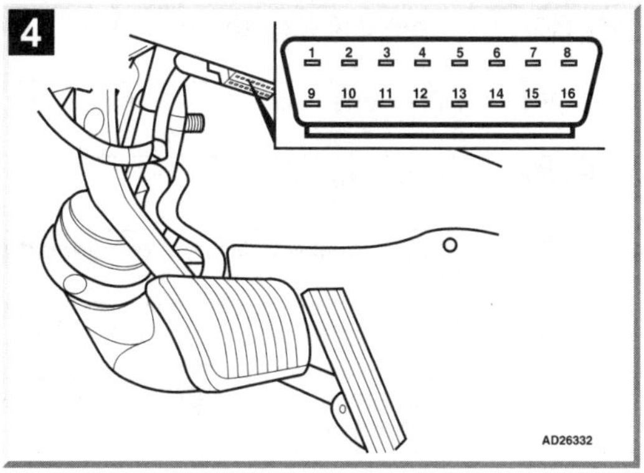

Erasing

- Trouble codes can only be erased by using a Scan Tool connected to the data link connector (DLC).

Trouble code identification

Scan code	Fault location	Probable cause
P0000-P0999	Refer to OBD-II trouble code table at the front of this manual	–
P1105	Barometric pressure (BARO) read solenoid – circuit malfunction	**Wiring, BARO solenoid**
P1115	Engine coolant temperature (ECT) sensor/ intake air temperature (IAT) sensor – signal variation	**Wiring, ECT sensor, IAT sensor, ECM**
P1192	Intake air temperature (IAT) sensor – voltage low	**Wiring, IAT sensor**
P1193	Intake air temperature (IAT) sensor – voltage high	**Wiring, IAT sensor**
P1195	Heated oxygen sensor (H02S) 1, bank 1 – slow response	**Wiring, heating inoperative, HO2S**
P1196	Heated oxygen sensor (HO2S) 1, bank 2 – slow response	**Wiring, heating inoperative, HO2S**

CHRYSLER

Cirrus 2.5L • Concorde 2.7/3.2/3.5L • LHS 3.5L • Prowler 3.5L • PT Cruiser 2.4L • Sebring Convertible 2.4L • Sebring Convertible 2.4/2.7L • Sebring Sedan 2.4L • Sebring Sedan 2.7L • 300M 3.5L

Scan code	Fault location	Probable cause
P1197	Heated oxygen sensor (HO2S) 2, bank 1 – slow response	Wiring, heating inoperative, HO2S
P1281	Engine temperature – low for long period	Wiring, cooling system, coolant thermostat, ECT sensor
P1282	Fuel pump relay	Wiring open circuit/short circuit, fuel pump relay
P1288	Intake manifold air control solenoid (short runner)	Wiring open circuit/short circuit, intake manifold air control solenoid (short runner)
P1289	Intake manifold air control solenoid	Wiring open circuit/short circuit, intake manifold air control solenoid
P1294	Idle speed control (ISC) system – rpm lower/ higher than expected	Wiring, air leak, throttle valve tight/sticking, IAC valve
P1295	Throttle position (TP) sensor – no 5 volt supply	Wiring, TP sensor, ECM
P1296	Manifold absolute pressure (MAP) sensor – no 5 volt supply	Wiring, MAP sensor, ECM
P1297	Manifold absolute pressure (MAP) sensor	Wiring, hose connection(s), MAP sensor
P1298	System too lean – at prolonged wide open throttle	Intake/exhaust leak, MAP sensor, fuel pressure/pump, injector(s), HO2S
P1299	Manifold absolute pressure (MAP) sensor/ throttle position (TP) sensor	Intake leak, hose connection(s), MAP sensor, TP sensor
P1388	Engine control relay	Wiring open circuit/short circuit, engine control relay
P1389	Engine control relay	Wiring open circuit, engine control relay
P1390	Crankshaft position (CKP) sensor/camshaft position (CMP) sensor – signal synchronisation	Timing belt loose/jumped position, CKP sensor, CMP sensor
P1391	Crankshaft position (CKP) sensor/camshaft position (CMP) sensor – intermittent signal	Wiring, poor connection, air gap, CKP sensor, CMP sensor
P1398	Engine control module (ECM) – misfire detection limit reached	Wiring, poor connection, CKP sensor
P1403	Exhaust gas recirculation (EGR) valve position sensor – no 5 volt supply	Wiring, ECM
P1478	Ambient/battery temperature sensor	Wiring, ambient/battery temperature sensor
P1479	Transmission cooling fan relay – circuit malfunction	Wiring, transmission fluid level low, transmission cooling fan, transmission cooling fan relay
P1486	Evaporative emission (EVAP) system – blockage detected	Hose connection(s), hose blocked/trapped, EVAP canister
P1488	Engine control module (ECM) – auxiliary 5 volt output low	Wiring, ECM
P1489	Engine coolant blower motor relay – high speed	Wiring open circuit/short circuit, engine coolant blower motor relay – high speed
P1490	Engine coolant blower motor relay – low speed	Wiring open circuit/short circuit, engine coolant blower motor relay – low speed

Scan code	Fault location	Probable cause
P1491	Engine coolant blower motor relay – circuit malfunction	**Wiring open circuit/short circuit, engine coolant blower motor relay**
P1492	Ambient/battery temperature sensor – voltage high	**Wiring, ambient/battery temperature sensor**
P1493	Ambient/battery temperature sensor – voltage low	**Wiring, ambient/battery temperature sensor**
P1494	Evaporative emission (EVAP) leak detection pump – pressure switch	**Wiring, hose leak/blockage, EVAP leak detection pump**
P1495	Evaporative emission (EVAP) leak detection pump solenoid	**Wiring open circuit/short circuit, hose leak/ blockage, EVAP leak detection pump solenoid**
P1496	Engine control module (ECM) – 5 volt output low	**ECM**
P1594	Alternator – voltage high	**Wiring, battery, alternator**
P1595	Cruise control	**Wiring open circuit/short circuit, hoses, cruise control master switch, cruise control vacuum switching valve**
P1596	Cruise control master switch – high voltage	**Wiring, cruise control master switch**
P1597	Cruise control master switch – low voltage	**Wiring, cruise control master switch**
P1598	A/C refrigerant pressure sensor – high input	**Wiring, A/C refrigerant pressure too high (cooling fault/incorrectly charged), A/C refrigerant pressure sensor**
P1599	A/C refrigerant pressure sensor – low input	**Wiring, A/C refrigerant pressure too low (incorrectly charged), A/C refrigerant pressure sensor**
P1602	Engine control module (ECM) – not programmed	**New ECM not programmed, incorrectly programmed, ECM**
P1603	Engine control module (ECM) – RAM link error	**ECM**
P1604	Engine control module (ECM) – RAM read/ write error	**ECM**
P1607	Engine control module (ECM) – shut down timer error	**ECM**
P1680	Clutch pedal position (CPP) switch – circuit malfunction	**Wiring, CPP switch, ECM**
P1681	CAN data bus, instrumentation – no signal	**Wiring, instrumentation control module**
P1682	Alternator – voltage low	**Wiring, battery, alternator drive belt, alternator**
P1683	Cruise control	**Wiring, cruise control vacuum switching valve**
P1684	Engine control module (ECM) – supply voltage disconnected within last 50 ignition switch ON-OFF cycles	**Wiring, battery**
P1685	Key not programmed into immobilizer control module	**Wiring, ignition key not matched/damaged, immobilizer control module**

Scan code	Fault location	Probable cause
P1686	CAN data bus, Immobilizer control module – no signal	**Wiring, immobilizer control module**
P1687	CAN data bus, instrumentation – no signal	**Wiring, instrumentation control module (mechanical)**
P1695	CAN data bus, body control module (BCM) – no signal	**Wiring, BCM**
P1696	Engine control module (ECM) – EEPROM error	**ECM**
P1697	Engine control module (ECM) – EEPROM error	**ECM**
P1698	CAN data bus, transmission control module (TCM) – no signal	**Wiring, TCM**
P1765	Transmission system relay	**Wiring open circuit/short circuit, transmission system relay, ECM/TCM**
P1899	Park/neutral position (PNP) switch – input circuit malfunction	**Wiring, PNP switch, ECM/TCM**
P2008	Intake manifold air control solenoid (short runner)	**Wiring open circuit/short circuit, intake manifold air control solenoid (short runner)**
P2302	Ignition coil 1, secondary – circuit malfunction	**Wiring, ignition coil**
P2305	Ignition coil 2, secondary – circuit malfunction	**Wiring, ignition coil**
P2308	Ignition coil 3, secondary – circuit malfunction	**Wiring, ignition coil**
P2311	Ignition coil 4, secondary – circuit malfunction	**Wiring, ignition coil**
P2314	Ignition coil 5, secondary – circuit malfunction	**Wiring, ignition coil**
P2317	Ignition coil 6, secondary – circuit malfunction	**Wiring, ignition coil**
P2503	Charging system voltage – low	**Wiring, poor connection, battery, alternator**

CHRYSLER

Model:	Engine identification:	Year:
Sebring 2.0L	**420A**	**1995-96**
Sebring 2.5L	**6G73**	**1996**

System: **SBEC III/Mitsubishi MPI**

Self-diagnosis

General information

- Refer to the front of this manual for general test conditions, terminology, detailed descriptions of wiring faults and a general trouble shooter for electrical and mechanical faults.
- Malfunction indicator lamp (MIL) will illuminate or flash if certain faults are recorded.
- ECM operates in backup mode if sensors fail, to enable vehicle to be driven to repair shop.
- Trouble codes can be displayed by the MIL or by using a Scan Tool connected to the data link connector(s) (DLC).

Accessing

Flash codes

- Within 5 seconds switch the ignition key ON-OFF-ON-OFF-ON.
- Count MIL flashes. Note trouble codes.
- Each trouble code consists of two groups **1** [**A**].
- The first group of flashes indicate the 'tens' of the trouble code.
- The second group of flashes indicate the 'units' of the trouble code.
- A short pause separates 'tens' and 'units' **1** [**B**].
- A long pause separates each trouble code **1** [**C**].
- For example: Trouble code 21 displayed **1**.

P codes & DRB codes

- Trouble codes can be displayed by using a Scan Tool connected to the data link connector(s) (DLC) **2**.

1

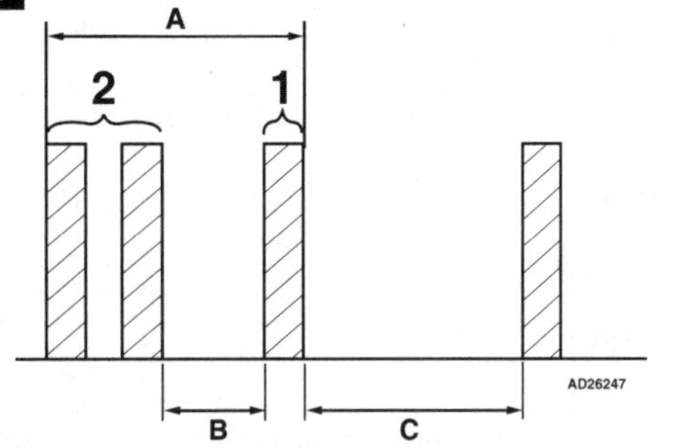

2

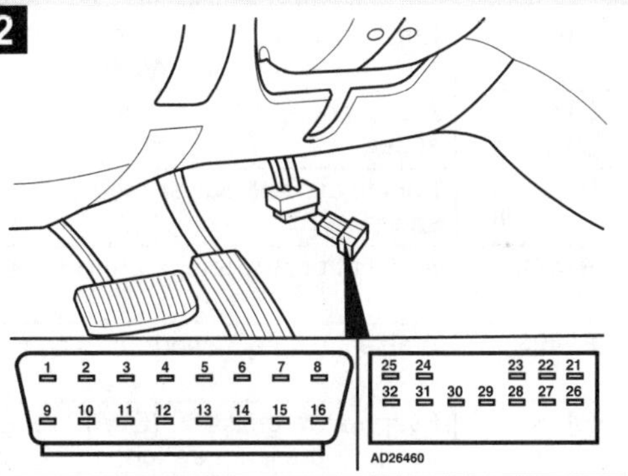

Erasing

NOTE: ***Disconnecting battery ground lead may cause driveability problems until the engine management system relearns basic values.***

- Ensure ignition switched OFF.
- Disconnect battery ground cable for at least 5 minutes.
- Reconnect battery ground cable.
- Repeat accessing procedure to ensure no trouble codes remain in ECM fault memory.

NOTE: ***Disconnecting battery ground lead may also erase memory from electronic units, such as radio.***

- Trouble codes can also be erased by using a Scan Tool connected to the data link connector(s) (DLC).

Trouble code identification

Scan code P type	Fault location	Probable cause
P0000-P0999	Refer to OBD-II trouble code table at the front of this manual	–
P1294	Idle speed control (ISC) system – rpm lower/ higher than expected	**Wiring, air leak, throttle valve tight/sticking, IAC valve**
P1295	Throttle position (TP) sensor – no 5 volt supply	**Wiring, TP sensor, ECM**
P1296	Manifold absolute pressure (MAP) sensor – no 5 volt supply	**Wiring, ECM**
P1297	Manifold absolute pressure (MAP) sensor	**Wiring, hose connection(s), MAP sensor**
P1390	Timing belt jumped 1 tooth or more	**Timing belt loose/jumped position**
P1391	Crankshaft position (CKP) sensor/camshaft position (CMP) sensor – intermittent signal	**Wiring, poor connection, air gap, CKP sensor, CMP sensor**
P1486	Evaporative emission (EVAP) system – blockage detected	**Hose connection(s), hose blocked/trapped, EVAP canister**
P1487	Engine coolant blower motor relay – high speed – circuit malfunction	**Wiring open circuit/short circuit, engine coolant blower motor relay – high speed**
P1489	A/C condenser blower motor relay – high speed	**Wiring open/short circuit, engine coolant blower motor relay – high speed**
P1490	Engine coolant blower motor relay – low speed	**Wiring open/short circuit, engine coolant blower motor relay – low speed**
P1492	Ambient/battery temperature sensor – voltage high	**Wiring, ambient/battery temperature sensor**
P1493	Ambient/battery temperature sensor – voltage low	**Wiring, ambient/battery temperature sensor**
P1494	Evaporative emission (EVAP) leak detection pump – pressure switch	**Wiring, hose leak/blockage, EVAP leak detection pump**
P1495	Evaporative emission (EVAP) leak detection pump solenoid	**Wiring open/short circuit, hose leak/ blockage, EVAP leak detection pump solenoid**
P1496	Manifold absolute pressure (MAP) sensor – 5 volt output low	**Wiring, MAP sensor**
P1696	Engine control module (ECM) – EEPROM error	**ECM**
P1697	Engine control module (ECM) – EEPROM error	**ECM**
P1698	CAN data bus, transmission control module (TCM) – no signal	**Wiring, TCM**
P1899	Park/neutral position (PNP) switch – input circuit malfunction	**Wiring, PNP switch, ECM/TCM**

Flash code	Scan code DRB III type	Fault location	Probable cause
11		Crankshaft position (CKP) sensor/camshaft position (CMP) sensor	**Wiring, poor connection, air gap, timing belt, CKP sensor, CMP sensor**
	133	Timing belt jumped 1 tooth or more	**Timing belt loose/jumped position**
	157	Crankshaft position (CKP) sensor/camshaft position (CMP) sensor – intermittent signal	**Wiring, poor connection, air gap, CKP sensor, CMP sensor**
	40	Crankshaft position (CKP) sensor – no signal	**Wiring, air gap, CKP sensor**
13	39	Manifold absolute pressure (MAP) sensor	**Wiring, hose connection(s), MAP sensor**
14		Manifold absolute pressure (MAP) sensor – input low/high	**Wiring, MAP sensor**
	36	Manifold absolute pressure (MAP) sensor – low input	**Wiring short circuit to ground, MAP sensor**
	37	Manifold absolute pressure (MAP) sensor – high input	**Wiring short circuit to positive, MAP sensor**
	135	Manifold absolute pressure (MAP) sensor – no 5 volt supply	**Wiring, ECM**
	146	Manifold absolute pressure (MAP) sensor – 5 volt output low	**Wiring, MAP sensor**
15	35	Vehicle speed sensor (VSS) – signal	**Wiring, VSS**
16	59	Knock sensor (KS) 1 – no signal	**Wiring, KS incorrectly tightened, KS**
17		Engine temperature – low for long period	**Wiring, cooling system, coolant thermostat, ECT sensor**
	33	Engine temperature – low for long period	**Wiring, cooling system, coolant thermostat, ECT sensor**
	128	Insufficient coolant temperature for closed loop fuel control	**Wiring, cooling system, coolant thermostat, ECT sensor**
21		Heated oxygen sensor (HO2S) 1/2, bank 1/2 – circuit malfunction/incorrect signal/voltage low/high	**Wiring, fuel pressure, injector control, EVAP canister purge valve, compression, air filter blocked, HO2S**
	155	Heated oxygen sensor (HO2S) 1, bank 1 – low voltage	**Wiring short circuit to ground, exhaust leak, HO2S**
	62	Heated oxygen sensor (HO2S) 1, bank 1 – high voltage	**Wiring short circuit to positive, fuel pressure, injector control, EVAP canister purge valve, air filter blocked, HO2S**
	102	Heated oxygen sensor (H02S) 1, bank 1 – slow response	**Wiring, heating inoperative, HO2S**
	32	Heated oxygen sensor (H02S) 1, bank 1 – no activity detected	**Wiring open circuit, heating inoperative, HO2S**
	103	Heated oxygen sensor (HO2S) 1, bank 1 – heater control – circuit malfunction	**Wiring, HO2S**

Flash code	Scan code DRB III type	Fault location	Probable cause
	156	Heated oxygen sensor (HO2S) 2, bank 1 – low voltage	**Wiring short circuit to ground, exhaust leak, HO2S**
	126	Heated oxygen sensor (HO2S) 2, bank 1 – high voltage	**Wiring short circuit to positive, catalytic converter, HO2S**
	129	Heated oxygen sensor (HO2S) 2, bank 1 – no activity detected	**Wiring, heating inoperative, HO2S**
	105	Heated oxygen sensor (HO2S) 2, bank 1, heater control – circuit malfunction	**Wiring, HO2S**
	104	Heated oxygen sensor (HO2S) 2, bank 1 – slow response	**Wiring, heating inoperative, HO2S**
	66	Heated oxygen sensor (HO2S) 1, bank 2 – high voltage	**Wiring short to positive, HO2S, ECM**
	65	Heated oxygen sensor (HO2S) 1, bank 2 – no activity detected	**Wiring, HO2S, ECM**
	122	Heated oxygen sensor (HO2S) 1, bank 2 – slow response	**Wiring, heating inoperative, HO2S**
	124	Heated oxygen sensor (HO2S) 1, bank 2, heater control – circuit malfunction	**Wiring, HO2S, ECM**
	125	Heated oxygen sensor (HO2S) 2, bank 2, heater control – circuit malfunction	**Wiring, HO2S, ECM**
	127	Heated oxygen sensor (HO2S) 2, bank 2 – high voltage	**Wiring short to positive, HO2S, ECM**
	130	Heated oxygen sensor (HO2S) 2, bank 2 – no activity detected	**Wiring, HO2S, ECM**
	181	Heated oxygen sensor (HO2S) 1, bank 2 – low voltage	**Wiring short to ground, exhaust leak, HO2S, ECM**
	182	Heated oxygen sensor (HO2S) 2, bank 2 – low voltage	**Wiring short to ground, exhaust leak, HO2S, ECM**
22		Engine coolant temperature (ECT) sensor – voltage low/high	**Wiring, coolant thermostat, ECT sensor**
	30	Engine coolant temperature (ECT) sensor – low input	**Wiring short circuit to ground, coolant thermostat, ECT sensor**
	31	Engine coolant temperature (ECT) sensor – high input	**Wiring open circuit/short circuit to positive, coolant thermostat, ground wire defective, ECT sensor**
23		Intake air temperature (IAT) sensor – voltage low/high	**Wiring, IAT sensor**
	57	Intake air temperature (IAT) sensor – low input	**Wiring short circuit to ground, IAT sensor**
	58	Intake air temperature (IAT) sensor – high input	**Wiring open circuit/short circuit to positive, ground wire defective, IAT sensor**
24		Throttle position (TP) sensor – voltage low/high	**Wiring, accelerator cable adjustment, TP sensor**

Flash code	Scan code DRB III type	Fault location	Probable cause
	132	Throttle position (TP) sensor – range/ performance problem	Wiring, accelerator cable adjustment, TP sensor
	26	Throttle position (TP) sensor – low input	Wiring short circuit to ground, TP sensor
	27	Throttle position (TP) sensor – high input	Wiring short circuit to positive, TP sensor
	136	Throttle position (TP) sensor – no 5 volt supply	Wiring, TP sensor, ECM
25		Idle air control (IAC) valve	Wiring open/short circuit, air leak, IAC valve
	25	Idle speed control (ISC) system – malfunction	Wiring open/short circuit, IAC valve
	138	Idle speed control (ISC) system – rpm lower/higher than expected	Wiring, air leak, throttle valve tight/ sticking, IAC valve
27		Injector – circuit malfunction	Wiring, injector
	21	Injector 1 – circuit malfunction	Wiring, injector
	20	Injector 2 – circuit malfunction	Wiring, injector
	19	Injector 3 – circuit malfunction	Wiring, injector
	61	Injector 4 – circuit malfunction	Wiring, injector
	69	Injector 5 – circuit malfunction	Wiring, injector
	70	Injector 6 – circuit malfunction	Wiring, injector
31		Evaporative emission (EVAP) canister purge valve – open/short circuit	Wiring, EVAP canister purge valve
	113	Evaporative emission (EVAP) system – incorrect flow detected	Hose connection(s), intake leak, EVAP canister purge valve
	160	Evaporative emission (EVAP) system – small leak detected	Hose connection(s), intake leak, EVAP canister, EVAP canister purge valve
	18	Evaporative emission (EVAP) canister purge valve – circuit malfunction	Wiring open/short circuit, EVAP canister purge valve
	161	Evaporative emission (EVAP) system – large leak detected	Hose connection(s), intake leak, EVAP canister, EVAP canister purge valve
	187	Evaporative emission (EVAP) system – blockage detected	Hose connection(s), hose blocked/ trapped, EVAP canister
	184	Evaporative emission (EVAP) leak detection pump – pressure switch	Wiring, hose leak/blockage, EVAP leak detection pump
	183	Evaporative emission (EVAP) leak detection pump solenoid	Wiring open/short circuit, hose leak/ blockage, EVAP leak detection pump solenoid
32		Exhaust gas recirculation (EGR) solenoid/ system – circuit malfunction	Wiring, hoses, EGR solenoid
	46	Exhaust gas recirculation (EGR) system – insufficient flow detected	Hose leak/blockage, wiring, EGR valve, EGR solenoid

Flash code	Scan code DRB III type	Fault location	Probable cause
	17	Exhaust gas recirculation (EGR) – circuit malfunction	**Wiring open/short circuit, EGR solenoid**
33	16	A/C compressor clutch relay – voltage low/high	**Wiring open/short circuit, A/C compressor clutch relay**
35		Engine coolant blower motor relay/A/C condenser blower motor relay – low/high speed	**Wiring open/short circuit, engine coolant blower motor relay – low/high speed**
	139	Engine coolant blower motor relay – high speed – circuit malfunction	**Wiring open circuit/short circuit, engine coolant blower motor relay – high speed**
	93	A/C condenser blower motor relay – high speed	**Wiring open/short circuit, engine coolant blower motor relay – high speed**
	92	Engine coolant blower motor relay – low speed	**Wiring open/short circuit, engine coolant blower motor relay – low speed**
36		Secondary air injection (AIR) system – incorrect flow detected	**AIR pump, AIR solenoid, AIR hose(s)**
	111	Secondary air injection (AIR) system – incorrect flow detected	**AIR pump, AIR solenoid, AIR hose(s)**
	51	Secondary air injection (AIR) solenoid – circuit malfunction	**Wiring, AIR solenoid, ECM**
37	114	Park/neutral position (PNP) switch – input circuit malfunction	**Wiring, PNP switch, ECM/TCM**
41	11	Alternator, field control – circuit malfunction	**Wiring open/short circuit, alternator, battery**
42		Fuel pump relay/engine control relay/fuel level sensor – circuit malfunction	**Wiring open/short circuit, fuel pump relay, engine control relay, fuel level sensor**
	101	Fuel pump relay	**Wiring open/short circuit, fuel pump relay**
	10	Engine control relay	**Wiring open/short circuit, engine control relay**
	44	Engine control relay	**Wiring open circuit, engine control relay**
	149	Fuel level sensor	**Wiring open circuit, fuel level sensor**
	150	Fuel level sensor	**Wiring short circuit to positive, fuel level sensor**
	151	Fuel level sensor	**Wiring, fuel level sensor**
43		Random/multiple cylinder(s) – misfire detected	**Wiring, spark plug(s), HT lead(s), injector(s), ignition coil(s), low compression**
	106	Random/multiple cylinder(s) – misfire detected	**Wiring, spark plug(s), HT lead(s), injector(s), ignition coil(s), low compression**

Flash code	Scan code DRB III type	Fault location	Probable cause
	107	Cylinder 1 – misfire detected	Wiring, engine mechanical fault, ignition/fuel system, injector
	108	Cylinder 2 – misfire detected	Wiring, engine mechanical fault, ignition/fuel system, injector
	109	Cylinder 3 – misfire detected	Wiring, engine mechanical fault, ignition/fuel system, injector
	110	Cylinder 4 – misfire detected	Wiring, engine mechanical fault, ignition/fuel system, injector
	174	Cylinder 5 – misfire detected	Wiring, engine mechanical fault, ignition/fuel system, injector
	175	Cylinder 6 – misfire detected	Wiring, engine mechanical fault, ignition/fuel system, injector
	43	Ignition coil A, primary/secondary – circuit malfunction	Wiring, ignition coil
	42	Ignition coil B, primary/secondary – circuit malfunction	Wiring, ignition coil
44		Ambient/battery temperature sensor – voltage low/high	Wiring, ambient/battery temperature sensor
	154	Ambient/battery temperature sensor – voltage high	Wiring, ambient/battery temperature sensor
	153	Ambient/battery temperature sensor – voltage low	Wiring, ambient/battery temperature sensor
	71	Ambient/battery temperature sensor – range/performance problem	Wiring, ambient/battery temperature sensor
45	137	Transmission control system – malfunction	Wiring, TCM
46	06	Alternator – voltage high	Wiring, battery, alternator
47	05	Alternator – voltage low	Wiring, battery, alternator drive belt, alternator
51		System too lean, bank 1/2	Intake/exhaust leak, MAP sensor, fuel pressure/pump, ECT sensor, injector(s), HO2S
	119	System too lean, bank 1	Intake/exhaust leak, MAP sensor, fuel pressure/pump, ECT sensor, injector(s), HO2S
	121	System too lean, bank 2	Intake/exhaust leak, MAP sensor, fuel pressure/pump, ECT sensor, injector(s), HO2S
52		System too rich, bank 1/2	Intake blocked, EVAP canister purge valve, fuel pressure, EGR system, injector(s), HO2S
	118	System too rich, bank 1	Intake blocked, EVAP canister purge valve, fuel pressure, EGR system, injector(s), HO2S

Flash code	Scan code DRB III type	Fault location	Probable cause
	120	System too rich, bank 2	**Intake blocked, EVAP canister purge valve, fuel pressure, EGR system, injector(s), HO2S**
53		Engine control module (ECM)/CAN data bus – internal fault/malfunction	**Wiring, connected system, ECM**
	02	Engine control module (ECM) – internal fault	**ECM**
	68	CAN data bus – malfunction	**Wiring, connected system, ECM**
54	01	Camshaft position (CMP) sensor – no signal	**Wiring, CMP sensor**
62	48	Engine control module (ECM) – EEPROM error	**ECM**
63	49	Engine control module (ECM) – EEPROM error	**ECM**
64	112	Catalytic converter, bank 1 – efficiency below threshold	**Wiring, exhaust leak, catalytic converter, HO2S 2**
65	115	Power steering pressure (PSP) switch – range/performance problem	**Wiring, PAS system, PSP switch**
66	96	CAN data bus, TCM – no signal	**Wiring, TCM**
72	180	Catalytic converter, bank 2 – efficiency below threshold	**Wiring, exhaust leak, catalytic converter, HO2S 2**

CHRYSLER

Model:	Engine identification:	Year:
Sebring 2.0L	420A	1997-99
Sebring 2.5L	6G73	1997-00
Sebring Convertible 2.5L	6G73	1996-00

System: **SBEC III/Mitsubishi MPI**

Self-diagnosis

General information

- Refer to the front of this manual for general test conditions, terminology, detailed descriptions of wiring faults and a general trouble shooter for electrical and mechanical faults.
- Malfunction indicator lamp (MIL) will illuminate or flash if certain faults are recorded.
- ECM operates in backup mode if sensors fail, to enable vehicle to be driven to repair shop.
- Trouble codes can be displayed by the MIL or by using a Scan Tool connected to the data link connector(s) (DLC).

Accessing

Flash codes

- Within 5 seconds switch the ignition key ON-OFF-ON-OFF-ON.
- Count MIL flashes. Note trouble codes.
- Each trouble code consists of two groups **1** [**A**].
- The first group of flashes indicate the 'tens' of the trouble code.
- The second group of flashes indicate the 'units' of the trouble code.
- A short pause separates 'tens' and 'units' **1** [**B**].
- A long pause separates each trouble code **1** [**C**].
- For example: Trouble code 21 displayed **1**.

P codes, Hex codes & DRB codes

- Trouble codes can be displayed by using a Scan Tool connected to the data link connector(s) (DLC) **2**.

1

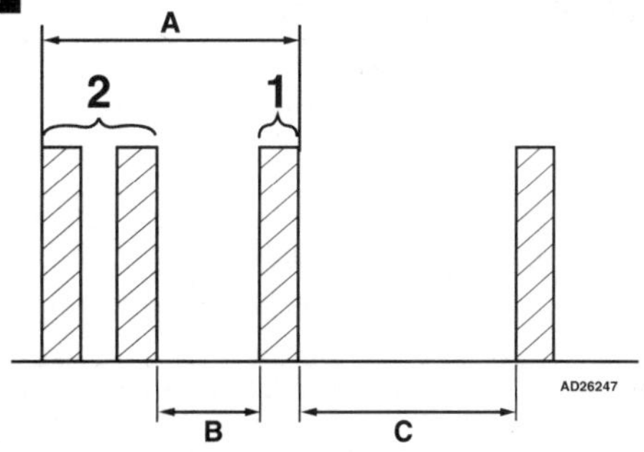

2

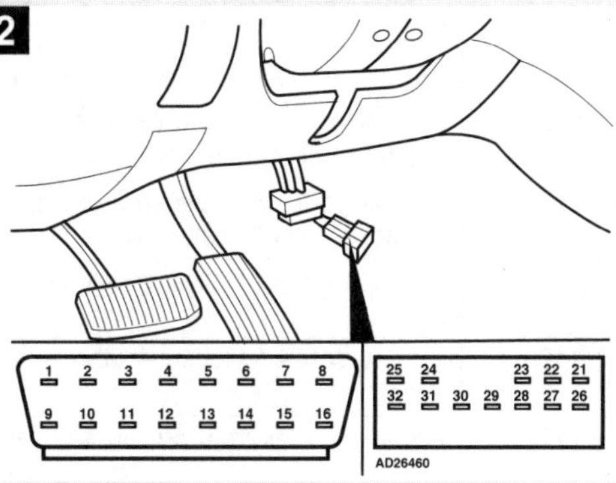

Erasing

NOTE: ***Disconnecting battery ground lead may cause driveability problems until the engine management system relearns basic values.***

- Ensure ignition switched OFF.
- Disconnect battery ground cable for at least 5 minutes.
- Reconnect battery ground cable.
- Repeat accessing procedure to ensure no trouble codes remain in ECM fault memory.

NOTE: ***Disconnecting battery ground lead may also erase memory from electronic units, such as radio.***

- Trouble codes can also be erased by using a Scan Tool connected to the data link connector(s) (DLC).

Trouble code identification

Scan code P type	Fault location	Probable cause
P0000-P0999	Refer to OBD-II trouble code table at the front of this manual	–
P1294	Idle speed control (ISC) system – rpm lower/higher than expected	**Wiring, air leak, throttle valve tight/sticking, IAC valve**
P1295	Throttle position (TP) sensor – no 5 volt supply	**Wiring, TP sensor, ECM**
P1296	Manifold absolute pressure (MAP) sensor – no 5 volt supply	**Wiring, ECM**
P1297	Manifold absolute pressure (MAP) sensor	**Wiring, hose connection(s), MAP sensor**
P1390	Timing belt jumped 1 tooth or more	**Timing belt loose/jumped position**
P1391	Crankshaft position (CKP) sensor/camshaft position (CMP) sensor – intermittent signal	**Wiring, poor connection, air gap, CKP sensor, CMP sensor**
P1486	Evaporative emission (EVAP) system – blockage detected	**Hose connection(s), hose blocked/trapped, EVAP canister**
P1487	Engine coolant blower motor relay – high speed – circuit malfunction	**Wiring open circuit/short circuit, engine coolant blower motor relay – high speed**
P1489	A/C condenser blower motor relay – high speed	**Wiring open/short circuit, engine coolant blower motor relay – high speed**
P1490	Engine coolant blower motor relay – low speed	**Wiring open/short circuit, engine coolant blower motor relay – low speed**
P1492	Ambient/battery temperature sensor – voltage high	**Wiring, ambient/battery temperature sensor**
P1493	Ambient/battery temperature sensor – voltage low	**Wiring, ambient/battery temperature sensor**
P1494	Evaporative emission (EVAP) leak detection pump – pressure switch	**Wiring, hose leak/blockage, EVAP leak detection pump**
P1495	Evaporative emission (EVAP) leak detection pump solenoid	**Wiring open/short circuit, hose leak/blockage, EVAP leak detection pump solenoid**
P1496	Manifold absolute pressure (MAP) sensor – 5 volt output low	**Wiring, MAP sensor**
P1696	Engine control module (ECM) – EEPROM error	**ECM**

Scan code P type	Fault location	Probable cause
P1697	Engine control module (ECM) – EEPROM error	**ECM**
P1698	CAN data bus, TCM – no signal	**Wiring, TCM**
P1899	Park/neutral position (PNP) switch – input circuit malfunction	**Wiring, PNP switch, ECM/TCM**

Flash code	Scan code Hex type	Scan code DRB III type	Fault location	Probable cause
11			Crankshaft position (CKP) sensor/ camshaft position (CMP) sensor	**Wiring, poor connection, air gap, timing belt, CKP sensor, CMP sensor**
	85	133	Timing belt jumped 1 tooth or more	**Timing belt loose/jumped position**
	9D	157	Crankshaft position (CKP) sensor/ camshaft position (CMP) sensor – intermittent signal	**Wiring, poor connection, air gap, CKP sensor, CMP sensor**
	28	40	Crankshaft position (CKP) sensor – no signal	**Wiring, air gap, CKP sensor**
13	27	39	Manifold absolute pressure (MAP) sensor	**Wiring, hose connection(s), MAP sensor**
14			Manifold absolute pressure (MAP) sensor – input low/high	**Wiring, MAP sensor**
	24	36	Manifold absolute pressure (MAP) sensor – low input	**Wiring short circuit to ground, MAP sensor**
	25	37	Manifold absolute pressure (MAP) sensor – high input	**Wiring short circuit to positive, MAP sensor**
	87	135	Manifold absolute pressure (MAP) sensor – no 5 volt supply	**Wiring, ECM**
	92	146	Manifold absolute pressure (MAP) sensor – 5 volt output low	**Wiring, MAP sensor**
15	23	35	Vehicle speed sensor (VSS) – signal	**Wiring, VSS**
16	3B	59	Knock sensor (KS) 1 – no signal	**Wiring, KS incorrectly tightened, KS**
17			Engine temperature – low for long period	**Wiring, cooling system, coolant thermostat, ECT sensor**
	21	33	Engine temperature – low for long period	**Wiring, cooling system, coolant thermostat, ECT sensor**
	80	128	Insufficient coolant temperature for closed loop fuel control	**Wiring, cooling system, coolant thermostat, ECT sensor**
21			Heated oxygen sensor (HO2S) 1/2, bank 1/2 – circuit malfunction/ incorrect signal/voltage low/high	**Wiring, fuel pressure, injector control, EVAP canister purge valve, compression, air filter blocked, HO2S**
	9B	155	Heated oxygen sensor (HO2S) 1, bank 1 – low voltage	**Wiring short circuit to ground, exhaust leak, HO2S**

Flash code	Scan code Hex type	Scan code DRB III type	Fault location	Probable cause
	3E	62	Heated oxygen sensor (HO2S) 1, bank 1 – high voltage	Wiring short circuit to positive, fuel pressure, injector control, EVAP canister purge valve, air filter blocked, HO2S
	66	102	Heated oxygen sensor (H02S) 1, bank 1 – slow response	Wiring, heating inoperative, HO2S
	20	32	Heated oxygen sensor (H02S) 1, bank 1 – no activity detected	Wiring open circuit, heating inoperative, HO2S
	67	103	Heated oxygen sensor (HO2S) 1, bank 1 – heater control – circuit malfunction	Wiring, HO2S
	9C	156	Heated oxygen sensor (HO2S) 2, bank 1 – low voltage	Wiring short circuit to ground, exhaust leak, HO2S
	7E	126	Heated oxygen sensor (HO2S) 2, bank 1 – high voltage	Wiring short circuit to positive, catalytic converter, HO2S
	81	129	Heated oxygen sensor (HO2S) 2, bank 1 – no activity detected	Wiring, heating inoperative, HO2S
	69	105	Heated oxygen sensor (HO2S) 2, bank 1, heater control – circuit malfunction	Wiring, HO2S
	42	66	Heated oxygen sensor (HO2S) 1, bank 2 – high voltage	Wiring short to positive, HO2S, ECM
	41	65	Heated oxygen sensor (HO2S) 1, bank 2 – no activity detected	Wiring, HO2S, ECM
	7A	122	Heated oxygen sensor (HO2S) 1, bank 2 – slow response	Wiring, heating inoperative, HO2S
	7C	124	Heated oxygen sensor (HO2S) 1, bank 2, heater control – circuit malfunction	Wiring, HO2S, ECM
	7D	125	Heated oxygen sensor (HO2S) 2, bank 2, heater control – circuit malfunction	Wiring, HO2S, ECM
	7F	127	Heated oxygen sensor (HO2S) 2, bank 2 – high voltage	Wiring short to positive, HO2S, ECM
	82	130	Heated oxygen sensor (HO2S) 2, bank 2 – no activity detected	Wiring, HO2S, ECM
	B5	181	Heated oxygen sensor (HO2S) 1, bank 2 – low voltage	Wiring short to ground, exhaust leak, HO2S, ECM
	B6	182	Heated oxygen sensor (HO2S) 2, bank 2 – low voltage	Wiring short to ground, exhaust leak, HO2S, ECM
22			Engine coolant temperature (ECT) sensor – voltage low/high	Wiring, coolant thermostat, ECT sensor
	1E	30	Engine coolant temperature (ECT) sensor – low input	Wiring short circuit to ground, coolant thermostat, ECT sensor

Flash code	Scan code Hex type	Scan code DRB III type	Fault location	Probable cause
	1F	31	Engine coolant temperature (ECT) sensor – high input	Wiring open circuit/short circuit to positive, coolant thermostat, ground wire defective, ECT sensor
23			Intake air temperature (IAT) sensor – voltage low/high	Wiring, IAT sensor
	39	57	Intake air temperature (IAT) sensor – low input	Wiring short circuit to ground, IAT sensor
	3A	58	Intake air temperature (IAT) sensor – high input	Wiring open circuit/short circuit to positive, ground wire defective, IAT sensor
24			Throttle position (TP) sensor – voltage low/high	Wiring, accelerator cable adjustment, TP sensor
	84	132	Throttle position (TP) sensor – range/ performance problem	Wiring, accelerator cable adjustment, TP sensor
	1A	26	Throttle position (TP) sensor – low input	Wiring short circuit to ground, TP sensor
	1B	27	Throttle position (TP) sensor – high input	Wiring short circuit to positive, TP sensor
	88	136	Throttle position (TP) sensor – no 5 volt supply	Wiring, TP sensor, ECM
25			Idle air control (IAC) valve	Wiring open/short circuit, air leak, IAC valve
	19	25	Idle speed control (ISC) system – malfunction	Wiring open/short circuit, IAC valve
	8A	138	Idle speed control (ISC) system – rpm lower/higher than expected	Wiring, air leak, throttle valve tight/ sticking, IAC valve
27			Injector – circuit malfunction	Wiring, injector
	15	21	Injector 1 – circuit malfunction	Wiring, injector
	14	20	Injector 2 – circuit malfunction	Wiring, injector
	13	19	Injector 3 – circuit malfunction	Wiring, injector
	3D	61	Injector 4 – circuit malfunction	Wiring, injector
	45	69	Injector 5 – circuit malfunction	Wiring, injector
	46	70	Injector 6 – circuit malfunction	Wiring, injector
31			Evaporative emission (EVAP) canister purge valve – open/short circuit	Wiring, EVAP canister purge valve
	71	113	Evaporative emission (EVAP) system – incorrect flow detected	Hose connection(s), intake leak, EVAP canister purge valve
	A0	160	Evaporative emission (EVAP) system – small leak detected	Hose connection(s), intake leak, EVAP canister, EVAP canister purge valve
	12	18	Evaporative emission (EVAP) canister purge valve – circuit malfunction	Wiring open/short circuit, EVAP canister purge valve

Flash code	Scan code Hex type	Scan code DRB III type	Fault location	Probable cause
	A1	161	Evaporative emission (EVAP) system – large leak detected	**Hose connection(s), intake leak, EVAP canister, EVAP canister purge valve**
	BB	187	Evaporative emission (EVAP) system – blockage detected	**Hose connection(s), hose blocked/ trapped, EVAP canister**
	B8	184	Evaporative emission (EVAP) leak detection pump – pressure switch	**Wiring, hose leak/blockage, EVAP leak detection pump**
	B7	183	Evaporative emission (EVAP) leak detection pump solenoid	**Wiring open/short circuit, hose leak/ blockage, EVAP leak detection pump solenoid**
32			Exhaust gas recirculation (EGR) solenoid/system – circuit malfunction	**Wiring, hoses, EGR solenoid**
	2E	46	Exhaust gas recirculation (EGR) system – insufficient flow detected	**Hose leak/blockage, wiring, EGR valve, EGR solenoid**
	11	17	Exhaust gas recirculation (EGR) – circuit malfunction	**Wiring open/short circuit, EGR solenoid**
33	10	16	A/C compressor clutch relay – voltage low/high	**Wiring open/short circuit, A/C compressor clutch relay**
35			Engine coolant blower motor relay/A/C condenser blower motor relay – low/ high speed	**Wiring open/short circuit, engine coolant blower motor relay – low/ high speed**
	8B	139	Engine coolant blower motor relay – high speed – circuit malfunction	**Wiring open circuit/short circuit, engine coolant blower motor relay – high speed**
	5D	93	A/C condenser blower motor relay – high speed	**Wiring open/short circuit, engine coolant blower motor relay – high speed**
	5C	92	Engine coolant blower motor relay – low speed	**Wiring open/short circuit, engine coolant blower motor relay – low speed**
37	72	114	Park/neutral position (PNP) switch – input circuit malfunction	**Wiring, PNP switch, ECM/TCM**
41	0B	11	Alternator, field control – circuit malfunction	**Wiring open/short circuit, alternator, battery**
42			Fuel pump relay/engine control relay/ fuel level sensor – circuit malfunction	**Wiring open/short circuit, fuel pump relay, engine control relay, fuel level sensor**
	65	101	Fuel pump relay	**Wiring open/short circuit, fuel pump relay**
	0A	10	Engine control relay	**Wiring open/short circuit, engine control relay**
	2C	44	Engine control relay	**Wiring open circuit, engine control relay**
	95	149	Fuel level sensor	**Wiring open circuit, fuel level sensor**

Flash code	Scan code Hex type	Scan code DRB III type	Fault location	Probable cause
	96	150	Fuel level sensor	**Wiring short circuit to positive, fuel level sensor**
	97	151	Fuel level sensor	**Wiring, fuel level sensor**
43			Random/multiple cylinder(s) – misfire detected	**Wiring, spark plug(s), HT lead(s), injector(s), ignition coil(s), low compression**
	6A	106	Random/multiple cylinder(s) – misfire detected	**Wiring, spark plug(s), HT lead(s), injector(s), ignition coil(s), low compression**
	6B	107	Cylinder 1 – misfire detected	**Wiring, engine mechanical fault, ignition/fuel system, injector**
	6C	108	Cylinder 2 – misfire detected	**Wiring, engine mechanical fault, ignition/fuel system, injector**
	6D	109	Cylinder 3 – misfire detected	**Wiring, engine mechanical fault, ignition/fuel system, injector**
	6E	110	Cylinder 4 – misfire detected	**Wiring, engine mechanical fault, ignition/fuel system, injector**
	AE	174	Cylinder 5 – misfire detected	**Wiring, engine mechanical fault, ignition/fuel system, injector**
	AF	175	Cylinder 6 – misfire detected	**Wiring, engine mechanical fault, ignition/fuel system, injector**
	2B	43	Ignition coil A, primary/secondary – circuit malfunction	**Wiring, ignition coil**
	2A	42	Ignition coil B, primary/secondary – circuit malfunction	**Wiring, ignition coil**
44			Ambient/battery temperature sensor – voltage low/high	**Wiring, ambient/battery temperature sensor**
	9A	154	Ambient/battery temperature sensor – voltage high	**Wiring, ambient/battery temperature sensor**
	99	153	Ambient/battery temperature sensor – voltage low	**Wiring, ambient/battery temperature sensor**
	47	71	Ambient/battery temperature sensor – range/performance problem	**Wiring, ambient/battery temperature sensor**
45	89	137	Transmission control system – malfunction	**Wiring, TCM**
46	06	06	Alternator – voltage high	**Wiring, battery, alternator**
47	05	05	Alternator – voltage low	**Wiring, battery, alternator drive belt, alternator**
51			System too lean, bank 1/2	**Intake/exhaust leak, MAP sensor, fuel pressure/pump, ECT sensor, injector(s), HO2S**
	77	119	System too lean, bank 1	**Intake/exhaust leak, MAP sensor, fuel pressure/pump, ECT sensor, injector(s), HO2S**

Flash code	Scan code Hex type	Scan code DRB III type	Fault location	Probable cause
	79	121	System too lean, bank 2	Intake/exhaust leak, MAP sensor, fuel pressure/pump, ECT sensor, injector(s), HO2S
52			System too rich, bank 1/2	Intake blocked, EVAP canister purge valve, fuel pressure, EGR system, injector(s), HO2S
	76	118	System too rich, bank 1	Intake blocked, EVAP canister purge valve, fuel pressure, EGR system, injector(s), HO2S
	78	120	System too rich, bank 2	Intake blocked, EVAP canister purge valve, fuel pressure, EGR system, injector(s), HO2S
53			Engine control module (ECM)/CAN data bus – internal fault/malfunction	Wiring, connected system, ECM
	02	02	Engine control module (ECM) – internal fault	ECM
	44	68	CAN data bus – malfunction	Wiring, connected system, ECM
54	01	01	Camshaft position (CMP) sensor – no signal	Wiring, CMP sensor
62	30	48	Engine control module (ECM) – EEPROM error	ECM
63	31	49	Engine control module (ECM) – EEPROM error	ECM
64	70	112	Catalytic converter, bank 1 – efficiency below threshold	Wiring, exhaust leak, catalytic converter, HO2S 2
65	73	115	Power steering pressure (PSP) switch – range/performance problem	Wiring, PAS system, PSP switch
66	60	96	CAN data bus, TCM – no signal	Wiring, TCM
72	B4	180	Catalytic converter, bank 2 – efficiency below threshold	Wiring, exhaust leak, catalytic converter, HO2S 2

CHRYSLER

Model:	Engine identification:	Year:
Sebring Coupe 2.4L	4G64	2001-02
Sebring Coupe 3.0L	6G72	2001-02

System: **Mitsubishi MPI**

Self-diagnosis

General information

- Refer to the front of this manual for general test conditions, terminology, detailed descriptions of wiring faults and a general trouble shooter for electrical and mechanical faults.
- Service engine soon/malfunction indicator lamp (MIL) will illuminate or flash if certain faults are recorded.
- ECM operates in backup mode if sensors fail, to enable vehicle to be driven to repair shop.

Accessing

- Trouble codes can only be displayed by using a Scan Tool connected to the data link connector (DLC) **1**.

1

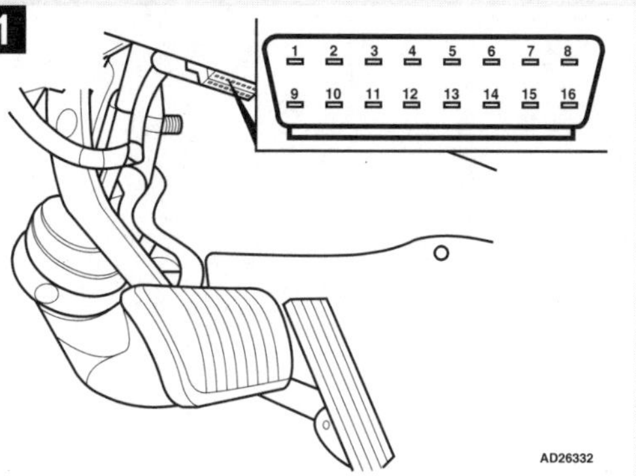

Erasing

NOTE: ***Disconnecting battery ground lead may cause driveability problems until the engine management system relearns basic values.***

- Ensure ignition switched OFF.
- Disconnect battery ground cable for at least 15 seconds.
- Reconnect battery ground lead.
- Repeat checking procedure to ensure no data remains in ECM fault memory.
- Trouble codes can also be erased by using a Scan Tool connected to the data link connector (DLC).

NOTE: ***Disconnecting battery ground lead may erase memory from electronic units such as radio.***

Trouble code identification

Scan code	Fault location	Probable cause
P0000-P0999	Refer to OBD-II trouble code table at the front of this manual	–
P1400	Manifold differential pressure (MDP) sensor, EGR system – circuit malfunction	**Wiring, MDP sensor, ECM**
P1500	Alternator – circuit malfunction	**Wiring, battery, alternator, ECM**
P1751	Transmission system relay – circuit malfunction	**Wiring, transmission system relay, ECM**
P1795	Throttle position (TP) sensor – circuit malfunction	**Wiring, TP sensor, ECM**

CHRYSLER

Model: **Sebring 2.5L**

Engine identification: **6G73**

Year: **1995**

System: **Mitsubishi MPI**

Self-diagnosis

General information

- Refer to the front of this manual for general test conditions, terminology, detailed descriptions of wiring faults and a general trouble shooter for electrical and mechanical faults.
- Malfunction indicator lamp (MIL) will illuminate if certain faults are recorded.
- ECM operates in backup mode if sensors fail, to enable vehicle to be driven to repair shop.
- Trouble codes can be displayed by the MIL or by using a Scan Tool connected to the data link connector(s) (DLC).

Accessing

- Within 5 seconds switch the ignition key ON-OFF-ON-OFF-ON.
- Count MIL flashes. Note trouble codes.
- Each trouble code consists of two groups **1** [**A**].
- The first group of flashes indicate the 'tens' of the trouble code.
- The second group of flashes indicate the 'units' of the trouble code.
- A short pause separates 'tens' and 'units' **1** [**B**].
- A long pause separates each trouble code **1** [**C**].
- For example: Trouble code 21 displayed **1**.
- Trouble codes can also be displayed by using a Scan Tool connected to the data link connector(s) (DLC) **2**.

1

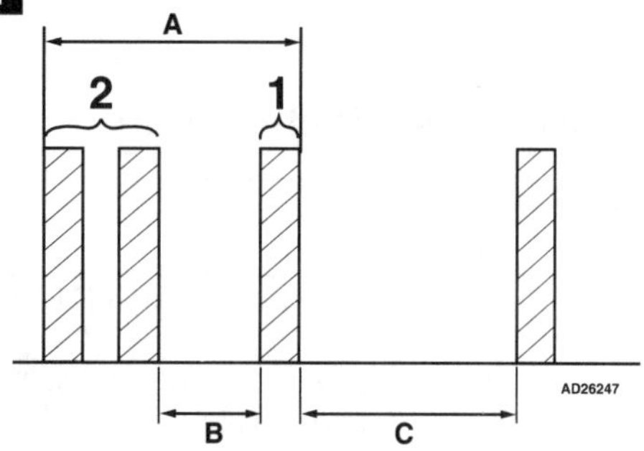

2

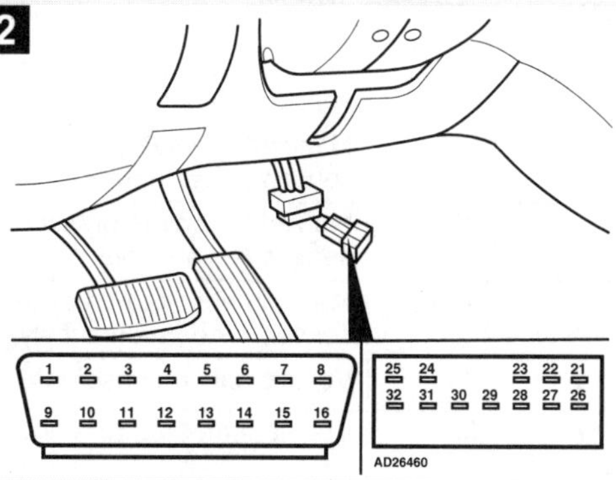

Erasing

NOTE: ***Disconnecting battery ground lead may cause driveability problems until the engine management system relearns basic values.***

- Ensure ignition switched OFF.
- Disconnect battery ground cable for at least 5 minutes.
- Reconnect battery ground cable.
- Repeat accessing procedure to ensure no trouble codes remain in ECM fault memory.

NOTE: ***Disconnecting battery ground lead may also erase memory from electronic units, such as radio.***

- Trouble codes can also be erased by using a Scan Tool connected to the data link connector(s) (DLC).

Trouble code identification

Flash code	Scan code DRB III type	Fault location	Probable cause
11		Crankshaft position (CKP) sensor/camshaft position (CMP) sensor	**Wiring, poor connection, air gap, timing belt, CKP sensor, CMP sensor**
	157	Crankshaft position (CKP) sensor/camshaft position (CMP) sensor – intermittent signal	**Wiring, poor connection, air gap, CKP sensor, CMP sensor**
	40	Crankshaft position (CKP) sensor – no signal	**Wiring, air gap, CKP sensor**
13	39	Manifold absolute pressure (MAP) sensor	**Wiring, hose connection(s), MAP sensor**
14		Manifold absolute pressure (MAP) sensor – input low/high	**Wiring, MAP sensor**
	36	Manifold absolute pressure (MAP) sensor – low input	**Wiring short circuit to ground, MAP sensor**
	37	Manifold absolute pressure (MAP) sensor – high input	**Wiring short circuit to positive, MAP sensor**
	135	Manifold absolute pressure (MAP) sensor – no 5 volt supply	**Wiring, ECM**
15	35	Vehicle speed sensor (VSS) – signal	**Wiring, VSS**
17	33	Engine temperature – low for long period	**Wiring, cooling system, coolant thermostat, ECT sensor**
21		Heated oxygen sensor (HO2S) 1/2, bank 1/2 – circuit malfunction/incorrect signal/voltage low/high	**Wiring, fuel pressure, injector control, EVAP canister purge valve, compression, air filter blocked, HO2S**
	32	Heated oxygen sensor (H02S) 1, bank 1 – no activity detected	**Wiring open circuit, heating inoperative, HO2S**
	41	Heated oxygen sensor (HO2S) 1, bank 2 – no activity detected	**Wiring open circuit, heating inoperative, HO2S**
	42	Heated oxygen sensor (HO2S) 1, bank 2 – high voltage	**Wiring short circuit to positive, fuel pressure, injector control, EVAP canister purge valve, air filter blocked, HO2S**

Flash code	Scan code DRB III type	Fault location	Probable cause
	62	Heated oxygen sensor (HO2S) 1, bank 1 – high voltage	**Wiring short circuit to positive, fuel pressure, injector control, EVAP canister purge valve, air filter blocked, HO2S**
22		Engine coolant temperature (ECT) sensor – voltage low/high	**Wiring, coolant thermostat, ECT sensor**
	30	Engine coolant temperature (ECT) sensor – low input	**Wiring short circuit to ground, coolant thermostat, ECT sensor**
	31	Engine coolant temperature (ECT) sensor – high input	**Wiring open circuit/short circuit to positive, coolant thermostat, ground wire defective, ECT sensor**
23		Intake air temperature (IAT) sensor – voltage low/high	**Wiring, IAT sensor**
	57	Intake air temperature (IAT) sensor – low input	**Wiring short circuit to ground, IAT sensor**
	58	Intake air temperature (IAT) sensor – high input	**Wiring open circuit/short circuit to positive, ground wire defective, IAT sensor**
24		Throttle position (TP) sensor – voltage low/high	**Wiring, accelerator cable adjustment, TP sensor**
	26	Throttle position (TP) sensor – low input	**Wiring short circuit to ground, TP sensor**
	27	Throttle position (TP) sensor – high input	**Wiring short circuit to positive, TP sensor**
	136	Throttle position (TP) sensor – no 5 volt supply	**Wiring, TP sensor, ECM**
25	25	Idle speed control (ISC) system – malfunction	**Wiring open/short circuit, IAC valve**
27		Injector – circuit malfunction	**Wiring, injector**
	21	Injector 1 – circuit malfunction	**Wiring, injector**
	20	Injector 2 – circuit malfunction	**Wiring, injector**
	19	Injector 3 – circuit malfunction	**Wiring, injector**
	61	Injector 4 – circuit malfunction	**Wiring, injector**
	69	Injector 5 – circuit malfunction	**Wiring, injector**
	70	Injector 6 – circuit malfunction	**Wiring, injector**
31	18	Evaporative emission (EVAP) canister purge valve – circuit malfunction	**Wiring open/short circuit, EVAP canister purge valve**
32		Exhaust gas recirculation (EGR) solenoid/system – circuit malfunction	**Wiring, hoses, EGR solenoid**
	46	Exhaust gas recirculation (EGR) system – insufficient flow detected	**Hose leak/blockage, wiring, EGR valve, EGR solenoid**
	17	Exhaust gas recirculation (EGR) – circuit malfunction	**Wiring open/short circuit, EGR solenoid**

Flash code	Scan code DRB III type	Fault location	Probable cause
33	16	A/C compressor clutch relay – voltage low/high	**Wiring open/short circuit, A/C compressor clutch relay**
35		Engine coolant blower motor relay/A/C condenser blower motor relay – low/high speed	**Wiring open/short circuit, engine coolant blower motor relay – low/high speed**
	139	Engine coolant blower motor relay – high speed – circuit malfunction	**Wiring open circuit/short circuit, engine coolant blower motor relay – high speed**
	93	A/C condenser blower motor relay – high speed	**Wiring open/short circuit, engine coolant blower motor relay – high speed**
	92	Engine coolant blower motor relay – low speed	**Wiring open/short circuit, engine coolant blower motor relay – low speed**
41	11	Alternator, field control – circuit malfunction	**Wiring open/short circuit, alternator, battery**
42		Fuel pump relay/engine control relay – circuit malfunction	**Wiring open/short circuit, fuel pump relay, engine control relay**
	101	Fuel pump relay	**Wiring open/short circuit, fuel pump relay**
	10	Engine control relay	**Wiring open/short circuit, engine control relay**
	44	Engine control relay	**Wiring open circuit, engine control relay**
44	71	Ambient/battery temperature sensor – range/performance problem	**Wiring, ambient/battery temperature sensor**
46	06	Alternator – voltage high	**Wiring, battery, alternator**
47	05	Alternator – voltage low	**Wiring, battery, alternator drive belt, alternator**
51		System too lean, bank 1/2	**Intake/exhaust leak, MAP sensor, fuel pressure/pump, ECT sensor, injector(s), HO2S**
	04	System too lean, bank 1	**Intake/exhaust leak, MAP sensor, fuel pressure/pump, ECT sensor, injector(s), HO2S**
	09	System too lean, bank 2	**Intake/exhaust leak, MAP sensor, fuel pressure/pump, ECT sensor, injector(s), HO2S**
52		System too rich, bank 1/2	**Intake blocked, EVAP canister purge valve, fuel pressure, EGR system, injector(s), HO2S**
	03	System too rich, bank 1	**Intake blocked, EVAP canister purge valve, fuel pressure, EGR system, injector(s), HO2S**
	08	System too rich, bank 2	**Intake blocked, EVAP canister purge valve, fuel pressure, EGR system, injector(s), HO2S**

Flash code	Scan code DRB III type	Fault location	Probable cause
53		Engine control module (ECM)/CAN data bus – internal fault/malfunction	**Wiring, connected system, ECM**
	02	Engine control module (ECM) – internal fault	**ECM**
	68	CAN data bus – malfunction	**Wiring, connected system, ECM**
54	01	Camshaft position (CMP) sensor – no signal	**Wiring, CMP sensor**
62	48	Engine control module (ECM) – EEPROM error	**ECM**
63	49	Engine control module (ECM) – EEPROM error	**ECM**
66	96	CAN data bus, transmission control module (TCM) – no signal	**Wiring, TCM**

CHRYSLER

Model:	Engine identification:	Year:
Town & Country 3.3L	EGA/EGM	1992-95
Town & Country 3.8L	EGH	1992-95

System: **SBEC II**

Self-diagnosis

General information

- Refer to the front of this manual for general test conditions, terminology, detailed descriptions of wiring faults and a general trouble shooter for electrical and mechanical faults.
- Malfunction indicator lamp (MIL) will illuminate if certain faults are recorded.
- ECM operates in backup mode if sensors fail, to enable vehicle to be driven to repair shop.
- Trouble codes can be displayed by the MIL or by using a Scan Tool connected to the data link connector (DLC).

Accessing

- Within 5 seconds switch the ignition key ON-OFF-ON-OFF-ON.
- Count MIL flashes. Note trouble codes.
- Each trouble code consists of two groups **1** [**A**].
- The first group of flashes indicate the 'tens' of the trouble code.
- The second group of flashes indicate the 'units' of the trouble code.
- A short pause separates 'tens' and 'units' **1** [**B**].
- A long pause separates each trouble code **1** [**C**].
- For example: Trouble code 21 displayed **1**.
- End of test sequence indicated by trouble code 55.
- Trouble codes can also be displayed by using a Scan Tool connected to the data link connector (DLC) 1992-93 **2** & 1994-95 **3**.

1

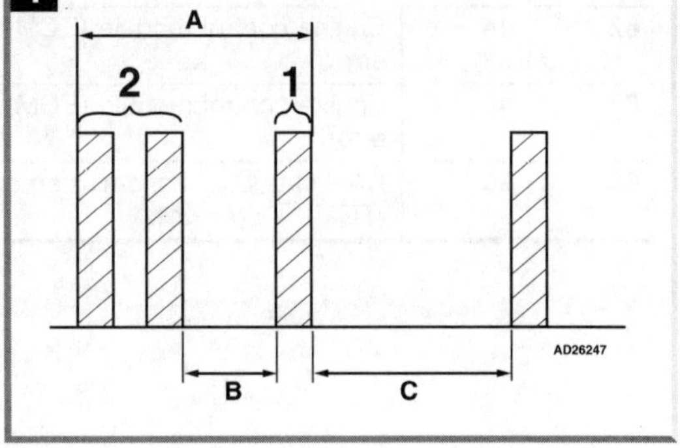

2

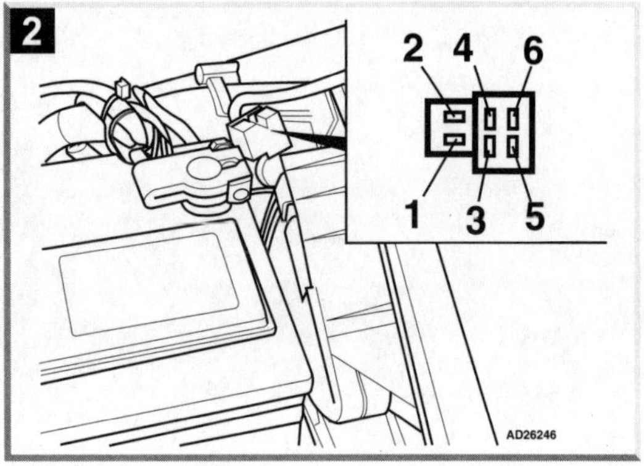

3

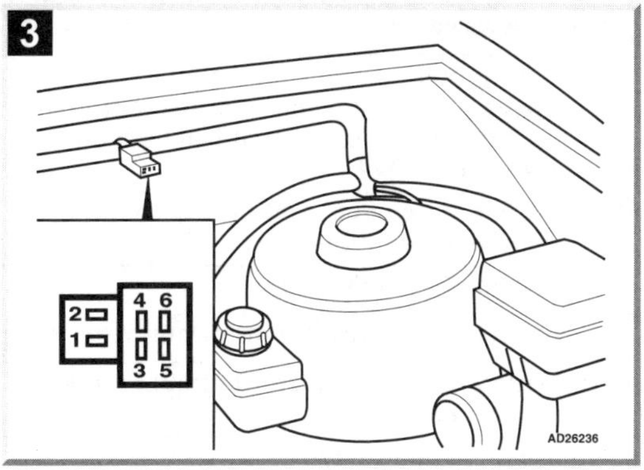

Erasing

NOTE: ***Disconnecting battery ground lead may cause driveability problems until the engine management system relearns basic values.***

- Ensure ignition switched OFF.
- Disconnect battery ground cable for at least 30 seconds.
- Reconnect battery ground cable.
- Repeat accessing procedure to ensure no trouble codes remain in ECM fault memory.

NOTE: ***Disconnecting battery ground lead may also erase memory from electronic units, such as radio.***

- Trouble codes can also be erased by using a Scan Tool connected to the data link connector (DLC).

Trouble code identification

Flash code	Fault location	Probable cause
11	Crankshaft position (CKP) sensor	**Wiring, air gap, CKP sensor**
12	Engine control module (ECM) – supply voltage disconnected within last 50 ignition switch ON-OFF cycles	**Wiring, battery**
13	Manifold absolute pressure (MAP) sensor	**Wiring, MAP sensor**
14	Manifold absolute pressure (MAP) sensor – voltage low/high	**Wiring, MAP sensor**
15	Vehicle speed sensor (VSS) – signal	**Wiring, VSS**
17	Engine temperature – low for long period	**Wiring, coolant thermostat, ECT sensor**
21	Heated oxygen sensor (HO2S) – incorrect signal/voltage high	**Wiring, fuel pressure, injector control, EVAP canister purge valve, compression, air filter blocked, HO2S**
22	Engine coolant temperature (ECT) sensor – voltage low/high	**Wiring, ECT sensor**
24	Throttle position (TP) sensor	**Wiring, accelerator cable adjustment, TP sensor**
25	Idle air control (IAC) valve	**Wiring, IAC valve**
27	Injector(s) – circuit malfunction	**Wiring, injector**
31	Evaporative emission (EVAP) canister purge valve – open/short circuit	**Wiring, EVAP canister purge valve**
32	Exhaust gas recirculation (EGR) solenoid/ system – circuit malfunction	**Wiring, hoses, EGR solenoid**
33	A/C compressor clutch relay	**Wiring, A/C compressor clutch relay**

Flash code	Fault location	Probable cause
34	Cruise control	**Wiring open circuit/short circuit, hoses, voltage low/high, cruise control master switch, cruise control vacuum switching valve**
35	Engine coolant blower motor relay – low/high speed	**Wiring, engine coolant blower motor relay – low/high speed**
37	Torque converter clutch (TCC) solenoid	**Wiring, TCC solenoid**
41	Alternator	**Wiring open circuit/short circuit, alternator**
42	Engine control relay	**Wiring open circuit/short circuit, engine control relay**
43	Ignition coil – primary circuit	**Wiring, ignition coil**
44	Battery temperature sensor/engine coolant temperature (ECT) sensor	**Wiring open circuit/short circuit, battery temperature sensor, ECT sensor, ECM**
46	Alternator – voltage high	**Wiring, battery, alternator**
47	Alternator – voltage low	**Wiring, battery, alternator drive belt, alternator**
51	System too lean	**Wiring, intake/exhaust leak, fuel pressure, injector control, EVAP canister purge valve, HO2S**
52	System too rich	**Wiring, exhaust blockage, fuel pressure, injector control, EVAP canister purge valve, compression, air filter blocked, ignition system, HO2S**
53	Engine control module (ECM) – internal fault – defective	**ECM**
54	Camshaft position (CMP) sensor – no signal	**Wiring, CMP sensor**
55	End of MIL code output	**–**
62	Engine control module (ECM) – EEPROM error	**ECM**
63	Engine control module (ECM) – EEPROM error	**ECM**

CHRYSLER

Model:	Engine identification:	Year:
Town & Country 3.3L	EGA/EGM	1996-97
Town & Country 3.8L	EGH	1996-97

System: **SBEC III**

Self-diagnosis

General information

- Refer to the front of this manual for general test conditions, terminology, detailed descriptions of wiring faults and a general trouble shooter for electrical and mechanical faults.
- Malfunction indicator lamp (MIL) will illuminate or flash if certain faults are recorded.
- ECM operates in backup mode if sensors fail, to enable vehicle to be driven to repair shop.
- Trouble codes can be displayed by the MIL or by using a Scan Tool connected to the data link connector (DLC).

Accessing

Flash codes

- Within 5 seconds switch the ignition key ON-OFF-ON-OFF-ON.
- Count MIL flashes. Note trouble codes.
- Each trouble code consists of two groups **1** [**A**].
- The first group of flashes indicate the 'tens' of the trouble code.
- The second group of flashes indicate the 'units' of the trouble code.
- A short pause separates 'tens' and 'units' **1** [**B**].
- A long pause separates each trouble code **1** [**C**].
- For example: Trouble code 21 displayed **1**.
- End of test sequence indicated by trouble code 55.

P codes & Hex codes

- Trouble codes can be displayed by using a Scan Tool connected to the data link connector (DLC) **2**.

1

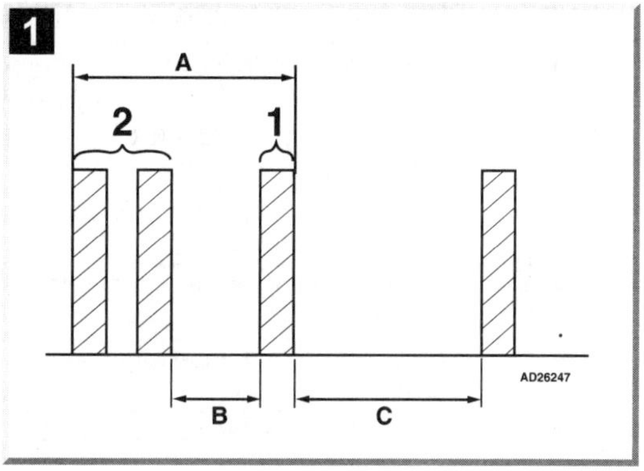

2

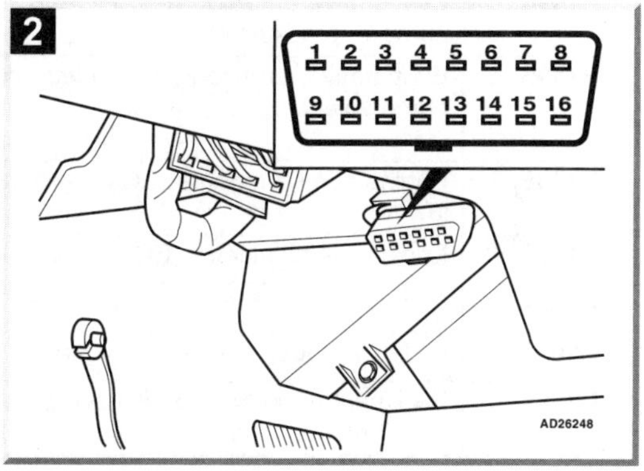

Erasing

- Trouble codes can only be erased by using a Scan Tool connected to the data link connector (DLC).

Trouble code identification

Scan code P type	Fault location	Probable cause
P0000-P0999	Refer to OBD-II trouble code table at the front of this manual	–
P1290	Compressed natural gas pressure sensor – pressure high	**Compressed natural gas system, wiring, compressed natural gas pressure sensor**
P1294	Idle speed control (ISC) system – rpm lower/higher than expected	**Wiring, air leak, throttle valve tight/sticking, IAC valve**
P1297	Manifold absolute pressure (MAP) sensor	**Wiring, hose connection(s), MAP sensor**
P1391	Crankshaft position (CKP) sensor/camshaft position (CMP) sensor – intermittent signal	**Wiring, poor connection, air gap, CKP sensor, CMP sensor**
P1398	Engine control module (ECM) – misfire detection limit reached	**Wiring, poor connection, engine mechanical fault, ignition/fuel system, CKP sensor**
P1486	Evaporative emission (EVAP) system – blockage detected	**Hose connection(s), hose blocked/trapped, EVAP canister**
P1491	Engine coolant blower motor relay – low/high speed	**Wiring open circuit/short circuit, engine coolant blower motor relay – low/high speed**
P1492	Ambient/battery temperature sensor – voltage high	**Wiring, ambient/battery temperature sensor**
P1493	Ambient/battery temperature sensor – voltage low	**Wiring, ambient/battery temperature sensor**
P1494	Evaporative emission (EVAP) leak detection pump – pressure switch	**Wiring, hose leak/blockage, EVAP leak detection pump**
P1495	Evaporative emission (EVAP) leak detection pump solenoid	**Wiring open circuit/short circuit, hose leak/blockage, EVAP leak detection pump solenoid**
P1496	Engine control module (ECM) – 5 volt output low	**ECM**
P1696	Engine control module (ECM) – EEPROM error	**ECM**
P1697	Engine control module (ECM) – EEPROM error	**ECM**
P1698	CAN data bus, TCM – no signal	**Wiring, TCM**
P1899	Park/neutral position (PNP) switch – input circuit malfunction	**Wiring, PNP switch, ECM/TCM**

Trouble code identification

Flash code	Scan code Hex type	Fault location	Probable cause
11		Crankshaft position (CKP) sensor/camshaft position (CMP) sensor	**Wiring, poor connection, air gap, CKP sensor, CMP sensor**
	9D	Crankshaft position (CKP) sensor/camshaft position (CMP) sensor – intermittent signal	**Wiring, poor connection, air gap, CKP sensor, CMP sensor**

Flash code	Scan code Hex type	Fault location	Probable cause
	28	Crankshaft position (CKP) sensor – no signal	**Wiring, air gap, CKP sensor**
	BA	Engine control module (ECM) – misfire detection limit reached	**Wiring, poor connection, engine mechanical fault, ignition/fuel system, CKP sensor**
12		Engine control module (ECM) – supply voltage disconnected within last 50 ignition switch ON-OFF cycles	**Wiring, battery**
13	27	Manifold absolute pressure (MAP) sensor	**Wiring, hose connection(s), MAP sensor**
14		Manifold absolute pressure (MAP) sensor – input low/high	**Wiring, MAP sensor**
	24	Manifold absolute pressure (MAP) sensor – low input	**Wiring short circuit to ground, MAP sensor**
	25	Manifold absolute pressure (MAP) sensor – high input	**Wiring short circuit to positive, MAP sensor**
15	23	Vehicle speed sensor (VSS) – signal	**Wiring, VSS**
17		Engine temperature – low for long period	**Wiring, cooling system, coolant thermostat, ECT sensor**
	21	Engine temperature – low for long period	**Wiring, cooling system, coolant thermostat, ECT sensor**
	80	Insufficient coolant temperature for closed loop fuel control	**Wiring, cooling system, coolant thermostat, ECT sensor**
21		Heated oxygen sensor (HO2S) 1/2, bank 1 – circuit malfunction/incorrect signal/voltage low/high	**Wiring, fuel pressure, injector control, EVAP canister purge valve, compression, air filter blocked, HO2S**
	9B	Heated oxygen sensor (HO2S) 1, bank 1 – low voltage	**Exhaust leak, wiring short circuit to ground, HO2S**
	3E	Heated oxygen sensor (HO2S) 1, bank 1 – high voltage	**Wiring short circuit to positive, fuel pressure, injector control, EVAP canister purge valve, air filter blocked, HO2S**
	66	Heated oxygen sensor (H02S) 1, bank 1 – slow response	**Heating inoperative, wiring, HO2S**
	20	Heated oxygen sensor (H02S) 1, bank 1 – no activity detected	**Wiring open circuit, heating inoperative, HO2S**
	67	Heated oxygen sensor (HO2S) 1, bank 1 – heater control – circuit malfunction	**Wiring, HO2S**
	9C	Heated oxygen sensor (HO2S) 2, bank 1 – low voltage	**Exhaust leak, wiring short circuit to ground, HO2S**
	7E	Heated oxygen sensor (HO2S) 2, bank 1 – high voltage	**Wiring short circuit to positive, catalytic converter, HO2S**
	81	Heated oxygen sensor (HO2S) 2, bank 1 – no activity detected	**Wiring, heating inoperative, HO2S**
	69	Heated oxygen sensor (HO2S) 2, bank 1, heater control – circuit malfunction	**Wiring, HO2S**

Flash code	Scan code Hex type	Fault location	Probable cause
	C0	Heated oxygen sensor (HO2S) 1, bank 1 – slow response	**Heating inoperative, wiring, HO2S**
22		Engine coolant temperature (ECT) sensor – voltage low/high	**Wiring, coolant thermostat, ECT sensor**
	1E	Engine coolant temperature (ECT) sensor – low input	**Coolant thermostat, wiring short circuit to ground, ECT sensor**
	1F	Engine coolant temperature (ECT) sensor – high input	**Coolant thermostat, wiring open circuit/ short circuit to positive, ground wire defective, ECT sensor**
23		Intake air temperature (IAT) sensor – voltage low/high	**Wiring, IAT sensor**
	39	Intake air temperature (IAT) sensor – low input	**Wiring short circuit to ground, IAT sensor**
	3A	Intake air temperature (IAT) sensor – high input	**Wiring open circuit/short circuit to positive, ground wire defective, IAT sensor**
24		Throttle position (TP) sensor – voltage low/ high	**Wiring, accelerator cable adjustment, TP sensor**
	84	Throttle position (TP) sensor – range/ performance problem	**Wiring, accelerator cable adjustment, TP sensor**
	1A	Throttle position (TP) sensor – low input	**Wiring short circuit to ground, TP sensor**
	1B	Throttle position (TP) sensor – high input	**Wiring short circuit to positive, TP sensor**
25		Idle air control (IAC) valve	**Wiring open circuit/short circuit, IAC valve**
	19	Idle speed control (ISC) system – malfunction	**Wiring open circuit/short circuit, IAC valve**
	8A	Idle speed control (ISC) system – rpm lower/higher than expected	**Wiring, air leak, throttle valve tight/ sticking, IAC valve**
27		Injector(s) – circuit malfunction	**Wiring, injector**
	15	Injector 1 – circuit malfunction	**Wiring, injector**
	14	Injector 2 – circuit malfunction	**Wiring, injector**
	13	Injector 3 – circuit malfunction	**Wiring, injector**
	3D	Injector 4 – circuit malfunction	**Wiring, injector**
	45	Injector 5 – circuit malfunction	**Wiring, injector**
	46	Injector 6 – circuit malfunction	**Wiring, injector**
31		Evaporative emission (EVAP) canister purge valve – open/short circuit	**Wiring, EVAP canister purge valve**
	71	Evaporative emission (EVAP) system – incorrect flow detected	**Hose connection(s), intake leak, EVAP canister purge valve**
	A0	Evaporative emission (EVAP) system – small leak detected	**Hose connection(s), intake leak, EVAP canister, EVAP canister purge valve**

Flash code	Scan code Hex type	Fault location	Probable cause
	12	Evaporative emission (EVAP) canister purge valve – circuit malfunction	**Wiring open circuit/short circuit, EVAP canister purge valve**
	A1	Evaporative emission (EVAP) system – large leak detected	**Hose connection(s), intake leak, EVAP canister, EVAP canister purge valve**
	BB	Evaporative emission (EVAP) system – blockage detected	**Hose connection(s), hose blocked/ trapped, EVAP canister**
	B8	Evaporative emission (EVAP) leak detection pump – pressure switch	**Wiring, hose leak/blockage, EVAP leak detection pump**
	B7	Evaporative emission (EVAP) leak detection pump solenoid	**Wiring open circuit/short circuit, hose leak/blockage, EVAP leak detection pump solenoid**
32		Exhaust gas recirculation (EGR) solenoid/ system – circuit malfunction	**Wiring, hoses, EGR solenoid**
	2E	Exhaust gas recirculation (EGR) system – insufficient flow detected	**Hose leak/blockage, wiring, EGR valve, EGR solenoid**
	11	Exhaust gas recirculation (EGR) – circuit malfunction	**Wiring open circuit/short circuit, EGR solenoid**
33		A/C compressor clutch relay/A/C refrigerant pressure sensor – voltage low/high	**A/C refrigerant pressure too low/high (cooling fault/incorrectly charged), wiring open circuit/short circuit, A/C compressor clutch relay, A/C refrigerant pressure sensor**
	10	A/C compressor clutch relay	**Wiring open circuit/short circuit, A/C compressor clutch relay**
	5A	A/C refrigerant pressure sensor – high input	**A/C refrigerant pressure too high (cooling fault/incorrectly charged), wiring, A/C refrigerant pressure sensor**
	5B	A/C refrigerant pressure sensor – low input	**A/C refrigerant pressure too low (incorrectly charged), wiring, A/C refrigerant pressure sensor**
34	0F	Cruise control	**Wiring open circuit/short circuit, hoses, voltage low/high, cruise control master switch, cruise control vacuum switching valve**
35	0E	Engine coolant blower motor relay – low/ high speed	**Wiring open circuit/short circuit, engine coolant blower motor relay – low/high speed**
37		Torque converter clutch (TCC) solenoid/ park/neutral position (PNP) switch – circuit malfunction	**Wiring, TCC solenoid**
	72	Park/neutral position (PNP) switch – input circuit malfunction	**Wiring, PNP switch, ECM/TCM**
	94	Torque converter clutch (TCC) solenoid – circuit malfunction	**Wiring, TCC solenoid**
	0C	Torque converter clutch (TCC) solenoid – electrical	**Wiring open circuit/short circuit, TCC solenoid**

Flash code	Scan code Hex type	Fault location	Probable cause
41	0B	Alternator, field control – circuit malfunction	**Wiring open circuit/short circuit, alternator, battery**
42		Fuel pump relay/engine control relay/fuel level sensor – circuit malfunction	**Wiring open circuit/short circuit, fuel pump relay, engine control relay, fuel level sensor**
	65	Fuel pump relay	**Wiring open circuit/short circuit, fuel pump relay**
	0A	Engine control relay	**Wiring open circuit/short circuit, engine control relay**
	2C	Engine control relay	**Wiring open circuit, engine control relay**
	95	Fuel level sensor – low input	**Wiring open circuit, fuel level sensor**
	96	Fuel level sensor – high input	**Wiring short circuit to positive, fuel level sensor**
	97	Fuel level sensor – malfunction	**Wiring, fuel level sensor**
43		Random/multiple cylinder(s) – misfire detected	**Wiring, spark plug(s), HT lead(s), injector(s), ignition coil(s), low compression**
	6A	Random/multiple cylinder(s) – misfire detected	**Wiring, spark plug(s), HT lead(s), injector(s), ignition coil(s), low compression**
	6B	Cylinder 1 – misfire detected	**Wiring, engine mechanical fault, ignition/fuel system, injector**
	6C	Cylinder 2 – misfire detected	**Wiring, engine mechanical fault, ignition/fuel system, injector**
	6D	Cylinder 3 – misfire detected	**Wiring, engine mechanical fault, ignition/fuel system, injector**
	6E	Cylinder 4 – misfire detected	**Wiring, engine mechanical fault, ignition/fuel system, injector**
	AE	Cylinder 5 – misfire detected	**Wiring, engine mechanical fault, ignition/fuel system, injector**
	AF	Cylinder 6 – misfire detected	**Wiring, engine mechanical fault, ignition/fuel system, injector**
	2B	Ignition coil A, primary/secondary – circuit malfunction	**Wiring, ignition coil**
	2A	Ignition coil B, primary/secondary – circuit malfunction	**Wiring, ignition coil**
	29	Ignition coil C, primary/secondary – circuit malfunction	**Wiring, ignition coil**
44		Ambient/battery temperature sensor – voltage low/high	**Wiring, ambient/battery temperature sensor**
	9A	Ambient/battery temperature sensor – voltage high	**Wiring, ambient/battery temperature sensor**
	99	Ambient/battery temperature sensor – voltage low	**Wiring, ambient/battery temperature sensor**

Flash code	Scan code Hex type	Fault location	Probable cause
45	89	Transmission control system – malfunction	Wiring, ECM/TCM
46	06	Alternator – voltage high	Wiring, battery, alternator
47	05	Alternator – voltage low	Wiring, battery, alternator drive belt, alternator
51	77	System too lean	Wiring, intake/exhaust leak, MAP sensor, fuel pressure/pump, EVAP canister purge valve, injector(s), HO2S
52	76	System too rich	Wiring, intake blocked, EVAP canister purge valve, fuel pressure, EGR system, injector(s), air filter blocked, ignition system, HO2S
53		Engine control module (ECM) – internal fault	ECM
	02	Engine control module (ECM) – internal fault	ECM
	44	Engine control module (ECM) – internal fault	ECM
54	01	Camshaft position (CMP) sensor – no signal	Wiring, CMP sensor
55		End of MIL code output	–
61	3C	Manifold absolute pressure (MAP) sensor – barometric pressure range/ performance problem	Intake/exhaust leak, wiring, MAP sensor
62	30	Engine control module (ECM) – EEPROM error	ECM
63	31	Engine control module (ECM) – EEPROM error	ECM
64	BE	Compressed natural gas pressure sensor – pressure high	Compressed natural gas system, wiring, compressed natural gas pressure sensor
65	98	Brake pedal position (BPP) switch – circuit malfunction	Wiring, BPP switch
66		CAN data bus, transmission control module (TCM)/body control module (BCM) – no signal	Wiring, TCM, BCM
	60	CAN data bus, transmission control module (TCM) – no signal	Wiring, TCM
	61	CAN data bus, body control module (BCM) – no signal	Wiring, BCM
71	92	Engine control module (ECM) – 5 volt output low	ECM
72	70	Catalytic converter system, bank 1 – efficiency below threshold	Catalytic converter, wiring, HO2S 2
77	52	Cruise control	Wiring, cruise control vacuum switching valve

CHRYSLER

Model:	Engine identification:	Year:
Town & Country 3.3L	EGA/EGM	1998
Town & Country 3.8L	EGH	1998

System: **SBEC III**

Self-diagnosis

General information

- Refer to the front of this manual for general test conditions, terminology, detailed descriptions of wiring faults and a general trouble shooter for electrical and mechanical faults.
- Malfunction indicator lamp (MIL) will illuminate or flash if certain faults are recorded.
- ECM operates in backup mode if sensors fail, to enable vehicle to be driven to repair shop.

Accessing

- Trouble codes can only be displayed by using a Scan Tool connected to the data link connector (DLC) **1**.

1

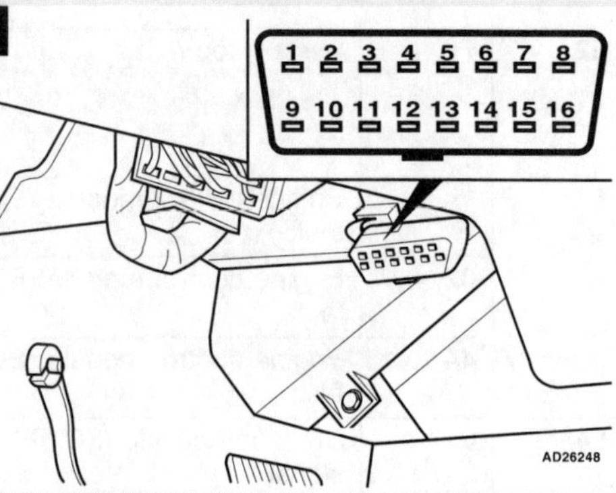

Erasing

- Trouble codes can only be erased by using a Scan Tool connected to the data link connector (DLC).

Trouble code identification

Scan code P type	Fault location	Probable cause
P0000-P0999	Refer to OBD-II trouble code table at the front of this manual	–
P1290	Compressed natural gas pressure sensor – pressure high	**Compressed natural gas system, wiring, compressed natural gas pressure sensor**
P1294	Idle speed control (ISC) system – rpm lower/higher than expected	**Wiring, air leak, throttle valve tight/sticking, IAC valve**
P1391	Crankshaft position (CKP) sensor/camshaft position (CMP) sensor – intermittent signal	**Wiring, poor connection, air gap, CKP sensor, CMP sensor**
P1398	Engine control module (ECM) – misfire detection limit reached	**Wiring, poor connection, CKP sensor**
P1486	Evaporative emission (EVAP) system – blockage detected	**Hose connection(s), hose blocked/trapped, EVAP canister**
P1491	Engine coolant blower motor relay – low/high speed	**Wiring open circuit/short circuit, engine coolant blower motor relay – low/high speed**
P1492	Ambient/battery temperature sensor – voltage high	**Wiring, ambient/battery temperature sensor**
P1493	Ambient/battery temperature sensor – voltage low	**Wiring, ambient/battery temperature sensor**

Scan code P type	Fault location	Probable cause
P1494	Evaporative emission (EVAP) leak detection pump – pressure switch	Wiring, hose leak/blockage, EVAP leak detection pump
P1495	Evaporative emission (EVAP) leak detection pump solenoid	Wiring open circuit/short circuit, hose leak/ blockage, EVAP leak detection pump solenoid
P1496	Engine control module (ECM) – 5 volt output low	ECM
P1696	Engine control module (ECM) – EEPROM error	ECM
P1697	Engine control module (ECM) – EEPROM error	ECM
P1698	CAN data bus, transmission control module (TCM) – no signal	Wiring, TCM
P1899	Park/neutral position (PNP) switch – input circuit malfunction	Wiring, PNP switch, ECM/TCM

Scan code Hex type	Fault location	Probable cause
01	Camshaft position (CMP) sensor – no signal	Wiring, CMP sensor
02	Engine control module (ECM) – internal fault	ECM
05	Alternator – voltage low	Wiring, battery, alternator drive belt, alternator
06	Alternator – voltage high	Wiring, battery, alternator
0A	Engine control relay	Wiring open circuit/short circuit, engine control relay
0B	Alternator, field control – circuit malfunction	Wiring open circuit/short circuit, alternator, battery
0C	Torque converter clutch (TCC) solenoid – electrical	Wiring open circuit/short circuit, TCC solenoid
0E	Engine coolant blower motor relay – low/high speed	Wiring open circuit/short circuit, engine coolant blower motor relay – low/high speed
0F	Cruise control	Wiring open circuit/short circuit, hoses, voltage low/high, cruise control master switch, cruise control vacuum switching valve
10	A/C compressor clutch relay	Wiring open circuit/short circuit, A/C compressor clutch relay
11	Exhaust gas recirculation (EGR) – circuit malfunction	Wiring open circuit/short circuit, EGR solenoid
12	Evaporative emission (EVAP) canister purge valve – circuit malfunction	Wiring open circuit/short circuit, EVAP canister purge valve
13	Injector 3 – circuit malfunction	Wiring, injector
14	Injector 2 – circuit malfunction	Wiring, injector
15	Injector 1 – circuit malfunction	Wiring, injector
19	Idle speed control (ISC) system – malfunction	Wiring open circuit/short circuit, IAC valve
1A	Throttle position (TP) sensor – low input	Wiring short circuit to ground, TP sensor
1B	Throttle position (TP) sensor – high input	Wiring short circuit to positive, TP sensor

Scan code Hex type	Fault location	Probable cause
1E	Engine coolant temperature (ECT) sensor – low input	Coolant thermostat, wiring short circuit to ground, ECT sensor
1F	Engine coolant temperature (ECT) sensor – high input	Coolant thermostat, wiring open circuit/ short circuit to positive, ground wire defective, ECT sensor
20	Heated oxygen sensor (H02S) 1, bank 1 – no activity detected	Wiring open circuit, heating inoperative, HO2S
21	Engine temperature – low for long period	Wiring, cooling system, coolant thermostat, ECT sensor
23	Vehicle speed sensor (VSS) – signal	Wiring, VSS
24	Manifold absolute pressure (MAP) sensor – low input	Wiring short circuit to ground, MAP sensor
25	Manifold absolute pressure (MAP) sensor – high input	Wiring short circuit to positive, MAP sensor
27	Manifold absolute pressure (MAP) sensor	Wiring, hose connection(s), MAP sensor
28	Crankshaft position (CKP) sensor – no signal	Wiring, air gap, CKP sensor
29	Ignition coil C, primary/secondary – circuit malfunction	Wiring, ignition coil
2A	Ignition coil B, primary/secondary – circuit malfunction	Wiring, ignition coil
2B	Ignition coil A, primary/secondary – circuit malfunction	Wiring, ignition coil
2C	Engine control relay	Wiring open circuit, engine control relay
2E	Exhaust gas recirculation (EGR) system – insufficient flow detected	Hose leak/blockage, wiring, EGR valve, EGR solenoid
30	Engine control module (ECM) – EEPROM error	ECM
31	Engine control module (ECM) – EEPROM error	ECM
39	Intake air temperature (IAT) sensor – low input	Wiring short circuit to ground, IAT sensor
3A	Intake air temperature (IAT) sensor – high input	Wiring open circuit/short circuit to positive, ground wire defective, IAT sensor
3C	Manifold absolute pressure (MAP) sensor – barometric pressure range/performance problem	Intake/exhaust leak, wiring, MAP sensor
3D	Injector 4 – circuit malfunction	Wiring, injector
3E	Heated oxygen sensor (HO2S) 1, bank 1 – high voltage	Wiring short circuit to positive, fuel pressure, injector control, EVAP canister purge valve, air filter blocked, HO2S
44	Engine control module (ECM) – internal fault	ECM
45	Injector 5 – circuit malfunction	Wiring, injector
46	Injector 6 – circuit malfunction	Wiring, injector
52	Cruise control	Wiring, cruise control vacuum switching valve

Scan code Hex type	Fault location	Probable cause
5A	A/C refrigerant pressure sensor – high input	**A/C refrigerant pressure too high (cooling fault/incorrectly charged), wiring, A/C refrigerant pressure sensor**
5B	A/C refrigerant pressure sensor – low input	**A/C refrigerant pressure too low (incorrectly charged), wiring, A/C refrigerant pressure sensor**
60	CAN data bus, transmission control module (TCM) – no signal	**Wiring, TCM**
61	CAN data bus, body control module (BCM) – no signal	**Wiring, BCM**
65	Fuel pump relay	**Wiring open circuit/short circuit, fuel pump relay**
66	Heated oxygen sensor (H02S) 1, bank 1 – slow response	**Heating inoperative, wiring, HO2S**
67	Heated oxygen sensor (HO2S) 1, bank 1 – heater control – circuit malfunction	**Wiring, HO2S**
69	Heated oxygen sensor (HO2S) 2, bank 1, heater control – circuit malfunction	**Wiring, HO2S**
6A	Random/multiple cylinder(s) – misfire detected	**Wiring, spark plug(s), HT lead(s), injector(s), ignition coil(s), low compression**
6B	Cylinder 1 – misfire detected	**Wiring, engine mechanical fault, ignition/fuel system, injector**
6C	Cylinder 2 – misfire detected	**Wiring, engine mechanical fault, ignition/fuel system, injector**
6D	Cylinder 3 – misfire detected	**Wiring, engine mechanical fault, ignition/fuel system, injector**
6E	Cylinder 4 – misfire detected	**Wiring, engine mechanical fault, ignition/fuel system, injector**
70	Catalytic converter system, bank 1 – efficiency below threshold	**Catalytic converter, wiring, heated oxygen sensor (HO2S) 2**
71	Evaporative emission (EVAP) system – incorrect flow detected	**Hose connection(s), intake leak, EVAP canister purge valve**
72	Park/neutral position (PNP) switch – input circuit malfunction	**Wiring, PNP switch, ECM/TCM**
76	System too rich	**Wiring, intake blocked, EVAP canister purge valve, fuel pressure, EGR system, injector(s), air filter blocked, ignition system, HO2S**
77	System too lean	**Wiring, intake/exhaust leak, MAP sensor, fuel pressure/pump, EVAP canister purge valve, injector(s), HO2S**
7E	Heated oxygen sensor (HO2S) 2, bank 1 – high voltage	**Wiring short circuit to positive, catalytic converter, HO2S**
80	Insufficient coolant temperature for closed loop fuel control	**Wiring, cooling system, coolant thermostat, ECT sensor**
81	Heated oxygen sensor (HO2S) 2, bank 1 – no activity detected	**Wiring, heating inoperative, HO2S**

Scan code Hex type	Fault location	Probable cause
84	Throttle position (TP) sensor – range/ performance problem	**Wiring, accelerator cable adjustment, TP sensor**
89	Transmission control system – malfunction	**Wiring, ECM/TCM**
8A	Idle speed control (ISC) system – rpm lower/ higher than expected	**Wiring, air leak, throttle valve tight/sticking, IAC valve**
92	Engine control module (ECM) – 5 volt output low	**ECM**
94	Torque converter clutch (TCC) solenoid – circuit malfunction	**Wiring, TCC solenoid**
95	Fuel level sensor – low input	**Wiring open circuit, fuel level sensor**
96	Fuel level sensor – high input	**Wiring short circuit to positive, fuel level sensor**
97	Fuel level sensor – malfunction	**Wiring, fuel level sensor**
98	Brake pedal position (BPP) switch – circuit malfunction	**Wiring, BPP switch**
99	Ambient/battery temperature sensor – voltage low	**Wiring, ambient/battery temperature sensor**
9A	Ambient/battery temperature sensor – voltage high	**Wiring, ambient/battery temperature sensor**
9B	Heated oxygen sensor (HO2S) 1, bank 1 – low voltage	**Exhaust leak, wiring short circuit to ground, HO2S**
9C	Heated oxygen sensor (HO2S) 2, bank 1 – low voltage	**Exhaust leak, wiring short circuit to ground, HO2S**
9D	Crankshaft position (CKP) sensor/camshaft position (CMP) sensor – intermittent signal	**Wiring, poor connection, air gap, CKP sensor, CMP sensor**
A0	Evaporative emission (EVAP) system – small leak detected	**Hose connection(s), intake leak, EVAP canister, EVAP canister purge valve**
A1	Evaporative emission (EVAP) system – large leak detected	**Hose connection(s), intake leak, EVAP canister, EVAP canister purge valve**
AE	Cylinder 5 – misfire detected	**Wiring, engine mechanical fault, ignition/fuel system, injector**
AF	Cylinder 6 – misfire detected	**Wiring, engine mechanical fault, ignition/fuel system, injector**
B7	Evaporative emission (EVAP) leak detection pump solenoid	**Wiring open circuit/short circuit, hose leak/ blockage, EVAP leak detection pump solenoid**
B8	Evaporative emission (EVAP) leak detection pump – pressure switch	**Wiring, hose leak/blockage, EVAP leak detection pump**
BA	Engine control module (ECM) – misfire detection limit reached	**Wiring, poor connection, CKP sensor**
BB	Evaporative emission (EVAP) system – blockage detected	**Hose connection(s), hose blocked/trapped, EVAP canister**
BE	Compressed natural gas pressure sensor – pressure high	**Compressed natural gas system, wiring, compressed natural gas pressure sensor**
C0	Heated oxygen sensor (HO2S) 1, bank 1 – slow response	**Heating inoperative, wiring, HO2S**

Model:	Engine identification:	Year:
Voyager 2.4L	EDZ	2000-02
Voyager 3.3L	EGA/EGM	2000-02
Town & Country 3.3L	EGA/EGM	1999-02
Town & Country 3.8L	EGH	1999-02

System: **SBEC III**

Self-diagnosis

General information

- Refer to the front of this manual for general test conditions, terminology, detailed descriptions of wiring faults and a general trouble shooter for electrical and mechanical faults.
- Malfunction indicator lamp (MIL) will illuminate or flash if certain faults are recorded.
- ECM operates in backup mode if sensors fail, to enable vehicle to be driven to repair shop.

Accessing

1999-01

- Trouble codes can only be displayed by using a Scan Tool connected to the data link connector (DLC) 1999-00 **1** & 2000→ **2**.

1

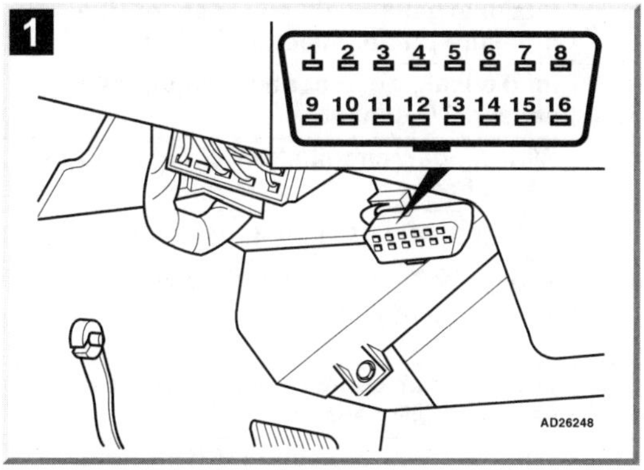

2

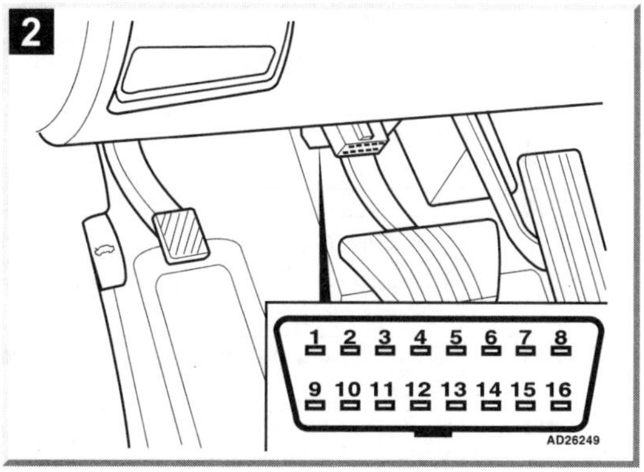

2001→

- Within 5 seconds switch the ignition key ON-OFF-ON-OFF-ON.
- Odometer will display trouble codes.
- If no trouble codes are stored in ECM, odometer will display 'P1684' and 'done' or only 'done'.
- Trouble codes can also be displayed by using a Scan Tool connected to the data link connector (DLC) **2**.

Erasing

- Trouble codes can only be erased by using a Scan Tool connected to the data link connector (DLC).

Trouble code identification

Scan code	Fault location	Probable cause
P0000-P0999	Refer to OBD-II trouble code table at the front of this manual	–
P1192	Intake air temperature (IAT) sensor – voltage low	**Wiring, IAT sensor**
P1193	Intake air temperature (IAT) sensor – voltage high	**Wiring, IAT sensor**
P1195	Heated oxygen sensor (H02S) 1, bank 1 – slow response	**Heating inoperative, wiring, HO2S**
P1196	Heated oxygen sensor (H02S) 2, bank 1 – slow response	**Heating inoperative, wiring, HO2S**
P1281	Engine temperature – low for long period	**Wiring, cooling system, coolant thermostat, ECT sensor**
P1282	Fuel pump relay	**Wiring open circuit/short circuit, fuel pump relay**
P1290	Compressed natural gas pressure sensor – pressure high	**Compressed natural gas system, wiring, compressed natural gas pressure sensor**
P1292	Compressed natural gas pressure sensor – voltage high	**Wiring, compressed natural gas pressure sensor**
P1293	Compressed natural gas pressure sensor – voltage low	**Wiring, compressed natural gas pressure sensor**
P1294	Idle speed control (ISC) system – rpm lower/ higher than expected	**Wiring, air leak, throttle valve tight/sticking, IAC valve**
P1295	Throttle position (TP) sensor – supply voltage	**Wiring, TP sensor, ECM**
P1296	Manifold absolute pressure (MAP) sensor – supply voltage	**Wiring, MAP sensor, ECM**
P1297	Manifold absolute pressure (MAP) sensor	**Wiring, hose connection(s), MAP sensor**
P1298	System too lean – at prolonged wide open throttle	**Intake/exhaust leak, MAP sensor, fuel pressure/pump, injector(s), HO2S**
P1299	Manifold absolute pressure (MAP) sensor/ throttle position (TP) sensor	**Intake leak, hose connection(s), MAP sensor, TP sensor**
P1388	Engine control relay	**Wiring open circuit/short circuit, engine control relay**
P1389	Engine control relay	**Wiring open circuit, engine control relay**
P1390	Crankshaft position (CKP) sensor/camshaft position (CMP) sensor – signal synchronisation	**Timing belt loose/jumped position, CKP sensor, CMP sensor**
P1391	Crankshaft position (CKP) sensor/camshaft position (CMP) sensor – intermittent signal	**Wiring, poor connection, air gap, CKP sensor, CMP sensor**
P1398	Engine control module (ECM) – misfire detection limit reached	**Wiring, poor connection, CKP sensor**
P1486	Evaporative emission (EVAP) system – blockage detected	**Hose connection(s), hose blocked/trapped, EVAP canister**
P1488	Engine control module (ECM) – auxiliary 5 volt output low	**Wiring, ECM**

Scan code	Fault location	Probable cause
P1489	Engine coolant blower motor relay	Wiring open circuit/short circuit, engine coolant blower motor relay – high speed
P1490	Engine coolant blower motor relay	Wiring open circuit/short circuit, engine coolant blower motor relay – low speed
P1491	Engine coolant blower motor relay – circuit malfunction	Wiring open circuit/short circuit, engine coolant blower motor relay
P1492	Ambient/battery temperature sensor – high input	Wiring short circuit to positive, ambient/ battery temperature sensor, ECM
P1493	Ambient/battery temperature sensor – voltage low	Wiring, ambient/battery temperature sensor
P1494	Evaporative emission (EVAP) leak detection pump – pressure switch	Wiring, hose leak/blockage, EVAP leak detection pump
P1495	Evaporative emission (EVAP) leak detection pump solenoid	Wiring open circuit/short circuit, hose leak/ blockage, EVAP leak detection pump solenoid
P1496	Engine control module (ECM) – 5 volt output low	ECM
P1594	Alternator – voltage high	Wiring, battery, alternator
P1595	Cruise control	Wiring open circuit/short circuit, hoses, cruise control master switch, cruise control vacuum switching valve
P1596	Cruise control master switch – high voltage	Wiring, cruise control master switch
P1597	Cruise control master switch – low voltage	Wiring, cruise control master switch
P1598	A/C refrigerant pressure sensor – high input	Wiring, A/C refrigerant pressure too high (cooling fault/incorrectly charged), A/C refrigerant pressure sensor
P1599	A/C refrigerant pressure sensor – low input	Wiring, A/C refrigerant pressure too low (incorrectly charged), A/C refrigerant pressure sensor
P1602	Engine control module (ECM) – not programmed	New ECM not programmed, incorrectly programmed, ECM
P1603	Engine control module (ECM) – RAM link error	ECM
P1604	Engine control module (ECM) – RAM read/ write error	ECM
P1607	Engine control module (ECM) – shut down timer error	ECM
P1681	CAN data bus, instrumentation – no signal	Wiring, instrumentation control module
P1682	Alternator – voltage low	Wiring, battery, alternator drive belt, alternator
P1683	Cruise control	Wiring, cruise control vacuum switching valve
P1684	Engine control module (ECM) – supply voltage disconnected within last 50 ignition switch ON-OFF cycles	Wiring, battery

Scan code	Fault location	Probable cause
P1685	Key not programmed into immobilizer control module	**Wiring, ignition key not matched/damaged, immobilizer control module**
P1686	CAN data bus, Immobilizer control module – no signal	**Wiring, immobilizer control module**
P1687	CAN data bus, instrumentation – no signal	**Wiring, instrumentation control module mechanical**
P1695	CAN data bus, body control module (BCM) – no signal	**Wiring, BCM**
P1696	Engine control module (ECM) – EEPROM error	**ECM**
P1697	Engine control module (ECM) – EEPROM error	**ECM**
P1698	CAN data bus, transmission control module (TCM) – no signal	**Wiring, TCM**
P1765	Transmission system relay	**Wiring open circuit/short circuit, transmission system relay, ECM/TCM**
P1899	Park/neutral position (PNP) switch – input circuit malfunction	**Wiring, PNP switch, ECM/TCM**
P2503	Charging system voltage – low	**Wiring, poor connection, battery, alternator**

DODGE

Model:	Engine identification:	Year:
Avenger 2.0L	**420A**	**1995-96**
Avenger 2.5L	**6G73**	**1996**

System: **SBEC III/Mitsubishi MPI**

Self-diagnosis

General information

- Refer to the front of this manual for general test conditions, terminology, detailed descriptions of wiring faults and a general trouble shooter for electrical and mechanical faults.
- Malfunction indicator lamp (MIL) will illuminate or flash if certain faults are recorded.
- ECM operates in backup mode if sensors fail, to enable vehicle to be driven to repair shop.
- Trouble codes can be displayed by the MIL or by using a Scan Tool connected to the data link connector(s) (DLC).

Accessing

Flash codes

- Within 5 seconds switch the ignition key ON-OFF-ON-OFF-ON.
- Count MIL flashes. Note trouble codes.
- Each trouble code consists of two groups **1** [**A**].
- The first group of flashes indicate the 'tens' of the trouble code.
- The second group of flashes indicate the 'units' of the trouble code.
- A short pause separates 'tens' and 'units' **1** [**B**].
- A long pause separates each trouble code **1** [**C**].
- For example: Trouble code 21 displayed **1**.

P codes & DRB codes

- Trouble codes can be displayed by using a Scan Tool connected to the data link connector(s) (DLC) **2**.

1

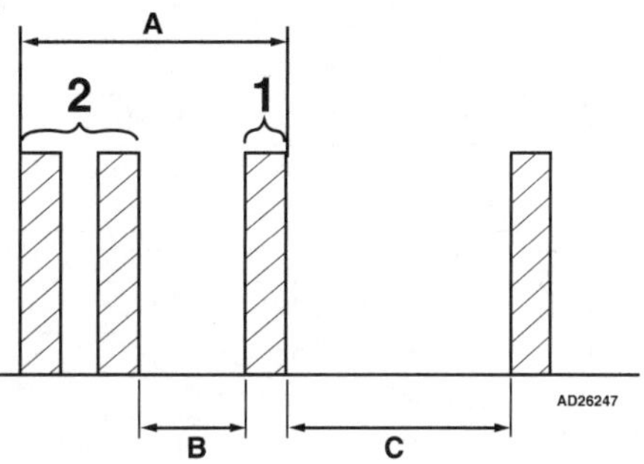

2

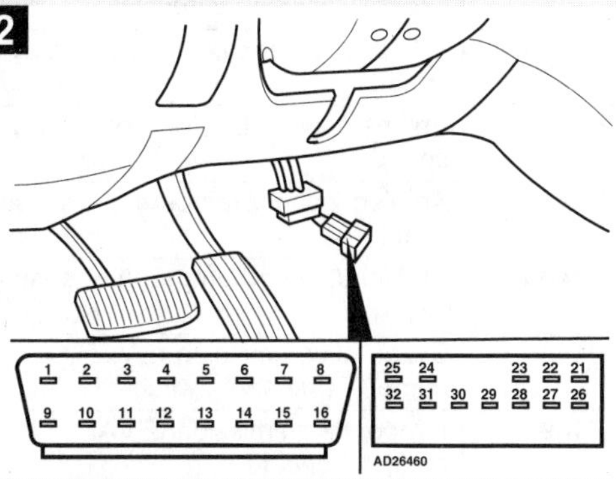

Erasing

NOTE: ***Disconnecting battery ground lead may cause driveability problems until the engine management system relearns basic values.***

- Ensure ignition switched OFF.
- Disconnect battery ground cable for at least 5 minutes.
- Reconnect battery ground cable.
- Repeat accessing procedure to ensure no trouble codes remain in ECM fault memory.

NOTE: ***Disconnecting battery ground lead may also erase memory from electronic units, such as radio.***

- Trouble codes can also be erased by using a Scan Tool connected to the data link connector(s) (DLC).

Trouble code identification

Scan code P type	Fault location	Probable cause
P0000-P0999	Refer to OBD-II trouble code table at front of this manual	–
P1294	Idle speed control (ISC) system – rpm lower/ higher than expected	**Wiring, air leak, throttle valve tight/sticking, IAC valve**
P1295	Throttle position (TP) sensor – no 5 volt supply	**Wiring, TP sensor, ECM**
P1296	Manifold absolute pressure (MAP) sensor – no 5 volt supply	**Wiring, ECM**
P1297	Manifold absolute pressure (MAP) sensor	**Wiring, hose connection(s), MAP sensor**
P1390	Timing belt jumped 1 tooth or more	**Timing belt loose/jumped position**
P1391	Crankshaft position (CKP) sensor/camshaft position (CMP) sensor – intermittent signal	**Wiring, poor connection, air gap, CKP sensor, CMP sensor**
P1486	Evaporative emission (EVAP) system – blockage detected	**Hose connection(s), hose blocked/trapped, EVAP canister**
P1487	Engine coolant blower motor relay – high speed – circuit malfunction	**Wiring open circuit/short circuit, engine coolant blower motor relay – high speed**
P1489	A/C condenser blower motor relay – high speed	**Wiring open/short circuit, engine coolant blower motor relay – high speed**
P1490	Engine coolant blower motor relay – low speed	**Wiring open/short circuit, engine coolant blower motor relay – low speed**
P1492	Ambient/battery temperature sensor – voltage high	**Wiring, ambient/battery temperature sensor**
P1493	Ambient/battery temperature sensor – voltage low	**Wiring, ambient/battery temperature sensor**
P1494	Evaporative emission (EVAP) leak detection pump – pressure switch	**Wiring, hose leak/blockage, EVAP leak detection pump**
P1495	Evaporative emission (EVAP) leak detection pump solenoid	**Wiring open/short circuit, hose leak/ blockage, EVAP leak detection pump solenoid**
P1496	Manifold absolute pressure (MAP) sensor – 5 volt output low	**Wiring, MAP sensor**
P1696	Engine control module (ECM) – EEPROM error	**ECM**
P1697	Engine control module (ECM) – EEPROM error	**ECM**
P1698	CAN data bus, TCM – no signal	**Wiring, TCM**
P1899	Park/neutral position (PNP) switch – input circuit malfunction	**Wiring, PNP switch, ECM/TCM**

Flash code	Scan code DRB III type	Fault location	Probable cause
11		Crankshaft position (CKP) sensor/camshaft position (CMP) sensor	**Wiring, poor connection, air gap, timing belt, CKP sensor, CMP sensor**
	133	Timing belt jumped 1 tooth or more	**Timing belt loose/jumped position**
	157	Crankshaft position (CKP) sensor/camshaft position (CMP) sensor – intermittent signal	**Wiring, poor connection, air gap, CKP sensor, CMP sensor**
	40	Crankshaft position (CKP) sensor – no signal	**Wiring, air gap, CKP sensor**
13	39	Manifold absolute pressure (MAP) sensor	**Wiring, hose connection(s), MAP sensor**
14		Manifold absolute pressure (MAP) sensor – input low/high	**Wiring, MAP sensor**
	36	Manifold absolute pressure (MAP) sensor – low input	**Wiring short circuit to ground, MAP sensor**
	37	Manifold absolute pressure (MAP) sensor – high input	**Wiring short circuit to positive, MAP sensor**
	135	Manifold absolute pressure (MAP) sensor – no 5 volt supply	**Wiring, ECM**
	146	Manifold absolute pressure (MAP) sensor – 5 volt output low	**Wiring, MAP sensor**
15	35	Vehicle speed sensor (VSS) – signal	**Wiring, VSS**
16	59	Knock sensor (KS) 1 – no signal	**Wiring, KS incorrectly tightened, KS**
17		Engine temperature – low for long period	**Wiring, cooling system, coolant thermostat, ECT sensor**
	33	Engine temperature – low for long period	**Wiring, cooling system, coolant thermostat, ECT sensor**
	128	Insufficient coolant temperature for closed loop fuel control	**Wiring, cooling system, coolant thermostat, ECT sensor**
21		Heated oxygen sensor (HO2S) 1/2, bank 1/2 – circuit malfunction/incorrect signal/ voltage low/high	**Wiring, fuel pressure, injector control, EVAP canister purge valve, compression, air filter blocked, HO2S**
	155	Heated oxygen sensor (HO2S) 1, bank 1 – low voltage	**Wiring short circuit to ground, exhaust leak, HO2S**
	62	Heated oxygen sensor (HO2S) 1, bank 1 – high voltage	**Wiring short circuit to positive, fuel pressure, injector control, EVAP canister purge valve, air filter blocked, HO2S**
	102	Heated oxygen sensor (H02S) 1, bank 1 – slow response	**Wiring, heating inoperative, HO2S**
	32	Heated oxygen sensor (H02S) 1, bank 1 – no activity detected	**Wiring open circuit, heating inoperative, HO2S**
	103	Heated oxygen sensor (HO2S) 1, bank 1 – heater control – circuit malfunction	**Wiring, HO2S**
	156	Heated oxygen sensor (HO2S) 2, bank 1 – low voltage	**Wiring short circuit to ground, exhaust leak, HO2S**
	126	Heated oxygen sensor (HO2S) 2, bank 1 – high voltage	**Wiring short circuit to positive, catalytic converter, HO2S**
	129	Heated oxygen sensor (HO2S) 2, bank 1 – no activity detected	**Wiring, heating inoperative, HO2S**

Flash code	Scan code DRB III type	Fault location	Probable cause
	105	Heated oxygen sensor (HO2S) 2, bank 1, heater control – circuit malfunction	**Wiring, HO2S**
	104	Heated oxygen sensor (HO2S) 2, bank 1 – slow response	**Wiring, heating inoperative, HO2S**
	66	Heated oxygen sensor (HO2S) 1, bank 2 – high voltage	**Wiring short to positive, HO2S, ECM**
	65	Heated oxygen sensor (HO2S) 1, bank 2 – no activity detected	**Wiring, HO2S, ECM**
	122	Heated oxygen sensor (HO2S) 1, bank 2 – slow response	**Wiring, heating inoperative, HO2S**
	124	Heated oxygen sensor (HO2S) 1, bank 2, heater control – circuit malfunction	**Wiring, HO2S, ECM**
	125	Heated oxygen sensor (HO2S) 2, bank 2, heater control – circuit malfunction	**Wiring, HO2S, ECM**
	127	Heated oxygen sensor (HO2S) 2, bank 2 – high voltage	**Wiring short to positive, HO2S, ECM**
	130	Heated oxygen sensor (HO2S) 2, bank 2 – no activity detected	**Wiring, HO2S, ECM**
	181	Heated oxygen sensor (HO2S) 1, bank 2 – low voltage	**Wiring short to ground, exhaust leak, HO2S, ECM**
	182	Heated oxygen sensor (HO2S) 2, bank 2 – low voltage	**Wiring short to ground, exhaust leak, HO2S, ECM**
22		Engine coolant temperature (ECT) sensor – voltage low/high	**Wiring, coolant thermostat, ECT sensor**
	30	Engine coolant temperature (ECT) sensor – low input	**Wiring short circuit to ground, coolant thermostat, ECT sensor**
	31	Engine coolant temperature (ECT) sensor – high input	**Wiring open circuit/short circuit to positive, coolant thermostat, ground wire defective, ECT sensor**
23		Intake air temperature (IAT) sensor – voltage low/high	**Wiring, IAT sensor**
	57	Intake air temperature (IAT) sensor – low input	**Wiring short circuit to ground, IAT sensor**
	58	Intake air temperature (IAT) sensor – high input	**Wiring open circuit/short circuit to positive, ground wire defective, IAT sensor**
24		Throttle position (TP) sensor – voltage low/high	**Wiring, accelerator cable adjustment, TP sensor**
	132	Throttle position (TP) sensor – range/performance problem	**Wiring, accelerator cable adjustment, TP sensor**
	26	Throttle position (TP) sensor – low input	**Wiring short circuit to ground, TP sensor**
	27	Throttle position (TP) sensor – high input	**Wiring short circuit to positive, TP sensor**
	136	Throttle position (TP) sensor – no 5 volt supply	**Wiring, TP sensor, ECM**
25		Idle air control (IAC) valve	**Wiring open/short circuit, air leak, IAC valve**

Flash code	Scan code DRB III type	Fault location	Probable cause
	25	Idle speed control (ISC) system – malfunction	**Wiring open/short circuit, IAC valve**
	138	Idle speed control (ISC) system – rpm lower/higher than expected	**Wiring, air leak, throttle valve tight/ sticking, IAC valve**
27		Injector – circuit malfunction	**Wiring, injector**
	21	Injector 1 – circuit malfunction	**Wiring, injector**
	20	Injector 2 – circuit malfunction	**Wiring, injector**
	19	Injector 3 – circuit malfunction	**Wiring, injector**
	61	Injector 4 – circuit malfunction	**Wiring, injector**
	69	Injector 5 – circuit malfunction	**Wiring, injector**
	70	Injector 6 – circuit malfunction	**Wiring, injector**
31		Evaporative emission (EVAP) canister purge valve – open/short circuit	**Wiring, EVAP canister purge valve**
	113	Evaporative emission (EVAP) system – incorrect flow detected	**Hose connection(s), intake leak, EVAP canister purge valve**
	160	Evaporative emission (EVAP) system – small leak detected	**Hose connection(s), intake leak, EVAP canister, EVAP canister purge valve**
	18	Evaporative emission (EVAP) canister purge valve – circuit malfunction	**Wiring open/short circuit, EVAP canister purge valve**
	161	Evaporative emission (EVAP) system – large leak detected	**Hose connection(s), intake leak, EVAP canister, EVAP canister purge valve**
	187	Evaporative emission (EVAP) system – blockage detected	**Hose connection(s), hose blocked/ trapped, EVAP canister**
	184	Evaporative emission (EVAP) leak detection pump – pressure switch	**Wiring, hose leak/blockage, EVAP leak detection pump**
	183	Evaporative emission (EVAP) leak detection pump solenoid	**Wiring open/short circuit, hose leak/ blockage, EVAP leak detection pump solenoid**
32		Exhaust gas recirculation (EGR) solenoid/ system – circuit malfunction	**Wiring, hoses, EGR solenoid**
	46	Exhaust gas recirculation (EGR) system – insufficient flow detected	**Hose leak/blockage, wiring, EGR valve, EGR solenoid**
	17	Exhaust gas recirculation (EGR) – circuit malfunction	**Wiring open/short circuit, EGR solenoid**
33	16	A/C compressor clutch relay – voltage low/ high	**Wiring open/short circuit, A/C compressor clutch relay**
35		Engine coolant blower motor relay/A/C condenser blower motor relay – low/high speed	**Wiring open/short circuit, engine coolant blower motor relay – low/high speed**
	139	Engine coolant blower motor relay – high speed – circuit malfunction	**Wiring open circuit/short circuit, engine coolant blower motor relay – high speed**
	93	A/C condenser blower motor relay – high speed	**Wiring open/short circuit, engine coolant blower motor relay – high speed**
	92	Engine coolant blower motor relay – low speed	**Wiring open/short circuit, engine coolant blower motor relay – low speed**

Flash code	Scan code DRB III type	Fault location	Probable cause
36		Secondary air injection (AIR) system – incorrect flow detected	**AIR pump, AIR solenoid, AIR hose(s)**
	111	Secondary air injection (AIR) system – incorrect flow detected	**AIR pump, AIR solenoid, AIR hose(s)**
	51	Secondary air injection (AIR) solenoid – circuit malfunction	**Wiring, AIR solenoid, ECM**
37	114	Park/neutral position (PNP) switch – input circuit malfunction	**Wiring, PNP switch, ECM/TCM**
41	11	Alternator, field control – circuit malfunction	**Wiring open/short circuit, alternator, battery**
42		Fuel pump relay/engine control relay/fuel level sensor – circuit malfunction	**Wiring open/short circuit, fuel pump relay, engine control relay, fuel level sensor**
	101	Fuel pump relay	**Wiring open/short circuit, fuel pump relay**
	10	Engine control relay	**Wiring open/short circuit, engine control relay**
	44	Engine control relay	**Wiring open circuit, engine control relay**
	149	Fuel level sensor	**Wiring open circuit, fuel level sensor**
	150	Fuel level sensor	**Wiring short circuit to positive, fuel level sensor**
	151	Fuel level sensor	**Wiring, fuel level sensor**
43		Random/multiple cylinder(s) – misfire detected	**Wiring, spark plug(s), HT lead(s), injector(s), ignition coil(s), low compression**
	106	Random/multiple cylinder(s) – misfire detected	**Wiring, spark plug(s), HT lead(s), injector(s), ignition coil(s), low compression**
	107	Cylinder 1 – misfire detected	**Wiring, engine mechanical fault, ignition/ fuel system, injector**
	108	Cylinder 2 – misfire detected	**Wiring, engine mechanical fault, ignition/ fuel system, injector**
	109	Cylinder 3 – misfire detected	**Wiring, engine mechanical fault, ignition/ fuel system, injector**
	110	Cylinder 4 – misfire detected	**Wiring, engine mechanical fault, ignition/ fuel system, injector**
	174	Cylinder 5 – misfire detected	**Wiring, engine mechanical fault, ignition/ fuel system, injector**
	175	Cylinder 6 – misfire detected	**Wiring, engine mechanical fault, ignition/ fuel system, injector**
	43	Ignition coil A, primary/secondary – circuit malfunction	**Wiring, ignition coil**
	42	Ignition coil B, primary/secondary – circuit malfunction	**Wiring, ignition coil**
44		Ambient/battery temperature sensor – voltage low/high	**Wiring, ambient/battery temperature sensor**

Flash code	Scan code DRB III type	Fault location	Probable cause
	154	Ambient/battery temperature sensor – voltage high	Wiring, ambient/battery temperature sensor
	153	Ambient/battery temperature sensor – voltage low	Wiring, ambient/battery temperature sensor
	71	Ambient/battery temperature sensor – range/performance problem	Wiring, ambient/battery temperature sensor
45	137	Transmission control system – malfunction	Wiring, TCM
46	06	Alternator – voltage high	Wiring, battery, alternator
47	05	Alternator – voltage low	Wiring, battery, alternator drive belt, alternator
51		System too lean, bank 1/2	Intake/exhaust leak, MAP sensor, fuel pressure/pump, ECT sensor, injector(s), HO2S
	119	System too lean, bank 1	Intake/exhaust leak, MAP sensor, fuel pressure/pump, ECT sensor, injector(s), HO2S
	121	System too lean, bank 2	Intake/exhaust leak, MAP sensor, fuel pressure/pump, ECT sensor, injector(s), HO2S
52		System too rich, bank 1/2	Intake blocked, EVAP canister purge valve, fuel pressure, EGR system, injector(s), HO2S
	118	System too rich, bank 1	Intake blocked, EVAP canister purge valve, fuel pressure, EGR system, injector(s), HO2S
	120	System too rich, bank 2	Intake blocked, EVAP canister purge valve, fuel pressure, EGR system, injector(s), HO2S
53		Engine control module (ECM)/CAN data bus – internal fault/malfunction	Wiring, connected system, ECM
	02	Engine control module (ECM) – internal fault	ECM
	68	CAN data bus – malfunction	Wiring, connected system, ECM
54	01	Camshaft position (CMP) sensor – no signal	Wiring, CMP sensor
62	48	Engine control module (ECM) – EEPROM error	ECM
63	49	Engine control module (ECM) – EEPROM error	ECM
64	112	Catalytic converter, bank 1 – efficiency below threshold	Wiring, exhaust leak, catalytic converter, HO2S 2
65	115	Power steering pressure (PSP) switch – range/performance problem	Wiring, PAS system, PSP switch
66	96	CAN data bus, transmission control module (TCM) – no signal	Wiring, TCM
72	180	Catalytic converter, bank 2 – efficiency below threshold	Wiring, exhaust leak, catalytic converter, HO2S 2

DODGE

Model:	Engine identification:	Year:
Avenger 2.0L	420A	1997-99
Avenger 2.5L	6G73	1997-00

System: **SBEC III/Mitsubishi MPI**

Self-diagnosis

General information

- Refer to the front of this manual for general test conditions, terminology, detailed descriptions of wiring faults and a general trouble shooter for electrical and mechanical faults.
- Malfunction indicator lamp (MIL) will illuminate or flash if certain faults are recorded.
- ECM operates in backup mode if sensors fail, to enable vehicle to be driven to repair shop.
- Trouble codes can be displayed by the MIL or by using a Scan Tool connected to the data link connector(s) (DLC).

Accessing

Flash codes

- Within 5 seconds switch the ignition key ON-OFF-ON-OFF-ON.
- Count MIL flashes. Note trouble codes.
- Each trouble code consists of two groups **1** [**A**].
- The first group of flashes indicate the 'tens' of the trouble code.
- The second group of flashes indicate the 'units' of the trouble code.
- A short pause separates 'tens' and 'units' **1** [**B**].
- A long pause separates each trouble code **1** [**C**].
- For example: Trouble code 21 displayed **1**.

P codes, Hex codes & DRB codes

- Trouble codes can also be displayed by using a Scan Tool connected to the data link connector(s) (DLC) **2**.

1

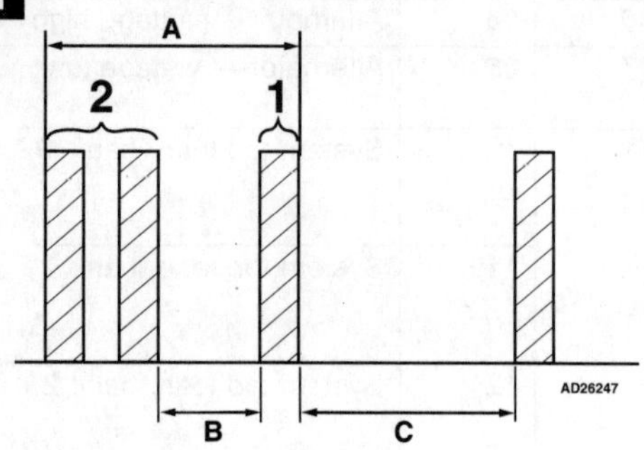

2

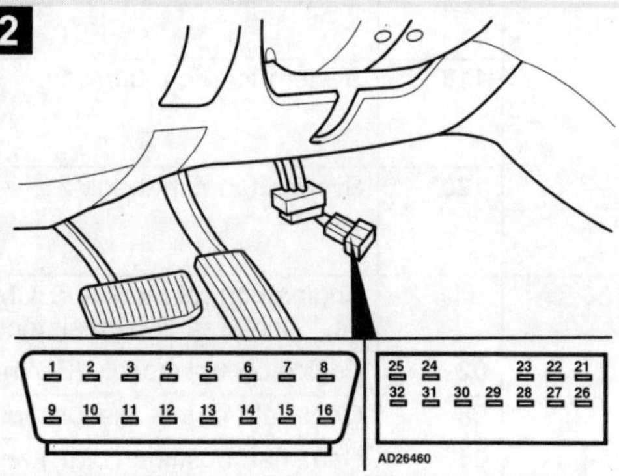

Erasing

NOTE: ***Disconnecting battery ground lead may cause driveability problems until the engine management system relearns basic values.***

- Ensure ignition switched OFF.
- Disconnect battery ground cable for at least 5 minutes.
- Reconnect battery ground cable.
- Repeat accessing procedure to ensure no trouble codes remain in ECM fault memory.

NOTE: ***Disconnecting battery ground lead may also erase memory from electronic units, such as radio.***

- Trouble codes can also be erased by using a Scan Tool connected to the data link connector(s) (DLC).

Trouble code identification

Scan code P type	Fault location	Probable cause
P0000-P0999	Refer to OBD-II trouble code table at the front of this manual	–
P1294	Idle speed control (ISC) system – rpm lower/higher than expected	**Wiring, air leak, throttle valve tight/sticking, IAC valve**
P1295	Throttle position (TP) sensor – no 5 volt supply	**Wiring, TP sensor, ECM**
P1296	Manifold absolute pressure (MAP) sensor – no 5 volt supply	**Wiring, ECM**
P1297	Manifold absolute pressure (MAP) sensor	**Wiring, hose connection(s), MAP sensor**
P1390	Timing belt jumped 1 tooth or more	**Timing belt loose/jumped position**
P1391	Crankshaft position (CKP) sensor/camshaft position (CMP) sensor – intermittent signal	**Wiring, poor connection, air gap, CKP sensor, CMP sensor**
P1486	Evaporative emission (EVAP) system – blockage detected	**Hose connection(s), hose blocked/trapped, EVAP canister**
P1487	Engine coolant blower motor relay – high speed – circuit malfunction	**Wiring open circuit/short circuit, engine coolant blower motor relay – high speed**
P1489	A/C condenser blower motor relay – high speed	**Wiring open/short circuit, engine coolant blower motor relay – high speed**
P1490	Engine coolant blower motor relay – low speed	**Wiring open/short circuit, engine coolant blower motor relay – low speed**
P1492	Ambient/battery temperature sensor – voltage high	**Wiring, ambient/battery temperature sensor**
P1493	Ambient/battery temperature sensor – voltage low	**Wiring, ambient/battery temperature sensor**
P1494	Evaporative emission (EVAP) leak detection pump – pressure switch	**Wiring, hose leak/blockage, EVAP leak detection pump**
P1495	Evaporative emission (EVAP) leak detection pump solenoid	**Wiring open/short circuit, hose leak/blockage, EVAP leak detection pump solenoid**
P1496	Manifold absolute pressure (MAP) sensor – 5 volt output low	**Wiring, MAP sensor**
P1696	Engine control module (ECM) – EEPROM error	**ECM**
P1697	Engine control module (ECM) – EEPROM error	**ECM**
P1698	CAN data bus, transmission control module (TCM) – no signal	**Wiring, TCM**
P1899	Park/neutral position (PNP) switch – input circuit malfunction	**Wiring, PNP switch, ECM/TCM**

Flash code	Scan code Hex type	Scan code DRB III type	Fault location	Probable cause
11			Crankshaft position (CKP) sensor/ camshaft position (CMP) sensor	**Wiring, poor connection, air gap, timing belt, CKP sensor, CMP sensor**
	85	133	Timing belt jumped 1 tooth or more	**Timing belt loose/jumped position**
	9D	157	Crankshaft position (CKP) sensor/ camshaft position (CMP) sensor – intermittent signal	**Wiring, poor connection, air gap, CKP sensor, CMP sensor**
	28	40	Crankshaft position (CKP) sensor – no signal	**Wiring, air gap, CKP sensor**
13	27	39	Manifold absolute pressure (MAP) sensor	**Wiring, hose connection(s), MAP sensor**
14			Manifold absolute pressure (MAP) sensor – input low/high	**Wiring, MAP sensor**
	24	36	Manifold absolute pressure (MAP) sensor – low input	**Wiring short circuit to ground, MAP sensor**
	25	37	Manifold absolute pressure (MAP) sensor – high input	**Wiring short circuit to positive, MAP sensor**
	87	135	Manifold absolute pressure (MAP) sensor – no 5 volt supply	**Wiring, ECM**
	92	146	Manifold absolute pressure (MAP) sensor – 5 volt output low	**Wiring, MAP sensor**
15	23	35	Vehicle speed sensor (VSS) – signal	**Wiring, VSS**
16	3B	59	Knock sensor (KS) 1 – no signal	**Wiring, KS incorrectly tightened, KS**
17			Engine temperature – low for long period	**Wiring, cooling system, coolant thermostat, ECT sensor**
	21	33	Engine temperature – low for long period	**Wiring, cooling system, coolant thermostat, ECT sensor**
	80	128	Insufficient coolant temperature for closed loop fuel control	**Wiring, cooling system, coolant thermostat, ECT sensor**
21			Heated oxygen sensor (HO2S) 1/2, bank 1/2 – circuit malfunction/ incorrect signal/voltage low/high	**Wiring, fuel pressure, injector control, EVAP canister purge valve, compression, air filter blocked, HO2S**
	9B	155	Heated oxygen sensor (HO2S) 1, bank 1 – low voltage	**Wiring short circuit to ground, exhaust leak, HO2S**
	3E	62	Heated oxygen sensor (HO2S) 1, bank 1 – high voltage	**Wiring short circuit to positive, fuel pressure, injector control, EVAP canister purge valve, air filter blocked, HO2S**
	66	102	Heated oxygen sensor (H02S) 1, bank 1 – slow response	**Wiring, heating inoperative, HO2S**
	20	32	Heated oxygen sensor (H02S) 1, bank 1 – no activity detected	**Wiring open circuit, heating inoperative, HO2S**

Flash code	Scan code Hex type	Scan code DRB III type	Fault location	Probable cause
	67	103	Heated oxygen sensor (HO2S) 1, bank 1 – heater control – circuit malfunction	**Wiring, HO2S**
	9C	156	Heated oxygen sensor (HO2S) 2, bank 1 – low voltage	**Wiring short circuit to ground, exhaust leak, HO2S**
	7E	126	Heated oxygen sensor (HO2S) 2, bank 1 – high voltage	**Wiring short circuit to positive, catalytic converter, HO2S**
	81	129	Heated oxygen sensor (HO2S) 2, bank 1 – no activity detected	**Wiring, heating inoperative, HO2S**
	69	105	Heated oxygen sensor (HO2S) 2, bank 1, heater control – circuit malfunction	**Wiring, HO2S**
	42	66	Heated oxygen sensor (HO2S) 1, bank 2 – high voltage	**Wiring short to positive, HO2S, ECM**
	41	65	Heated oxygen sensor (HO2S) 1, bank 2 – no activity detected	**Wiring, HO2S, ECM**
	7A	122	Heated oxygen sensor (HO2S) 1, bank 2 – slow response	**Wiring, heating inoperative, HO2S**
	7C	124	Heated oxygen sensor (HO2S) 1, bank 2, heater control – circuit malfunction	**Wiring, HO2S, ECM**
	7D	125	Heated oxygen sensor (HO2S) 2, bank 2, heater control – circuit malfunction	**Wiring, HO2S, ECM**
	7F	127	Heated oxygen sensor (HO2S) 2, bank 2 – high voltage	**Wiring short to positive, HO2S, ECM**
	82	130	Heated oxygen sensor (HO2S) 2, bank 2 – no activity detected	**Wiring, HO2S, ECM**
	B5	181	Heated oxygen sensor (HO2S) 1, bank 2 – low voltage	**Wiring short to ground, exhaust leak, HO2S, ECM**
	B6	182	Heated oxygen sensor (HO2S) 2, bank 2 – low voltage	**Wiring short to ground, exhaust leak, HO2S, ECM**
22			Engine coolant temperature (ECT) sensor – voltage low/high	**Wiring, coolant thermostat, ECT sensor**
	1E	30	Engine coolant temperature (ECT) sensor – low input	**Wiring short circuit to ground, coolant thermostat, ECT sensor**
	1F	31	Engine coolant temperature (ECT) sensor – high input	**Wiring open circuit/short circuit to positive, coolant thermostat, ground wire defective, ECT sensor**
23			Intake air temperature (IAT) sensor – voltage low/high	**Wiring, IAT sensor**
	39	57	Intake air temperature (IAT) sensor – low input	**Wiring short circuit to ground, IAT sensor**

Flash code	Scan code Hex type	Scan code DRB III type	Fault location	Probable cause
	3A	58	Intake air temperature (IAT) sensor – high input	**Wiring open circuit/short circuit to positive, ground wire defective, IAT sensor**
24			Throttle position (TP) sensor – voltage low/high	**Wiring, accelerator cable adjustment, TP sensor**
	84	132	Throttle position (TP) sensor – range/ performance problem	**Wiring, accelerator cable adjustment, TP sensor**
	1A	26	Throttle position (TP) sensor – low input	**Wiring short circuit to ground, TP sensor**
	1B	27	Throttle position (TP) sensor – high input	**Wiring short circuit to positive, TP sensor**
	88	136	Throttle position (TP) sensor – no 5 volt supply	**Wiring, TP sensor, ECM**
25			Idle air control (IAC) valve	**Wiring open/short circuit, air leak, IAC valve**
	19	25	Idle speed control (ISC) system – malfunction	**Wiring open/short circuit, IAC valve**
	8A	138	Idle speed control (ISC) system – rpm lower/higher than expected	**Wiring, air leak, throttle valve tight/ sticking, IAC valve**
27			Injector – circuit malfunction	**Wiring, injector**
	15	21	Injector 1 – circuit malfunction	**Wiring, injector**
	14	20	Injector 2 – circuit malfunction	**Wiring, injector**
	13	19	Injector 3 – circuit malfunction	**Wiring, injector**
	3D	61	Injector 4 – circuit malfunction	**Wiring, injector**
	45	69	Injector 5 – circuit malfunction	**Wiring, injector**
	46	70	Injector 6 – circuit malfunction	**Wiring, injector**
31			Evaporative emission (EVAP) canister purge valve – open/short circuit	**Wiring, EVAP canister purge valve**
	71	113	Evaporative emission (EVAP) system – incorrect flow detected	**Hose connection(s), intake leak, EVAP canister purge valve**
	A0	160	Evaporative emission (EVAP) system – small leak detected	**Hose connection(s), intake leak, EVAP canister, EVAP canister purge valve**
	12	18	Evaporative emission (EVAP) canister purge valve – circuit malfunction	**Wiring open/short circuit, EVAP canister purge valve**
	A1	161	Evaporative emission (EVAP) system – large leak detected	**Hose connection(s), intake leak, EVAP canister, EVAP canister purge valve**
	BB	187	Evaporative emission (EVAP) system – blockage detected	**Hose connection(s), hose blocked/ trapped, EVAP canister**
	B8	184	Evaporative emission (EVAP) leak detection pump – pressure switch	**Wiring, hose leak/blockage, EVAP leak detection pump**

Flash code	Scan code Hex type	Scan code DRB III type	Fault location	Probable cause
	B7	183	Evaporative emission (EVAP) leak detection pump solenoid	Wiring open/short circuit, hose leak/ blockage, EVAP leak detection pump solenoid
32			Exhaust gas recirculation (EGR) solenoid/system – circuit malfunction	Wiring, hoses, EGR solenoid
	2E	46	Exhaust gas recirculation (EGR) system – insufficient flow detected	Hose leak/blockage, wiring, EGR valve, EGR solenoid
	11	17	Exhaust gas recirculation (EGR) – circuit malfunction	Wiring open/short circuit, EGR solenoid
33	10	16	A/C compressor clutch relay – voltage low/high	Wiring open/short circuit, A/C compressor clutch relay
35			Engine coolant blower motor relay/A/C condenser blower motor relay – low/ high speed	Wiring open/short circuit, engine coolant blower motor relay – low/ high speed
	8B	139	Engine coolant blower motor relay – high speed – circuit malfunction	Wiring open circuit/short circuit, engine coolant blower motor relay – high speed
	5D	93	A/C condenser blower motor relay – high speed	Wiring open/short circuit, engine coolant blower motor relay – high speed
	5C	92	Engine coolant blower motor relay – low speed	Wiring open/short circuit, engine coolant blower motor relay – low speed
37	72	114	Park/neutral position (PNP) switch – input circuit malfunction	Wiring, PNP switch, ECM/TCM
41	0B	11	Alternator, field control – circuit malfunction	Wiring open/short circuit, alternator, battery
42			Fuel pump relay/engine control relay/ fuel level sensor – circuit malfunction	Wiring open/short circuit, fuel pump relay, engine control relay, fuel level sensor
	65	101	Fuel pump relay	Wiring open/short circuit, fuel pump relay
	0A	10	Engine control relay	Wiring open/short circuit, engine control relay
	2C	44	Engine control relay	Wiring open circuit, engine control relay
	95	149	Fuel level sensor	Wiring open circuit, fuel level sensor
	96	150	Fuel level sensor	Wiring short circuit to positive, fuel level sensor
	97	151	Fuel level sensor	Wiring, fuel level sensor
43			Random/multiple cylinder(s) – misfire detected	Wiring, spark plug(s), HT lead(s), injector(s), ignition coil(s), low compression

Flash code	Scan code Hex type	Scan code DRB III type	Fault location	Probable cause
	6A	106	Random/multiple cylinder(s) – misfire detected	**Wiring, spark plug(s), HT lead(s), injector(s), ignition coil(s), low compression**
	6B	107	Cylinder 1 – misfire detected	**Wiring, engine mechanical fault, ignition/fuel system, injector**
	6C	108	Cylinder 2 – misfire detected	**Wiring, engine mechanical fault, ignition/fuel system, injector**
	6D	109	Cylinder 3 – misfire detected	**Wiring, engine mechanical fault, ignition/fuel system, injector**
	6E	110	Cylinder 4 – misfire detected	**Wiring, engine mechanical fault, ignition/fuel system, injector**
	AE	174	Cylinder 5 – misfire detected	**Wiring, engine mechanical fault, ignition/fuel system, injector**
	AF	175	Cylinder 6 – misfire detected	**Wiring, engine mechanical fault, ignition/fuel system, injector**
	2B	43	Ignition coil A, primary/secondary – circuit malfunction	**Wiring, ignition coil**
	2A	42	Ignition coil B, primary/secondary – circuit malfunction	**Wiring, ignition coil**
44			Ambient/battery temperature sensor – voltage low/high	**Wiring, ambient/battery temperature sensor**
	9A	154	Ambient/battery temperature sensor – voltage high	**Wiring, ambient/battery temperature sensor**
	99	153	Ambient/battery temperature sensor – voltage low	**Wiring, ambient/battery temperature sensor**
	47	71	Ambient/battery temperature sensor – range/performance problem	**Wiring, ambient/battery temperature sensor**
45	89	137	Transmission control system – malfunction	**Wiring, TCM**
46	06	06	Alternator – voltage high	**Wiring, battery, alternator**
47	05	05	Alternator – voltage low	**Wiring, battery, alternator drive belt, alternator**
51			System too lean, bank 1/2	**Intake/exhaust leak, MAP sensor, fuel pressure/pump, ECT sensor, injector(s), HO2S**
	77	119	System too lean, bank 1	**Intake/exhaust leak, MAP sensor, fuel pressure/pump, ECT sensor, injector(s), HO2S**
	79	121	System too lean, bank 2	**Intake/exhaust leak, MAP sensor, fuel pressure/pump, ECT sensor, injector(s), HO2S**
52			System too rich, bank 1/2	**Intake blocked, EVAP canister purge valve, fuel pressure, EGR system, injector(s), HO2S**

Flash code	Scan code Hex type	Scan code DRB III type	Fault location	Probable cause
	76	118	System too rich, bank 1	Intake blocked, EVAP canister purge valve, fuel pressure, EGR system, injector(s), HO2S
	78	120	System too rich, bank 2	Intake blocked, EVAP canister purge valve, fuel pressure, EGR system, injector(s), HO2S
53			Engine control module (ECM)/CAN data bus – internal fault/malfunction	Wiring, connected system, ECM
	02	02	Engine control module (ECM) – internal fault	ECM
	44	68	CAN data bus – malfunction	Wiring, connected system, ECM
54	01	01	Camshaft position (CMP) sensor – no signal	Wiring, CMP sensor
62	30	48	Engine control module (ECM) – EEPROM error	ECM
63	31	49	Engine control module (ECM) – EEPROM error	ECM
64	70	112	Catalytic converter, bank 1 – efficiency below threshold	Wiring, exhaust leak, catalytic converter, HO2S 2
65	73	115	Power steering pressure (PSP) switch – range/performance problem	Wiring, PAS system, PSP switch
66	60	96	CAN data bus, transmission control module (TCM) – no signal	Wiring, TCM
72	B4	180	Catalytic converter, bank 2 – efficiency below threshold	Wiring, exhaust leak, catalytic converter, HO2S 2

DODGE

Model: **Avenger 2.5L**

Engine identification: **6G73**

Year: **1995**

System: **Mitsubishi MPI**

Self-diagnosis

General information

- Refer to the front of this manual for general test conditions, terminology, detailed descriptions of wiring faults and a general trouble shooter for electrical and mechanical faults.
- Malfunction indicator lamp (MIL) will illuminate if certain faults are recorded.
- ECM operates in backup mode if sensors fail, to enable vehicle to be driven to repair shop.
- Trouble codes can be displayed by the MIL or by using a Scan Tool connected to the data link connector(s) (DLC).

Accessing

- Within 5 seconds switch the ignition key ON-OFF-ON-OFF-ON.
- Count MIL flashes. Note trouble codes.
- Each trouble code consists of two groups **1** [**A**].
- The first group of flashes indicate the 'tens' of the trouble code.
- The second group of flashes indicate the 'units' of the trouble code.
- A short pause separates 'tens' and 'units' **1** [**B**].
- A long pause separates each trouble code **1** [**C**].
- For example: Trouble code 21 displayed **1**.
- Trouble codes can also be displayed by using a Scan Tool connected to the data link connector(s) (DLC) **2**.

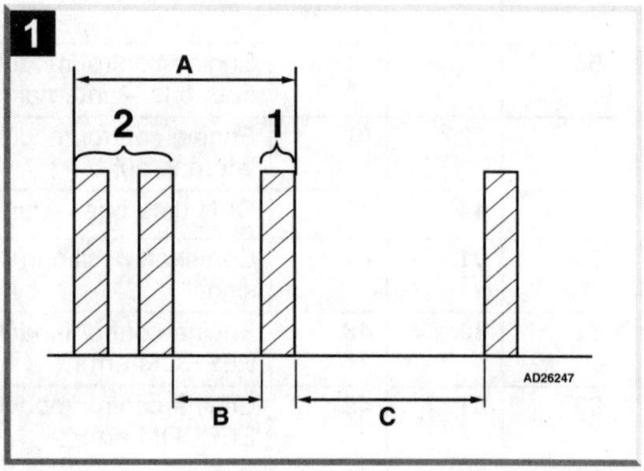

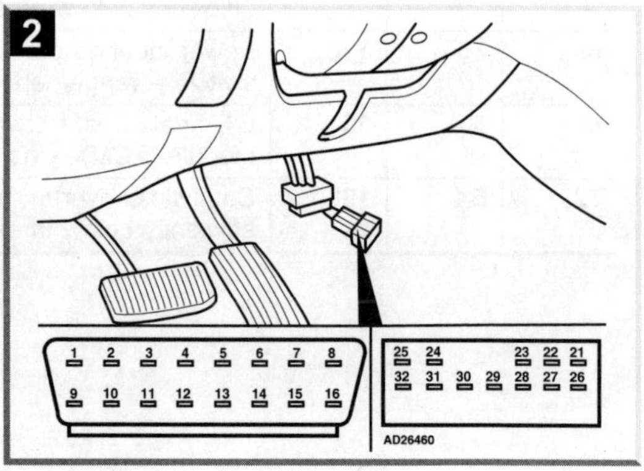

Erasing

NOTE: ***Disconnecting battery ground lead may cause driveability problems until the engine management system relearns basic values.***

- Ensure ignition switched OFF.
- Disconnect battery ground cable for at least 5 minutes.
- Reconnect battery ground cable.
- Repeat accessing procedure to ensure no trouble codes remain in ECM fault memory.

NOTE: ***Disconnecting battery ground lead may also erase memory from electronic units, such as radio.***

- Trouble codes can also be erased by using a Scan Tool connected to the data link connector(s) (DLC).

Trouble code identification

Flash code	Scan code DRB III type	Fault location	Probable cause
11		Crankshaft position (CKP) sensor/camshaft position (CMP) sensor	**Wiring, poor connection, air gap, timing belt, CKP sensor, CMP sensor**
	157	Crankshaft position (CKP) sensor/camshaft position (CMP) sensor – intermittent signal	**Wiring, poor connection, air gap, CKP sensor, CMP sensor**
	40	Crankshaft position (CKP) sensor – no signal	**Wiring, air gap, CKP sensor**
13	39	Manifold absolute pressure (MAP) sensor	**Wiring, hose connection(s), MAP sensor**
14		Manifold absolute pressure (MAP) sensor – input low/high	**Wiring, MAP sensor**
	36	Manifold absolute pressure (MAP) sensor – low input	**Wiring short circuit to ground, MAP sensor**
	37	Manifold absolute pressure (MAP) sensor – high input	**Wiring short circuit to positive, MAP sensor**
	135	Manifold absolute pressure (MAP) sensor – no 5 volt supply	**Wiring, ECM**
15	35	Vehicle speed sensor (VSS) – signal	**Wiring, VSS**
17	33	Engine temperature – low for long period	**Wiring, cooling system, coolant thermostat, ECT sensor**
21		Heated oxygen sensor (HO2S) 1/2, bank 1/2 – circuit malfunction/incorrect signal/voltage low/high	**Wiring, fuel pressure, injector control, EVAP canister purge valve, compression, air filter blocked, HO2S**
	32	Heated oxygen sensor (H02S) 1, bank 1 – no activity detected	**Wiring open circuit, heating inoperative, HO2S**
	41	Heated oxygen sensor (HO2S) 1, bank 2 – no activity detected	**Wiring open circuit, heating inoperative, HO2S**
	42	Heated oxygen sensor (HO2S) 1, bank 2 – high voltage	**Wiring short circuit to positive, fuel pressure, injector control, EVAP canister purge valve, air filter blocked, HO2S**
	62	Heated oxygen sensor (HO2S) 1, bank 1 – high voltage	**Wiring short circuit to positive, fuel pressure, injector control, EVAP canister purge valve, air filter blocked, HO2S**
22		Engine coolant temperature (ECT) sensor – voltage low/high	**Wiring, coolant thermostat, ECT sensor**
	30	Engine coolant temperature (ECT) sensor – low input	**Wiring short circuit to ground, coolant thermostat, ECT sensor**
	31	Engine coolant temperature (ECT) sensor – high input	**Wiring open circuit/short circuit to positive, coolant thermostat, ground wire defective, ECT sensor**

Flash code	Scan code DRB III type	Fault location	Probable cause
23		Intake air temperature (IAT) sensor – voltage low/high	**Wiring, IAT sensor**
	57	Intake air temperature (IAT) sensor – low input	**Wiring short circuit to ground, IAT sensor**
	58	Intake air temperature (IAT) sensor – high input	**Wiring open circuit/short circuit to positive, ground wire defective, IAT sensor**
24	26	Throttle position (TP) sensor – voltage low/high	**Wiring, accelerator cable adjustment, TP sensor**
	26	Throttle position (TP) sensor – low input	**Wiring short circuit to ground, TP sensor**
	27	Throttle position (TP) sensor – high input	**Wiring short circuit to positive, TP sensor**
	136	Throttle position (TP) sensor – no 5 volt supply	**Wiring, TP sensor, ECM**
25	25	Idle speed control (ISC) system – malfunction	**Wiring open/short circuit, IAC valve**
27		Injector – circuit malfunction	**Wiring, injector**
	21	Injector 1 – circuit malfunction	**Wiring, injector**
	20	Injector 2 – circuit malfunction	**Wiring, injector**
	19	Injector 3 – circuit malfunction	**Wiring, injector**
	61	Injector 4 – circuit malfunction	**Wiring, injector**
	69	Injector 5 – circuit malfunction	**Wiring, injector**
	70	Injector 6 – circuit malfunction	**Wiring, injector**
31	18	Evaporative emission (EVAP) canister purge valve – circuit malfunction	**Wiring open/short circuit, EVAP canister purge valve**
32		Exhaust gas recirculation (EGR) solenoid/system – circuit malfunction	**Wiring, hoses, EGR solenoid**
	46	Exhaust gas recirculation (EGR) system – insufficient flow detected	**Hose leak/blockage, wiring, EGR valve, EGR solenoid**
	17	Exhaust gas recirculation (EGR) – circuit malfunction	**Wiring open/short circuit, EGR solenoid**
33	16	A/C compressor clutch relay – voltage low/high	**Wiring open/short circuit, A/C compressor clutch relay**
35		Engine coolant blower motor relay/A/C condenser blower motor relay – low/high speed	**Wiring open/short circuit, engine coolant blower motor relay – low/high speed**
	139	Engine coolant blower motor relay – high speed – circuit malfunction	**Wiring open circuit/short circuit, engine coolant blower motor relay – high speed**
	93	A/C condenser blower motor relay – high speed	**Wiring open/short circuit, engine coolant blower motor relay – high speed**
	92	Engine coolant blower motor relay – low speed	**Wiring open/short circuit, engine coolant blower motor relay – low speed**

Flash code	Scan code DRB III type	Fault location	Probable cause
41	11	Alternator, field control – circuit malfunction	**Wiring open/short circuit, alternator, battery**
42		Fuel pump relay/engine control relay – circuit malfunction	**Wiring open/short circuit, fuel pump relay, engine control relay**
	101	Fuel pump relay	**Wiring open/short circuit, fuel pump relay**
	10	Engine control relay	**Wiring open/short circuit, engine control relay**
	44	Engine control relay	**Wiring open circuit, engine control relay**
44	71	Ambient/battery temperature sensor – range/performance problem	**Wiring, ambient/battery temperature sensor**
46	06	Alternator – voltage high	**Wiring, battery, alternator**
47	05	Alternator – voltage low	**Wiring, battery, alternator drive belt, alternator**
51		System too lean, bank 1/2	**Intake/exhaust leak, MAP sensor, fuel pressure/pump, ECT sensor, injector(s), HO2S**
	04	System too lean, bank 1	**Intake/exhaust leak, MAP sensor, fuel pressure/pump, ECT sensor, injector(s), HO2S**
	09	System too lean, bank 2	**Intake/exhaust leak, MAP sensor, fuel pressure/pump, ECT sensor, injector(s), HO2S**
52		System too rich, bank 1/2	**Intake blocked, EVAP canister purge valve, fuel pressure, EGR system, injector(s), HO2S**
	03	System too rich, bank 1	**Intake blocked, EVAP canister purge valve, fuel pressure, EGR system, injector(s), HO2S**
	08	System too rich, bank 2	**Intake blocked, EVAP canister purge valve, fuel pressure, EGR system, injector(s), HO2S**
53		Engine control module (ECM)/CAN data bus – internal fault/malfunction	**Wiring, connected system, ECM**
	02	Engine control module (ECM) – internal fault	**ECM**
	68	CAN data bus – malfunction	**Wiring, connected system, ECM**
54	01	Camshaft position (CMP) sensor – no signal	**Wiring, CMP sensor**
62	48	Engine control module (ECM) – EEPROM error	**ECM**
63	49	Engine control module (ECM) – EEPROM error	**ECM**
66	96	CAN data bus, transmission control module (TCM) – no signal	**Wiring, TCM**

DODGE

Model:	Engine identification:	Year:
Dakota 2.5/3.9L	EHC/EPE	1998
Dakota 5.2/5.9L	ELF/ELM	1998
Durango 3.9L	EHC	1998
Durango 5.2/5.9L	ELF/ELM	1998
Ram Truck 3.9/5.2L	EHC/ELF/ELN	1998
Ram Truck 5.9/8.0L	EML/EMM/EWA	1998
Ram Van/Wagon 3.9L	EHC	1998
Ram Van/Wagon 5.2/5.9L	ELF/ELN/EML	1998

System: **JTEC**

Self-diagnosis

General information

- Refer to the front of this manual for general test conditions, terminology, detailed descriptions of wiring faults and a general trouble shooter for electrical and mechanical faults.
- Malfunction indicator lamp (MIL) will illuminate or flash if certain faults are recorded.
- ECM operates in backup mode if sensors fail, to enable vehicle to be driven to repair shop.

Accessing

- Trouble codes can only be displayed by using a Scan Tool connected to the data link connector (DLC) **1**.

1

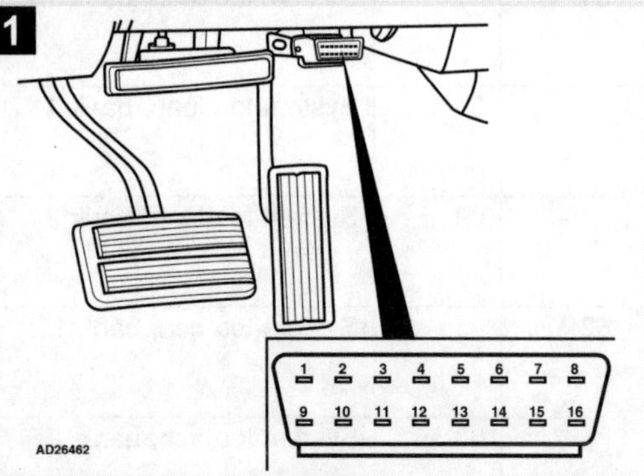

Erasing

- Trouble codes can only be erased by using a Scan Tool connected to the data link connector (DLC).

Trouble code identification

Scan code P type	Fault location	Probable cause
P0000-P0999	Refer to OBD-II trouble code table at the front of this manual	–
P1195	Heated oxygen sensor (HO2S) 1, bank 1 – slow response	**Wiring, heating inoperative, exhaust leak, HO2S**
P1196	Heated oxygen sensor (HO2S) 1, bank 2 – slow response	**Wiring, heating inoperative, exhaust leak, HO2S**

Scan code P type	Fault location	Probable cause
P1197	Heated oxygen sensor (HO2S) 2, bank 1 – slow response	**Wiring, heating inoperative, exhaust leak, HO2S**
P1281	Engine temperature – low for long period	**Wiring, cooling system, coolant thermostat, ECT sensor**
P1282	Fuel pump relay	**Wiring open circuit/short circuit, fuel pump relay**
P1294	Idle speed control (ISC) system – rpm lower/ higher than expected	**Wiring, air leak, throttle valve tight/sticking, IAC valve**
P1296	Manifold absolute pressure (MAP) sensor – no 5 volt supply	**Wiring, MAP sensor, ECM**
P1297	Manifold absolute pressure (MAP) sensor	**Wiring, hose connection(s), MAP sensor**
P1388	Engine control relay	**Wiring open circuit/short circuit, engine control relay**
P1389	Engine control relay	**Wiring open circuit, engine control relay**
P1391	Crankshaft position (CKP) sensor/camshaft position (CMP) sensor – intermittent signal	**Wiring, poor connection, air gap, CKP sensor, CMP sensor**
P1398	Engine control module (ECM) – misfire detection limit reached	**Wiring, poor connection, CKP sensor**
P1486	Evaporative emission (EVAP) system – blockage detected	**Hose connection(s), hose blocked/trapped, EVAP canister**
P1491	Engine coolant blower motor relay – circuit malfunction	**Wiring open circuit/short circuit, engine coolant blower motor**
P1492	Ambient/battery temperature sensor – voltage high	**Wiring, ambient/battery temperature sensor**
P1493	Ambient/battery temperature sensor – voltage low	**Wiring, ambient/battery temperature sensor**
P1494	Evaporative emission (EVAP) leak detection pump – pressure switch	**Wiring, hose leak/blockage, EVAP leak detection pump**
P1495	Evaporative emission (EVAP) leak detection pump solenoid	**Wiring open circuit/short circuit, hose leak/ blockage, EVAP leak detection pump solenoid**
P1594	Alternator – voltage high	**Wiring, battery, alternator**
P1595	Cruise control	**Wiring open circuit/short circuit, hoses, cruise control master switch, cruise control vacuum switching valve**
P1596	Cruise control master switch – high voltage	**Wiring, cruise control master switch**
P1597	Cruise control master switch – low voltage	**Wiring, cruise control master switch**
P1683	Cruise control	**Wiring, cruise control vacuum switching valve**
P1693	Engine control module (ECM)	**Companion module trouble code detected in ECM**
P1694	CAN data bus, engine control module (ECM) – no signal	**Wiring, ECM**
P1696	Engine control module (ECM) – EEPROM error	**ECM**

Scan code P type	Fault location	Probable cause
P1756	Governor pressure not equal to target (15-20 PSI) – mid pressure malfunction	**Signal wire open circuit/short circuit, valve body, sensor supply wire open circuit, transmission fluid level, governor pressure solenoid, transmission mechanical fault, ECM – no 5 volt supply**
P1757	Governor pressure not equal to target (15-20 PSI) – zero pressure malfunction	**Signal wire open circuit/short circuit, valve body, sensor supply wire open circuit, transmission fluid level, governor pressure solenoid, transmission mechanical fault, ECM – no 5 volt supply**
P1762	Governor pressure sensor – incorrect signal	**Signal wire short circuit to ground/positive, supply wire open circuit/short circuit, transmission fluid level, governor pressure solenoid, governor pressure sensor, ECM**
P1763	Governor pressure sensor – signal high	**Signal wire short circuit to positive, supply wire open circuit, transmission fluid level, governor pressure sensor, ECM**
P1764	Governor pressure sensor – signal low	**Signal wire short circuit to ground, supply wire open circuit, transmission fluid level, governor pressure sensor, ECM**
P1765	Transmission system relay	**Wiring open circuit/short circuit, transmission system relay, ECM**
P1899	Park/neutral position (PNP) switch – input circuit malfunction	**Wiring, PNP switch, ECM**

Scan code Hex type	Fault location	Probable cause
00	DTC error	–
01	Camshaft position (CMP) sensor – no signal	**Wiring, CMP sensor**
02	Engine control module (ECM) – internal fault	**ECM**
05	Alternator – voltage low	**Wiring, battery, alternator drive belt, alternator**
06	Alternator – voltage high	**Wiring, battery, alternator**
0A	Engine control relay	**Wiring open circuit/short circuit, engine control relay**
0B	Alternator, field control – circuit malfunction	**Wiring open circuit/short circuit, alternator, battery**
0C	Torque converter clutch (TCC) solenoid – electrical	**Wiring open circuit/short circuit, TCC solenoid**
0E	Engine coolant blower motor relay – low/high speed	**Wiring open circuit/short circuit, engine coolant blower motor relay – low/high speed**

Scan code Hex type	Fault location	Probable cause
0F	Cruise control	**Wiring open circuit/short circuit, hoses, voltage low/high, cruise control master switch, cruise control vacuum switching valve**
10	A/C compressor clutch relay	**Wiring open circuit/short circuit, A/C compressor clutch relay**
11	Exhaust gas recirculation (EGR) – circuit malfunction	**Wiring, EGR solenoid, ECM**
12	Evaporative emission (EVAP) canister purge valve – circuit malfunction	**Wiring open circuit/short circuit, EVAP canister purge valve**
13	Injector 3 – circuit malfunction	**Wiring, injector**
14	Injector 2 – circuit malfunction	**Wiring, injector**
15	Injector 1 – circuit malfunction	**Wiring, injector**
19	Idle speed control (ISC) system – malfunction	**Wiring open circuit/short circuit, IAC valve**
1A	Throttle position (TP) sensor – low input	**Wiring short circuit to ground, TP sensor**
1B	Throttle position (TP) sensor – high input	**Wiring short circuit to positive, TP sensor**
1E	Engine coolant temperature (ECT) sensor – low input	**Coolant thermostat, wiring short circuit to ground, ECT sensor**
1F	Engine coolant temperature (ECT) sensor – high input	**Coolant thermostat, wiring open circuit/ short circuit to positive, ground wire defective, ECT sensor**
21	Engine temperature – low for long period	**Wiring, cooling system, coolant thermostat, ECT sensor**
24	Manifold absolute pressure (MAP) sensor – low input	**Wiring short circuit to ground, MAP sensor**
25	Manifold absolute pressure (MAP) sensor – high input	**Wiring short circuit to positive, MAP sensor**
27	Manifold absolute pressure (MAP) sensor	**Wiring, hose connection(s), MAP sensor**
28	Crankshaft position (CKP) sensor – no signal	**Wiring, air gap, CKP sensor**
29	Ignition coil C, primary – circuit malfunction	**Wiring, ignition coil, ECM**
2A	Ignition coil B, primary – circuit malfunction	**Wiring, ignition coil, ECM**
2B	Ignition coil A, primary – circuit malfunction	**Wiring, ignition coil, ECM**
2C	Engine control relay	**Wiring open circuit, engine control relay**
31	Engine control module (ECM) – EEPROM error	**ECM**
32	Shift solenoid (SS) 3-4 – electrical	**Wiring, shift solenoid, ECM**
37	Injector 9 – circuit malfunction	**Wiring, injector**
38	Injector 10 – circuit malfunction	**Wiring, injector**
39	Intake air temperature (IAT) sensor – low input	**Wiring short circuit to ground, IAT sensor**

DODGE

Dakota 2.5/3.9L ● Dakota 5.2/5.9L ● Durango 3.9L ● Durango 5.2/5.9L
Ram Truck 3.9/5.2L ● Ram Truck 5.9/8.0L ● Ram Van/Wagon 3.9L
Ram Van/Wagon 5.2/5.9L

Scan code Hex type	Fault location	Probable cause
3A	Intake air temperature (IAT) sensor – high input	**Wiring open circuit/short circuit to positive, ground wire defective, IAT sensor**
3D	Injector 4 – circuit malfunction	**Wiring, injector**
3E	Heated oxygen sensor (HO2S) 1, bank 1 – high voltage	**Wiring short circuit to positive, fuel pressure, injector control, EVAP canister purge valve, air filter blocked, HO2S**
41	Heated oxygen sensor (HO2S) 1, bank 2 – no activity detected	**Wiring open circuit, heating inoperative, HO2S**
42	Heated oxygen sensor (HO2S) 1, bank 2 – high voltage	**Wiring short circuit to positive, fuel pressure, injector control, EVAP canister purge valve, air filter blocked, HO2S**
44	Engine control module (ECM) – internal fault	**ECM**
45	Injector 5 – circuit malfunction	**Wiring, injector**
46	Injector 6 – circuit malfunction	**Wiring, injector**
4A	Transmission fluid temperature (TFT) sensor – low input	**Wiring short circuit to ground, TFT sensor, ECM**
4B	Transmission fluid temperature (TFT) sensor – high input	**Wiring short circuit to positive, TFT sensor, ECM**
4C	Ignition coil D, primary – circuit malfunction	**Wiring, ignition coil, ECM**
4D	Ignition coil E, primary – circuit malfunction	**Wiring, ignition coil, ECM**
4F	Injector 7 – circuit malfunction	**Wiring, injector**
50	Injector 8 – circuit malfunction	**Wiring, injector**
52	Cruise control	**Wiring, cruise control vacuum switching valve**
56	Cruise control master switch – high voltage	**Wiring, cruise control master switch**
57	Cruise control master switch – low voltage	**Wiring, cruise control master switch**
65	Fuel pump relay	**Wiring open circuit/short circuit, fuel pump relay**
66	Heated oxygen sensor (H02S) 1, bank 1 – slow response	**Heating inoperative, wiring, HO2S**
67	Heated oxygen sensor (HO2S) 1, bank 1 – heater control – circuit malfunction	**Wiring, HO2S**
68	Heated oxygen sensor (H02S) 2, bank 1 – slow response	**Heating inoperative, wiring, exhaust leak, HO2S**
69	Heated oxygen sensor (HO2S) 2, bank 1, heater control – circuit malfunction	**Wiring, HO2S**
6A	Random/multiple cylinder(s) – misfire detected	**Wiring, spark plug(s), HT lead(s), injector(s), ignition coil(s), low compression**
6B	Cylinder 1 – misfire detected	**Wiring, engine mechanical fault, ignition/ fuel system, injector**
6C	Cylinder 2 – misfire detected	**Wiring, engine mechanical fault, ignition/ fuel system, injector**

Scan code Hex type	Fault location	Probable cause
6D	Cylinder 3 – misfire detected	**Wiring, engine mechanical fault, ignition/ fuel system, injector**
6E	Cylinder 4 – misfire detected	**Wiring, engine mechanical fault, ignition/ fuel system, injector**
70	Catalytic converter system, bank 1 – efficiency below threshold	**Catalytic converter, wiring, HO2S 2**
71	Evaporative emission (EVAP) system – incorrect flow detected	**Hose connection(s), intake leak, EVAP canister purge valve**
72	Park/neutral position (PNP) switch – input circuit malfunction	**Wiring, PNP switch, ECM**
73	Power steering pressure (PSP) switch – range/performance problem	**Wiring, PAS system, PSP switch**
76	System too rich, bank 1	**Wiring, intake blocked, EVAP canister purge valve, fuel pressure, EGR system, injector(s), air filter blocked, ignition system, HO2S**
77	System too lean, bank 1	**Wiring, intake/exhaust leak, MAP sensor, fuel pressure/pump, EVAP canister purge valve, injector(s), HO2S**
78	System too rich, bank 2	**Wiring, intake blocked, EVAP canister purge valve, fuel pressure, EGR system, injector(s), air filter blocked, ignition system, HO2S**
79	System too lean, bank 2	**Wiring, intake/exhaust leak, MAP sensor, fuel pressure/pump, EVAP canister purge valve, injector(s), HO2S**
7A	Heated oxygen sensor (HO2S) 1, bank 2 – slow response	**Heating inoperative, wiring, HO2S**
7B	Heated oxygen sensor (HO2S) 2, bank 2 – slow response	**Heating inoperative, wiring, exhaust leak, HO2S**
7C	Heated oxygen sensor (HO2S) 1, bank 2, heater control – circuit malfunction	**Wiring, HO2S, ECM**
7D	Heated oxygen sensor (HO2S) 2, bank 2, heater control – circuit malfunction	**Wiring, HO2S, ECM**
7E	Heated oxygen sensor (HO2S) 2, bank 1 – high voltage	**Wiring short circuit to positive, catalytic converter, HO2S, ECM**
7F	Heated oxygen sensor (HO2S) 2, bank 2 – high voltage	**Wiring short circuit to positive, catalytic converter, HO2S, ECM**
80	Insufficient coolant temperature for closed loop fuel control	**Wiring, cooling system, coolant thermostat, ECT sensor**
84	Throttle position (TP) sensor – range/ performance problem	**Wiring, accelerator cable adjustment, TP sensor**
87	Manifold absolute pressure (MAP) sensor – no 5 volt supply	**Wiring, MAP sensor, ECM**
8A	Idle speed control (ISC) system – rpm lower/ higher than expected	**Wiring, air leak, throttle valve tight/sticking, IAC valve**

DODGE

Dakota 2.5/3.9L • Dakota 5.2/5.9L • Durango 3.9L • Durango 5.2/5.9L
Ram Truck 3.9/5.2L • Ram Truck 5.9/8.0L • Ram Van/Wagon 3.9L
Ram Van/Wagon 5.2/5.9L

Scan code Hex type	Fault location	Probable cause
8D	Governor pressure not equal to target (15-20 PSI) – mid pressure malfunction	**Signal wire open circuit/short circuit, valve body, sensor supply wire open circuit, transmission fluid level, governor pressure solenoid, transmission mechanical fault, ECM – no 5 volt supply**
8E	Governor pressure not equal to target (15-20 PSI) – zero pressure malfunction	**Signal wire open circuit/short circuit, valve body, sensor supply wire open circuit, transmission fluid level, governor pressure solenoid, transmission mechanical fault, ECM – no 5 volt supply**
94	Torque converter clutch (TCC) solenoid – circuit malfunction	**Wiring, TCC solenoid**
95	Fuel level sensor – low input	**Wiring open circuit, fuel level sensor**
96	Fuel level sensor – high input	**Wiring short circuit to positive, fuel level sensor**
97	Fuel level sensor – malfunction	**Wiring, fuel level sensor**
99	Ambient/battery temperature sensor – voltage low	**Wiring, ambient/battery temperature sensor**
9A	Ambient/battery temperature sensor – voltage high	**Wiring, ambient/battery temperature sensor**
9B	Heated oxygen sensor (HO2S) 1, bank 1 – low voltage	**Exhaust leak, wiring short circuit to ground, HO2S**
9C	Heated oxygen sensor (HO2S) 2, bank 1 – low voltage	**Exhaust leak, wiring short circuit to ground, HO2S**
9D	Crankshaft position (CKP) sensor/camshaft position (CMP) sensor – intermittent signal	**Wiring, poor connection, air gap, CKP sensor, CMP sensor**
A0	Evaporative emission (EVAP) system – small leak detected	**Hose connection(s), intake leak, EVAP canister, EVAP canister purge valve**
A1	Evaporative emission (EVAP) system – large leak detected	**Hose connection(s), intake leak, EVAP canister, EVAP canister purge valve**
A4	Transmission fluid temperature (TFT) sensor – range/performance problem	**Wiring, TFT sensor**
A5	Gear selection, 3-4 – shift malfunction	**Wiring, TR sensor, shift solenoids, transmission mechanical fault**
A6	Output shaft speed (OSS) sensor – circuit malfunction	**Wiring, OSS sensor, ECM**
A7	Governor pressure sensor – signal low	**Signal wire short circuit to ground, supply wire open circuit, transmission fluid level, governor pressure sensor, ECM**
A8	Governor pressure sensor – signal high	**Signal wire short circuit to positive, supply wire open circuit, transmission fluid level, governor pressure sensor, ECM**
A9	Governor pressure sensor – incorrect signal	**Signal wire short circuit to ground/positive, supply wire open circuit/short circuit, transmission fluid level, governor pressure solenoid, governor pressure sensor, ECM**

Scan code Hex type	Fault location	Probable cause
AB	Transmission fluid pressure (TFP) solenoid – electrical	Wiring, TFP solenoid, ECM
AD	Transmission system relay	Wiring open circuit/short circuit, transmission system relay, ECM
AE	Cylinder 5 – misfire detected	Wiring, engine mechanical fault, ignition/ fuel system, injector
AF	Cylinder 6 – misfire detected	Wiring, engine mechanical fault, ignition/ fuel system, injector
B0	Cylinder 7 – misfire detected	Wiring, engine mechanical fault, ignition/ fuel system, injector
B1	Cylinder 8 – misfire detected	Wiring, engine mechanical fault, ignition/ fuel system, injector
B2	Cylinder 9 – misfire detected	Wiring, engine mechanical fault, ignition/ fuel system, injector
B3	Cylinder 10 – misfire detected	Wiring, engine mechanical fault, ignition/ fuel system, injector
B4	Catalytic converter system, bank 2 – efficiency below threshold	Catalytic converter, wiring, HO2S 2
B5	Heated oxygen sensor (HO2S) 1, bank 2 – low voltage	Exhaust leak, wiring short to ground, HO2S, ECM
B6	Heated oxygen sensor (HO2S) 2, bank 2 – low voltage	Exhaust leak, wiring short to ground, HO2S, ECM
B7	Evaporative emission (EVAP) leak detection pump solenoid	Wiring open circuit/short circuit, hose leak/ blockage, EVAP leak detection pump solenoid
B8	Evaporative emission (EVAP) leak detection pump – pressure switch	Wiring, hose leak/blockage, EVAP leak detection pump
BA	Engine control module (ECM) – misfire detection limit reached	Wiring, poor connection, CKP sensor
BB	Evaporative emission (EVAP) system – blockage detected	Hose connection(s), hose blocked/trapped, EVAP canister
BC	Overdrive solenoid/switch – performance problem	Wiring, overdrive solenoid/switch, transmission mechanical fault, ECM
BD	Heated oxygen sensor (HO2S) 3, bank 1, heater control – circuit malfunction	Wiring, HO2S, ECM
CO	Heated oxygen sensor (HO2S) 1, bank 1 – slow response	Heating inoperative, wiring, HO2S
C1	Heated oxygen sensor (HO2S) 1, bank 2 – slow response	Wiring, heating inoperative, exhaust leak, HO2S
C2	Heated oxygen sensor (HO2S) 2, bank 1 – slow response	Wiring, heating inoperative, exhaust leak, HO2S
DE	CAN data bus, engine control module (ECM) – no signal	Wiring, ECM
DF	Engine control module (ECM)	Companion module trouble code detected in ECM

DODGE

Model:	Engine identification:	Year:
Dakota 2.5/3.9L	EHC/EPE	1999-02
Dakota 4.7/5.2/5.9L	ELF/ELM/EVA	1999-02
Durango 3.9/4.7L	EHC/EVA	1999-02
Durango 5.2/5.9L	ELF/ELM	1999-02
Ram Truck 3.9/5.2L	EHC/ELF/ELN	1999-01
Ram Truck 3.7/4.7L	EKG/EVA	2002
Ram Truck 5.9/8.0L	EML/EMM/EWA	1999-02
Ram Van/Wagon 3.9/5.2/5.9L	EHC/ELF/ELN/ELP/EML	1999-02

System: **JTEC**

Self-diagnosis

General information

- Refer to the front of this manual for general test conditions, terminology, detailed descriptions of wiring faults and a general trouble shooter for electrical and mechanical faults.
- Malfunction indicator lamp (MIL) will illuminate or flash if certain faults are recorded.
- ECM operates in backup mode if sensors fail, to enable vehicle to be driven to repair shop.

Accessing

- Trouble codes can only be displayed by using a Scan Tool connected to the data link connector (DLC) **1**.

1

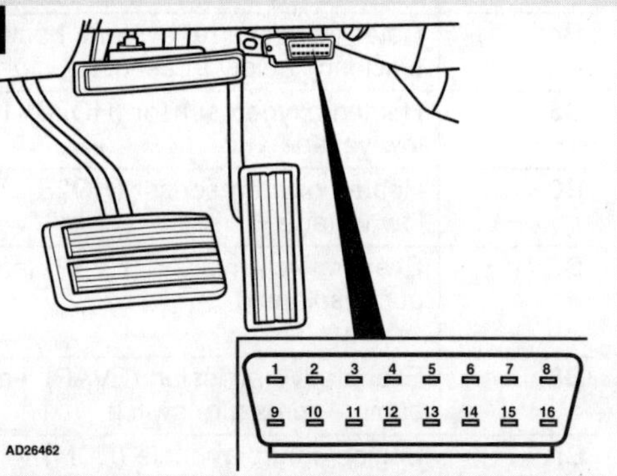

Erasing

- Trouble codes can only be erased by using a Scan Tool connected to the data link connector (DLC).

Trouble code identification

Scan code	Fault location	Probable cause
P0000-P0999	Refer to OBD-II trouble code table at the front of this manual	–
P1110	Decreased engine performance – high intake air temperature	**Wiring open circuit/short to positive, ground wire defective, IAT sensor, ECT sensor, cooling system, high engine bay temperature, ECM**
P1192	Intake air temperature (IAT) sensor – voltage low	**Wiring, IAT sensor, ECM**

Scan code	Fault location	Probable cause
P1193	Intake air temperature (IAT) sensor – voltage high	Wiring, IAT sensor, ECM
P1194	Heated oxygen sensor (HO2S), heater control – circuit malfunction	Wiring, HO2S, ECM
P1195	Heated oxygen sensor (HO2S) 1, bank 1 – slow response	Wiring, heating inoperative, exhaust leak, HO2S
P1196	Heated oxygen sensor (HO2S) 1, bank 2 – slow response	Wiring, heating inoperative, exhaust leak, HO2S
P1197	Heated oxygen sensor (HO2S) 2, bank 1 – slow response	Wiring, heating inoperative, exhaust leak, HO2S
P1280	Fuel pump relay – circuit malfunction	Wiring, fuel pump relay
P1281	Engine temperature – low for long period	Wiring, cooling system, coolant thermostat, ECT sensor
P1282	Fuel pump relay	Wiring open circuit/short circuit, fuel pump relay
P1290	Compressed natural gas pressure sensor – pressure high	Compressed natural gas system, wiring, compressed natural gas pressure sensor
P1292	Compressed natural gas pressure sensor – voltage high	Wiring, compressed natural gas pressure sensor
P1293	Compressed natural gas pressure sensor – voltage low	Wiring, compressed natural gas pressure sensor
P1294	Idle speed control (ISC) system – rpm lower/ higher than expected	Wiring, air leak, throttle valve tight/sticking, IAC valve
P1295	Throttle position (TP) sensor – no 5 volt supply	Wiring, TP sensor, ECM
P1296	Manifold absolute pressure (MAP) sensor – no 5 volt supply	Wiring, MAP sensor, ECM
P1297	Manifold absolute pressure (MAP) sensor	Wiring, hose connection(s), MAP sensor
P1298	System too lean – at prolonged wide open throttle	Intake/exhaust leak, MAP sensor, fuel pressure/pump, injector(s), HO2S
P1299	Manifold absolute pressure (MAP) sensor/ throttle position (TP) sensor	Intake leak, hose connection(s), MAP sensor, TP sensor
P1388	Engine control relay	Wiring open circuit/short circuit, engine control relay
P1389	Engine control relay	Wiring open circuit, engine control relay
P1391	Crankshaft position (CKP) sensor/camshaft position (CMP) sensor – intermittent signal	Wiring, poor connection, air gap, CKP sensor, CMP sensor
P1398	Engine control module (ECM) – misfire detection limit reached	Wiring, poor connection, CKP sensor
P1475	Engine control module (ECM) – auxiliary 5 volt output high	Wiring, ECM
P1476	Secondary air injection (AIR) system – insufficient flow detected	AIR pump, AIR valve, AIR hose(s)

Scan code	Fault location	Probable cause
P1477	Secondary air injection (AIR) system – excessive flow detected	**AIR pump, AIR valve, AIR hose(s), ECM**
P1478	Ambient/battery temperature sensor – range/ performance problem	**Wiring, ambient/battery temperature sensor**
P1479	Transmission cooling fan relay – circuit malfunction	**Wiring, transmission fluid level low, transmission cooling fan, transmission cooling fan relay**
P1486	Evaporative emission (EVAP) system – blockage detected	**Hose connection(s), hose blocked/trapped, EVAP canister**
P1488	Engine control module (ECM) – auxiliary 5 volt output low	**Wiring, ECM**
P1491	Engine coolant blower motor relay – circuit malfunction	**Wiring open circuit/short circuit, engine coolant blower motor**
P1492	Ambient/battery temperature sensor – voltage high	**Wiring, ambient/battery temperature sensor**
P1493	Ambient/battery temperature sensor – voltage low	**Wiring, ambient/battery temperature sensor**
P1494	Evaporative emission (EVAP) leak detection pump – pressure switch	**Wiring, hose leak/blockage, EVAP leak detection pump**
P1495	Evaporative emission (EVAP) leak detection pump solenoid	**Wiring open circuit/short circuit, hose leak/ blockage, EVAP leak detection pump solenoid**
P1496	Engine control module (ECM) – 5 volt output low	**ECM**
P1594	Alternator – voltage high	**Wiring, battery, alternator**
P1595	Cruise control	**Wiring open circuit/short circuit, hoses, cruise control master switch, cruise control vacuum switching valve**
P1596	Cruise control master switch – high voltage	**Wiring, cruise control master switch**
P1597	Cruise control master switch – low voltage	**Wiring, cruise control master switch**
P1598	A/C refrigerant pressure sensor – high input	**Wiring, A/C refrigerant pressure too high (cooling fault/incorrectly charged), A/C refrigerant pressure sensor**
P1599	A/C refrigerant pressure sensor – low input	**Wiring, A/C refrigerant pressure too low (incorrectly charged), A/C refrigerant pressure sensor**
P1602	Engine control module (ECM) – not programmed	**New ECM not programmed, incorrectly programmed, ECM**
P1680	Clutch pedal position (CPP) switch – circuit malfunction	**Wiring, CPP switch, ECM**
P1682	Alternator – voltage low	**Wiring, battery, alternator drive belt, alternator**
P1683	Cruise control	**Wiring, cruise control vacuum switching valve**

Scan code	Fault location	Probable cause
P1684	Engine control module (ECM) – supply voltage disconnected within last 50 ignition switch ON-OFF cycles	**Wiring, battery**
P1685	Key not programmed into immobilizer control module	**Wiring, ignition key not matched/damaged, immobilizer control module**
P1686	CAN data bus, Immobilizer control module – no signal	**Wiring, immobilizer control module**
P1687	CAN data bus, instrumentation – no signal	**Wiring, instrumentation control module (mechanical)**
P1693	Engine control module (ECM)	**Companion module trouble code detected in ECM**
P1694	CAN data bus, engine control module (ECM) – no signal	**Wiring, ECM**
P1696	Engine control module (ECM) – EEPROM error	**ECM**
P1698	CAN data bus, transmission control module (TCM) – no signal	**Wiring, TCM**
P1719	Skip shift solenoid – circuit malfunction	**Wiring open circuit/short circuit, skip shift solenoid**
P1740	Torque converter clutch (TCC)/overdrive solenoid – performance problem	**TCC, overdrive solenoid, transmission mechanical fault**
P1756	Governor pressure not equal to target (15-20 PSI) – mid pressure malfunction	**Signal wire open circuit/short circuit, valve body, sensor supply wire open circuit, transmission fluid level, governor pressure solenoid, transmission mechanical fault, ECM/TCM – no 5 volt supply**
P1757	Governor pressure not equal to target (15-20 PSI) – zero pressure malfunction	**Signal wire open circuit/short circuit, valve body, sensor supply wire open circuit, transmission fluid level, governor pressure solenoid, transmission mechanical fault, ECM/TCM – no 5 volt supply**
P1762	Governor pressure sensor – incorrect signal	**Signal wire short circuit to ground/positive, supply wire open circuit/short circuit, transmission fluid level, governor pressure solenoid, governor pressure sensor, ECM/TCM**
P1763	Governor pressure sensor – signal high	**Signal wire short circuit to positive, supply wire open circuit, transmission fluid level, governor pressure sensor, ECM/TCM**
P1764	Governor pressure sensor – signal low	**Signal wire short circuit to ground, supply wire open circuit, transmission fluid level, governor pressure sensor, ECM/TCM**
P1765	Transmission system relay	**Wiring open circuit/short circuit, transmission system relay, ECM/TCM**
P1899	Park/neutral position (PNP) switch – input circuit malfunction	**Wiring, PNP switch, ECM/TCM**

DODGE

Model:	Engine identification:	Year:
Daytona 2.2/2.5L	EDB/EDF/EDM/EDS/EDT	1992-93
Daytona 3.0L	EFA	1992-93
Dynasty 2.5/3.0L	EDM/EDB/EFA	1992-93
Intrepid 3.3/3.5L	EGB/EGE	1993-95
Monaco 3.0L	Z7X	1992
Shadow 2.2/2.5L	EDF/EDM/EDT/EDB	1992-94
Shadow 3.0L	EFA	1992-94
Spirit 2.2/2.5L	EDF/EDM/EDT/EDB/EDN	1992-95
Spirit 3.0L	EFA	1992-95
Stratus 2.4/2.5L	EDZ/EEB	1995

System: **SBEC II**

Self-diagnosis

General information

- Refer to the front of this manual for general test conditions, terminology, detailed descriptions of wiring faults and a general trouble shooter for electrical and mechanical faults.
- Malfunction indicator lamp (MIL) will illuminate if certain faults are recorded.
- ECM operates in backup mode if sensors fail, to enable vehicle to be driven to repair shop.
- Trouble codes can be displayed by the MIL or by using a Scan Tool connected to the data link connector (DLC).

Accessing

- Within 5 seconds switch the ignition key ON-OFF-ON-OFF-ON.
- Count MIL flashes. Note trouble codes.
- Each trouble code consists of two groups **1** [**A**].
- The first group of flashes indicate the 'tens' of the trouble code.
- The second group of flashes indicate the 'units' of the trouble code.
- A short pause separates 'tens' and 'units' **1** [**B**].
- A long pause separates each trouble code **1** [**C**].
- For example: Trouble code 21 displayed **1**.
- End of test sequence indicated by trouble code 55.
- Trouble codes can also be displayed by using a Scan Tool connected to the data link connector (DLC):
  - Intrepid **2**.
  - Stratus **3**.
  - Daytona **4**.
  - Shadow/Spirit **5**.
  - Dynasty **6**.
  - Monaco **7**.

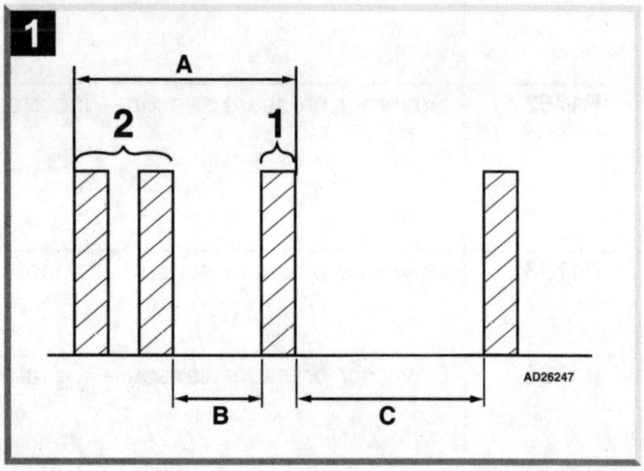

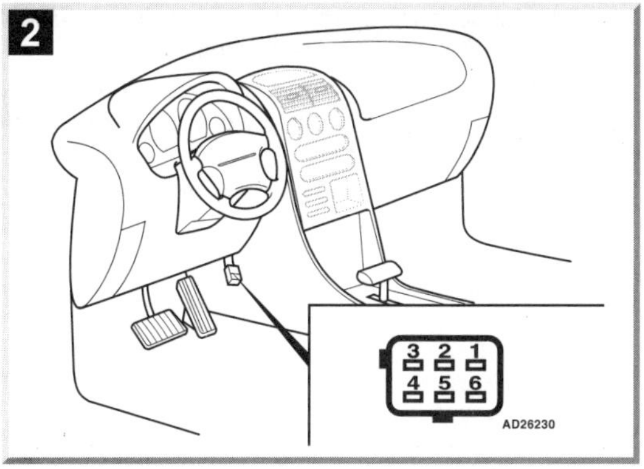
2
3 2 1
4 5 6
AD26230

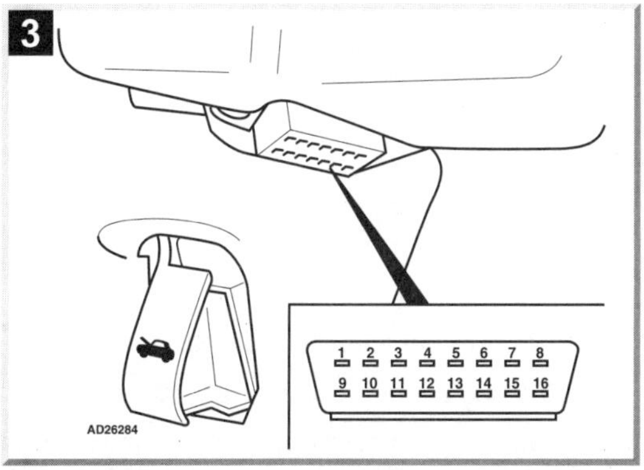
3
1 2 3 4 5 6 7 8
9 10 11 12 13 14 15 16
AD26284

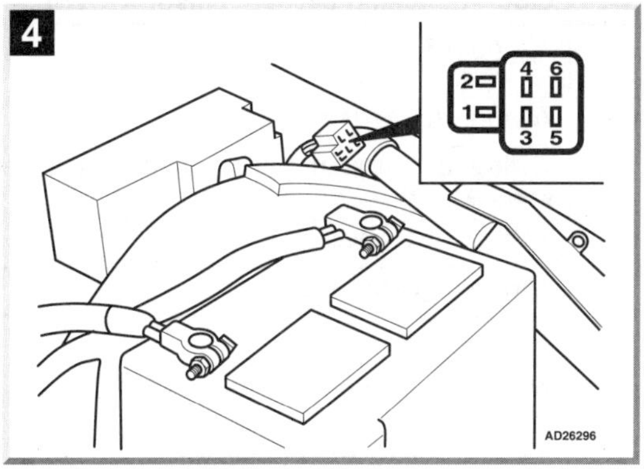
4
2 1 4 6 3 5
AD26296

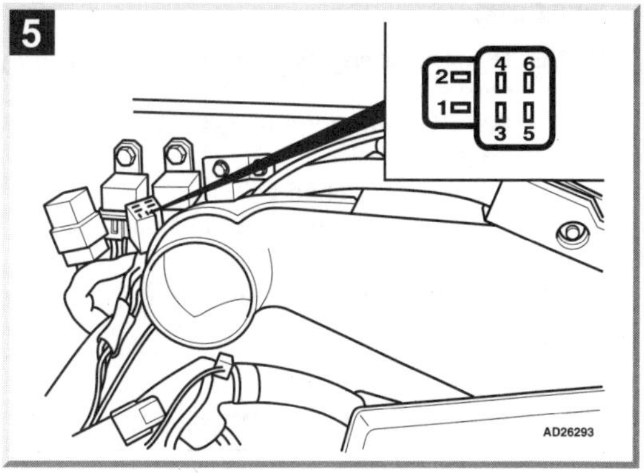
5
2 1 4 6 3 5
AD26293

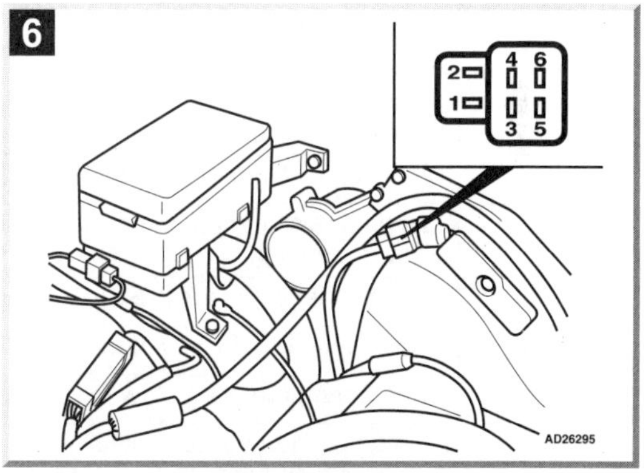
6
2 1 4 6 3 5
AD26295

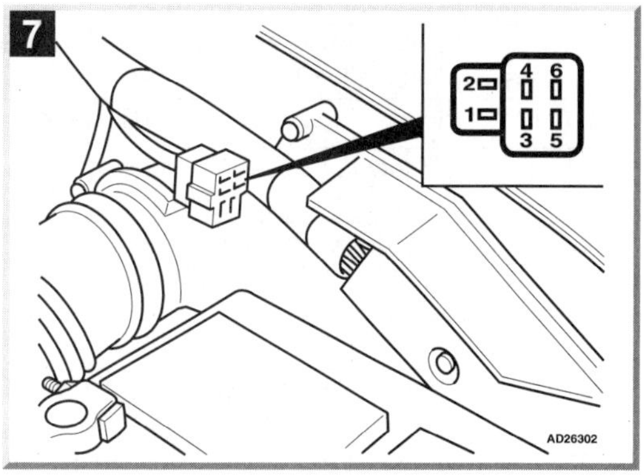
7
2 1 4 6 3 5
AD26302

Erasing

NOTE: ***Disconnecting battery ground lead may cause driveability problems until the engine management system relearns basic values.***

- Ensure ignition switched OFF.
- Disconnect battery ground cable for at least 30 seconds.
- Reconnect battery ground cable.
- Repeat accessing procedure to ensure no trouble codes remain in ECM fault memory.

NOTE: ***Disconnecting battery ground lead may also erase memory from electronic units, such as radio.***

- Trouble codes can also be erased by using a Scan Tool connected to the data link connector (DLC).

Trouble code identification

Flash code	Fault location	Probable cause
11	Crankshaft position (CKP) sensor	**Wiring, air gap, CKP sensor**
12	Engine control module (ECM) – supply voltage disconnected within last 50 ignition switch ON-OFF cycles	**Wiring, battery**
13	Manifold absolute pressure (MAP) sensor	**Wiring, MAP sensor**
14	Manifold absolute pressure (MAP) sensor – voltage low/high	**Wiring, MAP sensor**
15	Vehicle speed sensor (VSS) – signal	**Wiring, VSS**
16	Knock sensor (KS) – wiring	**Wiring open/short circuit, KS**
17	Engine temperature – low for long period	**Wiring, coolant thermostat, ECT sensor**
21	Heated oxygen sensor (HO2S) – incorrect signal/voltage high	**Wiring, fuel pressure, injector control, EVAP canister purge valve, compression, air filter blocked, heated oxygen sensor (HO2S)**
22	Engine coolant temperature (ECT) sensor – voltage low/high	**Wiring, ECT sensor**
23	Intake air temperature (IAT) sensor	**Wiring, IAT sensor**
24	Throttle position (TP) sensor	**Wiring, accelerator cable adjustment, TP sensor**
25	Idle air control (IAC) valve	**Wiring, IAC valve**
26	Injector – circuit low	**Wiring, injector, ECM**
27	Injector(s) – circuit malfunction	**Wiring, injector**
31	Evaporative emission (EVAP) canister purge valve – open/short circuit	**Wiring, EVAP canister purge valve**
32	Exhaust gas recirculation (EGR) solenoid/ system – circuit malfunction	**Wiring, hoses, EGR solenoid**
33	A/C compressor clutch relay/A/C refrigerant pressure sensor – voltage low/high	**Wiring, A/C compressor clutch relay, A/C refrigerant pressure sensor**
34	Cruise control	**Wiring open/short circuit, hoses, voltage low/high, cruise control master switch, cruise control vacuum switching valve**

Flash code	Fault location	Probable cause
35	Engine coolant blower motor relay – low/high speed	**Wiring, engine coolant blower motor relay – low/high speed**
41	Alternator	**Wiring open/short circuit, alternator**
42	Engine control relay	**Wiring open/short circuit, engine control relay**
43	Ignition coil – primary circuit	**Wiring, ignition coil**
44	Battery temperature sensor/engine coolant temperature (ECT) sensor	**Wiring open/short circuit, battery temperature sensor, ECT sensor, ECM**
46	Alternator – voltage high	**Wiring, battery, alternator**
47	Alternator – voltage low	**Wiring, battery, alternator drive belt, alternator**
51	System too lean	**Wiring, intake/exhaust leak, fuel pressure, injector control, EVAP canister purge valve, HO2S**
52	System too rich	**Wiring, exhaust blockage, fuel pressure, injector control, EVAP canister purge valve, compression, air filter blocked, ignition system, HO2S**
53	Engine control module (ECM) – internal fault	**ECM**
54	Camshaft position (CMP) sensor – no signal	**Wiring, CMP sensor**
55	End of MIL code output	–
62	Engine control module (ECM) – EEPROM error	**ECM**
63	Engine control module (ECM) – EEPROM error	**ECM**
65	Intake manifold air control solenoid	**Wiring open/short circuit, intake manifold air control solenoid**
66	CAN data bus, transmission control module (TCM)/body control module (BCM) – no signal	**Wiring, TCM, BCM**
71	Engine control module (ECM) – 5 volt output low	**ECM**
77	Cruise control system relay	**Wiring open/short circuit, cruise control system relay**

DODGE

Model:	Engine identification:	Year:
Caravan 2.4L	EDZ	1996-97
Caravan 3.0L	EFA	1996-97
Caravan 3.3L	EGA/EGM	1996-97
Caravan 3.8L	EGH	1996-97

System: **SBEC III**

Self-diagnosis

General information

- Refer to the front of this manual for general test conditions, terminology, detailed descriptions of wiring faults and a general trouble shooter for electrical and mechanical faults.
- Malfunction indicator lamp (MIL) will illuminate or flash if certain faults are recorded.
- ECM operates in backup mode if sensors fail, to enable vehicle to be driven to repair shop.
- Trouble codes can be displayed by the MIL or by using a Scan Tool connected to the data link connector (DLC).

Accessing

Flash codes

- Within 5 seconds switch the ignition key ON-OFF-ON-OFF-ON.
- Count MIL flashes. Note trouble codes.
- Each trouble code consists of two groups **1** [**A**].
- The first group of flashes indicate the 'tens' of the trouble code.
- The second group of flashes indicate the 'units' of the trouble code.
- A short pause separates 'tens' and 'units' **1** [**B**].
- A long pause separates each trouble code **1** [**C**].
- For example: Trouble code 21 displayed **1**.
- End of test sequence indicated by trouble code 55.

P codes & Hex codes

- Trouble codes can be displayed by using a Scan Tool connected to the data link connector (DLC) **2**.

1

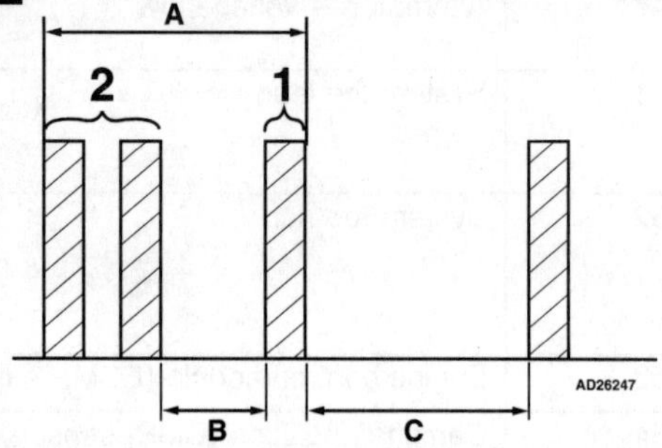

2

Erasing

- Trouble codes can only be erased by using a Scan Tool connected to the data link connector (DLC).

Trouble code identification

Scan code P type	Fault location	Probable cause
P0000-P0999	Refer to OBD-II trouble code table at the front of this manual	–
P1290	Compressed natural gas pressure sensor – pressure high	**Compressed natural gas system, wiring, compressed natural gas pressure sensor**
P1294	Idle speed control (ISC) system – rpm lower/higher than expected	**Wiring, air leak, throttle valve tight/sticking, IAC valve**
P1297	Manifold absolute pressure (MAP) sensor	**Wiring, hose connection(s), MAP sensor**
P1391	Crankshaft position (CKP) sensor/camshaft position (CMP) sensor – intermittent signal	**Wiring, poor connection, air gap, CKP sensor, CMP sensor**
P1398	Engine control module (ECM) – misfire detection limit reached	**Wiring, poor connection, CKP sensor**
P1486	Evaporative emission (EVAP) system – blockage detected	**Hose connection(s), hose blocked/trapped, EVAP canister**
P1491	Engine coolant blower motor relay – low/high speed	**Wiring open circuit/short circuit, engine coolant blower motor relay – low/high speed**
P1492	Ambient/battery temperature sensor – voltage high	**Wiring, ambient/battery temperature sensor**
P1493	Ambient/battery temperature sensor – voltage low	**Wiring, ambient/battery temperature sensor**
P1494	Evaporative emission (EVAP) leak detection pump – pressure switch	**Wiring, hose leak/blockage, EVAP leak detection pump**
P1495	Evaporative emission (EVAP) leak detection pump solenoid	**Wiring open circuit/short circuit, hose leak/blockage, EVAP leak detection pump solenoid**
P1496	Engine control module (ECM) – 5 volt output low	**ECM**
P1696	Engine control module (ECM) – EEPROM error	**ECM**
P1697	Engine control module (ECM) – EEPROM error	**ECM**
P1698	CAN data bus, transmission control module (TCM) – no signal	**Wiring, TCM**
P1899	Park/neutral position (PNP) switch – input circuit malfunction	**Wiring, PNP switch, ECM/TCM**

Flash code	Scan code Hex type	Fault location	Probable cause
11		Crankshaft position (CKP) sensor/camshaft position (CMP) sensor	**Wiring, poor connection, air gap, CKP sensor, CMP sensor**
	9D	Crankshaft position (CKP) sensor/camshaft position (CMP) sensor – intermittent signal	**Wiring, poor connection, air gap, CKP sensor, CMP sensor**

Flash code	Scan code Hex type	Fault location	Probable cause
	28	Crankshaft position (CKP) sensor – no signal	**Wiring, air gap, CKP sensor**
	BA	Engine control module (ECM) – misfire detection limit reached	**Wiring, poor connection, CKP sensor**
12		Engine control module (ECM) – supply voltage disconnected within last 50 ignition switch ON-OFF cycles	**Wiring, battery**
13	27	Manifold absolute pressure (MAP) sensor	**Wiring, hose connection(s), MAP sensor**
14		Manifold absolute pressure (MAP) sensor – input low/high	**Wiring, MAP sensor**
	24	Manifold absolute pressure (MAP) sensor – low input	**Wiring short circuit to ground, MAP sensor**
	25	Manifold absolute pressure (MAP) sensor – high input	**Wiring short circuit to positive, MAP sensor**
15	23	Vehicle speed sensor (VSS) – signal	**Wiring, VSS**
17		Engine temperature – low for long period	**Wiring, cooling system, coolant thermostat, ECT sensor**
	21	Engine temperature – low for long period	**Wiring, cooling system, coolant thermostat, ECT sensor**
	80	Insufficient coolant temperature for closed loop fuel control	**Wiring, cooling system, coolant thermostat, ECT sensor**
21		Heated oxygen sensor (HO2S) 1/2, bank 1 – circuit malfunction/incorrect signal/voltage low/high	**Wiring, fuel pressure, injector control, EVAP canister purge valve, compression, air filter blocked, HO2S**
	9B	Heated oxygen sensor (HO2S) 1, bank 1 – low voltage	**Exhaust leak, wiring short circuit to ground, HO2S**
	3E	Heated oxygen sensor (HO2S) 1, bank 1 – high voltage	**Wiring short circuit to positive, fuel pressure, injector control, EVAP canister purge valve, air filter blocked, HO2S**
	66	Heated oxygen sensor (H02S) 1, bank 1 – slow response	**Wiring, heating inoperative, HO2S**
	20	Heated oxygen sensor (H02S) 1, bank 1 – no activity detected	**Wiring open circuit, heating inoperative, HO2S**
	67	Heated oxygen sensor (HO2S) 1, bank 1 – heater control – circuit malfunction	**wiring, HO2S**
	9C	Heated oxygen sensor (HO2S) 2, bank 1 – low voltage	**Exhaust leak, wiring short circuit to ground, HO2S**
	7E	Heated oxygen sensor (HO2S) 2, bank 1 – high voltage	**Wiring short circuit to positive, catalytic converter, HO2S**
	81	Heated oxygen sensor (HO2S) 2, bank 1 – no activity detected	**Wiring, heating inoperative, HO2S**
	69	Heated oxygen sensor (HO2S) 2, bank 1, heater control – circuit malfunction	**Wiring, HO2S**

Flash code	Scan code Hex type	Fault location	Probable cause
	C0	Heated oxygen sensor (HO2S) 1, bank 1 – slow response	**Wiring, heating inoperative, HO2S**
22		Engine coolant temperature (ECT) sensor – voltage low/high	**Wiring, coolant thermostat, ECT sensor**
	1E	Engine coolant temperature (ECT) sensor – low input	**Coolant thermostat, wiring short circuit to ground, ECT sensor**
	1F	Engine coolant temperature (ECT) sensor – high input	**Coolant thermostat, wiring open circuit/ short circuit to positive, ground wire defective, ECT sensor**
23		Intake air temperature (IAT) sensor – voltage low/high	**Wiring, IAT sensor**
	39	Intake air temperature (IAT) sensor – low input	**Wiring short circuit to ground, IAT sensor**
	3A	Intake air temperature (IAT) sensor – high input	**Wiring open circuit/short circuit to positive, ground wire defective, IAT sensor**
24		Throttle position (TP) sensor – voltage low/ high	**Wiring, accelerator cable adjustment, TP sensor**
	84	Throttle position (TP) sensor – range/ performance problem	**Wiring, accelerator cable adjustment, TP sensor**
	1A	Throttle position (TP) sensor – low input	**Wiring short circuit to ground, TP sensor**
	1B	Throttle position (TP) sensor – high input	**Wiring short circuit to positive, TP sensor**
25		Idle air control (IAC) valve	**Wiring open circuit/short circuit, IAC valve**
	19	Idle speed control (ISC) system – malfunction	**Wiring open circuit/short circuit, IAC valve**
	8A	Idle speed control (ISC) system – rpm lower/higher than expected	**Wiring, air leak, throttle valve tight/ sticking, IAC valve**
27		Injector(s) – circuit malfunction	**Wiring, injector**
	15	Injector 1 – circuit malfunction	**Wiring, injector**
	14	Injector 2 – circuit malfunction	**Wiring, injector**
	13	Injector 3 – circuit malfunction	**Wiring, injector**
	3D	Injector 4 – circuit malfunction	**Wiring, injector**
	45	Injector 5 – circuit malfunction	**Wiring, injector**
	46	Injector 6 – circuit malfunction	**Wiring, injector**
31		Evaporative emission (EVAP) canister purge valve – open/short circuit	**Wiring, EVAP canister purge valve**
	71	Evaporative emission (EVAP) system – incorrect flow detected	**Hose connection(s), intake leak, EVAP canister purge valve**
	A0	Evaporative emission (EVAP) system – small leak detected	**Hose connection(s), intake leak, EVAP canister, EVAP canister purge valve**

Flash code	Scan code Hex type	Fault location	Probable cause
	12	Evaporative emission (EVAP) canister purge valve – circuit malfunction	**Wiring open circuit/short circuit, EVAP canister purge valve**
	A1	Evaporative emission (EVAP) system – large leak detected	**Hose connection(s), intake leak, EVAP canister, EVAP canister purge valve**
	BB	Evaporative emission (EVAP) system – blockage detected	**Hose connection(s), hose blocked/ trapped, EVAP canister**
	B8	Evaporative emission (EVAP) leak detection pump – pressure switch	**Wiring, hose leak/blockage, EVAP leak detection pump**
	B7	Evaporative emission (EVAP) leak detection pump solenoid	**Wiring open circuit/short circuit, hose leak/blockage, EVAP leak detection pump solenoid**
32		Exhaust gas recirculation (EGR) solenoid/ system – circuit malfunction	**Wiring, hoses, EGR solenoid**
	2E	Exhaust gas recirculation (EGR) system – insufficient flow detected	**Hose leak/blockage, wiring, EGR valve, EGR solenoid**
	11	Exhaust gas recirculation (EGR) – circuit malfunction	**Wiring open circuit/short circuit, EGR solenoid**
33		A/C compressor clutch relay/A/C refrigerant pressure sensor – voltage low/high	**A/C refrigerant pressure too low/high (cooling fault/incorrectly charged), wiring open circuit/short circuit, A/C compressor clutch relay, A/C refrigerant pressure sensor**
	10	A/C compressor clutch relay	**Wiring open circuit/short circuit, A/C compressor clutch relay**
	5A	A/C refrigerant pressure sensor – high input	**A/C refrigerant pressure too high (cooling fault/incorrectly charged), wiring, A/C refrigerant pressure sensor**
	5B	A/C refrigerant pressure sensor – low input	**A/C refrigerant pressure too low (incorrectly charged), wiring, A/C refrigerant pressure sensor**
34	0F	Cruise control	**Wiring open circuit/short circuit, hoses, voltage low/high, cruise control master switch, cruise control vacuum switching valve**
35	0E	Engine coolant blower motor relay – low/ high speed	**Wiring open circuit/short circuit, engine coolant blower motor relay – low/high speed**
37		Torque converter clutch (TCC) solenoid/ park/neutral position (PNP) switch – circuit malfunction	**Wiring, TCC solenoid**
	72	Park/neutral position (PNP) switch – input circuit malfunction	**Wiring, PNP switch, ECM/TCM**
	94	Torque converter clutch (TCC) solenoid – circuit malfunction	**Wiring, TCC solenoid**
	0C	Torque converter clutch (TCC) solenoid – electrical	**Wiring open circuit/short circuit, TCC solenoid**

Flash code	Scan code Hex type	Fault location	Probable cause
41	0B	Alternator, field control – circuit malfunction	Wiring open circuit/short circuit, alternator, battery
42		Fuel pump relay/engine control relay/fuel level sensor – circuit malfunction	Wiring open circuit/short circuit, Fuel pump relay, engine control relay, fuel level sensor
	65	Fuel pump relay	Wiring open circuit/short circuit, fuel pump relay
	0A	Engine control relay	Wiring open circuit/short circuit, engine control relay
	2C	Engine control relay	Wiring open circuit, engine control relay
	95	Fuel level sensor – low input	Wiring open circuit, fuel level sensor
	96	Fuel level sensor – high input	Wiring short circuit to positive, fuel level sensor
	97	Fuel level sensor – malfunction	Wiring, fuel level sensor
43		Random/multiple cylinder(s) – misfire detected	Wiring, spark plug(s), HT lead(s), injector(s), ignition coil(s), low compression
	6A	Random/multiple cylinder(s) – misfire detected	Wiring, spark plug(s), HT lead(s), injector(s), ignition coil(s), low compression
	6B	Cylinder 1 – misfire detected	Wiring, engine mechanical fault, ignition/fuel system, injector
	6C	Cylinder 2 – misfire detected	Wiring, engine mechanical fault, ignition/fuel system, injector
	6D	Cylinder 3 – misfire detected	Wiring, engine mechanical fault, ignition/fuel system, injector
	6E	Cylinder 4 – misfire detected	Wiring, engine mechanical fault, ignition/fuel system, injector
	AE	Cylinder 5 – misfire detected	Wiring, engine mechanical fault, ignition/fuel system, injector
	AF	Cylinder 6 – misfire detected	Wiring, engine mechanical fault, ignition/fuel system, injector
	2B	Ignition coil A, primary/secondary – circuit malfunction	Wiring, ignition coil
	2A	Ignition coil B, primary/secondary – circuit malfunction	Wiring, ignition coil
	29	Ignition coil C, primary/secondary – circuit malfunction	Wiring, ignition coil
44		Battery temperature sensor – voltage low/high	Wiring, battery temperature sensor
	9A	Ambient/battery temperature sensor – voltage high	Wiring, ambient/battery temperature sensor
	99	Ambient/battery temperature sensor – voltage low	Wiring, ambient/battery temperature sensor

Flash code	Scan code Hex type	Fault location	Probable cause
45	89	Transmission control system – malfunction	**Wiring, ECM/TCM**
46	06	Alternator – voltage high	**Wiring, battery, alternator**
47	05	Alternator – voltage low	**Wiring, battery, alternator drive belt, alternator**
51	77	System too lean	**Wiring, intake/exhaust leak, MAP sensor, fuel pressure/pump, EVAP canister purge valve, injector(s), HO2S**
52	76	System too rich	**Wiring, intake blocked, EVAP canister purge valve, fuel pressure, EGR system, injector(s), air filter blocked, ignition system, HO2S**
53		Engine control module (ECM) – internal fault	**ECM**
	02	Engine control module (ECM) – internal fault	**ECM**
	44	Engine control module (ECM) – internal fault	**ECM**
54	01	Camshaft position (CMP) sensor – no signal	**Wiring, CMP sensor**
55		End of MIL code output	–
61	3C	Manifold absolute pressure (MAP) sensor – barometric pressure range/ performance problem	**Intake/exhaust leak, wiring, MAP sensor**
62	30	Engine control module (ECM) – EEPROM error	**ECM**
63	31	Engine control module (ECM) – EEPROM error	**ECM**
64	BE	Compressed natural gas pressure sensor – pressure high	**Compressed natural gas system, wiring, compressed natural gas pressure sensor**
65	98	Brake pedal position (BPP) switch – circuit malfunction	**Wiring, BPP switch**
66		CAN data bus, transmission control module (TCM)/body control module (BCM) – no signal	**Wiring, TCM, BCM**
	60	CAN data bus, transmission control module (TCM) – no signal	**Wiring, TCM**
	61	CAN data bus, body control module (BCM) – no signal	**Wiring, BCM**
71	92	Engine control module (ECM) – 5 volt output low	**ECM**
72	70	Catalytic converter system, bank 1 – efficiency below threshold	**Catalytic converter, wiring, HO2S 2**
77	52	Cruise control	**Wiring, cruise control vacuum switching valve**

DODGE

Model:	Engine identification:	Year:
Caravan 2.4L	EDZ	1998
Caravan 3.0L	EFA	1998
Caravan 3.3L	EGA/EGM	1998
Caravan 3.8L	EGH	1998

System: **SBEC III**

Self-diagnosis

General information

- Refer to the front of this manual for general test conditions, terminology, detailed descriptions of wiring faults and a general trouble shooter for electrical and mechanical faults.
- Malfunction indicator lamp (MIL) will illuminate or flash if certain faults are recorded.
- ECM operates in backup mode if sensors fail, to enable vehicle to be driven to repair shop.

Accessing

- Trouble codes can only be displayed by using a Scan Tool connected to the data link connector (DLC) **1**.

1

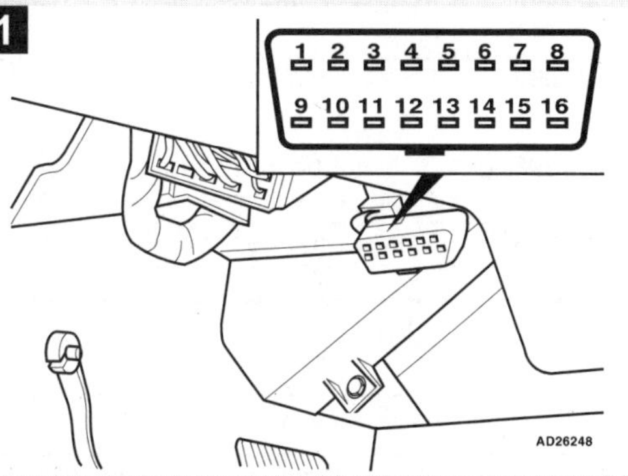

Erasing

- Trouble codes can only be erased by using a Scan Tool connected to the data link connector (DLC).

Trouble code identification

Scan code P type	Fault location	Probable cause
P0000-P0999	Refer to OBD-II trouble code table at the front of this manual	–
P1290	Compressed natural gas pressure sensor – pressure high	**Compressed natural gas system, wiring, compressed natural gas pressure sensor**
P1294	Idle speed control (ISC) system – rpm lower/ higher than expected	**Wiring, air leak, throttle valve tight/sticking, IAC valve**
P1297	Manifold absolute pressure (MAP) sensor	**Wiring, hose connection(s), MAP sensor**
P1391	Crankshaft position (CKP) sensor/camshaft position (CMP) sensor – intermittent signal	**Wiring, poor connection, air gap, CKP sensor, CMP sensor**
P1398	Engine control module (ECM) – misfire detection limit reached	**Wiring, poor connection, CKP sensor**
P1486	Evaporative emission (EVAP) system – blockage detected	**Hose connection(s), hose blocked/trapped, EVAP canister**

Scan code P type	Fault location	Probable cause
P1491	Engine coolant blower motor relay – low/high speed	**Wiring open circuit/short circuit, engine coolant blower motor relay – low/high speed**
P1492	Ambient/battery temperature sensor – voltage high	**Wiring, ambient/battery temperature sensor**
P1493	Ambient/battery temperature sensor – voltage low	**Wiring, ambient/battery temperature sensor**
P1494	Evaporative emission (EVAP) leak detection pump – pressure switch	**Wiring, hose leak/blockage, EVAP leak detection pump**
P1495	Evaporative emission (EVAP) leak detection pump solenoid	**Wiring open circuit/short circuit, hose leak/ blockage, EVAP leak detection pump solenoid**
P1496	Engine control module (ECM) – 5 volt output low	**ECM**
P1696	Engine control module (ECM) – EEPROM error	**ECM**
P1697	Engine control module (ECM) – EEPROM error	**ECM**
P1698	CAN data bus, transmission control module (TCM) – no signal	**Wiring, TCM**
P1899	Park/neutral position (PNP) switch – input circuit malfunction	**Wiring, PNP switch, ECM/TCM**

Scan code Hex type	Fault location	Probable cause
01	Camshaft position (CMP) sensor – no signal	**Wiring, CMP sensor**
02	Engine control module (ECM) – internal fault	**ECM**
05	Alternator – voltage low	**Wiring, battery, alternator drive belt, alternator**
06	Alternator – voltage high	**Wiring, battery, alternator**
0A	Engine control relay	**Wiring open circuit/short circuit, engine control relay**
0B	Alternator, field control – circuit malfunction	**Wiring open circuit/short circuit, alternator, battery**
0C	Torque converter clutch (TCC) solenoid – electrical	**Wiring open circuit/short circuit, TCC solenoid**
0E	Engine coolant blower motor relay – low/high speed	**Wiring open circuit/short circuit, engine coolant blower motor relay – low/high speed**
0F	Cruise control	**Wiring open circuit/short circuit, hoses, voltage low/high, cruise control master switch, cruise control vacuum switching valve**
10	A/C compressor clutch relay	**Wiring open circuit/short circuit, A/C compressor clutch relay**
11	Exhaust gas recirculation (EGR) – circuit malfunction	**Wiring open circuit/short circuit, EGR solenoid**
12	Evaporative emission (EVAP) canister purge valve – circuit malfunction	**Wiring open circuit/short circuit, EVAP canister purge valve**
13	Injector 3 – circuit malfunction	**Wiring, injector**
14	Injector 2 – circuit malfunction	**Wiring, injector**

Scan code Hex type	Fault location	Probable cause
15	Injector 1 – circuit malfunction	**Wiring, injector**
19	Idle speed control (ISC) system – malfunction	**Wiring open circuit/short circuit, IAC valve**
1A	Throttle position (TP) sensor – low input	**Wiring short circuit to ground, TP sensor**
1B	Throttle position (TP) sensor – high input	**Wiring short circuit to positive, TP sensor**
1E	Engine coolant temperature (ECT) sensor – low input	**Coolant thermostat, wiring short circuit to ground, ECT sensor**
1F	Engine coolant temperature (ECT) sensor – high input	**Coolant thermostat, wiring open circuit/ short circuit to positive, ground wire defective, ECT sensor**
20	Heated oxygen sensor (H02S) 1, bank 1 – no activity detected	**Wiring open circuit, heating inoperative, HO2S**
21	Engine temperature – low for long period	**Wiring, cooling system, coolant thermostat, ECT sensor**
23	Vehicle speed sensor (VSS) – signal	**Wiring, VSS**
24	Manifold absolute pressure (MAP) sensor – low input	**Wiring short circuit to ground, MAP sensor**
25	Manifold absolute pressure (MAP) sensor – high input	**Wiring short circuit to positive, MAP sensor**
27	Manifold absolute pressure (MAP) sensor	**Wiring, hose connection(s), MAP sensor**
28	Crankshaft position (CKP) sensor – no signal	**Wiring, air gap, CKP sensor**
29	Ignition coil C, primary/secondary – circuit malfunction	**Wiring, ignition coil**
2A	Ignition coil B, primary/secondary – circuit malfunction	**Wiring, ignition coil**
2B	Ignition coil A, primary/secondary – circuit malfunction	**Wiring, ignition coil**
2C	Engine control relay	**Wiring open circuit, engine control relay**
2E	Exhaust gas recirculation (EGR) system – insufficient flow detected	**Hose leak/blockage, wiring, EGR valve, EGR solenoid**
30	Engine control module (ECM) – EEPROM error	**ECM**
31	Engine control module (ECM) – EEPROM error	**ECM**
39	Intake air temperature (IAT) sensor – low input	**Wiring short circuit to ground, IAT sensor**
3A	Intake air temperature (IAT) sensor – high input	**Wiring open circuit/short circuit to positive, ground wire defective, IAT sensor**
3C	Manifold absolute pressure (MAP) sensor – barometric pressure range/performance problem	**Intake/exhaust leak, wiring, MAP sensor**
3D	Injector 4 – circuit malfunction	**Wiring, injector**
3E	Heated oxygen sensor (HO2S) 1, bank 1 – high voltage	**Wiring short circuit to positive, fuel pressure, injector control, EVAP canister purge valve, air filter blocked, HO2S**
44	Engine control module (ECM) – internal fault	**ECM**

Scan code Hex type	Fault location	Probable cause
45	Injector 5 – circuit malfunction	**Wiring, injector**
46	Injector 6 – circuit malfunction	**Wiring, injector**
52	Cruise control	**Wiring, cruise control vacuum switching valve**
5A	A/C refrigerant pressure sensor – high input	**A/C refrigerant pressure too high (cooling fault/incorrectly charged), wiring, A/C refrigerant pressure sensor**
5B	A/C refrigerant pressure sensor – low input	**A/C refrigerant pressure too low (incorrectly charged), wiring, A/C refrigerant pressure sensor**
60	CAN data bus, TCM – no signal	**Wiring, TCM**
61	CAN data bus, body control module (BCM)– no signal	**Wiring, BCM**
65	Fuel pump relay	**Wiring open circuit/short circuit, fuel pump relay**
66	Heated oxygen sensor (H02S) 1, bank 1 – slow response	**Wiring, heating inoperative, HO2S**
67	Heated oxygen sensor (HO2S) 1, bank 1 – heater control – circuit malfunction	**Wiring, HO2S**
69	Heated oxygen sensor (HO2S) 2, bank 1, heater control – circuit malfunction	**Wiring, HO2S**
6A	Random/multiple cylinder(s) – misfire detected	**Wiring, spark plug(s), HT lead(s), injector(s), ignition coil(s), low compression**
6B	Cylinder 1 – misfire detected	**Wiring, engine mechanical fault, ignition/fuel system, injector**
6C	Cylinder 2 – misfire detected	**Wiring, engine mechanical fault, ignition/fuel system, injector**
6D	Cylinder 3 – misfire detected	**Wiring, engine mechanical fault, ignition/fuel system, injector**
6E	Cylinder 4 – misfire detected	**Wiring, engine mechanical fault, ignition/fuel system, injector**
70	Catalytic converter system, bank 1 – efficiency below threshold	**Catalytic converter, wiring, HO2S 2**
71	Evaporative emission (EVAP) system – incorrect flow detected	**Hose connection(s), intake leak, EVAP canister purge valve**
72	Park/neutral position (PNP) switch – input circuit malfunction	**Wiring, PNP switch, ECM/TCM**
76	System too rich	**Wiring, intake blocked, EVAP canister purge valve, fuel pressure, EGR system, injector(s), air filter blocked, ignition system, HO2S**
77	System too lean	**Wiring, intake/exhaust leak, MAP sensor, fuel pressure/pump, EVAP canister purge valve, injector(s), HO2S**
7E	Heated oxygen sensor (HO2S) 2, bank 1 – high voltage	**Wiring short circuit to positive, catalytic converter, HO2S**
80	Insufficient coolant temperature for closed loop fuel control	**Wiring, cooling system, coolant thermostat, ECT sensor**

Scan code Hex type	Fault location	Probable cause
81	Heated oxygen sensor (HO2S) 2, bank 1 – no activity detected	Wiring, heating inoperative, HO2S
84	Throttle position (TP) sensor – range/ performance problem	Wiring, accelerator cable adjustment, TP sensor
89	Transmission control system – malfunction	Wiring, ECM/TCM
8A	Idle speed control (ISC) system – rpm lower/ higher than expected	Wiring, air leak, throttle valve tight/sticking, IAC valve
92	Engine control module (ECM) – 5 volt output low	ECM
94	Torque converter clutch (TCC) solenoid – circuit malfunction	Wiring, TCC solenoid
95	Fuel level sensor – low input	Wiring open circuit, fuel level sensor
96	Fuel level sensor – high input	Wiring short circuit to positive, fuel level sensor
97	Fuel level sensor – malfunction	Wiring, fuel level sensor
98	Brake pedal position (BPP) switch – circuit malfunction	Wiring, BPP switch
99	Ambient/battery temperature sensor – voltage low	Wiring, ambient/battery temperature sensor
9A	Ambient/battery temperature sensor – voltage high	Wiring, ambient/battery temperature sensor
9B	Heated oxygen sensor (HO2S) 1, bank 1 – low voltage	Exhaust leak, wiring short circuit to ground, HO2S
9C	Heated oxygen sensor (HO2S) 2, bank 1 – low voltage	Exhaust leak, wiring short circuit to ground, HO2S
9D	Crankshaft position (CKP) sensor/camshaft position (CMP) sensor – intermittent signal	Wiring, poor connection, air gap, CKP sensor, CMP sensor
A0	Evaporative emission (EVAP) system – small leak detected	Hose connection(s), intake leak, EVAP canister, EVAP canister purge valve
A1	Evaporative emission (EVAP) system – large leak detected	Hose connection(s), intake leak, EVAP canister, EVAP canister purge valve
AE	Cylinder 5 – misfire detected	Wiring, engine mechanical fault, ignition/fuel system, injector
AF	Cylinder 6 – misfire detected	Wiring, engine mechanical fault, ignition/fuel system, injector
B7	Evaporative emission (EVAP) leak detection pump solenoid	Wiring open circuit/short circuit, hose leak/ blockage, EVAP leak detection pump solenoid
B8	Evaporative emission (EVAP) leak detection pump – pressure switch	Wiring, hose leak/blockage, EVAP leak detection pump
BA	Engine control module (ECM) – misfire detection limit reached	Wiring, poor connection, CKP sensor
BB	Evaporative emission (EVAP) system – blockage detected	Hose connection(s), hose blocked/trapped, EVAP canister
BE	Compressed natural gas pressure sensor – pressure high	Compressed natural gas system, wiring, compressed natural gas pressure sensor
C0	Heated oxygen sensor (HO2S) 1, bank 1 – slow response	Wiring, heating inoperative, HO2S

DODGE

Model:	Engine identification:	Year:
Caravan 2.4L	**EDZ**	**1999-02**
Caravan 3.0L	**EFA**	**1999-02**
Caravan 3.3L	**EGA/EGM**	**1999-02**
Caravan 3.8L	**EGH**	**1999-02**

System: **SBEC III**

Self-diagnosis

General information

- Refer to the front of this manual for general test conditions, terminology, detailed descriptions of wiring faults and a general trouble shooter for electrical and mechanical faults.
- Malfunction indicator lamp (MIL) will illuminate or flash if certain faults are recorded.
- ECM operates in backup mode if sensors fail, to enable vehicle to be driven to repair shop.

Accessing

1999-01

- Trouble codes can only be displayed by using a Scan Tool connected to the data link connector (DLC) 1999-00 **1** & 2000→ **2**.

1

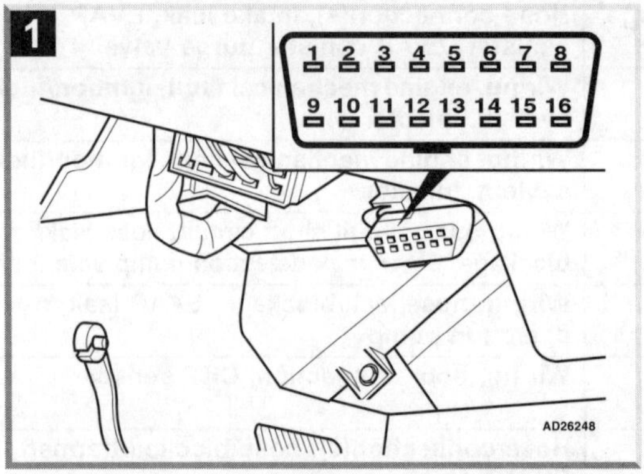

2

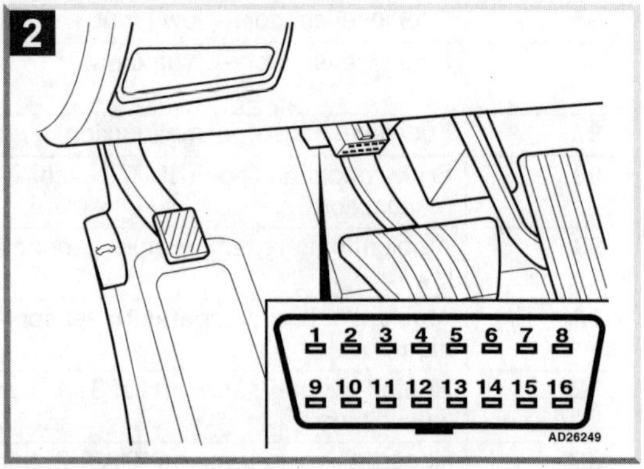

2001→

- Within 5 seconds switch the ignition key ON-OFF-ON-OFF-ON.
- Odometer will display trouble codes.
- If no trouble codes are stored in ECM, odometer will display 'P1684' and 'done' or only 'done'.
- Trouble codes can also be displayed by using a Scan Tool connected to the data link connector (DLC) **2**.

Erasing

- Trouble codes can only be erased by using a Scan Tool connected to the data link connector (DLC).

Trouble code identification

Scan code	Fault location	Probable cause
P0000-P0999	Refer to OBD-II trouble code table at the front of this manual	–
P1192	Intake air temperature (IAT) sensor – voltage low	**Wiring, IAT sensor**
P1193	Intake air temperature (IAT) sensor – voltage high	**Wiring, IAT sensor**
P1195	Heated oxygen sensor (H02S) 1, bank 1 – slow response	**Heating inoperative, wiring, HO2S**
P1196	Heated oxygen sensor (H02S) 2, bank 1 – slow response	**Heating inoperative, wiring, HO2S**
P1281	Engine temperature – low for long period	**Wiring, cooling system, coolant thermostat, ECT sensor**
P1282	Fuel pump relay	**Wiring open circuit/short circuit, fuel pump relay**
P1290	Compressed natural gas pressure sensor – pressure high	**Compressed natural gas system, wiring, compressed natural gas pressure sensor**
P1292	Compressed natural gas pressure sensor – voltage high	**Wiring, compressed natural gas pressure sensor**
P1293	Compressed natural gas pressure sensor – voltage low	**Wiring, compressed natural gas pressure sensor**
P1294	Idle speed control (ISC) system – rpm lower/higher than expected	**Wiring, air leak, throttle valve tight/sticking, IAC valve**
P1295	Throttle position (TP) sensor – supply voltage	**Wiring, TP sensor, ECM**
P1296	Manifold absolute pressure (MAP) sensor – supply voltage	**Wiring, MAP sensor, ECM**
P1297	Manifold absolute pressure (MAP) sensor	**Wiring, hose connection(s), MAP sensor**
P1298	System too lean – at prolonged wide open throttle	**Intake/exhaust leak, MAP sensor, fuel pressure/pump, injector(s), HO2S**
P1299	Manifold absolute pressure (MAP) sensor/throttle position (TP) sensor	**Intake leak, hose connection(s), MAP sensor, TP sensor**
P1388	Engine control relay	**Wiring open circuit/short circuit, engine control relay**
P1389	Engine control relay	**Wiring open circuit, engine control relay**
P1390	Crankshaft position (CKP) sensor/camshaft position (CMP) sensor – signal synchronisation	**Timing belt loose/jumped position, CKP sensor, CMP sensor**
P1391	Crankshaft position (CKP) sensor/camshaft position (CMP) sensor – intermittent signal	**Wiring, poor connection, air gap, CKP sensor, CMP sensor**
P1398	Engine control module (ECM) – misfire detection limit reached	**Wiring, poor connection, CKP sensor**
P1486	Evaporative emission (EVAP) system – blockage detected	**Hose connection(s), hose blocked/trapped, EVAP canister**
P1488	Engine control module (ECM) – auxiliary 5 volt output low	**Wiring, ECM**

Scan code	Fault location	Probable cause
P1489	Engine coolant blower motor relay	**Wiring open circuit/short circuit, engine coolant blower motor relay – high speed**
P1490	Engine coolant blower motor relay	**Wiring open circuit/short circuit, engine coolant blower motor relay – low speed**
P1491	Engine coolant blower motor relay – circuit malfunction	**Wiring open circuit/short circuit, engine coolant blower motor relay**
P1492	Ambient/battery temperature sensor – high input	**Wiring short circuit to positive, ambient/ battery temperature sensor, ECM**
P1493	Ambient/battery temperature sensor – voltage low	**Wiring, ambient/battery temperature sensor**
P1494	Evaporative emission (EVAP) leak detection pump – pressure switch	**Wiring, hose leak/blockage, EVAP leak detection pump**
P1495	Evaporative emission (EVAP) leak detection pump solenoid	**Wiring open circuit/short circuit, hose leak/ blockage, EVAP leak detection pump solenoid**
P1496	Engine control module (ECM) – 5 volt output low	**ECM**
P1594	Alternator – voltage high	**Wiring, battery, alternator**
P1595	Cruise control	**Wiring open circuit/short circuit, hoses, cruise control master switch, cruise control vacuum switching valve**
P1596	Cruise control master switch – high voltage	**Wiring, cruise control master switch**
P1597	Cruise control master switch – low voltage	**Wiring, cruise control master switch**
P1598	A/C refrigerant pressure sensor – high input	**Wiring, A/C refrigerant pressure too high (cooling fault/incorrectly charged), A/C refrigerant pressure sensor**
P1599	A/C refrigerant pressure sensor – low input	**Wiring, A/C refrigerant pressure too low (incorrectly charged), A/C refrigerant pressure sensor**
P1602	Engine control module (ECM) – not programmed	**New ECM not programmed, incorrectly programmed, ECM**
P1603	Engine control module (ECM) – RAM link error	**ECM**
P1604	Engine control module (ECM) – RAM read/ write error	**ECM**
P1607	Engine control module (ECM) – shut down timer error	**ECM**
P1681	CAN data bus, instrumentation – no signal	**Wiring, instrumentation control module**
P1682	Alternator – voltage low	**Wiring, battery, alternator drive belt, alternator**
P1683	Cruise control	**Wiring, cruise control vacuum switching valve**
P1684	Engine control module (ECM) – supply voltage disconnected within last 50 ignition switch ON-OFF cycles	**Wiring, battery**

Scan code	Fault location	Probable cause
P1685	Key not programmed into immobilizer control module	**Wiring, ignition key not matched/damaged, immobilizer control module**
P1686	CAN data bus, immobilizer control module – no signal	**Wiring, immobilizer control module**
P1687	CAN data bus, instrumentation – no signal	**Wiring, instrumentation control module mechanical**
P1695	CAN data bus, body control module (BCM) – no signal	**Wiring, BCM**
P1696	Engine control module (ECM) – EEPROM error	**ECM**
P1697	Engine control module (ECM) – EEPROM error	**ECM**
P1698	CAN data bus, transmission control module (TCM) – no signal	**Wiring, TCM**
P1765	Transmission system relay	**Wiring open circuit/short circuit, transmission system relay, ECM/TCM**
P1899	Park/neutral position (PNP) switch – input circuit malfunction	**Wiring, PNP switch, ECM/TCM**
P2503	Charging system voltage – low	**Wiring, poor connection, battery, alternator**

DODGE

Model:	Engine identification:	Year:
Caravan 2.5L	EDM	1992-95
Caravan 3.0L	EFA	1992-95
Caravan 3.3L	EGA/EGM	1992-95
Caravan 3.8L	EGH	1992-95

System: **SBEC II**

Self-diagnosis

General information

- Refer to the front of this manual for general test conditions, terminology, detailed descriptions of wiring faults and a general trouble shooter for electrical and mechanical faults.
- Malfunction indicator lamp (MIL) will illuminate if certain faults are recorded.
- ECM operates in backup mode if sensors fail, to enable vehicle to be driven to repair shop.
- Trouble codes can be displayed by the MIL or by using a Scan Tool connected to the data link connector (DLC).

Accessing

- Within 5 seconds switch the ignition key ON-OFF-ON-OFF-ON.
- Count MIL flashes. Note trouble codes.
- Each trouble code consists of two groups **1** [**A**].
- The first group of flashes indicate the 'tens' of the trouble code.
- The second group of flashes indicate the 'units' of the trouble code.
- A short pause separates 'tens' and 'units' **1** [**B**].
- A long pause separates each trouble code **1** [**C**].
- For example: Trouble code 21 displayed **1**.
- End of test sequence indicated by trouble code 55.
- Trouble codes can also be displayed by using a Scan Tool connected to the data link connector (DLC) 1992-93 **2** & 1994-95 **3**.

1

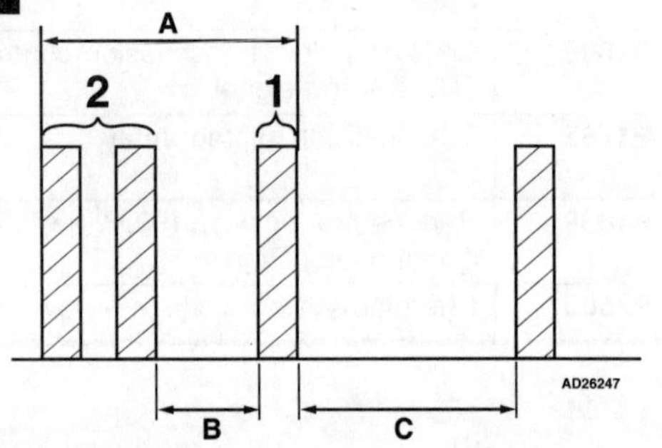

2

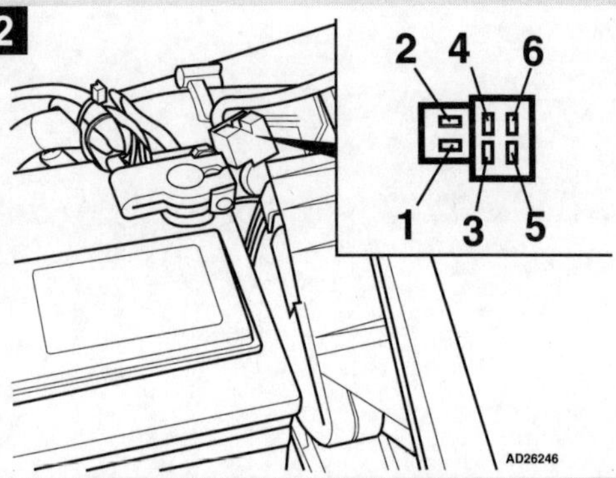

3

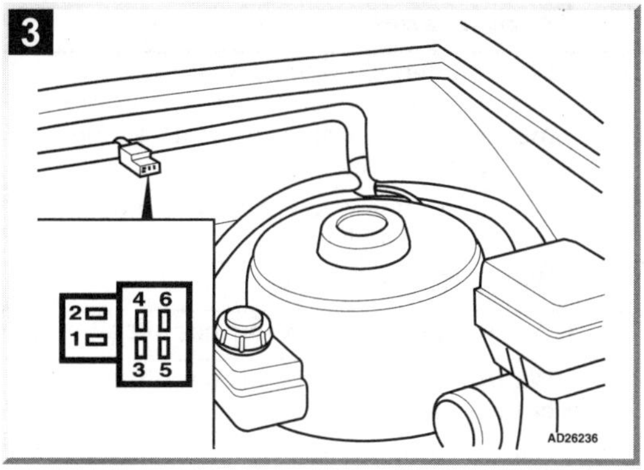

Erasing

NOTE: ***Disconnecting battery ground lead may cause driveability problems until the engine management system relearns basic values.***

- Ensure ignition switched OFF.
- Disconnect battery ground cable for at least 30 seconds.
- Reconnect battery ground cable.
- Repeat accessing procedure to ensure no trouble codes remain in ECM fault memory.

NOTE: ***Disconnecting battery ground lead may also erase memory from electronic units, such as radio.***

- Trouble codes can also be erased by using a Scan Tool connected to the data link connector (DLC).

Trouble code identification

Flash code	Fault location	Probable cause
11	Crankshaft position (CKP) sensor	**Wiring, air gap, CKP sensor**
12	Engine control module (ECM) – supply voltage disconnected within last 50 ignition switch ON-OFF cycles	**Wiring, battery**
13	Manifold absolute pressure (MAP) sensor	**Wiring, MAP sensor**
14	Manifold absolute pressure (MAP) sensor – voltage low/high	**Wiring, MAP sensor**
15	Vehicle speed sensor (VSS) – signal	**Wiring, VSS**
17	Engine temperature – low for long period	**Wiring, coolant thermostat, ECT sensor**
21	Heated oxygen sensor (HO2S) – incorrect signal/voltage high	**Wiring, fuel pressure, injector control, EVAP canister purge valve, compression, air filter blocked, HO2S**
22	Engine coolant temperature (ECT) sensor – voltage low/high	**Wiring, ECT sensor**
24	Throttle position (TP) sensor	**Wiring, accelerator cable adjustment, TP sensor**
25	Idle air control (IAC) valve	**Wiring, IAC valve**
27	Injector(s) – circuit malfunction	**Wiring, injector**
31	Evaporative emission (EVAP) canister purge valve – open/short circuit	**Wiring, EVAP canister purge valve**
32	Exhaust gas recirculation (EGR) solenoid/ system – circuit malfunction	**Wiring, hoses, EGR solenoid**
33	A/C compressor clutch relay	**Wiring, A/C compressor clutch relay**

Flash code	Fault location	Probable cause
34	Cruise control	**Wiring open circuit/short circuit, hoses, voltage low/high, cruise control master switch, cruise control vacuum switching valve**
35	Engine coolant blower motor relay – low/high speed	**Wiring, engine coolant blower motor relay – low/high speed**
37	Torque converter clutch (TCC) solenoid	**Wiring, TCC solenoid**
41	Alternator	**Wiring open circuit/short circuit, alternator**
42	Engine control relay	**Wiring open circuit/short circuit, engine control relay**
43	Ignition coil – primary circuit	**Wiring, ignition coil**
44	Battery temperature sensor/engine coolant temperature (ECT) sensor	**Wiring open circuit/short circuit, battery temperature sensor, ECT sensor, ECM**
46	Alternator – voltage high	**Wiring, battery, alternator**
47	Alternator – voltage low	**Wiring, battery, alternator drive belt, alternator**
51	System too lean	**Wiring, intake/exhaust leak, fuel pressure, injector control, EVAP canister purge valve, HO2S**
52	System too rich	**Wiring, exhaust blockage, fuel pressure, injector control, EVAP canister purge valve, compression, air filter blocked, ignition system, HO2S**
53	Engine control module (ECM) – internal fault – defective	**ECM**
54	Camshaft position (CMP) sensor – no signal	**Wiring, CMP sensor**
55	End of MIL code output	**–**
62	Engine control module (ECM) – EEPROM error	**ECM**
63	Engine control module (ECM) – EEPROM error	**ECM**

DODGE

Model:	Engine identification:	Year:
Colt 1.5L	**4G15**	**1992-93**
Colt 1.8L	**4G93**	**1992-93**
Colt 2.4L	**4G64**	**1992-93**
Stealth 3.0L	**6G72**	**1992-93**
Stealth 3.0L Turbo	**6G72**	**1992-93**

System: **Mitsubishi MPI**

Self-diagnosis

General information

- Refer to the front of this manual for general test conditions, terminology, detailed descriptions of wiring faults and a general trouble shooter for electrical and mechanical faults.
- Malfunction indicator lamp (MIL) will illuminate if certain faults are recorded.
- ECM operates in backup mode if sensors fail, to enable vehicle to be driven to repair shop.
- 1992: Trouble codes can be displayed by using an analogue voltmeter connected to the data link connector (DLC).
- 1993: Trouble codes can be displayed by using an analogue voltmeter or Scan Tool connected to the data link connector (DLC).

Accessing

- Ensure ignition switched OFF.
- Access data link connector (DLC) **1**.
- Connect an analogue voltmeter between data link connector (DLC) terminals 1 and 12 **2**.
- Switch ignition ON.
- Count voltmeter needle deflections. Note trouble codes. Compare with trouble code table.
- Each trouble code consists of two groups of one or more needle deflections **3** :
  - Long needle deflections indicate the 'tens' of the trouble code **3** [**A**].
  - Short needle deflections indicate the 'units' of the trouble code **3** [**D**].
  - A short pause separates each trouble code group **3** [**C**].
  - A long pause separates each trouble code **3** [**E**].
  - For example: Trouble code 22 displayed **3**.
- If a constant deflection is displayed: Suspect engine control module (ECM).
- If constant short deflections are displayed: No trouble code is stored.
- 1993: Trouble codes can also be displayed by using a Scan Tool connected to the data link connector (DLC).

1

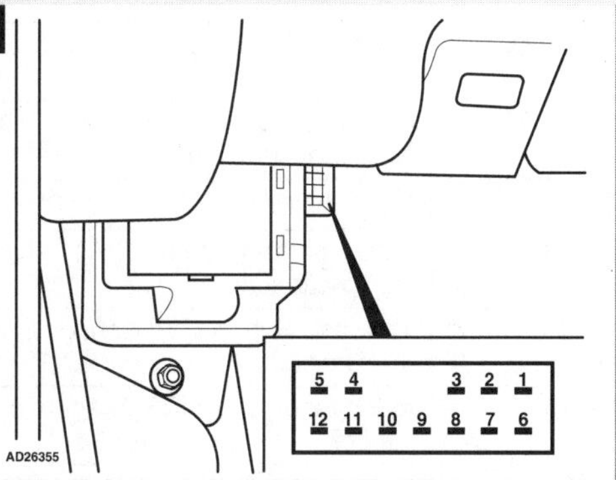

2

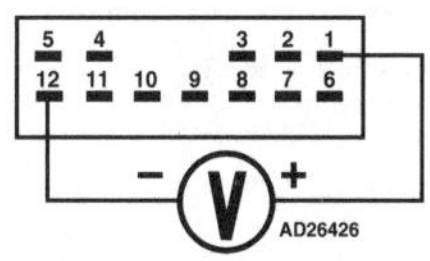

3

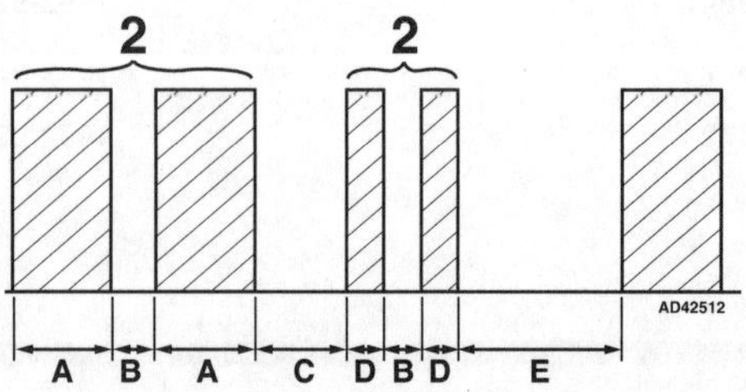

Erasing

NOTE: ***Disconnecting battery ground lead may cause driveability problems until the engine management system relearns basic values.***

- Ensure ignition switched OFF.
- Disconnect battery ground cable for at least 15 seconds.
- Reconnect battery ground lead.
- Repeat checking procedure to ensure no data remains in ECM fault memory.
- 1993: Trouble codes can also be erased by using a Scan Tool connected to the data link connector (DLC).

NOTE: ***Disconnecting battery ground lead may erase memory from electronic units such as radio.***

Trouble code identification

Flash code	Fault location	Probable cause
11 1	Heated oxygen sensor (HO2S)/oxygen sensor (O2S) 1, bank 1 – circuit malfunction	**Wiring, fuel pressure, injector(s), intake leak, HO2S/O2S, ECM**
11 2	Heated oxygen sensor (HO2S) 1, bank 2 – circuit malfunction	**Wiring, fuel pressure, injector(s), intake leak, HO2S, ECM**
12	Volume air flow (VAF) sensor – malfunction	**Wiring, VAF sensor, ECM**
13	Intake air temperature (IAT) sensor – circuit malfunction	**Wiring, IAT sensor, ECM**
14	Throttle position (TP) sensor – circuit malfunction	**Wiring, TP sensor, CTP switch, ECM**
15	Idle speed control (ISC) actuator position sensor – circuit malfunction	**Wiring, ISC actuator, ISC actuator position sensor, ECM**
21	Engine coolant temperature (ECT) sensor – circuit malfunction	**Wiring, ECT sensor, ECM**
22	Crankshaft position (CKP) sensor – circuit malfunction	**Wiring, CKP sensor, ECM**
23	Camshaft position (CMP) sensor – circuit malfunction	**Wiring, CMP sensor, ECM**
24	Vehicle speed sensor (VSS) – circuit malfunction	**Wiring, VSS, ECM**
25	Barometric pressure (BARO) sensor – circuit malfunction	**Wiring, BARO sensor, ECM**
31	Knock sensor (KS) – circuit malfunction	**Wiring, poor connection, KS, ECM**
32	Manifold absolute pressure (MAP) sensor – circuit malfunction	**Wiring, MAP sensor, ECM**
36	Ignition timing adjustment	**Wiring, ignition timing connector short to ground, ECM**

Flash code	Fault location	Probable cause
39 **2**	Heated oxygen sensor (HO2S) 1, bank 1 – circuit malfunction	**Wiring, fuel pressure, injector(s), intake leak, HO2S, ECM**
41	Injectors – circuit malfunction	**Wiring, injectors, ECM**
42	Fuel pump – circuit malfunction	**Wiring, engine control relay, fuel pump, ECM**
43	Exhaust gas recirculation (EGR) system – circuit malfunction	**Wiring, hose leak/blockage, EGR valve, EGR temperature sensor, EGR solenoid, ECM**
44	Ignition coil/ignition control module (ICM) cylinders 1 & 4 – circuit malfunction	**Wiring, ignition coil, ICM, ECM**
52	Ignition coil/ignition control module (ICM) cylinders 2 & 5 – circuit malfunction	**Wiring, ignition coil, ICM, ECM**
53	Ignition coil/ignition control module (ICM) cylinders 3 & 6 – circuit malfunction	**Wiring, ignition coil, ICM, ECM**
55	Idle air control (IAC) valve position sensor – circuit malfunction	**Wiring, IAC valve, IAC valve position sensor, ECM**
59	Heated oxygen sensor (HO2S) 2, bank 1 – circuit malfunction	**Wiring, HO2S, ECM**
61	Transmission control module (TCM)/engine control module (ECM), torque reduction signal – malfunction	**Wiring, TCM, ECM**
62	Intake manifold air control actuator position sensor – malfunction	**Wiring, intake manifold air control actuator, intake manifold air control actuator position sensor, ECM**

1 Except turbo
2 Turbo

DODGE

Model: **Colt 1.5L**

Engine identification: **4G15**

Year: **1994**

System: **Mitsubishi MPI**

Self-diagnosis

General information

- Refer to the front of this manual for general test conditions, terminology, detailed descriptions of wiring faults and a general trouble shooter for electrical and mechanical faults.
- Malfunction indicator lamp (MIL) will illuminate if certain faults are recorded.
- ECM operates in backup mode if sensors fail, to enable vehicle to be driven to repair shop.
- Trouble codes can be displayed by using an analogue voltmeter or Scan Tool connected to the data link connector(s) (DLC).

Accessing

- Ensure ignition switched OFF.
- Access 12 pin and 16 pin data link connectors (DLC) **1**.
- Connect an analogue voltmeter between 12 pin data link connector (DLC) terminal 5 and 16 pin data link connector (DLC) terminal 4 or 5 **2**
- Switch ignition ON.
- Count voltmeter needle deflections. Note trouble codes. Compare with trouble code table.
- Each trouble code consists of two groups of one or more needle deflections **3** :
  - Long needle deflections indicate the 'tens' of the trouble code **3** [**A**].
  - Short needle deflections indicate the 'units' of the trouble code **3** [**D**].
  - A short pause separates each trouble code group **3** [**C**].
  - A long pause separates each trouble code **3** [**E**].
  - For example: Trouble code 22 displayed **3**.
- If a constant deflection is displayed: Suspect engine control module (ECM).
- If constant short deflections are displayed: No trouble code is stored.
- Trouble codes can also be displayed by using a Scan Tool connected to the data link connector(s) (DLC).

1

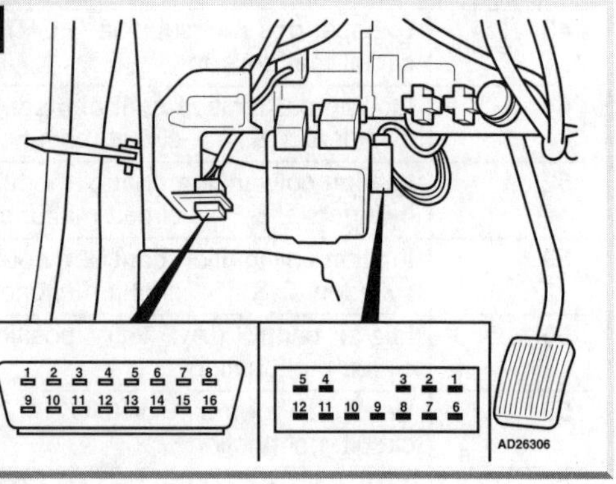

2

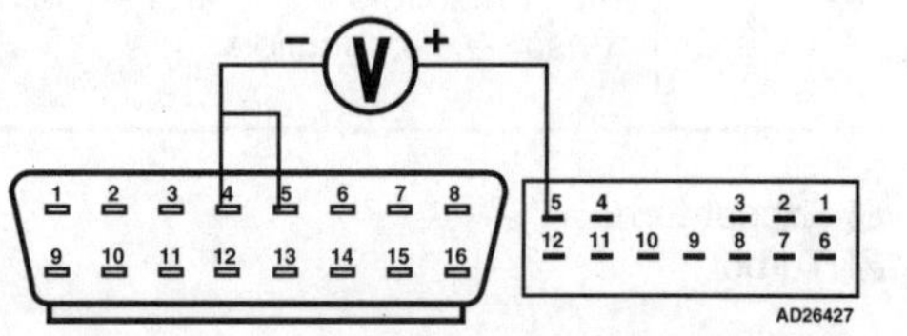

3

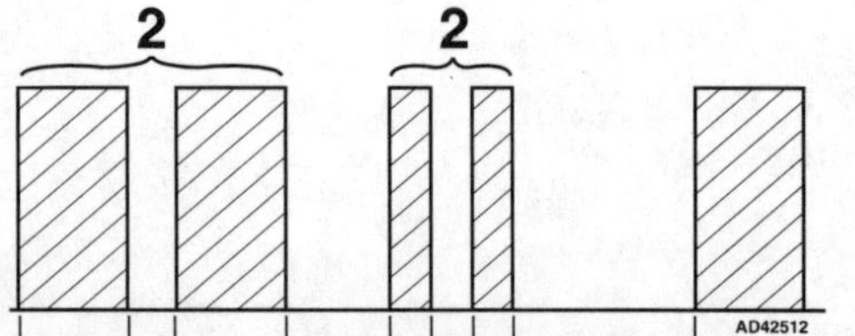

Erasing

NOTE: ***Disconnecting battery ground lead may cause driveability problems until the engine management system relearns basic values.***

- Ensure ignition switched OFF.
- Disconnect battery ground cable for at least 15 seconds.
- Reconnect battery ground lead.
- Repeat checking procedure to ensure no data remains in ECM fault memory.
- Trouble codes can also be erased by using a Scan Tool connected to the data link connector(s) (DLC).

NOTE: ***Disconnecting battery ground lead may erase memory from electronic units such as radio.***

Trouble code identification

Flash code	Fault location	Probable cause
11	Heated oxygen sensor (HO2S) 1, bank 1 – circuit malfunction	**Wiring, fuel pressure, injector(s), intake leak, HO2S, ECM**
13	Intake air temperature (IAT) sensor – circuit malfunction	**Wiring, IAT sensor, ECM**
14	Throttle position (TP) sensor – circuit malfunction	**Wiring, TP sensor, CTP switch, ECM**
21	Engine coolant temperature (ECT) sensor – circuit malfunction	**Wiring, ECT sensor, ECM**
22	Crankshaft position (CKP) sensor – circuit malfunction	**Wiring, CKP sensor, ECM**
23	Camshaft position (CMP) sensor – circuit malfunction	**Wiring, CMP sensor, ECM**
24	Vehicle speed sensor (VSS) – circuit malfunction	**Wiring, VSS, ECM**
32	Manifold absolute pressure (MAP) sensor – circuit malfunction	**Wiring, MAP sensor, ECM**
36	Ignition timing adjustment	**Wiring, ignition timing connector short to ground, ECM**
41	Injectors – circuit malfunction	**Wiring, injectors, ECM**
43	Exhaust gas recirculation (EGR) system – circuit malfunction	**Wiring, hose leak/blockage, EGR valve, EGR temperature sensor, EGR solenoid, ECM**
55	Idle air control (IAC) valve position sensor – circuit malfunction	**Wiring, IAC valve, IAC valve position sensor, ECM**
59	Heated oxygen sensor (HO2S) 2, bank 1 – circuit malfunction	**Wiring, HO2S, ECM**

DODGE

Model:	Engine identification:	Year:
Colt 1.8L	4G93	1994
Colt 2.4L	4G64	1994
Stealth 3.0L	6G72	1994-95
Stealth 3.0L Turbo	6G72	1994-95

System: **Mitsubishi MPI**

Self-diagnosis

General information

- Refer to the front of this manual for general test conditions, terminology, detailed descriptions of wiring faults and a general trouble shooter for electrical and mechanical faults.
- Malfunction indicator lamp (MIL) will illuminate if certain faults are recorded.
- ECM operates in backup mode if sensors fail, to enable vehicle to be driven to repair shop.
- Trouble codes can be displayed by the MIL or by using a Scan Tool connected to the data link connector(s) (DLC).

Accessing

- Ensure ignition switched OFF.
- Access 16 pin data link connector (DLC):
  - Colt **1**.
  - Stealth **2**.
- Jump data link connector (DLC) terminal 1 and ground **3**.
- Switch ignition ON.
- Count MIL flashes. Note trouble codes. Compare with trouble code table.
- Each trouble code consists of two groups **4** :
  - Long flashes indicate the 'tens' of the trouble code **4** [**A**].
  - Short flashes indicate the 'units' of the trouble code **4** [**D**].
  - A short pause separates each trouble code group **4** [**C**].
  - A long pause separates each trouble code **4** [**E**].
  - For example: Trouble code 22 displayed **4**.
- If MIL remains ON: Suspect faulty engine control module (ECM).
- If constant short flashes are displayed, no trouble code is stored.
- Trouble codes can also be displayed by using a Scan Tool connected to the data link connector(s) (DLC).

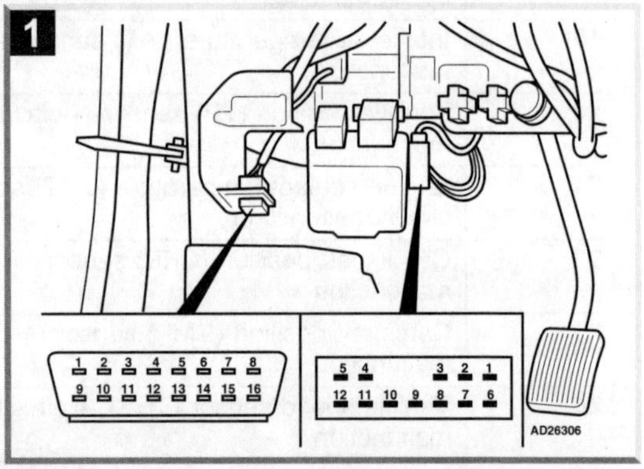

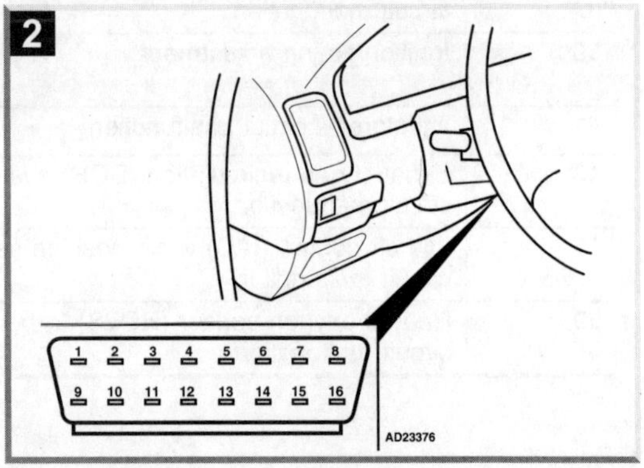

3

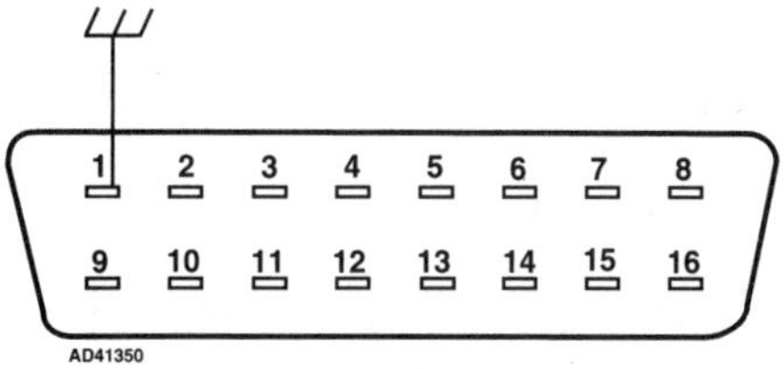

4

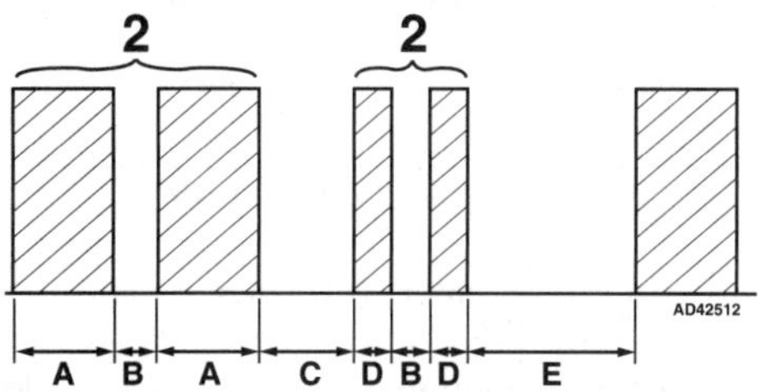

Erasing

NOTE: ***Disconnecting battery ground lead may cause driveability problems until the engine management system relearns basic values.***

- Ensure ignition switched OFF.
- Disconnect battery ground cable for at least 15 seconds.
- Reconnect battery ground lead.
- Repeat checking procedure to ensure no data remains in ECM fault memory.
- Trouble codes can also be erased by using a Scan Tool connected to the data link connector(s) (DLC).

NOTE: ***Disconnecting battery ground lead may erase memory from electronic units such as radio.***

Trouble code identification

Flash code	Fault location	Probable cause
11 1 2	Heated oxygen sensor (HO2S) 1, bank 1 – circuit malfunction	**Wiring, fuel pressure, injector(s), intake leak, HO2S, ECM**
11 3 4	Heated oxygen sensor (HO2S) 1, bank 2 – circuit malfunction	**Wiring, fuel pressure, injector(s), intake leak, HO2S, ECM**
12	Volume air flow (VAF) sensor – malfunction	**Wiring, VAF sensor, ECM**
13	Intake air temperature (IAT) sensor – circuit malfunction	**Wiring, IAT sensor, ECM**
14	Throttle position (TP) sensor – circuit malfunction	**Wiring, TP sensor, CTP switch, ECM**
21	Engine coolant temperature (ECT) sensor – circuit malfunction	**Wiring, ECT sensor, ECM**
22	Crankshaft position (CKP) sensor – circuit malfunction	**Wiring, CKP sensor, ECM**

Flash code	Fault location	Probable cause
23	Camshaft position (CMP) sensor – circuit malfunction	**Wiring, CMP sensor, ECM**
24	Vehicle speed sensor (VSS) – circuit malfunction	**Wiring, VSS, ECM**
25	Barometric pressure (BARO) sensor – circuit malfunction	**Wiring, BARO sensor, ECM**
31	Knock sensor (KS) – circuit malfunction	**Wiring, poor connection, KS, ECM**
36	Ignition timing adjustment	**Wiring, ignition timing connector short to ground, ECM**
39	Heated oxygen sensor (HO2S) 1, bank 1 – circuit malfunction	**Wiring, fuel pressure, injector(s), intake leak, HO2S, ECM**
41	Injectors – circuit malfunction	**Wiring, injectors, ECM**
43	Exhaust gas recirculation (EGR) system – circuit malfunction	**Wiring, hose leak/blockage, EGR valve, EGR temperature sensor, EGR solenoid, ECM**
44	Ignition coil/ignition control module (ICM) cylinders 1 & 4 – circuit malfunction	**Wiring, ignition coil, ICM, ECM**
52	Ignition coil/ignition control module (ICM) cylinders 2 & 5 – circuit malfunction	**Wiring, ignition coil, ICM, ECM**
53	Ignition coil/ignition control module (ICM) cylinders 3 & 6 – circuit malfunction	**Wiring, ignition coil, ICM, ECM**
55	Idle air control (IAC) valve position sensor – circuit malfunction	**Wiring, IAC valve, IAC valve position sensor, ECM**
59	Heated oxygen sensor (HO2S) 2, bank 1 – circuit malfunction	**Heating inoperative, wiring, HO2S, ECM**
61	Transmission control module (TCM)/engine control module (ECM), torque reduction signal – malfunction	**Wiring, TCM, ECM**
62	Intake manifold air control actuator position sensor – malfunction	**Wiring, intake manifold air control actuator, intake manifold air control actuator position sensor, ECM**
69	Heated oxygen sensor (HO2S) 2, bank 2 – circuit malfunction	**Heating inoperative, wiring, HO2S, ECM**

1 Colt
2 Stealth non-turbo – Federal
3 Stealth non-turbo – California
4 Stealth Turbo

DODGE

Model:	Engine identification:	Year:
Dakota 2.5/3.9/5.2L	**VIN position 8 = G/K/X/Y**	**1992-95**
Ramcharger 5.2/5.9L	**VIN position 8 = Y/Z/5**	**1992-93**
Ram Pickup D/W 3.9/5.2L	**VIN position 8 = X/Y**	**1992-93**
Ram Pickup/Chassis Cab D/W 5.9L	**VIN position 8 = Z/5**	**1992-93**
Ram Truck 3.9/5.2L	**VIN position 8 = X/Y**	**1994-95**
Ram Truck 5.9/8.0L	**VIN position 8 = W/Z/5**	**1994-95**
Ram Van/Wagon 3.9/5.2/5.9L	**VIN position 8 = A/T/X/Y/Z**	**1992-94**
Ram Van/Wagon 3.9/5.2/5.9L	**EHC/ELF/EML**	**1995**

System: **SBEC II**

Self-diagnosis

General information

- Refer to the front of this manual for general test conditions, terminology, detailed descriptions of wiring faults and a general trouble shooter for electrical and mechanical faults.
- Malfunction indicator lamp (MIL) will illuminate if certain faults are recorded.
- ECM operates in backup mode if sensors fail, to enable vehicle to be driven to repair shop.
- Trouble codes can be displayed by the MIL or by using a Scan Tool connected to the data link connector (DLC).

Accessing

- Within 3 seconds switch the ignition key ON-OFF-ON-OFF-ON.
- Count MIL flashes. Note trouble codes.
- Each trouble code consists of two groups **1** [**A**].
- The first group of flashes indicate the 'tens' of the trouble code.
- The second group of flashes indicate the 'units' of the trouble code.
- A short pause separates 'tens' and 'units' **1** [**B**].
- A long pause separates each trouble code **1** [**C**].
- For example: Trouble code 21 displayed **1**.
- End of test sequence indicated by trouble code 55.
- Trouble codes can also be displayed by using a Scan Tool connected to the data link connector (DLC):
  - Dakota **2**.
  - Ramcharger/Ram Pickup **3**.
  - Ram Truck **4**.
  - Ram Van/Wagon **5**.

1

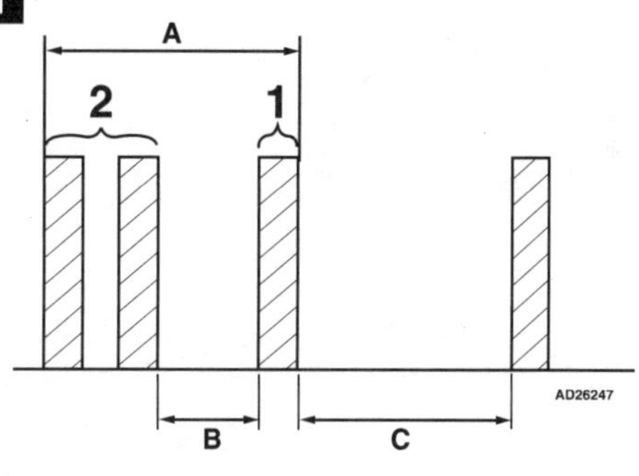

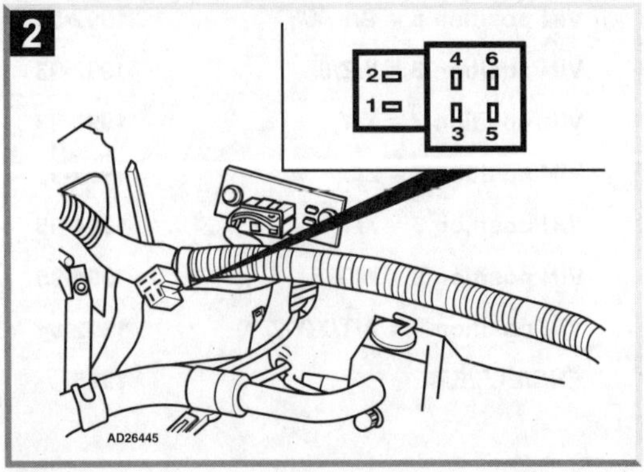

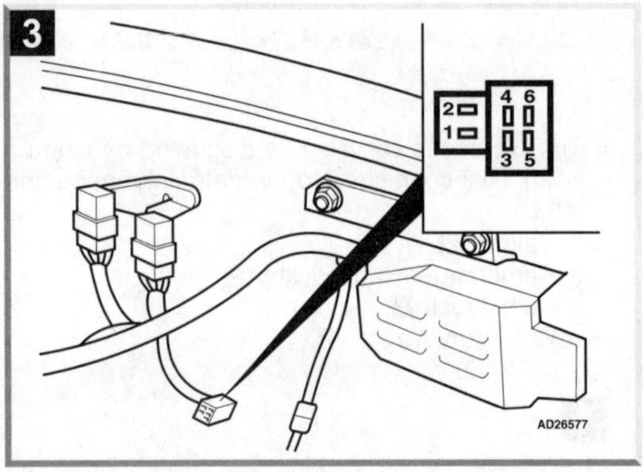

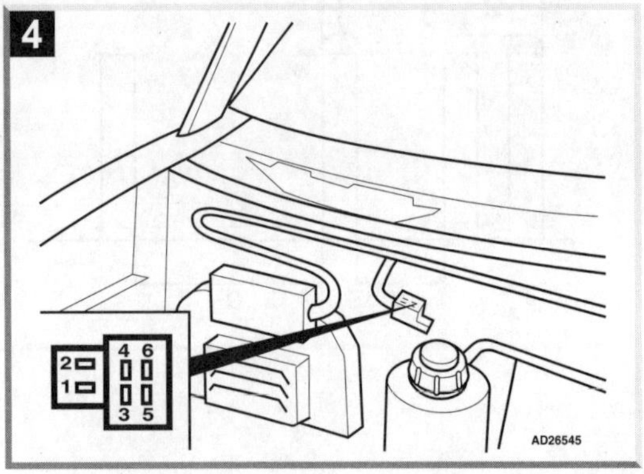

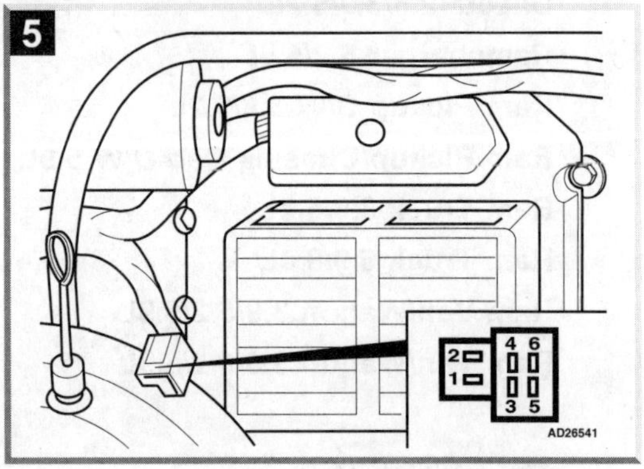

Erasing

NOTE: ***Disconnecting battery ground lead may cause driveability problems until the engine management system relearns basic values.***

- Ensure ignition switched OFF.
- Disconnect battery ground cable for at least 30 seconds.
- Reconnect battery ground cable.
- Repeat accessing procedure to ensure no trouble codes remain in ECM fault memory.

NOTE: ***Disconnecting battery ground lead may also erase memory from electronic units, such as radio.***

- Trouble codes can also be erased by using a Scan Tool connected to the data link connector (DLC).

Trouble code identification

Flash code	Fault location	Probable cause
11	Crankshaft position (CKP) sensor	**Wiring, air gap, CKP sensor**
12	Engine control module (ECM) – supply voltage disconnected within last 50 ignition switch ON-OFF cycles	**Wiring, battery**
13	Manifold absolute pressure (MAP) sensor	**Wiring, MAP sensor**
14	Manifold absolute pressure (MAP) sensor – voltage low/high	**Wiring, MAP sensor**
15	Vehicle speed sensor (VSS) – signal	**Wiring, VSS**
17	Engine temperature – low for long period	**Wiring, coolant thermostat, ECT sensor**
21	Heated oxygen sensor (HO2S) – incorrect signal/voltage high	**Wiring, fuel pressure, injector control, EVAP canister purge valve, compression, air filter blocked, HO2S**
22	Engine coolant temperature (ECT) sensor – voltage low/high	**Wiring, ECT sensor**
23	Intake air temperature (IAT) sensor – voltage low/high	**Wiring, IAT sensor**
24	Throttle position (TP) sensor	**Wiring, accelerator cable adjustment, TP sensor**
25	Idle air control (IAC) valve/idle speed control (ISC) actuator	**Wiring, IAC valve/ISC actuator**
27	Injector(s) – circuit malfunction	**Wiring, injector**
31	Evaporative emission (EVAP) canister purge valve – open/short circuit	**Wiring, EVAP canister purge valve**
32	Exhaust gas recirculation (EGR) solenoid/ system – circuit malfunction	**Wiring, hoses, EGR solenoid**
33	A/C compressor clutch relay	**Wiring, A/C compressor clutch relay**
34	Cruise control	**Wiring open circuit/short circuit, hoses, voltage low/high, cruise control master switch, cruise control vacuum switching valve**
35 [1]	Engine coolant blower motor relay – low/high speed	**Wiring, engine coolant blower motor relay – low/high speed**
35 [2]	Closed throttle position (CTP) switch	**Wiring open circuit/short circuit, ISC actuator, CTP switch**
36	Secondary air injection (AIR) solenoid	**Wiring open circuit/short circuit, AIR solenoid**
37	Torque converter clutch (TCC) solenoid/ transmission fluid temperature (TFT) sensor	**Wiring, transmission fluid level, TCC solenoid, TFT sensor, ECM**
41	Alternator	**Wiring open circuit/short circuit, alternator**
42	Engine control relay	**Wiring open circuit/short circuit, engine control relay**
43	Ignition coil(s)	**Wiring open circuit/short circuit, ignition coil(s)**

Flash code	Fault location	Probable cause
44	Battery temperature sensor/engine coolant temperature (ECT) sensor	**Wiring open circuit/short circuit, battery temperature sensor, ECT sensor, ECM**
45	Overdrive solenoid	**Wiring open circuit/short circuit, overdrive solenoid**
46	Alternator – voltage high	**Wiring, battery, alternator**
47	Alternator – voltage low	**Wiring, battery, alternator drive belt, alternator**
51	System too lean	**Wiring, intake/exhaust leak, fuel pressure, injector control, EVAP canister purge valve, HO2S**
52	System too rich	**Wiring, exhaust blockage, fuel pressure, injector control, EVAP canister purge valve, compression, air filter blocked, ignition system, HO2S**
53	Engine control module (ECM) – internal fault – defective	**ECM**
54	Camshaft position (CMP) sensor – no signal	**Wiring, CMP sensor**
55	End of malfunction indicator lamp (MIL) code output	**–**
62	Engine control module (ECM) – EEPROM error	**ECM**
63	Engine control module (ECM) – EEPROM error	**ECM**

1 2.5L
2 Except 2.5L

DODGE

Model:	Engine identification:	Year:
Dakota 2.5L	EPE	1996-97
Dakota 3.9L	EHC	1996-97
Dakota 5.2L	ELF	1996-97
Ram Truck 3.9L	EHC	1996-97
Ram Truck 5.2L	ELF/ELN	1996-97
Ram Truck 5.9L	EML/EMM	1996-97
Ram Truck 8.0L	EWA	1996-97
Ram Van/Wagon 3.9L	EHC	1996-97
Ram Van/Wagon 5.2L	ELF/ELN	1996-97
Ram Van/Wagon 5.9L	EML	1996-97

System: **JTEC**

Self-diagnosis

General information

- Refer to the front of this manual for general test conditions, terminology, detailed descriptions of wiring faults and a general trouble shooter for electrical and mechanical faults.
- Malfunction indicator lamp (MIL) will illuminate or flash if certain faults are recorded.
- ECM operates in backup mode if sensors fail, to enable vehicle to be driven to repair shop.
- Trouble codes can be displayed by the MIL or by using a Scan Tool connected to the data link connector (DLC).
- Dakota 1997: Trouble codes can also be displayed by the digital odometer.

Accessing

Flash codes

- Within 5 seconds switch the ignition key ON-OFF-ON-OFF-ON.
- Count MIL flashes. Note trouble codes.
- Each trouble code consists of two groups **1** [**A**].
- The first group of flashes indicate the 'tens' of the trouble code.
- The second group of flashes indicate the 'units' of the trouble code.
- A short pause separates 'tens' and 'units' **1** [**B**].
- A long pause separates each trouble code **1** [**C**].
- For example: Trouble code 21 displayed **1**.
- End of test sequence indicated by trouble code 55.
- Dakota 1997: Trouble codes can also be displayed by the digital odometer using the same key cycling procedure as above. After code 55 is displayed the odometer will return to its normal mode.

P codes & Hex codes

- Trouble codes can also be displayed by using a Scan Tool connected to the data link connector (DLC) **2**.

1

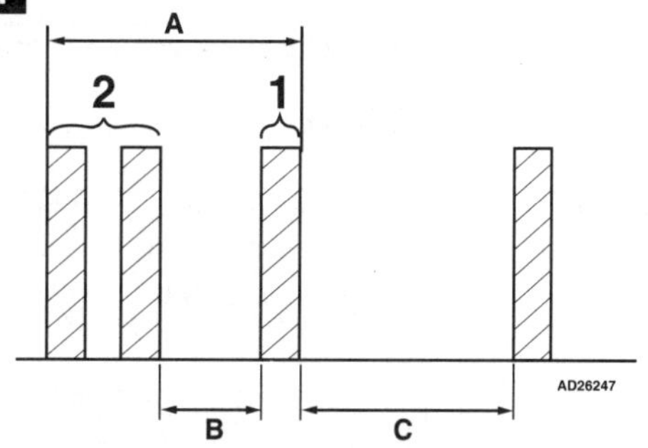

2

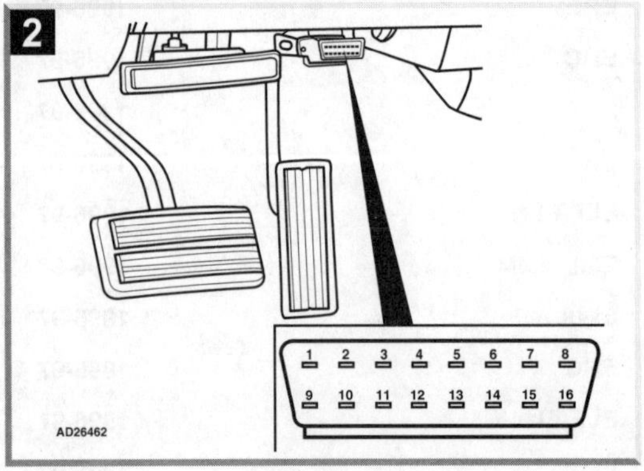

Erasing

- Trouble codes can only be erased by using a Scan Tool connected to the data link connector (DLC).

Trouble code identification

Scan code P type	Fault location	Probable cause
P0000-P0999	Refer to OBD-II trouble code table at the front of this manual	–
P1294	Idle speed control (ISC) system – rpm lower/higher than expected	**Wiring, air leak, throttle valve tight/sticking, IAC valve**
P1296	Manifold absolute pressure (MAP) sensor – supply voltage	**Wiring, MAP sensor, ECM**
P1297	Manifold absolute pressure (MAP) sensor	**Wiring, hose connection(s), MAP sensor**
P1391	Crankshaft position (CKP) sensor/camshaft position (CMP) sensor – intermittent signal	**Wiring, poor connection, air gap, CKP sensor, CMP sensor**
P1398	Engine control module (ECM) – misfire detection limit reached	**Wiring, poor connection, engine mechanical fault, ignition/fuel system, CKP sensor, ECM**
P1486	Evaporative emission (EVAP) system – blockage detected	**Hose connection(s), hose blocked/trapped, EVAP canister**
P1491	Engine coolant blower motor relay	**Wiring open circuit/short circuit, engine coolant blower motor relay – low/high speed**
P1492	Ambient/battery temperature sensor – voltage high	**Wiring, ambient/battery temperature sensor, ECM**
P1493	Ambient/battery temperature sensor – voltage low	**Wiring, ambient/battery temperature sensor, ECM**
P1494	Evaporative emission (EVAP) leak detection pump – pressure switch	**Wiring, hose leak/blockage, EVAP leak detection pump**
P1495	Evaporative emission (EVAP) leak detection pump solenoid	**Wiring open circuit/short circuit, hose leak/blockage, EVAP leak detection pump solenoid**
P1496	Engine control module (ECM) – 5 volt output low	**ECM**
P1696	Engine control module (ECM) – EEPROM error	**ECM**

Scan code P type	Fault location	Probable cause
P1756	Governor pressure not equal to target (15-20 PSI) – mid pressure malfunction	Signal wire open circuit/short circuit, valve body, sensor supply wire open circuit, transmission fluid level, governor pressure solenoid, transmission mechanical fault, ECM – no 5 volt supply
P1757	Governor pressure not equal to target – low pressure malfunction	Signal wire open circuit/short circuit, valve body, sensor supply wire open circuit, transmission fluid level, governor pressure solenoid, transmission mechanical fault, ECM – no 5 volt supply
P1762	Governor pressure sensor – incorrect signal	Signal wire short circuit to ground/positive, supply wire open circuit/short circuit, transmission fluid level, governor pressure solenoid, governor pressure sensor, ECM
P1763	Governor pressure sensor – signal high	Signal wire short circuit to positive, supply wire open circuit, transmission fluid level, governor pressure sensor, ECM
P1764	Governor pressure sensor – signal low	Signal wire short circuit to ground, supply wire open circuit, transmission fluid level, governor pressure sensor, ECM
P1765	Transmission system relay	Wiring open circuit/short circuit, transmission system relay, ECM
P1899	Park/neutral position (PNP) switch – input circuit malfunction	Wiring, PNP switch, ECM

Flash code	Scan code Hex type	Fault location	Probable cause
11		Crankshaft position (CKP) sensor/camshaft position (CMP) sensor	Wiring, poor connection, air gap, CKP sensor, CMP sensor
	28	Crankshaft position (CKP) sensor – no signal	Wiring, air gap, CKP sensor
	9D	Crankshaft position (CKP) sensor/camshaft position (CMP) sensor – intermittent signal	Wiring, poor connection, air gap, CKP sensor, CMP sensor
	BA	Engine control module (ECM) – misfire detection limit reached	Wiring, poor connection, engine mechanical fault, ignition/fuel system, CKP sensor, ECM
12		Engine control module (ECM) – supply voltage disconnected within last 50 ignition switch ON-OFF cycles	Wiring, battery
13	27	Manifold absolute pressure (MAP) sensor	Wiring, hose connection(s), MAP sensor
14		Manifold absolute pressure (MAP) sensor – input low/high	Wiring, MAP sensor
	24	Manifold absolute pressure (MAP) sensor – low input	Wiring short circuit to ground, MAP sensor
	25	Manifold absolute pressure (MAP) sensor – high input	Wiring short circuit to positive, MAP sensor

Flash code	Scan code Hex type	Fault location	Probable cause
	87	Manifold absolute pressure (MAP) sensor – supply voltage	Wiring, MAP sensor, ECM
15		Vehicle speed sensor (VSS) – signal	Wiring, VSS, ECM
	23	Vehicle speed sensor (VSS) – no signal	Wiring, VSS, ECM
	A6	Output shaft speed (OSS) sensor – circuit malfunction	Wiring, OSS sensor, ECM
17		Engine temperature – low for long period	Wiring, cooling system, coolant thermostat, ECT sensor
	21	Engine temperature – low for long period	Wiring, cooling system, coolant thermostat, ECT sensor
	80	Insufficient coolant temperature for closed loop fuel control	Wiring, cooling system, coolant thermostat, ECT sensor
21		Heated oxygen sensor (HO2S) 1/2/3, bank 1/2 – circuit/signal malfunction	Wiring, fuel pressure, injector control, EVAP canister purge valve, compression, air filter blocked, catalytic converter, HO2S
	3E	Heated oxygen sensor (HO2S) 1, bank 1 – high voltage	Wiring short circuit to positive, HO2S
	42	Heated oxygen sensor (HO2S) 1, bank 2 – high voltage	Wiring short to positive, HO2S
	66	Heated oxygen sensor (H02S) 1, bank 1 – slow response	Heating inoperative, wiring, HO2S
	67	Heated oxygen sensor (HO2S) 1, bank 1 – heater control – circuit malfunction	Wiring, HO2S
	68	Heated oxygen sensor (HO2S) 2, bank 1 – slow response	Heating inoperative, wiring, HO2S
	69	Heated oxygen sensor (HO2S) 2, bank 1, heater control – circuit malfunction	Wiring, HO2S
	7A	Heated oxygen sensor (HO2S) 1, bank 2 – slow response	Heating inoperative, wiring, HO2S
	7B	Heated oxygen sensor (HO2S) 2, bank 2 – slow response	Heating inoperative, wiring, HO2S
	7C	Heated oxygen sensor (HO2S) 1, bank 2, heater control – circuit malfunction	Wiring, HO2S, ECM
	7D	Heated oxygen sensor (HO2S) 2, bank 2, heater control – circuit malfunction	Wiring, HO2S, ECM
	7E	Heated oxygen sensor (HO2S) 2, bank 1 – high voltage	Wiring short circuit to positive, catalytic converter, HO2S
	7F	Heated oxygen sensor (HO2S) 2, bank 2 – high voltage	Wiring short to positive, HO2S, ECM
	90	Heated oxygen sensor (HO2S) 3, bank 1 – low voltage	Exhaust leak, wiring short to ground, HO2S, ECM
	93	Heated oxygen sensor (HO2S) 3, bank 1 – high voltage	Wiring short to positive, HO2S, ECM

Flash code	Scan code Hex type	Fault location	Probable cause
	9B	Heated oxygen sensor (HO2S) 1, bank 1 – low voltage	Exhaust leak, wiring short circuit to ground, HO2S
	9C	Heated oxygen sensor (HO2S) 2, bank 1 – low voltage	Exhaust leak, wiring short circuit to ground, HO2S
	9F	Heated oxygen sensor (HO2S) 3, bank 1 – slow response	Heating inoperative, wiring, HO2S
	B5	Heated oxygen sensor (HO2S) 1, bank 2 – low voltage	Exhaust leak, wiring short to ground, HO2S, ECM
	B6	Heated oxygen sensor (HO2S) 2, bank 2 – low voltage	Exhaust leak, wiring short to ground, HO2S, ECM
	BD	Heated oxygen sensor (HO2S) 3, bank 1, heater control – circuit malfunction	Wiring, HO2S, ECM
	CO	Heated oxygen sensor (HO2S) 1, bank 1 – slow response	Heating inoperative, wiring, HO2S
22		Engine coolant temperature (ECT) sensor – voltage low/high	Wiring, coolant thermostat, ECT sensor
	1E	Engine coolant temperature (ECT) sensor – low input	Coolant thermostat, wiring short circuit to ground, ECT sensor
	1F	Engine coolant temperature (ECT) sensor – high input	Coolant thermostat, wiring open circuit/ short circuit to positive, ground wire defective, ECT sensor
23		Intake air temperature (IAT) sensor – voltage low/high	Wiring, IAT sensor
	39	Intake air temperature (IAT) sensor – low input	Wiring short circuit to ground, IAT sensor
	3A	Intake air temperature (IAT) sensor – high input	Wiring open circuit/short circuit to positive, ground wire defective, IAT sensor
24		Throttle position (TP) sensor – voltage low/ high	Wiring, accelerator cable adjustment, TP sensor
	1A	Throttle position (TP) sensor – low input	Wiring short circuit to ground, TP sensor
	1B	Throttle position (TP) sensor – high input	Wiring short circuit to positive, TP sensor
	84	Throttle position (TP) sensor – range/ performance problem	Wiring, accelerator cable adjustment, TP sensor
25		Idle air control (IAC) valve	Wiring open circuit/short circuit, IAC valve
	19	Idle speed control (ISC) system – malfunction	Wiring open circuit/short circuit, IAC valve
	8A	Idle speed control (ISC) system – rpm lower/higher than expected	Wiring, air leak, throttle valve tight/ sticking, IAC valve
27		Injector(s) – circuit malfunction	Wiring, injector
	13	Injector 3 – circuit malfunction	Wiring, injector
	14	Injector 2 – circuit malfunction	Wiring, injector

DODGE

Dakota 2.5L ● Dakota 3.9L ● Dakota 5.2L ● Ram Truck 3.9L
Ram Truck 5.2L ● Ram Truck 5.9L ● Ram Truck 8.0L
Ram Van/Wagon 3.9L ● Ram Van/Wagon 5.2L ● Ram Van/Wagon 5.9L

Flash code	Scan code Hex type	Fault location	Probable cause
	15	Injector 1 – circuit malfunction	**Wiring, injector**
	37	Injector 9 – circuit malfunction	**Wiring, injector**
	38	Injector 10 – circuit malfunction	**Wiring, injector**
	3D	Injector 4 – circuit malfunction	**Wiring, injector**
	45	Injector 5 – circuit malfunction	**Wiring, injector**
	46	Injector 6 – circuit malfunction	**Wiring, injector**
	4F	Injector 7 – circuit malfunction	**Wiring, injector**
	50	Injector 8 – circuit malfunction	**Wiring, injector**
31		Evaporative emission (EVAP) canister purge valve/system	**Wiring, EVAP canister, EVAP canister purge valve**
	12	Evaporative emission (EVAP) canister purge valve – circuit malfunction	**Wiring open circuit/short circuit, EVAP canister purge valve**
	71	Evaporative emission (EVAP) system – incorrect flow detected	**Hose connection(s), intake leak, EVAP canister purge valve**
	A0	Evaporative emission (EVAP) system – small leak detected	**Hose connection(s), intake leak, EVAP canister, EVAP canister purge valve**
	A1	Evaporative emission (EVAP) system – large leak detected	**Hose connection(s), intake leak, EVAP canister, EVAP canister purge valve**
	BB	Evaporative emission (EVAP) system – blockage detected	**Hose connection(s), hose blocked/ trapped, EVAP canister**
	B8	Evaporative emission (EVAP) leak detection pump – pressure switch	**Wiring, hose leak/blockage, EVAP leak detection pump**
	B7	Evaporative emission (EVAP) leak detection pump solenoid	**Wiring open circuit/short circuit, hose leak/blockage, EVAP leak detection pump solenoid**
32	11	Exhaust gas recirculation (EGR) – circuit malfunction	**Wiring, EGR solenoid, ECM**
33	10	A/C compressor clutch relay	**Wiring open circuit/short circuit, A/C compressor clutch relay**
34		Cruise control	**Wiring open circuit/short circuit, hoses, voltage low/high, cruise control master switch, cruise control vacuum switching valve**
	0F	Cruise control	**Wiring open circuit/short circuit, hoses, voltage low/high, cruise control vacuum switching valve**
	56	Cruise control master switch – high voltage	**Wiring, cruise control master switch**
	57	Cruise control master switch – low voltage	**Wiring, cruise control master switch**
35	0E	Engine coolant blower motor relay	**Wiring open circuit/short circuit, engine coolant blower motor relay – low/high speed**

Flash code	Scan code Hex type	Fault location	Probable cause
37		Transmission control system – malfunction	Wiring, transmission fluid level, transmission mechanical fault, PNP switch, TCC solenoid, TFT sensor, TFP solenoid, shift solenoid, governor pressure solenoid, transmission system relay, ECM
	0C	Torque converter clutch (TCC) solenoid – electrical	Wiring open circuit/short circuit, TCC solenoid
	32	Shift solenoid (SS) 3-4 – electrical	Wiring, shift solenoid, ECM
	4A	Transmission fluid temperature (TFT) sensor – low input	Wiring short circuit to ground, TFT sensor, ECM
	4B	Transmission fluid temperature (TFT) sensor – high input	Wiring short circuit to positive, TFT sensor, ECM
	72	Park/neutral position (PNP) switch – input circuit malfunction	Wiring, PNP switch, ECM
	8D	Governor pressure not equal to target (15-20 PSI) – mid pressure malfunction	Signal wire open circuit/short circuit, valve body, sensor supply wire open circuit, transmission fluid level, governor pressure solenoid, transmission mechanical fault, ECM – no 5 volt supply
	8E	Governor pressure not equal to target – low pressure malfunction	Signal wire open circuit/short circuit, valve body, sensor supply wire open circuit, transmission fluid level, governor pressure solenoid, transmission mechanical fault, ECM – no 5 volt supply
	94	Torque converter clutch (TCC) solenoid – circuit malfunction	Wiring, TCC solenoid
	A4	Transmission fluid temperature (TFT) sensor – range/performance problem	Wiring, TFT sensor
	A5	Gear selection, 3-4 – shift malfunction	Wiring, TR sensor, shift solenoids, transmission mechanical fault
	AB	Transmission fluid pressure (TFP) solenoid – electrical	Wiring, TFP solenoid, ECM
	AD	Transmission system relay	Wiring open circuit/short circuit, transmission system relay, ECM
41	0B	Alternator, field control – circuit malfunction	Wiring open circuit/short circuit, alternator, battery
42		Fuel pump relay/engine control relay/fuel level sensor – circuit malfunction	Wiring open circuit/short circuit, fuel pump relay, engine control relay, fuel level sensor
	0A	Engine control relay	Wiring open circuit/short circuit, engine control relay
	2C	Engine control relay	Wiring open circuit, engine control relay
	65	Fuel pump relay	Wiring open circuit/short circuit, fuel pump relay

Flash code	Scan code Hex type	Fault location	Probable cause
	95	Fuel level sensor – low input	Wiring open circuit, fuel level sensor
	96	Fuel level sensor – high input	Wiring short circuit to positive, fuel level sensor
	97	Fuel level sensor – malfunction	Wiring, fuel level sensor
43		Random/multiple cylinder(s) – misfire detected	Wiring, spark plug(s), HT lead(s), injector(s), ignition coil(s), low compression
	29	Ignition coil C, primary – circuit malfunction	Wiring, ignition coil, ECM
	2A	Ignition coil B, primary – circuit malfunction	Wiring, ignition coil, ECM
	2B	Ignition coil A, primary – circuit malfunction	Wiring, ignition coil, ECM
	4C	Ignition coil D, primary – circuit malfunction	Wiring, ignition coil, ECM
	4D	Ignition coil E, primary – circuit malfunction	Wiring, ignition coil, ECM
	6A	Random/multiple cylinder(s) – misfire detected	Wiring, spark plug(s), HT lead(s), injector(s), ignition coil(s), low compression
	6B	Cylinder 1 – misfire detected	Wiring, engine mechanical fault, ignition/fuel system, injector
	6C	Cylinder 2 – misfire detected	Wiring, engine mechanical fault, ignition/fuel system, injector
	6D	Cylinder 3 – misfire detected	Wiring, engine mechanical fault, ignition/fuel system, injector
	6E	Cylinder 4 – misfire detected	Wiring, engine mechanical fault, ignition/fuel system, injector
	AE	Cylinder 5 – misfire detected	Wiring, engine mechanical fault, ignition/fuel system, injector
	AF	Cylinder 6 – misfire detected	Wiring, engine mechanical fault, ignition/fuel system, injector
	B0	Cylinder 7 – misfire detected	Wiring, engine mechanical fault, ignition/fuel system, injector
	B1	Cylinder 8 – misfire detected	Wiring, engine mechanical fault, ignition/fuel system, injector
	B2	Cylinder 9 – misfire detected	Wiring, engine mechanical fault, ignition/fuel system, injector
	B3	Cylinder 10 – misfire detected	Wiring, engine mechanical fault, ignition/fuel system, injector
44		Ambient/battery temperature sensor – voltage low/high	Wiring, ambient/battery temperature sensor, ECM
	9A	Ambient/battery temperature sensor – voltage high	Wiring, ambient/battery temperature sensor, ECM
	99	Ambient/battery temperature sensor – voltage low	Wiring, ambient/battery temperature sensor, ECM

Flash code	Scan code Hex type	Fault location	Probable cause
45		Transmission control system – malfunction	Wiring, transmission fluid level, transmission mechanical fault, PNP switch, TCC solenoid, TFT sensor, TFP solenoid, shift solenoid, overdrive solenoid/switch, governor pressure solenoid, transmission system relay, ECM
	32	Shift solenoid (SS) 3-4 – electrical	Wiring, shift solenoid, ECM
	4A	Transmission fluid temperature (TFT) sensor – low input	Wiring short circuit to ground, TFT sensor, ECM
	4B	Transmission fluid temperature (TFT) sensor – high input	Wiring short circuit to positive, TFT sensor, ECM
	8D	Governor pressure not equal to target (15-20 PSI) – mid pressure malfunction	Signal wire open circuit/short circuit, valve body, sensor supply wire open circuit, transmission fluid level, governor pressure solenoid, transmission mechanical fault, ECM – no 5 volt supply
	8E	Governor pressure not equal to target – low pressure malfunction	Signal wire open circuit/short circuit, valve body, sensor supply wire open circuit, transmission fluid level, governor pressure solenoid, transmission mechanical fault, ECM – no 5 volt supply
	A4	Transmission fluid temperature (TFT) sensor – range/performance problem	Wiring, TFT sensor
	A5	Gear selection, 3-4 – shift malfunction	Wiring, TR sensor, shift solenoids, transmission mechanical fault
	A7	Governor pressure sensor – signal low	Signal wire short circuit to ground, supply wire open circuit, transmission fluid level, governor pressure sensor, ECM
	A8	Governor pressure sensor – signal high	Signal wire short circuit to positive, supply wire open circuit, transmission fluid level, governor pressure sensor, ECM
	A9	Governor pressure sensor – incorrect signal	Signal wire short circuit to ground/ positive, supply wire open circuit/short circuit, transmission fluid level, governor pressure solenoid, governor pressure sensor, ECM
	AB	Transmission fluid pressure (TFP) solenoid – electrical	Wiring, TFP solenoid, ECM
	AD	Transmission system relay	Wiring open circuit/short circuit, transmission system relay, ECM
	BC	Overdrive solenoid/switch – performance problem	Wiring, overdrive solenoid/switch, transmission mechanical fault, ECM
46	06	Alternator – voltage high	Wiring, battery, alternator

Flash code	Scan code Hex type	Fault location	Probable cause
47	05	Alternator – voltage low	Wiring, battery, alternator drive belt, alternator
51		System too lean, bank 1/2	Wiring, intake/exhaust leak, MAP sensor, fuel pressure/pump, EVAP canister purge valve, injector(s), HO2S
	77	System too lean, bank 1	Intake/exhaust leak, MAP sensor, fuel pressure/pump, injector(s), HO2S
	79	System too lean, bank 2	Intake/exhaust leak, fuel pressure/pump, injector(s), hose connection(s)
52		System too rich, bank 1/2	Wiring, intake blocked, EVAP canister purge valve, fuel pressure, EGR system, injector(s), air filter blocked, ignition system, HO2S
	76	System too rich, bank 1	Intake blocked, EVAP canister purge valve, fuel pressure, EGR system, injector(s), HO2S
	78	System too rich, bank 2	Intake blocked, EVAP canister purge valve, fuel pressure, EGR system, injector(s), HO2S
53		Engine control module (ECM) – internal fault	ECM
	02	Engine control module (ECM) – internal fault	ECM
	44	Engine control module (ECM) – internal fault	ECM
54	01	Camshaft position (CMP) sensor – no signal	Wiring, CMP sensor
55		End of malfunction indicator lamp (MIL) code output	–
63	31	Engine control module (ECM) – EEPROM error	ECM
64	70	Catalytic converter system, bank 1 – efficiency below threshold	Catalytic converter, wiring, HO2S 2
65	73	Power steering pressure (PSP) switch – range/performance problem	Wiring, PAS system, PSP switch
71	B9	Engine control module (ECM) – 5 volt output low	ECM
72		Catalytic converter system, bank 1/2 – efficiency below threshold	Catalytic converter, wiring, HO2S 2
	70	Catalytic converter system, bank 1 – efficiency below threshold	Catalytic converter, wiring, HO2S 2
	B4	Catalytic converter system, bank 2 – efficiency below threshold	Catalytic converter, wiring, HO2S 2
77	52	Cruise control	Wiring, cruise control vacuum switching valve

DODGE

Model:	Engine identification:	Year:
Intrepid 2.7L	**EER**	**1998**
Intrepid 3.2L	**EGW**	**1998**
Neon 2.0L	**ECB/ECC**	**1998**
Stratus 2.0L	**ECB**	**1998**
Stratus 2.4L	**EDZ**	**1998**
Stratus 2.5L	**EEB**	**1998**

System: **SBEC III**

Self-diagnosis

General information

- Refer to the front of this manual for general test conditions, terminology, detailed descriptions of wiring faults and a general trouble shooter for electrical and mechanical faults.
- Malfunction indicator lamp (MIL) will illuminate or flash if certain faults are recorded.
- ECM operates in backup mode if sensors fail, to enable vehicle to be driven to repair shop.

Accessing

- Trouble codes can only be displayed by using a Scan Tool connected to the data link connector (DLC):
  - Intrepid **1**.
  - Neon **2**.
  - Stratus **3**.

1

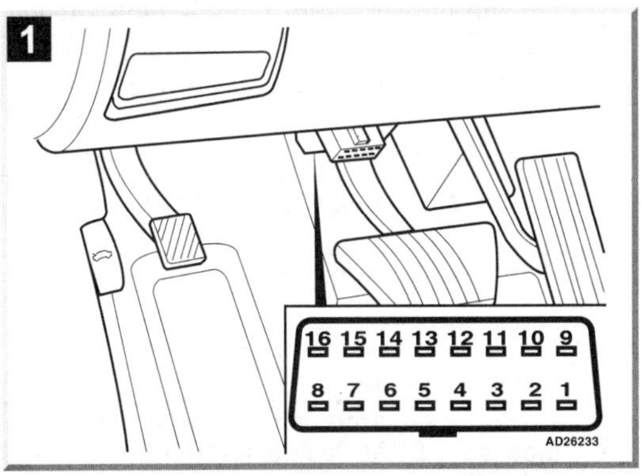

2

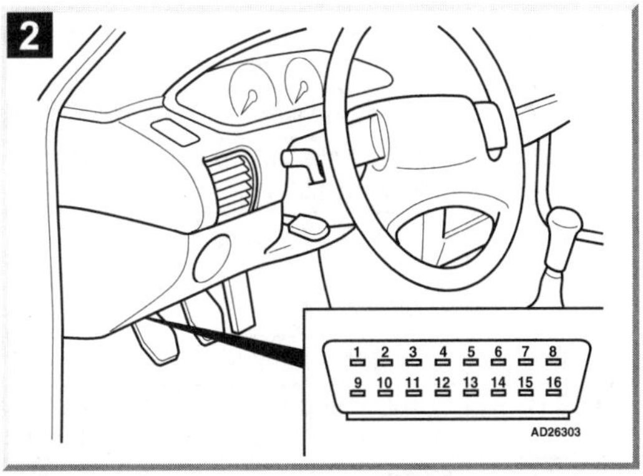

3

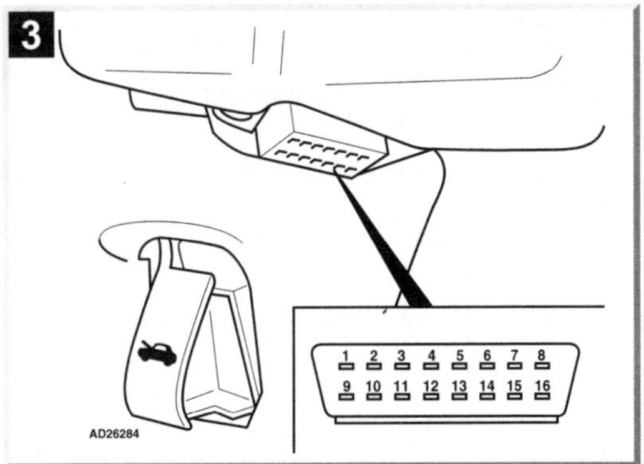

Erasing

- Trouble codes can only be erased by using a Scan Tool connected to the data link connector (DLC).

Trouble code identification

Scan code P type	Fault location	Probable cause
P0000-P0999	Refer to OBD-II trouble code table at the front of this manual	–
P1195	Heated oxygen sensor (HO2S) 1, bank 1 – slow response	**Wiring, heating inoperative, HO2S**
P1196	Heated oxygen sensor (HO2S) 1, bank 2 – slow response	**Wiring, heating inoperative, HO2S**
P1281	Engine temperature – low for long period	**Wiring, cooling system, coolant thermostat, ECT sensor**
P1282	Fuel pump relay	**Wiring open circuit/short circuit, fuel pump relay**
P1289	Intake manifold air control solenoid	**Wiring open circuit/short circuit, intake manifold air control solenoid**
P1294	Idle speed control (ISC) system – rpm lower/ higher than expected	**Wiring, air leak, throttle valve tight/sticking, IAC valve**
P1297	Manifold absolute pressure (MAP) sensor	**Wiring, hose connection(s), MAP sensor**
P1299	Manifold absolute pressure (MAP) sensor/ throttle position (TP) sensor	**Intake leak, hose connection(s), MAP sensor, TP sensor**
P1388	Engine control relay	**Wiring open circuit/short circuit, engine control relay**
P1389	Engine control relay	**Wiring open circuit, engine control relay**
P1390	Timing belt jumped 1 tooth or more	**Timing belt loose/jumped position**
P1391	Crankshaft position (CKP) sensor/camshaft position (CMP) sensor – intermittent signal	**Wiring, poor connection, air gap, CKP sensor, CMP sensor**
P1398	Engine control module (ECM) – misfire detection limit reached	**Wiring, poor connection, CKP sensor**
P1478	Ambient/battery temperature sensor	**Wiring, ambient/battery temperature sensor**
P1486	Evaporative emission (EVAP) system – blockage detected	**Hose connection(s), hose blocked/trapped, EVAP canister**
P1489	Engine coolant blower motor relay – high speed	**Wiring open circuit/short circuit, engine coolant blower motor relay – high speed**
P1490	Engine coolant blower motor relay – low speed	**Wiring open circuit/short circuit, engine coolant blower motor relay – low speed**
P1491	Engine coolant blower motor relay – low/high speed	**Wiring open circuit/short circuit, engine coolant blower motor relay – low/high speed**
P1492	Ambient/battery temperature sensor – voltage high	**Wiring, ambient/battery temperature sensor**
P1493	Ambient/battery temperature sensor – voltage low	**Wiring, ambient/battery temperature sensor**
P1494	Evaporative emission (EVAP) leak detection pump – pressure switch	**Wiring, hose leak/blockage, EVAP leak detection pump**
P1495	Evaporative emission (EVAP) leak detection pump solenoid	**Wiring open circuit/short circuit, hose leak/ blockage, EVAP leak detection pump solenoid**

Scan code P type	Fault location	Probable cause
P1496	Engine control module (ECM) – 5 volt output low	ECM
P1594	Alternator – voltage high	Wiring, battery, alternator
P1595	Cruise control	Wiring open circuit/short circuit, hoses, cruise control master switch, cruise control vacuum switching valve
P1596	Cruise control master switch – high voltage	Wiring, cruise control master switch
P1597	Cruise control master switch – low voltage	Wiring, cruise control master switch
P1598	A/C refrigerant pressure sensor – high input	Wiring, A/C refrigerant pressure too high (cooling fault/incorrectly charged), A/C refrigerant pressure sensor
P1599	A/C refrigerant pressure sensor – low input	Wiring, A/C refrigerant pressure too low (incorrectly charged), A/C refrigerant pressure sensor
P1682	Alternator – voltage low	Wiring, battery, alternator drive belt, alternator
P1683	Cruise control	Wiring, cruise control vacuum switching valve
P1685	Key not programmed into immobilizer control module	Wiring, ignition key not matched/damaged, immobilizer control module
P1686	CAN data bus, Immobilizer control module – no signal	Wiring, immobilizer control module
P1687	CAN data bus, instrumentation – no signal	Wiring, instrumentation control module (mechanical)
P1695	CAN data bus, body control module (BCM) – no signal	Wiring, BCM
P1696	Engine control module (ECM) – EEPROM error	ECM
P1697	Engine control module (ECM) – EEPROM error	ECM
P1698	CAN data bus, transmission control module (TCM) – no signal	Wiring, TCM
P1899	Park/neutral position (PNP) switch – input circuit malfunction	Wiring, PNP switch, ECM/TCM

Scan code Hex type	Fault location	Probable cause
01	Camshaft position (CMP) sensor – no signal	Wiring, CMP sensor
02	Engine control module (ECM) – internal fault	ECM
05	Alternator – voltage low	Wiring, battery, alternator drive belt, alternator
06	Alternator – voltage high	Wiring, battery, alternator
0A	Engine control relay	Wiring open circuit/short circuit, engine control relay

DODGE

Intrepid 2.7L • Intrepid 3.2L • Neon 2.0L • Stratus 2.0L
Stratus 2.4L • Stratus 2.5L

Scan code Hex type	Fault location	Probable cause
0B	Alternator, field control – circuit malfunction	**Wiring open circuit/short circuit, alternator, battery**
0C	Torque converter clutch (TCC) solenoid – electrical	**Wiring open circuit/short circuit, TCC solenoid**
0E	Engine coolant blower motor relay – low/high speed	**Wiring open circuit/short circuit, engine coolant blower motor relay – low/high speed**
0F	Cruise control	**Wiring open circuit/short circuit, hoses, cruise control master switch, cruise control vacuum switching valve**
10	A/C compressor clutch relay	**Wiring open circuit/short circuit, A/C compressor clutch relay**
11	Exhaust gas recirculation (EGR) – circuit malfunction	**Wiring open circuit/short circuit, EGR solenoid**
12	Evaporative emission (EVAP) canister purge valve – circuit malfunction	**Wiring open circuit/short circuit, EVAP canister purge valve**
13	Injector 3 – circuit malfunction	**Wiring, injector**
14	Injector 2 – circuit malfunction	**Wiring, injector**
15	Injector 1 – circuit malfunction	**Wiring, injector**
19	Idle speed control (ISC) system – malfunction	**Wiring open circuit/short circuit, IAC valve**
1A	Throttle position (TP) sensor – low input	**Wiring short circuit to ground, TP sensor**
1B	Throttle position (TP) sensor – high input	**Wiring short circuit to positive, TP sensor**
1E	Engine coolant temperature (ECT) sensor – low input	**Wiring short circuit to ground, coolant thermostat, ECT sensor**
1F	Engine coolant temperature (ECT) sensor – high input	**Wiring open circuit/short circuit to positive, coolant thermostat, ground wire defective, ECT sensor**
20	Heated oxygen sensor (H02S) 1, bank 1 – no activity detected	**Wiring open circuit, heating inoperative, HO2S**
21	Engine temperature – low for long period	**Wiring, cooling system, coolant thermostat, ECT sensor**
23	Vehicle speed sensor (VSS) – signal	**Wiring, VSS**
24	Manifold absolute pressure (MAP) sensor – low input	**Wiring short circuit to ground, MAP sensor**
25	Manifold absolute pressure (MAP) sensor – high input	**Wiring short circuit to positive, MAP sensor**
27	Manifold absolute pressure (MAP) sensor	**Wiring, hose connection(s), MAP sensor**
28	Crankshaft position (CKP) sensor – no signal	**Wiring, air gap, CKP sensor**
29	Ignition coil C, primary/secondary – circuit malfunction	**Wiring, ignition coil**
2A	Ignition coil B, primary/secondary – circuit malfunction	**Wiring, ignition coil**

Scan code Hex type	Fault location	Probable cause
2B	Ignition coil A, primary/secondary – circuit malfunction	**Wiring, ignition coil**
2C	Engine control relay	**Wiring open circuit, engine control relay**
2E	Exhaust gas recirculation (EGR) system – insufficient flow detected	**Wiring, hose leak/blockage, EGR valve, EGR solenoid**
30	Engine control module (ECM) – EEPROM error	**ECM**
31	Engine control module (ECM) – EEPROM error	**ECM**
39	Intake air temperature (IAT) sensor – low input	**Wiring short circuit to ground, IAT sensor**
3A	Intake air temperature (IAT) sensor – high input	**Wiring open circuit/short circuit to positive, ground wire defective, IAT sensor**
3B	Knock sensor (KS) 1 – no signal	**Wiring, KS incorrectly tightened, KS**
3C	Manifold absolute pressure (MAP) sensor – barometric pressure range/performance problem	**Intake/exhaust leak, wiring, MAP sensor**
3D	Injector 4 – circuit malfunction	**Wiring, injector**
3E	Heated oxygen sensor (HO2S) 1, bank 1 – high voltage	**Wiring short circuit to positive, fuel pressure, injector control, EVAP canister purge valve, air filter blocked, HO2S**
41	Heated oxygen sensor (HO2S) 1, bank 2 – no activity detected	**Wiring, HO2S**
42	Heated oxygen sensor (HO2S) 1, bank 2 – high voltage	**Wiring short circuit to positive, fuel pressure, injector control, EVAP canister purge valve, air filter blocked, HO2S**
44	Engine control module (ECM) – internal fault	**ECM**
45	Injector 5 – circuit malfunction	**Wiring, injector**
46	Injector 6 – circuit malfunction	**Wiring, injector**
47	Ambient/battery temperature sensor	**Wiring, ambient/battery temperature sensor**
4C	Ignition coil D, primary/secondary – circuit malfunction	**Wiring, ignition coil**
4D	Ignition coil E, primary/secondary – circuit malfunction	**Wiring, ignition coil**
52	Cruise control	**Wiring, cruise control vacuum switching valve**
56	Cruise control master switch – high voltage	**Wiring, cruise control master switch**
57	Cruise control master switch – low voltage	**Wiring, cruise control master switch**
58	Intake manifold air control solenoid	**Wiring open circuit/short circuit, intake manifold air control solenoid**
5A	A/C refrigerant pressure sensor – high input	**Wiring, A/C refrigerant pressure too high (cooling fault/incorrectly charged), A/C refrigerant pressure sensor**
5B	A/C refrigerant pressure sensor – low input	**Wiring, A/C refrigerant pressure too low (incorrectly charged), A/C refrigerant pressure sensor**

Scan code Hex type	Fault location	Probable cause
5C	Engine coolant blower motor relay – low speed	**Wiring open circuit/short circuit, engine coolant blower motor relay – low speed**
5D	Engine coolant blower motor relay – high speed	**Wiring open circuit/short circuit, engine coolant blower motor relay – high speed**
60	CAN data bus, transmission control module (TCM) – no signal	**Wiring, TCM**
61	CAN data bus, body control module (BCM) – no signal	**Wiring, BCM**
65	Fuel pump relay	**Wiring open circuit/short circuit, fuel pump relay**
66	Heated oxygen sensor (H02S) 1, bank 1 – slow response	**Wiring, heating inoperative, HO2S**
67	Heated oxygen sensor (HO2S) 1, bank 1 – heater control – circuit malfunction	**Wiring, HO2S**
69	Heated oxygen sensor (HO2S) 2, bank 1, heater control – circuit malfunction	**Wiring, HO2S**
6A	Random/multiple cylinder(s) – misfire detected	**Wiring, spark plug(s), HT lead(s), injector(s), ignition coil(s), low compression**
6B	Cylinder 1 – misfire detected	**Wiring, engine mechanical fault, ignition/ fuel system, injector**
6C	Cylinder 2 – misfire detected	**Wiring, engine mechanical fault, ignition/ fuel system, injector**
6D	Cylinder 3 – misfire detected	**Wiring, engine mechanical fault, ignition/ fuel system, injector**
6E	Cylinder 4 – misfire detected	**Wiring, engine mechanical fault, ignition/ fuel system, injector**
70	Catalytic converter system, bank 1 – efficiency below threshold	**Wiring, heated oxygen sensor (HO2S) 2, catalytic converter**
71	Evaporative emission (EVAP) system – incorrect flow detected	**Hose connection(s), intake leak, EVAP canister purge valve**
72	Park/neutral position (PNP) switch – input circuit malfunction	**Wiring, PNP switch, ECM/TCM**
73	Power steering pressure (PSP) switch – range/performance problem	**Wiring, PAS system, PSP switch**
76	System too rich	**Wiring, intake blocked, EVAP canister purge valve, fuel pressure, EGR system, injector(s), air filter blocked, ignition system, HO2S**
77	System too lean	**Wiring, intake/exhaust leak, MAP sensor, fuel pressure/pump, EVAP canister purge valve, injector(s), HO2S**
78	System too rich, bank 2	**Intake blocked, EVAP canister purge valve, fuel pressure, EGR system, injector(s), HO2S**
79	System too lean, bank 2	**Intake/exhaust leak, fuel pressure/pump, injector(s), AIR system, hose connection(s)**

Scan code Hex type	Fault location	Probable cause
7A	Heated oxygen sensor (HO2S) 1, bank 2 – slow response	**Wiring, heating inoperative, HO2S**
7C	Heated oxygen sensor (HO2S) 1, bank 2, heater control – circuit malfunction	**Wiring, HO2S**
7D	Heated oxygen sensor (HO2S) 2, bank 2, heater control – circuit malfunction	**Wiring, HO2S**
7E	Heated oxygen sensor (HO2S) 2, bank 1 – high voltage	**Wiring short circuit to positive, catalytic converter, HO2S**
7F	Heated oxygen sensor (HO2S) 2, bank 2 – high voltage	**Wiring short circuit to positive, catalytic converter, HO2S**
80	Insufficient coolant temperature for closed loop fuel control	**Wiring, cooling system, coolant thermostat, ECT sensor**
81	Heated oxygen sensor (HO2S) 2, bank 1 – no activity detected	**Wiring, heating inoperative, HO2S**
82	Heated oxygen sensor (HO2S) 2, bank 2 – no activity detected	**Wiring, HO2S**
84	Throttle position (TP) sensor – range/ performance problem	**Wiring, accelerator cable adjustment, TP sensor**
85	Timing belt jumped 1 tooth or more	**Timing belt loose/jumped position**
89	Transmission control system – malfunction	**Wiring, ECM/TCM**
8A	Idle speed control (ISC) system – rpm lower/ higher than expected	**Wiring, air leak, throttle valve tight/sticking, IAC valve**
91	Manifold absolute pressure (MAP) sensor/ throttle position (TP) sensor	**Intake leak, hose connection(s), MAP sensor, TP sensor**
92	Engine control module (ECM) – 5 volt output low	**ECM**
94	Torque converter clutch (TCC) solenoid – circuit malfunction	**Wiring, TCC solenoid**
95	Fuel level sensor – low input	**Wiring open circuit, fuel level sensor**
96	Fuel level sensor – high input	**Wiring short circuit to positive, fuel level sensor**
97	Fuel level sensor – malfunction	**Wiring, fuel level sensor**
98	Brake pedal position (BPP) switch – circuit malfunction	**Wiring, BPP switch**
99	Ambient/battery temperature sensor – voltage low	**Wiring, ambient/battery temperature sensor**
9A	Ambient/battery temperature sensor – voltage high	**Wiring, ambient/battery temperature sensor**
9B	Heated oxygen sensor (HO2S) 1, bank 1 – low voltage	**Wiring short circuit to ground, exhaust leak, HO2S**
9C	Heated oxygen sensor (HO2S) 2, bank 1 – low voltage	**Wiring short circuit to ground, exhaust leak, HO2S**
9D	Crankshaft position (CKP) sensor/camshaft position (CMP) sensor – intermittent signal	**Wiring, poor connection, air gap, CKP sensor, CMP sensor**

DODGE

Intrepid 2.7L • Intrepid 3.2L • Neon 2.0L • Stratus 2.0L
Stratus 2.4L • Stratus 2.5L

Scan code Hex type	Fault location	Probable cause
A0	Evaporative emission (EVAP) system – small leak detected	**Hose connection(s), intake leak, EVAP canister, EVAP canister purge valve**
A1	Evaporative emission (EVAP) system – large leak detected	**Hose connection(s), intake leak, EVAP canister, EVAP canister purge valve**
AE	Cylinder 5 – misfire detected	**Wiring, engine mechanical fault, ignition/ fuel system, injector**
AF	Cylinder 6 – misfire detected	**Wiring, engine mechanical fault, ignition/ fuel system, injector**
B5	Heated oxygen sensor (HO2S) 1, bank 2 – low voltage	**Wiring short circuit to ground, exhaust leak, HO2S**
B6	Heated oxygen sensor (HO2S) 2, bank 2 – low voltage	**Wiring short circuit to ground, exhaust leak, HO2S**
B7	Evaporative emission (EVAP) leak detection pump solenoid	**Wiring open circuit/short circuit, hose leak/ blockage, EVAP leak detection pump solenoid**
B8	Evaporative emission (EVAP) leak detection pump – pressure switch	**Wiring, hose leak/blockage, EVAP leak detection pump**
BA	Engine control module (ECM) – misfire detection limit reached	**Wiring, poor connection, CKP sensor**
BB	Evaporative emission (EVAP) system – blockage detected	**Hose connection(s), hose blocked/trapped, EVAP canister**
C0	Heated oxygen sensor (HO2S) 1, bank 1 – slow response	**Wiring, heating inoperative, HO2S**
C1	Heated oxygen sensor (HO2S) 1, bank 2 – slow response	**Wiring, heating inoperative, HO2S**
D4	Exhaust gas recirculation (EGR) valve position sensor A – low input	**Wiring short circuit to ground, EGR valve position sensor**
D5	Exhaust gas recirculation (EGR) valve position sensor A – high input	**Wiring short circuit to positive, EGR valve position sensor**
D7	Exhaust gas recirculation (EGR) system – range/performance problem	**Wiring, hose leak/blockage, EGR valve/ solenoid**
D8	Ignition coil F, primary/secondary – circuit malfunction	**Wiring, ignition coil**
E1	CAN data bus, instrumentation – no signal	**Wiring, instrumentation control module (mechanical)**
E2	CAN data bus, immobilizer control module – no signal	**Wiring, immobilizer control module**
E8	Key not programmed into immobilizer control module	**Wiring, ignition key not matched/damaged, immobilizer control module**

Model:	Engine identification:	Year:
Intrepid 2.7/3.2/3.5L	EER/EES/EGW/EGG/ECG/EGK/EGJ	1999-02
Neon 2.0L	ECB/ECC	1999-02
Stratus Sedan 2.0L	ECB	1999-02
Stratus Sedan 2.4L	EDZ	1999-02
Stratus Sedan 2.5L	EEB	1999-02
Stratus Sedan 2.7L	EER	1999-02

System: **SBEC III**

Self-diagnosis

General information

- Refer to the front of this manual for general test conditions, terminology, detailed descriptions of wiring faults and a general trouble shooter for electrical and mechanical faults.
- Malfunction indicator lamp (MIL) will illuminate or flash if certain faults are recorded.
- ECM operates in backup mode if sensors fail, to enable vehicle to be driven to repair shop.

Accessing

1999-2001

- Trouble codes can only be displayed by using a Scan Tool connected to the data link connector (DLC):
  - Intrepid **1**.
  - Neon **2**.
  - Stratus 1999-00 **3**.

1

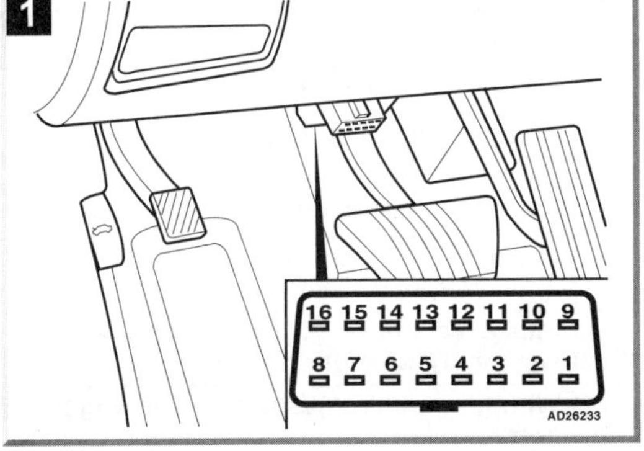

2

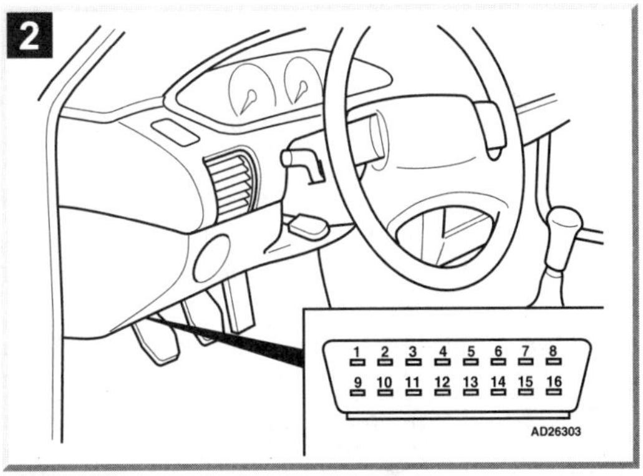

3

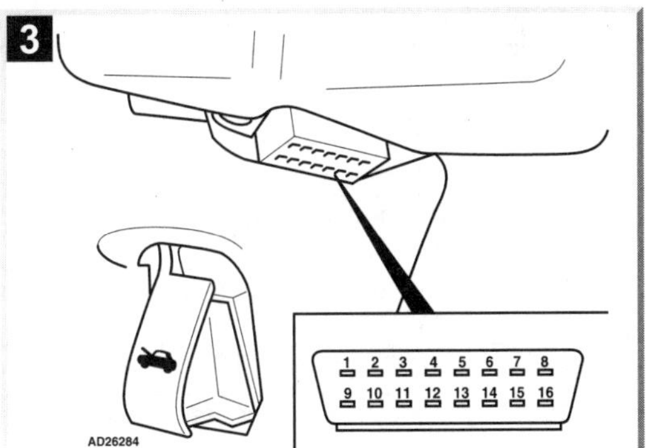

Accessing

2001→

- Within 5 seconds switch the ignition key ON-OFF-ON-OFF-ON.
- Odometer will display trouble codes.
- If no trouble codes are stored in ECM, odometer will display 'P1684' and 'done' or only 'done'.
- Trouble codes can also be displayed by using a Scan Tool connected to the data link connector (DLC):
  - Intrepid/Stratus 2001-02 **1**.
  - Neon **2**.

Erasing

- Trouble codes can only be erased by using a Scan Tool connected to the data link connector (DLC).

Trouble code identification

Scan code	Fault location	Probable cause
P0000-P0999	Refer to OBD-II trouble code table at the front of this manual	–
P1192	Intake air temperature (IAT) sensor – voltage low	**Wiring, IAT sensor**
P1193	Intake air temperature (IAT) sensor – voltage high	**Wiring, IAT sensor**
P1195	Heated oxygen sensor (H02S) 1, bank 1 – slow response	**Wiring, heating inoperative, HO2S**
P1196	Heated oxygen sensor (HO2S) 1, bank 2 – slow response	**Wiring, heating inoperative, HO2S**
P1197	Heated oxygen sensor (HO2S) 2, bank 1 – slow response	**Wiring, heating inoperative, HO2S**
P1281	Engine temperature – low for long period	**Wiring, cooling system, coolant thermostat, ECT sensor**
P1282	Fuel pump relay	**Wiring open circuit/short circuit, fuel pump relay**
P1288	Intake manifold air control solenoid (short runner)	**Wiring open circuit/short circuit, intake manifold air control solenoid (short runner)**
P1289	Intake manifold air control solenoid	**Wiring open circuit/short circuit, intake manifold air control solenoid**
P1294	Idle speed control (ISC) system – rpm lower/higher than expected	**Wiring, air leak, throttle valve tight/sticking, IAC valve**
P1295	Throttle position (TP) sensor – no 5 volt supply	**Wiring, TP sensor, ECM**
P1296	Manifold absolute pressure (MAP) sensor – no 5 volt supply	**Wiring, MAP sensor, ECM**
P1297	Manifold absolute pressure (MAP) sensor	**Wiring, hose connection(s), MAP sensor**
P1298	System too lean – at prolonged wide open throttle	**Intake/exhaust leak, MAP sensor, fuel pressure/pump, injector(s), HO2S**
P1299	Manifold absolute pressure (MAP) sensor/throttle position (TP) sensor	**Intake leak, hose connection(s), MAP sensor, TP sensor**
P1388	Engine control relay	**Wiring open circuit/short circuit, engine control relay**

Scan code	Fault location	Probable cause
P1389	Engine control relay	**Wiring open circuit, engine control relay**
P1390	Crankshaft position (CKP) sensor/camshaft position (CMP) sensor – signal synchronisation	**Timing belt loose/jumped position, CKP sensor, CMP sensor**
P1391	Crankshaft position (CKP) sensor/camshaft position (CMP) sensor – intermittent signal	**Wiring, poor connection, air gap, CKP sensor, CMP sensor**
P1398	Engine control module (ECM) – misfire detection limit reached	**Wiring, poor connection, CKP sensor**
P1403	Exhaust gas recirculation (EGR) valve position sensor – no 5 volt supply	**Wiring, ECM**
P1478	Ambient/battery temperature sensor	**Wiring, ambient/battery temperature sensor**
P1486	Evaporative emission (EVAP) system – blockage detected	**Hose connection(s), hose blocked/trapped, EVAP canister**
P1488	Engine control module (ECM) – auxiliary 5 volt output low	**Wiring, ECM**
P1489	Engine coolant blower motor relay – high speed	**Wiring open circuit/short circuit, engine coolant blower motor relay – high speed**
P1490	Engine coolant blower motor relay – low speed	**Wiring open circuit/short circuit, engine coolant blower motor relay – low speed**
P1491	Engine coolant blower motor relay – circuit malfunction	**Wiring open circuit/short circuit, engine coolant blower motor relay**
P1492	Ambient/battery temperature sensor – voltage high	**Wiring, ambient/battery temperature sensor**
P1493	Ambient/battery temperature sensor – voltage low	**Wiring, ambient/battery temperature sensor**
P1494	Evaporative emission (EVAP) leak detection pump – pressure switch	**Wiring, hose leak/blockage, EVAP leak detection pump**
P1495	Evaporative emission (EVAP) leak detection pump solenoid	**Wiring open circuit/short circuit, hose leak/blockage, EVAP leak detection pump solenoid**
P1496	Engine control module (ECM) – 5 volt output low	**ECM**
P1594	Alternator – voltage high	**Wiring, battery, alternator**
P1595	Cruise control	**Wiring open circuit/short circuit, hoses, cruise control master switch, cruise control vacuum switching valve**
P1596	Cruise control master switch – high voltage	**Wiring, cruise control master switch**
P1597	Cruise control master switch – low voltage	**Wiring, cruise control master switch**
P1598	A/C refrigerant pressure sensor – high input	**Wiring, A/C refrigerant pressure too high (cooling fault/incorrectly charged), A/C refrigerant pressure sensor**
P1599	A/C refrigerant pressure sensor – low input	**Wiring, A/C refrigerant pressure too low (incorrectly charged), A/C refrigerant pressure sensor**
P1602	Engine control module (ECM) – not programmed	**New ECM not programmed, incorrectly programmed, ECM**

Scan code	Fault location	Probable cause
P1603	Engine control module (ECM) – RAM link error	**ECM**
P1604	Engine control module (ECM) – RAM read/ write error	**ECM**
P1607	Engine control module (ECM) – shut down timer error	**ECM**
P1681	CAN data bus, instrumentation – no signal	**Wiring, instrumentation control module**
P1682	Alternator – voltage low	**Wiring, battery, alternator drive belt, alternator**
P1683	Cruise control	**Wiring, cruise control vacuum switching valve**
P1684	Engine control module (ECM) – supply voltage disconnected within last 50 ignition switch ON-OFF cycles	**Wiring, battery**
P1685	Key not programmed into immobilizer control module	**Wiring, ignition key not matched/damaged, immobilizer control module**
P1686	CAN data bus, immobilizer control module – no signal	**Wiring, immobilizer control module**
P1687	CAN data bus, instrumentation – no signal	**Wiring, instrumentation control module (mechanical)**
P1695	CAN data bus, body control module (BCM) – no signal	**Wiring, BCM**
P1696	Engine control module (ECM) – EEPROM error	**ECM**
P1697	Engine control module (ECM) – EEPROM error	**ECM**
P1698	CAN data bus, transmission control module (TCM) – no signal	**Wiring, TCM**
P1765	Transmission system relay	**Wiring open circuit/short circuit, transmission system relay, ECM/TCM**
P1899	Park/neutral position (PNP) switch – input circuit malfunction	**Wiring, PNP switch, ECM/TCM**
P2008	Intake manifold air control solenoid (short runner)	**Wiring open circuit/short circuit, intake manifold air control solenoid (short runner)**
P2302	Ignition coil 1, secondary – circuit malfunction	**Wiring, ignition coil**
P2305	Ignition coil 2, secondary – circuit malfunction	**Wiring, ignition coil**
P2308	Ignition coil 3, secondary – circuit malfunction	**Wiring, ignition coil**
P2311	Ignition coil 4, secondary – circuit malfunction	**Wiring, ignition coil**
P2314	Ignition coil 5, secondary – circuit malfunction	**Wiring, ignition coil**
P2317	Ignition coil 6, secondary – circuit malfunction	**Wiring, ignition coil**
P2503	Charging system voltage – low	**Wiring, poor connection, battery, alternator**

DODGE

Model:	Engine identification:	Year:
Intrepid 3.3L	EGB	1996-97
Intrepid 3.5L	EGE	1996-97
Neon 2.0L	ECB/ECC	1996-97
Stratus 2.0L	ECB	1996-97
Stratus 2.4L	EDZ	1996-97
Stratus 2.5L	EEB	1996-97

System: **SBEC III**

Self-diagnosis

General information

- Refer to the front of this manual for general test conditions, terminology, detailed descriptions of wiring faults and a general trouble shooter for electrical and mechanical faults.
- Malfunction indicator lamp (MIL) will illuminate or flash if certain faults are recorded.
- ECM operates in backup mode if sensors fail, to enable vehicle to be driven to repair shop.
- Trouble codes can be displayed by the MIL or by using a Scan Tool connected to the data link connector (DLC).

Accessing

Flash codes

- Within 5 seconds switch the ignition key ON-OFF-ON-OFF-ON.
- Count MIL flashes. Note trouble codes.
- Each trouble code consists of two groups **1** [**A**].
- The first group of flashes indicate the 'tens' of the trouble code.
- The second group of flashes indicate the 'units' of the trouble code.
- A short pause separates 'tens' and 'units' **1** [**B**].
- A long pause separates each trouble code **1** [**C**].
- For example: Trouble code 21 displayed **1**.
- End of test sequence indicated by trouble code 55.

P codes & Hex codes

- Trouble codes can be displayed by using a Scan Tool connected to the data link connector (DLC):
  - Intrepid **2**.
  - Neon **3**.
  - Stratus **4**.

1

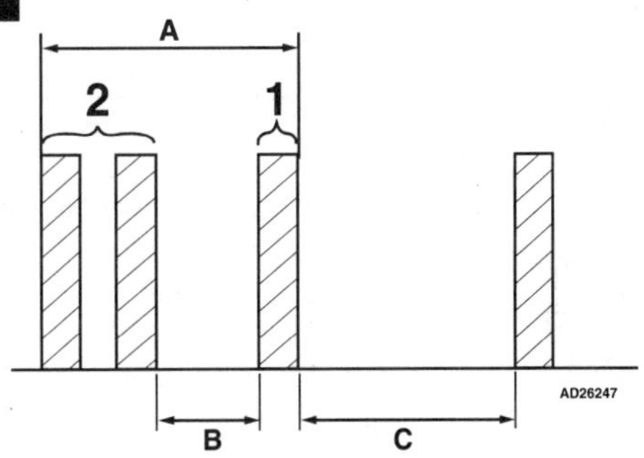

2

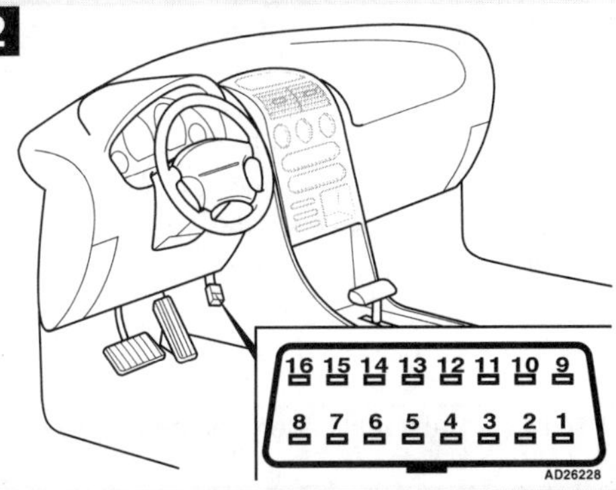

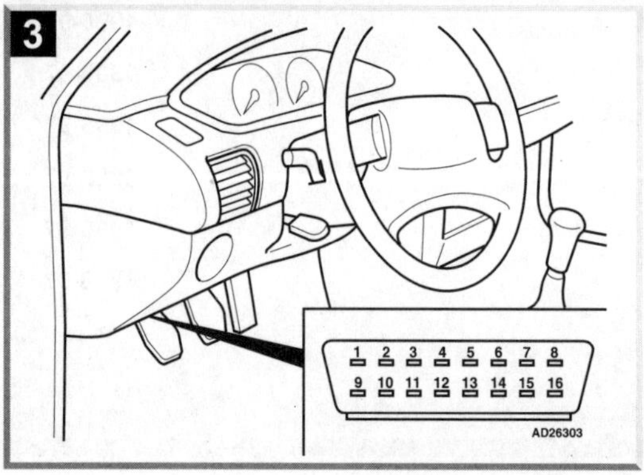

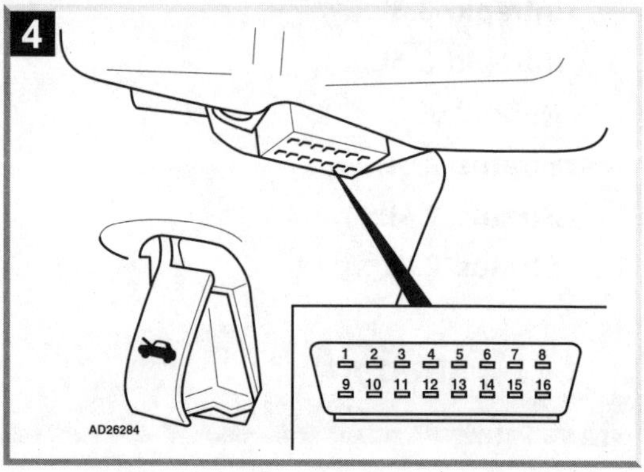

Erasing

- Trouble codes can only be erased by using a Scan Tool connected to the data link connector (DLC).

Trouble code identification

Scan code P type	Fault location	Probable cause
P0000-P0999	Refer to OBD-II trouble code table at the front of this manual	–
P1289	Intake manifold air control solenoid	**Wiring open/short circuit, intake manifold air control solenoid**
P1296	Manifold absolute pressure (MAP) sensor – no 5 volt supply	**Wiring, ECM**
P1297	Manifold absolute pressure (MAP) sensor	**Wiring, hose connection(s), MAP sensor**
P1299	Manifold absolute pressure (MAP) sensor/ throttle position (TP) sensor	**Intake leak, hose connection(s), MAP sensor, TP sensor**
P1390	Timing belt jumped 1 tooth or more	**Timing belt loose/jumped position**
P1391	Crankshaft position (CKP) sensor/camshaft position (CMP) sensor – intermittent signal	**Wiring, poor connection, air gap, CKP sensor, CMP sensor**
P1398	Engine control module (ECM) – misfire detection limit reached	**Wiring, poor connection, CKP sensor**
P1486	Evaporative emission (EVAP) system – blockage detected	**Hose connection(s), hose blocked/trapped, EVAP canister**
P1489	Engine coolant blower motor relay – high speed	**Wiring open/short circuit, engine coolant blower motor relay – high speed**
P1490	Engine coolant blower motor relay – low speed	**Wiring open/short circuit, engine coolant blower motor relay – low speed**
P1492	Ambient/battery temperature sensor – voltage high	**Wiring, ambient/battery temperature sensor**

Scan code P type	Fault location	Probable cause
P1493	Ambient/battery temperature sensor – voltage low	**Wiring, ambient/battery temperature sensor**
P1494	Evaporative emission (EVAP) leak detection pump – pressure switch	**Wiring, hose leak/blockage, EVAP leak detection pump**
P1495	Evaporative emission (EVAP) leak detection pump solenoid	**Wiring open/short circuit, hose leak/ blockage, EVAP leak detection pump solenoid**
P1496	Engine control module (ECM) – 5 volt output low	**ECM**
P1496	Manifold absolute pressure (MAP) sensor – 5 volt output low	**Wiring, MAP sensor**
P1696	Engine control module (ECM) – EEPROM error	**ECM**
P1697	Engine control module (ECM) – EEPROM error	**ECM**
P1698	CAN data bus, Transmission control module (TCM) – no signal	**Wiring, TCM**
P1899	Park/neutral position (PNP) switch – input circuit malfunction	**Wiring, PNP switch, ECM/TCM**

Flash code	Scan code Hex type	Fault location	Probable cause
11		Crankshaft position (CKP) sensor/camshaft position (CMP) sensor	**Wiring, poor connection, air gap, CKP sensor, CMP sensor**
	85	Timing belt jumped 1 tooth or more	**Timing belt loose/jumped position**
	9D	Crankshaft position (CKP) sensor/camshaft position (CMP) sensor – intermittent signal	**Wiring, poor connection, air gap, CKP sensor, CMP sensor**
	28	Crankshaft position (CKP) sensor – no signal	**Wiring, air gap, CKP sensor**
	BA	Engine control module (ECM) – misfire detection limit reached	**Wiring, poor connection, CKP sensor**
12		Engine control module (ECM) – supply voltage disconnected within last 50 ignition switch ON-OFF cycles	**Wiring, battery**
13	27	Manifold absolute pressure (MAP) sensor	**Wiring, hose connection(s), MAP sensor**
14		Manifold absolute pressure (MAP) sensor – input low/high	**Wiring, MAP sensor**
	24	Manifold absolute pressure (MAP) sensor – low input	**Wiring short circuit to ground, MAP sensor**
	25	Manifold absolute pressure (MAP) sensor – high input	**Wiring short circuit to positive, MAP sensor**
	87	Manifold absolute pressure (MAP) sensor – no 5 volt supply	**Wiring, ECM**

Flash code	Scan code Hex type	Fault location	Probable cause
	92	Manifold absolute pressure (MAP) sensor – 5 volt output low	**Wiring, MAP sensor**
15	23	Vehicle speed sensor (VSS) – signal	**Wiring, VSS**
16		Knock sensor (KS) 1/2 – no signal	**Wiring, KS incorrectly tightened, KS**
	3B	Knock sensor (KS) 1 – no signal	**Wiring, KS incorrectly tightened, KS**
	53	Knock sensor (KS) 2 – no signal	**Wiring, KS incorrectly tightened, KS**
17		Engine temperature – low for long period	**Wiring, cooling system, coolant thermostat, ECT sensor**
	21	Engine temperature – low for long period	**Wiring, cooling system, coolant thermostat, ECT sensor**
	80	Insufficient coolant temperature for closed loop fuel control	**Wiring, cooling system, coolant thermostat, ECT sensor**
21		Heated oxygen sensor (HO2S) 1/2, bank 1/2 – circuit malfunction/incorrect signal/ voltage low/high	**Wiring, fuel pressure, injector control, EVAP canister purge valve, compression, air filter blocked, HO2S**
	9B	Heated oxygen sensor (HO2S) 1, bank 1 – low voltage	**Wiring short circuit to ground, exhaust leak, HO2S**
	3E	Heated oxygen sensor (HO2S) 1, bank 1 – high voltage	**Wiring short circuit to positive, fuel pressure, injector control, EVAP canister purge valve, air filter blocked, HO2S**
	66	Heated oxygen sensor (H02S) 1, bank 1 – slow response	**Wiring, heating inoperative, HO2S**
	20	Heated oxygen sensor (H02S) 1, bank 1 – no activity detected	**Wiring open circuit, heating inoperative, HO2S**
	67	Heated oxygen sensor (HO2S) 1, bank 1 – heater control – circuit malfunction	**Wiring, HO2S**
	9C	Heated oxygen sensor (HO2S) 2, bank 1 – low voltage	**Wiring short circuit to ground, exhaust leak, HO2S**
	7E	Heated oxygen sensor (HO2S) 2, bank 1 – high voltage	**Wiring short circuit to positive, catalytic converter, HO2S**
	81	Heated oxygen sensor (HO2S) 2, bank 1 – no activity detected	**Wiring, heating inoperative, HO2S**
	69	Heated oxygen sensor (HO2S) 2, bank 1, heater control – circuit malfunction	**Wiring, HO2S**
	B5	Heated oxygen sensor (HO2S) 1, bank 2 – low voltage	**Wiring short circuit to ground, exhaust leak, HO2S**
	42	Heated oxygen sensor (HO2S) 1, bank 2 – high voltage	**Wiring short circuit to positive, fuel pressure, injector control, EVAP canister purge valve, air filter blocked, HO2S**
	7A	Heated oxygen sensor (HO2S) 1, bank 2 – slow response	**Wiring, heating inoperative, HO2S**

Flash code	Scan code Hex type	Fault location	Probable cause
	41	Heated oxygen sensor (HO2S) 1, bank 2 – no activity detected	Wiring, HO2S
	7C	Heated oxygen sensor (HO2S) 1, bank 2, heater control – circuit malfunction	Wiring, HO2S
	B6	Heated oxygen sensor (HO2S) 2, bank 2 – low voltage	Wiring short circuit to ground, exhaust leak, HO2S
	7F	Heated oxygen sensor (HO2S) 2, bank 2 – high voltage	Wiring short circuit to positive, catalytic converter, HO2S
	82	Heated oxygen sensor (HO2S) 2, bank 2 – no activity detected	Wiring, HO2S
	7D	Heated oxygen sensor (HO2S) 2, bank 2, heater control – circuit malfunction	Wiring, HO2S
22		Engine coolant temperature (ECT) sensor – voltage low/high	Wiring, coolant thermostat, ECT sensor
	1E	Engine coolant temperature (ECT) sensor – low input	Wiring short circuit to ground, coolant thermostat, ECT sensor
	1F	Engine coolant temperature (ECT) sensor – high input	Wiring open circuit/short circuit to positive, coolant thermostat, ground wire defective, ECT sensor
23		Intake air temperature (IAT) sensor – voltage low/high	Wiring, IAT sensor
	39	Intake air temperature (IAT) sensor – low input	Wiring short circuit to ground, IAT sensor
	3A	Intake air temperature (IAT) sensor – high input	Wiring open circuit/short circuit to positive, ground wire defective, IAT sensor
24		Throttle position (TP) sensor – voltage low/high	Wiring, accelerator cable adjustment, TP sensor
	84	Throttle position (TP) sensor – range/performance problem	Wiring, accelerator cable adjustment, TP sensor
	1A	Throttle position (TP) sensor – low input	Wiring short circuit to ground, TP sensor
	1B	Throttle position (TP) sensor – high input	Wiring short circuit to positive, TP sensor
25		Idle air control (IAC) valve	Wiring open/short circuit, air leak, IAC valve
	19	Idle speed control (ISC) system – malfunction	Wiring open/short circuit, IAC valve
	8A	Idle speed control (ISC) system – rpm lower/higher than expected	Wiring, air leak, throttle valve tight/sticking, IAC valve
	91	Manifold absolute pressure (MAP) sensor/throttle position (TP) sensor	Intake leak, hose connection(s), MAP sensor, TP sensor
27		Injector – circuit malfunction	Wiring, injector
	15	Injector 1 – circuit malfunction	Wiring, injector

Flash code	Scan code Hex type	Fault location	Probable cause
	14	Injector 2 – circuit malfunction	**Wiring, injector**
	13	Injector 3 – circuit malfunction	**Wiring, injector**
	3D	Injector 4 – circuit malfunction	**Wiring, injector**
	45	Injector 5 – circuit malfunction	**Wiring, injector**
	46	Injector 6 – circuit malfunction	**Wiring, injector**
31		Evaporative emission (EVAP) canister purge valve – open/short circuit	**Wiring, EVAP canister purge valve**
	71	Evaporative emission (EVAP) system – incorrect flow detected	**Hose connection(s), intake leak, EVAP canister purge valve**
	A0	Evaporative emission (EVAP) system – small leak detected	**Hose connection(s), intake leak, EVAP canister, EVAP canister purge valve**
	12	Evaporative emission (EVAP) canister purge valve – circuit malfunction	**Wiring open/short circuit, EVAP canister purge valve**
	A1	Evaporative emission (EVAP) system – large leak detected	**Hose connection(s), intake leak, EVAP canister, EVAP canister purge valve**
	BB	Evaporative emission (EVAP) system – blockage detected	**Hose connection(s), hose blocked/ trapped, EVAP canister**
	B8	Evaporative emission (EVAP) leak detection pump – pressure switch	**Wiring, hose leak/blockage, EVAP leak detection pump**
	B7	Evaporative emission (EVAP) leak detection pump solenoid	**Wiring open/short circuit, hose leak/ blockage, EVAP leak detection pump solenoid**
32		Exhaust gas recirculation (EGR) solenoid/ system – circuit malfunction	**Wiring, hoses, EGR solenoid**
	2E	Exhaust gas recirculation (EGR) system – insufficient flow detected	**Hose leak/blockage, wiring, EGR valve, EGR solenoid**
	11	Exhaust gas recirculation (EGR) – circuit malfunction	**Wiring open/short circuit, EGR solenoid**
33		A/C compressor clutch relay/A/C refrigerant pressure sensor – voltage low/high	**Wiring open/short circuit, A/C refrigerant pressure too low/high (cooling fault/incorrectly charged), A/C compressor clutch relay, A/C refrigerant pressure sensor**
	10	A/C compressor clutch relay	**Wiring open/short circuit, A/C compressor clutch relay**
	5A	A/C refrigerant pressure sensor – high input	**Wiring, A/C refrigerant pressure too high (cooling fault/incorrectly charged), A/C refrigerant pressure sensor**
	5B	A/C refrigerant pressure sensor – low input	**Wiring, A/C refrigerant pressure too low (incorrectly charged), A/C refrigerant pressure sensor**
	86	A/C – no 5 volt supply	**Wiring, ECM**

Flash code	Scan code Hex type	Fault location	Probable cause
34		Cruise control	Wiring open/short circuit, voltage low/ high, cruise control master switch, cruise control vacuum switching valve
	0F	Cruise control	Wiring open/short circuit, voltage low/ high, cruise control vacuum switching valve
	56	Cruise control master switch – high voltage	Wiring, cruise control master switch
	57	Cruise control master switch – low voltage	Wiring, cruise control master switch
35		Engine coolant blower motor relay – low/ high speed	Wiring open/short circuit, engine coolant blower motor relay – low/high speed
	5D	Engine coolant blower motor relay – high speed	Wiring open/short circuit, engine coolant blower motor relay – high speed
	5C	Engine coolant blower motor relay – low speed	Wiring open/short circuit, engine coolant blower motor relay – low speed
37		Torque converter clutch (TCC) solenoid/ park/neutral position (PNP) switch – circuit malfunction	Wiring, TCC solenoid
	94	Torque converter clutch (TCC) solenoid – circuit malfunction	Wiring, TCC solenoid
	0C	Torque converter clutch (TCC) solenoid – electrical	Wiring open/short circuit, TCC solenoid
	72	Park/neutral position (PNP) switch – input circuit malfunction	Wiring, PNP switch, ECM/TCM
41	0B	Alternator, field control – circuit malfunction	Wiring open/short circuit, alternator, battery
42		Fuel pump relay/engine control relay/fuel level sensor – circuit malfunction	Wiring open/short circuit, fuel pump relay, engine control relay, fuel level sensor
	65	Fuel pump relay	Wiring open/short circuit, fuel pump relay
	0A	Engine control relay	Wiring open/short circuit, engine control relay
	2C	Engine control relay	Wiring open circuit, engine control relay
	95	Fuel level sensor	Wiring open circuit, fuel level sensor
	96	Fuel level sensor	Wiring short circuit to positive, fuel level sensor
	97	Fuel level sensor	Wiring, fuel level sensor
43		Random/multiple cylinder(s) – misfire detected	Wiring, spark plug(s), HT lead(s), injector(s), ignition coil(s), low compression

Flash code	Scan code Hex type	Fault location	Probable cause
	6A	Random/multiple cylinder(s) – misfire detected	Wiring, spark plug(s), HT lead(s), injector(s), ignition coil(s), low compression
	6B	Cylinder 1 – misfire detected	Wiring, engine mechanical fault, ignition/fuel system, injector
	6C	Cylinder 2 – misfire detected	Wiring, engine mechanical fault, ignition/fuel system, injector
	6D	Cylinder 3 – misfire detected	Wiring, engine mechanical fault, ignition/fuel system, injector
	6E	Cylinder 4 – misfire detected	Wiring, engine mechanical fault, ignition/fuel system, injector
	AE	Cylinder 5 – misfire detected	Wiring, engine mechanical fault, ignition/fuel system, injector
	AF	Cylinder 6 – misfire detected	Wiring, engine mechanical fault, ignition/fuel system, injector
	2B	Ignition coil A, primary/secondary – circuit malfunction	Wiring, ignition coil
	2A	Ignition coil B, primary/secondary – circuit malfunction	Wiring, ignition coil
	29	Ignition coil C, primary/secondary – circuit malfunction	Wiring, ignition coil
44		Ambient/battery temperature sensor – voltage low/high	Wiring, ambient/battery temperature sensor
	9A	Ambient/battery temperature sensor – voltage high	Wiring, ambient/battery temperature sensor
	99	Ambient/battery temperature sensor – voltage low	Wiring, ambient/battery temperature sensor
45	89	Transmission control system – malfunction	Wiring, TCM
46	06	Alternator – voltage high	Wiring, battery, alternator
47	05	Alternator – voltage low	Wiring, battery, alternator drive belt, alternator
51		System too lean	Wiring, intake/exhaust leak, fuel pressure, injector control, EVAP canister purge valve, HO2S
	77	System too lean, bank 1	Intake/exhaust leak, MAP sensor, fuel pressure/pump, injector(s), HO2S
	79	System too lean, bank 2	Intake/exhaust leak, MAP sensor, fuel pressure/pump, injector(s), HO2S
52		System too rich	Wiring, exhaust blockage, fuel pressure, injector control, EVAP canister purge valve, compression, air filter blocked, ignition system, HO2S
	76	System too rich, bank 1	Intake blocked, EVAP canister purge valve, fuel pressure, EGR system, injector(s), HO2S

Flash code	Scan code Hex type	Fault location	Probable cause
	78	System too rich, bank 2	**Intake blocked, EVAP canister purge valve, fuel pressure, EGR system, injector(s), HO2S**
53		Engine control module (ECM) – internal fault	**ECM**
	02	Engine control module (ECM) – internal fault	**ECM**
	44	Engine control module (ECM) – internal fault	**ECM**
54	01	Camshaft position (CMP) sensor – no signal	**Wiring, CMP sensor**
55		End of MIL code output	–
61	3C	Manifold absolute pressure (MAP) sensor – barometric pressure range/ performance problem	**Wiring, intake/exhaust leak, MAP sensor**
62	30	Engine control module (ECM) – EEPROM error	**ECM**
63	31	Engine control module (ECM) – EEPROM error	**ECM**
65		Power steering pressure (PSP) switch/ brake pedal position (BPP) switch/Intake manifold air control solenoid	**Wiring, PAS system, PSP switch, BPP switch, intake manifold air control solenoid**
	73	Power steering pressure (PSP) switch – range/performance problem	**Wiring, PAS system, PSP switch**
	98	Brake pedal position (BPP) switch B – circuit malfunction	**Wiring, BPP switch**
	58	Intake manifold air control solenoid	**Wiring open/short circuit, intake manifold air control solenoid**
66		CAN data bus, transmission control module (TCM)/body control module (BCM) – no signal	**Wiring, TCM, BCM**
	60	CAN data bus, transmission control module (TCM) – no signal	**Wiring, TCM**
	61	CAN data bus, body control module (BCM) – no signal	**Wiring, BCM**
71	92	Engine control module (ECM) – 5 volt output low	**ECM**
72		Catalytic converter system, bank 1/2 – efficiency below threshold	**Wiring, catalytic converter, HO2S 2**
	70	Catalytic converter system, bank 1 – efficiency below threshold	**Wiring, catalytic converter, HO2S 2**
	B4	Catalytic converter system, bank 2 – efficiency below threshold	**Wiring, catalytic converter, HO2S 2**
77	52	Cruise control	**Wiring, cruise control vacuum switching valve**

DODGE

Model:	Engine identification:	Year:
Neon 2.0L	ECB/ECC	1995
Stratus 2.0L	ECB	1995

System: **SBEC II**

Self-diagnosis

General information

- Malfunction indicator lamp (MIL) will illuminate or flash if certain faults are recorded.
- ECM operates in backup mode if sensors fail, to enable vehicle to be driven to repair shop.
- Trouble codes can be displayed by the MIL or by using a Scan Tool connected to the data link connector (DLC).

Accessing

Flash codes

- Within 5 seconds switch the ignition key ON-OFF-ON-OFF-ON.
- Count MIL flashes. Note trouble codes.
- Each trouble code consists of two groups **1** [**A**].
- The first group of flashes indicate the 'tens' of the trouble code.
- The second group of flashes indicate the 'units' of the trouble code.
- A short pause separates 'tens' and 'units' **1** [**B**].
- A long pause separates each trouble code **1** [**C**].
- For example: Trouble code 21 displayed **1**.
- End of test sequence indicated by trouble code 55.

P codes

- Trouble codes can be displayed by using a Scan Tool connected to the data link connector (DLC):
  - Stratus **2**.
  - Neon **3**.

1

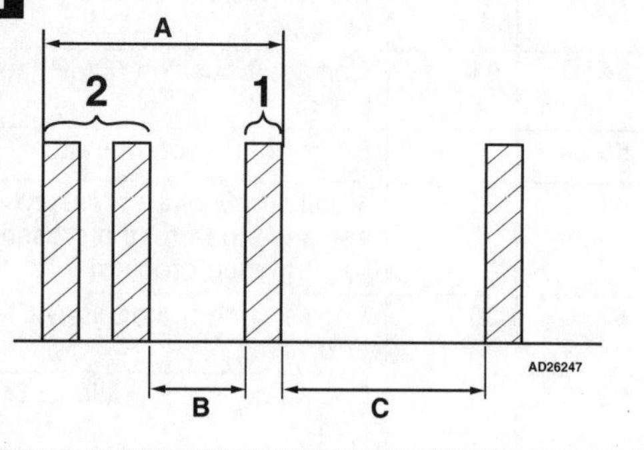

2

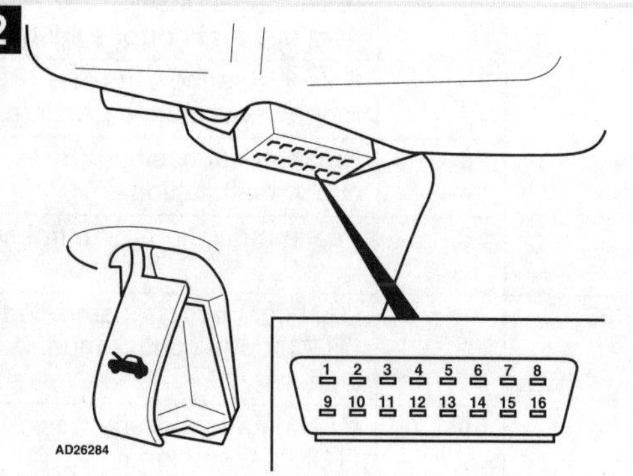

3

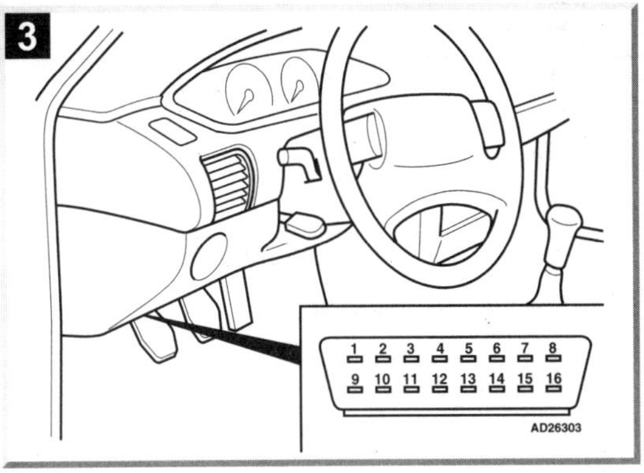

Erasing

NOTE: ***Disconnecting battery ground lead may cause driveability problems until the engine management system relearns basic values.***

- Ensure ignition switched OFF.
- Disconnect battery ground cable for at least 30 seconds.
- Reconnect battery ground cable.
- Repeat accessing procedure to ensure no trouble codes remain in ECM fault memory.

NOTE: ***Disconnecting battery ground lead may also erase memory from electronic units, such as radio.***

- Trouble codes can also be erased by using a Scan Tool connected to the data link connector (DLC).

Trouble code identification

Scan code P type	Fault location	Probable cause
P0000-P0999	Refer to OBD-II trouble code table at the front of this manual	–
P1294	Idle speed control (ISC) system – rpm lower/ higher than expected	**Wiring, air leak, throttle valve tight/sticking, IAC valve**
P1295	Throttle position (TP) sensor – supply voltage	**Wiring, TP sensor, ECM**
P1296	Manifold absolute pressure (MAP) sensor – supply voltage	**Wiring, MAP sensor, ECM**
P1297	Manifold absolute pressure (MAP) sensor	**Wiring, hose connection(s), MAP sensor**
P1299	Manifold absolute pressure (MAP) sensor/ throttle position (TP) sensor	**Intake leak, hose connection(s), MAP sensor, TP sensor, ECM**
P1390	Crankshaft position (CKP) sensor/camshaft position (CMP) sensor – signal synchronisation	**Timing belt loose/jumped position, CKP sensor, CMP sensor**
P1391	Crankshaft position (CKP) sensor/camshaft position (CMP) sensor – intermittent signal	**Wiring, poor connection, air gap, CKP sensor, CMP sensor**
P1489	Engine coolant blower motor relay – high speed	**Wiring open circuit/short circuit, engine coolant blower motor relay – high speed**
P1490	Engine coolant blower motor relay – low speed	**Wiring open circuit/short circuit, engine coolant blower motor relay – low speed**
P1491	Engine coolant blower motor relay – circuit malfunction	**Wiring open circuit/short circuit, engine coolant blower motor relay**
P1899	Park/neutral position (PNP) switch – input circuit malfunction	**Wiring, PNP switch, ECM/TCM**

Flash code	Fault location	Probable cause
11	Crankshaft position (CKP) sensor/camshaft position (CMP) sensor	**Wiring, poor connection, air gap, CKP sensor, CMP sensor**
12	Engine control module (ECM) – supply voltage disconnected within last 50 ignition switch ON-OFF cycles	**Wiring, battery**
13	Manifold absolute pressure (MAP) sensor	**Wiring, hose connection(s), MAP sensor**
14	Manifold absolute pressure (MAP) sensor – input low/high	**Wiring, MAP sensor**
15	Vehicle speed sensor (VSS) – signal	**Wiring, VSS**
16	Knock sensor (KS) – wiring	**Wiring open circuit/short circuit, KS**
17	Engine temperature – low for long period/ insufficient coolant temperature for closed loop fuel control	**Wiring, cooling system, coolant thermostat, ECT sensor**
21	Heated oxygen sensor (HO2S) 1/2, bank 1 – circuit malfunction/incorrect signal/voltage low/ high	**Wiring, fuel pressure, injector control, EVAP canister purge valve, compression, air filter blocked, HO2S**
22	Engine coolant temperature (ECT) sensor – voltage low/high	**Wiring, coolant thermostat, ECT sensor**
23	Intake air temperature (IAT) sensor – voltage low/high	**Wiring, IAT sensor**
24	Throttle position (TP) sensor – voltage low/ high	**Wiring, accelerator cable adjustment, TP sensor**
25	Idle air control (IAC) valve	**Wiring open circuit/short circuit, IAC valve**
27	Injector(s) – circuit malfunction	**Wiring, injector**
31	Evaporative emission (EVAP) solenoid/system	**Wiring, hose connection(s), intake leak, EVAP canister purge valve**
	Evaporative emission (EVAP) system – insufficient/excessive flow detected	**Hose leak/blockage, wiring, intake leak, EVAP canister purge valve**
	Evaporative emission (EVAP) solenoid – circuit malfunction	**Wiring open circuit/short circuit, EVAP canister purge valve**
32	Exhaust gas recirculation (EGR) solenoid/ system – circuit malfunction	**Wiring, hoses, EGR valve, EGR solenoid**
	Exhaust gas recirculation (EGR) system – insufficient flow detected	**Hose leak/blockage, wiring, EGR valve, EGR solenoid**
	Exhaust gas recirculation (EGR) solenoid – circuit malfunction	**Wiring open circuit/short circuit, EGR solenoid**
33	A/C compressor clutch relay/A/C refrigerant pressure sensor – voltage low/high	**A/C refrigerant pressure too low/high (cooling fault/incorrectly charged), wiring open circuit/short circuit, A/C compressor clutch relay, A/C refrigerant pressure sensor**
34	Cruise control	**Wiring open circuit/short circuit, hoses, voltage low/high, cruise control master switch, cruise control vacuum switching valve**

Flash code	Fault location	Probable cause
35	Engine coolant blower motor relay – low/high speed	Wiring open circuit/short circuit, engine coolant blower motor relay – low/high speed
37	Torque converter clutch (TCC) solenoid/park/neutral position (PNP) switch – circuit malfunction	Wiring, TCC solenoid
41	Alternator, field control – circuit malfunction	Wiring open circuit/short circuit, alternator, battery
42	Fuel pump relay/engine control relay/fuel level sensor – circuit malfunction	Wiring open circuit/short circuit, Fuel pump relay, engine control relay, fuel level sensor
43	Random/multiple cylinder(s) – misfire detected	Wiring, spark plug(s), HT lead(s), injector(s), ignition coil(s), low compression
44	Ambient/battery temperature sensor – voltage low/high	Wiring, ambient/battery temperature sensor
46	Alternator – voltage high	Wiring, battery, alternator
47	Alternator – voltage low	Wiring, battery, alternator drive belt, alternator
51	System too lean	Wiring, intake/exhaust leak, MAP sensor, fuel pressure/pump, EVAP canister purge valve, injector(s), HO2S
52	System too rich	Wiring, intake blocked, EVAP canister purge valve, fuel pressure, EGR system, injector(s), air filter blocked, ignition system, HO2S
53	Engine control module (ECM) – internal fault	ECM
54	Camshaft position (CMP) sensor – no signal	Wiring, CMP sensor
55	End of malfunction indicator lamp (MIL) code output	–
62	Engine control module (ECM) – EEPROM error	ECM
63	Engine control module (ECM) – EEPROM error	ECM
64	Catalytic converter system – efficiency below threshold	Catalytic converter, wiring, HO2S
65	Power steering pressure (PSP) switch – circuit malfunction	Wiring, PAS system, PSP switch, ECM
66	CAN data bus, body control module (BCM) – no signal	Wiring, BCM
71	Engine control module (ECM) – 5 volt output low	ECM
77	Cruise control	Wiring, cruise control vacuum switching valve

DODGE

Model:	Engine identification:	Year:
Ram 50 2.4L	4G64	1992-93
Ram 50 3.0L	6G72	1992

System: **Mitsubishi MPI**

Self-diagnosis

General information

- Refer to the front of this manual for general test conditions, terminology, detailed descriptions of wiring faults and a general trouble shooter for electrical and mechanical faults.
- Malfunction indicator lamp (MIL) will illuminate if certain faults are recorded.
- ECM operates in backup mode if sensors fail, to enable vehicle to be driven to repair shop.
- 1992: Trouble codes can be displayed by using an analogue voltmeter connected to the data link connector (DLC).
- 1993: Trouble codes can be displayed by using an analogue voltmeter or Scan Tool connected to the data link connector (DLC).

Accessing

- Ensure ignition switched OFF.
- Access data link connector (DLC) **1**.
- Connect an analogue voltmeter between data link connector (DLC) terminals 1 and 12 **2**.
- Switch ignition ON.
- Count voltmeter needle deflections. Note trouble codes. Compare with trouble code table.
- Each trouble code consists of two groups of one or more needle deflections **3** :
  - Long needle deflections indicate the 'tens' of the trouble code **3** [**A**].
  - Short needle deflections indicate the 'units' of the trouble code **3** [**D**].
  - A short pause separates each trouble code group **3** [**C**].
  - A long pause separates each trouble code **3** [**E**].
  - For example: Trouble code 22 displayed **3**.
- If a constant deflection is displayed: Suspect engine control module (ECM).
- If constant short deflections are displayed: No trouble code is stored.
- 1993: Trouble codes can also be displayed by using a Scan Tool connected to the data link connector (DLC).

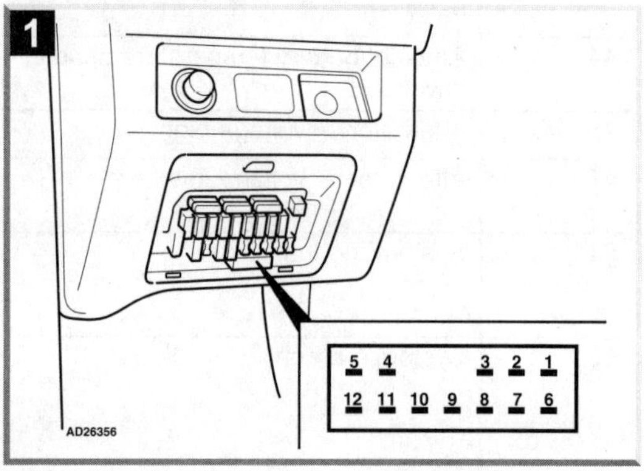

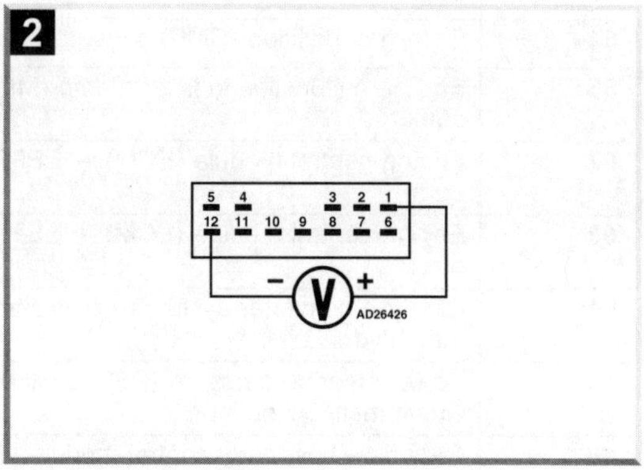

3

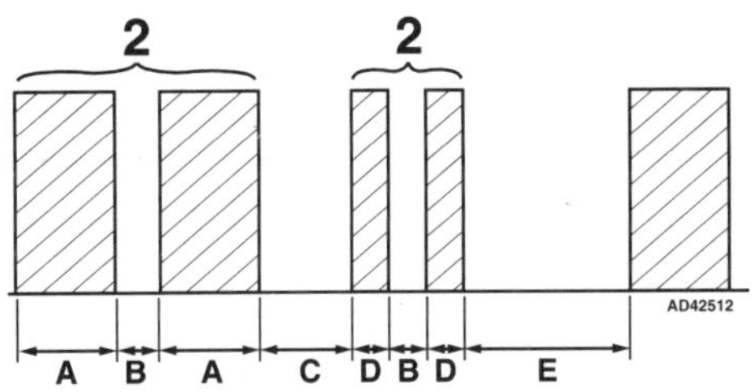

Erasing

NOTE: ***Disconnecting battery ground lead may cause driveability problems until the engine management system relearns basic values.***

- Ensure ignition switched OFF.
- Disconnect battery ground cable for at least 15 seconds.
- Reconnect battery ground lead.
- Repeat checking procedure to ensure no data remains in ECM fault memory.
- 1993: Trouble codes can also be erased by using a Scan Tool connected to the data link connector (DLC).

NOTE: ***Disconnecting battery ground lead may erase memory from electronic units such as radio.***

Trouble code identification

Flash code	Fault location	Probable cause
11	Heated oxygen sensor (HO2S)/oxygen sensor (O2S) 1, bank 1 – circuit malfunction	**Wiring, fuel pressure, injector(s), intake leak, HO2S/O2S, ECM**
12	Volume air flow (VAF) sensor – malfunction	**Wiring, VAF sensor, ECM**
13	Intake air temperature (IAT) sensor – circuit malfunction	**Wiring, IAT sensor, ECM**
14	Throttle position (TP) sensor – circuit malfunction	**Wiring, TP sensor, CTP switch, ECM**
15	Idle speed control (ISC) actuator position sensor – circuit malfunction	**Wiring, ISC actuator, ISC actuator position sensor, TP sensor, ECM**
21	Engine coolant temperature (ECT) sensor – circuit malfunction	**Wiring, ECT sensor, ECM**
22	Crankshaft position (CKP) sensor – circuit malfunction	**Wiring, CKP sensor, ECM**
23	Camshaft position (CMP) sensor – circuit malfunction	**Wiring, CMP sensor, ECM**
24	Vehicle speed sensor (VSS) – circuit malfunction	**Wiring, VSS, ECM**
25	Barometric pressure (BARO) sensor – circuit malfunction	**Wiring, BARO sensor, ECM**
36	Ignition timing adjustment	**Wiring, ignition timing connector short to ground, ECM**
41	Injectors – circuit malfunction	**Wiring, injectors, ECM**
42	Fuel pump – circuit malfunction	**Wiring, engine control relay, fuel pump, ECM**
43	Exhaust gas recirculation (EGR) system – circuit malfunction	**Wiring, Hose leak/blockage, EGR valve, EGR temperature sensor, EGR solenoid, ECM**
55	Idle air control (IAC) valve position sensor – circuit malfunction	**Wiring, IAC valve, IAC valve position sensor, ECM**
59	Heated oxygen sensor (HO2S) 2, bank 1 – circuit malfunction	**Wiring, HO2S, ECM**

DODGE

Model:	Engine identification:	Year:
Ram Pickup/Chassis Cab D/W 5.9L Turbo Diesel	**ETA**	**1992-93**
Ram Truck 5.9L Turbo Diesel	**ETB**	**1994-95**

System: **SBEC II**

Self-diagnosis

General information

- Refer to the front of this manual for general test conditions, terminology, detailed descriptions of wiring faults and a general trouble shooter for electrical and mechanical faults.
- Malfunction indicator lamp (MIL) will illuminate if certain faults are recorded.
- Trouble codes can be displayed by the MIL or by using a Scan Tool connected to the data link connector (DLC).

Accessing

- Within 3 seconds switch the ignition key ON-OFF-ON-OFF-ON.
- Count MIL flashes. Note trouble codes.
- Each trouble code consists of two groups **1** [**A**].
- The first group of flashes indicate the 'tens' of the trouble code.
- The second group of flashes indicate the 'units' of the trouble code.
- A short pause separates 'tens' and 'units' **1** [**D**].
- A long pause separates each trouble code **1** [**E**].
- For example: Trouble code 44 displayed **1**.
- End of test sequence indicated by trouble code 55.
- Trouble codes can also be displayed by using a Scan Tool connected to the data link connector (DLC):
  - Ram Pickup **2**.
  - Ram Truck **3**.

1

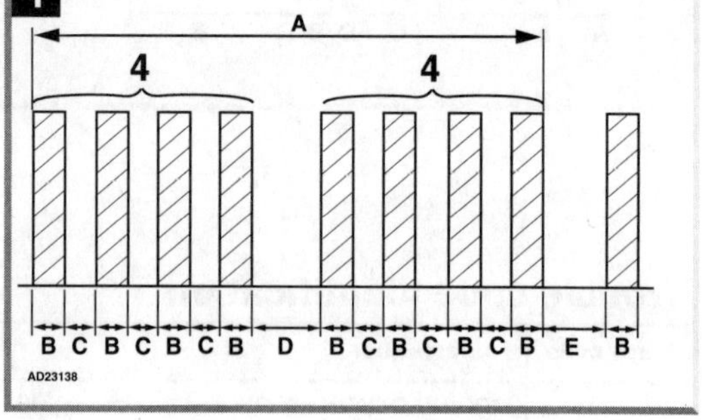

2

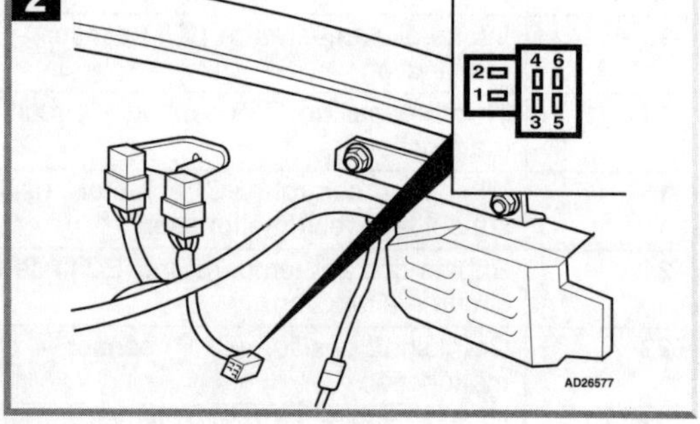

3

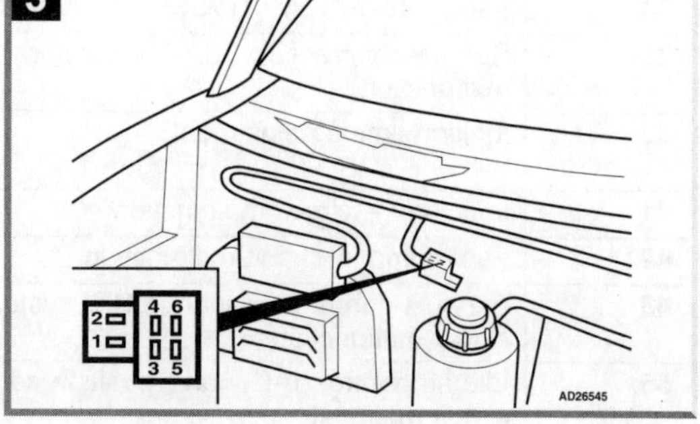

Erasing

- Ensure ignition switched OFF.
- Disconnect battery ground cable for at least 30 seconds.
- Reconnect battery ground cable.
- Repeat accessing procedure to ensure no trouble codes remain in ECM fault memory.

NOTE: ***Disconnecting battery ground lead may also erase memory from electronic units, such as radio.***

- Trouble codes can also be erased by using a Scan Tool connected to the data link connector (DLC).

Trouble code identification

Flash code	Fault location	Probable cause
11	Crankshaft position (CKP) sensor	**Wiring, air gap, CKP sensor**
12	Engine control module (ECM) – supply voltage disconnected within last 50 ignition switch ON-OFF cycles	**Wiring, battery**
15	Vehicle speed sensor (VSS) – signal	**Wiring, VSS**
22	Engine coolant temperature (ECT) sensor – voltage low/high	**Wiring, ECT sensor**
23	Intake air temperature (IAT) sensor – voltage low/high	**Wiring, IAT sensor**
24	Throttle position (TP) sensor	**Wiring, accelerator cable adjustment, TP sensor**
33	A/C compressor clutch relay	**Wiring, A/C compressor clutch relay**
34	Cruise control	**Wiring open circuit/short circuit, hoses, voltage low/high, cruise control master switch, cruise control vacuum switching valve**
37	Torque converter clutch (TCC) solenoid/ transmission fluid temperature (TFT) sensor	**Wiring, transmission fluid level, torque converter clutch (TCC) solenoid, TFT sensor, ECM**
41	Alternator	**Wiring open circuit/short circuit, alternator**
42	Engine control relay	**Wiring open circuit/short circuit, engine control relay**
44	Battery temperature sensor/engine coolant temperature (ECT) sensor	**Wiring open circuit/short circuit, battery temperature sensor, ECT sensor, ECM**
45	Overdrive solenoid	**Wiring open circuit/short circuit, overdrive solenoid**
46	Alternator – voltage high	**Wiring, battery, alternator**
47	Alternator – voltage low	**Wiring, battery, alternator drive belt, alternator**
53	Engine control module (ECM) – internal fault – defective	**ECM**
55	End of MIL code output	**–**
62	Engine control module (ECM) – EEPROM error	**ECM**
63	Engine control module (ECM) – EEPROM error	**ECM**

DODGE

Model:	Engine identification:	Year:
Stealth 3.0L	6G72	1996
Stealth 3.0L Turbo	6G72	1996
Stratus Coupe 2.4L	4G64	2001-02
Stratus Coupe 3.0L	6G72	2001-02

System: **Mitsubishi MPI**

Self-diagnosis

General information

- Refer to the front of this manual for general test conditions, terminology, detailed descriptions of wiring faults and a general trouble shooter for electrical and mechanical faults.
- Service engine soon/malfunction indicator lamp (MIL) will illuminate or flash if certain faults are recorded.
- ECM operates in backup mode if sensors fail, to enable vehicle to be driven to repair shop.

Accessing

- Trouble codes can only be displayed by using a Scan Tool connected to the data link connector (DLC) **1**.

1

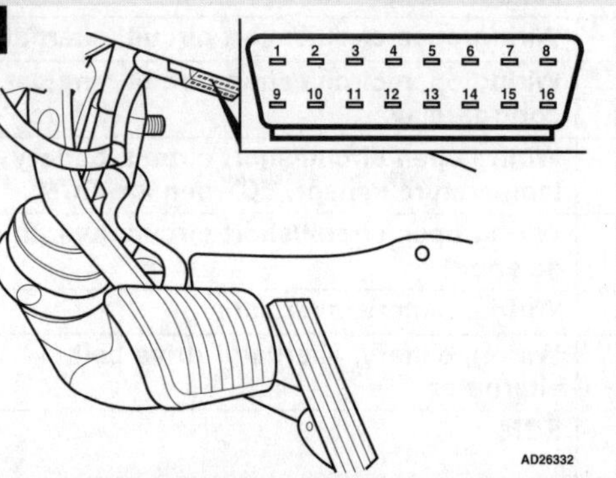

Erasing

NOTE: ***Disconnecting battery ground lead may cause driveability problems until the engine management system relearns basic values.***

- Ensure ignition switched OFF.
- Disconnect battery ground cable for at least 15 seconds.
- Reconnect battery ground lead.
- Repeat checking procedure to ensure no data remains in ECM fault memory.
- Trouble codes can also be erased by using a Scan Tool connected to the data link connector (DLC).

NOTE: ***Disconnecting battery ground lead may erase memory from electronic units such as radio.***

Trouble code identification

Scan code	Fault location	Probable cause
P0000-P0999	Refer to OBD-II trouble code table at the front of this manual	–
P1103	Turbocharger (TC) wastegate actuator – malfunction	Hose connection(s), wiring, TC wastegate regulating valve, TC wastegate, ECM
P1104	Turbocharger (TC) wastegate regulating valve – circuit malfunction	Wiring, TC wastegate regulating valve, ECM
P1105	Fuel pressure regulator control solenoid – circuit malfunction	Wiring, fuel pressure regulator control solenoid, ECM
P1300	Ignition timing adjustment	Wiring, ignition timing connector short to ground, ECM
P1400	Manifold differential pressure (MDP) sensor, exhaust gas recirculation (EGR) system – circuit malfunction	Wiring, MDP sensor, ECM
P1500	Alternator – circuit malfunction	Wiring, battery, alternator, ECM
P1715	Transmission speed sensor A/B – circuit malfunction	Wiring, transmission speed sensor A/B, ECM
P1750	Torque converter clutch (TCC) solenoid/ transmission shift control valve/transmission pressure control solenoid – circuit malfunction	Wiring, TCC solenoid/transmission shift control valve/transmission pressure control solenoid, TCM, ECM
P1751	Transmission system relay – circuit malfunction	Wiring, transmission system relay, ECM
P1795	Throttle position (TP) sensor – circuit malfunction	Wiring, TP sensor, ECM

DODGE

Model: **Ram Truck 5.9L Turbo Diesel**

Engine identification: **ETB**

Year: **1996-97**

System: **JTEC**

Self-diagnosis

General information

- Refer to the front of this manual for general test conditions, terminology, detailed descriptions of wiring faults and a general trouble shooter for electrical and mechanical faults.
- Malfunction indicator lamp (MIL) will illuminate or flash if certain faults are recorded.
- Trouble codes can be displayed by the MIL or by using a Scan Tool connected to the data link connector (DLC).

Accessing

Flash codes

- Within 5 seconds switch the ignition key ON-OFF-ON-OFF-ON.
- Count MIL flashes. Note trouble codes.
- Each trouble code consists of two groups **1** [**A**].
- The first group of flashes indicate the 'tens' of the trouble code.
- The second group of flashes indicate the 'units' of the trouble code.
- A short pause separates 'tens' and 'units' **1** [**D**].
- A long pause separates each trouble code **1** [**E**].
- For example: Trouble code 42 displayed **1**.
- End of test sequence indicated by trouble code 55.

P codes & Hex codes

- Trouble codes can also be displayed by using a Scan Tool connected to the data link connector (DLC) **2**.

1

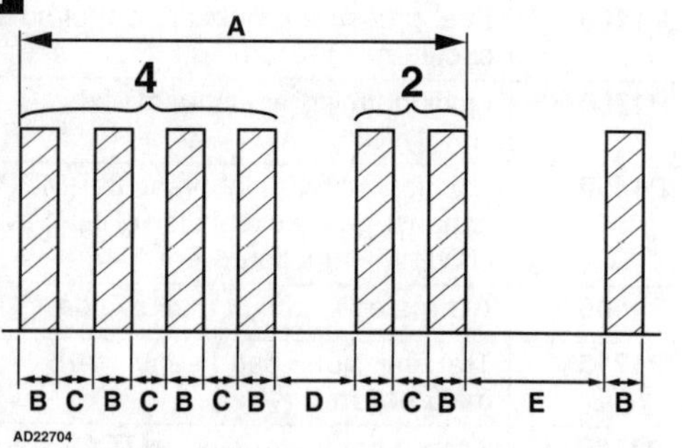

2

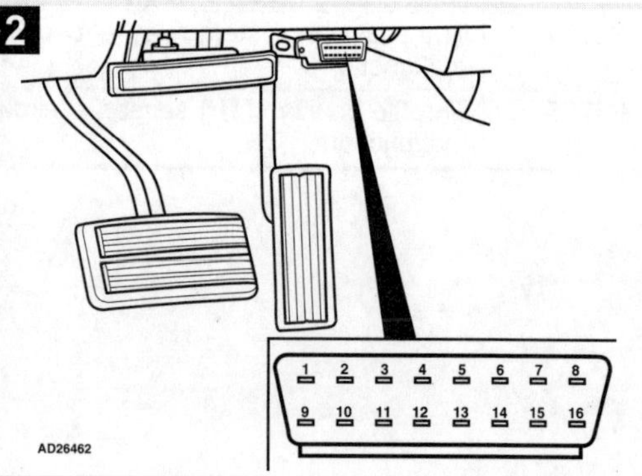

Erasing

- Trouble codes can only be erased by using a Scan Tool connected to the data link connector (DLC).

Trouble code identification

Scan code P type	Fault location	Probable cause
P0000-P0999	Refer to OBD-II trouble code table at the front of this manual	–
P1492	Ambient/battery temperature sensor – voltage high	**Wiring, ambient/battery temperature sensor, ECM**
P1493	Ambient/battery temperature sensor – voltage low	**Wiring, ambient/battery temperature sensor, ECM**
P1496	Engine control module (ECM) – 5 volt output low	**ECM**
P1696	Engine control module (ECM) – EEPROM error	**ECM**
P1756	Governor pressure not equal to target (15-20 PSI) – mid pressure malfunction	**Signal wire open circuit/short circuit, valve body, sensor supply wire open circuit, transmission fluid level, governor pressure solenoid, transmission mechanical fault, ECM – no 5 volt supply**
P1757	Governor pressure not equal to target – low pressure malfunction	**Signal wire open circuit/short circuit, valve body, sensor supply wire open circuit, transmission fluid level, governor pressure solenoid, transmission mechanical fault, ECM – no 5 volt supply**
P1762	Governor pressure sensor – incorrect signal	**Signal wire short circuit to ground/positive, supply wire open circuit/short circuit, transmission fluid level, governor pressure solenoid, governor pressure sensor, ECM**
P1763	Governor pressure sensor – signal high	**Signal wire short circuit to positive, supply wire open circuit, transmission fluid level, governor pressure sensor, ECM**
P1764	Governor pressure sensor – signal low	**Signal wire short circuit to ground, supply wire open circuit, transmission fluid level, governor pressure sensor, ECM**
P1765	Transmission system relay	**Wiring open circuit/short circuit, transmission system relay, ECM**
P1899	Park/neutral position (PNP) switch – input circuit malfunction	**Wiring, PNP switch, ECM**

Flash code	Scan code Hex type	Fault location	Probable cause
11	28	Crankshaft position (CKP) sensor	**Wiring, air gap, CKP sensor**
12		Engine control module (ECM) – supply voltage disconnected within last 50 ignition switch ON-OFF cycles	**Wiring, battery**
15		Vehicle speed sensor (VSS) – signal	**Wiring, VSS, ECM**
	23	Vehicle speed sensor (VSS) – no signal	**Wiring, VSS, ECM**

Flash code	Scan code Hex type	Fault location	Probable cause
	A6	Output shaft speed (OSS) sensor – circuit malfunction	**Wiring, OSS sensor, ECM**
22		Engine coolant temperature (ECT) sensor – voltage low/high	**Wiring, coolant thermostat, ECT sensor**
	1E	Engine coolant temperature (ECT) sensor – low input	**Coolant thermostat, wiring short circuit to ground, ECT sensor**
	1F	Engine coolant temperature (ECT) sensor – high input	**Coolant thermostat, wiring open circuit/ short circuit to positive, ground wire defective, ECT sensor**
23		Intake air temperature (IAT) sensor – voltage low/high	**Wiring, IAT sensor**
	39	Intake air temperature (IAT) sensor – low input	**Wiring short circuit to ground, IAT sensor**
	3A	Intake air temperature (IAT) sensor – high input	**Wiring open circuit/short circuit to positive, ground wire defective, IAT sensor**
24		Throttle position (TP) sensor – voltage low/ high	**Wiring, accelerator cable adjustment, TP sensor**
	1A	Throttle position (TP) sensor – low input	**Wiring short circuit to ground, TP sensor**
	1B	Throttle position (TP) sensor – high input	**Wiring short circuit to positive, TP sensor**
	84	Throttle position (TP) sensor – range/ performance problem	**Wiring, accelerator cable adjustment, TP sensor**
32	11	Exhaust gas recirculation (EGR) – circuit malfunction	**Wiring, EGR solenoid, EGR control module, ECM**
33	10	A/C compressor clutch relay	**Wiring open circuit/short circuit, A/C compressor clutch relay**
34		Cruise control	**Wiring open circuit/short circuit, hoses, voltage low/high, cruise control master switch, cruise control vacuum switching valve**
	0F	Cruise control	**Wiring open circuit/short circuit, hoses, voltage low/high, cruise control vacuum switching valve**
	56	Cruise control master switch – high voltage	**Wiring, cruise control master switch**
	57	Cruise control master switch – low voltage	**Wiring, cruise control master switch**
37		Transmission control system – malfunction	**Wiring, transmission fluid level, transmission mechanical fault, PNP switch, TCC solenoid, TFT sensor, TFP solenoid, SS, governor pressure solenoid, transmission system relay, ECM**
	0C	Torque converter clutch (TCC) solenoid – electrical	**Wiring open circuit/short circuit, TCC solenoid**

Flash code	Scan code Hex type	Fault location	Probable cause
	32	Shift solenoid (SS) 3-4 – electrical	Wiring, SS, ECM
	4A	Transmission fluid temperature (TFT) sensor – low input	Wiring short circuit to ground, TFT sensor, ECM
	4B	Transmission fluid temperature (TFT) sensor – high input	Wiring short circuit to positive, TFT sensor, ECM
	72	Park/neutral position (PNP) switch – input circuit malfunction	Wiring, PNP switch, ECM
	8D	Governor pressure not equal to target (15-20 PSI) – mid pressure malfunction	Signal wire open circuit/short circuit, valve body, sensor supply wire open circuit, transmission fluid level, governor pressure solenoid, transmission mechanical fault, ECM – no 5 volt supply
	8E	Governor pressure not equal to target – low pressure malfunction	Signal wire open circuit/short circuit, valve body, sensor supply wire open circuit, transmission fluid level, governor pressure solenoid, transmission mechanical fault, ECM – no 5 volt supply
	94	Torque converter clutch (TCC) solenoid – circuit malfunction	Wiring, TCC solenoid
	A4	Transmission fluid temperature (TFT) sensor – range/performance problem	Wiring, TFT sensor
	A5	Gear selection, 3-4 – shift malfunction	Wiring, TR sensor, shift solenoids, transmission mechanical fault
	AB	Transmission fluid pressure (TFP) solenoid – electrical	Wiring, TFP solenoid, ECM
	AD	Transmission system relay	Wiring open circuit/short circuit, transmission system relay, ECM
41	0B	Alternator, field control – circuit malfunction	Wiring open circuit/short circuit, alternator, battery
42		Engine control relay/fuel level sensor – circuit malfunction	Wiring open circuit/short circuit, engine control relay, fuel level sensor
	0A	Engine control relay	Wiring open circuit/short circuit, engine control relay
	2C	Engine control relay	Wiring open circuit, engine control relay
	95	Fuel level sensor – low input	Wiring open circuit, fuel level sensor
	96	Fuel level sensor – high input	Wiring short circuit to positive, fuel level sensor
	97	Fuel level sensor – malfunction	Wiring, fuel level sensor
44		Ambient/battery temperature sensor – voltage low/high	Wiring, ambient/battery temperature sensor, ECM
	9A	Ambient/battery temperature sensor – voltage high	Wiring, ambient/battery temperature sensor, ECM

Flash code	Scan code Hex type	Fault location	Probable cause
	99	Ambient/battery temperature sensor – voltage low	**Wiring, ambient/battery temperature sensor, ECM**
45		Transmission control system – malfunction	**Wiring, transmission fluid level, transmission mechanical fault, PNP switch, TCC solenoid, TFT sensor, TFP solenoid, shift solenoid, overdrive solenoid/switch, governor pressure solenoid, transmission system relay, ECM**
	32	Shift solenoid (SS) 3-4 – electrical	**Wiring, SS, ECM**
	4A	Transmission fluid temperature (TFT) sensor – low input	**Wiring short circuit to ground, TFT sensor, ECM**
	4B	Transmission fluid temperature (TFT) sensor – high input	**Wiring short circuit to positive, TFT sensor, ECM**
	8D	Governor pressure not equal to target (15-20 PSI) – mid pressure malfunction	**Signal wire open circuit/short circuit, valve body, sensor supply wire open circuit, transmission fluid level, governor pressure solenoid, transmission mechanical fault, ECM – no 5 volt supply**
	8E	Governor pressure not equal to target – low pressure malfunction	**Signal wire open circuit/short circuit, valve body, sensor supply wire open circuit, transmission fluid level, governor pressure solenoid, transmission mechanical fault, ECM – no 5 volt supply**
	A4	Transmission fluid temperature (TFT) sensor – range/performance problem	**Wiring, TFT sensor**
	A5	Gear selection, 3-4 – shift malfunction	**Wiring, TR sensor, shift solenoids, transmission mechanical fault**
	A7	Governor pressure sensor – signal low	**Signal wire short circuit to ground, supply wire open circuit, transmission fluid level, governor pressure sensor, ECM**
	A8	Governor pressure sensor – signal high	**Signal wire short circuit to positive, supply wire open circuit, transmission fluid level, governor pressure sensor, ECM**
	A9	Governor pressure sensor – incorrect signal	**Signal wire short circuit to ground/ positive, supply wire open circuit/short circuit, transmission fluid level, governor pressure solenoid, governor pressure sensor, ECM**
	AB	Transmission fluid pressure (TFP) solenoid – electrical	**Wiring, TFP solenoid, ECM**
	AD	Transmission system relay	**Wiring open circuit/short circuit, transmission system relay, ECM**

Flash code	Scan code Hex type	Fault location	Probable cause
	BC	Overdrive solenoid/switch – performance problem	**Wiring, overdrive solenoid/switch, transmission mechanical fault, ECM**
46	06	Alternator – voltage high	**Wiring, battery, alternator**
47	05	Alternator – voltage low	**Wiring, battery, alternator drive belt, alternator**
53		Engine control module (ECM) – internal fault	**ECM**
	02	Engine control module (ECM) – internal fault	**ECM**
	44	Engine control module (ECM) – internal fault	**ECM**
55		End of MIL lamp code output	–
63	31	Engine control module (ECM) – EEPROM error	**ECM**
65	73	Power steering pressure (PSP) switch – range/performance problem	**Wiring, PAS system, PSP switch**
68	8C	Exhaust gas recirculation (EGR) system – flow malfunction	**Hose leak/blockage, wiring, EGR valve, EGR solenoid, EGR control module, ECM**
71	B9	Engine control module (ECM) – 5 volt output low	**ECM**
77	52	Cruise control	**Wiring, cruise control vacuum switching valve**

DODGE

Model:	Engine identification:	Year:
Ram Truck 2500/3500 5.9L Turbo Diesel	**ETB**	**1998**

System: **JTEC**

Self-diagnosis

General information

- Refer to the front of this manual for general test conditions, terminology, detailed descriptions of wiring faults and a general trouble shooter for electrical and mechanical faults.
- Malfunction indicator lamp (MIL) will illuminate or flash if certain faults are recorded.

Accessing

- Trouble codes can only be displayed by using a Scan Tool connected to the data link connector (DLC) **1**.

1

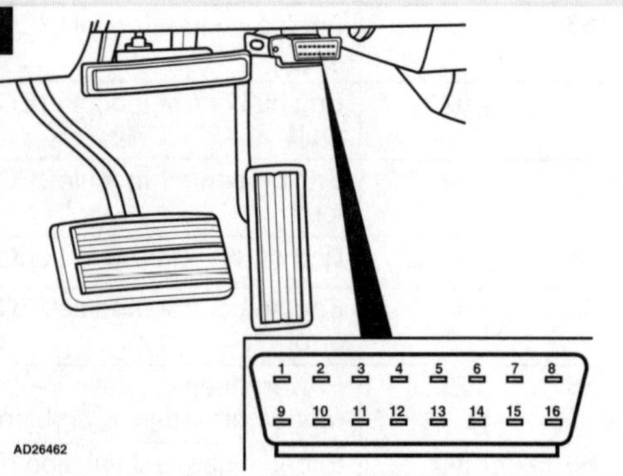

Erasing

- Trouble codes can only be erased by using a Scan Tool connected to the data link connector (DLC).

Trouble code identification

Scan code P type	Fault location	Probable cause
P0000-P0999	Refer to OBD-II trouble code table at the front of this manual	–
P1281	Engine temperature – low for long period	**Wiring, cooling system, coolant thermostat, ECT sensor**
P1388	Engine control relay	**Wiring open circuit/short circuit, engine control relay**
P1389	Engine control relay	**Wiring open circuit, engine control relay**
P1492	Ambient/battery temperature sensor – voltage high	**Wiring, ambient/battery temperature sensor, ECM**
P1493	Ambient/battery temperature sensor – voltage low	**Wiring, ambient/battery temperature sensor, ECM**
P1496	Engine control module (ECM) – 5 volt output low	**ECM**
P1594	Alternator – voltage high	**Wiring, battery, alternator**

Scan code P type	Fault location	Probable cause
P1595	Cruise control	Wiring open circuit/short circuit, hoses, cruise control master switch, cruise control vacuum switching valve
P1596	Cruise control master switch – high voltage	Wiring, cruise control master switch
P1597	Cruise control master switch – low voltage	Wiring, cruise control master switch
P1693	Engine control module (ECM)	Companion module trouble code detected in ECM
P1694	CAN data bus, engine control module (ECM) – no signal	Wiring, ECM
P1696	Engine control module (ECM) – EEPROM error	ECM
P1756	Governor pressure not equal to target (15-20 PSI) – mid pressure malfunction	Signal wire open circuit/short circuit, valve body, sensor supply wire open circuit, transmission fluid level, governor pressure solenoid, transmission mechanical fault, ECM – no 5 volt supply
P1757	Governor pressure not equal to target – low pressure malfunction	Signal wire open circuit/short circuit, valve body, sensor supply wire open circuit, transmission fluid level, governor pressure solenoid, transmission mechanical fault, ECM – no 5 volt supply
P1762	Governor pressure sensor – incorrect signal	Signal wire short circuit to ground/positive, supply wire open circuit/short circuit, transmission fluid level, governor pressure solenoid, governor pressure sensor, ECM
P1763	Governor pressure sensor – signal high	Signal wire short circuit to positive, supply wire open circuit, transmission fluid level, governor pressure sensor, ECM
P1764	Governor pressure sensor – signal low	Signal wire short circuit to ground, supply wire open circuit, transmission fluid level, governor pressure sensor, ECM
P1765	Transmission system relay	Wiring open circuit/short circuit, transmission system relay, ECM
P1899	Park/neutral position (PNP) switch – input circuit malfunction	Wiring, PNP switch, ECM

Scan code Hex type	Fault location	Probable cause
00	DTC error	–
02	Engine control module (ECM) – internal fault	ECM
05	Alternator – voltage low	Wiring, battery, alternator drive belt, alternator
06	Alternator – voltage high	Wiring, battery, alternator
0A	Engine control relay	Wiring open circuit/short circuit, engine control relay

DODGE

Ram Truck 2500/3500 5.9L Turbo Diesel

Scan code Hex type	Fault location	Probable cause
0B	Alternator, field control – circuit malfunction	**Wiring open circuit/short circuit, alternator, battery**
0C	Torque converter clutch (TCC) solenoid – electrical	**Wiring open circuit/short circuit, TCC solenoid**
0F	Cruise control	**Wiring open circuit/short circuit, hoses, voltage low/high, cruise control master switch, cruise control vacuum switching valve**
10	A/C compressor clutch relay	**Wiring open circuit/short circuit, A/C compressor clutch relay**
11	Exhaust gas recirculation (EGR) – circuit malfunction	**Wiring, EGR solenoid, EGR control module, ECM**
12	Evaporative emission (EVAP) canister purge valve – circuit malfunction	**Wiring open circuit/short circuit, EVAP canister purge valve**
13	Injector 3 – circuit malfunction	**Wiring, injector**
14	Injector 2 – circuit malfunction	**Wiring, injector**
15	Injector 1 – circuit malfunction	**Wiring, injector**
19	Idle speed control (ISC) system – malfunction	**Wiring open circuit/short circuit, IAC valve**
1A	Throttle position (TP) sensor – low input	**Wiring short circuit to ground, TP sensor**
1B	Throttle position (TP) sensor – high input	**Wiring short circuit to positive, TP sensor**
1E	Engine coolant temperature (ECT) sensor – low input	**Coolant thermostat, wiring short circuit to ground, ECT sensor**
1F	Engine coolant temperature (ECT) sensor – high input	**Coolant thermostat, wiring open circuit/ short circuit to positive, ground wire defective, ECT sensor**
21	Engine temperature – low for long period	**Wiring, cooling system, coolant thermostat, ECT sensor**
24	Manifold absolute pressure (MAP) sensor – low input	**Wiring short circuit to ground, MAP sensor**
25	Manifold absolute pressure (MAP) sensor – high input	**Wiring short circuit to positive, MAP sensor**
27	Manifold absolute pressure (MAP) sensor	**Wiring, hose connection(s), MAP sensor**
28	Crankshaft position (CKP) sensor – no signal	**Wiring, air gap, CKP sensor**
29	Ignition coil C, primary – circuit malfunction	**Wiring, ignition coil, ECM**
2A	Ignition coil B, primary – circuit malfunction	**Wiring, ignition coil, ECM**
2B	Ignition coil A, primary – circuit malfunction	**Wiring, ignition coil**
2C	Engine control relay	**Wiring open circuit, engine control relay**
31	Engine control module (ECM) – EEPROM error	**ECM**
32	Shift solenoid (SS) 3-4 – electrical	**Wiring, SS, ECM**
37	Injector 9 – circuit malfunction	**Wiring, injector**
38	Injector 10 – circuit malfunction	**Wiring, injector**

Scan code Hex type	Fault location	Probable cause
39	Intake air temperature (IAT) sensor – low input	**Wiring short circuit to ground, IAT sensor**
3A	Intake air temperature (IAT) sensor – high input	**Wiring open circuit/short circuit to positive, ground wire defective, IAT sensor**
3D	Injector 4 – circuit malfunction	**Wiring, injector**
3E	Heated oxygen sensor (HO2S) 1, bank 1 – high voltage	**Wiring short circuit to positive, fuel pressure, injector control, EVAP canister purge valve, air filter blocked, HO2S**
41	Heated oxygen sensor (HO2S) 1, bank 2 – no activity detected	**Wiring open circuit, heating inoperative, HO2S**
42	Heated oxygen sensor (HO2S) 1, bank 2 – high voltage	**Wiring short circuit to positive, fuel pressure, injector control, EVAP canister purge valve, air filter blocked, HO2S**
44	Engine control module (ECM) – internal fault	**ECM**
45	Injector 5 – circuit malfunction	**Wiring, injector**
46	Injector 6 – circuit malfunction	**Wiring, injector**
4A	Transmission fluid temperature (TFT) sensor – low input	**Wiring short circuit to ground, TFT sensor, ECM**
4B	Transmission fluid temperature (TFT) sensor – high input	**Wiring short circuit to positive, TFT sensor, ECM**
4C	Ignition coil D, primary – circuit malfunction	**Wiring, ignition coil, ECM**
4D	Ignition coil E, primary – circuit malfunction	**Wiring, ignition coil, ECM**
4F	Injector 7 – circuit malfunction	**Wiring, injector**
50	Injector 8 – circuit malfunction	**Wiring, injector**
52	Cruise control	**Wiring, cruise control vacuum switching valve**
56	Cruise control master switch – high voltage	**Wiring, cruise control master switch**
57	Cruise control master switch – low voltage	**Wiring, cruise control master switch**
65	Fuel pump relay	**Wiring open circuit/short circuit, fuel pump relay**
66	Heated oxygen sensor (H02S) 1, bank 1 – slow response	**Heating inoperative, wiring, HO2S**
67	Heated oxygen sensor (HO2S) 1, bank 1 – heater control – circuit malfunction	**Wiring, HO2S**
68	Heated oxygen sensor (H02S) 2, bank 1 – slow response	**Heating inoperative, wiring, exhaust leak, HO2S**
69	Heated oxygen sensor (HO2S) 2, bank 1, heater control – circuit malfunction	**Wiring, HO2S**
6A	Random/multiple cylinder(s) – misfire detected	**Wiring, spark plug(s), HT lead(s), injector(s), ignition coil(s), low compression**
6B	Cylinder 1 – misfire detected	**Wiring, engine mechanical fault, ignition/ fuel system, injector**
6C	Cylinder 2 – misfire detected	**Wiring, engine mechanical fault, ignition/ fuel system, injector**

Scan code Hex type	Fault location	Probable cause
6D	Cylinder 3 – misfire detected	**Wiring, engine mechanical fault, ignition/ fuel system, injector**
6E	Cylinder 4 – misfire detected	**Wiring, engine mechanical fault, ignition/ fuel system, injector**
70	Catalytic converter system, bank 1 – efficiency below threshold	**Catalytic converter, wiring, HO2S 2**
71	Evaporative emission (EVAP) system – incorrect flow detected	**Hose connection(s), intake leak, EVAP canister purge valve**
72	Park/neutral position (PNP) switch – input circuit malfunction	**Wiring, PNP switch, ECM**
73	Power steering pressure (PSP) switch – range/performance problem	**Wiring, PAS system, PSP switch**
76	System too rich, bank 1	**Wiring, intake blocked, EVAP canister purge valve, fuel pressure, EGR system, injector(s), air filter blocked, ignition system, HO2S**
77	System too lean, bank 1	**Wiring, intake/exhaust leak, MAP sensor, fuel pressure/pump, EVAP canister purge valve, injector(s), HO2S**
78	System too rich, bank 2	**Wiring, intake blocked, EVAP canister purge valve, fuel pressure, EGR system, injector(s), air filter blocked, ignition system, HO2S**
79	System too lean, bank 2	**Wiring, intake/exhaust leak, MAP sensor, fuel pressure/pump, EVAP canister purge valve, injector(s), HO2S**
7A	Heated oxygen sensor (HO2S) 1, bank 2 – slow response	**Heating inoperative, wiring, HO2S**
7B	Heated oxygen sensor (HO2S) 2, bank 2 – slow response	**Heating inoperative, wiring, exhaust leak, HO2S**
7C	Heated oxygen sensor (HO2S) 1, bank 2, heater control – circuit malfunction	**Wiring, HO2S, ECM**
7D	Heated oxygen sensor (HO2S) 2, bank 2, heater control – circuit malfunction	**Wiring, HO2S, ECM**
7E	Heated oxygen sensor (HO2S) 2, bank 1 – high voltage	**Wiring short circuit to positive, catalytic converter, HO2S, ECM**
7F	Heated oxygen sensor (HO2S) 2, bank 2 – high voltage	**Wiring short circuit to positive, catalytic converter, HO2S, ECM**
80	Insufficient coolant temperature for closed loop fuel control	**Wiring, cooling system, coolant thermostat, ECT sensor**
84	Throttle position (TP) sensor – range/ performance problem	**Wiring, accelerator cable adjustment, TP sensor**
87	Manifold absolute pressure (MAP) sensor – no 5 volt supply	**Wiring, MAP sensor, ECM**
8A	Idle speed control (ISC) system – rpm lower/ higher than expected	**Wiring, air leak, throttle valve tight/sticking, IAC valve**
8C	Exhaust gas recirculation (EGR) system – flow malfunction	**Hose leak/blockage, wiring, EGR valve, EGR solenoid, ECM**

Scan code Hex type	Fault location	Probable cause
8D	Governor pressure not equal to target (15-20 PSI) – mid pressure malfunction	**Signal wire open circuit/short circuit, valve body, sensor supply wire open circuit, transmission fluid level, governor pressure solenoid, transmission mechanical fault, ECM – no 5 volt supply**
8E	Governor pressure not equal to target (15-20 PSI) – zero pressure malfunction	**Signal wire open circuit/short circuit, valve body, sensor supply wire open circuit, transmission fluid level, governor pressure solenoid, transmission mechanical fault, ECM – no 5 volt supply**
94	Torque converter clutch (TCC) solenoid – circuit malfunction	**Wiring, TCC solenoid**
95	Fuel level sensor – low input	**Wiring open circuit, fuel level sensor**
96	Fuel level sensor – high input	**Wiring short circuit to positive, fuel level sensor**
97	Fuel level sensor – malfunction	**Wiring, fuel level sensor**
99	Ambient/battery temperature sensor – voltage low	**Wiring, ambient/battery temperature sensor**
9A	Ambient/battery temperature sensor – voltage high	**Wiring, ambient/battery temperature sensor**
9B	Heated oxygen sensor (HO2S) 1, bank 1 – low voltage	**Exhaust leak, wiring short circuit to ground, HO2S**
9C	Heated oxygen sensor (HO2S) 2, bank 1 – low voltage	**Exhaust leak, wiring short circuit to ground, HO2S**
9D	Crankshaft position (CKP) sensor/camshaft position (CMP) sensor – intermittent signal	**Wiring, poor connection, air gap, CKP sensor, CMP sensor**
A0	Evaporative emission (EVAP) system – small leak detected	**Hose connection(s), intake leak, EVAP canister, EVAP canister purge valve**
A1	Evaporative emission (EVAP) system – large leak detected	**Hose connection(s), intake leak, EVAP canister, EVAP canister purge valve**
A4	Transmission fluid temperature (TFT) sensor – range/performance problem	**Wiring, TFT sensor**
A5	Gear selection, 3-4 – shift malfunction	**Wiring, TR sensor, shift solenoids, transmission mechanical fault**
A6	Output shaft speed (OSS) sensor – circuit malfunction	**Wiring, OSS sensor, ECM**
A7	Governor pressure sensor – signal low	**Signal wire short circuit to ground, supply wire open circuit, transmission fluid level, governor pressure sensor, ECM**
A8	Governor pressure sensor – signal high	**Signal wire short circuit to positive, supply wire open circuit, transmission fluid level, governor pressure sensor, ECM**
A9	Governor pressure sensor – incorrect signal	**Signal wire short circuit to ground/positive, supply wire open circuit/short circuit, transmission fluid level, governor pressure solenoid, governor pressure sensor, ECM**

Scan code Hex type	Fault location	Probable cause
AB	Transmission fluid pressure (TFP) solenoid – electrical	**Wiring, TFP solenoid, ECM**
AD	Transmission system relay	**Wiring open circuit/short circuit, transmission system relay, ECM**
AE	Cylinder 5 – misfire detected	**Wiring, engine mechanical fault, ignition/ fuel system, injector**
AF	Cylinder 6 – misfire detected	**Wiring, engine mechanical fault, ignition/ fuel system, injector**
B0	Cylinder 7 – misfire detected	**Wiring, engine mechanical fault, ignition/ fuel system, injector**
B1	Cylinder 8 – misfire detected	**Wiring, engine mechanical fault, ignition/ fuel system, injector**
B2	Cylinder 9 – misfire detected	**Wiring, engine mechanical fault, ignition/ fuel system, injector**
B3	Cylinder 10 – misfire detected	**Wiring, engine mechanical fault, ignition/ fuel system, injector**
B4	Catalytic converter system, bank 2 – efficiency below threshold	**Catalytic converter, wiring, HO2S 2**
B5	Heated oxygen sensor (HO2S) 1, bank 2 – low voltage	**Exhaust leak, wiring short to ground, HO2S, ECM**
B6	Heated oxygen sensor (HO2S) 2, bank 2 – low voltage	**Exhaust leak, wiring short to ground, HO2S, ECM**
B7	Evaporative emission (EVAP) leak detection pump solenoid	**Wiring open circuit/short circuit, hose leak/ blockage, EVAP leak detection pump solenoid**
B8	Evaporative emission (EVAP) leak detection pump – pressure switch	**Wiring, hose leak/blockage, EVAP leak detection pump**
BA	Engine control module (ECM) – misfire detection limit reached	**Wiring, poor connection, CKP sensor**
BB	Evaporative emission (EVAP) system – blockage detected	**Hose connection(s), hose blocked/trapped, EVAP canister**
BC	Overdrive solenoid/switch – performance problem	**Wiring, overdrive solenoid/switch, transmission mechanical fault, ECM**
BD	Heated oxygen sensor (HO2S) 3, bank 1, heater control – circuit malfunction	**Wiring, HO2S, ECM**
CO	Heated oxygen sensor (HO2S) 1, bank 1 – slow response	**Heating inoperative, wiring, HO2S**
C1	Heated oxygen sensor (HO2S) 1, bank 2 – slow response	**Wiring, heating inoperative, exhaust leak, HO2S**
C2	Heated oxygen sensor (HO2S) 2, bank 1 – slow response	**Wiring, heating inoperative, exhaust leak, HO2S**
DE	CAN data bus, engine control module (ECM) – no signal	**Wiring, ECM**
DF	Engine control module (ECM)	**Companion module trouble code detected in ECM**

DODGE

Model: Ram Truck 5.9L Turbo Diesel

Engine identification: ETC

Year: 1999-02

System: **Cummins CM551/A**

Self-diagnosis

General information

- Refer to the front of this manual for general test conditions, terminology, detailed descriptions of wiring faults and a general trouble shooter for electrical and mechanical faults.
- Malfunction indicator lamp (MIL) will illuminate or flash if certain faults are recorded.
- ECM operates in backup mode if sensors fail, to enable vehicle to be driven to repair shop.

Accessing

- Trouble codes can only be displayed by using a Scan Tool connected to the data link connector (DLC) **1**.

1

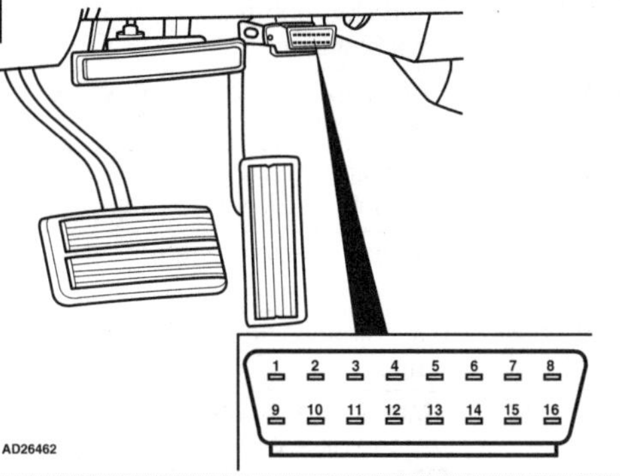

Erasing

- Trouble codes can only be erased by using a Scan Tool connected to the data link connector (DLC).

Trouble code identification

Scan code	Fault location	Probable cause
P0000-P0999	Refer to OBD-II trouble code table at the front of this manual	–
P1110	Decreased engine performance – high intake air temperature	**Wiring open circuit/short to positive, ground wire defective, IAT sensor, ECT sensor, cooling system, high engine bay temperature, intake manifold heater relay(s), intake manifold heater(s), ECM 1**
P1180	Decreased engine performance – high injection pump fuel temperature	**Wiring open circuit/short to positive, ground wire defective, fuel temperature sensor, fuel temperature sensor, mechanical fault, high engine bay temperature, ECM 1**
P1192	Intake air temperature (IAT) sensor – voltage low	**Wiring, IAT sensor, ECM 1**
P1193	Intake air temperature (IAT) sensor – voltage high	**Wiring, IAT sensor, ECM 1**

Scan code	Fault location	Probable cause
P1281	Engine temperature – low for long period	**Wiring, cooling system, coolant thermostat, ECT sensor**
P1283 [1]	Engine control module (ECM) 1 – idle signal malfunction	**Wiring, injection pump control module [3], ECM 1**
P1284 [3]	Injection pump control module – battery voltage out of range	**Wiring, injection pump control module, ECM 1**
P1285	Injection pump control module relay – circuit malfunction	**Wiring, injection pump control module relay,**
P1286	Accelerator pedal position (APP) sensor – high input	**Wiring short to positive, APP sensor, ECM 1**
P1287 [1] [3]	Injection pump control module/engine control module (ECM) 1 – supply voltage low	**Wiring, battery, alternator**
P1291	Intake manifold heaters – no temperature rise	**Wiring, intake manifold heater relay(s), intake manifold heater(s), ECM 1**
P1295	Accelerator pedal position (APP) sensor – low input	**Signal wire short circuit to ground, supply wire open circuit, APP sensor, ECM 1**
P1296	Manifold absolute pressure (MAP) sensor – no 5 volt supply	**Wiring, MAP sensor, ECM 1**
P1297	Manifold absolute pressure (MAP) sensor	**Wiring, hose connection(s), MAP sensor**
P1388	Engine control relay	**Wiring open circuit/short circuit, engine control relay, ECM 2**
P1389	Engine control relay	**Wiring open circuit, engine control relay, ECM 2**
P1399	Glow plug warning lamp – circuit malfunction	**Wiring, glow plug warning lamp, ECM 1**
P1475 [1]	Engine control module (ECM) 1 – auxiliary 5 volt output high	**Wiring, ECM 1**
P1478	Ambient/battery temperature sensor – range/ performance problem	**Wiring, ambient/battery temperature sensor**
P1488 [1]	Engine control module (ECM) 1 – auxiliary 5 volt output low	**Wiring, ECM 1**
P1492	Ambient/battery temperature sensor – voltage high	**Wiring, ambient/battery temperature sensor**
P1493	Ambient/battery temperature sensor – voltage low	**Wiring, ambient/battery temperature sensor**
P1496 [1]	Engine control module (ECM) 1 – 5 volt output low	**ECM 1**
P1594	Alternator – voltage high	**Wiring, battery, alternator**
P1595	Cruise control	**Wiring open circuit/short circuit, hoses, cruise control master switch, cruise control vacuum switching valve, ECM 2**
P1596	Cruise control master switch – high voltage	**Wiring, cruise control master switch, ECM 2**
P1597	Cruise control master switch – low voltage	**Wiring, cruise control master switch, ECM 2**

Scan code	Fault location	Probable cause
P1602 1	Engine control module (ECM) 1 – not programmed	**New ECM not programmed, incorrectly programmed, ECM 1**
P1682	Alternator – voltage low	**Wiring, battery, alternator drive belt, alternator**
P1683	Cruise control	**Wiring, cruise control vacuum switching valve, ECM 2**
P1684 1	Engine control module (ECM) 1 – supply voltage disconnected within last 50 ignition switch ON-OFF cycles	**Wiring, battery**
P1688 3	Injection pump/injection pump control module – internal fault	**Injection pump/injection pump control module**
P1689 1 3	Injection pump control module communication to engine control module (ECM) 1 – malfunction	**Wiring, poor connection, injection pump control module, ECM 1**
P1690	Crankshaft position/injection pump camshaft position sensor – correlation	**Wiring, CKP sensor, injection pump CMP sensor, mechanical fault, ECM 1**
P1691	Injection pump – internal fault	**Injection pump/injection pump control module 3**
P1693 1	Engine control module (ECM) 1	**Companion module trouble code detected in ECM 1**
P1694 1	CAN data bus, engine control module (ECM) 1 – no signal	**Wiring, ECM 1**
P1696 1	Engine control module (ECM) 1 – EEPROM error	**ECM 1**
P1698 2	CAN data bus, engine control module (ECM) 2 – no signal	**Wiring, ECM 2**
P1719	Skip shift solenoid – circuit malfunction	**Wiring open circuit/short circuit, skip shift solenoid**
P1740	Torque converter clutch (TCC)/overdrive solenoid – performance problem	**TCC, overdrive solenoid, transmission mechanical fault**
P1756	Governor pressure not equal to target (15-20 PSI) – mid pressure malfunction	**Signal wire open circuit/short circuit, valve body, sensor supply wire open circuit, transmission fluid level, governor pressure solenoid, transmission mechanical fault, ECM 2 – no 5 volt supply 2**
P1757	Governor pressure not equal to target (15-20 PSI) – zero pressure malfunction	**Signal wire open circuit/short circuit, valve body, sensor supply wire open circuit, transmission fluid level, governor pressure solenoid, transmission mechanical fault, ECM 2 – no 5 volt supply 2**
P1762	Governor pressure sensor – incorrect signal	**Signal wire short circuit to ground/positive, supply wire open circuit/short circuit, transmission fluid level, governor pressure solenoid, governor pressure sensor, ECM 2 2**

Scan code	Fault location	Probable cause
P1763	Governor pressure sensor – signal high	**Signal wire short circuit to positive, supply wire open circuit, transmission fluid level, governor pressure sensor, ECM 2 2**
P1764	Governor pressure sensor – signal low	**Signal wire short circuit to ground, supply wire open circuit, transmission fluid level, governor pressure sensor, ECM 2 2**
P1765	Transmission system relay	**Wiring open circuit/short circuit, transmission system relay, ECM 2 2**
P1899	Park/neutral position (PNP) switch – input circuit malfunction	**Wiring, PNP switch, ECM 2 2**

1 ECM 1 = Engine control module mounted on engine near injection pump.
2 ECM 2 = Powertrain control module mounted in engine bay right rear. Does not control the fuel system.
3 Fuel injection pump control module in fuel injection pump (unserviceable).

EAGLE

Model:	Engine identification:	Year:
Premier 3.0L	**Z7X**	**1992**
Vision 3.3L	**EGB**	**1993-95**
Vision 3.5L	**EGE**	**1993-95**

System: **SBEC II**

Self-diagnosis

General information

- Refer to the front of this manual for general test conditions, terminology, detailed descriptions of wiring faults and a general trouble shooter for electrical and mechanical faults.
- Malfunction indicator lamp (MIL) will illuminate if certain faults are recorded.
- ECM operates in backup mode if sensors fail, to enable vehicle to be driven to repair shop.
- Trouble codes can be displayed by the MIL or by using a Scan Tool connected to the data link connector (DLC).

Accessing

- Within 5 seconds switch the ignition key ON-OFF-ON-OFF-ON.
- Count MIL flashes. Note trouble codes.
- Each trouble code consists of two groups **1** [**A**].
- The first group of flashes indicate the 'tens' of the trouble code.
- The second group of flashes indicate the 'units' of the trouble code.
- A short pause separates 'tens' and 'units' **1** [**B**].
- A long pause separates each trouble code **1** [**C**].
- For example: Trouble code 21 displayed **1**
- End of test sequence indicated by trouble code 55.
- Trouble codes can also be displayed by using a Scan Tool connected to the data link connector (DLC):
  - Vision **2**.
  - Premier **3**.

1

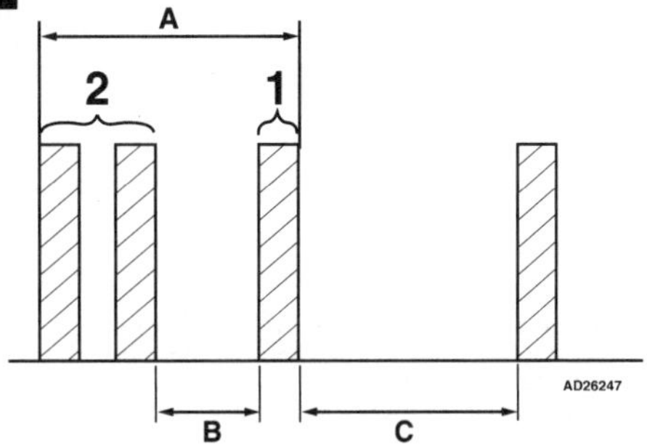

2

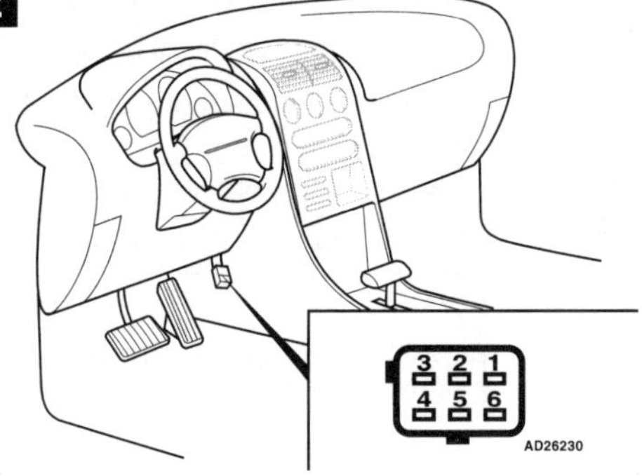

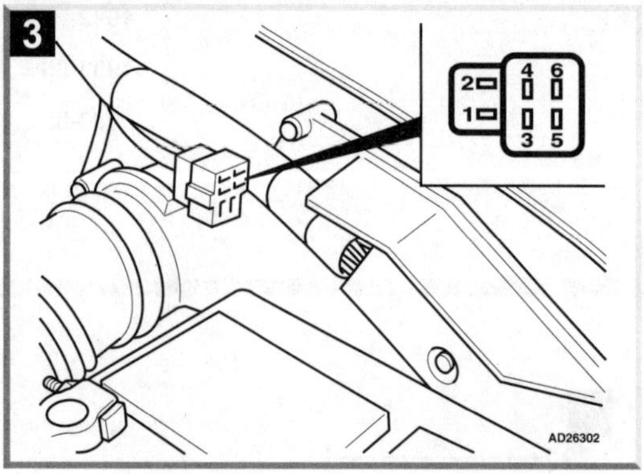

Erasing

NOTE: ***Disconnecting battery ground lead may cause driveability problems until the engine management system relearns basic values.***

- Ensure ignition switched OFF.
- Disconnect battery ground cable for at least 30 seconds.
- Reconnect battery ground cable.
- Repeat accessing procedure to ensure no trouble codes remain in ECM fault memory.

NOTE: ***Disconnecting battery ground lead may also erase memory from electronic units, such as radio.***

- Trouble codes can also be erased by using a Scan Tool connected to the data link connector (DLC).

Trouble code identification

Flash code	Fault location	Probable cause
11	Crankshaft position (CKP) sensor	**Wiring, air gap, CKP sensor**
12	Engine control module (ECM) – supply voltage disconnected within last 50 ignition switch ON-OFF cycles	**Wiring, battery**
13	Manifold absolute pressure (MAP) sensor	**Wiring, MAP sensor**
14	Manifold absolute pressure (MAP) sensor – voltage low/high	**Wiring, MAP sensor**
15	Vehicle speed sensor (VSS) – signal	**Wiring, VSS**
16	Knock sensor (KS) – wiring	**Wiring open/short circuit, KS**
17	Engine temperature – low for long period	**Wiring, coolant thermostat, ECT sensor**
21	Heated oxygen sensor (HO2S) – incorrect signal/voltage high	**Wiring, fuel pressure, injector control, EVAP canister purge valve, compression, air filter blocked, heated oxygen sensor (HO2S)**
22	Engine coolant temperature (ECT) sensor – voltage low/high	**Wiring, ECT sensor**
23	Intake air temperature (IAT) sensor	**Wiring, IAT sensor**
24	Throttle position (TP) sensor	**Wiring, accelerator cable adjustment, TP sensor**
25	Idle air control (IAC) valve	**Wiring, IAC valve**
26	Injector – circuit low	**Wiring, injector, ECM**
27	Injector – circuit malfunction	**Wiring, injector**
31	Evaporative emission (EVAP) canister purge valve – open/short circuit	**Wiring, EVAP canister purge valve**

Flash code	Fault location	Probable cause
32	Exhaust gas recirculation (EGR) solenoid/ system – circuit malfunction	**Wiring, hoses, EGR solenoid**
33	A/C compressor clutch relay/A/C refrigerant pressure sensor – voltage low/high	**Wiring, A/C compressor clutch relay, A/C refrigerant pressure sensor**
34	Cruise control	**Wiring open/short circuit, hoses, voltage low/high, cruise control master switch, cruise control vacuum switching valve**
35	Engine coolant blower motor relay – low/high speed	**Wiring, engine coolant blower motor relay – low/high speed**
41	Alternator	**Wiring open/short circuit, alternator**
42	Engine control relay	**Wiring open/short circuit, engine control relay**
43	Ignition coil – primary circuit	**Wiring, ignition coil**
44	Battery temperature sensor/engine coolant temperature (ECT) sensor	**Wiring open/short circuit, battery temperature sensor, ECT sensor, ECM**
46	Alternator – voltage high	**Wiring, battery, alternator**
47	Alternator – voltage low	**Wiring, battery, alternator drive belt, alternator**
51	System too lean	**Wiring, intake/exhaust leak, fuel pressure, injector control, EVAP canister purge valve, heated oxygen sensor (HO2S)**
52	System too rich	**Wiring, exhaust blockage, fuel pressure, injector control, EVAP canister purge valve, compression, air filter blocked, ignition system, heated oxygen sensor (HO2S)**
53	Engine control module (ECM) – internal fault	**ECM**
54	Camshaft position (CMP) sensor – no signal	**Wiring, CMP sensor**
55	End of MIL code output	**–**
62	Engine control module (ECM) – EEPROM error	**ECM**
63	Engine control module (ECM) – EEPROM error	**ECM**
65	Intake manifold air control solenoid	**Wiring open/short circuit, intake manifold air control solenoid**
66	CAN data bus, transmission control module (TCM)/body control module (BCM) – no signal	**Wiring, TCM, BCM**
77	Cruise control system relay	**Wiring open/short circuit, cruise control system relay**

EAGLE

Model:	Engine identification:	Year:
Summit 1.5L	4G15	1992-93
Summit 1.8L	4G93	1992-93
Summit Wagon 1.8L	4G93	1992-93
Summit Wagon 2.4L	4G64	1992-93
Talon 1.8L	4G37	1993-94
Talon 2.0L	4G63	1992-94
Talon 2.0L Turbo	4G63	1992-94

System: **Mitsubishi MPI**

Self-diagnosis

General information

- Refer to the front of this manual for general test conditions, terminology, detailed descriptions of wiring faults and a general trouble shooter for electrical and mechanical faults.
- Malfunction indicator lamp (MIL) will illuminate if certain faults are recorded.
- ECM operates in backup mode if sensors fail, to enable vehicle to be driven to repair shop.
- 1992: Trouble codes can be displayed by using an analogue voltmeter connected to the data link connector (DLC).
- 1993 →: Trouble codes can be displayed by using an analogue voltmeter or Scan Tool connected to the data link connector (DLC).

Accessing

- Ensure ignition switched OFF.
- Access data link connector (DLC) **1**.
- Connect an analogue voltmeter between data link connector (DLC) terminals 1 and 12 **2**.
- Switch ignition ON.
- Count voltmeter needle deflections. Note trouble codes. Compare with trouble code table.
- Each trouble code consists of two groups of one or more needle deflections **3** :
  - Long needle deflections indicate the 'tens' of the trouble code **3** [**A**].
  - Short needle deflections indicate the 'units' of the trouble code **3** [**D**].
  - A short pause separates each trouble code group **3** [**C**].
  - A long pause separates each trouble code **3** [**E**].
  - For example: Trouble code 22 displayed **3**.
- If a constant deflection is displayed: Suspect engine control module (ECM).
- If constant short deflections are displayed: No trouble code is stored.
- 1993 →: Trouble codes can also be displayed by using a Scan Tool connected to the data link connector (DLC).

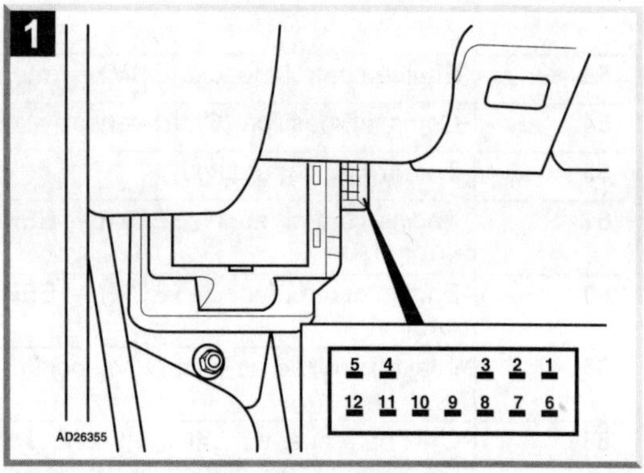

2

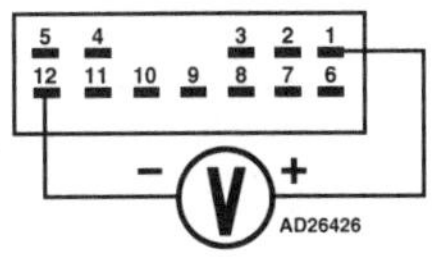

3

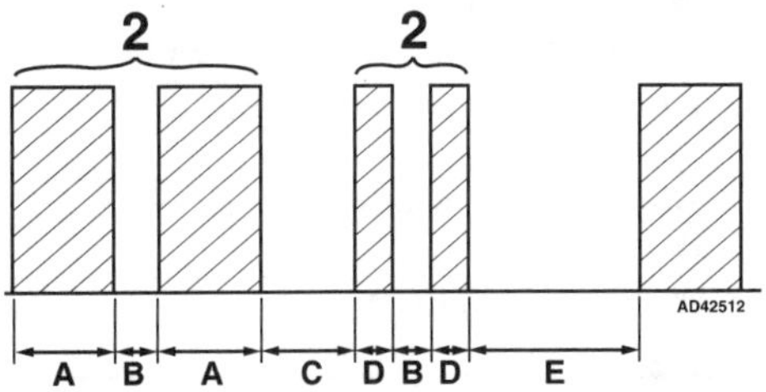

Erasing

NOTE: ***Disconnecting battery ground lead may cause driveability problems until the engine management system relearns basic values.***

- Ensure ignition switched OFF.
- Disconnect battery ground cable for at least 15 seconds.
- Reconnect battery ground lead.
- Repeat checking procedure to ensure no data remains in ECM fault memory.
- 1993 →: Trouble codes can also be erased by using a Scan Tool connected to the data link connector (DLC).

NOTE: ***Disconnecting battery ground lead may erase memory from electronic units such as radio.***

Trouble code identification

Flash code	Fault location	Probable cause
11	Heated oxygen sensor (HO2S)/oxygen sensor (O2S) 1, bank 1 – circuit malfunction	**Wiring, fuel pressure, injector(s), intake leak, HO2S/O2S, ECM**
12	Volume air flow (VAF) sensor – malfunction	**Wiring, VAF sensor, ECM**
13	Intake air temperature (IAT) sensor – circuit malfunction	**Wiring, IAT sensor, ECM**
14	Throttle position (TP) sensor – circuit malfunction	**Wiring, TP sensor, CTP switch, ECM**
15	Idle speed control (ISC) actuator position sensor – circuit malfunction	**Wiring, ISC actuator, ISC actuator position sensor, ECM**
21	Engine coolant temperature (ECT) sensor – circuit malfunction	**Wiring, ECT sensor, ECM**
22	Crankshaft position (CKP) sensor – circuit malfunction	**Wiring, CKP sensor, ECM**

Flash code	Fault location	Probable cause
23	Camshaft position (CMP) sensor – circuit malfunction	**Wiring, CMP sensor, ECM**
24	Vehicle speed sensor (VSS) – circuit malfunction	**Wiring, VSS, ECM**
25	Barometric pressure (BARO) sensor – circuit malfunction	**Wiring, BARO sensor, ECM**
31	Knock sensor (KS) – circuit malfunction	**Wiring, KS, ECM**
32	Manifold absolute pressure (MAP) sensor – circuit malfunction	**Wiring, MAP sensor, ECM**
36	Ignition timing adjustment	**Wiring, ignition timing connector short to ground, ECM**
41	Injectors – circuit malfunction	**Wiring, injectors, ECM**
42	Fuel pump – circuit malfunction	**Wiring, engine control relay, fuel pump, ECM**
43	Exhaust gas recirculation (EGR) system – circuit malfunction	**Wiring, hose leak/blockage, EGR valve, EGR temperature sensor, EGR solenoid, ECM**
44	Ignition coil/ignition control module (ICM) – circuit malfunction	**Wiring, ignition coil/ICM, ECM**
55	Idle air control (IAC) valve position sensor – circuit malfunction	**Wiring, IAC valve, IAC valve position sensor, ECM**
59	Heated oxygen sensor (HO2S) 2, bank 1 – circuit malfunction	**Wiring, HO2S, ECM**

EAGLE

Model: **Summit 1.5L** | Engine identification: **4G15** | Year: **1994**

System: **Mitsubishi MPI**

Self-diagnosis

General information

- Refer to the front of this manual for general test conditions, terminology, detailed descriptions of wiring faults and a general trouble shooter for electrical and mechanical faults.
- Malfunction indicator lamp (MIL) will illuminate if certain faults are recorded.
- ECM operates in backup mode if sensors fail, to enable vehicle to be driven to repair shop.
- Trouble codes can be displayed by using an analogue voltmeter or Scan Tool connected to the data link connector(s) (DLC).

Accessing

- Ensure ignition switched OFF.
- Access 12 pin and 16 pin data link connectors (DLC) **1**.
- Connect an analogue voltmeter between 12 pin data link connector (DLC) terminal 5 and 16 pin data link connector (DLC) terminal 4 or 5 **2**
- Switch ignition ON.
- Count voltmeter needle deflections. Note trouble codes. Compare with trouble code table.
- Each trouble code consists of two groups of one or more needle deflections **3** :
  - Long needle deflections indicate the 'tens' of the trouble code **3** [**A**].
  - Short needle deflections indicate the 'units' of the trouble code **3** [**D**].
  - A short pause separates each trouble code group **3** [**C**].
  - A long pause separates each trouble code **3** [**E**].
  - For example: Trouble code 22 displayed **3**.
- If a constant deflection is displayed: Suspect engine control module (ECM).
- If constant short deflections are displayed: No trouble code is stored.
- Trouble codes can also be displayed by using a Scan Tool connected to the data link connector(s) (DLC).

1

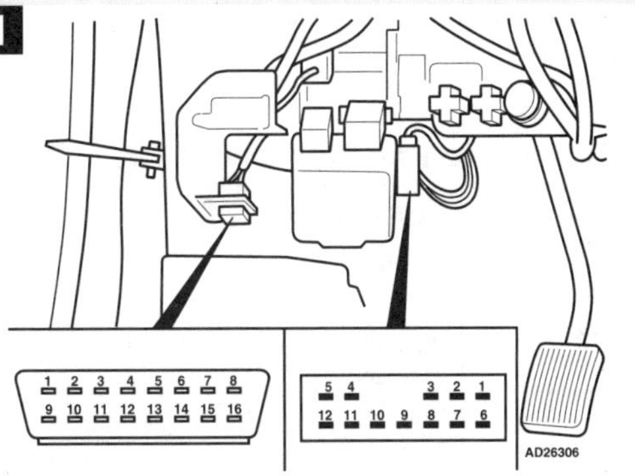

2

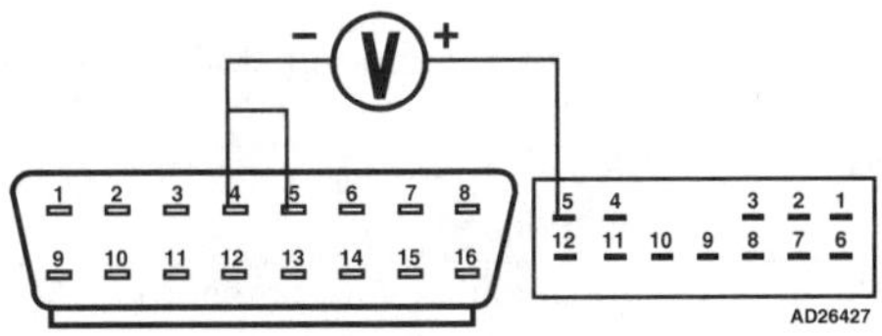

3

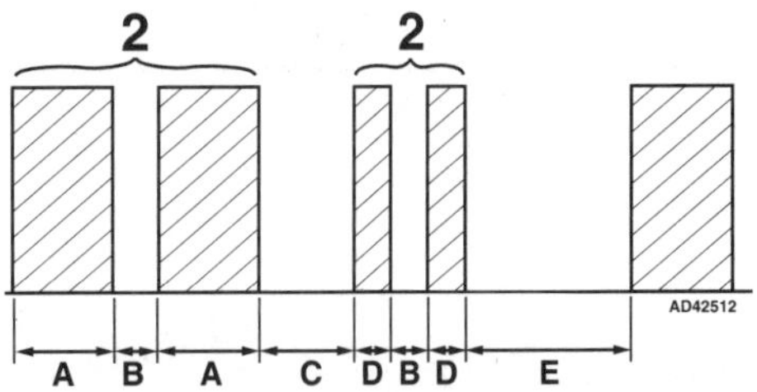

Erasing

NOTE: ***Disconnecting battery ground lead may cause driveability problems until the engine management system relearns basic values.***

- Ensure ignition switched OFF.
- Disconnect battery ground cable for at least 15 seconds.
- Reconnect battery ground lead.
- Repeat checking procedure to ensure no data remains in ECM fault memory.
- Trouble codes can also be erased by using a Scan Tool connected to the data link connector(s) (DLC).

NOTE: ***Disconnecting battery ground lead may erase memory from electronic units such as radio.***

Trouble code identification

Flash code	Fault location	Probable cause
11	Heated oxygen sensor (HO2S) 1, bank 1 – circuit malfunction	**Wiring, fuel pressure, injector(s), intake leak, HO2S, ECM**
13	Intake air temperature (IAT) sensor – circuit malfunction	**Wiring, IAT sensor, ECM**
14	Throttle position (TP) sensor – circuit malfunction	**Wiring, TP sensor, CTP switch, ECM**
21	Engine coolant temperature (ECT) sensor – circuit malfunction	**Wiring, ECT sensor, ECM**
22	Crankshaft position (CKP) sensor – circuit malfunction	**Wiring, CKP sensor, ECM**
23	Camshaft position (CMP) sensor – circuit malfunction	**Wiring, CMP sensor, ECM**
24	Vehicle speed sensor (VSS) – circuit malfunction	**Wiring, VSS, ECM**
32	Manifold absolute pressure (MAP) sensor – circuit malfunction	**Wiring, MAP sensor, ECM**
36	Ignition timing adjustment	**Wiring, ignition timing connector short to ground, ECM**
41	Injectors – circuit malfunction	**Wiring, injectors, ECM**
43	Exhaust gas recirculation (EGR) system – circuit malfunction	**Wiring, Hose leak/blockage, EGR valve, EGR temperature sensor, EGR solenoid, ECM**
55	Idle air control (IAC) valve position sensor – circuit malfunction	**Wiring, IAC valve, IAC valve position sensor, ECM**
59	Heated oxygen sensor (HO2S) 2, bank 1 – circuit malfunction	**Wiring, HO2S, ECM**

EAGLE

Model:	Engine identification:	Year:
Summit 1.8L	**4G93**	**1994**
Summit Wagon 1.8L FWD – California	**4G93**	**1994**
Summit Wagon 2.4L FWD – California	**4G64**	**1994**
Summit Wagon 2.4L AWD – California	**4G64**	**1994-95**
Summit Wagon 1.8L – Federal	**4G93**	**1994-95**
Summit Wagon 2.4L – Federal	**4G64**	**1994-95**

System: **Mitsubishi MPI**

Self-diagnosis

General information

- Refer to the front of this manual for general test conditions, terminology, detailed descriptions of wiring faults and a general trouble shooter for electrical and mechanical faults.
- Malfunction indicator lamp (MIL) will illuminate if certain faults are recorded.
- ECM operates in backup mode if sensors fail, to enable vehicle to be driven to repair shop.
- Trouble codes can be displayed by the MIL or by using a Scan Tool connected to the data link connector(s) (DLC).

Accessing

- Ensure ignition switched OFF.
- Access 16 pin data link connector (DLC) **1**.
- Jump data link connector (DLC) terminal 1 and ground **2**.
- Switch ignition ON.
- Count MIL flashes. Note trouble codes. Compare with trouble code table.
- Each trouble code consists of two groups **3** :
  - Long flashes indicate the 'tens' of the trouble code **3** [**A**].
  - Short flashes indicate the 'units' of the trouble code **3** [**D**].
  - A short pause separates each trouble code group **3** [**C**].
  - A long pause separates each trouble code **3** [**E**].
  - For example: Trouble code 22 displayed **3**.
- If MIL remains ON: Suspect faulty engine control module (ECM).
- If constant short flashes are displayed, no trouble code is stored.
- Trouble codes can also be displayed by using a Scan Tool connected to the data link connector(s) (DLC).

1

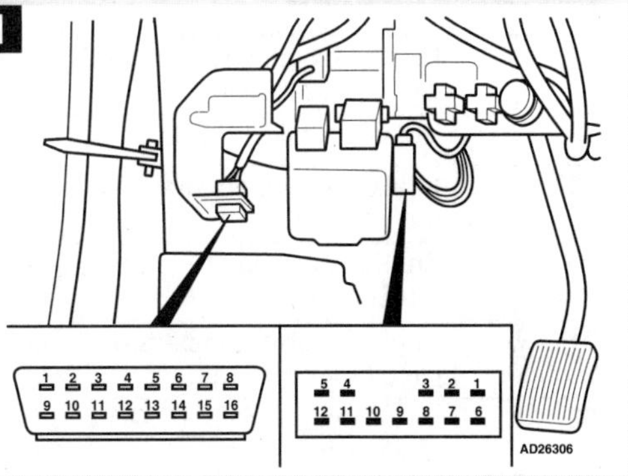

2

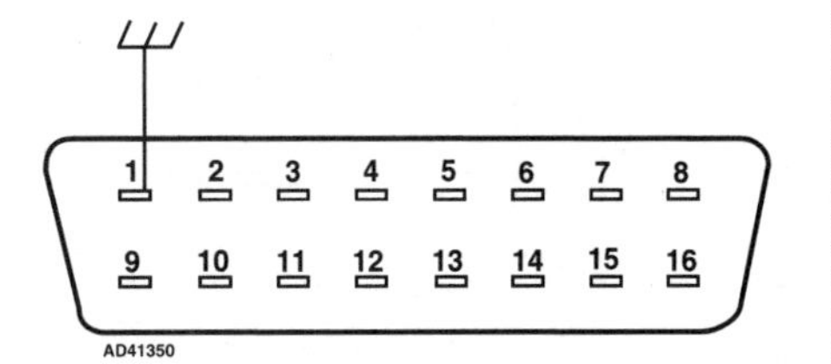

3

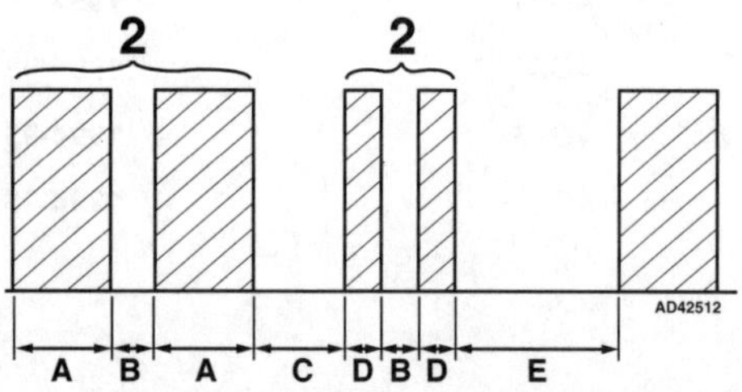

Erasing

NOTE: ***Disconnecting battery ground lead may cause driveability problems until the engine management system relearns basic values.***

- Ensure ignition switched OFF.
- Disconnect battery ground cable for at least 15 seconds.
- Reconnect battery ground lead.
- Repeat checking procedure to ensure no data remains in ECM fault memory.
- Trouble codes can also be erased by using a Scan Tool connected to the data link connector(s) (DLC).

NOTE: ***Disconnecting battery ground lead may erase memory from electronic units such as radio.***

Trouble code identification

Flash code	Fault location	Probable cause
11	Heated oxygen sensor (HO2S) 1, bank 1 – circuit malfunction	**Wiring, fuel pressure, injector(s), intake leak, HO2S, ECM**
12	Volume air flow (VAF) sensor – malfunction	**Wiring, VAF sensor, ECM**
13	Intake air temperature (IAT) sensor – circuit malfunction	**Wiring, IAT sensor, ECM**
14	Throttle position (TP) sensor – circuit malfunction	**Wiring, TP sensor, CTP switch, ECM**
21	Engine coolant temperature (ECT) sensor – circuit malfunction	**Wiring, ECT sensor, ECM**
22	Crankshaft position (CKP) sensor – circuit malfunction	**Wiring, CKP sensor, ECM**
23	Camshaft position (CMP) sensor – circuit malfunction	**Wiring, CMP sensor, ECM**
24	Vehicle speed sensor (VSS) – circuit malfunction	**Wiring, VSS, ECM**
25	Barometric pressure (BARO) sensor – circuit malfunction	**Wiring, BARO sensor, ECM**
36	Ignition timing adjustment	**Wiring, ignition timing connector short to ground, ECM**
41	Injectors – circuit malfunction	**Wiring, injectors, ECM**
43	Exhaust gas recirculation (EGR) system – circuit malfunction	**Wiring, Hose leak/blockage, EGR valve, EGR temperature sensor, EGR solenoid, ECM**
55	Idle air control (IAC) valve position sensor – circuit malfunction	**Wiring, IAC valve, IAC valve position sensor, ECM**
59	Heated oxygen sensor (HO2S) 2, bank 1 – circuit malfunction	**Wiring, HO2S, ECM**

EAGLE

Model:	Engine identification:	Year:
Summit 1.5L	**4G15**	**1995-96**
Summit 1.8L	**4G93**	**1995-96**
Summit Wagon 1.8L FWD – California	**4G93**	**1995-96**
Summit Wagon 2.4L FWD – California	**4G64**	**1995-96**
Summit Wagon 2.4L AWD – California	**4G64**	**1996**
Summit Wagon 1.8L – Federal	**4G93**	**1996**
Summit Wagon 2.4L – Federal	**4G64**	**1996**
Talon 2.0L Turbo	**4G63**	**1995-98**

System: **Mitsubishi MPI**

Self-diagnosis

General information

- Refer to the front of this manual for general test conditions, terminology, detailed descriptions of wiring faults and a general trouble shooter for electrical and mechanical faults.
- Malfunction indicator lamp (MIL) will illuminate or flash if certain faults are recorded.
- ECM operates in backup mode if sensors fail, to enable vehicle to be driven to repair shop.
- 1995 Summit – Federal: Trouble codes can be displayed by the MIL or by using a Scan Tool connected to the data link connector(s) (DLC).

Accessing

Summit – 1995 Federal

Flash codes

- Ensure ignition switched OFF.
- Access 16 pin data link connector (DLC) **1**.
- Jump data link connector (DLC) terminal 1 and ground **2**.
- Switch ignition ON.
- Count MIL flashes. Note trouble codes. Compare with trouble code table.
- Each trouble code consists of two groups **3** :
  - Long flashes indicate the 'tens' of the trouble code **3** [**A**].
  - Short flashes indicate the 'units' of the trouble code **3** [**D**].
  - A short pause separates each trouble code group **3** [**C**].
  - A long pause separates each trouble code **3** [**E**].
  - For example: Trouble code 22 displayed **3**.
- If MIL remains ON: Suspect faulty engine control module (ECM).
- If constant short flashes are displayed, no trouble code is stored.

P codes

- Trouble codes can also be displayed by using a Scan Tool connected to the data link connector(s) (DLC).

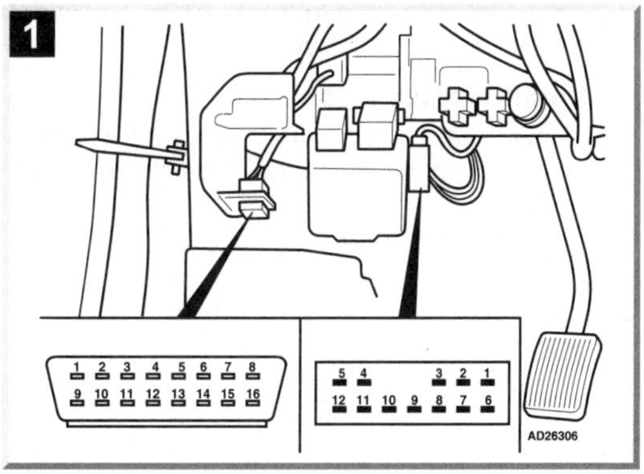

2

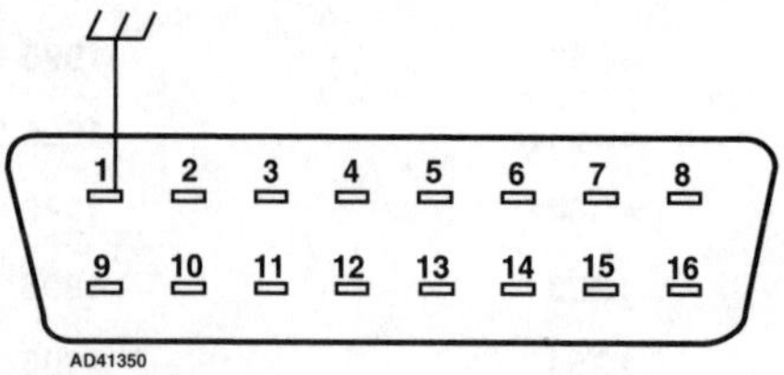

4

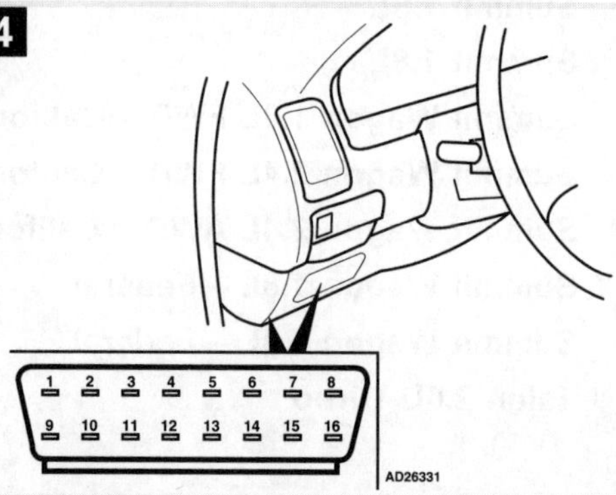

3

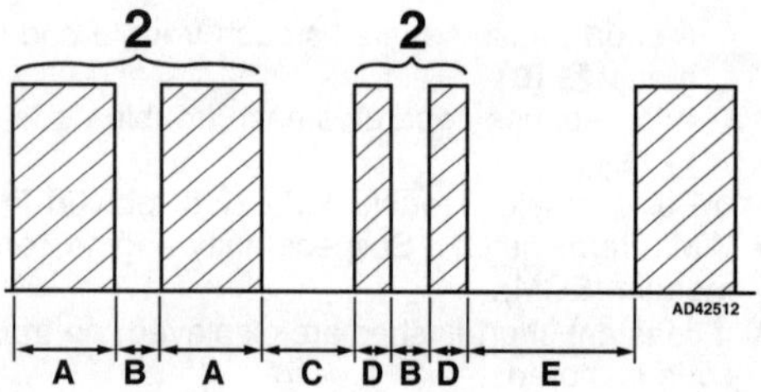

5

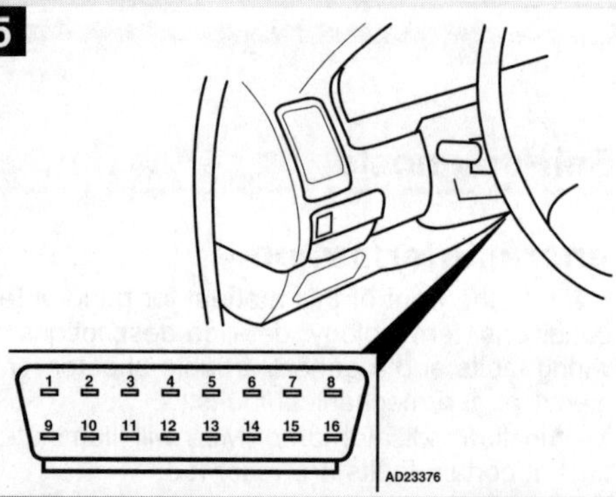

Summit - except 1995 Federal

- Trouble codes can only be displayed by using a Scan Tool connected to the 12 pin or 16 pin data link connectors (DLC) **1**.

Summit Wagon/Talon

- Trouble codes can only be displayed by using a Scan Tool connected to the data link connector (DLC):
  - Summit Wagon **4**.
  - Talon **5**.

Erasing

NOTE: ***Disconnecting battery ground lead may cause driveability problems until the engine management system relearns basic values.***

- Ensure ignition switched OFF.
- Disconnect battery ground cable for at least 15 seconds.
- Reconnect battery ground lead.
- Repeat checking procedure to ensure no data remains in ECM fault memory.
- Trouble codes can also be erased by using a Scan Tool connected to the data link connector(s) (DLC).

NOTE: ***Disconnecting battery ground lead may erase memory from electronic units such as radio.***

Trouble code identification

Scan code P type	Fault location	Probable cause
P0000-P0999	Refer to OBD-II trouble code table at the front of this manual	–
P1103	Turbocharger (TC) wastegate actuator – malfunction	**Hose connection(s), wiring, TC wastegate regulating valve, TC wastegate, ECM**
P1104	Turbocharger (TC) wastegate regulating valve – circuit malfunction	**Wiring, TC wastegate regulating valve, ECM**
P1105	Fuel pressure regulator control solenoid – circuit malfunction	**Wiring, fuel pressure regulator control solenoid, ECM**
P1300	Ignition timing adjustment	**Wiring, ignition timing connector short to ground, ECM**
P1400	Manifold differential pressure (MDP) sensor, EGR system – circuit malfunction	**Wiring, MDP sensor, ECM**
P1443	Evaporative emission (EVAP) canister purge valve 2 – circuit malfunction	**Wiring, EVAP canister purge valve 2, ECM**
P1500	Alternator – circuit malfunction	**Wiring, battery, alternator, ECM**
P1600	CAN data bus, TCM – circuit malfunction	**Wiring, TCM, ECM**
P1715	Transmission speed sensor A/B – circuit malfunction	**Wiring, transmission speed sensor A/B, ECM**
P1750	Torque converter clutch (TCC) solenoid/ transmission shift control valve/transmission pressure control solenoid – circuit malfunction	**Wiring, TCC solenoid/transmission shift control valve/transmission pressure control solenoid, TCM, ECM**
P1791	CAN data bus, TCM – coolant temperature signal	**Wiring, TCM, ECM**

Flash code	Fault location	Probable cause
11	Heated oxygen sensor (HO2S) 1, bank 1 – circuit malfunction	**Wiring, fuel pressure, injector(s), intake leak, HO2S, ECM**
12	Volume air flow (VAF) sensor – malfunction	**Wiring, VAF sensor, ECM**
13	Intake air temperature (IAT) sensor – circuit malfunction	**Wiring, IAT sensor, ECM**
14	Throttle position (TP) sensor – circuit malfunction	**Wiring, TP sensor, CTP switch, ECM**
21	Engine coolant temperature (ECT) sensor – circuit malfunction	**Wiring, ECT sensor, ECM**
22	Crankshaft position (CKP) sensor – circuit malfunction	**Wiring, CKP sensor, ECM**
23	Camshaft position (CMP) sensor – circuit malfunction	**Wiring, CMP sensor, ECM**
24	Vehicle speed sensor (VSS) – circuit malfunction	**Wiring, VSS, ECM**
25	Barometric pressure (BARO) sensor – circuit malfunction	**Wiring, BARO sensor, ECM**

Flash code	Fault location	Probable cause
32	Manifold absolute pressure (MAP) sensor – circuit malfunction	Wiring, MAP sensor, ECM
36	Ignition timing adjustment	Wiring, ignition timing connector short to ground, ECM
41	Injectors – circuit malfunction	Wiring, injectors, ECM
43	Exhaust gas recirculation (EGR) system – circuit malfunction	Wiring, hose leak/blockage, EGR valve, EGR temperature sensor, EGR solenoid, ECM
55	Idle air control (IAC) valve position sensor – circuit malfunction	Wiring, IAC valve, IAC valve position sensor, ECM
59	Heated oxygen sensor (HO2S) 2, bank 1 – circuit malfunction	Wiring, HO2S, ECM

EAGLE

Model: **Talon 2.0L**

Engine identification: **420A**

Year: **1995-96**

System: **SBEC III**

Self-diagnosis

General information

- Refer to the front of this manual for general test conditions, terminology, detailed descriptions of wiring faults and a general trouble shooter for electrical and mechanical faults.
- Malfunction indicator lamp (MIL) will illuminate or flash if certain faults are recorded.
- ECM operates in backup mode if sensors fail, to enable vehicle to be driven to repair shop.
- Trouble codes can be displayed by the MIL or by using a Scan Tool connected to the data link connector(s) (DLC).

Accessing

Flash codes

- Within 5 seconds switch the ignition key ON-OFF-ON-OFF-ON.
- Count MIL flashes. Note trouble codes.
- Each trouble code consists of two groups **1** [**A**].
- The first group of flashes indicate the 'tens' of the trouble code.
- The second group of flashes indicate the 'units' of the trouble code.
- A short pause separates 'tens' and 'units' **1** [**B**].
- A long pause separates each trouble code **1** [**C**].
- For example: Trouble code 21 displayed **1**.

P codes & DRB codes

- Trouble codes can also be displayed by using a Scan Tool connected to the data link connector(s) (DLC) **2**.

1

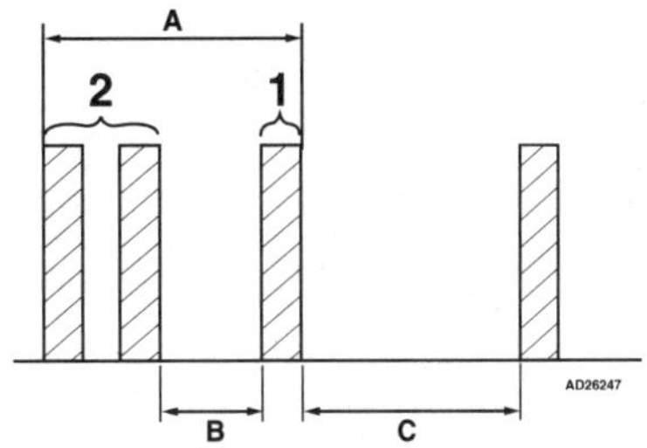

2

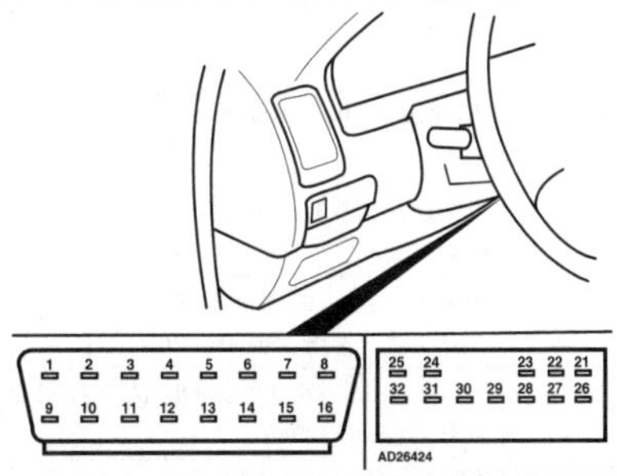

Erasing

NOTE: ***Disconnecting battery ground lead may cause driveability problems until the engine management system relearns basic values.***

- Ensure ignition switched OFF.
- Disconnect battery ground cable for at least 5 minutes.
- Reconnect battery ground cable.
- Repeat accessing procedure to ensure no trouble codes remain in ECM fault memory.

NOTE: ***Disconnecting battery ground lead may also erase memory from electronic units, such as radio.***

- Trouble codes can also be erased by using a Scan Tool connected to the data link connector(s) (DLC).

Trouble code identification

Scan code P type	Fault location	Probable cause
P0000-P0999	Refer to OBD-II trouble code table at the front of this manual	–
P1294	Idle speed control (ISC) system – rpm lower/higher than expected	**Wiring, air leak, throttle valve tight/sticking, IAC valve**
P1295	Throttle position (TP) sensor – no 5 volt supply	**Wiring, TP sensor, ECM**
P1296	Manifold absolute pressure (MAP) sensor – no 5 volt supply	**Wiring, ECM**
P1297	Manifold absolute pressure (MAP) sensor	**Wiring, hose connection(s), MAP sensor**
P1390	Timing belt jumped 1 tooth or more	**Timing belt loose/jumped position**
P1391	Crankshaft position (CKP) sensor/camshaft position (CMP) sensor – intermittent signal	**Wiring, poor connection, air gap, CKP sensor, CMP sensor**
P1486	Evaporative emission (EVAP) system – blockage detected	**Hose connection(s), hose blocked/trapped, EVAP canister**
P1487	Engine coolant blower motor relay – high speed – circuit malfunction	**Wiring open circuit/short circuit, engine coolant blower motor relay – high speed**
P1489	A/C condenser blower motor relay – high speed	**Wiring open/short circuit, engine coolant blower motor relay – high speed**
P1490	Engine coolant blower motor relay – low speed	**Wiring open/short circuit, engine coolant blower motor relay – low speed**
P1492	Ambient/battery temperature sensor – voltage high	**Wiring, ambient/battery temperature sensor**
P1493	Ambient/battery temperature sensor – voltage low	**Wiring, ambient/battery temperature sensor**
P1494	Evaporative emission (EVAP) leak detection pump – pressure switch	**Wiring, hose leak/blockage, EVAP leak detection pump**
P1495	Evaporative emission (EVAP) leak detection pump solenoid	**Wiring open/short circuit, hose leak/blockage, EVAP leak detection pump solenoid**
P1496	Manifold absolute pressure (MAP) sensor – 5 volt output low	**Wiring, MAP sensor**
P1696	Engine control module (ECM) – EEPROM error	**ECM**
P1697	Engine control module (ECM) – EEPROM error	**ECM**
P1698	CAN data bus, transmission control module (TCM) – no signal	**Wiring, TCM**
P1899	Park/neutral position (PNP) switch – input circuit malfunction	**Wiring, PNP switch, ECM/TCM**

Flash code	Scan code DRB III type	Fault location	Probable cause
11		Crankshaft position (CKP) sensor/camshaft position (CMP) sensor	**Wiring, poor connection, air gap, timing belt,CKP sensor, CMP sensor**
	133	Timing belt jumped 1 tooth or more	**Timing belt loose/jumped position**
	157	Crankshaft position (CKP) sensor/camshaft position (CMP) sensor – intermittent signal	**Wiring, poor connection, air gap, CKP sensor, CMP sensor**
	40	Crankshaft position (CKP) sensor – no signal	**Wiring, air gap, CKP sensor**

Flash code	Scan code DRB III type	Fault location	Probable cause
13	39	Manifold absolute pressure (MAP) sensor	**Wiring, hose connection(s), MAP sensor**
14		Manifold absolute pressure (MAP) sensor – input low/high	**Wiring, MAP sensor**
	36	Manifold absolute pressure (MAP) sensor – low input	**Wiring short circuit to ground, MAP sensor**
	37	Manifold absolute pressure (MAP) sensor – high input	**Wiring short circuit to positive, MAP sensor**
	135	Manifold absolute pressure (MAP) sensor – no 5 volt supply	**Wiring, ECM**
	146	Manifold absolute pressure (MAP) sensor – 5 volt output low	**Wiring, MAP sensor**
15	35	Vehicle speed sensor (VSS) – signal	**Wiring, VSS**
16	59	Knock sensor (KS) 1 – no signal	**Wiring, KS incorrectly tightened,KS**
17		Engine temperature – low for long period	**Wiring, cooling system, coolant thermostat, ECT sensor**
	33	Engine temperature – low for long period	**Wiring, cooling system, coolant thermostat, ECT sensor**
	128	Insufficient coolant temperature for closed loop fuel control	**Wiring, cooling system, coolant thermostat, ECT sensor**
21		Heated oxygen sensor (HO2S) 1/2, bank 1 – circuit malfunction/incorrect signal/ voltage low/high	**Wiring, fuel pressure, injector control, EVAP canister purge valve, compression, air filter blocked, HO2S**
	155	Heated oxygen sensor (HO2S) 1, bank 1 – low voltage	**Wiring short circuit to ground, exhaust leak, HO2S**
	62	Heated oxygen sensor (HO2S) 1, bank 1 – high voltage	**Wiring short circuit to positive, fuel pressure, injector control, EVAP canister purge valve, air filter blocked, HO2S**
	102	Heated oxygen sensor (H02S) 1, bank 1 – slow response	**Wiring, heating inoperative, HO2S**
	32	Heated oxygen sensor (H02S) 1, bank 1 – no activity detected	**Wiring open circuit, heating inoperative, HO2S**
	103	Heated oxygen sensor (HO2S) 1, bank 1 – heater control – circuit malfunction	**Wiring, HO2S**
	156	Heated oxygen sensor (HO2S) 2, bank 1 – low voltage	**Wiring short circuit to ground, exhaust leak, HO2S**
	126	Heated oxygen sensor (HO2S) 2, bank 1 – high voltage	**Wiring short circuit to positive, catalytic converter, HO2S**
	129	Heated oxygen sensor (HO2S) 2, bank 1 – no activity detected	**Wiring, heating inoperative, HO2S**
	105	Heated oxygen sensor (HO2S) 2, bank 1, heater control – circuit malfunction	**Wiring, HO2S**
	104	Heated oxygen sensor (HO2S) 2, bank 1 – slow response	**Wiring, heating inoperative, HO2S**
22		Engine coolant temperature (ECT) sensor – voltage low/high	**Wiring, coolant thermostat, ECT sensor**

Flash code	Scan code DRB III type	Fault location	Probable cause
	30	Engine coolant temperature (ECT) sensor – low input	**Wiring short circuit to ground, coolant thermostat, ECT sensor**
	31	Engine coolant temperature (ECT) sensor – high input	**Wiring open circuit/short circuit to positive, coolant thermostat, ground wire defective, ECT sensor**
23		Intake air temperature (IAT) sensor – voltage low/high	**Wiring, IAT sensor**
	57	Intake air temperature (IAT) sensor – low input	**Wiring short circuit to ground, IAT sensor**
	58	Intake air temperature (IAT) sensor – high input	**Wiring open circuit/short circuit to positive, ground wire defective, IAT sensor**
24		Throttle position (TP) sensor – voltage low/ high	**Wiring, accelerator cable adjustment, TP sensor**
	132	Throttle position (TP) sensor – range/ performance problem	**Wiring, accelerator cable adjustment, TP sensor**
	26	Throttle position (TP) sensor – low input	**Wiring short circuit to ground, TP sensor**
	27	Throttle position (TP) sensor – high input	**Wiring short circuit to positive, TP sensor**
	136	Throttle position (TP) sensor – no 5 volt supply	**Wiring, TP sensor, ECM**
25		Idle air control (IAC) valve	**Wiring open/short circuit, air leak, IAC valve**
	25	Idle speed control (ISC) system – malfunction	**Wiring open/short circuit, IAC valve**
	138	Idle speed control (ISC) system – rpm lower/ higher than expected	**Wiring, air leak, throttle valve tight/ sticking, IAC valve**
27		Injector – circuit malfunction	**Wiring, injector**
	21	Injector 1 – circuit malfunction	**Wiring, injector**
	20	Injector 2 – circuit malfunction	**Wiring, injector**
	19	Injector 3 – circuit malfunction	**Wiring, injector**
	61	Injector 4 – circuit malfunction	**Wiring, injector**
31		Evaporative emission (EVAP) canister purge valve – open/short circuit	**Wiring, EVAP canister purge valve**
	113	Evaporative emission (EVAP) system – incorrect flow detected	**Hose connection(s), intake leak, EVAP canister purge valve**
	160	Evaporative emission (EVAP) system – small leak detected	**Hose connection(s), intake leak, EVAP canister, EVAP canister purge valve**
	18	Evaporative emission (EVAP) canister purge valve – circuit malfunction	**Wiring open/short circuit, EVAP canister purge valve**
	161	Evaporative emission (EVAP) system – large leak detected	**Hose connection(s), intake leak, EVAP canister, EVAP canister purge valve**
	187	Evaporative emission (EVAP) system – blockage detected	**Hose connection(s), hose blocked/ trapped, EVAP canister**
	184	Evaporative emission (EVAP) leak detection pump – pressure switch	**Wiring, hose leak/blockage, EVAP leak detection pump**

Flash code	Scan code DRB III type	Fault location	Probable cause
	183	Evaporative emission (EVAP) leak detection pump solenoid	**Wiring open/short circuit, hose leak/ blockage, EVAP leak detection pump solenoid**
32		Exhaust gas recirculation (EGR) solenoid/ system – circuit malfunction	**Wiring, hoses, EGR solenoid**
	46	Exhaust gas recirculation (EGR) system – insufficient flow detected	**Hose leak/blockage, wiring, EGR valve, EGR solenoid**
	17	Exhaust gas recirculation (EGR) – circuit malfunction	**Wiring open/short circuit, EGR solenoid**
33	16	A/C compressor clutch relay – voltage low/ high	**Wiring open/short circuit, A/C compressor clutch relay**
35		Engine coolant blower motor relay/A/C condenser blower motor relay – low/high speed	**Wiring open/short circuit, engine coolant blower motor relay – low/high speed**
	139	Engine coolant blower motor relay – high speed – circuit malfunction	**Wiring open circuit/short circuit, engine coolant blower motor relay – high speed**
	93	A/C condenser blower motor relay – high speed	**Wiring open/short circuit, engine coolant blower motor relay – high speed**
	92	Engine coolant blower motor relay – low speed	**Wiring open/short circuit, engine coolant blower motor relay – low speed**
37	114	Park/neutral position (PNP) switch – input circuit malfunction	**Wiring, PNP switch, ECM/TCM**
41	11	Alternator, field control – circuit malfunction	**Wiring open/short circuit, alternator, battery**
42		Fuel pump relay/engine control relay/fuel level sensor – circuit malfunction	**Wiring open/short circuit, fuel pump relay, engine control relay, fuel level sensor**
	101	Fuel pump relay	**Wiring open/short circuit, fuel pump relay**
	10	Engine control relay	**Wiring open/short circuit, engine control relay**
	44	Engine control relay	**Wiring open circuit, engine control relay**
	149	Fuel level sensor	**Wiring open circuit, fuel level sensor**
	150	Fuel level sensor	**Wiring short circuit to positive, fuel level sensor**
	151	Fuel level sensor	**Wiring, fuel level sensor**
43		Random/multiple cylinder(s) – misfire detected	**Wiring, spark plug(s), HT lead(s), injector(s), ignition coil(s), low compression**
	106	Random/multiple cylinder(s) – misfire detected	**Wiring, spark plug(s), HT lead(s), injector(s), ignition coil(s), low compression**
	107	Cylinder 1 – misfire detected	**Wiring, engine mechanical fault, ignition/ fuel system, injector**
	108	Cylinder 2 – misfire detected	**Wiring, engine mechanical fault, ignition/ fuel system, injector**
	109	Cylinder 3 – misfire detected	**Wiring, engine mechanical fault, ignition/ fuel system, injector**

Flash code	Scan code DRB III type	Fault location	Probable cause
	110	Cylinder 4 – misfire detected	**Wiring, engine mechanical fault, ignition/ fuel system, injector**
	43	Ignition coil A, primary/secondary – circuit malfunction	**Wiring, ignition coil**
	42	Ignition coil B, primary/secondary – circuit malfunction	**Wiring, ignition coil**
44		Ambient/battery temperature sensor – voltage low/high	**Wiring, ambient/battery temperature sensor**
	154	Ambient/battery temperature sensor – voltage high	**Wiring, ambient/battery temperature sensor**
	153	Ambient/battery temperature sensor – voltage low	**Wiring, ambient/battery temperature sensor**
	71	Ambient/battery temperature sensor – range/performance problem	**Wiring, ambient/battery temperature sensor**
45	137	Transmission control system – malfunction	**Wiring, transmission control module (TCM)**
46	06	Alternator – voltage high	**Wiring, battery, alternator**
47	05	Alternator – voltage low	**Wiring, battery, alternator drive belt, alternator**
51	119	System too lean, bank 1	**Intake/exhaust leak, MAP sensor, fuel pressure/pump, injector(s), heated oxygen sensor (HO2S)**
52	118	System too rich, bank 1	**Intake blocked, EVAP canister purge valve, fuel pressure, EGR system, injector(s), heated oxygen sensor (HO2S)**
53		Engine control module (ECM)/CAN data bus – internal fault/malfunction	**Wiring, connected system, ECM**
	02	Engine control module (ECM) – internal fault	**ECM**
	68	CAN data bus – malfunction	**Wiring, connected system, ECM**
54	01	Camshaft position (CMP) sensor – no signal	**Wiring, CMP sensor**
62	48	Engine control module (ECM) – EEPROM error	**ECM**
63	49	Engine control module (ECM) – EEPROM error	**ECM**
64	112	Catalytic converter, bank 1 – efficiency below threshold	**Wiring, exhaust leak, catalytic converter, HO2S 2**
65	115	Power steering pressure (PSP) switch – range/performance problem	**Wiring, PAS system, PSP switch**
66	96	CAN data bus, transmission control module (TCM) – no signal	**Wiring, TCM**

Self-diagnosis

General information

- Refer to the front of this manual for general test conditions, terminology, detailed descriptions of wiring faults and a general trouble shooter for electrical and mechanical faults.
- Malfunction indicator lamp (MIL) will illuminate or flash if certain faults are recorded.
- ECM operates in backup mode if sensors fail, to enable vehicle to be driven to repair shop.
- Trouble codes can be displayed by the MIL or by using a Scan Tool connected to the data link connector(s) (DLC).

Accessing

Flash codes

- Within 5 seconds switch the ignition key ON-OFF-ON-OFF-ON.
- Count MIL flashes. Note trouble codes.
- Each trouble code consists of two groups **1** [**A**].
- The first group of flashes indicate the 'tens' of the trouble code.
- The second group of flashes indicate the 'units' of the trouble code.
- A short pause separates 'tens' and 'units' **1** [**B**].
- A long pause separates each trouble code **1** [**C**].
- For example: Trouble code 21 displayed **1**.

P codes, Hex codes & DRB codes

- Trouble codes can also be displayed by using a Scan Tool connected to the data link connector(s) (DLC) **2**.

1

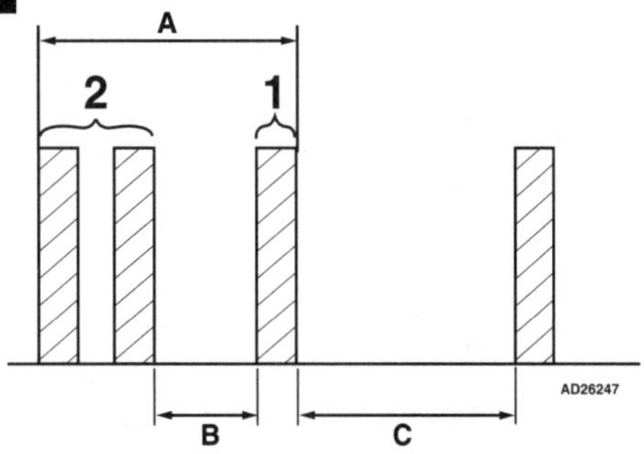

2

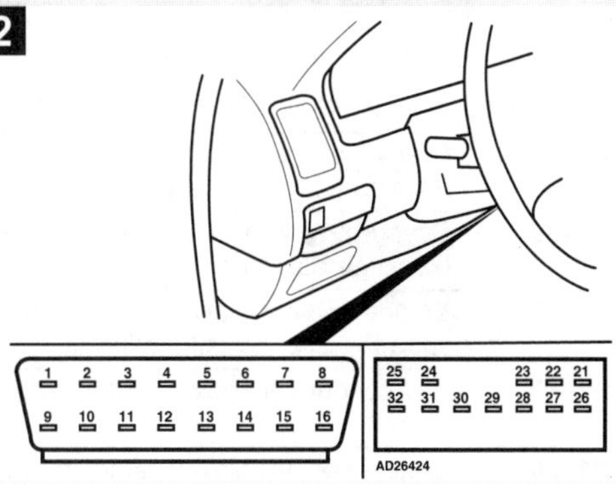

Erasing

NOTE: ***Disconnecting battery ground lead may cause driveability problems until the engine management system relearns basic values.***

- Ensure ignition switched OFF.
- Disconnect battery ground cable for at least 5 minutes.
- Reconnect battery ground cable.
- Repeat accessing procedure to ensure no trouble codes remain in ECM fault memory.

NOTE: ***Disconnecting battery ground lead may also erase memory from electronic units, such as radio.***

- Trouble codes can also be erased by using a Scan Tool connected to the data link connector(s) (DLC).

Trouble code identification

Scan code P type	Fault location	Probable cause
P0000-P0999	Refer to OBD-II trouble code table at the front of this manual	–
P1294	Idle speed control (ISC) system – rpm lower/higher than expected	**Wiring, air leak, throttle valve tight/sticking, IAC valve**
P1295	Throttle position (TP) sensor – no 5 volt supply	**Wiring, TP sensor, ECM**
P1296	Manifold absolute pressure (MAP) sensor – no 5 volt supply	**Wiring, ECM**
P1297	Manifold absolute pressure (MAP) sensor	**Wiring, hose connection(s), MAP sensor**
P1390	Timing belt jumped 1 tooth or more	**Timing belt loose/jumped position**
P1391	Crankshaft position (CKP) sensor/camshaft position (CMP) sensor – intermittent signal	**Wiring, poor connection, air gap, CKP sensor, CMP sensor**
P1486	Evaporative emission (EVAP) system – blockage detected	**Hose connection(s), hose blocked/trapped, EVAP canister**
P1487	Engine coolant blower motor relay – high speed – circuit malfunction	**Wiring open circuit/short circuit, engine coolant blower motor relay – high speed**
P1489	A/C condenser blower motor relay – high speed	**Wiring open/short circuit, engine coolant blower motor relay – high speed**
P1490	Engine coolant blower motor relay – low speed	**Wiring open/short circuit, engine coolant blower motor relay – low speed**
P1492	Ambient/battery temperature sensor – voltage high	**Wiring, ambient/battery temperature sensor**
P1493	Ambient/battery temperature sensor – voltage low	**Wiring, ambient/battery temperature sensor**
P1494	Evaporative emission (EVAP) leak detection pump – pressure switch	**Wiring, hose leak/blockage, EVAP leak detection pump**
P1495	Evaporative emission (EVAP) leak detection pump solenoid	**Wiring open/short circuit, hose leak/blockage, EVAP leak detection pump solenoid**
P1496	Manifold absolute pressure (MAP) sensor – 5 volt output low	**Wiring, MAP sensor**
P1696	Engine control module (ECM) – EEPROM error	**ECM**

Scan code P type	Fault location	Probable cause
P1697	Engine control module (ECM) – EEPROM error	ECM
P1698	CAN data bus, transmission control module (TCM) – no signal	Wiring, TCM
P1899	Park/neutral position (PNP) switch – input circuit malfunction	Wiring, PNP switch, ECM/TCM

Flash code	Scan code Hex type	Scan code DRB III type	Fault location	Probable cause
11			Crankshaft position (CKP) sensor/ camshaft position (CMP) sensor	Wiring, poor connection, air gap, timing belt, CKP sensor, CMP sensor
	85	133	Timing belt jumped 1 tooth or more	Timing belt loose/jumped position
	9D	157	Crankshaft position (CKP) sensor/ camshaft position (CMP) sensor – intermittent signal	Wiring, poor connection, air gap, CKP sensor, CMP sensor
	28	40	Crankshaft position (CKP) sensor – no signal	Wiring, air gap, CKP sensor
13	27	39	Manifold absolute pressure (MAP) sensor	Wiring, hose connection(s), MAP sensor
14			Manifold absolute pressure (MAP) sensor – input low/high	Wiring, MAP sensor
	24	36	Manifold absolute pressure (MAP) sensor – low input	Wiring short circuit to ground, MAP sensor
	25	37	Manifold absolute pressure (MAP) sensor – high input	Wiring short circuit to positive, MAP sensor
	87	135	Manifold absolute pressure (MAP) sensor – no 5 volt supply	Wiring, ECM
	92	146	Manifold absolute pressure (MAP) sensor – 5 volt output low	Wiring, MAP sensor
15	23	35	Vehicle speed sensor (VSS) – signal	Wiring, VSS
16	3B	59	Knock sensor (KS) 1 – no signal	Wiring, KS incorrectly tightened, KS
17			Engine temperature – low for long period	Wiring, cooling system, coolant thermostat, ECT sensor
	21	33	Engine temperature – low for long period	Wiring, cooling system, coolant thermostat, ECT sensor
	80	128	Insufficient coolant temperature for closed loop fuel control	Wiring, cooling system, coolant thermostat, ECT sensor
21			Heated oxygen sensor (HO2S) 1/2, bank 1 – circuit malfunction/incorrect signal/voltage low/high	Wiring, fuel pressure, injector control, EVAP canister purge valve, compression, air filter blocked, HO2S

Flash code	Scan code Hex type	Scan code DRB III type	Fault location	Probable cause
	9B	155	Heated oxygen sensor (HO2S) 1, bank 1 – low voltage	**Wiring short circuit to ground, exhaust leak, HO2S**
	3E	62	Heated oxygen sensor (HO2S) 1, bank 1 – high voltage	**Wiring short circuit to positive, fuel pressure, injector control, EVAP canister purge valve, air filter blocked, HO2S**
	66	102	Heated oxygen sensor (H02S) 1, bank 1 – slow response	**Wiring, heating inoperative, HO2S**
	20	32	Heated oxygen sensor (H02S) 1, bank 1 – no activity detected	**Wiring open circuit, heating inoperative, HO2S**
	67	103	Heated oxygen sensor (HO2S) 1, bank 1 – heater control – circuit malfunction	**Wiring, HO2S**
	9C	156	Heated oxygen sensor (HO2S) 2, bank 1 – low voltage	**Wiring short circuit to ground, exhaust leak, HO2S**
	7E	126	Heated oxygen sensor (HO2S) 2, bank 1 – high voltage	**Wiring short circuit to positive, catalytic converter, HO2S**
	81	129	Heated oxygen sensor (HO2S) 2, bank 1 – no activity detected	**Wiring, heating inoperative, HO2S**
	69	105	Heated oxygen sensor (HO2S) 2, bank 1, heater control – circuit malfunction	**Wiring, HO2S**
22			Engine coolant temperature (ECT) sensor – voltage low/high	**Wiring, coolant thermostat, ECT sensor**
	1E	30	Engine coolant temperature (ECT) sensor – low input	**Wiring short circuit to ground, coolant thermostat, ECT sensor**
	1F	31	Engine coolant temperature (ECT) sensor – high input	**Wiring open circuit/short circuit to positive, coolant thermostat, ground wire defective, ECT sensor**
23			Intake air temperature (IAT) sensor – voltage low/high	**Wiring, IAT sensor**
	39	57	Intake air temperature (IAT) sensor – low input	**Wiring short circuit to ground, IAT sensor**
	3A	58	Intake air temperature (IAT) sensor – high input	**Wiring open circuit/short circuit to positive, ground wire defective, IAT sensor**
24			Throttle position (TP) sensor – voltage low/high	**Wiring, accelerator cable adjustment, TP sensor**
	84	132	Throttle position (TP) sensor – range/ performance problem	**Wiring, accelerator cable adjustment, TP sensor**
	1A	26	Throttle position (TP) sensor – low input	**Wiring short circuit to ground, TP sensor**
	1B	27	Throttle position (TP) sensor – high input	**Wiring short circuit to positive, TP sensor**

Flash code	Scan code Hex type	Scan code DRB III type	Fault location	Probable cause
	88	136	Throttle position (TP) sensor – no 5 volt supply	**Wiring, TP sensor, ECM**
25			Idle air control (IAC) valve	**Wiring open/short circuit, air leak, IAC valve**
	19	25	Idle speed control (ISC) system – malfunction	**Wiring open/short circuit, IAC valve**
	8A	138	Idle speed control (ISC) system – rpm lower/higher than expected	**Wiring, air leak, throttle valve tight/ sticking, IAC valve**
27			Injector – circuit malfunction	**Wiring, injector**
	15	21	Injector 1 – circuit malfunction	**Wiring, injector**
	14	20	Injector 2 – circuit malfunction	**Wiring, injector**
	13	19	Injector 3 – circuit malfunction	**Wiring, injector**
	3D	61	Injector 4 – circuit malfunction	**Wiring, injector**
31			Evaporative emission (EVAP) canister purge valve – open/short circuit	**Wiring, EVAP canister purge valve**
	71	113	Evaporative emission (EVAP) system – incorrect flow detected	**Hose connection(s), intake leak, EVAP canister purge valve**
	A0	160	Evaporative emission (EVAP) system – small leak detected	**Hose connection(s), intake leak, EVAP canister, EVAP canister purge valve**
	12	18	Evaporative emission (EVAP) canister purge valve – circuit malfunction	**Wiring open/short circuit, EVAP canister purge valve**
	A1	161	Evaporative emission (EVAP) system – large leak detected	**Hose connection(s), intake leak, EVAP canister, EVAP canister purge valve**
	BB	187	Evaporative emission (EVAP) system – blockage detected	**Hose connection(s), hose blocked/ trapped, EVAP canister**
	B8	184	Evaporative emission (EVAP) leak detection pump – pressure switch	**Wiring, hose leak/blockage, EVAP leak detection pump**
	B7	183	Evaporative emission (EVAP) leak detection pump solenoid	**Wiring open/short circuit, hose leak/ blockage, EVAP leak detection pump solenoid**
32			Exhaust gas recirculation (EGR) solenoid/system – circuit malfunction	**Wiring, hoses, EGR solenoid**
	2E	46	Exhaust gas recirculation (EGR) system – insufficient flow detected	**Hose leak/blockage, wiring, EGR valve, EGR solenoid**
	11	17	Exhaust gas recirculation (EGR) – circuit malfunction	**Wiring open/short circuit, EGR solenoid**
33	10	16	A/C compressor clutch relay – voltage low/high	**Wiring open/short circuit, A/C compressor clutch relay**
35			Engine coolant blower motor relay/A/C condenser blower motor relay – low/ high speed	**Wiring open/short circuit, engine coolant blower motor relay – low/ high speed**

Flash code	Scan code Hex type	Scan code DRB III type	Fault location	Probable cause
	8B	139	Engine coolant blower motor relay – high speed – circuit malfunction	**Wiring open circuit/short circuit, engine coolant blower motor relay – high speed**
	5D	93	A/C condenser blower motor relay – high speed	**Wiring open/short circuit, engine coolant blower motor relay – high speed**
	5C	92	Engine coolant blower motor relay – low speed	**Wiring open/short circuit, engine coolant blower motor relay – low speed**
37	72	114	Park/neutral position (PNP) switch – input circuit malfunction	**Wiring, PNP switch, ECM/TCM**
41	0B	11	Alternator, field control – circuit malfunction	**Wiring open/short circuit, alternator, battery**
42			Fuel pump relay/engine control relay/ fuel level sensor – circuit malfunction	**Wiring open/short circuit, Fuel pump relay, engine control relay, fuel level sensor**
	65	101	Fuel pump relay	**Wiring open/short circuit, fuel pump relay**
	0A	10	Engine control relay	**Wiring open/short circuit, engine control relay**
	2C	44	Engine control relay	**Wiring open circuit, engine control relay**
	95	149	Fuel level sensor	**Wiring open circuit, fuel level sensor**
	96	150	Fuel level sensor	**Wiring short circuit to positive, fuel level sensor**
	97	151	Fuel level sensor	**Wiring, fuel level sensor**
43			Random/multiple cylinder(s) – misfire detected	**Wiring, spark plug(s), HT lead(s), injector(s), ignition coil(s), low compression**
	6A	106	Random/multiple cylinder(s) – misfire detected	**Wiring, spark plug(s), HT lead(s), injector(s), ignition coil(s), low compression**
	6B	107	Cylinder 1 – misfire detected	**Wiring, engine mechanical fault, ignition/fuel system, injector**
	6C	108	Cylinder 2 – misfire detected	**Wiring, engine mechanical fault, ignition/fuel system, injector**
	6D	109	Cylinder 3 – misfire detected	**Wiring, engine mechanical fault, ignition/fuel system, injector**
	6E	110	Cylinder 4 – misfire detected	**Wiring, engine mechanical fault, ignition/fuel system, injector**
	2B	43	Ignition coil A, primary/secondary – circuit malfunction	**Wiring, ignition coil**
	2A	42	Ignition coil B, primary/secondary – circuit malfunction	**Wiring, ignition coil**

Flash code	Scan code Hex type	Scan code DRB III type	Fault location	Probable cause
44			Ambient/battery temperature sensor – voltage low/high	Wiring, ambient/battery temperature sensor
	9A	154	Ambient/battery temperature sensor – voltage high	Wiring, ambient/battery temperature sensor
	99	153	Ambient/battery temperature sensor – voltage low	Wiring, ambient/battery temperature sensor
	47	71	Ambient/battery temperature sensor – range/performance problem	Wiring, ambient/battery temperature sensor
45	89	137	Transmission control system – malfunction	Wiring, transmission control module (TCM)
46	06	06	Alternator – voltage high	Wiring, battery, alternator
47	05	05	Alternator – voltage low	Wiring, battery, alternator drive belt, alternator
51	77	119	System too lean, bank 1	Intake/exhaust leak, MAP sensor, fuel pressure/pump, injector(s), heated oxygen sensor (HO2S)
52	76	118	System too rich, bank 1	Intake blocked, EVAP canister purge valve, fuel pressure, EGR system, injector(s), heated oxygen sensor (HO2S)
53			Engine control module (ECM)/CAN data bus – internal fault/malfunction	Wiring, connected system, ECM
	02	02	Engine control module (ECM) – internal fault	ECM
	44	68	CAN data bus – malfunction	Wiring, connected system, ECM
54	01	01	Camshaft position (CMP) sensor – no signal	Wiring, CMP sensor
62	30	48	Engine control module (ECM) – EEPROM error	ECM
63	31	49	Engine control module (ECM) – EEPROM error	ECM
64	70	112	Catalytic converter, bank 1 – efficiency below threshold	Wiring, exhaust leak, catalytic converter, HO2S 2
65	73	115	Power steering pressure (PSP) switch – range/performance problem	Wiring, PAS system, PSP switch
66	60	96	CAN data bus, transmission control module (TCM) – no signal	Wiring, TCM

EAGLE

Model:	Engine identification:	Year:
Vision 3.3L	EGB	1996-97
Vision 3.5L	EGE	1996-97

System: **SBEC III**

Self-diagnosis

General information

- Refer to the front of this manual for general test conditions, terminology, detailed descriptions of wiring faults and a general trouble shooter for electrical and mechanical faults.
- Malfunction indicator lamp (MIL) will illuminate or flash if certain faults are recorded.
- ECM operates in backup mode if sensors fail, to enable vehicle to be driven to repair shop.
- Trouble codes can be displayed by the MIL or by using a Scan Tool connected to the data link connector (DLC).

Accessing

Flash codes

- Within 5 seconds switch the ignition key ON-OFF-ON-OFF-ON.
- Count MIL flashes. Note trouble codes.
- Each trouble code consists of two groups **1** [**A**].
- The first group of flashes indicate the 'tens' of the trouble code.
- The second group of flashes indicate the 'units' of the trouble code.
- A short pause separates 'tens' and 'units' **1** [**B**].
- A long pause separates each trouble code **1** [**C**].
- For example: Trouble code 21 displayed **1**.
- End of test sequence indicated by trouble code 55.

P codes & Hex codes

- Trouble codes can be displayed by using a Scan Tool connected to the data link connector (DLC) **2**.

1

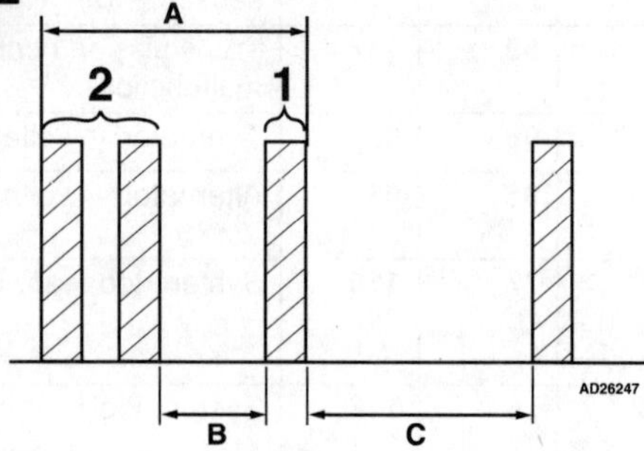

2

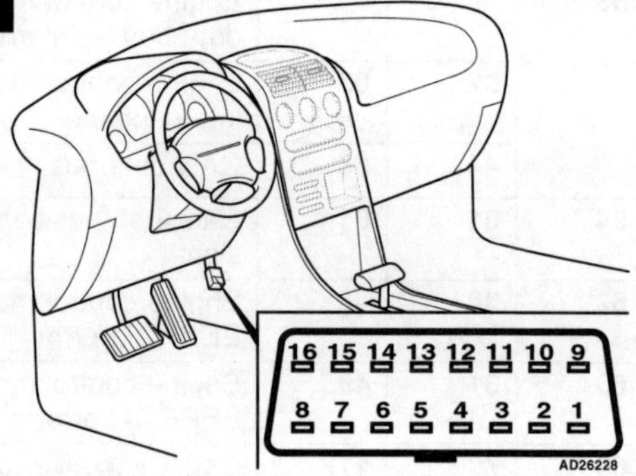

Erasing

- Trouble codes can only be erased by using a Scan Tool connected to the data link connector (DLC).

Trouble code identification

P type	Fault location	Probable cause
P0000-P0999	Refer to OBD-II trouble code table at the front of this manual	–
P1289	Intake manifold air control solenoid	**Wiring open circuit/short circuit, intake manifold air control solenoid**
P1294	Idle speed control (ISC) system – rpm lower/ higher than expected	**Wiring, air leak, throttle valve tight/sticking, IAC valve**
P1296	Manifold absolute pressure (MAP) sensor – no 5 volt supply	**Wiring, ECM**
P1297	Manifold absolute pressure (MAP) sensor	**Wiring, hose connection(s), MAP sensor**
P1299	Manifold absolute pressure (MAP) sensor/ throttle position (TP) sensor	**Intake leak, hose connection(s), MAP sensor, TP sensor**
P1391	Crankshaft position (CKP) sensor/camshaft position (CMP) sensor – intermittent signal	**Wiring, poor connection, air gap, CKP sensor, CMP sensor**
P1398	Engine control module (ECM) – misfire detection limit reached	**Wiring, poor connection, CKP sensor**
P1489	Engine coolant blower motor relay – high speed	**Wiring open circuit/short circuit, engine coolant blower motor relay – high speed**
P1490	Engine coolant blower motor relay – low speed	**Wiring open circuit/short circuit, engine coolant blower motor relay – low speed**
P1492	Ambient/battery temperature sensor – voltage high	**Wiring, ambient/battery temperature sensor**
P1493	Ambient/battery temperature sensor – voltage low	**Wiring, ambient/battery temperature sensor**
P1494	Evaporative emission (EVAP) leak detection pump – pressure switch	**Wiring, hose leak/blockage, EVAP leak detection pump**
P1495	Evaporative emission (EVAP) leak detection pump solenoid	**Wiring open circuit/short circuit, hose leak/ blockage, EVAP leak detection pump solenoid**
P1496	Manifold absolute pressure (MAP) sensor – 5 volt output low	**Wiring, MAP sensor**
P1496	Engine control module (ECM) – 5 volt output low	**ECM**
P1696	Engine control module (ECM) – EEPROM error	**ECM**
P1697	Engine control module (ECM) – EEPROM error	**ECM**
P1698	CAN data bus, transmission control module (TCM) – no signal	**Wiring, TCM**
P1899	Park/neutral position (PNP) switch – input circuit malfunction	**Wiring, PNP switch, ECM/TCM**

Flash code	Scan code Hex type	Fault location	Probable cause
11		Crankshaft position (CKP) sensor/camshaft position (CMP) sensor	**Wiring, poor connection, air gap, CKP sensor, CMP sensor**
	9D	Crankshaft position (CKP) sensor/camshaft position (CMP) sensor – intermittent signal	**Wiring, poor connection, air gap, CKP sensor, CMP sensor**

Flash code	Scan code Hex type	Fault location	Probable cause
	28	Crankshaft position (CKP) sensor – no signal	**Wiring, air gap, CKP sensor**
	BA	Engine control module (ECM) – misfire detection limit reached	**Wiring, poor connection, CKP sensor**
12		Engine control module (ECM) – supply voltage disconnected within last 50 ignition switch ON-OFF cycles	**Wiring, battery**
13	27	Manifold absolute pressure (MAP) sensor	**Wiring, hose connection(s), MAP sensor**
14		Manifold absolute pressure (MAP) sensor – input low/high	**Wiring, MAP sensor**
	24	Manifold absolute pressure (MAP) sensor – low input	**Wiring short circuit to ground, MAP sensor**
	25	Manifold absolute pressure (MAP) sensor – high input	**Wiring short circuit to positive, MAP sensor**
	87	Manifold absolute pressure (MAP) sensor – no 5 volt supply	**Wiring, ECM**
	92	Manifold absolute pressure (MAP) sensor – 5 volt output low	**Wiring, MAP sensor**
15	23	Vehicle speed sensor (VSS) – signal	**Wiring, VSS**
16		Knock sensor (KS) 1/2 – no signal	**Wiring, KS incorrectly tightened, KS**
	3B	Knock sensor (KS) 1 – no signal	**Wiring, KS incorrectly tightened, KS**
	53	Knock sensor (KS) 2 – no signal	**Wiring, KS incorrectly tightened, KS**
17		Engine temperature – low for long period	**Wiring, cooling system, coolant thermostat, ECT sensor**
	21	Engine temperature – low for long period	**Wiring, cooling system, coolant thermostat, ECT sensor**
	80	Insufficient coolant temperature for closed loop fuel control	**Wiring, cooling system, coolant thermostat, ECT sensor**
21		Heated oxygen sensor (HO2S) 1/2, bank 1/2 – circuit malfunction/incorrect signal/ voltage low/high	**Wiring, fuel pressure, injector control, EVAP canister purge valve, compression, air filter blocked, HO2S**
	9B	Heated oxygen sensor (HO2S) 1, bank 1 – low voltage	**Exhaust leak, wiring short circuit to ground, HO2S**
	3E	Heated oxygen sensor (HO2S) 1, bank 1 – high voltage	**Wiring short circuit to positive, fuel pressure, injector control, EVAP canister purge valve, air filter blocked, HO2S**
	66	Heated oxygen sensor (H02S) 1, bank 1 – slow response	**Wiring, heating inoperative, HO2S**
	20	Heated oxygen sensor (H02S) 1, bank 1 – no activity detected	**Wiring open circuit, heating inoperative, HO2S**
	67	Heated oxygen sensor (HO2S) 1, bank 1 – heater control – circuit malfunction	**Wiring, HO2S**
	9C	Heated oxygen sensor (HO2S) 2, bank 1 – low voltage	**Exhaust leak, wiring short circuit to ground, HO2S**
	7E	Heated oxygen sensor (HO2S) 2, bank 1 – high voltage	**Wiring short circuit to positive, catalytic converter, HO2S**

Flash code	Scan code Hex type	Fault location	Probable cause
	81	Heated oxygen sensor (HO2S) 2, bank 1 – no activity detected	Wiring, heating inoperative, HO2S
	69	Heated oxygen sensor (HO2S) 2, bank 1, heater control – circuit malfunction	Wiring, HO2S
	B5	Heated oxygen sensor (HO2S) 1, bank 2 – low voltage	Exhaust leak, wiring short circuit to ground, HO2S
	42	Heated oxygen sensor (HO2S) 1, bank 2 – high voltage	Wiring short circuit to positive, fuel pressure, injector control, EVAP canister purge valve, air filter blocked, HO2S
	7A	Heated oxygen sensor (HO2S) 1, bank 2 – slow response	Wiring, heating inoperative, HO2S
	41	Heated oxygen sensor (HO2S) 1, bank 2 – no activity detected	Wiring, HO2S
	7C	Heated oxygen sensor (HO2S) 1, bank 2, heater control – circuit malfunction	Wiring, HO2S
	B6	Heated oxygen sensor (HO2S) 2, bank 2 – low voltage	Exhaust leak, wiring short circuit to ground, HO2S
	7F	Heated oxygen sensor (HO2S) 2, bank 2 – high voltage	Wiring short circuit to positive, catalytic converter, HO2S
	82	Heated oxygen sensor (HO2S) 2, bank 2 – no activity detected	Wiring, HO2S
	7D	Heated oxygen sensor (HO2S) 2, bank 2, heater control – circuit malfunction	Wiring, HO2S
22		Engine coolant temperature (ECT) sensor – voltage low/high	Wiring, coolant thermostat, ECT sensor
	1E	Engine coolant temperature (ECT) sensor – low input	Coolant thermostat, wiring short circuit to ground, ECT sensor
	1F	Engine coolant temperature (ECT) sensor – high input	Coolant thermostat, wiring open circuit/ short circuit to positive, ground wire defective, ECT sensor
23		Intake air temperature (IAT) sensor – voltage low/high	Wiring, IAT sensor
	39	Intake air temperature (IAT) sensor – low input	Wiring short circuit to ground, IAT sensor
	3A	Intake air temperature (IAT) sensor – high input	Wiring open circuit/short circuit to positive, ground wire defective, IAT sensor
24		Throttle position (TP) sensor – voltage low/ high	Wiring, accelerator cable adjustment, TP sensor
	84	Throttle position (TP) sensor – range/ performance problem	Wiring, accelerator cable adjustment, TP sensor
	1A	Throttle position (TP) sensor – low input	Wiring short circuit to ground, TP sensor
	1B	Throttle position (TP) sensor – high input	Wiring short circuit to positive, TP sensor
25		Idle air control (IAC) valve	Wiring open circuit/short circuit, air leak, IAC valve
	19	Idle speed control (ISC) system – malfunction	Wiring open circuit/short circuit, IAC valve

Flash code	Scan code Hex type	Fault location	Probable cause
	8A	Idle speed control (ISC) system – rpm lower/higher than expected	**Wiring, air leak, throttle valve tight/ sticking, IAC valve**
	91	Manifold absolute pressure (MAP) sensor/ throttle position (TP) sensor	**Intake leak, hose connection(s), MAP sensor, TP sensor**
27		Injector – circuit malfunction	**Wiring, injector**
	15	Injector 1 – circuit malfunction	**Wiring, injector**
	14	Injector 2 – circuit malfunction	**Wiring, injector**
	13	Injector 3 – circuit malfunction	**Wiring, injector**
	3D	Injector 4 – circuit malfunction	**Wiring, injector**
	45	Injector 5 – circuit malfunction	**Wiring, injector**
	46	Injector 6 – circuit malfunction	**Wiring, injector**
31		Evaporative emission (EVAP) canister purge valve – open/short circuit	**Wiring, EVAP canister purge valve**
	71	Evaporative emission (EVAP) system – incorrect flow detected	**Hose connection(s), intake leak, EVAP canister purge valve**
	A0	Evaporative emission (EVAP) system – small leak detected	**Hose connection(s), intake leak, EVAP canister, EVAP canister purge valve**
	12	Evaporative emission (EVAP) canister purge valve – circuit malfunction	**Wiring open circuit/short circuit, EVAP canister purge valve**
	A1	Evaporative emission (EVAP) system – large leak detected	**Hose connection(s), intake leak, EVAP canister, EVAP canister purge valve**
	B8	Evaporative emission (EVAP) leak detection pump – pressure switch	**Wiring, hose leak/blockage, EVAP leak detection pump**
	B7	Evaporative emission (EVAP) leak detection pump solenoid	**Wiring open circuit/short circuit, hose leak/blockage, EVAP leak detection pump solenoid**
32		Exhaust gas recirculation (EGR) solenoid/ system – circuit malfunction	**Wiring, hoses, EGR solenoid**
	2E	Exhaust gas recirculation (EGR) system – insufficient flow detected	**Hose leak/blockage, wiring, EGR valve, EGR solenoid**
	11	Exhaust gas recirculation (EGR) – circuit malfunction	**Wiring open circuit/short circuit, EGR solenoid**
33		A/C compressor clutch relay/A/C refrigerant pressure sensor – voltage low/high	**A/C refrigerant pressure too low/high (cooling fault/incorrectly charged), wiring open circuit/short circuit, A/C compressor clutch relay, A/C refrigerant pressure sensor**
	10	A/C compressor clutch relay	**Wiring open circuit/short circuit, A/C compressor clutch relay**
	5A	A/C refrigerant pressure sensor – high input	**A/C refrigerant pressure too high (cooling fault/incorrectly charged), wiring, A/C refrigerant pressure sensor**
	5B	A/C refrigerant pressure sensor – low input	**A/C refrigerant pressure too low (incorrectly charged), wiring, A/C refrigerant pressure sensor**

Flash code	Scan code Hex type	Fault location	Probable cause
	86	Air conditioning (A/C) – no 5 volt supply	**Wiring, ECM**
34		Cruise control	**Wiring open circuit/short circuit, voltage low/high, cruise control master switch, cruise control vacuum switching valve**
	0F	Cruise control	**Wiring open circuit/short circuit, voltage low/high, cruise control vacuum switching valve**
	56	Cruise control master switch – high voltage	**Wiring, cruise control master switch**
	57	Cruise control master switch – low voltage	**Wiring, cruise control master switch**
35		Engine coolant blower motor relay – low/ high speed	**Wiring open circuit/short circuit, engine coolant blower motor relay – low/high speed**
	5D	Engine coolant blower motor relay – high speed	**Wiring open circuit/short circuit, engine coolant blower motor relay – high speed**
	5C	Engine coolant blower motor relay – low speed	**Wiring open circuit/short circuit, engine coolant blower motor relay – low speed**
37		Torque converter clutch (TCC) solenoid/ park/neutral position (PNP) switch – circuit malfunction	**Wiring, PNP switch, TCC solenoid**
	94	Torque converter clutch (TCC) solenoid – circuit malfunction	**Wiring, TCC solenoid**
	0C	Torque converter clutch (TCC) solenoid – electrical	**Wiring open circuit/short circuit, TCC solenoid**
	72	Park/neutral position (PNP) switch – input circuit malfunction	**Wiring, PNP switch, ECM/TCM**
41	0B	Alternator, field control – circuit malfunction	**Wiring open circuit/short circuit, alternator, battery**
42		Fuel pump relay/engine control relay/fuel level sensor – circuit malfunction	**Wiring open circuit/short circuit, fuel pump relay, engine control relay, fuel level sensor**
	65	Fuel pump relay	**Wiring open circuit/short circuit, fuel pump relay**
	0A	Engine control relay	**Wiring open circuit/short circuit, engine control relay**
	2C	Engine control relay	**Wiring open circuit, engine control relay**
	95	Fuel level sensor	**Wiring open circuit, fuel level sensor**
	96	Fuel level sensor	**Wiring short circuit to positive, fuel level sensor**
	97	Fuel level sensor	**Wiring, fuel level sensor**
43		Random/multiple cylinder(s) – misfire detected	**Wiring, spark plug(s), HT lead(s), injector(s), ignition coil(s), low compression**
	6A	Random/multiple cylinder(s) – misfire detected	**Wiring, spark plug(s), HT lead(s), injector(s), ignition coil(s), low compression**

Flash code	Scan code Hex type	Fault location	Probable cause
	6B	Cylinder 1 – misfire detected	**Wiring, engine mechanical fault, ignition/ fuel system, injector**
	6C	Cylinder 2 – misfire detected	**Wiring, engine mechanical fault, ignition/ fuel system, injector**
	6D	Cylinder 3 – misfire detected	**Wiring, engine mechanical fault, ignition/ fuel system, injector**
	6E	Cylinder 4 – misfire detected	**Wiring, engine mechanical fault, ignition/ fuel system, injector**
	AE	Cylinder 5 – misfire detected	**Wiring, engine mechanical fault, ignition/ fuel system, injector**
	AF	Cylinder 6 – misfire detected	**Wiring, engine mechanical fault, ignition/ fuel system, injector**
	2B	Ignition coil A, primary/secondary – circuit malfunction	**Wiring, ignition coil**
	2A	Ignition coil B, primary/secondary – circuit malfunction	**Wiring, ignition coil**
	29	Ignition coil C, primary/secondary – circuit malfunction	**Wiring, ignition coil**
44		Ambient/battery temperature sensor – voltage low/high	**Wiring, ambient/battery temperature sensor**
	9A	Ambient/battery temperature sensor – voltage high	**Wiring, ambient/battery temperature sensor**
	99	Ambient/battery temperature sensor – voltage low	**Wiring, ambient/battery temperature sensor**
45	89	Transmission control system – malfunction	**Wiring, transmission control module (TCM)**
46	06	Alternator – voltage high	**Wiring, battery, alternator**
47	05	Alternator – voltage low	**Wiring, battery, alternator drive belt, alternator**
51		System too lean	**Wiring, intake/exhaust leak, fuel pressure, injector control, EVAP canister purge valve, heated oxygen sensor (HO2S)**
	77	System too lean, bank 1	**Intake/exhaust leak, MAP sensor, fuel pressure/pump, injector(s), heated oxygen sensor (HO2S)**
	79	System too lean, bank 2	**Intake/exhaust leak, MAP sensor, fuel pressure/pump, injector(s), heated oxygen sensor (HO2S)**
52		System too rich	**Wiring, exhaust blockage, fuel pressure, injector control, EVAP canister purge valve, compression, air filter blocked, ignition system, heated oxygen sensor (HO2S)**
	76	System too rich, bank 1	**Intake blocked, EVAP canister purge valve, fuel pressure, EGR system, injector(s), heated oxygen sensor (HO2S)**

Flash code	Scan code Hex type	Fault location	Probable cause
	78	System too rich, bank 2	Intake blocked, EVAP canister purge valve, fuel pressure, EGR system, injector(s), heated oxygen sensor (HO2S)
53		Engine control module (ECM) – internal fault	ECM
	02	Engine control module (ECM) – internal fault	ECM
	44	Engine control module (ECM) – internal fault	ECM
54	01	Camshaft position (CMP) sensor – no signal	Wiring, CMP sensor
55		End of MIL code output	–
61	3C	Manifold absolute pressure (MAP) sensor – barometric pressure range/ performance problem	Intake/exhaust leak, wiring, MAP sensor
62	30	Engine control module (ECM) – EEPROM error	ECM
63	31	Engine control module (ECM) – EEPROM error	ECM
65		Power steering pressure (PSP) switch/ brake pedal position (BPP) switch/Intake manifold air control solenoid	Wiring, PAS system, PSP switch, BPP switch, intake manifold air control solenoid
	73	Power steering pressure (PSP) switch – range/performance problem	Wiring, PAS system, PSP switch
	98	Brake pedal position (BPP) switch B – circuit malfunction	Wiring, BPP switch
	58	Intake manifold air control solenoid	Wiring open circuit/short circuit, intake manifold air control solenoid
66		CAN data bus, transmission control module (TCM)/body control module (BCM) – no signal	Wiring, TCM, BCM
	60	CAN data bus, transmission control module (TCM) – no signal	Wiring, TCM
	61	CAN data bus, body control module (BCM) – no signal	Wiring, BCM
71	92	Engine control module (ECM) – 5 volt output low	ECM
72		Catalytic converter system, bank 1/2 – efficiency below threshold	Catalytic converter, wiring, heated oxygen sensor (HO2S) 2
	70	Catalytic converter system, bank 1 – efficiency below threshold	Catalytic converter, wiring, heated oxygen sensor (HO2S) 2
	B4	Catalytic converter system, bank 2 – efficiency below threshold	Catalytic converter, wiring, heated oxygen sensor (HO2S) 2
77	52	Cruise control	Wiring, cruise control vacuum switching valve

FORD

Model:	Engine identification:	Year:
Aerostar/Bronco 3.0/4.0/5.0/5.8L	**VIN position 8 = H/N/U/X**	**1996-97**
Econoline/Club Wagon 4.2/4.6/4.9/ 5.0/5.4/6.8L	**VIN position 8 = L/M/N/S/W/Y/Z/2/6**	**1996-02**
Econoline/Club Wagon/F-Series 5.8/7.5L (with 16 pin DLC)	**VIN position 8 = G/H**	**1996-97**
Econoline/Club Wagon/F-Series 7.3L Turbo Diesel	**VIN position 8 = F**	**1995-02**
Escape/Excursion 2.0/3.0/5.4/6.8L	**VIN position 8 = B/L/S/1**	**2000-02**
Excursion 7.3L Turbo Diesel	**VIN position 8 = F**	**2000-02**
Expedition/Explorer 4.0/4.6/5.0/5.4L	**VIN position 8 = A/E/K/L/M/P/W/X/6**	**1996-02**
F-Series 4.2/4.6/4.9/5.0/5.4/6.8L	**VIN position 8 = L/M/N/S/W/Y/Z/2/3/6/9**	**1996-02**
Ranger 2.3/2.5/3.0/4.0L	**VIN position 8 = A/C/E/U/V/X**	**1995-02**
Windstar 3.0/3.8L	**VIN position 8 = U/4**	**1995-02**

System: **Ford EEC-V**

Self-diagnosis

General information

- Refer to the front of this manual for general test conditions, terminology, detailed descriptions of wiring faults and a general trouble shooter for electrical and mechanical faults.
- Malfunction indicator lamp (MIL) will illuminate or flash if certain faults are recorded.
- ECM operates in backup mode if sensors fail, to enable vehicle to be driven to repair shop.

Accessing

- Trouble codes can only be displayed by using a Scan Tool connected to the data link connector (DLC):
  - Bronco **1**.
  - Econoline/Club Wagon/Escape/Expedition/Explorer/F-Series 4.2L/4.6L/5.4L SOHC/Ranger/Windstar **2**.
  - Aerostar/Excursion/F-Series 4.9L/5.0L/5.8L/6.8L/7.5L/7.3L Turbo Diesel **3**.

1

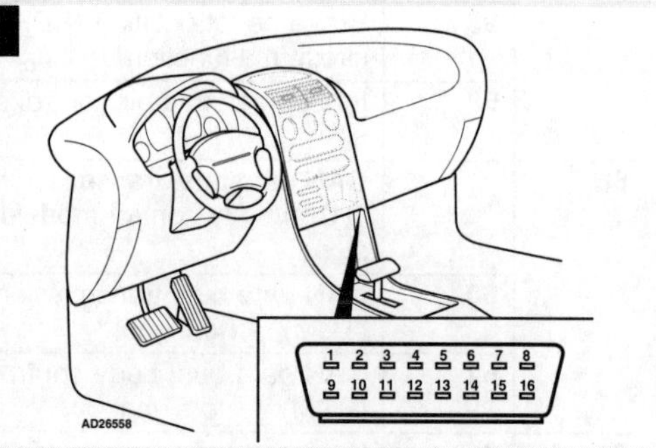

2

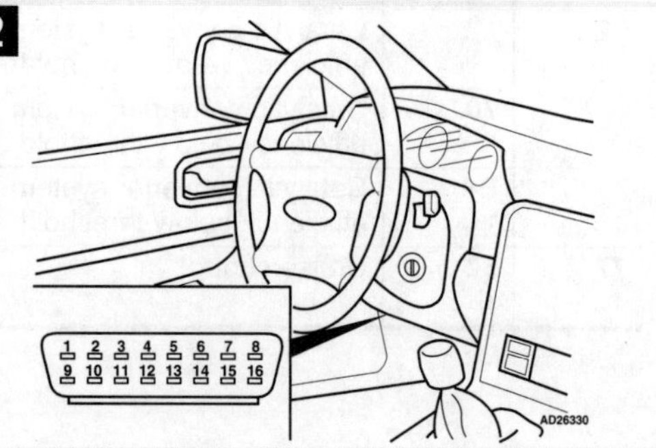

3

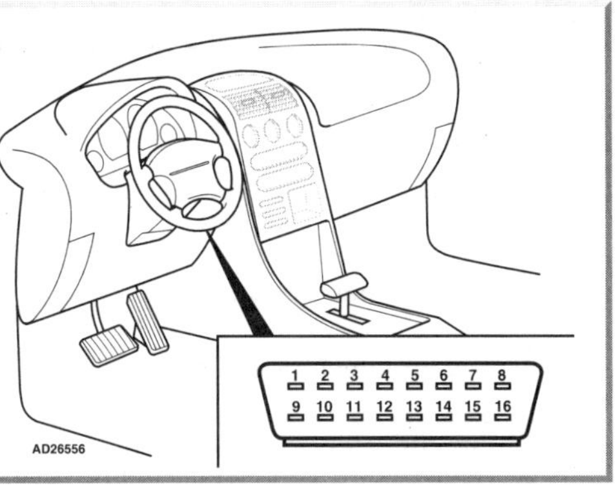

Erasing

- Trouble codes can only be erased by using a Scan Tool connected to the data link connector (DLC).

Trouble code identification

Scan code	Fault location	Probable cause
P0000-P0999	Refer to OBD-II trouble code table at the front of this manual	–
P1000	System readiness test not complete	–
P1001	Self Test	–
P1098	Engine oil temperature (EOT) sensor – low input	**Wiring open circuit/short circuit to ground, EOT sensor, ECM**
P1100	Mass air flow (MAF) sensor – circuit intermittent	**Wiring, MAF sensor**
P1101	Mass air flow (MAF) sensor – signal malfunction	**Wiring, battery, MAF sensor, ECM**
P1109	Intake air temperature (IAT) sensor 2 – circuit intermittent	**Wiring, IAT sensor, ECM**
P1111	No fault found	–
P1112	Intake air temperature (IAT) – circuit intermittent	**Wiring, IAT sensor, ECM**
P1114	Intake air temperature (IAT) sensor 2 – low input	**Wiring short circuit to ground, IAT sensor, ECM**
P1115	Intake air temperature (IAT) sensor 2 – high input	**Wiring, IAT sensor, ECM**
P1116 [1]	Cylinder head temperature (CHT) sensor – out of self-test range	**Wiring, cooling system fault, CHT sensor**
P1116 [2]	Engine coolant temperature (ECT) sensor – signal malfunction	**Wiring, coolant thermostat, cooling system, ECT sensor, ECM**
P1117 [1]	Cylinder head temperature (CHT) sensor – out of self-test range	**Wiring, cooling system fault, CHT sensor**
P1117 [2]	Engine coolant temperature (ECT) sensor – circuit intermittent	**Wiring, ECT sensor, cooling system, ECM**
P1120	Throttle position (TP) sensor – signal malfunction, low	**Wiring, TP sensor, ECM**
P1121	Throttle position (TP) sensor – inconsistent with MAF sensor signal	**Wiring, MAF sensor, TP sensor, air leak**

Scan code	Fault location	Probable cause
P1124	Throttle position (TP) sensor – signal malfunction	Mechanical fault, TP sensor, ECM
P1125	Throttle position (TP) sensor – circuit intermittent	Wiring, TP sensor
P1127	Self test not completed – heated oxygen sensor (HO2S), not tested	Engine not running long enough prior to self-test
P1128	Heated oxygen sensor (HO2S) 1, bank 1/bank 2 – interchanged	Wiring, connectors swapped
P1129	Heated oxygen sensor (HO2S) 2, bank 1/bank 2 – interchanged	Wiring, connectors swapped
P1130	Heated oxygen sensor (HO2S) 1, bank 1 – fuel trim (FT), at limit	Wiring, intake/fuel system, HO2S, EGR system, oil level, camshaft timing, cylinder compression, ECM
P1131	Heated oxygen sensor (HO2S) 1, bank 1 – not switching, fuel trim (FT) weak mixture	Wiring, intake/fuel system, HO2S, EGR system, oil level, camshaft timing, cylinder compression, ECM
P1132	Heated oxygen sensor (HO2S) 1, bank 1 – not switching, fuel trim (FT) rich mixture	Wiring, intake/fuel system, HO2S, EGR system, oil level, camshaft timing, cylinder compression, ECM
P1137	Heated oxygen sensor (HO2S) 2, bank 1 – not switching, fuel trim (FT) weak mixture	Wiring, exhaust system leaking, HO2S
P1138	Heated oxygen sensor (HO2S) 2, bank 1 – not switching, fuel trim (FT) rich mixture	Wiring, exhaust system leaking, HO2S
P1150	Heated oxygen sensor (HO2S) 1, bank 2 – fuel trim (FT), at limit	Wiring, intake/fuel system, HO2S, EGR system, oil level, camshaft timing, cylinder compression, ECM
P1151	Heated oxygen sensor (HO2S) 1, bank 2 – not switching, fuel trim (FT) weak mixture	Wiring, intake/fuel system, HO2S, EGR system, oil level, camshaft timing, cylinder compression, ECM
P1152	Heated oxygen sensor (HO2S) 1, bank 2 – not switching, fuel trim (FT) rich mixture	Wiring, intake/fuel system, HO2S, EGR system, oil level, camshaft timing, cylinder compression, ECM
P1157	Heated oxygen sensor (HO2S) 2, bank 2 – not switching, fuel trim (FT) weak mixture	Wiring, exhaust system leaking, HO2S
P1158	Heated oxygen sensor (HO2S) 2, bank 2 – not switching, fuel trim (FT) rich mixture	Wiring, exhaust system leaking, HO2S
P1168	Fuel rail pressure sensor – signal low	Wiring, fuel level, low fuel pressure, fuel rail pressure sensor
P1169	Fuel rail pressure sensor – signal high	Wiring, high fuel pressure, fuel rail pressure sensor
P1180	Fuel delivery system – pressure low	Fuel filter blocked, fuel supply blocked
P1181	Fuel delivery system – pressure high	Fuel pressure regulator
P1183	Engine oil temperature sensor – circuit malfunction	Wiring, engine oil temperature sensor, ECM
P1184	Engine oil temperature sensor – out of self-test range	Wiring, engine oil temperature sensor, ECM

Scan code	Fault location	Probable cause
P1211	Injector control pressure – too high/low	**Injector(s), high pressure fuel pump, fuel pressure regulator control solenoid**
P1212	Fuel pressure sensor – signal too high/low	**Wiring, fuel pressure sensor**
P1218	Cylinder identification – signal failure	**Injector control module, ECM**
P1219	Cylinder identification – signal failure	**Injector control module, ECM**
P1220	Throttle control system (traction control) – malfunction	**Wiring, throttle position motor, mechanical fault, TP sensor, ECM**
P1224	Throttle position (TP) sensor (traction control), voltage out of range	**Wiring, mechanical fault, TP sensor**
P1229	Supercharger (SC) intercooler pump	**Wiring, SC intercooler pump, SC intercooler pump relay, ECM**
P1230	Fuel pump/variable load control module – circuit intermittent	**Wiring**
P1231	Fuel pump – circuit malfunction	**Wiring, fuel pump relay**
P1232	Low speed fuel pump – malfunction	**Wiring, low speed fuel pump relay, ECM**
P1233	Fuel system – malfunction	**Wiring, IFS switch, fuel pump, fuel pump control module, ECM**
P1234	Fuel system – malfunction	**Wiring, IFS switch, fuel pump, fuel pump control module, ECM**
P1235	Fuel pump control – signal malfunction	**Wiring, fuel pump, fuel pump control module, ECM**
P1236	Fuel pump control – signal malfunction	**Wiring, fuel pump, fuel pump control module, ECM**
P1237	Fuel pump – circuit malfunction	**Wiring, fuel pump, fuel pump relay, fuel pump control module**
P1238	Fuel pump – circuit malfunction	**Wiring, fuel pump, fuel pump relay, fuel pump control module**
P1244	Alternator – low input	**Wiring, alternator, drive belt, ECM**
P1245	Alternator – high input	**Wiring, alternator, ECM**
P1246	Alternator	**Wiring, alternator, drive belt, ECM**
P1260	Theft detected – vehicle immobilized	**–**
P1261	Injector circuit, cylinder 1 – high to low side short	**Wiring short circuit**
P1262	Injector circuit, cylinder 2 – high to low side short	**Wiring short circuit**
P1263	Injector circuit, cylinder 3 – high to low side short	**Wiring short circuit**
P1264	Injector circuit, cylinder 4 – high to low side short	**Wiring short circuit**
P1265	Injector circuit, cylinder 5 – high to low side short	**Wiring short circuit**
P1266	Injector circuit, cylinder 6 – high to low side short	**Wiring short circuit**
P1267	Injector circuit, cylinder 7 – high to low side short	**Wiring short circuit**
P1268	Injector circuit, cylinder 8 – high to low side short	**Wiring short circuit**
P1270	Maximum engine RPM or vehicle speed limit reached	**Engine over revved in neutral, wheel slippage**
P1271	Injector circuit, cylinder 1 – high to low side open	**Wiring open circuit**
P1272	Injector circuit, cylinder 2 – high to low side open	**Wiring open circuit**

Scan code	Fault location	Probable cause
P1273	Injector circuit, cylinder 3 – high to low side open	**Wiring open circuit**
P1274	Injector circuit, cylinder 4 – high to low side open	**Wiring open circuit**
P1275	Injector circuit, cylinder 5 – high to low side open	**Wiring open circuit**
P1276	Injector circuit, cylinder 6 – high to low side open	**Wiring open circuit**
P1277	Injector circuit, cylinder 7 – high to low side open	**Wiring open circuit**
P1278	Injector circuit, cylinder 8 – high to low side open	**Wiring open circuit**
P1280	Fuel pressure sensor – low input	**Wiring, biased sensor, ECM**
P1281	Fuel pressure sensor – high input	**Wiring, biased sensor, ECM**
P1282	Injector control pressure – too high	**Wiring, fuel pressure regulator control solenoid**
P1283	Fuel pressure regulator control solenoid – circuit malfunction	**Wiring open circuit/short circuit to positive**
P1284	Self test not complete – fuel pressure sensor failure	**Wiring, biased sensor, ECM**
P1285	Cylinder head temperature (CHT) sensor – overheat condition	**Wiring, cooling system fault, CHT sensor**
P1288	Cylinder head temperature (CHT) sensor – out of self-test range	**Wiring, cooling system fault, CHT sensor**
P1289	Cylinder head temperature (CHT) sensor – high input	**Wiring, cooling system fault, CHT sensor**
P1290	Cylinder head temperature (CHT) sensor – low input	**Wiring, cooling system fault, CHT sensor**
P1291	Injector circuit, right cylinder bank – circuit malfunction	**High side short circuit to ground/positive**
P1292	Injector circuit, left cylinder bank – circuit malfunction	**High side short circuit to ground/positive**
P1293	Injector circuit, left cylinder bank – circuit malfunction	**High side open circuit**
P1294	Injector circuit, right cylinder bank – circuit malfunction	**High side open circuit**
P1295	Multiple faults, right cylinder bank	**Wiring short circuit to ground**
P1296	Multiple faults, left cylinder bank	**Wiring short circuit to ground**
P1297	Injector circuit – malfunction	**Wiring short circuit, injector control module**
P1298	Injector control module – internal fault	**Injector control module**
P1299	Cylinder head temperature (CHT) sensor – over temperature condition	**Wiring, cooling system fault, CHT sensor**
P1309	Misfire monitor disabled	**CMP sensor, CKP sensor, ECT sensor, MAF sensor, ECM**
P1351 3	Ignition system diagnostic monitor – input circuit malfunction	**Wiring, ICM, ECM**
P1351 4	Ignition system diagnostic monitor – input circuit malfunction	**Wiring, ECM**
P1352	Ignition coil 1, primary – circuit malfunction	**Wiring, spark plug(s), ignition coil, ICM**
P1353	Ignition coil 2, primary – circuit malfunction	**Wiring, spark plug(s), ignition coil, ICM**

Scan code	Fault location	Probable cause
P1354	Ignition coil 3, primary – circuit malfunction	**Wiring, spark plug(s), ignition coil, ICM**
P1355	Ignition coil 4, primary – circuit malfunction	**Wiring, spark plug(s), ignition coil, ICM**
P1356	Ignition system diagnostic monitor – indicates engine not turning	**Wiring, ignition coil, ICM, ECM**
P1358 3	Ignition system diagnostic monitor – out of self-test range	**Wiring, ignition coil, ICM**
P1358 4	Ignition system diagnostic monitor – out of self-test range	**Wiring, ignition coil, ECM**
P1359 3	Ignition system – spark output (SPOUT) signal – circuit malfunction	**Wiring, ICM, ECM**
P1359 4	Ignition system – spark output (SPOUT) signal – circuit malfunction	**Wiring, ECM**
P1364	Ignition coil, primary – circuit malfunction	**Wiring, spark plug(s), ignition coil, ICM**
P1380	Camshaft position (CMP) actuator – circuit malfunction	**Wiring, CMP actuator, ECM**
P1381	Camshaft position (CMP) actuator, bank 1 – timing over advanced	**Wiring, timing belt/chain, CMP actuator, low oil pressure**
P1383	Camshaft position (CMP) actuator, bank 1 – timing over retarded	**Wiring, timing belt/chain, CMP actuator**
P1390	Octane adjustment	**Octane adjustment connector, ECM**
P1400 5	Exhaust gas recirculation (EGR) valve position sensor – low input	**Wiring, EGR solenoid, EGR valve position sensor, ECM**
P1400 6	Exhaust gas recirculation (EGR) pressure sensor – circuit malfunction, low input	**Wiring, EGR pressure sensor, ECM**
P1401 5	Exhaust gas recirculation (EGR) valve position sensor – high input	**Wiring, EGR solenoid, EGR valve position sensor, ECM**
P1401 6	Exhaust gas recirculation (EGR) pressure sensor – circuit malfunction, high input	**Wiring, EGR pressure sensor, ECM**
P1403	Exhaust gas recirculation (EGR) pressure sensor	**Pressure hoses reversed**
P1405	Exhaust gas recirculation (EGR) system – upstream hose off or plugged	**EGR hoses**
P1406	Exhaust gas recirculation (EGR) system – downstream hose off or blocked	**EGR hoses**
P1407	Exhaust gas recirculation (EGR) solenoid	**Wiring, hose leak/blockage, EGR solenoid, EGR valve position sensor, ECM**
P1408	Exhaust gas recirculation (EGR) system – out of self-test range	**Wiring, EGR hoses, EGR valve, EGR solenoid, ECM**
P1409	Exhaust gas recirculation (EGR) solenoid – circuit malfunction	**Wiring, EGR solenoid, ECM**
P1411	Secondary air injection (AIR) system – malfunction	**AIR hoses, AIR pump, AIR solenoid**
P1413	Secondary air injection monitor circuit – low input	**Wiring, AIR solenoid, AIR pump, ECM**
P1414	Secondary air injection monitor circuit – high input	**Wiring, AIR solenoid, ECM**

Scan code	Fault location	Probable cause
P1442	Evaporative emission (EVAP) system – small leak detected	Hose connection(s), intake leak, EVAP canister, EVAP canister purge valve
P1443	Evaporative emission (EVAP) system – malfunction	Wiring, EVAP system hoses, EVAP canister, EVAP canister purge valve
P1444	Evaporative emission (EVAP) system – low input	Wiring, EVAP pressure sensor, ECM
P1445	Evaporative emission (EVAP) system – high input	Wiring, EVAP pressure sensor, ECM
P1450	Evaporative emission (EVAP) system, vent control – circuit malfunction	Wiring, EVAP system hoses, EVAP canister purge valve, fuel filler cap blocked
P1451	Evaporative emission (EVAP) canister purge valve – circuit malfunction	Wiring, EVAP canister purge valve, ECM
P1452	Evaporative emission (EVAP) system – unable to bleed fuel tank vacuum	Wiring, EVAP system hoses, EVAP canister purge valve, fuel filler cap blocked
P1455	Evaporative emission (EVAP) system – malfunction	Fuel filler cap blocked, EVAP system hoses
P1460	Wide open throttle (WOT) relay, A/C – circuit malfunction	Wiring, WOT relay, ECM
P1461	A/C refrigerant pressure sensor – high voltage	Wiring, A/C refrigerant pressure sensor, ECM
P1462	A/C refrigerant pressure sensor – low voltage	Wiring, A/C refrigerant pressure sensor, ECM
P1463	A/C refrigerant pressure sensor – insufficient pressure change	Wiring, A/C system – mechanical fault, A/C refrigerant pressure sensor, A/C compressor clutch
P1464	A/C demand out	Wiring, A/C ON – during self-test, WOT relay – (A/C cut-out), ECM
P1469	Frequent A/C compressor clutch cycling	Wiring, A/C system – mechanical fault
P1473	Engine coolant blower motor – circuit malfunction	Wiring open circuit/short circuit to positive, engine coolant blower motor, variable load control module
P1474	Engine coolant blower motor (low speed) – control circuit malfunction	Wiring, engine coolant blower motor relay, ECM
P1477	Engine coolant blower motor (medium speed) – control circuit malfunction	Wiring, engine coolant blower motor relay, ECM
P1479	Engine coolant blower motor (high speed) – control circuit malfunction	Wiring, engine coolant blower motor relay, ECM
P1480	Engine coolant blower motor (low speed) – with low speed fan ON	Wiring, engine coolant blower motor relay
P1481	Engine coolant blower motor (low speed) – with high speed fan ON	Wiring, engine coolant blower motor relay
P1483	Engine coolant blower motor – intermittent fault	Wiring, engine coolant blower motor, variable load control module
P1484	Variable load control module – malfunction	Wiring open circuit, variable load control module
P1500	Vehicle speed sensor (VSS) – circuit intermittent	Wiring, VSS, ECM

Scan code	Fault location	Probable cause
P1501 7	Vehicle moved during self-test	**Repeat test**
P1501 8	Vehicle speed sensor (VSS) – out of self-test range	**Electro-magnetic interference**
P1502	Vehicle speed sensor (VSS) – intermittent signal	**Wiring, VSS/ABS wheel speed sensors**
P1504	Idle air control (IAC) valve – circuit malfunction	**Wiring, IAC valve, intake system leaking, ECM**
P1505	Idle air control (IAC) system – air trim, at limit	**Intake system leaking, air filter blocked, throttle body, IAC valve**
P1506	Idle air control (IAC) valve – over-speed error	**Wiring, IAC valve, vacuum leaks, ECM**
P1507	Idle air control (IAC) valve – under-speed error	**Wiring, IAC valve, hose blocked, ECM**
P1512	Intake manifold runner control (IMRC) module – malfunction, stuck closed –	**Mechanical fault**
P1513	Intake manifold runner control (IMRC) module – malfunction, stuck closed –	**Mechanical fault**
P1516	Intake manifold runner control (IMRC) module – input error (bank 1)	**Wiring, mechanical fault**
P1517	Intake manifold runner control (IMRC) module – input error (bank 2)	**Wiring, mechanical fault**
P1518	Intake manifold runner control (IMRC) module – malfunction, stuck open	**Wiring, IMRC actuator, ECM**
P1519	Intake manifold runner control (IMRC) module – malfunction, stuck closed	**Wiring, IMRC actuator, ECM**
P1520	Intake manifold runner control (IMRC) module – circuit malfunction	**Mechanical fault, vacuum hoses, IMRC actuator, ECM**
P1530	A/C compressor clutch– circuit malfunction	**Wiring open circuit/short circuit to positive, A/C compressor clutch, variable load control module**
P1531	Accelerator pedal moved during self-test	**Repeat test**
P1536	Parking brake switch – circuit malfunction	**Wiring short circuit to ground/positive, wiring open circuit, low brake fluid**
P1537	Intake manifold runner control (IMRC) module – malfunction, stuck closed	**Wiring, mechanical fault, ECM**
P1538	Intake manifold runner control (IMRC) module – malfunction, stuck closed	**Wiring, mechanical fault, ECM**
P1539	A/C compressor clutch – circuit malfunction	**Wiring, A/C compressor clutch, variable load control module**
P1549	Intake manifold air control system – malfunction	**Wiring, intake manifold air control solenoid, ECM**
P1550	Power steering pressure (PSP) sensor – malfunction	**Wiring, PSP sensor, ECM**
P1565	Cruise control system	**Wiring, CPP switch, BPP switch, cruise control module, cruise control master switch, VSS**
P1566	Cruise control system	**Wiring, CPP switch, BPP switch, cruise control module, cruise control master switch, VSS**

Scan code	Fault location	Probable cause
P1567	Cruise control system	**Wiring, CPP switch, BPP switch, cruise control module, cruise control master switch, VSS**
P1568	Cruise control system	**Wiring, CPP switch, BPP switch, cruise control module, cruise control master switch, VSS**
P1572	Brake pedal position (BPP) switch – out of self-test range	**Wiring, lights control module, BPP switch, ECM**
P1605	'Keep alive' memory test failure	**Wiring, battery terminal corrosion, loose battery connection, ECM**
P1625	Variable load control module, power supply to engine coolant blower motor – circuit malfunction	**Wiring, variable load control module**
P1626	Variable load control module, power supply to A/C system – circuit malfunction	**Wiring, variable load control module**
P1633	'Keep alive' power supply – voltage too low	**Wiring, ECM**
P1635	Tire/axle ratio out of range	**Incorrect tire size, incorrect axle ratio, incorrect vehicle ID configuration**
P1636	Engine control module (ECM) – internal fault	**ECM**
P1640	Diagnostic trouble codes stored	**P-codes stored in another control module**
P1641	Fuel pump – circuit malfunction	**Wiring, fuel pump relay, fuel pump, ECM**
P1650	Power steering pressure (PSP) switch – malfunction	**Wiring, PSP switch, steering wheel turned during self-test, ECM**
P1651	Power steering pressure (PSP) switch – input	**Wiring, vehicle towed with engine running, PSP switch, DTC not erased, ECM**
P1662	Injector control module relay – circuit malfunction	**Wiring, fuses, injector control module relay**
P1663	Fuel delivery command signal – malfunction	**Wiring open circuit/short circuit to ground, injector control module, ECM**
P1667	Cylinder identification – circuit malfunction	**Wiring open circuit/short circuit to ground, injector control module relay, ECM**
P1668	Engine control module (ECM)/injector control module – communication error	**Wiring open circuit/short circuit to ground, injector control module, ECM**
P1700	Transmission system – malfunction	**Transmission**
P1701	Transmission system – reverse engagement error	**Transmission**
P1702	Transmission range (TR) sensor – circuit intermittent	**Wiring, TR sensor**
P1703	Brake pedal position (BPP) switch – out of self-test range	**Wiring, brake pedal depressed during self-test, rear electronic module/lights control module, BPP switch, ECM**
P1704	Transmission range (TR) sensor – malfunction	**Wiring, TR sensor**
P1705 8	Park/neutral position (PNP) switch/clutch pedal position (CPP) switch	**Wiring, PNP switch, CPP switch, ECM**
P1706	Transmission range (TR) sensor – vehicle speed in park detected	**Wiring, mechanical fault, ECM**

Scan code	Fault location	Probable cause
P1707	Transmission range (TR) sensor – low input	**Wiring, mechanical fault, ECM**
P1708	Transmission range (TR) sensor – high input	**Wiring, mechanical fault, ECM**
P1709	Park/neutral position (PNP) switch/clutch pedal position (CPP) switch – out of self-test range	**Wiring, PNP switch/CPP switch, ECM**
P1710	Transmission fluid temperature sensor – mid range failure	**Wiring, transmission fluid temperature sensor**
P1711	Transmission fluid temperature sensor – out of self-test range	**Wiring, transmission fluid temperature sensor, fluid level, ECM**
P1713	Transmission fluid temperature sensor – low range failure	**Wiring, transmission fluid temperature sensor**
P1714	Shift solenoid (SS) A – inductive malfunction	**Wiring, SS, ECM/TCM**
P1715	Shift solenoid (SS) B – inductive malfunction	**Wiring, SS, ECM/TCM**
P1716	Shift solenoid (SS) C – inductive malfunction	**Wiring, SS, ECM/TCM**
P1717	Shift solenoid (SS) D – inductive malfunction	**Wiring, SS, ECM/TCM**
P1718	Transmission fluid temperature sensor – high range failure	**Wiring, transmission fluid temperature sensor**
P1719	Overdrive speed sensor – circuit malfunction	**Wiring, overdrive speed sensor, TCM**
P1728	Torque converter clutch (TCC) – slip fault	**TCC**
P1729	4x4L switch – malfunction	**Wiring, four wheel drive shift control module, ECM**
P1740	Torque converter clutch (TCC) solenoid – circuit malfunction	**Wiring, TCC solenoid, ECM/TCM**
P1741	Torque converter clutch (TCC) – control error	**Wiring, TCC, TCM**
P1742	Torque converter clutch (TCC) solenoid	**Wiring, TCC solenoid**
P1743	Torque converter clutch (TCC) solenoid	**Wiring, TCC solenoid**
P1744	Torque converter clutch (TCC) – malfunction	**Wiring, TCC, TCM**
P1745	Transmission fluid pressure (TFP) solenoid – circuit malfunction	**Wiring, TFP solenoid, TCM, ECM**
P1746	Transmission fluid pressure (TFP) solenoid – circuit malfunction	**Wiring, TFP solenoid, TCM, ECM**
P1747 7	Transmission fluid pressure (TFP) solenoid – short circuit	**Wiring, TFP solenoid, ECM**
P1747 8	Transmission fluid pressure (TFP) solenoid – circuit malfunction	**Wiring short circuit, TFP solenoid, TCM**
P1748	Transmission fluid pressure (TFP) solenoid – malfunction	**Wiring short circuit to ground/positive, wiring open circuit, TFP solenoid, ECM**
P1749	Transmission fluid pressure (TFP) solenoid – malfunction	**Wiring, TFP solenoid, TCM**
P1751	Shift solenoid (SS) A – performance	**Wiring, SS**
P1754	Coast/converter clutch solenoid – circuit malfunction	**Wiring, coast clutch solenoid**
P1756	Shift solenoid (SS) B – performance	**Wiring, SS**
P1760	Transmission fluid pressure (TFP) solenoid – short circuit	**Wiring, TFP solenoid, TCM**

Scan code	Fault location	Probable cause
P1761	Shift solenoid (SS) C – performance	Wiring, SS
P1762	Overdrive band failed/off	Transmission
P1767	Torque converter clutch (TCC) – circuit malfunction	Wiring, TCC, TCM
P1779	Transmission warning lamp – out of self-test range	Wiring, transmission warning lamp
P1780	Overdrive selection switch	Wiring, transmission control switch, ECM
P1781	4x4L switch – out of self-test range	Wiring, four wheel drive shift control module, 4x4L switch, ECM
P1783	Transmission fluid temperature sensor (TFT) – overheating condition	Wiring, TFT sensor, TCM, ECM
P1784	Transmission, reverse/first gear – mechanical fault	Transmission
P1785	Transmission, first/second gear – mechanical fault	Transmission
P1786	Gear selection, 3-2 – downshift malfunction	Wiring, TR sensor, shift solenoids, transmission mechanical fault
P1787	Gear selection, 2-1 – downshift malfunction	Wiring, TR sensor, shift solenoids, transmission mechanical fault
P1788	Transmission shift control valve, 3-2 downshift/ coastdown downshift – open circuit	Wiring, transmission shift control valve, TCM
P1789	Transmission shift control valve, 3-2 downshift/ coastdown downshift – short circuit	Wiring, transmission shift control valve, TCM
P1900	Output shaft speed (OSS) sensor – circuit intermittent	Wiring, poor connection, OSS sensor, ECM/ TCM
P1901	Turbine shaft speed (TSS) sensor – circuit intermittent	Wiring, poor connection, TSS sensor, ECM/ TCM
U1011 9	CAN data bus, air intake – invalid or missing data	Wiring, ECM (alternative fuel), ECM
U1020	CAN data bus, A/C – A/C compressor clutch invalid or missing data	Wiring, A/C trouble code(s) stored
U1021	CAN data bus, A/C – communication error	Wiring, A/C trouble code(s) stored, ECM
U1039	CAN data bus, antilock brake system (ABS) – vehicle speed invalid or missing data	Wiring, ABS trouble code(s) stored
U1041	CAN data bus, antilock brake system (ABS) – vehicle speed invalid or missing data	Wiring, ABS trouble code(s) stored
U1051	CAN data bus, engine control module (ECM) – brake signal invalid or missing data	Wiring, BPP switch, bulb(s)
U1071 9	CAN data bus, engine sensor data – invalid or missing data	Wiring, ECM (alternative fuel), ECM
U1073 9	CAN data bus, cooling system – communication error	Wiring, variable load control module, ECM (alternative fuel) – if fitted, ECM
U1089	CAN data bus, active suspension – invalid or missing data	Wiring, active suspension trouble code(s) stored
U1098	CAN data bus, antilock brake system (ABS) – vehicle speed invalid or missing data	Wiring, ABS trouble code(s) stored

Scan code	Fault location	Probable cause
U1130	CAN data bus, fuel pump – communication error	**Wiring, instrument panel, ECM**
U1131	CAN data bus, fuel pump – communication error	**Wiring, instrument panel, ECM**
U1135	CAN data bus, engine control module (ECM) – ignition switch invalid or missing data	**Wiring**
U1147	CAN data bus, alarm	**Wiring, alarm system trouble code(s) stored**
U1243	CAN data bus, A/C – invalid or missing data	**Wiring, A/C system trouble code(s) stored**
U1256	CAN data bus – communication error	**Wiring, trouble code(s) stored in other control modules**
U1260 9	CAN data bus – communication error	**Wiring, ECM (alternative fuel)**
U1261 9	CAN data bus – communication error	**Wiring, ECM (alternative fuel)**
U1262 9	CAN data bus – communication error	**Wiring, ECM (alternative fuel), ECM**
U1262 10	CAN data bus – communication error	**Wiring, trouble code(s) stored in other control modules**
U1341	CAN data bus, A/C – vehicle speed invalid or missing data	**Wiring, A/C trouble code(s) stored**
U1451	CAN data bus, alarm	**Wiring, alarm system trouble code(s) stored**
U2015 9 11	Signal link fault	**Compuvalve/ECM (alternative fuel)**
U2195	CAN data bus, lights control module – invalid or missing data	**Wiring, lights control module trouble code(s) stored**
U2243	CAN data bus, A/C	**Wiring, A/C system trouble code(s) stored**

1 Expedition 4.6L, Econoline/Club Wagon/F-Series 5.4L, Econoline/Club Wagon 6.8L
2 Except Expedition 4.6L, Econoline/Club Wagon/F-Series 5.4L, Econoline/Club Wagon 6.8L
3 Windstar
4 Except Windstar
5 Explorer 5.0L
6 Except Explorer 5.0L
7 1995 Econoline/Club Wagon 7.3L Turbo Diesel
8 Except 1995 Econoline/Club Wagon 7.3L Turbo Diesel
9 Econoline Bi-Fuel
10 Except Econoline Bi-Fuel
11 Compuvalve incorporates ECM (alternative fuel) for Econoline

FORD

Model:	Engine identification:	Year:
Aerostar Van/Wagon 3.0/4.0L	VIN position 8 = U/X	1992-95
Bronco 5.0/5.8L	VIN position 8 = H/N/R	1992-95
Econoline/Club Wagon 4.9/5.0/5.8/7.5L	VIN position 8 = G/H/N/R/Y/Z	1992-95
Econoline/Club Wagon 5.8/7.5L (without 16 pin DLC)	VIN position 8 = G/H	1996
Explorer 4.0L	VIN position 8 = X	1992-95
F-Series 4.9/5.0/5.8/7.5L	VIN position 8 = G/H/N/R/Y	1992-95
F-Series 5.8/7.5L (without 16 pin DLC)	VIN position 8 = G/H	1996-97
Ranger 2.3/3.0/4.0L	VIN position 8 = A/U/X	1992-94

System: **Ford EEC-IV**

Self-diagnosis

General information

- Refer to the front of this manual for general test conditions, terminology, detailed descriptions of wiring faults and a general trouble shooter for electrical and mechanical faults.
- Malfunction indicator lamp (MIL) will illuminate if certain faults are recorded.
- ECM operates in backup mode if sensors fail, to enable vehicle to be driven to repair shop.
- Trouble codes can be displayed by the MIL or by using an analogue voltmeter or Scan Tool connected to the data link connectors (DLC) 1 and 2:
  - Aerostar **1**.
  - Bronco/F-Series **2**.
  - Econoline/Club Wagon **3**.
  - Explorer **4**.
  - Ranger **5**.

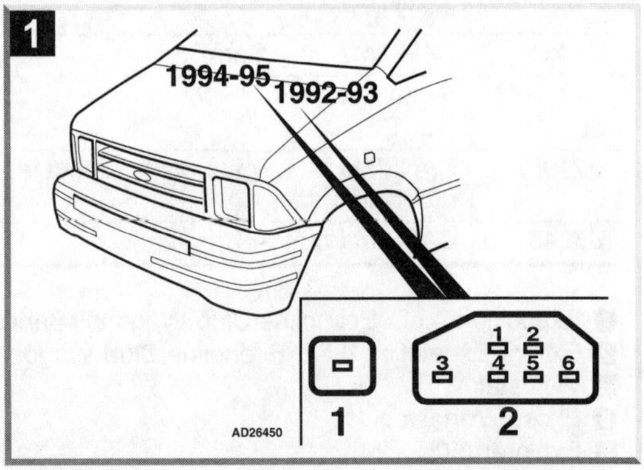

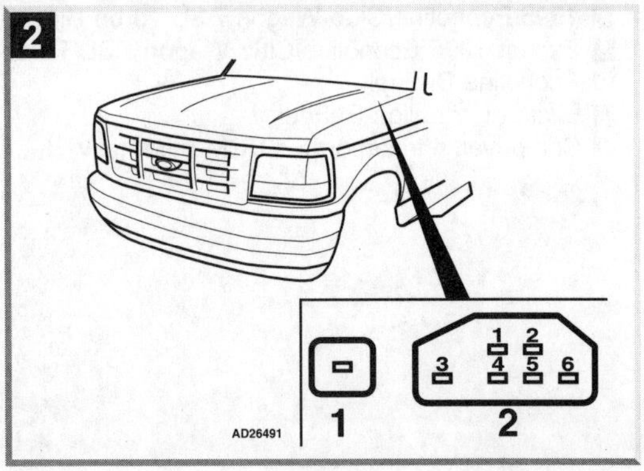

3

4

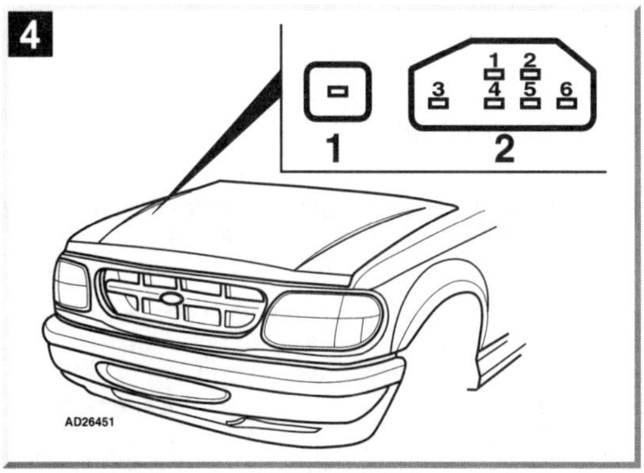

5

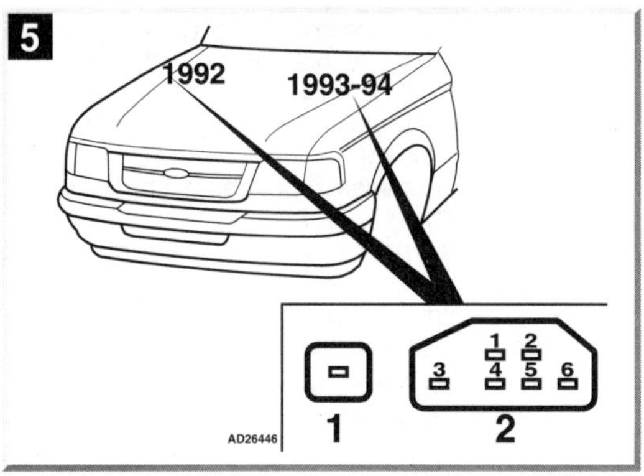

Preparatory conditions

- Automatic transmission in 'PARK'.
- All auxiliary equipment, including air conditioning, switched OFF.

Accessing – Key On Engine Off (KOEO) test

Using malfunction indicator lamp (MIL)

- Start engine.
- Allow to idle until engine at normal operating temperature.
- Switch ignition OFF.
- Wait 10 seconds.
- If engine does not start or stalls: Continue with KOEO test.
- If fitted: Remove 'EEC TEST' cover to access data link connector (DLC).
- 4.9L M/T: Depress clutch pedal for entire accessing procedure.
- Jump DLC 1 terminal (STI) to DLC 2 terminal 2 (SIG RTN) **6**.
- Switch ignition ON.

NOTE: ***MIL may flash briefly while fast DTCs are issued for Scan Tool followed by slow DTC output.***

- Count MIL flashes. Note trouble codes. Compare with trouble code table.
- Each trouble code is displayed twice. Firstly for present faults followed by those for memorized faults as follows:
  - Trouble codes are displayed by illuminated MIL switching OFF.
  - Each trouble code consists of two or three digits.
  - A group of 0.5 second flashes indicate each digit **7** [**A**].
  - A 0.5 second pause separates each flash **7** [**B**].
  - A 2 second pause separates each trouble code digit **7** [**C**].
  - A 4 second pause separates each trouble code **7** [**D**].
  - For example: Three digit trouble code 122 displayed **7**. Throttle position (TP) sensor.

NOTE: ***Two digit trouble codes displayed in a similar pattern.***

  - A 6-9 second pause **8** [**B**] indicates the end of trouble code output for present faults **8** [**A**].
  - A 0.5 second flash (separator code) **8** [**C**], followed by a 6-9 second pause **8** [**D**] indicates the start of trouble code output for memorized faults **8** [**E**].
  - Trouble codes for memorized faults displayed in the same manner as those for present faults.

NOTE: ***Memorized trouble codes refer to faults that have occurred in the last 80 warm-up cycles.***

- Switch ignition OFF.
- Remove jump lead.
- 4.9L M/T: Release clutch pedal.
- Trouble codes can also be displayed by using a Scan Tool connected to the data link connectors (DLC) 1 and 2.

6

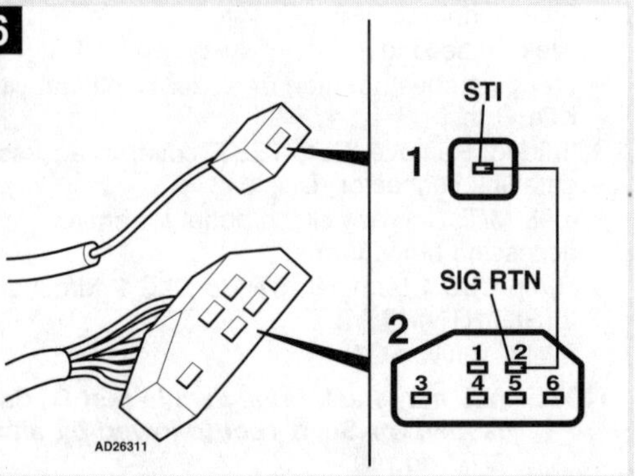

7

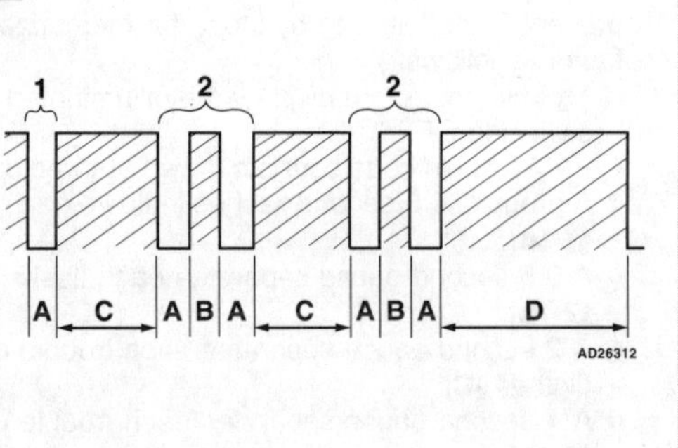

8

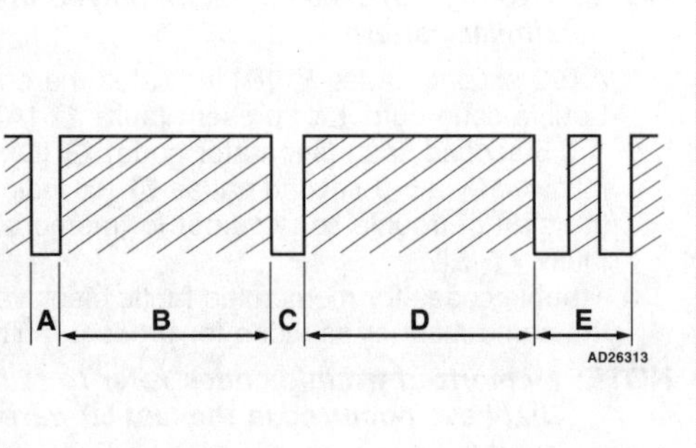

Using analogue voltmeter

- Ensure ignition switched OFF.
- If fitted: Remove 'EEC TEST' cover to access data link connector (DLC).
- Connect an analogue voltmeter between data link connector (DLC) 2 terminal 4 (STO) and battery (12 volt) positive **9**.
- Set analogue voltmeter to a DC voltage range of 0-20 V.
- Start engine.
- Allow to idle until engine at normal operating temperature.
- Switch ignition OFF.
- Wait 10 seconds.
- If engine does not start or stalls: Continue with KOEO test.
- 4.9L M/T: Depress clutch pedal for entire accessing procedure.
- Jump DLC 1 terminal (STI) to DLC 2 terminal 2 (SIG RTN) **9**.
- Switch ignition ON.

NOTE: ***Voltmeter needle may briefly deflect while fast DTCs are issued for Scan Tool followed by slow DTC output.***

- Count voltmeter needle deflections. Note trouble codes. Compare with trouble code table.
- Each trouble code is displayed twice. Firstly for present faults followed by those for memorized faults as follows:
  - Trouble codes are displayed by voltmeter needle deflecting to zero.
  - Each trouble code consists of two or three digits.
  - A group of 0.5 second needle deflections indicate each digit **7** [**A**].
  - A 0.5 second pause separates each needle deflection **7** [**B**].
  - A 2 second pause separates each trouble code digit **7** [**C**].
  - A 4 second pause separates each trouble code **7** [**D**].
  - For example: Three digit trouble code 122 displayed **7**. Throttle position (TP) sensor.

NOTE: ***Two digit trouble codes displayed in a similar pattern.***

  - A 6-9 second pause **8** [**B**] indicates the end of trouble code output for present faults **8** [**A**].
  - A 0.5 second needle deflection (separator code) **8** [**C**], followed by a 6-9 second pause **8** [**D**] indicates the start of trouble code output for memorized faults **8** [**E**].
  - Trouble codes for memorized faults displayed in the same manner as those for present faults.

NOTE: ***Memorized trouble codes refer to faults that have occurred in the last 80 warm-up cycles.***

- Switch ignition OFF.
- Remove jump lead and voltmeter.
- 4.9L M/T: Release clutch pedal.
- Trouble codes can also be displayed by using a Scan Tool connected to the data link connectors (DLC) 1 and 2.

9

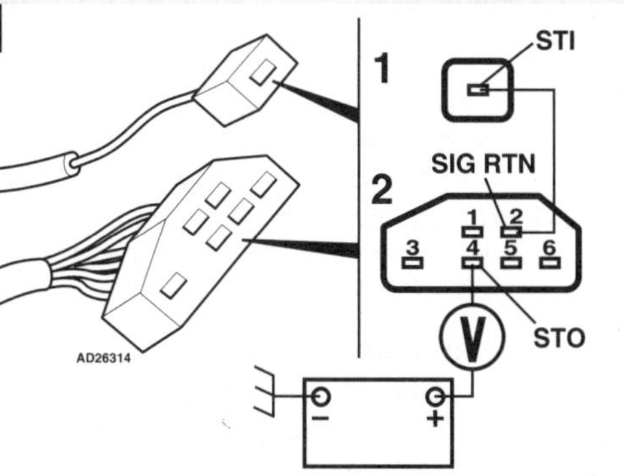

Accessing – Key On Engine Running (KOER) test

Using malfunction indicator lamp (MIL)

- Start engine.
- Allow to idle until engine at normal operating temperature.
- Increase engine speed to over 2000 rpm for 2 minutes.
- Switch ignition OFF.
- Wait 10 seconds.
- If fitted: Remove 'EEC TEST' cover to access data link connector (DLC).
- Jump DLC 1 terminal (STI) to DLC 2 terminal 2 (SIG RTN) 6.
- Start engine.
- Engine identification code displayed as follows:
  - 4 cylinder: Two 0.5 second flashes.
  - 6 cylinder: Three 0.5 second flashes.
  - 8 cylinder: Four 0.5 second flashes.
  - For example: 6 cylinder engine identified 10 [**A**].
- With PSP switch: Within 1-2 seconds, turn steering wheel at least ½ a turn and release.
- With BPP switch: Depress and release brake pedal.
- If fitted: Operate overdrive selection switch twice.
- If after 6-20 seconds 10 [**B**], a single 0.5 second flash is displayed (dynamic response signal) 10 [**C**]: Briefly fully open throttle.

NOTE: ***DO NOT fully open throttle during KOER test if no dynamic response signal is displayed.***

- Trouble codes displayed for present faults after 4-15 second pause 10 [**D**].
- Fast DTCs are issued for Scan Tool during 4-15 second pause 10 [**E**].
- Count MIL flashes. Note trouble codes. Compare with trouble code table.

NOTE: ***Each trouble code is displayed twice.***

- Trouble codes for KOER test are displayed in the same manner as those for KOEO test 10 [**F**].
- For example: Three digit trouble code 111 is displayed 10. No fault found.
- Switch ignition OFF.
- Remove jump lead.
- Trouble codes can also be displayed by using a Scan Tool connected to the data link connectors (DLC) 1 and 2.

10

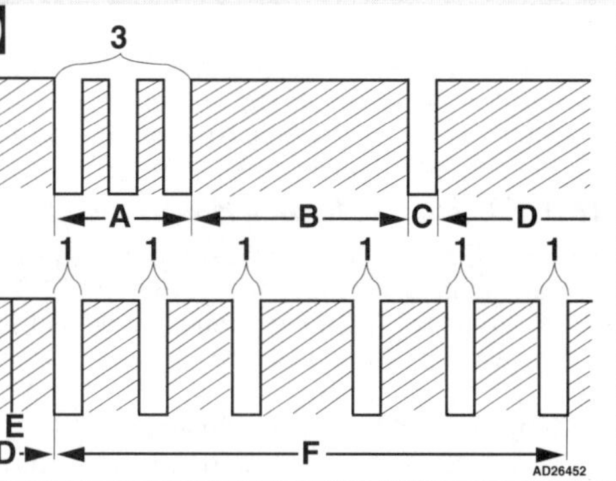

Using analogue voltmeter

- Ensure ignition switched OFF.
- If fitted: Remove 'EEC TEST' cover to access data link connector (DLC).
- Connect an analogue voltmeter between data link connector (DLC) 2 terminal 4 (STO) and battery (12 volt) positive 9.
- Set analogue voltmeter to a DC voltage range of 0-20 V.
- Start engine.
- Allow to idle until engine at normal operating temperature.

- Increase engine speed to over 2000 rpm for 2 minutes.
- Switch ignition OFF.
- Wait 10 seconds.
- Jump DLC 1 terminal (STI) to DLC 2 terminal 2 (SIG RTN) **9**.
- Start engine.
- Engine identification code displayed as follows:
  - 4 cylinder: Two 0.5 second needle deflections.
  - 6 cylinder: Three 0.5 second needle deflections.
- 8 cylinder: Four 0.5 second needle deflections.
  - For example: 6 cylinder engine identified **10** [**A**].
- With PSP switch: Within 1-2 seconds, turn steering wheel at least ½ a turn and release.
- With BPP switch: Depress and release brake pedal.
- If fitted: Operate overdrive selection switch twice.
- If after 6-20 seconds **10** [**B**], a single 0.5 second needle deflection is displayed (dynamic response signal) **10** [**C**]: Briefly fully open throttle.

NOTE: ***DO NOT fully open throttle during KOER test if no dynamic response signal is displayed.***

- Trouble codes displayed for present faults after 4-15 second pause **10** [**D**].
- Fast DTCs are issued for Scan Tool during 4-15 second pause **10** [**E**].
- Count needle deflections. Note trouble codes. Compare with trouble code table.

NOTE: ***Each trouble code is displayed twice.***

- Trouble codes for KOER test are displayed in the same manner as those for KOEO test **10** [**F**].
- For example: Three digit trouble code 111 is displayed **10**. No fault found.
- Switch ignition OFF.
- Remove jump lead and voltmeter.
- Trouble codes can also be displayed by using a Scan Tool connected to the data link connectors (DLC) 1 and 2.

Erasing

NOTE: ***Disconnecting battery ground lead may cause driveability problems until the engine management system relearns basic values.***

Using malfunction indicator lamp (MIL)

- Ensure ignition switched OFF.
- If fitted: Remove 'EEC TEST' cover to access data link connector (DLC).
- Jump DLC 1 terminal (STI) to DLC 2 terminal 2 (SIG RTN) **6**.
- Switch ignition ON.
- When trouble codes start to be displayed: Remove jump lead.
- Switch ignition OFF.
- Repeat accessing procedure to ensure no memorized trouble codes remain in ECM fault memory.
- Trouble codes can also be erased by using a Scan Tool connected to the data link connectors (DLC) 1 and 2.
- If emission related component(s) have been replaced:
  - Disconnect battery ground cable for at least 5 minutes to erase adaptive memory from ECM.
  - If required: Drive vehicle for 10 miles or more to allow ECM to relearn basic values.

NOTE: ***Disconnecting battery ground lead may also erase memory from electronic units, such as radio.***

Using analogue voltmeter

- Ensure ignition switched OFF.
- If fitted: Remove 'EEC TEST' cover to access data link connector (DLC).
- Connect an analogue voltmeter between data link connector (DLC) 2 terminal 4 (STO) and battery (12 volt) positive **9**.
- Set analogue voltmeter to a DC voltage range of 0-20 V.
- Jump DLC 1 terminal (STI) to DLC 2 terminal 2 (SIG RTN) **9**.
- Switch ignition ON.
- When trouble codes start to be displayed: Remove jump lead.
- Switch ignition OFF.
- Repeat accessing procedure to ensure no memorized trouble codes remain in ECM fault memory.
- Trouble codes can also be erased by using a Scan Tool connected to the data link connectors (DLC) 1 and 2.
- If emission related component(s) have been replaced:
  - Disconnect battery ground cable for at least 5 minutes to erase adaptive memory from ECM.
  - If required: Drive vehicle for 10 miles or more to allow ECM to relearn basic values.

NOTE: ***Disconnecting battery ground lead may also erase memory from electronic units, such as radio.***

Trouble code identification

Two digit

Flash code	Fault location	Probable cause
11	No fault found	–
12	Idle speed, high RPM check – unable to control during KOER test	**Intake system, throttle tight/sticking, IAC valve/ISC actuator, ECM**
13	Idle speed, low RPM check – unable to control during KOER test	**Intake system, throttle tight/sticking, IAC valve/ISC actuator, ECM**
14	Engine speed (RPM) sensor/profile ignition pickup (PIP) circuit – intermittent signal	**Incorrectly installed RPM sensor, wiring, poor connection, electrical interference, RPM sensor, ECM**
15	Engine control module (ECM) – KAM/ROM fault	**Battery disconnected, wiring, fuses, ECM**
16	Engine RPM too low for heated oxygen sensor (HO2S) test	**Intake system, throttle tight/sticking, IAC valve/ISC actuator, ECM**
	Ignition diagnostic monitor (IDM) circuit – no signal during KOEO test	**Wiring, ICM, ECM**
18	Spark output (SPOUT) circuit open	**Wiring, ICM, ECM**
	Ignition diagnostic monitor (IDM) circuit/spark output (SPOUT) circuit grounded	**Wiring, ICM, ECM**
19	Engine control module (ECM) – internal voltage fault	**ECM**
21	Engine coolant temperature (ECT) sensor – incorrect signal	**Wiring, ECT sensor, ECM**
22	Manifold absolute pressure (MAP) sensor/ barometric pressure (BARO) sensor – incorrect signal	**Vacuum hose damaged, trapped vacuum, high atmospheric pressure, excessive EGR, wiring, MAP/BARO sensor, ECM**
23	Throttle position (TP) sensor – incorrect signal	**Throttle valve tight/sticking, wiring, TP sensor, ECM**
24	Intake air temperature (IAT) sensor – incorrect signal	**Wiring, IAT sensor, ECM**
25	Knock sensor (KS) – no signal during dynamic response test	**Wiring, KS, ECM**
26 [2]	Mass air flow (MAF) sensor – incorrect signal	**Intake leak/blockage, wiring, MAF sensor, ECM**
26 [1] [3]	Transmission fluid temperature (TFT) sensor – signal high/low	**ATF level, wiring, TFT sensor, ECM**
29	Vehicle speed signal – incorrect signal	**Wiring, rear ABS system, vehicle speed interface module, ECM**
31	Exhaust gas recirculation (EGR) valve position sensor – below minimum voltage	**Wiring short circuit to ground, EGR valve/ position sensor, ECM**
32	Exhaust gas recirculation (EGR) valve position sensor – below closed limit	**Wiring short circuit to ground, EGR valve/ position sensor, ECM**
33	Exhaust gas recirculation (EGR) valve – not opening	**Wiring, EGR valve tight/sticking, ECM**

Flash code	Fault location	Probable cause
34	Exhaust gas recirculation (EGR) valve position sensor – above closed limit	**Wiring short circuit to positive, EGR valve/ position sensor, ECM**
35	Exhaust gas recirculation (EGR) valve position sensor – above maximum voltage	**Wiring short circuit to positive, EGR valve/ position sensor, ECM**
41	Heated oxygen sensor (HO2S), bank 1 – mixture lean/not switching	**Intake leak, fuel pressure low, fuel filter/ pump, injectors, HO2S, AIR system, MAF/ MAP sensor, ECM**
42	Heated oxygen sensor (HO2S), bank 1 – mixture rich/not switching	**Intake blocked, excessive fuel in engine oil, fuel pressure high, breather system, injectors, HO2S, AIR system, EVAP system, MAF/MAP sensor, ignition system, ECM**
43	Throttle position (TP) sensor – incorrect signal at idle	**Incorrectly adjusted, TP sensor, ECM**
44	Secondary air injection (AIR) system, bank 1 – inoperative	**Hose connection, wiring, AIR pump motor, AIR solenoid, ECM**
45	Secondary air injection (AIR) – upstream during KOER test	**Hose connection, wiring, AIR solenoid, ECM**
46	Secondary air injection (AIR) – not bypassed during KOER test	**Hose connection, wiring, AIR solenoid, ECM**
47	Four wheel drive selector switch – closed during KOEO test	**Wiring short circuit, four wheel drive selector switch, internal damage to transfer case, ECM**
49 **1**	Gear shift, 1-2 – malfunction	**ATF level/condition, poor connection, wiring, TR sensor, SS, clutch slipping, ECM**
51	Engine coolant temperature (ECT) sensor – open circuit	**Wiring open circuit, ECT sensor, ECM**
52	Power steering pressure (PSP) switch – open circuit/no signal change	**Wiring open circuit, PSP switch, ECM**
53	Throttle position (TP) sensor – above maximum voltage	**TP sensor loose, wiring short circuit to positive, TP sensor, ECM**
54	Intake air temperature (IAT) sensor – open circuit	**Wiring open circuit, IAT sensor, ECM**
56 **2**	Mass air flow (MAF) sensor – above maximum voltage	**Wiring short circuit to positive, MAF sensor, ECM**
56 **1** **3**	Transmission fluid temperature (TFT) sensor – above maximum voltage	**ATF level, wiring open circuit, TFT sensor, ECM**
59 **1**	Gear shift, 2-3 – malfunction	**ATF level/condition, poor connection, wiring, TR sensor, SS, clutch slipping, ECM**
61	Engine coolant temperature (ECT) sensor – short circuit to ground	**Wiring short circuit to ground, ECT sensor, ECM**
62 **1**	Torque converter clutch (TCC) – excessive clutch slip	**ATF level/condition, TCC slipping, ECM**
63	Throttle position (TP) sensor – below minimum voltage	**TP sensor loose, wiring open circuit/short circuit to ground, TP sensor, ECM**
64	Intake air temperature (IAT) sensor – short circuit to ground	**Wiring short circuit to ground, IAT sensor, ECM**
65 **1**	Overdrive selection switch – no signal change during KOEO test	**Incorrect switch operation, wiring, fuse, overdrive selection switch, ECM**

Flash code	Fault location	Probable cause
66 **2**	Mass air flow (MAF) sensor – below minimum voltage	**Wiring short circuit to ground, MAF sensor, ECM**
66 **1** **3**	Transmission fluid temperature (TFT) sensor – below minimum voltage	**ATF level, wiring short circuit to ground, TFT sensor, ECM**
67	Neutral position (NP) switch – open circuit	**Incorrect adjustment, wiring open circuit, NP switch, ECM**
	Clutch pedal position (CPP) switch – circuit malfunction	**Incorrect adjustment, wiring, CPP switch, ECM**
	A/C – on during KOEO test	**Wiring, A/C switch, ECM**
67 **1**	Transmission range (TR) sensor – voltage high/low	**Incorrect adjustment, wiring, TR sensor, ECM**
68 **1**	Automatic transmission fluid (ATF) – temperature high	**ATF level, transmission cooling system, towing weight limit exceeded, clutch slipping, wiring, TFT sensor, ECM**
69 **1**	Gear shift, 3-4 – malfunction	**ATF level/condition, poor connection, wiring, TR sensor, SS, clutch slipping, ECM**
72	Manifold absolute pressure (MAP) sensor/ mass air flow (MAF) sensor – incorrect signal change during dynamic response test	**Intake leak, wiring, MAP sensor, ECM**
73	Throttle position (TP) sensor – incorrect signal change during dynamic response test	**Incorrect throttle operation, wiring, TP sensor, ECM**
74	Brake pedal position (BPP) switch – open circuit/no signal change during KOER test	**Incorrect brake pedal operation, wiring, BPP switch, ECM**
75	Brake pedal position (BPP) switch – closed circuit	**Wiring, BPP switch, ECM**
77	Wide open throttle (WOT) switch – no signal change during dynamic response test	**Incorrect throttle operation, wiring, WOT switch, ECM**
79	A/C – on during KOEO test	**Wiring, A/C switch, ECM**
81	Secondary air injection (AIR) divert solenoid – circuit malfunction during KOEO test	**Wiring, AIR divert solenoid, ECM**
82	Secondary air injection (AIR) bypass solenoid – circuit malfunction during KOEO test	**Wiring, AIR bypass solenoid, ECM**
84	Exhaust gas recirculation (EGR) solenoid – circuit malfunction during KOEO test	**Wiring, EGR solenoid, ECM**
85	Evaporative emission (EVAP) canister purge valve – circuit malfunction during KOEO test	**Wiring, EVAP canister purge valve, ECM**
86	AT – shift solenoid	**Wiring, shift solenoid, ECM**
87	Fuel pump, primary circuit – circuit malfunction	**IFS switch, wiring, fuel pump, ECM**
89	Torque converter clutch (TCC) lock-up solenoid	**Wiring, TCC lock-up solenoid, ECM**
91	Heated oxygen sensor (HO2S), bank 2 – mixture lean/not switching during KOER test	**Intake leak, fuel pressure low, fuel filter/ pump, injectors, HO2S, AIR system, MAF/ MAP sensor, ECM**
91 **1**	Shift solenoid (SS) 1 – circuit malfunction during KOEO test	**Wiring, SS, ECM**

Flash code	Fault location	Probable cause
92	Heated oxygen sensor (HO2S), bank 1 – mixture rich/not switching during KOER test	**Intake blocked, excessive fuel in engine oil, fuel pressure high, breather system, injectors, HO2S, AIR system, EVAP system, MAF/MAP sensor, ignition system, ECM**
92 [1]	Shift solenoid (SS) 2 – circuit malfunction during KOEO test	**Wiring, SS, ECM**
93 [1]	Transmission coast clutch solenoid – circuit malfunction during KOEO test	**Wiring, transmission coast clutch solenoid, ECM**
94	Secondary air injection (AIR) system, bank 2 – inoperative during KOER test	**Hose connection, wiring, AIR pump motor, AIR solenoid, ECM**
94 [1]	Torque converter clutch (TCC) solenoid – circuit malfunction during KOEO test	**Wiring, TCC solenoid, ECM**
95	Fuel pump, secondary circuit – circuit malfunction	**IFS switch, wiring, fuel pump, ECM**
96	Fuel pump, secondary circuit – circuit malfunction	**IFS switch, wiring open circuit to battery, fuel pump, ECM**
97 [1]	Overdrive indicator lamp – circuit malfunction	**Bulb, wiring, fuse, ECM**
98	Engine control module (ECM) – hard fault detected during KOER test	**HO2S bank 2 – indicates lean, ECM**
98 [1]	Engine control module (ECM), transmission fluid pressure (TFP) control – open driver circuit during KOEO test	**ECM**
99	Engine control module – idle speed not learnt during KOER test	**Idle speed adjusted, ECM**
99 [1]	Transmission fluid pressure (TFP) solenoid – circuit malfunction during KOEO test	**Wiring, poor connection, TFP solenoid, ECM**

[1] ECM incorporates transmission control function
[2] Except 5.0/5.8/7.5L
[3] 5.0/5.8/7.5L

Three digit

Flash code	Fault location	Probable cause
111	No fault found	–
112	Intake air temperature (IAT) sensor – below minimum voltage	**Wiring short circuit to ground, poor connection, IAT sensor, ECM**
113	Intake air temperature (IAT) sensor – above maximum voltage	**Wiring open circuit, poor connection, IAT sensor, ECM**
114	Intake air temperature (IAT) sensor – incorrect signal	**Air temperature below 10°C (50°F), poor connection, IAT sensor, ECM**
116	Engine coolant temperature (ECT) sensor – incorrect signal	**Coolant level low, poor connection, ECT sensor, ECM**
117	Engine coolant temperature (ECT) sensor – below minimum voltage	**Wiring short circuit to ground, poor connection, ECT sensor, ECM**

Flash code	Fault location	Probable cause
118	Engine coolant temperature (ECT) sensor – above maximum voltage	Wiring open circuit, poor connection, IAT sensor, ECM
121	Throttle position (TP) sensor – closed throttle voltage high/low	Throttle valve tight/sticking, TP sensor loose/damaged, ECM
	Throttle position (TP) sensor – voltage inconsistent with MAF sensor signal	Throttle valve tight/sticking, TP sensor loose/damaged, ECM
122	Throttle position (TP) sensor – below minimum voltage	TP sensor loose, wiring open circuit/short circuit to ground, TP sensor, ECM
123	Throttle position (TP) sensor – above maximum voltage	TP sensor loose, wiring short circuit to positive, TP sensor, ECM
124	Throttle position (TP) sensor – voltage high	Throttle tight/sticking, intake leak, MAP/BARO sensor, ECM
125	Throttle position (TP) sensor – voltage low	Throttle tight/sticking, intake leak, MAP/BARO sensor, ECM
126	Manifold absolute pressure (MAP) sensor/barometric pressure (BARO) sensor – voltage high/low	Vacuum hose damaged/trapped, high atmospheric pressure, excessive EGR, wiring, MAP/BARO sensor, ECM
128	Manifold absolute pressure (MAP) sensor – loss of vacuum	Vacuum hose damaged/disconnected, MAP sensor, ECM
129	Manifold absolute pressure (MAP) sensor – insufficient voltage change during KOER dynamic response test	Incorrect throttle operation, vacuum hose damaged/disconnected, MAP sensor, IAC system, ECM
	Mass air flow (MAF) sensor – insufficient voltage change during KOER dynamic response test	Incorrect throttle operation, intake leak, wiring, MAF sensor, IAC system, ECM
136	Heated oxygen sensor (HO2S), bank 2 – mixture lean during KOER test	Intake leak, fuel pressure low, fuel filter/pump, injectors, HO2S, AIR system, MAP sensor, ECM
137	Heated oxygen sensor (HO2S), bank 2 – mixture rich during KOER test	Intake blocked, excessive fuel in engine oil, fuel pressure high, breather system, injectors, HO2S, AIR system, EVAP system, MAP sensor, ignition system, ECM
138	Cold start injector – insufficient flow during KOER test	Fuel pressure, wiring, cold start injector, ECM
139	Heated oxygen sensor (HO2S), bank 2 – not switching	Intake leak/blocked, injectors, HO2S, AIR system, breather system, EVAP system, MAP sensor, ignition system, ECM
141	Fuel system indicates lean	Intake leak, fuel pressure low, fuel filter/pump, injectors, HO2S, ECM
144	Heated oxygen sensor (HO2S), bank 1 – not switching	Intake leak/blocked, injectors, HO2S, AIR system, breather system, EVAP system, MAP sensor, ignition system, ECM
157	Mass air flow (MAF) sensor – below minimum voltage	Wiring short circuit to ground, poor connection, MAF sensor, IAC system, ECM
158	Mass air flow (MAF) sensor – above maximum voltage	Wiring short circuit to positive, MAF sensor screen blocked, MAF sensor, ECM
159 **1**	Mass air flow (MAF) sensor – voltage high/low	Intake leak, IAC valve, MAF sensor, ECM

Flash code	Fault location	Probable cause
167	Throttle position (TP) sensor – insufficient voltage change during KOER dynamic response test	Incorrect throttle operation, TP sensor, ECM
171	Fuel system control, bank 1 – adaptive limits reached	Intake leak/blocked, breather system, fuel pressure, fuel filter/pump, injectors, HO2S, AIR system, EVAP system, MAP sensor, ignition system, ECM
172	Heated oxygen sensor (HO2S), bank 1 – mixture lean	Intake leak, fuel pressure low, fuel filter/pump, injectors, HO2S, AIR system, MAP sensor, ECM
173	Heated oxygen sensor (HO2S), bank 1 – mixture rich	Intake blocked, excessive fuel in engine oil, fuel pressure high, breather system, injectors, HO2S, AIR system, EVAP system, MAP sensor, ignition system, ECM
174	Heated oxygen sensor (HO2S), bank 1 – slow response	Intake leak, excessive fuel in engine oil, fuel pressure, fuel filter/pump, injectors, HO2S, AIR system, MAF/MAP sensor, ECM
175	Fuel system control, bank 2 – adaptive limits reached	Intake leak/blocked, excessive fuel in engine oil, breather system, fuel pressure, fuel filter/pump, injectors, HO2S, AIR system, EVAP system, MAP sensor, ignition system, ECM
176	Heated oxygen sensor (HO2S), bank 2 – mixture lean	Intake leak, fuel pressure low, fuel filter/pump, injectors, HO2S, AIR system, MAP sensor, ECM
177	Heated oxygen sensor (HO2S), bank 2 – mixture rich	Intake blocked, fuel pressure high, breather system, injectors, HO2S, AIR system, EVAP system, MAP sensor, ignition system, ECM
178	Heated oxygen sensor (HO2S), bank 2 – slow response	Intake leak, excessive fuel in engine oil, fuel pressure, fuel filter/pump, injectors, HO2S, AIR system, MAF/MAP sensor, ECM
179	Fuel control system, bank 1 – lean adaption limit reached at part throttle, mixture rich	Intake leak, fuel pressure low, fuel filter blocked, wiring, injector, MAP sensor, ECM
181	Fuel control system, bank 1 – rich adaption limit reached at part throttle, mixture lean	Intake blocked, fuel pressure high, wiring, injector, MAP sensor, ECM
182	Fuel control system, bank 1 – lean adaption limit reached at idle, mixture rich	Intake leak, fuel pressure low, fuel filter blocked, wiring, injector, MAP sensor, ECM
183	Fuel control system, bank 1 – rich adaption limit reached at idle, mixture lean	Intake blocked, fuel pressure high, wiring, injector, MAP sensor, ECM
184	Mass air flow (MAF) sensor – signal high	MAF sensor screen blocked, MAF sensor, ECM
185	Mass air flow (MAF) sensor – signal low	Intake leak, breather system, MAF sensor, ECM
186	Injector – pulse width long/mass air flow (MAF) sensor – signal low	Intake leak, fuel pressure low, MAF sensor, ECM
187	Injector – pulse width short/mass air flow (MAF) sensor – signal high	Intake blocked, fuel pressure high, MAF sensor, ECM
188	Fuel system control, bank 2 – lean adaption limit reached at part throttle, mixture rich	Intake leak, fuel pressure low, fuel filter blocked, wiring, injector, MAP sensor, ECM

Flash code	Fault location	Probable cause
189	Fuel system control, bank 2 – rich adaption limit reached at part throttle, mixture lean	**Intake blocked, fuel pressure high, wiring, injector, MAP sensor, ECM**
191	Fuel system control, bank 2 – lean adaption limit reached at idle, mixture rich	**Intake leak, fuel pressure low, fuel filter blocked, wiring, injector, MAP sensor, ECM**
192	Fuel system control, bank 2 – rich adaption limit reached at idle, mixture lean	**Intake blocked, fuel pressure high, wiring, injector, MAP sensor, ECM**
193	Fuel quality sensor – circuit malfunction	**Wiring, fuel separation, fuel quality sensor, ECM**
194	Cylinder balance test – malfunction	**Wiring, injector, ECM**
195	Cylinder balance test – malfunction	**Wiring, injector, ECM**
211	Profile ignition pickup (PIP) – circuit malfunction	**Wiring, poor connection, arcing secondary ignition component – coil/cap/rotor/HT leads/plugs, electrical interference, ECM**
212	Ignition control module (ICM) – ignition diagnostic monitor (IDM) signal missing	**Wiring, ICM, ECM**
213	Spark output (SPOUT) – open circuit	**Wiring, ICM, ECM**
214 2	Camshaft position (CMP) sensor – circuit malfunction	**Wiring, CMP sensor, ICM, ECM**
214 3	Crankshaft position (CKP) sensor, CID signal – circuit malfunction	**Wiring, CKP sensor, ICM, ECM**
215	Ignition coil 1 – primary circuit malfunction	**Wiring, ignition coil, ICM, ECM**
216	Ignition coil 2 – primary circuit malfunction	**Wiring, ignition coil, ICM, ECM**
217	Ignition coil 3 – primary circuit malfunction	**Wiring, ignition coil, ICM, ECM**
218	Ignition diagnostic monitor (IDM), left side – signal missing	**Wiring, ignition coil, ECM**
219	Ignition timing, defaulted to 10° – spark output (SPOUT) circuit open	**Wiring open circuit, ICM, ECM**
221	Ignition timing error	**Wiring, ICM, ECM**
222	Ignition diagnostic monitor (IDM), right side – signal missing	**Wiring, ignition coil, ECM**
223	Loss of dual plug inhibit (DPI) control	**Wiring, ICM, ECM**
224	Ignition coil 1, 2, 3 or 4 – primary circuit malfunction	**Wiring, ignition coil, ICM, ECM**
225	Knock sensor (KS) – no signal during KOER dynamic response test	**Wiring, KS, ECM**
226	Ignition control module (ICM) – ignition diagnostic monitor (IDM) signal missing	**Wiring open circuit, ICM, ECM**
232	Ignition coil 1, 2, 3 or 4 – primary circuit malfunction	**Wiring, ignition coil, ICM, ECM**
238	Ignition coil 4 – primary circuit malfunction	**Wiring, ignition coil, ICM, ECM**
241	Ignition control module (ICM) – ignition diagnostic monitor (IDM) signal malfunction	**Wiring, ICM, ECM**
244	Cylinder identification (CID) fault when cylinder balance test requested	**Wiring, CKP sensor, CMP sensor, ICM, ECM**
311	Secondary air injection (AIR) system, bank 1 – inoperative during KOER test	**Hose blocked/leaking, AIR bypass/divert solenoid, AIR pump, ECM**

Flash code	Fault location	Probable cause
312	Secondary air injection (AIR) system – misdirected during KOER test	**Hose connections**
313	Secondary air injection (AIR) system – not bypassed during KOER test	**Hose blocked/leaking, AIR bypass solenoid, AIR pump**
314	Secondary air injection (AIR) system, bank 2 – inoperative during KOER test	**Hose blocked/leaking, AIR bypass/divert solenoid, AIR pump, ECM**
326	Exhaust gas pressure sensor – voltage low	**Hose blocked/leaking, wiring, exhaust gas pressure sensor, ECM**
327 [4]	Exhaust gas recirculation (EGR) valve position sensor – minimum voltage exceeded	**Wiring open circuit/short circuit to ground, EGR valve/position sensor, ECM**
327 [5]	Exhaust gas pressure sensor – minimum voltage exceeded	**Hose blocked/leaking, wiring open circuit/ short circuit to ground, exhaust gas pressure sensor, ECM**
328	Exhaust gas recirculation (EGR) valve position sensor, closed position – voltage low	**Wiring, EGR valve/position sensor, ECM**
332 [4]	Exhaust gas recirculation (EGR) system – insufficient flow detected	**Hose blocked/leaking, wiring, EGR solenoid filter, EGR solenoid, EGR valve/position sensor, ECM**
332 [5]	Exhaust gas recirculation (EGR) system – insufficient flow detected	**Hose blocked/leaking, EGR solenoid, exhaust gas pressure sensor, EGR valve, ECM**
334	Exhaust gas recirculation (EGR) valve position sensor, closed position – voltage high	**Wiring, EGR valve/position sensor, EGR solenoid, ECM**
335 [1]	Exhaust gas pressure sensor – voltage high/ low during KOEO test	**Exhaust gas pressure sensor, blocked inlet hose, ECM**
336	Exhaust gas pressure sensor – voltage high	**Exhaust system restricted, wiring, EGR solenoid, exhaust gas pressure sensor, EGR valve, hose routing, ECM**
337 [4]	Exhaust gas recirculation (EGR) position sensor – above maximum voltage	**Wiring open circuit/short circuit to positive, EGR valve/position sensor, ECM**
337 [5]	Exhaust gas pressure sensor – above maximum voltage	**Incorrect hose routing, wiring short circuit, signal wire open circuit, exhaust gas pressure sensor, EGR solenoid/valve, ECM**
338	Engine coolant temperature (ECT) – too low	**Thermostat stuck open, coolant leak, ECM**
339	Engine coolant temperature (ECT) – too high	**Coolant level low, thermostat stuck closed, cooling system blocked, coolant blower motor worn/damaged, water pump defective, coolant leak, ECM**
341	Octane adjuster – open circuit	**Wiring, octane adjuster, ECM**
381	A/C compressor clutch – excessive cycling	**A/C system incorrectly charged, cycling switch defective, A/C circuit fault, wiring intermittent open circuit to A/C system, ECM**
411	Idle air control (IAC) valve – no control during KOER low RPM check	**Idle speed adjustment, intake leak, throttle valve tight/sticking, ignition timing incorrect, wiring short circuit to ground, IAC valve, ECM**

Flash code	Fault location	Probable cause
412	Idle air control (IAC) valve – no control during KOER high RPM check	Idle speed adjustment, throttle valve tight/sticking, wiring open circuit/short circuit to ground, IAC valve, ECM
415	Idle air control (IAC) valve – lower limit reached	Intake leak, throttle valve tight/sticking, IAC valve, ECM
416	Idle air control (IAC) valve – upper limit reached	Air filter blocked, throttle body dirty/damaged, IAC valve, ECM
452 [6]	Vehicle speed signal – incorrect signal	Wiring, rear ABS system, vehicle speed interface module, ECM
452 [7]	Vehicle speed sensor (VSS) – incorrect signal	Wiring, VSS, ECM
453	Servo leaking down – cruise control KOER test	Servo defective, ECM
454	Servo leaking up – cruise control KOER test	Servo defective, ECM
455	Insufficient RPM increase – cruise control KOER test	Intake system, ECM
456	Insufficient RPM decrease – cruise control KOER test	Intake system, ECM
457	Cruise control master/selector switch – circuit malfunction during KOEO test	Wiring, cruise control master/selector switch, ECM
458	Cruise control master/selector switch – short circuit to ground during KOEO test	Wiring short circuit to ground, cruise control master/selector switch, ECM
459	Cruise control system – ground open circuit during KOEO test	Wiring open circuit, ECM
511	Engine control module (ECM) – ROM fault during KOEO test	ECM
512	Engine control module (ECM) – KAM fault	Battery disconnected, wiring, fuses, ECM
513	Engine control module (ECM) – internal voltage fault	ECM
519	Power steering pressure (PSP) switch – open circuit during KOEO test	PSP switch not fitted – disregard trouble code, wiring open circuit, PSP switch, ECM
521	Power steering pressure (PSP) switch – no signal change during KOEO test	PSP switch not fitted – disregard trouble code, wiring, PSP switch, ECM
522 [8]	Park/neutral position (PNP) switch/neutral position (NP) switch – not in park/neutral during KOEO test/open circuit	Incorrect gear lever position, incorrect adjustment, wiring open circuit, starter motor relay disconnected, PNP/NP switch, ECM
	Transmission range (TR) sensor – not in park/neutral during KOEO test	Incorrect shift lever position, incorrect adjustment, wiring, TR sensor, ECM
524	Fuel pump (FP) – engine does not start	Wiring open circuit, fuel pump (FP) relay, ECM
	Fuel pump (FP) – engine starts	Left/front HO2S short circuit to positive, ECM
525	Vehicle in gear – during KOEO test	Incorrect shift lever position, incorrect adjustment, PNP switch open, ECM
	A/C – ON during KOEO test	A/C ON, A/C wiring short circuit to positive, ECM
526	Neutral position (NP) switch – closed circuit during KOEO test	Wiring short circuit, NP switch, ECM
	A/C – ON during KOEO test	A/C ON, A/C wiring short circuit to positive, A/C switch, ECM

Flash code	Fault location	Probable cause
527 8	Park/neutral position (PNP) switch/neutral position switch – open circuit during KOEO test	Wiring open circuit, starter motor relay disconnected, PNP/NP switch, ECM
	A/C – ON during KOEO test	A/C ON, A/C wiring short circuit to positive, A/C switch, ECM
528	Clutch pedal position (CPP) switch – circuit malfunction	Wiring open circuit, CPP switch, ECM
529	Data link/engine control module (ECM) – circuit malfunction	Wiring, ECM
532	Instrument panel – circuit malfunction	Wiring, instrument panel, ECM
533	Data link/instrument panel – circuit malfunction	Wiring, instrument panel, ECM
536	Brake pedal position (BPP) switch – circuit malfunction/no signal change during KOER test	Incorrect brake pedal operation, wiring, BPP switch, ECM
538 2	KOER dynamic response test – insufficient rpm change	Incorrect throttle operation, ECM
	KOER cylinder balance test – invalid test	Throttle operated during test, CMP sensor, ECM
538 3	KOER dynamic response test – insufficient rpm change	Incorrect throttle operation, ECM
	KOER cylinder balance test – invalid test	Throttle operated during test, CKP sensor – CID signal, ECM
539	A/C – ON during KOEO test	Wiring, A/C switch, ECM
542	Fuel pump (FP) secondary circuit – engine does not start	IFS switch, supply wire open circuit, FP ground wire, FP, ECM
	Fuel pump (FP) secondary circuit – engine starts	Supply wire short circuit to positive, FP relay, FP to ECM wire open circuit, left/front HO2S short circuit to positive, ECM
543	Fuel pump (FP) secondary circuit – engine does not start	Supply wire open circuit, fuel pump relay, ECM
	Fuel pump (FP) secondary circuit – engine starts	Left/front HO2S short circuit to positive, ECM
551	Idle air control (IAC) valve – circuit malfunction during KOEO test	Wiring, IAC valve, ECM
552	Secondary air injection (AIR) bypass solenoid – circuit malfunction during KOEO test	Wiring, AIR bypass solenoid, ECM
553	Secondary air injection (AIR) divert solenoid – circuit malfunction during KOEO test	Wiring, AIR divert solenoid, ECM
554	Fuel pressure control solenoid – circuit malfunction	Hose connection, wiring, fuel pressure control solenoid, ECM
556	Fuel pump (FP) primary circuit	IFS switch, wiring, fuel pump relay, constant control relay module, ECM
557	Fuel pump (FP) primary circuit, low speed	Wiring, FP relay/constant control relay module, ECM
558	Exhaust gas recirculation (EGR) solenoid – circuit malfunction during KOEO test	Hose damaged, wiring, EGR solenoid, ECM
559	A/C relay, primary circuit – circuit malfunction	Wiring, A/C relay, ECM

Flash code	Fault location	Probable cause
562	Engine coolant blower motor relay – circuit malfunction	**Wiring, engine coolant blower motor relay, ECM**
563	Engine coolant blower motor relay, high speed – circuit malfunction	**Wiring, engine coolant blower motor relay/ constant control relay module, ECM**
564	Engine coolant blower motor relay, low speed – circuit malfunction	**Wiring, engine coolant blower motor relay/ constant control relay module, ECM**
565	Evaporative emission (EVAP) canister purge valve 1 – malfunction	**Hose damaged, wiring, EVAP canister purge valve, ECM**
566 8	Shift solenoid (SS) 3-4 – circuit malfunction	**Wiring, SS, ECM**
567	Cruise control vent valve – circuit malfunction	**Wiring, cruise control vent valve, ECM**
568	Cruise control vacuum valve – circuit malfunction	**Wiring, cruise control vacuum valve, ECM**
569	Evaporative emission (EVAP) canister purge valve 2 – malfunction	**Hose damaged, wiring, EVAP canister purge valve, ECM**
571	Exhaust gas recirculation (EGR) atmospheric solenoid – circuit malfunction during KOEO	**Wiring, EGR atmospheric solenoid, ECM**
572	Exhaust gas recirculation (EGR) vacuum solenoid – circuit malfunction during KOEO	**Wiring, EGR vacuum solenoid, ECM**
578	A/C pressure sensor – short circuit	**Supply wire short ground, A/C pressure sensor, variable control relay module, ECM**
579	A/C pressure sensor – insufficient pressure change	**Wiring, A/C fault, A/C pressure sensor, variable control relay module, wiring to A/C compressor clutch open circuit, ECM**
581	Engine coolant blower motor – excessive current draw	**Wiring short circuit to ground, engine coolant blower motor, variable control relay module, ECM**
582	Engine coolant blower motor – open circuit/ short circuit to positive	**Supply wire open circuit/short circuit to positive, ground wire open circuit, engine coolant blower motor, variable control relay module, ECM**
583	Fuel pump – excessive current draw	**Wiring short circuit to ground, fuel pump, ECM**
584	Variable control relay module – ground open circuit	**Wiring open circuit, variable control relay module, ECM**
585	A/C compressor clutch – excessive current draw	**Wiring short circuit to ground, A/C compressor clutch, variable control relay module, ECM**
586	A/C compressor clutch – open circuit/short circuit to positive	**Wiring open circuit/short circuit to positive, A/C compressor clutch, variable control relay module, ECM**
587	Variable control relay module – communication error	**Wiring, variable control relay module, ECM**
593	Heated oxygen sensor (HO2S) – heater circuit malfunction	**Wiring, HO2S heater relay, HO2S, ECM**
617 8	Gear shift, 1-2 – malfunction	**ATF level/condition, poor connection, wiring, TR sensor, SS, clutch slipping, ECM**
618 8	Gear shift, 2-3 – malfunction	**ATF level/condition, poor connection, wiring, TR sensor, SS, clutch slipping, ECM**

Flash code	Fault location	Probable cause
619 8	Gear shift, 3-4 – malfunction	**ATF level/condition, poor connection, wiring, TR sensor, SS, clutch slipping, ECM**
621 8	Shift solenoid (SS) 1 – circuit malfunction during KOEO	**Wiring, SS, ECM**
622 8	Shift solenoid (SS) 2 – circuit malfunction during KOEO	**Wiring, SS, ECM**
623 8	Overdrive indicator lamp – circuit malfunction	**Bulb, wiring, fuse, ECM**
624 8	Transmission fluid pressure (TFP) solenoid – circuit malfunction	**Wiring, poor connection, TFP solenoid, ECM**
625 8	Engine control module (ECM), TFP control – open driver circuit	**ECM**
626 8	Transmission coast clutch solenoid – circuit malfunction during KOEO test	**Wiring, transmission coast clutch solenoid, ECM**
627 8	Torque converter clutch (TCC) solenoid – circuit malfunction	**Wiring, TCC solenoid, ECM**
628 8	Torque converter clutch (TCC) – excessive clutch slip	**ATF level/condition, TCC slipping, ECM**
629 8	Torque converter clutch (TCC) solenoid – circuit malfunction	**Wiring, TCC solenoid, ECM**
631 8	Overdrive indicator lamp – circuit malfunction during KOEO test	**Bulb, wiring, fuse, ECM**
632 8	Overdrive selection switch – no signal change during KOEO test	**Incorrect switch operation, wiring, fuse, overdrive selection switch, ECM**
633	Four wheel drive selector switch – closed during KOEO test	**Wiring short circuit, four wheel drive selector switch, internal damage to transfer case, ECM**
634 8	Transmission range (TR) sensor – voltage high/low	**Incorrect adjustment, wiring, TR sensor, ECM**
636 8	Transmission fluid temperature (TFT) sensor – signal high/low	**ATF level, wiring, TFT sensor, ECM**
637 8	Transmission fluid temperature (TFT) sensor – above maximum voltage	**ATF level, wiring open circuit, TFT sensor, ECM**
638 8	Transmission fluid temperature (TFT) sensor – below minimum voltage	**ATF level, wiring short circuit to ground, TFT sensor, ECM**
639 8	Turbine shaft speed (TSS) sensor/output shaft speed (OSS) sensor – signal low	**Wiring, TSS/OSS sensor, ECM**
641 8	Shift solenoid (SS) 3 – circuit malfunction during KOEO	**Wiring, SS, ECM**
643 8	Torque converter clutch (TCC) solenoid – circuit malfunction during KOEO test	**Wiring, TCC solenoid, ECM**
645 8	Incorrect gear ratio, 1st gear	**ATF level/condition, poor connection, wiring, TR sensor, SS, clutch slipping, ECM**
646 8	Incorrect gear ratio, 2nd gear	**ATF level/condition, poor connection, wiring, TR sensor, SS, clutch slipping, ECM**
647 8	Incorrect gear ratio, 3rd gear	**ATF level/condition, poor connection, wiring, TR sensor, SS, clutch slipping, ECM**

Flash code	Fault location	Probable cause
648 8	Incorrect gear ratio, 4th gear	**ATF level/condition, poor connection, wiring, TR sensor, SS, clutch slipping, ECM**
649 8	Transmission fluid pressure (TFP) solenoid – signal high/low	**Wiring, poor connection, TFP solenoid, ECM**
651 8	Transmission fluid pressure (TFP) solenoid – circuit malfunction	**Wiring, poor connection, TFP solenoid, ECM**
652 8	Torque converter clutch (TCC) solenoid – circuit malfunction	**Wiring, TCC solenoid, ECM**
653 8	Overdrive selection switch – no signal change during KOEO test	**Incorrect switch operation, wiring, fuse, overdrive selection switch, ECM**
654 8	Transmission range (TR) sensor – 'P' signal during KOEO test	**Transmission not in 'P' during KOEO test, incorrect adjustment, ECM**
655 8	Transmission range (TR) sensor – 'N' signal during KOEO test	**Transmission not in 'N' during KOEO test, incorrect adjustment, ECM**
656 8	Torque converter clutch (TCC) – continuous clutch slip	**ATF level/condition, TCC slipping, ECM**
657 8	Automatic transmission fluid (ATF) – temperature high	**ATF level, transmission cooling system, towing weight limit exceeded, clutch slipping, wiring, TFT sensor, ECM**
659 8	High vehicle speed – transmission in 'P'	**TR sensor/PNP switch, ECM**
667 8	Transmission range (TR) sensor – below minimum voltage	**Wiring open circuit/short circuit to ground, TR sensor, ECM**
668 8	Transmission range (TR) sensor – above maximum voltage	**Wiring open circuit/short circuit to positive, TR sensor, ECM**
675 8	Transmission range (TR) sensor – out of range	**Wiring, TR sensor, ECM**
691	Four wheel drive selector fault – incorrect engine RPM/vehicle speed	**Wiring short circuit, four wheel drive selector switch, internal damage to transfer case, ECM**
692 8	Transmission state does not match calculated ratio	**Transmission fault, ECM**
998	Engine control module (ECM) – hard fault present	**ECM**

1 Trouble code may be generated if an efficient exhaust extraction system used during KOER test – vehicles with pressure feedback EGR
2 Aerostar, Bronco, Econoline/Club Wagon, Explorer, F-Series, Ranger 2.3L (California), Ranger 3.0/4.0L
3 Ranger 2.3L (except California)
4 Bronco (except 5.8L SFI), Econoline/Club Wagon/F-Series (except 4.9/5.8L SFI)
5 Aerostar, Bronco 5.8L SFI, Econoline/Club Wagon/F-Series 4.9/5.8L SFI, Explorer, Ranger
6 Aerostar, Bronco, Econoline/Club Wagon, F-Series
7 Explorer, Ranger
8 ECM incorporates transmission control function

FORD

Model:	Engine identification:	Year:
Aspire 1.3L	VIN position 8 = H	1996-97
Contour 2.0/2.5L	VIN position 8 = L/3	1996-00
Crown Victoria 4.6L SOHC	VIN position 8 = W	1995-02
Escort 1.8/1.9/2.0L SOHC/DOHC	VIN position 8 = J/P/3/8	1996-02
Focus 2.0L SOHC/DOHC	VIN position 8 = P/3	2000-02
Mustang/GT/Cobra 3.8/4.6L	VIN position 8 = V/W/X/4	1994-02
Probe 2.0/2.5L	VIN position 8 = A/B	1996-97
Taurus 3.0L/24V	VIN position 8 = U/S/1/2	1996-02
Taurus 3.4L SHO	VIN position 8 = N	1996-99
Thunderbird 3.8/4.6L	VIN position 8 = W/4	1994-97

System: **Ford EEC-V**

Self-diagnosis

General information

- Refer to the front of this manual for general test conditions, terminology, detailed descriptions of wiring faults and a general trouble shooter for electrical and mechanical faults.
- Malfunction indicator lamp (MIL) will illuminate or flash if certain faults are recorded.
- ECM operates in backup mode if sensors fail, to enable vehicle to be driven to repair shop.

Accessing

- Trouble codes can only be displayed by using a Scan Tool connected to the data link connector (DLC):
  - Aspire/Mustang/Crown Victoria **1**.
  - Contour/Taurus **2**.
  - Escort/Focus/Probe **3**.
  - Thunderbird **4**.

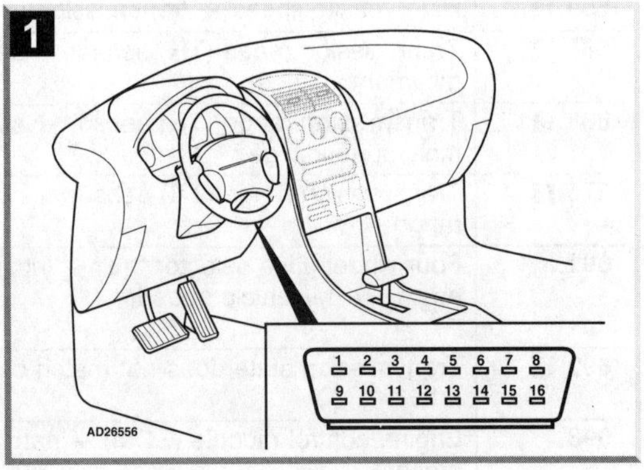

3

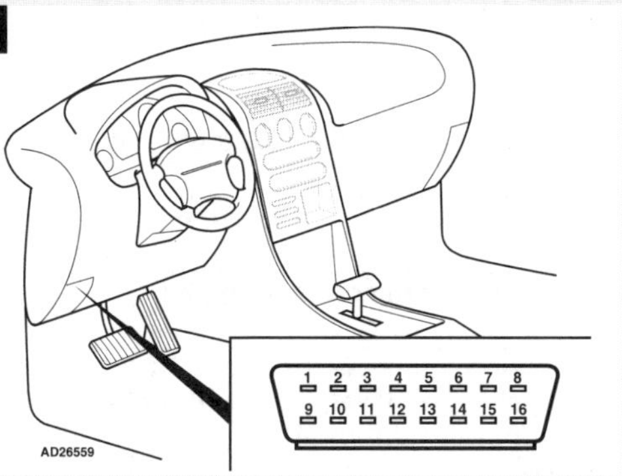

4

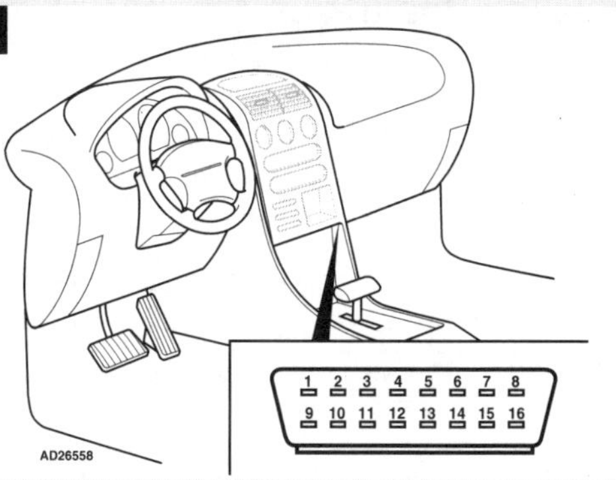

Erasing

- Trouble codes can only be erased by using a Scan Tool connected to the data link connector (DLC).

Trouble code identification

Scan code	Fault location	Probable cause
P0000-P0999	Refer to OBD-II trouble code table at the front of this manual	–
P1000	System readiness test not complete	–
P1001	Self Test	–
P1100	Mass air flow (MAF) sensor – circuit intermittent	**Wiring, MAF sensor**
P1101	Mass air flow (MAF) sensor – signal malfunction	**Wiring, battery, MAF sensor, ECM**
P1109	Intake air temperature (IAT) sensor 2 – circuit intermittent	**Wiring, IAT sensor, ECM**
P1112	Intake air temperature (IAT) – circuit intermittent	**Wiring, IAT sensor, ECM**
P1114	Intake air temperature (IAT) sensor 2 – low input	**Wiring short circuit to ground, IAT sensor, ECM**
P1115	Intake air temperature (IAT) sensor 2 – high input	**Wiring, IAT sensor, ECM**
P1116	Engine coolant temperature (ECT) sensor – signal malfunction	**Wiring, coolant thermostat, cooling system, ECT sensor, ECM**
P1117	Engine coolant temperature (ECT) sensor – circuit intermittent	**Wiring, ECT sensor, cooling system, ECM**
P1120	Throttle position (TP) sensor – signal malfunction, low	**Wiring, TP sensor**
P1121	Throttle position (TP) sensor – inconsistent with mass air flow (MAF) sensor signal	**Wiring, MAF sensor, TP sensor, air leak**
P1124	Throttle position (TP) sensor – signal malfunction	**Mechanical fault, TP sensor, ECM**

Scan code	Fault location	Probable cause
P1125	Throttle position (TP) sensor – circuit intermittent	**Wiring, TP sensor**
P1127	Self test not completed – heated oxygen sensor (HO2S), not tested	**Engine not running long enough prior to self-test**
P1128	Heated oxygen sensor (HO2S) 1, bank 1/bank 2 – interchanged	**Wiring, connectors swapped**
P1129	Heated oxygen sensor (HO2S) 2, bank 1/bank 2 – interchanged	**Wiring, connectors swapped**
P1130	Heated oxygen sensor (HO2S) 1, bank 1 – fuel trim (FT), at limit	**Wiring, intake/fuel system, HO2S, EGR system, oil level, camshaft timing, cylinder compression, ECM**
P1131	Heated oxygen sensor (HO2S) 1, bank 1 – not switching, fuel trim (FT) weak mixture	**Wiring, intake/fuel system, HO2S, EGR system, oil level, camshaft timing, cylinder compression, ECM**
P1132	Heated oxygen sensor (HO2S) 1, bank 1 – not switching, fuel trim (FT) rich mixture	**Wiring, intake/fuel system, HO2S, EGR system, oil level, camshaft timing, cylinder compression, ECM**
P1137	Heated oxygen sensor (HO2S) 2, bank 1 – not switching, fuel trim (FT) weak mixture	**Wiring, exhaust system leaking, HO2S**
P1138	Heated oxygen sensor (HO2S) 2, bank 1 – not switching, fuel trim (FT) rich mixture	**Wiring, exhaust system leaking, HO2S**
P1150	Heated oxygen sensor (HO2S) 1, bank 2 – fuel trim (FT), at limit	**Wiring, intake/fuel system, HO2S, EGR system, oil level, camshaft timing, cylinder compression, ECM**
P1151	Heated oxygen sensor (HO2S) 1, bank 2 – not switching, fuel trim (FT) weak mixture	**Wiring, intake/fuel system, HO2S, EGR system, oil level, camshaft timing, cylinder compression, ECM**
P1152	Heated oxygen sensor (HO2S) 1, bank 2 – not switching, fuel trim (FT) rich mixture	**Wiring, intake/fuel system, HO2S, EGR system, oil level, camshaft timing, cylinder compression, ECM**
P1157	Heated oxygen sensor (HO2S) 2, bank 2 – not switching, fuel trim (FT) weak mixture	**Wiring, exhaust system leaking, HO2S**
P1158	Heated oxygen sensor (HO2S) 2, bank 2 – not switching, fuel trim (FT) rich mixture	**Wiring, exhaust system leaking, HO2S**
P1168	Fuel rail pressure sensor – signal low	**Wiring, fuel level, low fuel pressure, fuel rail pressure sensor**
P1169	Fuel rail pressure sensor – signal high	**Wiring, high fuel pressure, fuel rail pressure sensor**
P1170	Oxygen sensor (O2S)/Heated oxygen sensor (HO2S) 1, bank 1 – no activity detected	**Wiring open circuit, heating inoperative, O2S/HO2S**
P1173	Oxygen sensor (O2S)/Heated oxygen sensor (HO2S) 1, bank 2 – no activity detected	**Wiring, O2S/HO2S, ECM**
P1180	Fuel delivery system – pressure low	**Fuel filter blocked, fuel supply blocked**
P1181	Fuel delivery system – pressure high	**Fuel pressure regulator**
P1183	Engine oil temperature sensor – circuit malfunction	**Wiring, engine oil temperature sensor, ECM**

Scan code	Fault location	Probable cause
P1184	Engine oil temperature sensor – out of self-test range	Wiring, engine oil temperature sensor, ECM
P1195	Exhaust gas pressure sensor – circuit malfunction	Wiring, exhaust gas pressure sensor, exhaust gas pressure control solenoid, ECM
P1196	Starter switch – circuit malfunction	Wiring, starter motor, starter switch, ECM
P1220	Throttle control system (traction control) – malfunction	Wiring, throttle position motor, mechanical fault, throttle position (TP) sensor, ECM
P1224	Throttle position (TP) sensor (traction control), voltage out of range	Wiring, mechanical fault, TP sensor
P1229	Supercharger (SC) intercooler pump	Wiring, SC intercooler pump, SC intercooler pump relay, ECM
P1230	Fuel pump/variable load control module – circuit intermittent	Wiring
P1231	Fuel pump – circuit malfunction	Wiring, fuel pump relay
P1232	Low speed fuel pump – malfunction	Wiring, low speed fuel pump relay, ECM
P1233	Fuel system – malfunction	Wiring, IFS switch, fuel pump, fuel pump control module, ECM
P1234	Fuel system – malfunction	Wiring, IFS switch, fuel pump, fuel pump control module, ECM
P1235	Fuel pump control – signal malfunction	Wiring, fuel pump, fuel pump control module, ECM
P1236	Fuel pump control – signal malfunction	Wiring, fuel pump, fuel pump control module, ECM
P1237	Fuel pump – circuit malfunction	Wiring, fuel pump, fuel pump relay, fuel pump control module
P1238	Fuel pump – circuit malfunction	Wiring, fuel pump, fuel pump relay, fuel pump control module
P1244	Alternator – low input	Wiring, drive belt, alternator, ECM
P1245	Alternator – high input	Wiring, alternator, ECM
P1246	Alternator	Wiring, alternator, drive belt, ECM
P1250	Fuel pressure regulator control solenoid	Wiring, fuel pressure regulator control solenoid, vacuum hoses, ECM
P1260	Theft detected – vehicle immobilized	–
P1270	Maximum engine RPM or vehicle speed limit reached	Engine over revved in neutral, wheel slippage
P1285	Cylinder head temperature (CHT) sensor – overheat condition	Wiring, cooling system fault, CHT sensor
P1288	Cylinder head temperature (CHT) sensor – out of self-test range	Wiring, cooling system fault, CHT sensor
P1289	Cylinder head temperature (CHT) sensor – high input	Wiring, cooling system fault, CHT sensor
P1290	Cylinder head temperature (CHT) sensor – low input	Wiring, cooling system fault, CHT sensor
P1299	Cylinder head temperature (CHT) sensor – over temperature condition	Wiring, cooling system fault, CHT sensor

Scan code	Fault location	Probable cause
P1309	Misfire monitor disabled	**CMP sensor, CKP sensor, ECT sensor, MAF sensor, ECM**
P1345	Camshaft position (CMP) sensor – circuit malfunction	**Wiring, CMP sensor, ECM**
P1351	Ignition system diagnostic monitor – input circuit malfunction	**Wiring, ECM**
P1352	Ignition coil 1, primary – circuit malfunction	**Wiring, spark plug(s), ignition coil, ICM**
P1353	Ignition coil 2, primary – circuit malfunction	**Wiring, spark plug(s), ignition coil, ICM**
P1354	Ignition coil 3, primary – circuit malfunction	**Wiring, spark plug(s), ignition coil, ICM**
P1355	Ignition coil 4, primary – circuit malfunction	**Wiring, spark plug(s), ignition coil, ICM**
P1358	Ignition system diagnostic monitor – out of self-test range	**Wiring, ignition coil, ECM**
P1359	Ignition system – spark output (SPOUT) signal – circuit malfunction	**Wiring, ECM**
P1364	Ignition coil, primary – circuit malfunction	**Wiring, spark plug(s), ignition coil, ICM**
P1380	Camshaft position (CMP) actuator – circuit malfunction	**Wiring, CMP actuator, ECM**
P1381	Camshaft position (CMP) actuator, bank 1 – timing over advanced	**Wiring, timing belt/chain, CMP actuator, low oil pressure**
P1383	Camshaft position (CMP) actuator, bank 1 – timing over retarded	**Wiring, timing belt/chain, CMP actuator**
P1390	Octane adjustment	**Octane adjustment connector, ECM**
P1400 1	Exhaust gas recirculation (EGR) valve position sensor – low input	**Wiring, EGR solenoid, EGR valve position sensor, ECM**
P1400 2	Exhaust gas recirculation (EGR) pressure sensor – circuit malfunction, low input	**Wiring, EGR pressure sensor, ECM**
P1401 1	Exhaust gas recirculation (EGR) valve position sensor – high input	**Wiring, EGR solenoid, EGR valve position sensor, ECM**
P1401 2	Exhaust gas recirculation (EGR) pressure sensor – circuit malfunction, high input	**Wiring, EGR pressure sensor, ECM**
P1402	Exhaust gas recirculation (EGR) valve position sensor – circuit malfunction	**Wiring, EGR solenoid, EGR valve position sensor, ECM**
P1403	Exhaust gas recirculation (EGR) pressure sensor	**Pressure hoses reversed**
P1405	Exhaust gas recirculation (EGR) system – upstream hose off or plugged	**EGR hoses**
P1406	Exhaust gas recirculation (EGR) system – downstream hose off or blocked	**EGR hoses**
P1407	Exhaust gas recirculation (EGR) solenoid	**Wiring, hose leak/blockage, EGR solenoid, EGR valve position sensor, ECM**
P1408 3	Exhaust gas recirculation (EGR) solenoid	**Wiring, hose leak/blockage, EGR solenoid, EGR valve position sensor, ECM**
P1408 4	Exhaust gas recirculation (EGR) system – out of self-test range	**Wiring, EGR hoses, EGR valve, EGR solenoid, ECM**
P1409	Exhaust gas recirculation (EGR) solenoid – circuit malfunction	**Wiring, EGR solenoid, ECM**

Scan code	Fault location	Probable cause
P1410	Fuel pressure regulator control solenoid/ exhaust gas recirculation (EGR) control solenoid	**Wiring, connectors swapped**
P1411	Secondary air injection (AIR) system – malfunction	**AIR hoses, AIR pump, AIR solenoid**
P1413	Secondary air injection monitor circuit – low input	**Wiring, AIR solenoid, AIR pump, ECM**
P1414	Secondary air injection monitor circuit – high input	**Wiring, AIR solenoid, ECM**
P1442	Evaporative emission (EVAP) system – small leak detected	**Hose connection(s), intake leak, EVAP canister, EVAP canister purge valve**
P1443 5	Evaporative emission (EVAP) system – malfunction	**Wiring, EVAP system hoses, EVAP canister, EVAP canister purge valve**
P1443 6	Idle air control (IAC) valve	**Wiring, IAC valve, ECM**
P1444	Evaporative emission (EVAP) system – low input	**Wiring, EVAP pressure sensor, ECM**
P1445	Evaporative emission (EVAP) system – high input	**Wiring, EVAP pressure sensor, ECM**
P1450	Evaporative emission (EVAP) system, vent control – circuit malfunction	**Wiring, EVAP system hoses, EVAP canister purge valve, fuel filler cap blocked**
P1451	Evaporative emission (EVAP) canister purge valve – circuit malfunction	**Wiring, EVAP canister purge valve, ECM**
P1452	Evaporative emission (EVAP) system – unable to bleed fuel tank vacuum	**Wiring, EVAP system hoses, EVAP canister purge valve, fuel filler cap blocked**
P1455	Evaporative emission (EVAP) system – malfunction	**Fuel filler cap blocked, EVAP system hoses**
P1460	Wide open throttle (WOT) relay, A/C – circuit malfunction	**Wiring, WOT relay, ECM**
P1461	A/C refrigerant pressure sensor – high voltage	**Wiring, A/C refrigerant pressure sensor, ECM**
P1462	A/C refrigerant pressure sensor – low voltage	**Wiring, A/C refrigerant pressure sensor, ECM**
P1463	A/C refrigerant pressure sensor – insufficient pressure change	**Wiring, A/C system – mechanical fault, A/C refrigerant pressure sensor, A/C compressor clutch**
P1464	A/C demand out	**Wiring, A/C ON – during self-test, WOT relay – (A/C cut-out), ECM**
P1469	Frequent A/C compressor clutch cycling	**Wiring, A/C system – mechanical fault**
P1473	Engine coolant blower motor – circuit malfunction	**Wiring, engine coolant blower motor, variable load control module**
P1474	Engine coolant blower motor (low speed) – control circuit malfunction	**Wiring, engine coolant blower motor relay, ECM**
P1477	Engine coolant blower motor (medium speed) – control circuit malfunction	**Wiring, engine coolant blower motor relay, ECM**
P1479	Engine coolant blower motor (high speed) – control circuit malfunction	**Wiring, engine coolant blower motor relay, ECM**

Scan code	Fault location	Probable cause
P1480	Engine coolant blower motor (low speed) – with low speed fan ON	**Wiring, engine coolant blower motor relay**
P1481	Engine coolant blower motor (low speed) – with high speed fan ON	**Wiring, engine coolant blower motor relay**
P1483	Engine coolant blower motor – intermittent fault	**Wiring, engine coolant blower motor, variable load control module**
P1484	Variable load control module – malfunction	**Wiring open circuit, variable load control module**
P1485	Exhaust gas recirculation (EGR) control solenoid – circuit malfunction	**Wiring, hose leak/blockage, EGR control solenoid, ECM**
P1486	Exhaust gas recirculation (EGR) vent solenoid – circuit malfunction	**Wiring, hose leak/blockage, EGR vent solenoid, ECM**
P1487	Exhaust gas recirculation (EGR) control solenoid – circuit malfunction	**Wiring, hose leak/blockage, EGR control solenoid, EGR valve position sensor, EGR solenoid, exhaust system blocked, ECM**
P1500	Vehicle speed sensor (VSS) – circuit intermittent	**Wiring, VSS, ECM**
P1501	Vehicle speed sensor (VSS) – out of self-test range	**Electro-magnetic interference**
P1502	Vehicle speed sensor (VSS) – intermittent signal	**Wiring, VSS/ABS wheel speed sensors**
P1504	Idle air control (IAC) valve – circuit malfunction	**Wiring, IAC valve, ECM**
P1505	Idle air control (IAC) system – air trim, at limit	**Intake system leaking, air filter blocked, throttle body, IAC valve**
P1506	Idle air control (IAC) valve – over-speed error	**Wiring, IAC valve, vacuum leaks, ECM**
P1507	Idle air control (IAC) valve – under-speed error	**Wiring, IAC valve, hose blocked, ECM**
P1512	Intake manifold runner control (IMRC) module – malfunction, stuck closed –	**Mechanical fault**
P1513	Intake manifold runner control (IMRC) module – malfunction, stuck closed	**Mechanical fault**
P1516	Intake manifold runner control (IMRC) module – input error (bank 1)	**Wiring, mechanical fault**
P1517	Intake manifold runner control (IMRC) module – input error (bank 2)	**Wiring, mechanical fault**
P1518	Intake manifold runner control (IMRC) module – malfunction, stuck open	**Wiring, IMRC actuator, ECM**
P1519	Intake manifold runner control (IMRC) module – malfunction, stuck closed	**Wiring, IMRC actuator, ECM**
P1520	Intake manifold runner control (IMRC) module – circuit malfunction	**Mechanical fault, vacuum hoses, IMRC actuator, ECM**
P1521	Intake manifold air control solenoid 1 – circuit malfunction	**Wiring, intake manifold air control solenoid 1, ECM**
P1522	Intake manifold air control solenoid 2 – circuit malfunction	**Wiring, intake manifold air control solenoid 2, ECM**

Scan code	Fault location	Probable cause
P1523	Intake manifold air control solenoid, high speed – circuit malfunction	**Wiring, intake manifold air control solenoid, ECM**
P1530	A/C compressor clutch – circuit malfunction	**Wiring open circuit/short circuit to positive, A/C compressor clutch, variable load control module**
P1537	Intake manifold runner control (IMRC) module – malfunction, stuck closed	**Wiring, mechanical fault, ECM**
P1538	Intake manifold runner control (IMRC) module – malfunction, stuck closed	**Wiring, mechanical fault, ECM**
P1539	A/C compressor clutch – circuit malfunction	**Wiring, A/C compressor clutch, variable load control module**
P1549	Intake manifold air control system – malfunction	**Wiring, intake manifold air control solenoid, ECM**
P1550	Power steering pressure (PSP) sensor – malfunction	**Wiring, PSP sensor, ECM**
P1565	Cruise control system	**Wiring, CPP switch, BPP switch, cruise control module, cruise control master switch, VSS**
P1566	Cruise control system	**Wiring, CPP switch, BPP switch, cruise control module, cruise control master switch, VSS**
P1567	Cruise control system	**Wiring, CPP switch, BPP switch, cruise control module, cruise control master switch, VSS**
P1568	Cruise control system	**Wiring, CPP switch, BPP switch, cruise control module, cruise control master switch, VSS**
P1605	'Keep alive' memory test failure	**Wiring, battery terminal corrosion, loose battery connection, ECM**
P1609	Engine control module (ECM) – internal fault	**ECM**
P1625	Variable load control module, power supply to engine coolant blower motor – circuit malfunction	**Wiring, variable load control module**
P1626	Variable load control module, power supply to A/C system – circuit malfunction	**Wiring, variable load control module**
P1633	'Keep alive' power supply – voltage too low	**Wiring, ECM**
P1635	Tire/axle ratio out of range	**Incorrect tire size, incorrect axle ratio, incorrect vehicle ID configuration**
P1636	Engine control module (ECM) – internal fault	**ECM**
P1640	Diagnostic trouble codes stored	**P-codes stored in another control module**
P1641	Fuel pump – circuit malfunction	**Wiring, fuel pump relay, fuel pump, ECM**
P1650	Power steering pressure (PSP) switch – malfunction	**Wiring, PSP switch, steering wheel turned during self-test, ECM**
P1651	Power steering pressure (PSP) switch – input	**Wiring, vehicle towed with engine running, PSP switch, DTC not erased, ECM**
P1700	Transmission system – malfunction	**Transmission**

Scan code	Fault location	Probable cause
P1701	Transmission system – reverse engagement error	**Transmission**
P1702	Transmission range (TR) sensor – circuit intermittent	**Wiring, TR sensor**
P1703	Brake pedal position (BPP) switch – out of self-test range	**Wiring, brake pedal depressed during self-test, rear electronic module/lights control module, BPP switch, ECM**
P1704	Transmission range (TR) sensor – malfunction	**Wiring, TR sensor**
P1705 7	Transmission range (TR) sensor – out of self-test range	**Wiring, TR sensor**
P1705 8	Park/neutral position (PNP) switch/clutch pedal position (CPP) switch	**Wiring, PNP switch, CPP switch, ECM**
P1709 9	Throttle position (TP) sensor – malfunction	**Wiring, mechanical fault, TP sensor, ECM**
P1709 10	Park/neutral position (PNP) switch/clutch pedal position (CPP) switch – out of self-test range	**Wiring, PNP switch/CPP switch, ECM**
P1710	Transmission fluid temperature (TFT) sensor – mid range failure	**Wiring, TFT sensor**
P1711	Transmission fluid temperature (TFT) sensor – out of self-test range	**Wiring, TFT sensor, ECM**
P1713	Transmission fluid temperature (TFT) sensor – low range failure	**Wiring, TFT sensor**
P1714	Shift solenoid (SS) A – inductive malfunction	**Wiring, SS, ECM/TCM**
P1715	Shift solenoid (SS) B – inductive malfunction	**Wiring, SS, ECM/TCM**
P1716	Shift solenoid (SS) C – inductive malfunction	**Wiring, SS, ECM/TCM**
P1717	Shift solenoid (SS) D – inductive malfunction	**Wiring, SS, ECM/TCM**
P1718	Transmission fluid temperature sensor – high range failure	**Wiring, transmission fluid temperature sensor**
P1719	Overdrive speed sensor – circuit malfunction	**Wiring, overdrive speed sensor, TCM**
P1720	Vehicle speed sensor (VSS) – circuit malfunction	**Wiring, VSS, vehicle speedometer, ECM**
P1728	Torque converter clutch (TCC) – slip fault	**TCC**
P1729	4x4L switch – malfunction	**Wiring, four wheel drive shift control module, ECM**
P1740	Torque converter clutch (TCC) solenoid – circuit malfunction	**Wiring, TCC solenoid, ECM/TCM**
P1741	Torque converter clutch (TCC) – control error	**Wiring, TCC, TCM**
P1742	Torque converter clutch (TCC) solenoid	**Wiring, TCC solenoid**
P1743	Torque converter clutch (TCC) solenoid	**Wiring, TCC solenoid**
P1744	Torque converter clutch (TCC) – malfunction	**Wiring, TCC, TCC solenoid, TCM**
P1745	Transmission fluid pressure (TFP) solenoid – circuit malfunction	**Wiring, TFP solenoid, TCM, ECM**
P1746	Transmission fluid pressure (TFP) solenoid – circuit malfunction	**Wiring, TFP solenoid, TCM, ECM**

Scan code	Fault location	Probable cause
P1747	Transmission fluid pressure (TFP) solenoid – circuit malfunction	Wiring short circuit, TFP solenoid, TCM
P1749	Transmission fluid pressure (TFP) solenoid – malfunction	Wiring, TFP solenoid, TCM
P1751	Shift solenoid (SS) A – performance	Wiring, SS
P1754	Coast clutch solenoid – circuit malfunction	Wiring, coast clutch solenoid
P1756	Shift solenoid (SS) B – performance	Wiring, SS
P1760	Transmission fluid pressure (TFP) solenoid – short circuit	Wiring, TFP solenoid, TCM
P1761	Shift solenoid (SS) C – performance	Wiring, SS
P1762	Overdrive band failed/off	Transmission
P1765	Shift solenoid (SS), downshift 3-2 – malfunction	Wiring, SS, ECM/TCM
P1767	Torque converter clutch (TCC) – circuit malfunction	Wiring, TCC, TCM
P1780	Overdrive selection switch	Wiring, transmission control switch, ECM
P1781	4x4L switch – out of self-test range	Wiring, four wheel drive shift control module, 4x4L switch, ECM
P1783	Transmission fluid temperature (TFT) sensor – overheating condition	Wiring, TFT sensor, TCM, ECM
P1784	Transmission, reverse/first gear – mechanical fault	Transmission
P1785	Transmission, first/second gear – mechanical fault	Transmission
P1786	Gear selection, 3-2 – downshift malfunction	Wiring, TR sensor, shift solenoids, transmission mechanical fault
P1787	Gear selection, 2-1 – downshift malfunction	Wiring, TR sensor, shift solenoids, transmission mechanical fault
P1788	Transmission shift control valve, 3-2 downshift/ coastdown downshift – open circuit	Wiring, transmission shift control valve, TCM
P1789	Transmission shift control valve, 3-2 downshift/ coastdown downshift – short circuit	Wiring, transmission shift control valve, TCM
P1792	Closed throttle position (CTP) switch – circuit malfunction	Wiring, CTP switch, ECM
P1794	Battery voltage – low	Wiring, engine control relay, battery
P1795	Exhaust gas pressure sensor – circuit malfunction	Wiring, exhaust gas pressure sensor, exhaust gas pressure control solenoid, ECM
P1797	Park/neutral position (PNP) switch/clutch pedal position (CPP) switch	Wiring, PNP switch, CPP switch, ECM
P1900	Output shaft speed (OSS) sensor – circuit intermittent	Wiring, poor connection, OSS sensor, ECM/ TCM
P1901	Turbine shaft speed (TSS) sensor – circuit intermittent	Wiring, poor connection, TSS sensor, ECM/ TCM
U1020	CAN data bus, A/C – A/C compressor clutch invalid or missing data	Wiring, A/C trouble code(s) stored

Scan code	Fault location	Probable cause
U1021	CAN data bus, A/C – communication error	Wiring, A/C trouble code(s) stored, ECM
U1039	CAN data bus, antilock brake system (ABS) – vehicle speed invalid or missing data	Wiring, ABS trouble code(s) stored
U1041	CAN data bus, antilock brake system (ABS) – vehicle speed invalid or missing data	Wiring, ABS trouble code(s) stored
U1051	CAN data bus, engine control module (ECM) – brake signal invalid or missing data	Wiring, BPP switch, bulb(s)
U1073	CAN data bus, cooling system – communication error	Wiring, variable load control module, ECM (alternative fuel) – if fitted, ECM
U1089	CAN data bus, active suspension – invalid or missing data	Wiring, active suspension trouble code(s) stored
U1098	CAN data bus, antilock brake system (ABS) – vehicle speed invalid or missing data	Wiring, ABS trouble code(s) stored
U1130	CAN data bus, fuel pump – communication error	Wiring, instrument panel, ECM
U1131	CAN data bus, fuel pump – communication error	Wiring, instrument panel, ECM
U1135	CAN data bus, engine control module (ECM) – ignition switch invalid or missing data	Wiring
U1147	CAN data bus, alarm	Wiring, alarm system trouble code(s) stored
U1243	CAN data bus, A/C – invalid or missing data	Wiring, A/C system trouble code(s) stored
U1256	CAN data bus – communication error	Wiring, trouble code(s) stored in other control modules
U1262	CAN data bus – communication error	Wiring, trouble code(s) stored in other control modules
U1341	CAN data bus, A/C – vehicle speed invalid or missing data	Wiring, A/C trouble code(s) stored
U1451	CAN data bus, alarm	Wiring, alarm system trouble code(s) stored
U2195	CAN data bus, lights control module – invalid or missing data	Wiring, lights control module trouble code(s) stored
U2243	CAN data bus, A/C	Wiring, A/C system trouble code(s) stored

1 Probe
2 Except Probe
3 Probe
4 Except Probe
5 Crown Victoria/Escort/Mustang/Thunderbird
6 Aspire/Contour/Focus/Probe/Taurus
7 Mustang/Thunderbird/Crown Victoria
8 Except Mustang/Thunderbird/Crown Victoria
9 Aspire
10 Except Aspire

FORD

Model:	Engine identification:	Year:
Contour 2.0L	VIN position 8 = 3	1995
Contour 2.5L	VIN position 8 = L	1995
Escort 1.9L/HO	VIN position 8 = J/K	1992-95
Mustang 2.3L	VIN position 8 = M	1992-93
Probe 2.0L	VIN position 8 = A	1993-95
Taurus 3.0L	VIN position 8 = U	1992-95
Taurus 3.0L SHO	VIN position 8 = Y	1992-95
Taurus 3.2L SHO	VIN position 8 = P	1992-95
Tempo 2.3L	VIN position 8 = X	1992-94
Tempo 3.0L	VIN position 8 = U	1992-94

System: **Ford EEC-IV**

Self-diagnosis

General information

- Refer to the front of this manual for general test conditions, terminology, detailed descriptions of wiring faults and a general trouble shooter for electrical and mechanical faults.
- Malfunction indicator lamp (MIL) will illuminate if certain faults are recorded.
- ECM operates in backup mode if sensors fail, to enable vehicle to be driven to repair shop.
- Trouble codes can be displayed by the MIL or by using an analogue voltmeter or Scan Tool connected to the data link connectors (DLC) 1 and 2:
  - Contour **1**.
  - Escort **2**.
  - Mustang **3**.
  - Probe **4**.
  - Taurus **5**.
  - Tempo **6**.

1

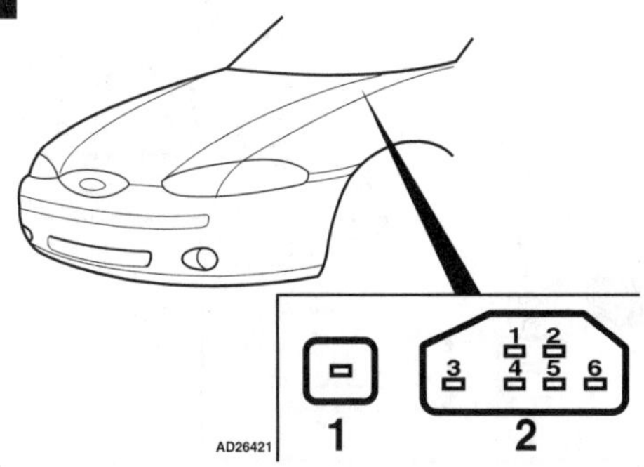

2

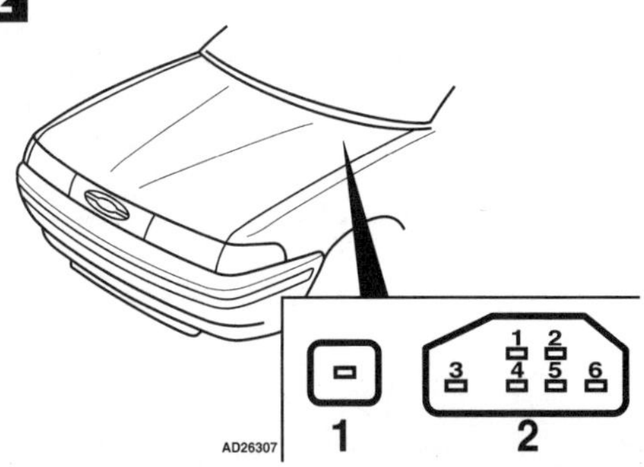

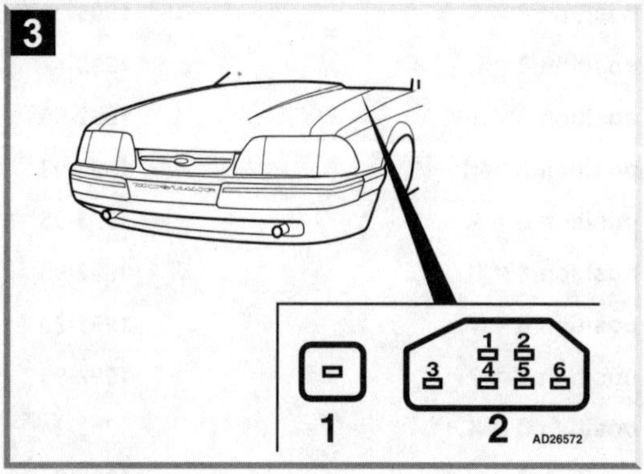

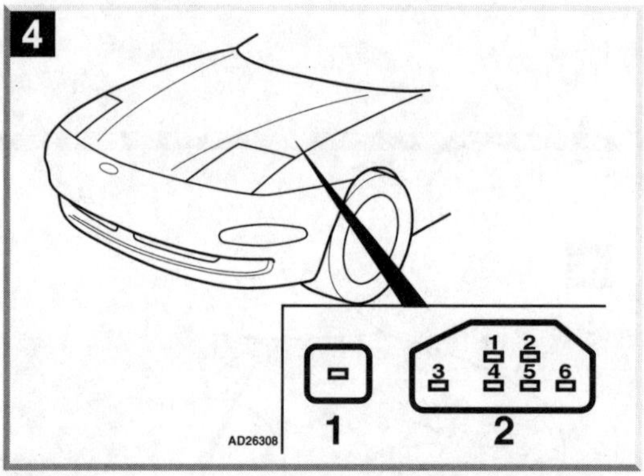

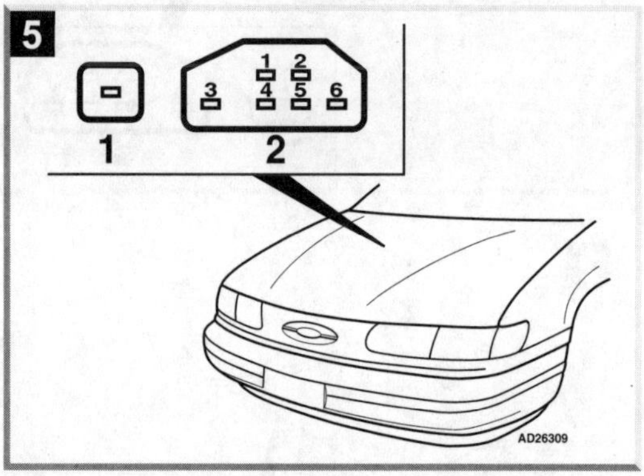

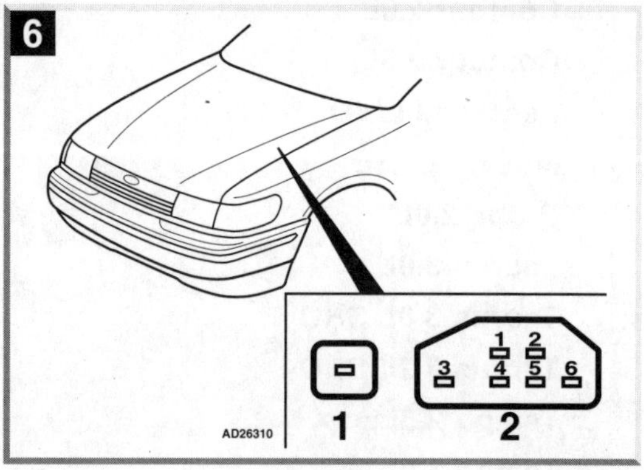

Preparatory conditions

- Automatic transmission in 'PARK'.
- All auxiliary equipment, including air conditioning, switched OFF.

Accessing – Key On Engine Off (KOEO) test

Using malfunction indicator lamp (MIL)

- Start engine.
- Allow to idle until engine at normal operating temperature.
- Switch ignition OFF.
- Wait 10 seconds.
- If engine does not start or stalls: Continue with KOEO test.
- If fitted: Remove 'EEC TEST' cover to access data link connector (DLC).
- Jump DLC 1 terminal (STI) to DLC 2 terminal 2 (SIG RTN) **7**.
- Switch ignition ON.

NOTE: ***MIL may flash briefly while fast DTCs are issued for Scan Tool followed by slow DTC output.***

- Count MIL flashes. Note trouble codes. Compare with trouble code table.
- Each trouble code is displayed twice. Firstly for present faults followed by those for memorized faults as follows:
  - Trouble codes are displayed by illuminated MIL switching OFF.
  - Each trouble code consists of two or three digits.
  - A group of 0.5 second flashes indicate each digit **8** [**A**].
  - A 0.5 second pause separates each flash **8** [**B**].

- ○ A 2 second pause separates each trouble code digit 8 [**C**].
- ○ A 4 second pause separates each trouble code 8 [**D**].
- ○ For example: Three digit trouble code 122 displayed 8. Throttle position (TP) sensor.

NOTE: ***Two digit trouble codes displayed in a similar pattern.***

- ○ A 6-9 second pause 9 [**B**] indicates the end of trouble code output for present faults 9 [**A**].
- ○ A 0.5 second flash (separator code) 9 [**C**], followed by a 6-9 second pause 9 [**D**] indicates the start of trouble code output for memorized faults 9 [**E**].
- ○ Trouble codes for memorized faults displayed in the same manner as those for present faults.

NOTE: ***Memorized trouble codes refer to faults that have occurred in the last 80 warm-up cycles.***

- Switch ignition OFF.
- Remove jump lead.
- Trouble codes can also be displayed by using a Scan Tool connected to the data link connectors (DLC) 1 and 2.

7

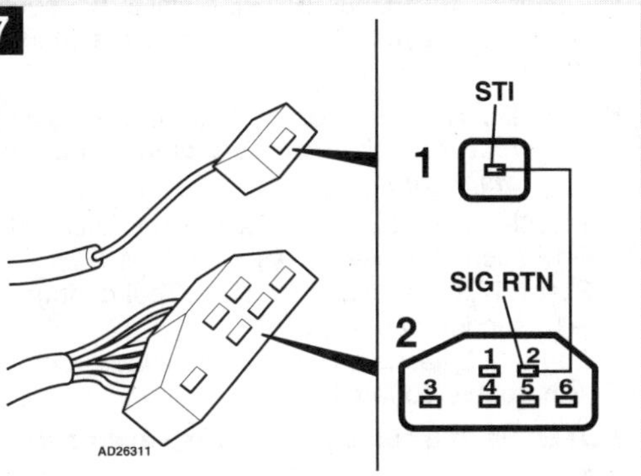

8

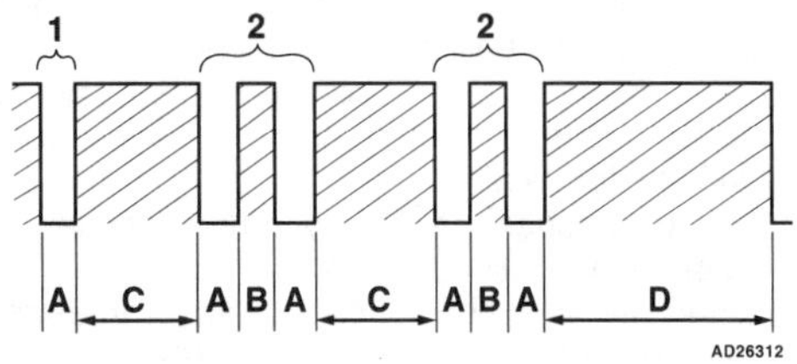

9

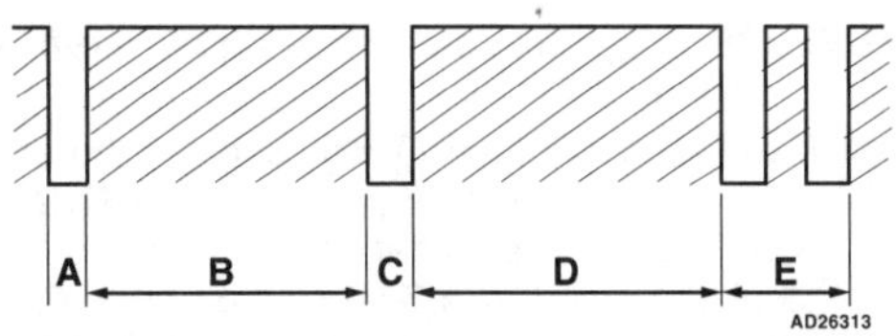

Using analogue voltmeter

- Ensure ignition switched OFF.
- If fitted: Remove 'EEC TEST' cover to access data link connector (DLC).
- Connect an analogue voltmeter between data link connector (DLC) 2 terminal 4 (STO) and battery (12 volt) positive 10.
- Set analogue voltmeter to a DC voltage range of 0-20 V.
- Start engine.
- Allow to idle until engine at normal operating temperature.
- Switch ignition OFF.
- Wait 10 seconds.
- If engine does not start or stalls: Continue with KOEO test.
- Jump DLC 1 terminal (STI) to DLC 2 terminal 2 (SIG RTN) 10.
- Switch ignition ON.

NOTE: ***Voltmeter needle may briefly deflect while fast DTCs are issued for Scan Tool followed by slow DTC output.***

- Count voltmeter needle deflections. Note trouble codes. Compare with trouble code table.
- Each trouble code is displayed twice. Firstly for present faults followed by those for memorized faults as follows:
  - Trouble codes are displayed by voltmeter needle deflecting to zero.
  - Each trouble code consists of two or three digits.
  - A group of 0.5 second needle deflections indicate each digit **8** [**A**].
  - A 0.5 second pause separates each needle deflection **8** [**B**].
  - A 2 second pause separates each trouble code digit **8** [**C**].
  - A 4 second pause separates each trouble code **8** [**D**].
  - For example: Three digit trouble code 122 displayed **8**. Throttle position (TP) sensor.

NOTE: ***Two digit trouble codes displayed in a similar pattern.***

  - A 6-9 second pause **9** [**B**] indicates the end of trouble code output for present faults **9** [**A**].
  - A 0.5 second needle deflection (separator code) **9** [**C**], followed by a 6-9 second pause **9** [**D**] indicates the start of trouble code output for memorized faults **9** [**E**].
  - Trouble codes for memorized faults displayed in the same manner as those for present faults.

NOTE: ***Memorized trouble codes refer to faults that have occurred in the last 80 warm-up cycles.***

- Switch ignition OFF.
- Remove jump lead and voltmeter.
- Trouble codes can also be displayed by using a Scan Tool connected to the data link connectors (DLC) 1 and 2.

10

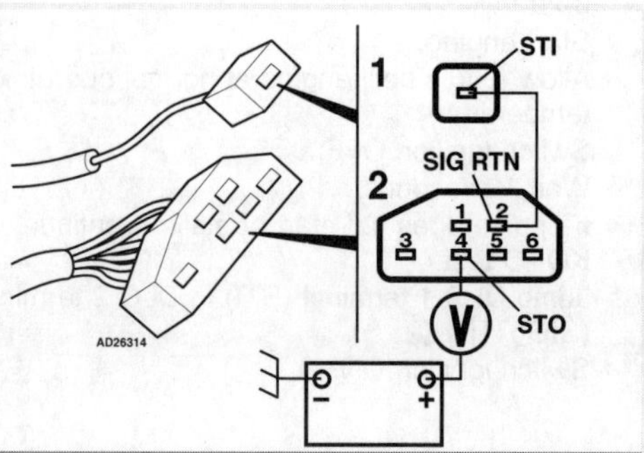

Accessing – Key On Engine Running (KOER) test

Using malfunction indicator lamp (MIL)

- Start engine.
- Allow to idle until engine at normal operating temperature.
- Increase engine speed to over 2000 rpm for 2 minutes.
- Switch ignition OFF.
- Wait 10 seconds.
- If fitted: Remove 'EEC TEST' cover to access data link connector (DLC).
- Jump DLC 1 terminal (STI) to DLC 2 terminal 2 (SIG RTN) **7**.
- Start engine.
- Engine identification code displayed as follows:
  - 4 cylinder: Two 0.5 second flashes.
  - 6 cylinder: Three 0.5 second flashes.
  - For example: 6 cylinder engine identified **11** [**A**].
- With PSP switch: Within 1-2 seconds, turn steering wheel at least ½ a turn and release.
- With BPP switch: Depress and release brake pedal.
- If fitted: Operate overdrive selection switch twice.
- If after 6-20 seconds **11** [**B**], a single 0.5 second flash is displayed (dynamic response signal) **11** [**C**]: Briefly fully open throttle.

NOTE: ***DO NOT fully open throttle during KOER test if no dynamic response signal is displayed.***

- Trouble codes displayed for present faults after 4-15 second pause **11** [**D**].
- Fast DTCs are issued for Scan Tool during 4-15 second pause **11** [**E**].
- Count MIL flashes. Note trouble codes. Compare with trouble code table.

NOTE: ***Each trouble code is displayed twice.***

- Trouble codes for KOER test are displayed in the same manner as those for KOEO test **11** [**F**].
- For example: Three digit trouble code 111 is displayed **11**. No fault found.
- Switch ignition OFF.
- Remove jump lead.
- Trouble codes can also be displayed by using a Scan Tool connected to the data link connectors (DLC) 1 and 2.

11

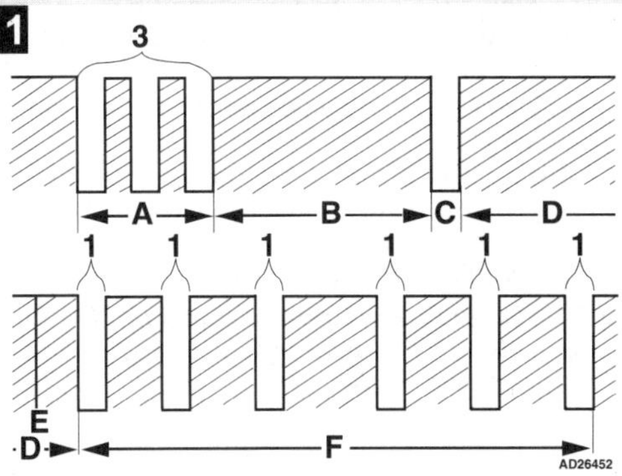

Using analogue voltmeter

- Ensure ignition switched OFF.
- If fitted: Remove 'EEC TEST' cover to access data link connector (DLC).
- Connect an analogue voltmeter between data link connector (DLC) 2 terminal 4 (STO) and battery (12 volt) positive 10.
- Set analogue voltmeter to a DC voltage range of 0-20 V.
- Start engine.
- Allow to idle until engine at normal operating temperature.
- Increase engine speed to over 2000 rpm for 2 minutes.
- Switch ignition OFF.
- Wait 10 seconds.
- Jump DLC 1 terminal (STI) to DLC 2 terminal 2 (SIG RTN) 10.
- Start engine.
- Engine identification code displayed as follows:
  - 4 cylinder: Two 0.5 second needle deflections.
  - 6 cylinder: Three 0.5 second needle deflections.
  - For example: 6 cylinder engine identified 11 [**A**].
- With PSP switch: Within 1-2 seconds, turn steering wheel at least ½ a turn and release.
- With BPP switch: Depress and release brake pedal.
- If fitted: Operate overdrive selection switch twice.
- If after 6-20 seconds 11 [**B**], a single 0.5 second needle deflection is displayed (dynamic response signal) 11 [**C**]: Briefly fully open throttle.

NOTE: ***DO NOT fully open throttle during KOER test if no dynamic response signal is displayed.***

- Trouble codes displayed for present faults after 4-15 second pause 11 [**D**].
- Fast DTCs are issued for Scan Tool during 4-15 second pause 11 [**E**].
- Count needle deflections. Note trouble codes. Compare with trouble code table.

NOTE: ***Each trouble code is displayed twice.***

- Trouble codes for KOER test are displayed in the same manner as those for KOEO test 11 [**F**].
- For example: Three digit trouble code 111 is displayed 11. No fault found.
- Switch ignition OFF.
- Remove jump lead and voltmeter.
- Trouble codes can also be displayed by using a Scan Tool connected to the data link connectors (DLC) 1 and 2.

Erasing

NOTE: ***Disconnecting battery ground lead may cause driveability problems until the engine management system relearns basic values.***

Using malfunction indicator lamp (MIL)

- Ensure ignition switched OFF.
- If fitted: Remove 'EEC TEST' cover to access data link connector (DLC).
- Jump DLC 1 terminal (STI) to DLC 2 terminal 2 (SIG RTN) 7.
- Switch ignition ON.
- When trouble codes start to be displayed: Remove jump lead.
- Switch ignition OFF.
- Repeat accessing procedure to ensure no memorized trouble codes remain in ECM fault memory.
- Trouble codes can also be erased by using a Scan Tool connected to the data link connectors (DLC) 1 and 2.
- If emission related component(s) have been replaced:
  - Disconnect battery ground cable for at least 5 minutes to erase adaptive memory from ECM.
  - If required: Drive vehicle for 10 miles or more to allow ECM to relearn basic values.

NOTE: ***Disconnecting battery ground lead may also erase memory from electronic units, such as radio.***

Using analogue voltmeter

- Ensure ignition switched OFF.
- If fitted: Remove 'EEC TEST' cover to access data link connector (DLC).

- Connect an analogue voltmeter between data link connector (DLC) 2 terminal 4 (STO) and battery (12 volt) positive 10.
- Set analogue voltmeter to a DC voltage range of 0-20 V.
- Jump DLC 1 terminal (STI) to DLC 2 terminal 2 (SIG RTN) 10.
- Switch ignition ON.
- When trouble codes start to be displayed: Remove jump lead.
- Switch ignition OFF.
- Repeat accessing procedure to ensure no memorized trouble codes remain in ECM fault memory.
- Trouble codes can also be erased by using a Scan Tool connected to the data link connectors (DLC) 1 and 2.
- If emission related component(s) have been replaced:
  - Disconnect battery ground cable for at least 5 minutes to erase adaptive memory from ECM.
  - If required: Drive vehicle for 10 miles or more to allow ECM to relearn basic values.

NOTE: ***Disconnecting battery ground lead may also erase memory from electronic units, such as radio.***

Trouble code identification

Two digit

Flash code	Fault location	Probable cause
11	No fault found	–
12	Idle speed, high RPM check – unable to control during KOER test	**Intake system, throttle tight/sticking, IAC valve/ISC actuator, ECM**
13	Idle speed, low RPM check – unable to control during KOER test	**Intake system, throttle tight/sticking, IAC valve/ISC actuator, ECM**
14	Engine speed (RPM) sensor/profile ignition pickup (PIP) circuit – intermittent signal	**Incorrectly installed RPM sensor, wiring, poor connection, electrical interference, RPM sensor, ECM**
15	Engine control module (ECM) – KAM/ROM fault	**Battery disconnected, wiring, fuses, ECM**
16	Engine RPM too low for heated oxygen sensor (HO2S) test	**Intake system, throttle tight/sticking, IAC valve/ISC actuator, ECM**
	Ignition diagnostic monitor (IDM) circuit – no signal during KOEO test	**Wiring, ICM, ECM**
18	Spark output (SPOUT) circuit open	**Wiring, ICM, ECM**
	Ignition diagnostic monitor (IDM) circuit/spark output (SPOUT) circuit grounded	**Wiring, ICM, ECM**
19	Engine control module (ECM) – internal voltage fault	**ECM**
21	Engine coolant temperature (ECT) sensor – incorrect signal	**Wiring, ECT sensor, ECM**
22	Manifold absolute pressure (MAP) sensor/ barometric pressure (BARO) sensor – incorrect signal	**Vacuum hose damaged, trapped vacuum, high atmospheric pressure, excessive EGR, wiring, MAP/BARO sensor, ECM**
23	Throttle position (TP) sensor – incorrect signal	**Throttle valve tight/sticking, wiring, TP sensor, ECM**
24	Intake air temperature (IAT) sensor – incorrect signal	**Wiring, IAT sensor, ECM**

Flash code	Fault location	Probable cause
25	Knock sensor (KS) – no signal during dynamic response test	**Wiring, KS, ECM**
26	Mass air flow (MAF) sensor – incorrect signal	**Intake leak/blockage, wiring, MAF sensor, ECM**
29	Vehicle speed signal – incorrect signal	**Wiring, vehicle speed interface module, VSS, ECM**
31	Exhaust gas recirculation (EGR) valve position sensor – below minimum voltage	**Wiring short circuit to ground, EGR valve/ position sensor, ECM**
32	Exhaust gas recirculation (EGR) valve position sensor – below closed limit	**Wiring short circuit to ground, EGR valve/ position sensor, ECM**
33	Exhaust gas recirculation (EGR) valve – not opening	**Wiring, EGR valve tight/sticking, ECM**
34	Exhaust gas recirculation (EGR) valve position sensor – above closed limit	**Wiring short circuit to positive, EGR valve/ position sensor, ECM**
35	Exhaust gas recirculation (EGR) valve position sensor – above maximum voltage	**Wiring short circuit to positive, EGR valve/ position sensor, ECM**
41	Heated oxygen sensor (HO2S), bank 1 – mixture lean/not switching	**Intake leak, fuel pressure low, fuel filter/ pump, injectors, HO2S, AIR system, MAF/ MAP sensor, ECM**
42	Heated oxygen sensor (HO2S), bank 1 – mixture rich/not switching	**Intake blocked, excessive fuel in engine oil, fuel pressure high, breather system, injectors, HO2S, AIR system, EVAP system, MAF/MAP sensor, ignition system, ECM**
43	Throttle position (TP) sensor – incorrect signal at idle	**Incorrectly adjusted, TP sensor, ECM**
44	Secondary air injection (AIR) system, bank 1 – inoperative	**Hose connection, wiring, AIR pump motor, AIR solenoid, ECM**
45	Secondary air injection (AIR) – upstream during KOER test	**Hose connection, wiring, AIR solenoid, ECM**
46	Secondary air injection (AIR) – not bypassed during KOER test	**Hose connection, wiring, AIR solenoid, ECM**
47	Four wheel drive selector switch – closed during KOEO test	**Wiring short circuit, four wheel drive selector switch, internal damage to transfer case, ECM**
49 **1**	Gear shift, 1-2 – malfunction	**ATF level/condition, poor connection, wiring, TR sensor, SS, clutch slipping, ECM**
51	Engine coolant temperature (ECT) sensor – open circuit	**Wiring open circuit, ECT sensor, ECM**
52	Power steering pressure (PSP) switch – open circuit/no signal change	**Wiring open circuit, PSP switch, ECM**
53	Throttle position (TP) sensor – above maximum voltage	**TP sensor loose, wiring short circuit to positive, TP sensor, ECM**
54	Intake air temperature (IAT) sensor – open circuit	**Wiring open circuit, IAT sensor, ECM**
56	Mass air flow (MAF) sensor – above maximum voltage	**Wiring short circuit to positive, MAF sensor, ECM**
59 **1**	Gear shift, 2-3 – malfunction	**ATF level/condition, poor connection, wiring, TR sensor, SS, clutch slipping, ECM**

Flash code	Fault location	Probable cause
61	Engine coolant temperature (ECT) sensor – short circuit to ground	**Wiring short circuit to ground, ECT sensor, ECM**
62 **1**	Torque converter clutch (TCC) – excessive clutch slip	**ATF level/condition, TCC slipping, ECM**
63	Throttle position (TP) sensor – below minimum voltage	**TP sensor loose, wiring open circuit/short circuit to ground, TP sensor, ECM**
64	Intake air temperature (IAT) sensor – short circuit to ground	**Wiring short circuit to ground, IAT sensor, ECM**
65 **1**	Overdrive selection switch – no signal change during KOEO test	**Incorrect switch operation, wiring, fuse, overdrive selection switch, ECM**
66	Mass air flow (MAF) sensor – below minimum voltage	**Wiring short circuit to ground, MAF sensor, ECM**
67	Neutral position (NP) switch – open circuit	**Incorrect adjustment, wiring open circuit, NP switch, ECM**
	Clutch pedal position (CPP) switch – circuit malfunction	**Incorrect adjustment, wiring, CPP switch, ECM**
	A/C – on during KOEO test	**Wiring, A/C switch, ECM**
67 **1**	Transmission range (TR) sensor – voltage high/low	**Incorrect adjustment, wiring, TR sensor, ECM**
68 **1**	Automatic transmission fluid (ATF) – temperature high	**ATF level, transmission cooling system, towing weight limit exceeded, clutch slipping, wiring, TFT sensor, ECM**
69 **1**	Gear shift, 3-4 – malfunction	**ATF level/condition, poor connection, wiring, TR sensor, SS, clutch slipping, ECM**
72	Manifold absolute pressure (MAP) sensor/ mass air flow (MAF) sensor – incorrect signal change during dynamic response test	**Intake leak, wiring, MAP sensor, ECM**
73	Throttle position (TP) sensor – incorrect signal change during dynamic response test	**Incorrect throttle operation, wiring, TP sensor, ECM**
74	Brake pedal position (BPP) switch – open circuit/no signal change during KOER test	**Incorrect brake pedal operation, wiring, BPP switch, ECM**
75	Brake pedal position (BPP) switch – closed circuit	**Wiring, BPP switch, ECM**
77	Wide open throttle (WOT) switch – no signal change during dynamic response test	**Incorrect throttle operation, wiring, WOT switch, ECM**
79	A/C – on during KOEO test	**Wiring, A/C switch, ECM**
81	Secondary air injection (AIR) divert solenoid – circuit malfunction during KOEO test	**Wiring, AIR divert solenoid, ECM**
82	Secondary air injection (AIR) bypass solenoid – circuit malfunction during KOEO test	**Wiring, AIR bypass solenoid, ECM**
84	Exhaust gas recirculation (EGR) solenoid – circuit malfunction during KOEO test	**Wiring, EGR solenoid, ECM**
85	Evaporative emission (EVAP) canister purge valve – circuit malfunction during KOEO test	**Wiring, EVAP canister purge valve, ECM**
86	AT – shift solenoid	**Wiring, shift solenoid, ECM**

Flash code	Fault location	Probable cause
87	Fuel pump, primary circuit – circuit malfunction	**IFS switch, wiring, fuel pump, ECM**
89	Torque converter clutch (TCC) lock-up solenoid	**Wiring, TCC lock-up solenoid, ECM**
91	Heated oxygen sensor (HO2S), bank 2 – mixture lean/not switching during KOER test	**Intake leak, fuel pressure low, fuel filter/ pump, injectors, HO2S, AIR system, MAF/ MAP sensor, ECM**
91 [1]	Shift solenoid (SS) 1 – circuit malfunction during KOEO test	**Wiring, SS, ECM**
92	Heated oxygen sensor (HO2S), bank 1 – mixture rich/not switching during KOER test	**Intake blocked, excessive fuel in engine oil, fuel pressure high, breather system, injectors, HO2S, AIR system, EVAP system, MAF/MAP sensor, ignition system, ECM**
92 [1]	Shift solenoid (SS) 2 – circuit malfunction during KOEO test	**Wiring, SS, ECM**
93 [1]	Transmission coast clutch solenoid – circuit malfunction during KOEO test	**Wiring, transmission coast clutch solenoid, ECM**
94	Secondary air injection (AIR) system, bank 2 – inoperative during KOER test	**Hose connection, wiring, AIR pump motor, AIR solenoid, ECM**
94 [1]	Torque converter clutch (TCC) solenoid – circuit malfunction during KOEO test	**Wiring, TCC solenoid, ECM**
95	Fuel pump, secondary circuit – circuit malfunction	**IFS switch, wiring, fuel pump, ECM**
96	Fuel pump, secondary circuit – circuit malfunction	**IFS switch, wiring open circuit to battery, fuel pump, ECM**
97 [1]	Overdrive indicator lamp – circuit malfunction	**Bulb, wiring, fuse, ECM**
98	Engine control module (ECM) – hard fault detected during KOER test	**HO2S bank 2 – indicates lean, ECM**
98 [1]	Engine control module (ECM), transmission fluid pressure (TFP) control – open driver circuit during KOEO test	**ECM**
99	Engine control module – idle speed not learnt during KOER test	**Idle speed adjusted, ECM**
99 [1]	Transmission fluid pressure (TFP) solenoid – circuit malfunction during KOEO test	**Wiring, poor connection, TFP solenoid, ECM**

[1] ECM incorporates transmission control function

Three digit

Flash code	Fault location	Probable cause
111	No fault found	–
112	Intake air temperature (IAT) sensor – below minimum voltage	**Wiring short circuit to ground, poor connection, IAT sensor, ECM**
113	Intake air temperature (IAT) sensor – above maximum voltage	**Wiring open circuit, poor connection, IAT sensor, ECM**

Flash code	Fault location	Probable cause
114 **1**	Intake air temperature (IAT) sensor – incorrect signal	**Air temperature below 10°C (50°F), poor connection, IAT sensor, ECM**
116	Engine coolant temperature (ECT) sensor – incorrect signal	**Coolant level low, poor connection, ECT sensor, ECM**
117	Engine coolant temperature (ECT) sensor – below minimum voltage	**Wiring short circuit to ground, poor connection, ECT sensor, ECM**
118	Engine coolant temperature (ECT) sensor – above maximum voltage	**Wiring open circuit, poor connection, IAT sensor, ECM**
121	Throttle position (TP) sensor – closed throttle voltage high/low	**Throttle valve tight/sticking, TP sensor loose/damaged, ECM**
	Throttle position (TP) sensor – voltage inconsistent with MAF sensor signal	**Throttle valve tight/sticking, TP sensor loose/damaged, ECM**
122	Throttle position (TP) sensor – below minimum voltage	**TP sensor loose, wiring open circuit/short circuit to ground, TP sensor, ECM**
123	Throttle position (TP) sensor – above maximum voltage	**TP sensor loose, wiring short circuit to positive, TP sensor, ECM**
124	Throttle position (TP) sensor – voltage high	**Throttle tight/sticking, intake leak, MAP/BARO sensor, ECM**
125	Throttle position (TP) sensor – voltage low	**Throttle tight/sticking, intake leak, MAP/BARO sensor, ECM**
126	Manifold absolute pressure (MAP) sensor/barometric pressure (BARO) sensor – voltage high/low	**Vacuum hose damaged/trapped, high atmospheric pressure, excessive EGR, wiring, MAP/BARO sensor, ECM**
128	Manifold absolute pressure (MAP) sensor – loss of vacuum	**Vacuum hose damaged/disconnected, MAP sensor, ECM**
129	Manifold absolute pressure (MAP) sensor – insufficient voltage change during KOER dynamic response test	**Incorrect throttle operation, vacuum hose damaged/disconnected, MAP sensor, IAC system, ECM**
	Mass air flow (MAF) sensor – insufficient voltage change during KOER dynamic response test	**Incorrect throttle operation, intake leak, wiring, MAF sensor, IAC system, ECM**
136	Heated oxygen sensor (HO2S), bank 2 – mixture lean during KOER test	**Intake leak, fuel pressure low, fuel filter/pump, injectors, HO2S, AIR system, MAP sensor, ECM**
137	Heated oxygen sensor (HO2S), bank 2 – mixture rich during KOER test	**Intake blocked, excessive fuel in engine oil, fuel pressure high, breather system, injectors, HO2S, AIR system, EVAP system, MAP sensor, ignition system, ECM**
138	Cold start injector – insufficient flow during KOER test	**Fuel pressure, wiring, cold start injector, ECM**
139	Heated oxygen sensor (HO2S), bank 2 – not switching	**Intake leak/blocked, injectors, HO2S, AIR system, breather system, EVAP system, MAP sensor, ignition system, ECM**
141	Fuel system indicates lean	**Intake leak, fuel pressure low, fuel filter/pump, injectors, HO2S, ECM**
144	Heated oxygen sensor (HO2S), bank 1 – not switching	**Intake leak/blocked, injectors, HO2S, AIR system, breather system, EVAP system, MAP sensor, ignition system, ECM**

Flash code	Fault location	Probable cause
157	Mass air flow (MAF) sensor – below minimum voltage	**Wiring short circuit to ground, poor connection, MAF sensor, IAC system, ECM**
158	Mass air flow (MAF) sensor – above maximum voltage	**Wiring short circuit to positive, MAF sensor screen blocked, MAF sensor, ECM**
159 **2**	Mass air flow (MAF) sensor – voltage high/low	**Intake leak, IAC valve, MAF sensor, ECM**
167	Throttle position (TP) sensor – insufficient voltage change during KOER dynamic response test	**Incorrect throttle operation, TP sensor, ECM**
171	Fuel system control, bank 1 – adaptive limits reached	**Intake leak/blocked, breather system, fuel pressure, fuel filter/pump, injectors, HO2S, AIR system, EVAP system, MAP sensor, ignition system, ECM**
172	Heated oxygen sensor (HO2S), bank 1 – mixture lean	**Intake leak, fuel pressure low, fuel filter/pump, injectors, HO2S, AIR system, MAP sensor, ECM**
173	Heated oxygen sensor (HO2S), bank 1 – mixture rich	**Intake blocked, excessive fuel in engine oil, fuel pressure high, breather system, injectors, HO2S, AIR system, EVAP system, MAP sensor, ignition system, ECM**
174	Heated oxygen sensor (HO2S), bank 1 – slow response	**Intake leak, excessive fuel in engine oil, fuel pressure, fuel filter/pump, injectors, HO2S, AIR system, MAF/MAP sensor, ECM**
175	Fuel system control, bank 2 – adaptive limits reached	**Intake leak/blocked, breather system, excessive fuel in engine oil, fuel pressure, fuel filter/pump, injectors, HO2S, AIR system, EVAP system, MAP sensor, ignition system, ECM**
176	Heated oxygen sensor (HO2S), bank 2 – mixture lean	**Intake leak, fuel pressure low, fuel filter/pump, injectors, HO2S, AIR system, MAP sensor, ECM**
177	Heated oxygen sensor (HO2S), bank 2 – mixture rich	**Intake blocked, fuel pressure high, breather system, injectors, HO2S, AIR system, EVAP system, MAP sensor, ignition system, ECM**
178	Heated oxygen sensor (HO2S), bank 2 – slow response	**Intake leak, excessive fuel in engine oil, fuel pressure, fuel filter/pump, injectors, HO2S, AIR system, MAF/MAP sensor, ECM**
179	Fuel control system, bank 1 – lean adaption limit reached at part throttle, mixture rich	**Intake leak, fuel pressure low, fuel filter blocked, wiring, injector, MAP sensor, ECM**
181	Fuel control system, bank 1 – rich adaption limit reached at part throttle, mixture lean	**Intake blocked, fuel pressure high, wiring, injector, MAP sensor, ECM**
182	Fuel control system, bank 1 – lean adaption limit reached at idle, mixture rich	**Intake leak, fuel pressure low, fuel filter blocked, wiring, injector, MAP sensor, ECM**
183	Fuel control system, bank 1 – rich adaption limit reached at idle, mixture lean	**Intake blocked, fuel pressure high, wiring, injector, MAP sensor, ECM**
184	Mass air flow (MAF) sensor – signal high	**MAF sensor screen blocked, MAF sensor, ECM**
185	Mass air flow (MAF) sensor – signal low	**Intake leak, breather system, MAF sensor, ECM**

Flash code	Fault location	Probable cause
186	Injector – pulse width long/mass air flow (MAF) sensor – signal low	**Intake leak, fuel pressure low, MAF sensor, ECM**
187	Injector – pulse width short/mass air flow (MAF) sensor – signal high	**Intake blocked, fuel pressure high, MAF sensor, ECM**
188	Fuel system control, bank 2 – lean adaption limit reached at part throttle, mixture rich	**Intake leak, fuel pressure low, fuel filter blocked, wiring, injector, MAP sensor, ECM**
189	Fuel system control, bank 2 – rich adaption limit reached at part throttle, mixture lean	**Intake blocked, fuel pressure high, wiring, injector, MAP sensor, ECM**
191	Fuel system control, bank 2 – lean adaption limit reached at idle, mixture rich	**Intake leak, fuel pressure low, fuel filter blocked, wiring, injector, MAP sensor, ECM**
192	Fuel system control, bank 2 – rich adaption limit reached at idle, mixture lean	**Intake blocked, fuel pressure high, wiring, injector, MAP sensor, ECM**
193	Fuel quality sensor – circuit malfunction	**Wiring, fuel separation, fuel quality sensor, ECM**
194	Cylinder balance test – malfunction	**Wiring, injector, ECM**
195	Cylinder balance test – malfunction	**Wiring, injector, ECM**
211	Profile ignition pickup (PIP) – circuit malfunction	**Wiring, poor connection, arcing secondary ignition component – coil/cap/rotor/HT leads/plugs, electrical interference, ECM**
212	Ignition control module (ICM) – ignition diagnostic monitor (IDM) signal missing	**Wiring, ICM, ECM**
213	Spark output (SPOUT) – open circuit	**Wiring, ICM, ECM**
214	Camshaft position (CMP) sensor – circuit malfunction	**Wiring, CMP sensor, ICM, ECM**
215	Ignition coil 1 – primary circuit malfunction	**Wiring, ignition coil, ICM, ECM**
216	Ignition coil 2 – primary circuit malfunction	**Wiring, ignition coil, ICM, ECM**
217	Ignition coil 3 – primary circuit malfunction	**Wiring, ignition coil, ICM, ECM**
218	Ignition diagnostic monitor (IDM), left side – signal missing	**Wiring, ignition coil, ECM**
219	Ignition timing, defaulted to 10° – spark output (SPOUT) circuit open	**Wiring open circuit, ICM, ECM**
221	Ignition timing error	**Wiring, ICM, ECM**
222	Ignition diagnostic monitor (IDM), right side – signal missing	**Wiring, ignition coil, ECM**
223	Loss of dual plug inhibit (DPI) control	**Wiring, ICM, ECM**
224	Ignition coil 1, 2, 3 or 4 – primary circuit malfunction	**Wiring, ignition coil, ICM, ECM**
225	Knock sensor (KS) – no signal during KOER dynamic response test	**Wiring, KS, ECM**
226	Ignition control module (ICM) – ignition diagnostic monitor (IDM) signal missing	**Wiring open circuit, ICM, ECM**
232	Ignition coil 1, 2, 3 or 4 – primary circuit malfunction	**Wiring, ignition coil, ICM, ECM**
238	Ignition coil 4 – primary circuit malfunction	**Wiring, ignition coil, ICM, ECM**
241	Ignition control module (ICM) – ignition diagnostic monitor (IDM) signal malfunction	**Wiring, ICM, ECM**

Flash code	Fault location	Probable cause
244	Cylinder identification (CID) fault when cylinder balance test requested	**Wiring, CKP sensor, CMP sensor, ICM, ECM**
311	Secondary air injection (AIR) system, bank 1 – inoperative during KOER test	**Hose blocked/leaking, AIR bypass/divert solenoid, AIR pump, ECM**
312	Secondary air injection (AIR) system – misdirected during KOER test	**Hose connections**
313	Secondary air injection (AIR) system – not bypassed during KOER test	**Hose blocked/leaking, AIR bypass solenoid, AIR pump**
314	Secondary air injection (AIR) system, bank 2 – inoperative during KOER test	**Hose blocked/leaking, AIR bypass/divert solenoid, AIR pump, ECM**
326 [3]	Exhaust gas recirculation (EGR) system – flow at idle detected	**EGR valve stuck open, wiring, EGRT sensor damaged, ECM**
326 [4]	Exhaust gas pressure sensor – voltage low	**Hose blocked/leaking, wiring, exhaust gas pressure sensor, ECM**
326 [2] [5]	Exhaust gas pressure sensor – voltage low	**Hose blocked/leaking, EGR solenoid filter, EGR solenoid, wiring, EGR valve, exhaust gas pressure sensor, ECM**
327 [6]	Exhaust gas recirculation (EGR) valve position sensor – minimum voltage exceeded	**Wiring open circuit/short circuit to ground, EGR valve/position sensor, ECM**
327 [7]	Exhaust gas recirculation temperature (EGRT) sensor – minimum voltage exceeded	**Wiring short circuit to ground, EGRT sensor, ECM**
327 [8]	Exhaust gas recirculation (EGR) valve position sensor – minimum voltage exceeded	**Supply wire open circuit, signal wire open circuit/short circuit to ground, EGR valve/ position sensor, ECM**
327 [9]	Exhaust gas pressure sensor – minimum voltage exceeded	**Hose blocked/leaking, wiring open circuit/ short circuit to ground, exhaust gas pressure sensor, ECM**
328	Exhaust gas recirculation (EGR) valve position sensor, closed position – voltage low	**Wiring, EGR valve/position sensor, ECM**
332 [6]	Exhaust gas recirculation (EGR) system – insufficient flow detected	**Hose blocked/leaking, wiring, EGR solenoid filter, EGR solenoid, EGR valve/position sensor, ECM**
332 [7]	Exhaust gas recirculation (EGR) system – insufficient flow detected	**Hose blocked/leaking, EGR valve, wiring, exhaust gas pressure sensor, exhaust system, EGRT sensor, ECM**
332 [8]	Exhaust gas recirculation (EGR) system – insufficient flow detected	**Hose blocked/leaking, EGR solenoid, EGR valve/position sensor, EGR atmospheric solenoid, ECM**
332 [9]	Exhaust gas recirculation (EGR) system – insufficient flow detected	**Hose blocked/leaking, EGR solenoid, exhaust gas pressure sensor, EGR valve, ECM**
334	Exhaust gas recirculation (EGR) valve position sensor, closed position – voltage high	**Wiring, EGR valve/position sensor, EGR solenoid, ECM**
335 [2]	Exhaust gas pressure sensor – voltage high/ low during KOEO test	**Exhaust gas pressure sensor, blocked inlet hose, ECM**
336	Exhaust gas pressure sensor – voltage high	**Exhaust system restricted, wiring, EGR solenoid, exhaust gas pressure sensor, EGR valve, hose routing, ECM**

Flash code	Fault location	Probable cause
337 [6]	Exhaust gas recirculation (EGR) position sensor – above maximum voltage	**Wiring open circuit/short circuit to positive, EGR valve/position sensor, ECM**
337 [7]	Exhaust gas recirculation temperature (EGRT) sensor – above maximum voltage	**Wiring open circuit, EGRT sensor, ECM**
337 [8]	Exhaust gas recirculation (EGR) position sensor – above maximum voltage	**Incorrect hose routing, Wiring open circuit/ short circuit to positive, EGR valve/position sensor, EGR solenoid, ECM**
337 [4]	Exhaust gas pressure sensor – above maximum voltage	**Incorrect hose routing, wiring short circuit, signal wire open circuit, exhaust gas pressure sensor, EGR solenoid/valve, ECM**
337 [5]	Exhaust gas pressure sensor – above maximum voltage	**Exhaust system restricted, wiring short circuit, signal wire open circuit, exhaust gas pressure sensor, ECM**
338	Engine coolant temperature (ECT) – too low	**Thermostat stuck open, coolant leak, ECM**
339	Engine coolant temperature (ECT) – too high	**Coolant level low, thermostat stuck closed, cooling system blocked, coolant blower motor worn/damaged, water pump defective, coolant leak, ECM**
341	Octane adjuster – open circuit	**Wiring, octane adjuster, ECM**
381	A/C compressor clutch – excessive cycling	**A/C system incorrectly charged, cycling switch defective, A/C circuit fault, wiring intermittent open circuit to A/C system, ECM**
411	Idle air control (IAC) valve – no control during KOER low RPM check	**Idle speed adjustment, intake leak, throttle valve tight/sticking, ignition timing incorrect, wiring short circuit to ground, IAC valve, ECM**
412	Idle air control (IAC) valve – no control during KOER high RPM check	**Idle speed adjustment, throttle valve tight/ sticking, wiring open circuit/short circuit to ground, IAC valve, ECM**
415	Idle air control (IAC) valve – lower limit reached	**Intake leak, throttle valve tight/sticking, IAC valve, ECM**
416	Idle air control (IAC) valve – upper limit reached	**Air filter blocked, throttle body dirty/ damaged, IAC valve, ECM**
452	Vehicle speed sensor (VSS) – incorrect signal	**Wiring, VSS, ECM**
453	Servo leaking down – cruise control KOER test	**Servo defective, ECM**
454	Servo leaking up – cruise control KOER test	**Servo defective, ECM**
455	Insufficient RPM increase – cruise control KOER test	**Intake system, ECM**
456	Insufficient RPM decrease – cruise control KOER test	**Intake system, ECM**
457	Cruise control master/selector switch – circuit malfunction during KOEO test	**Wiring, cruise control master/selector switch, ECM**
458	Cruise control master/selector switch – short circuit to ground during KOEO test	**Wiring short circuit to ground, cruise control master/selector switch, ECM**
459	Cruise control system – ground open circuit during KOEO test	**Wiring open circuit, ECM**

Flash code	Fault location	Probable cause
511	Engine control module (ECM) – ROM fault during KOEO test	**ECM**
512	Engine control module (ECM) – KAM fault	**Battery disconnected, wiring, fuses, ECM**
513	Engine control module (ECM) – internal voltage fault	**ECM**
519	Power steering pressure (PSP) switch – open circuit during KOEO test	**PSP switch not fitted – disregard trouble code, wiring open circuit, PSP switch, ECM**
521	Power steering pressure (PSP) switch – no signal change during KOEO test	**PSP switch not fitted – disregard trouble code, wiring, PSP switch, ECM**
522 [13]	Park/neutral position (PNP) switch/neutral position (NP) switch – not in park/neutral during KOEO test/open circuit	**Incorrect gear lever position, incorrect adjustment, wiring open circuit, starter motor relay disconnected, PNP/NP switch, ECM**
	Transmission range (TR) sensor – not in park/neutral during KOEO test	**Incorrect shift lever position, incorrect adjustment, wiring, TR sensor, ECM**
524 [10]	Fuel pump (FP) – engine does not start	**Wiring open circuit, fuel pump (FP) relay, ECM**
	Fuel pump (FP) – engine starts	**Left/front HO2S short circuit to positive, ECM**
524 [11]	Fuel pump (FP), low speed	**Wiring open circuit, FP resistor, FP relay, ECM**
525	Vehicle in gear – during KOEO test	**Incorrect shift lever position, incorrect adjustment, PNP switch open, ECM**
	A/C – ON during KOEO test	**A/C ON, A/C wiring short circuit to positive, ECM**
526	Neutral position (NP) switch – closed circuit during KOEO test	**Wiring short circuit, NP switch, ECM**
	A/C – ON during KOEO test	**A/C ON, A/C wiring short circuit to positive, A/C switch, ECM**
527 [13]	Park/neutral position (PNP) switch/neutral position switch – open circuit during KOEO test	**Wiring open circuit, starter motor relay disconnected, PNP/NP switch, ECM**
	A/C – ON during KOEO test	**A/C ON, A/C wiring short circuit to positive, A/C switch, ECM**
528	Clutch pedal position (CPP) switch – circuit malfunction	**Wiring open circuit, CPP switch, ECM**
529	Data link/engine control module (ECM) – circuit malfunction	**Wiring, ECM**
532	Instrument panel – circuit malfunction	**Wiring, instrument panel, ECM**
533	Data link/instrument panel – circuit malfunction	**Wiring, instrument panel, ECM**
536	Brake pedal position (BPP) switch – circuit malfunction/no signal change during KOER test	**Incorrect brake pedal operation, wiring, BPP switch, ECM**
538	KOER dynamic response test – insufficient rpm change	**Incorrect throttle operation, ECM**
	KOER cylinder balance test – invalid test	**Throttle operated during test, CMP sensor, ECM**

Flash code	Fault location	Probable cause
539	A/C – ON during KOEO test	**Wiring, A/C switch, ECM**
542	Fuel pump (FP) secondary circuit – engine does not start	**IFS switch, supply wire open circuit, FP ground wire, FP, ECM**
	Fuel pump (FP) secondary circuit – engine starts	**Supply wire short circuit to positive, FP relay, FP to ECM wire open circuit, left/front HO2S short circuit to positive, ECM**
543	Fuel pump (FP) secondary circuit – engine does not start	**Supply wire open circuit, fuel pump relay, ECM**
	Fuel pump (FP) secondary circuit – engine starts	**Left/front HO2S short circuit to positive, ECM**
551	Idle air control (IAC) valve – circuit malfunction during KOEO test	**Wiring, IAC valve, ECM**
552	Secondary air injection (AIR) bypass solenoid – circuit malfunction during KOEO test	**Wiring, AIR bypass solenoid, ECM**
553	Secondary air injection (AIR) divert solenoid – circuit malfunction during KOEO test	**Wiring, AIR divert solenoid, ECM**
554	Fuel pressure control solenoid – circuit malfunction	**Hose connection, wiring, fuel pressure control solenoid, ECM**
556 [10]	Fuel pump (FP) primary circuit, high speed	**Wiring, FP relay/constant control relay module, ECM**
556 [12]	Fuel pump (FP) primary circuit	**IFS switch, wiring, fuel pump relay, constant control relay module, ECM**
557	Fuel pump (FP) primary circuit, low speed	**Wiring, FP relay/constant control relay module, ECM**
558	Exhaust gas recirculation (EGR) solenoid – circuit malfunction during KOEO test	**Hose damaged, wiring, EGR solenoid, ECM**
559	A/C relay, primary circuit – circuit malfunction	**Wiring, A/C relay, ECM**
562	Engine coolant blower motor relay – circuit malfunction	**Wiring, engine coolant blower motor relay, ECM**
563	Engine coolant blower motor relay, high speed – circuit malfunction	**Wiring, engine coolant blower motor relay/ constant control relay module, ECM**
564	Engine coolant blower motor relay, low speed – circuit malfunction	**Wiring, engine coolant blower motor relay/ constant control relay module, ECM**
565	Evaporative emission (EVAP) canister purge valve 1 – malfunction	**Hose damaged, wiring, EVAP canister purge valve, ECM**
566 [13]	Shift solenoid (SS) 3-4 – circuit malfunction	**Wiring, SS, ECM**
567	Cruise control vent valve – circuit malfunction	**Wiring, cruise control vent valve, ECM**
568	Cruise control vacuum valve – circuit malfunction	**Wiring, cruise control vacuum valve, ECM**
569	Evaporative emission (EVAP) canister purge valve 2 – malfunction	**Hose damaged, wiring, EVAP canister purge valve, ECM**
571	Exhaust gas recirculation (EGR) atmospheric solenoid – circuit malfunction during KOEO	**Wiring, EGR atmospheric solenoid, ECM**
572	Exhaust gas recirculation (EGR) vacuum solenoid – circuit malfunction during KOEO	**Wiring, EGR vacuum solenoid, ECM**

Flash code	Fault location	Probable cause
578	A/C pressure sensor – short circuit	**Supply wire short ground, A/C pressure sensor, variable control relay module, ECM**
579	A/C pressure sensor – insufficient pressure change	**Wiring, A/C fault, A/C pressure sensor, variable control relay module, wiring to A/C compressor clutch open circuit, ECM**
581	Engine coolant blower motor – excessive current draw	**Wiring short circuit to ground, engine coolant blower motor, variable control relay module, ECM**
582	Engine coolant blower motor – open circuit/ short circuit to positive	**Supply wire open circuit/short circuit to positive, ground wire open circuit, engine coolant blower motor, variable control relay module, ECM**
583	Fuel pump – excessive current draw	**Wiring short circuit to ground, fuel pump, ECM**
584	Variable control relay module – ground open circuit	**Wiring open circuit, variable control relay module, ECM**
585	A/C compressor clutch – excessive current draw	**Wiring short circuit to ground, A/C compressor clutch, variable control relay module, ECM**
586	A/C compressor clutch – open circuit/short circuit to positive	**Wiring open circuit/short circuit to positive, A/C compressor clutch, variable control relay module, ECM**
587	Variable control relay module – communication error	**Wiring, variable control relay module, ECM**
593	Heated oxygen sensor (HO2S) – heater circuit malfunction	**Wiring, HO2S heater relay, HO2S, ECM**
617 [13]	Gear shift, 1-2 – malfunction	**ATF level/condition, poor connection, wiring, TR sensor, SS, clutch slipping, ECM**
618 [13]	Gear shift, 2-3 – malfunction	**ATF level/condition, poor connection, wiring, TR sensor, SS, clutch slipping, ECM**
619 [13]	Gear shift, 3-4 – malfunction	**ATF level/condition, poor connection, wiring, TR sensor, SS, clutch slipping, ECM**
621 [13]	Shift solenoid (SS) 1 – circuit malfunction during KOEO	**Wiring, SS, ECM**
622 [13]	Shift solenoid (SS) 2 – circuit malfunction during KOEO	**Wiring, SS, ECM**
623 [13]	Overdrive indicator lamp – circuit malfunction	**Bulb, wiring, fuse, ECM**
624 [13]	Transmission fluid pressure (TFP) solenoid – circuit malfunction	**Wiring, poor connection, TFP solenoid, ECM**
625 [13]	Engine control module (ECM), TFP control – open driver circuit	**ECM**
626 [13]	Transmission coast clutch solenoid – circuit malfunction during KOEO test	**Wiring, transmission coast clutch solenoid, ECM**
627 [13]	Torque converter clutch (TCC) solenoid – circuit malfunction	**Wiring, TCC solenoid, ECM**
628 [13]	Torque converter clutch (TCC) – excessive clutch slip	**ATF level/condition, TCC slipping, ECM**

Flash code	Fault location	Probable cause
629 [13]	Torque converter clutch (TCC) solenoid – circuit malfunction	**Wiring, TCC solenoid, ECM**
631 [13]	Overdrive indicator lamp – circuit malfunction during KOEO test	**Bulb, wiring, fuse, ECM**
632 [13]	Overdrive selection switch – no signal change during KOEO test	**Incorrect switch operation, wiring, fuse, overdrive selection switch, ECM**
633	Four wheel drive selector switch – closed during KOEO test	**Wiring short circuit, four wheel drive selector switch, internal damage to transfer case, ECM**
634 [13] [14]	Transmission range (TR) sensor – voltage high/low	**Incorrect adjustment, wiring, TR sensor, ECM**
634 [13] [15]	Park/neutral position (PNP) switch – incorrect signal	**Wiring, PNP switch, ECM**
636 [13]	Transmission fluid temperature (TFT) sensor – signal high/low	**ATF level, wiring, TFT sensor, ECM**
637 [13]	Transmission fluid temperature (TFT) sensor – above maximum voltage	**ATF level, wiring open circuit, TFT sensor, ECM**
638 [13]	Transmission fluid temperature (TFT) sensor – below minimum voltage	**ATF level, wiring short circuit to ground, TFT sensor, ECM**
639 [13] [14]	Turbine shaft speed (TSS) sensor/output shaft speed (OSS) sensor – signal low	**Wiring, TSS/OSS sensor, ECM**
639 [13] [15]	Transmission speed sensor – circuit malfunction	**Wiring, transmission speed sensor, ECM**
641 [13]	Shift solenoid (SS) 3 – circuit malfunction during KOEO	**Wiring, SS, ECM**
643 [13]	Torque converter clutch (TCC) solenoid – circuit malfunction during KOEO test	**Wiring, TCC solenoid, ECM**
645 [13]	Incorrect gear ratio, 1st gear	**ATF level/condition, poor connection, wiring, TR sensor, SS, clutch slipping, ECM**
646 [13]	Incorrect gear ratio, 2nd gear	**ATF level/condition, poor connection, wiring, TR sensor, SS, clutch slipping, ECM**
647 [13]	Incorrect gear ratio, 3rd gear	**ATF level/condition, poor connection, wiring, TR sensor, SS, clutch slipping, ECM**
648 [13]	Incorrect gear ratio, 4th gear	**ATF level/condition, poor connection, wiring, TR sensor, SS, clutch slipping, ECM**
649 [13]	Transmission fluid pressure (TFP) solenoid – signal high/low	**Wiring, poor connection, TFP solenoid, ECM**
651 [13]	Transmission fluid pressure (TFP) solenoid – circuit malfunction	**Wiring, poor connection, TFP solenoid, ECM**
652 [13]	Torque converter clutch (TCC) solenoid – circuit malfunction	**Wiring, TCC solenoid, ECM**
653 [13]	Overdrive selection switch – no signal change during KOEO test	**Incorrect switch operation, wiring, fuse, overdrive selection switch, ECM**
654 [13]	Transmission range (TR) sensor – 'P' signal during KOEO test	**Transmission not in 'P' during KOEO test, incorrect adjustment, ECM**
655 [13]	Transmission range (TR) sensor – 'N' signal during KOEO test	**Transmission not in 'N' during KOEO test, incorrect adjustment, ECM**

Flash code	Fault location	Probable cause
656 [13]	Torque converter clutch (TCC) – continuous clutch slip	**ATF level/condition, TCC slipping, ECM**
657 [13]	Automatic transmission fluid (ATF) – temperature high	**ATF level, transmission cooling system, towing weight limit exceeded, clutch slipping, wiring, TFT sensor, ECM**
659 [13]	High vehicle speed – transmission in 'P'	**TR sensor/PNP switch, ECM**
667 [13]	Transmission range (TR) sensor – below minimum voltage	**Wiring open circuit/short circuit to ground, TR sensor, ECM**
668 [13]	Transmission range (TR) sensor – above maximum voltage	**Wiring open circuit/short circuit to positive, TR sensor, ECM**
675 [13]	Transmission range (TR) sensor – out of range	**Wiring, TR sensor, ECM**
691	Four wheel drive selector fault – incorrect engine RPM/vehicle speed	**Wiring short circuit, four wheel drive selector switch, internal damage to transfer case, ECM**
692 [13]	Transmission state does not match calculated ratio	**Transmission fault, ECM**
998	Engine control module (ECM) – hard fault present	**ECM**

[1] Tempo: Additional warm-up time may be required due to IAT location. Run engine until under hood temperature warm and stabilized
[2] Trouble code may be generated if an efficient exhaust extraction system used during KOER test – vehicles with pressure feedback EGR
[3] Probe
[4] Contour, Taurus 3.0L Flex fuel, Taurus 3.2L SHO
[5] Escort, Tempo, Taurus (except 3.0L Flex fuel/3.2L SHO)
[6] Mustang
[7] Probe (except California)
[8] Probe (California)
[9] Contour, Escort, Tempo, Taurus
[10] Taurus 3.0L SHO
[11] Taurus 3.0L Flex fuel
[12] Except Taurus 3.0L SHO
[13] ECM incorporates transmission control function (AT)
[14] Except Escort
[15] Escort

FORD

Model:	Engine identification:	Year:
Crown Victoria 4.6L	VIN position 8 = W	1992-94
Taurus 3.8L	VIN position 8 = 4	1992-95
Thunderbird 3.8L	VIN position 8 = 4	1992-95
Thunderbird SC 3.8L	VIN position 8 = C/R	1992-95
Thunderbird 5.0L	VIN position 8 = T	1992-93
Mustang LX/GT 5.0L	VIN position 8 = E/T	1992-95
Mustang Cobra 5.0L	VIN position 8 = D	1993-95

System: **Ford EEC-IV**

Self-diagnosis

General information

- Refer to the front of this manual for general test conditions, terminology, detailed descriptions of wiring faults and a general trouble shooter for electrical and mechanical faults.
- Malfunction indicator lamp (MIL) will illuminate if certain faults are recorded.
- ECM operates in backup mode if sensors fail, to enable vehicle to be driven to repair shop.
- Trouble codes can be displayed by the MIL or by using an analogue voltmeter or Scan Tool connected to the data link connectors (DLC) 1 and 2:
  - Crown Victoria **1**.
  - Mustang 1992-93 **2**.
  - Mustang 1994-95 **3**.
  - Taurus **4**.
  - Thunderbird **5**.

1

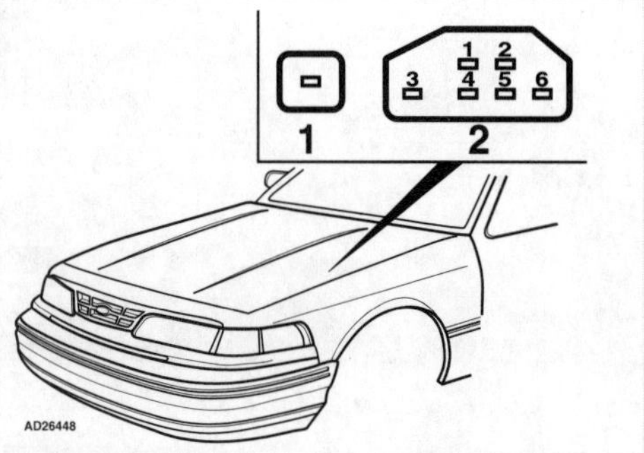

2

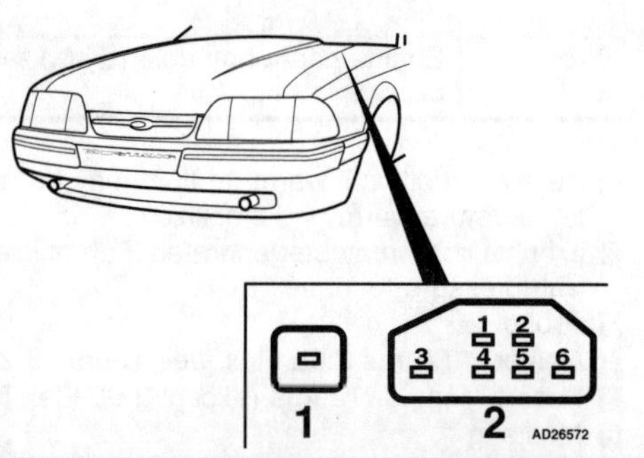

3

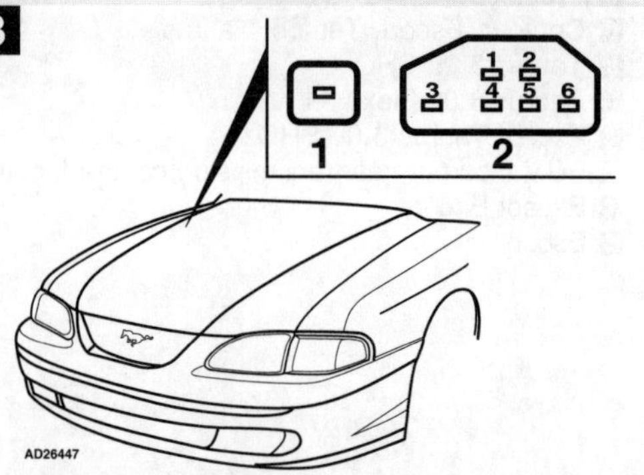

4

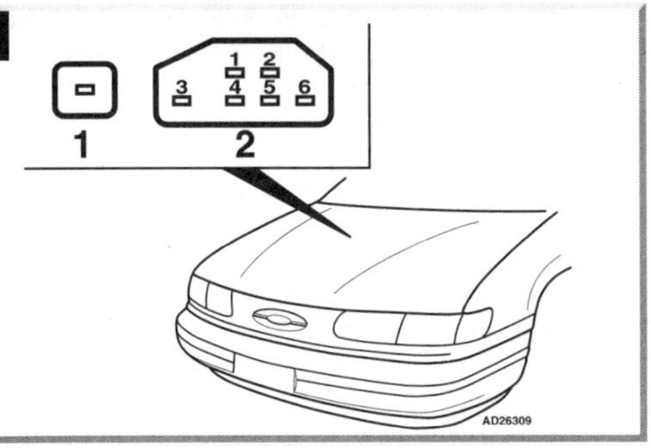

5

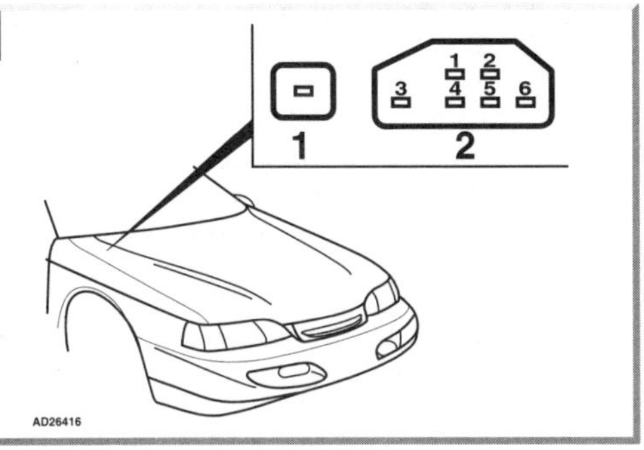

Preparatory conditions

- Automatic transmission in 'PARK'.
- All auxiliary equipment, including air conditioning, switched OFF.

Accessing – Key On Engine Off (KOEO) test

Using malfunction indicator lamp (MIL)

- Start engine.
- Allow to idle until engine at normal operating temperature.
- Switch ignition OFF.
- Wait 10 seconds.
- If engine does not start or stalls: Continue with KOEO test.
- If fitted: Remove 'EEC TEST' cover to access data link connector (DLC).
- Jump DLC 1 terminal (STI) to DLC 2 terminal 2 (SIG RTN) **6**.
- Switch ignition ON.

NOTE: ***MIL may flash briefly while fast DTCs are issued for Scan Tool followed by slow DTC output.***

- Count MIL flashes. Note trouble codes. Compare with trouble code table.
- Each trouble code is displayed twice. Firstly for present faults followed by those for memorized faults as follows:
  - Trouble codes are displayed by illuminated MIL switching OFF.
  - Each trouble code consists of two or three digits.
  - A group of 0.5 second flashes indicate each digit **7** [**A**].
  - A 0.5 second pause separates each flash **7** [**B**].
  - A 2 second pause separates each trouble code digit **6** [**C**].
  - A 4 second pause separates each trouble code **7** [**D**].
  - For example: Three digit trouble code 122 displayed **6**. Throttle position (TP) sensor.

NOTE: ***Two digit trouble codes displayed in a similar pattern.***

  - A 6-9 second pause **8** [**B**] indicates the end of trouble code output for present faults **8** [**A**].
  - A 0.5 second flash (separator code) **8** [**C**], followed by a 6-9 second pause **8** [**D**] indicates the start of trouble code output for memorized faults **8** [**E**].
  - Trouble codes for memorized faults displayed in the same manner as those for present faults.

NOTE: ***Memorized trouble codes refer to faults that have occurred in the last 80 warm-up cycles.***

- Switch ignition OFF.
- Remove jump lead.
- Trouble codes can also be displayed by using a Scan Tool connected to the data link connectors (DLC) 1 and 2.

6

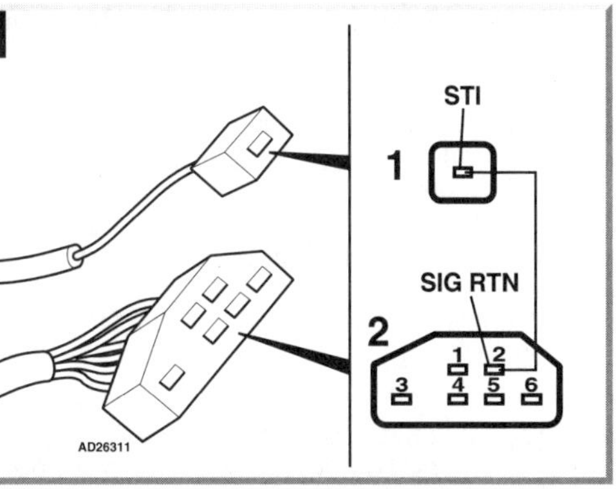

7

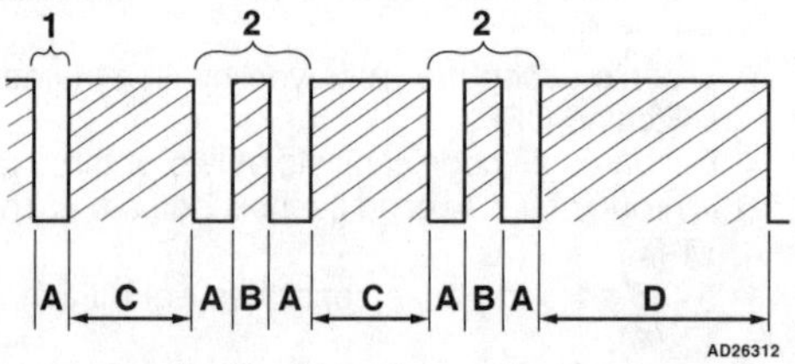

8

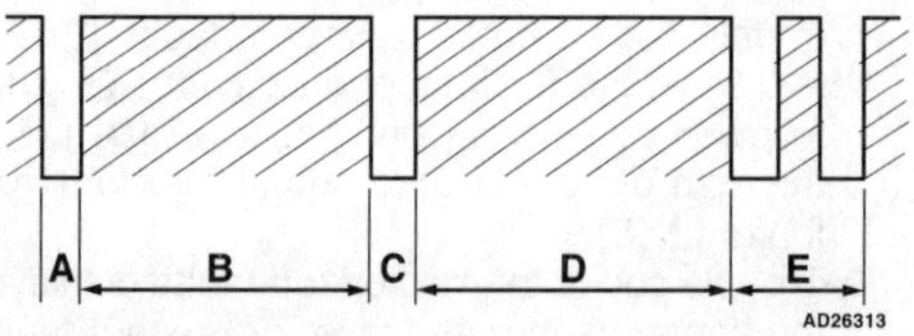

Using analogue voltmeter

- Ensure ignition switched OFF.
- If fitted: Remove 'EEC TEST' cover to access data link connector (DLC).
- Connect an analogue voltmeter between data link connector (DLC) 2 terminal 4 (STO) and battery (12 volt) positive **9**.
- Set analogue voltmeter to a DC voltage range of 0-20 V.
- Start engine.
- Allow to idle until engine at normal operating temperature.
- Switch ignition OFF.
- Wait 10 seconds.
- If engine does not start or stalls: Continue with KOEO test.
- Jump DLC 1 terminal (STI) to DLC 2 terminal 2 (SIG RTN) **9**.
- Switch ignition ON.

NOTE: ***Voltmeter needle may briefly deflect while fast DTCs are issued for Scan Tool followed by slow DTC output.***

- Count voltmeter needle deflections. Note trouble codes. Compare with trouble code table.
- Each trouble code is displayed twice. Firstly for present faults followed by those for memorized faults as follows:
  - Trouble codes are displayed by voltmeter needle deflecting to zero.
  - Each trouble code consists of two or three digits.
  - A group of 0.5 second needle deflections indicate each digit **7** [**A**].
  - A 0.5 second pause separates each needle deflection **7** [**B**].
  - A 2 second pause separates each trouble code digit **7** [**C**].
  - A 4 second pause separates each trouble code **7** [**D**].
  - For example: Three digit trouble code 122 displayed **7**. Throttle position (TP) sensor.

NOTE: ***Two digit trouble codes displayed in a similar pattern.***

  - A 6-9 second pause **8** [**B**] indicates the end of trouble code output for present faults **8** [**A**].
  - A 0.5 second needle deflection (separator code) **8** [**C**], followed by a 6-9 second pause **8** [**D**] indicates the start of trouble code output for memorized faults **8** [**E**].
  - Trouble codes for memorized faults displayed in the same manner as those for present faults.

NOTE: ***Memorized trouble codes refer to faults that have occurred in the last 80 warm-up cycles.***

- Switch ignition OFF.
- Remove jump lead and voltmeter.
- Trouble codes can also be displayed by using a Scan Tool connected to the data link connectors (DLC) 1 and 2.

9

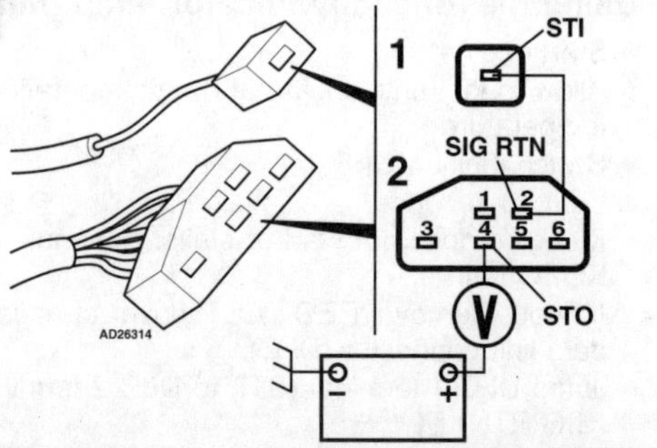

Accessing – Key On Engine Running (KOER) test

Using malfunction indicator lamp (MIL)

- Start engine.
- Allow to idle until engine at normal operating temperature.
- Increase engine speed to over 2000 rpm for 2 minutes.
- Switch ignition OFF.
- Wait 10 seconds.
- If fitted: Remove 'EEC TEST' cover to access data link connector (DLC).
- Jump DLC 1 terminal (STI) to DLC 2 terminal 2 (SIG RTN) **5**.
- Start engine.
- Engine identification code displayed as follows:
  - 6 cylinder: Three 0.5 second flashes.
  - 8 cylinder: Four 0.5 second flashes.
  - For example: 6 cylinder engine identified **10** [**A**].
- With PSP switch: Within 1-2 seconds, turn steering wheel at least ½ a turn and release.
- With BPP switch: Depress and release brake pedal.
- If fitted: Operate overdrive selection switch twice.
- If after 6-20 seconds **10** [**B**], a single 0.5 second flash is displayed (dynamic response signal) **9** [**C**]: Briefly fully open throttle.

NOTE: ***DO NOT fully open throttle during KOER test if no dynamic response signal is displayed.***

- Trouble codes displayed for present faults after 4-15 second pause **10** [**D**].
- Fast DTCs are issued for Scan Tool during 4-15 second pause **10** [**E**].
- Count MIL flashes. Note trouble codes. Compare with trouble code table.

NOTE: ***Each trouble code is displayed twice.***

- Trouble codes for KOER test are displayed in the same manner as those for KOEO test **10** [**F**].
- For example: Three digit trouble code 111 is displayed **10** . No fault found.
- Switch ignition OFF.
- Remove jump lead.
- Trouble codes can also be displayed by using a Scan Tool connected to the data link connectors (DLC) 1 and 2.

10

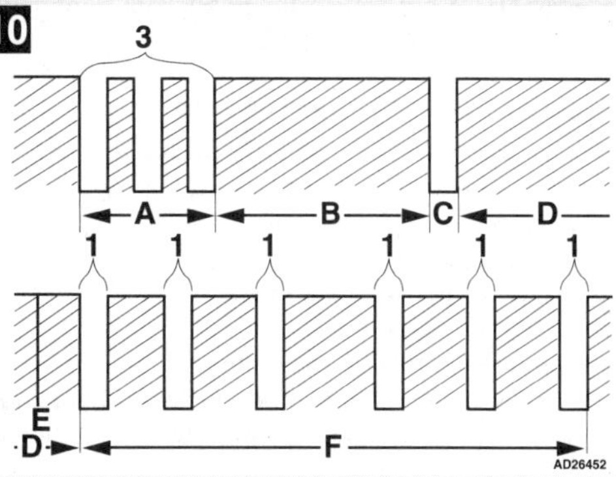

Using analogue voltmeter

- Ensure ignition switched OFF.
- If fitted: Remove 'EEC TEST' cover to access data link connector (DLC).
- Connect an analogue voltmeter between data link connector (DLC) 2 terminal 4 (STO) and battery (12 volt) positive **9**.
- Set analogue voltmeter to a DC voltage range of 0-20 V.
- Start engine.
- Allow to idle until engine at normal operating temperature.
- Increase engine speed to over 2000 rpm for 2 minutes.
- Switch ignition OFF.
- Wait 10 seconds.
- Jump DLC 1 terminal (STI) to DLC 2 terminal 2 (SIG RTN) **9**.
- Start engine.
- Engine identification code displayed as follows:
  - 6 cylinder: Three 0.5 second needle deflections.
  - 8 cylinder: Four 0.5 second needle deflections.
  - For example: 6 cylinder engine identified **10** [**A**].
- With PSP switch: Within 1-2 seconds, turn steering wheel at least ½ a turn and release.
- With BPP switch: Depress and release brake pedal.
- If fitted: Operate overdrive selection switch twice.
- If after 6-20 seconds **10** [**B**], a single 0.5 second needle deflection is displayed (dynamic response signal) **10** [**C**]: Briefly fully open throttle.

NOTE: ***DO NOT fully open throttle during KOER test if no dynamic response signal is displayed.***

- Trouble codes displayed for present faults after 4-15 second pause **10** [**D**].

- Fast DTCs are issued for Scan Tool during 4-15 second pause **10** [**E**].
- Count needle deflections. Note trouble codes. Compare with trouble code table.

NOTE: *Each trouble code is displayed twice.*

- Trouble codes for KOER test are displayed in the same manner as those for KOEO test **10** [**F**].
- For example: Three digit trouble code 111 is displayed **10** . No fault found.
- Switch ignition OFF.
- Remove jump lead and voltmeter.
- Trouble codes can also be displayed by using a Scan Tool connected to the data link connectors (DLC) 1 and 2.

Erasing

NOTE: *Disconnecting battery ground lead may cause driveability problems until the engine management system relearns basic values.*

Using malfunction indicator lamp (MIL)

- Ensure ignition switched OFF.
- If fitted: Remove 'EEC TEST' cover to access data link connector (DLC).
- Jump DLC 1 terminal (STI) to DLC 2 terminal 2 (SIG RTN) **5**.
- Switch ignition ON.
- When trouble codes start to be displayed: Remove jump lead.
- Switch ignition OFF.
- Repeat accessing procedure to ensure no memorized trouble codes remain in ECM fault memory.
- Trouble codes can also be erased by using a Scan Tool connected to the data link connectors (DLC) 1 and 2.
- If emission related component(s) have been replaced:
  - Disconnect battery ground cable for at least 5 minutes to erase adaptive memory from ECM.
  - If required: Drive vehicle for 10 miles or more to allow ECM to relearn basic values.

NOTE: *Disconnecting battery ground lead may also erase memory from electronic units, such as radio.*

Using analogue voltmeter

- Ensure ignition switched OFF.
- If fitted: Remove 'EEC TEST' cover to access data link connector (DLC).
- Connect an analogue voltmeter between data link connector (DLC) 2 terminal 4 (STO) and battery (12 volt) positive **9**.
- Set analogue voltmeter to a DC voltage range of 0-20 V.
- Jump DLC 1 terminal (STI) to DLC 2 terminal 2 (SIG RTN) **9**.
- Switch ignition ON.
- When trouble codes start to be displayed: Remove jump lead.
- Switch ignition OFF.
- Repeat accessing procedure to ensure no memorized trouble codes remain in ECM fault memory.
- Trouble codes can also be erased by using a Scan Tool connected to the data link connectors (DLC) 1 and 2.
- If emission related component(s) have been replaced:
  - Disconnect battery ground cable for at least 5 minutes to erase adaptive memory from ECM.
  - If required: Drive vehicle for 10 miles or more to allow ECM to relearn basic values.

NOTE: *Disconnecting battery ground lead may also erase memory from electronic units, such as radio.*

Trouble code identification

Two digit

Flash code	Fault location	Probable cause
11	No fault found	–
12	Idle speed, high RPM check – unable to control during KOER test	**Intake system, throttle tight/sticking, IAC valve/ISC actuator, ECM**
13	Idle speed, low RPM check – unable to control during KOER test	**Intake system, throttle tight/sticking, IAC valve/ISC actuator, ECM**

Flash code	Fault location	Probable cause
14	Engine speed (RPM) sensor/profile ignition pickup (PIP) circuit – intermittent signal	**Incorrectly installed RPM sensor, wiring, poor connection, electrical interference, RPM sensor, ECM**
15	Engine control module (ECM) – KAM/ROM fault	**Battery disconnected, wiring, fuses, ECM**
16	Engine RPM too low for heated oxygen sensor (HO2S) test	**Intake system, throttle tight/sticking, IAC valve/ISC actuator, ECM**
	Ignition diagnostic monitor (IDM) circuit – no signal during KOEO test	**Wiring, ICM, ECM**
18	Spark output (SPOUT) circuit open	**Wiring, ICM, ECM**
	Ignition diagnostic monitor (IDM) circuit/spark output (SPOUT) circuit grounded	**Wiring, ICM, ECM**
19	Engine control module (ECM) – internal voltage fault	**ECM**
21	Engine coolant temperature (ECT) sensor – incorrect signal	**Wiring, ECT sensor, ECM**
22	Manifold absolute pressure (MAP) sensor/ barometric pressure (BARO) sensor – incorrect signal	**Vacuum hose damaged, trapped vacuum, high atmospheric pressure, excessive EGR, wiring, MAP/BARO sensor, ECM**
23	Throttle position (TP) sensor – incorrect signal	**Throttle valve tight/sticking, wiring, TP sensor, ECM**
24	Intake air temperature (IAT) sensor – incorrect signal	**Wiring, IAT sensor, ECM**
25	Knock sensor (KS) – no signal during dynamic response test	**Wiring, KS, ECM**
26	Mass air flow (MAF) sensor – incorrect signal	**Intake leak/blockage, wiring, MAF sensor, ECM**
29	Vehicle speed signal – incorrect signal	**Wiring, vehicle speed interface module, VSS, ECM**
31	Exhaust gas recirculation (EGR) valve position sensor – below minimum voltage	**Wiring short circuit to ground, EGR valve/ position sensor, ECM**
32	Exhaust gas recirculation (EGR) valve position sensor – below closed limit	**Wiring short circuit to ground, EGR valve/ position sensor, ECM**
33	Exhaust gas recirculation (EGR) valve – not opening	**Wiring, EGR valve tight/sticking, ECM**
34	Exhaust gas recirculation (EGR) valve position sensor – above closed limit	**Wiring short circuit to positive, EGR valve/ position sensor, ECM**
35	Exhaust gas recirculation (EGR) valve position sensor – above maximum voltage	**Wiring short circuit to positive, EGR valve/ position sensor, ECM**
41	Heated oxygen sensor (HO2S), bank 1 – mixture lean/not switching	**Intake leak, fuel pressure low, fuel filter/pump, injectors, HO2S, AIR system, MAF/MAP sensor, ECM**
42	Heated oxygen sensor (HO2S), bank 1 – mixture rich/not switching	**Intake blocked, excessive fuel in engine oil, fuel pressure high, breather system, injectors, HO2S, AIR system, EVAP system, MAF/MAP sensor, ignition system, ECM**
43	Throttle position (TP) sensor – incorrect signal at idle	**Incorrectly adjusted, TP sensor, ECM**
44	Secondary air injection (AIR) system, bank 1 – inoperative	**Hose connection, wiring, AIR pump motor, AIR solenoid, ECM**

Flash code	Fault location	Probable cause
45	Secondary air injection (AIR) – upstream during KOER test	**Hose connection, wiring, AIR solenoid, ECM**
46	Secondary air injection (AIR) – not bypassed during KOER test	**Hose connection, wiring, AIR solenoid, ECM**
47	Four wheel drive selector switch – closed during KOEO test	**Wiring short circuit, four wheel drive selector switch, internal damage to transfer case, ECM**
49 [1]	Gear shift, 1-2 – malfunction	**ATF level/condition, poor connection, wiring, TR sensor, SS, clutch slipping, ECM**
51	Engine coolant temperature (ECT) sensor – open circuit	**Wiring open circuit, ECT sensor, ECM**
52	Power steering pressure (PSP) switch – open circuit/no signal change	**Wiring open circuit, PSP switch, ECM**
53	Throttle position (TP) sensor – above maximum voltage	**TP sensor loose, wiring short circuit to positive, TP sensor, ECM**
54	Intake air temperature (IAT) sensor – open circuit	**Wiring open circuit, IAT sensor, ECM**
56	Mass air flow (MAF) sensor – above maximum voltage	**Wiring short circuit to positive, MAF sensor, ECM**
59 [1]	Gear shift, 2-3 – malfunction	**ATF level/condition, poor connection, wiring, TR sensor, SS, clutch slipping, ECM**
61	Engine coolant temperature (ECT) sensor – short circuit to ground	**Wiring short circuit to ground, ECT sensor, ECM**
62 [1]	Torque converter clutch (TCC) – excessive clutch slip	**ATF level/condition, TCC slipping, ECM**
63	Throttle position (TP) sensor – below minimum voltage	**TP sensor loose, wiring open circuit/short circuit to ground, TP sensor, ECM**
64	Intake air temperature (IAT) sensor – short circuit to ground	**Wiring short circuit to ground, IAT sensor, ECM**
65 [1]	Overdrive selection switch – no signal change during KOEO test	**Incorrect switch operation, wiring, fuse, overdrive selection switch, ECM**
66	Mass air flow (MAF) sensor – below minimum voltage	**Wiring short circuit to ground, MAF sensor, ECM**
67	Neutral position (NP) switch – open circuit	**Incorrect adjustment, wiring open circuit, NP switch, ECM**
	Clutch pedal position (CPP) switch – circuit malfunction	**Incorrect adjustment, wiring, CPP switch, ECM**
	Air conditioning (A/C) – on during KOEO test	**Wiring, A/C switch, ECM**
67 [1]	Transmission range (TR) sensor – voltage high/low	**Incorrect adjustment, wiring, TR sensor, ECM**
68 [1]	Automatic transmission fluid (ATF) – temperature high	**ATF level, transmission cooling system, towing weight limit exceeded, clutch slipping, wiring, TFT sensor, ECM**
69 [1]	Gear shift, 3-4 – malfunction	**ATF level/condition, poor connection, wiring, TR sensor, SS, clutch slipping, ECM**
72	Manifold absolute pressure (MAP) sensor/mass air flow (MAF) sensor – incorrect signal change during dynamic response test	**Intake leak, wiring, MAP sensor, ECM**
73	Throttle position (TP) sensor – incorrect signal change during dynamic response test	**Incorrect throttle operation, wiring, TP sensor, ECM**

Flash code	Fault location	Probable cause
74	Brake pedal position (BPP) switch – open circuit/ no signal change during KOER test	**Incorrect brake pedal operation, wiring, BPP switch, ECM**
75	Brake pedal position (BPP) switch – closed circuit	**Wiring, BPP switch, ECM**
77	Wide open throttle (WOT) switch – no signal change during dynamic response test	**Incorrect throttle operation, wiring, WOT switch, ECM**
79	Air conditioning (A/C) – on during KOEO test	**Wiring, A/C switch, ECM**
81	Secondary air injection (AIR) divert solenoid – circuit malfunction during KOEO test	**Wiring, AIR divert solenoid, ECM**
82	Secondary air injection (AIR) bypass solenoid – circuit malfunction during KOEO test	**Wiring, AIR bypass solenoid, ECM**
84	Exhaust gas recirculation (EGR) solenoid – circuit malfunction during KOEO test	**Wiring, EGR solenoid, ECM**
85	Evaporative emission (EVAP) canister purge valve – circuit malfunction during KOEO test	**Wiring, EVAP canister purge valve, ECM**
86	AT – shift solenoid	**Wiring, shift solenoid, ECM**
87	Fuel pump, primary circuit – circuit malfunction	**IFS switch, wiring, fuel pump, ECM**
89	Torque converter clutch (TCC) lock-up solenoid	**Wiring, TCC lock-up solenoid, ECM**
91	Heated oxygen sensor (HO2S), bank 2 – mixture lean/not switching during KOER test	**Intake leak, fuel pressure low, fuel filter/pump, injectors, HO2S, AIR system, MAF/MAP sensor, ECM**
91 1	Shift solenoid (SS) 1 – circuit malfunction during KOEO test	**Wiring, SS, ECM**
92	Heated oxygen sensor (HO2S), bank 1 – mixture rich/not switching during KOER test	**Intake blocked, excessive fuel in engine oil, fuel pressure high, breather system, injectors, HO2S, AIR system, EVAP system, MAF/MAP sensor, ignition system, ECM**
92 1	Shift solenoid (SS) 2 – circuit malfunction during KOEO test	**Wiring, SS, ECM**
93 1	Transmission coast clutch solenoid – circuit malfunction during KOEO test	**Wiring, transmission coast clutch solenoid, ECM**
94	Secondary air injection (AIR) system, bank 2 – inoperative during KOER test	**Hose connection, wiring, AIR pump motor, AIR solenoid, ECM**
94 1	Torque converter clutch (TCC) solenoid – circuit malfunction during KOEO test	**Wiring, TCC solenoid, ECM**
95	Fuel pump, secondary circuit – circuit malfunction	**IFS switch, wiring, fuel pump, ECM**
96	Fuel pump, secondary circuit – circuit malfunction	**IFS switch, wiring open circuit to battery, fuel pump, ECM**
97 1	Overdrive indicator lamp – circuit malfunction	**Bulb, wiring, fuse, ECM**
98	Engine control module (ECM) – hard fault detected during KOER test	**HO2S bank 2 – indicates lean, ECM**
98 1	Engine control module (ECM), transmission fluid pressure (TFP) control – open driver circuit during KOEO test	**ECM**

Flash code	Fault location	Probable cause
99	Engine control module – idle speed not learnt during KOER test	**Idle speed adjusted, ECM**
99 [1]	Transmission fluid pressure (TFP) solenoid – circuit malfunction during KOEO test	**Wiring, poor connection, TFP solenoid, ECM**

[1] ECM incorporates transmission control function

Three digit

Flash code	Fault location	Probable cause
111	No fault found	–
112	Intake air temperature (IAT) sensor – below minimum voltage	**Wiring short circuit to ground, poor connection, IAT sensor, ECM**
113	Intake air temperature (IAT) sensor – above maximum voltage	**Wiring open circuit, poor connection, IAT sensor, ECM**
114	Intake air temperature (IAT) sensor – incorrect signal	**Air temperature below 10°C (50°F), poor connection, IAT sensor, ECM**
116	Engine coolant temperature (ECT) sensor – incorrect signal	**Coolant level low, poor connection, ECT sensor, ECM**
117	Engine coolant temperature (ECT) sensor – below minimum voltage	**Wiring short circuit to ground, poor connection, ECT sensor, ECM**
118	Engine coolant temperature (ECT) sensor – above maximum voltage	**Wiring open circuit, poor connection, IAT sensor, ECM**
121	Throttle position (TP) sensor – closed throttle voltage high/low	**Throttle valve tight/sticking, TP sensor loose/damaged, ECM**
	Throttle position (TP) sensor – voltage inconsistent with MAF sensor signal	**Throttle valve tight/sticking, TP sensor loose/damaged, ECM**
122	Throttle position (TP) sensor – below minimum voltage	**TP sensor loose, wiring open circuit/short circuit to ground, TP sensor, ECM**
123	Throttle position (TP) sensor – above maximum voltage	**TP sensor loose, wiring short circuit to positive, TP sensor, ECM**
124	Throttle position (TP) sensor – voltage high	**Throttle tight/sticking, intake leak, MAP/BARO sensor, ECM**
125	Throttle position (TP) sensor – voltage low	**Throttle tight/sticking, intake leak, MAP/BARO sensor, ECM**
126	Manifold absolute pressure (MAP) sensor/ barometric pressure (BARO) sensor – voltage high/low	**Vacuum hose damaged/trapped, high atmospheric pressure, excessive EGR, wiring, MAP/BARO sensor, ECM**
128	Manifold absolute pressure (MAP) sensor – loss of vacuum	**Vacuum hose damaged/disconnected, MAP sensor, ECM**
129	Manifold absolute pressure (MAP) sensor – insufficient voltage change during KOER dynamic response test	**Incorrect throttle operation, vacuum hose damaged/disconnected, MAP sensor, IAC system, ECM**
	Mass air flow (MAF) sensor – insufficient voltage change during KOER dynamic response test	**Incorrect throttle operation, intake leak, wiring, MAF sensor, IAC system, ECM**

Flash code	Fault location	Probable cause
136	Heated oxygen sensor (HO2S), bank 2 – mixture lean during KOER test	**Intake leak, fuel pressure low, fuel filter/pump, injectors, HO2S, AIR system, MAP sensor, ECM**
137	Heated oxygen sensor (HO2S), bank 2 – mixture rich during KOER test	**Intake blocked, excessive fuel in engine oil, fuel pressure high, breather system, injectors, HO2S, AIR system, EVAP system, MAP sensor, ignition system, ECM**
138	Cold start injector – insufficient flow during KOER test	**Fuel pressure, wiring, cold start injector, ECM**
139	Heated oxygen sensor (HO2S), bank 2 – not switching	**Intake leak/blocked, injectors, HO2S, AIR system, breather system, EVAP system, MAP sensor, ignition system, ECM**
141	Fuel system indicates lean	**Intake leak, fuel pressure low, fuel filter/pump, injectors, HO2S, ECM**
144	Heated oxygen sensor (HO2S), bank 1 – not switching	**Intake leak/blocked, injectors, HO2S, AIR system, breather system, EVAP system, MAP sensor, ignition system, ECM**
157	Mass air flow (MAF) sensor – below minimum voltage	**Wiring short circuit to ground, poor connection, MAF sensor, IAC system, ECM**
158	Mass air flow (MAF) sensor – above maximum voltage	**Wiring short circuit to positive, MAF sensor screen blocked, MAF sensor, ECM**
159 1	Mass air flow (MAF) sensor – voltage high/low	**Intake leak, IAC valve, MAF sensor, ECM**
167	Throttle position (TP) sensor – insufficient voltage change during KOER dynamic response test	**Incorrect throttle operation, TP sensor, ECM**
171	Fuel system control, bank 1 – adaptive limits reached	**Intake leak/blocked, breather system, fuel pressure, fuel filter/pump, injectors, HO2S, AIR system, EVAP system, MAP sensor, ignition system, ECM**
172	Heated oxygen sensor (HO2S), bank 1 – mixture lean	**Intake leak, fuel pressure low, fuel filter/pump, injectors, HO2S, AIR system, MAP sensor, ECM**
173	Heated oxygen sensor (HO2S), bank 1 – mixture rich	**Intake blocked, excessive fuel in engine oil, fuel pressure high, breather system, injectors, HO2S, AIR system, EVAP system, MAP sensor, ignition system, ECM**
174	Heated oxygen sensor (HO2S), bank 1 – slow response	**Intake leak, excessive fuel in engine oil, fuel pressure, fuel filter/pump, injectors, HO2S, AIR system, MAF/MAP sensor, ECM**
175	Fuel system control, bank 2 – adaptive limits reached	**Intake leak/blocked, excessive fuel in engine oil, breather system, fuel pressure, fuel filter/pump, injectors, HO2S, AIR system, EVAP system, MAP sensor, ignition system, ECM**
176	Heated oxygen sensor (HO2S), bank 2 – mixture lean	**Intake leak, fuel pressure low, fuel filter/pump, injectors, HO2S, AIR system, MAP sensor, ECM**
177	Heated oxygen sensor (HO2S), bank 2 – mixture rich	**Intake blocked, fuel pressure high, breather system, injectors, HO2S, AIR system, EVAP system, MAP sensor, ignition system, ECM**

Flash code	Fault location	Probable cause
178	Heated oxygen sensor (HO2S), bank 2 – slow response	Intake leak, excessive fuel in engine oil, fuel pressure, fuel filter/pump, injectors, HO2S, AIR system, MAF/MAP sensor, ECM
179	Fuel control system, bank 1 – lean adaption limit reached at part throttle, mixture rich	Intake leak, fuel pressure low, fuel filter blocked, wiring, injector, MAP sensor, ECM
181	Fuel control system, bank 1 – rich adaption limit reached at part throttle, mixture lean	Intake blocked, fuel pressure high, wiring, injector, MAP sensor, ECM
182	Fuel control system, bank 1 – lean adaption limit reached at idle, mixture rich	Intake leak, fuel pressure low, fuel filter blocked, wiring, injector, MAP sensor, ECM
183	Fuel control system, bank 1 – rich adaption limit reached at idle, mixture lean	Intake blocked, fuel pressure high, wiring, injector, MAP sensor, ECM
184	Mass air flow (MAF) sensor – signal high	MAF sensor screen blocked, MAF sensor, ECM
185	Mass air flow (MAF) sensor – signal low	Intake leak, breather system, MAF sensor, ECM
186	Injector – pulse width long/mass air flow (MAF) sensor – signal low	Intake leak, fuel pressure low, MAF sensor, ECM
187	Injector – pulse width short/mass air flow (MAF) sensor – signal high	Intake blocked, fuel pressure high, MAF sensor, ECM
188	Fuel system control, bank 2 – lean adaption limit reached at part throttle, mixture rich	Intake leak, fuel pressure low, fuel filter blocked, wiring, injector, MAP sensor, ECM
189	Fuel system control, bank 2 – rich adaption limit reached at part throttle, mixture lean	Intake blocked, fuel pressure high, wiring, injector, MAP sensor, ECM
191	Fuel system control, bank 2 – lean adaption limit reached at idle, mixture rich	Intake leak, fuel pressure low, fuel filter blocked, wiring, injector, MAP sensor, ECM
192	Fuel system control, bank 2 – rich adaption limit reached at idle, mixture lean	Intake blocked, fuel pressure high, wiring, injector, MAP sensor, ECM
193	Fuel quality sensor – circuit malfunction	Wiring, fuel separation, fuel quality sensor, ECM
194	Cylinder balance test – malfunction	Wiring, injector, ECM
195	Cylinder balance test – malfunction	Wiring, injector, ECM
211	Profile ignition pickup (PIP) – circuit malfunction	Wiring, poor connection, arcing secondary ignition component – coil/cap/rotor/HT leads/plugs, electrical interference, ECM
212	Ignition control module (ICM) – ignition diagnostic monitor (IDM) signal missing	Wiring, ICM, ECM
213	Spark output (SPOUT) – open circuit	Wiring, ICM, ECM
214	Camshaft position (CMP) sensor – circuit malfunction	Wiring, CMP sensor, ICM, ECM
215	Ignition coil 1 – primary circuit malfunction	Wiring, ignition coil, ICM, ECM
216	Ignition coil 2 – primary circuit malfunction	Wiring, ignition coil, ICM, ECM
217	Ignition coil 3 – primary circuit malfunction	Wiring, ignition coil, ICM, ECM
218	Ignition diagnostic monitor (IDM), left side – signal missing	Wiring, ignition coil, ECM
219	Ignition timing, defaulted to 10° – spark output (SPOUT) circuit open	Wiring open circuit, ICM, ECM
221	Ignition timing error	Wiring, ICM, ECM
222	Ignition diagnostic monitor (IDM), right side – signal missing	Wiring, ignition coil, ECM

Flash code	Fault location	Probable cause
223	Loss of dual plug inhibit (DPI) control	**Wiring, ICM, ECM**
224	Ignition coil 1, 2, 3 or 4 – primary circuit malfunction	**Wiring, ignition coil, ICM, ECM**
225	Knock sensor (KS) – no signal during KOER dynamic response test	**Wiring, KS, ECM**
226	Ignition control module (ICM) – ignition diagnostic monitor (IDM) signal missing	**Wiring open circuit, ICM, ECM**
232	Ignition coil 1, 2, 3 or 4 – primary circuit malfunction	**Wiring, ignition coil, ICM, ECM**
238	Ignition coil 4 – primary circuit malfunction	**Wiring, ignition coil, ICM, ECM**
241	Ignition control module (ICM) – ignition diagnostic monitor (IDM) signal malfunction	**Wiring, ICM, ECM**
244	Cylinder identification (CID) fault when cylinder balance test requested	**Wiring, CKP sensor, CMP sensor, ICM, ECM**
311	Secondary air injection (AIR) system, bank 1 – inoperative during KOER test	**Hose blocked/leaking, AIR bypass/divert solenoid, AIR pump, ECM**
312	Secondary air injection (AIR) system – misdirected during KOER test	**Hose connections**
313	Secondary air injection (AIR) system – not bypassed during KOER test	**Hose blocked/leaking, AIR bypass solenoid, AIR pump**
314	Secondary air injection (AIR) system, bank 2 – inoperative during KOER test	**Hose blocked/leaking, AIR bypass/divert solenoid, AIR pump, ECM**
326 [2]	Exhaust gas pressure sensor – voltage low	**Hose blocked/leaking, wiring, exhaust gas pressure sensor, ECM**
326 [1] [3]	Exhaust gas pressure sensor – voltage low	**Hose blocked/leaking, EGR solenoid filter, EGR solenoid, wiring, EGR valve, exhaust gas pressure sensor, ECM**
327 [4]	Exhaust gas recirculation (EGR) valve position sensor – minimum voltage exceeded	**Wiring open circuit/short circuit to ground, EGR valve/position sensor, ECM**
327 [5]	Exhaust gas pressure sensor – minimum voltage exceeded	**Hose blocked/leaking, wiring open circuit/ short circuit to ground, exhaust gas pressure sensor, ECM**
328	Exhaust gas recirculation (EGR) valve position sensor, closed position – voltage low	**Wiring, EGR valve/position sensor, ECM**
332 [4]	Exhaust gas recirculation (EGR) system – insufficient flow detected	**Hose blocked/leaking, wiring, EGR solenoid filter, EGR solenoid, EGR valve/position sensor, ECM**
332 [5]	Exhaust gas recirculation (EGR) system – insufficient flow detected	**Hose blocked/leaking, EGR solenoid, exhaust gas pressure sensor, EGR valve, ECM**
334	Exhaust gas recirculation (EGR) valve position sensor, closed position – voltage high	**Wiring, EGR valve/position sensor, EGR solenoid, ECM**
335 [1]	Exhaust gas pressure sensor – voltage high/low during KOEO test	**Exhaust gas pressure sensor, blocked inlet hose, ECM**
336	Exhaust gas pressure sensor – voltage high	**Exhaust system restricted, wiring, EGR solenoid, exhaust gas pressure sensor, EGR valve, hose routing, ECM**
337 [4]	Exhaust gas recirculation (EGR) position sensor – above maximum voltage	**Wiring open circuit/short circuit to positive, EGR valve/position sensor, ECM**

Flash code	Fault location	Probable cause
337 **2**	Exhaust gas pressure sensor – above maximum voltage	**Incorrect hose routing, wiring short circuit, signal wire open circuit, exhaust gas pressure sensor, EGR solenoid/valve, ECM**
337 **3**	Exhaust gas pressure sensor – above maximum voltage	**Exhaust system restricted, wiring short circuit, signal wire open circuit, exhaust gas pressure sensor, ECM**
338	Engine coolant temperature (ECT) – too low	**Thermostat stuck open, coolant leak, ECM**
339	Engine coolant temperature (ECT) – too high	**Coolant level low, thermostat stuck closed, cooling system blocked, coolant blower motor worn/damaged, water pump defective, coolant leak, ECM**
341	Octane adjuster – open circuit	**Wiring, octane adjuster, ECM**
381	A/C compressor clutch – excessive cycling	**A/C system incorrectly charged, cycling switch defective, A/C circuit fault, wiring intermittent open circuit to A/C system, ECM**
411	Idle air control (IAC) valve – no control during KOER low RPM check	**Idle speed adjustment, intake leak, throttle valve tight/sticking, ignition timing incorrect, wiring short circuit to ground, IAC valve, ECM**
412	Idle air control (IAC) valve – no control during KOER high RPM check	**Idle speed adjustment, throttle valve tight/sticking, wiring open circuit/short circuit to ground, IAC valve, ECM**
415	Idle air control (IAC) valve – lower limit reached	**Intake leak, throttle valve tight/sticking, IAC valve, ECM**
416	Idle air control (IAC) valve – upper limit reached	**Air filter blocked, throttle body dirty/damaged, IAC valve, ECM**
452	Vehicle speed signal – incorrect signal	**Wiring, rear ABS system, vehicle speed interface module, ECM**
452	Vehicle speed sensor (VSS) – incorrect signal	**Wiring, VSS, ECM**
453	Servo leaking down – cruise control KOER test	**Servo defective, ECM**
454	Servo leaking up – cruise control KOER test	**Servo defective, ECM**
455	Insufficient RPM increase – cruise control KOER test	**Intake system, ECM**
456	Insufficient RPM decrease – cruise control KOER test	**Intake system, ECM**
457	Cruise control master/selector switch – circuit malfunction during KOEO test	**Wiring, cruise control master/selector switch, ECM**
458	Cruise control master/selector switch – short circuit to ground during KOEO test	**Wiring short circuit to ground, cruise control master/selector switch, ECM**
459	Cruise control system – ground open circuit during KOEO test	**Wiring open circuit, ECM**
511	Engine control module (ECM) – ROM fault during KOEO test	**ECM**
512	Engine control module (ECM) – KAM fault	**Battery disconnected, wiring, fuses, ECM**
513	Engine control module (ECM) – internal voltage fault	**ECM**
519	Power steering pressure (PSP) switch – circuit malfunction during KOEO test	**PSP switch not fitted – disregard trouble code, wiring, PSP switch, ECM**

Flash code	Fault location	Probable cause
521	Power steering pressure (PSP) switch – no signal change during KOEO test	**PSP switch not fitted – disregard trouble code, wiring, PSP switch, ECM**
522 [6]	Park/neutral position (PNP) switch/neutral position (NP) switch – not in park/neutral during KOEO test/open circuit	**Incorrect gear lever position, incorrect adjustment, wiring open circuit, starter motor relay disconnected, PNP/NP switch, ECM**
	Transmission range (TR) sensor – not in park/ neutral during KOEO test	**Incorrect shift lever position, incorrect adjustment, wiring, TR sensor, ECM**
525	Vehicle in gear – during KOEO test	**Incorrect shift lever position, incorrect adjustment, PNP switch open, ECM**
	A/C – ON during KOEO test	**A/C ON, A/C wiring short circuit to positive, ECM**
526	Neutral position switch (NP) – closed circuit during KOEO test	**Wiring short circuit, NP switch, ECM**
	A/C – ON during KOEO test	**A/C ON, A/C wiring short circuit to positive, A/C switch, ECM**
527 [6]	Park/neutral position (PNP) switch/neutral position switch – open circuit during KOEO test	**Wiring open circuit, starter motor relay disconnected, PNP/NP switch, ECM**
	A/C – ON during KOEO test	**A/C ON, A/C wiring short circuit to positive, A/C switch, ECM**
528	Clutch pedal position (CPP) switch – circuit malfunction	**Wiring open circuit, CPP switch, ECM**
529	Data link/engine control module (ECM) – circuit malfunction	**Wiring, ECM**
532	Instrument panel – circuit malfunction	**Wiring, instrument panel, ECM**
533	Data link/instrument panel – circuit malfunction	**Wiring, instrument panel, ECM**
536	Brake pedal position (BPP) switch – circuit malfunction/no signal change during KOER test	**Incorrect brake pedal operation, wiring, BPP switch, ECM**
538	KOER dynamic response test – insufficient rpm change	**Incorrect throttle operation, ECM**
	KOER cylinder balance test – invalid test	**Throttle operated during test, CMP sensor, ECM**
539	A/C – ON during KOEO test	**Wiring, A/C switch, ECM**
542	Fuel pump (FP) secondary circuit – engine does not start	**IFS switch, supply wire open circuit, FP ground wire, FP, ECM**
	Fuel pump (FP) secondary circuit – engine starts	**Supply wire short circuit to positive, FP relay, FP to ECM wire open circuit, left/front HO2S short circuit to positive, ECM**
543	Fuel pump (FP) secondary circuit – engine does not start	**Supply wire open circuit, fuel pump relay, ECM**
	Fuel pump (FP) secondary circuit – engine starts	**Left/front HO2S short circuit to positive, ECM**
551	Idle air control (IAC) valve – circuit malfunction during KOEO test	**Wiring, IAC valve, ECM**
552	Secondary air injection (AIR) bypass solenoid – circuit malfunction during KOEO test	**Wiring, AIR bypass solenoid, ECM**
553	Secondary air injection (AIR) divert solenoid – circuit malfunction during KOEO test	**Wiring, AIR divert solenoid, ECM**

Flash code	Fault location	Probable cause
554	Fuel pressure control solenoid – circuit malfunction	**Hose connection, wiring, fuel pressure control solenoid, ECM**
556	Fuel pump (FP) primary circuit	**IFS switch, wiring, fuel pump relay, constant control relay module, ECM**
557	Fuel pump (FP) primary circuit, low speed	**Wiring, FP relay/constant control relay module, ECM**
558	Exhaust gas recirculation (EGR) solenoid – circuit malfunction during KOEO test	**Hose damaged, wiring, EGR solenoid, ECM**
559	A/C relay, primary circuit – circuit malfunction	**Wiring, A/C relay, ECM**
562	Engine coolant blower motor relay – circuit malfunction	**Wiring, engine coolant blower motor relay, ECM**
563	Engine coolant blower motor relay, high speed – circuit malfunction	**Wiring, engine coolant blower motor relay/ constant control relay module, ECM**
564	Engine coolant blower motor relay, low speed – circuit malfunction	**Wiring, engine coolant blower motor relay/ constant control relay module, ECM**
565	Evaporative emission (EVAP) canister purge valve 1 – malfunction	**Hose damaged, wiring, EVAP canister purge valve, ECM**
566 6	Shift solenoid (SS) 3-4 – circuit malfunction	**Wiring, SS, ECM**
567	Cruise control vent valve – circuit malfunction	**Wiring, cruise control vent valve, ECM**
568	Cruise control vacuum valve – circuit malfunction	**Wiring, cruise control vacuum valve, ECM**
569	Evaporative emission (EVAP) canister purge valve 2 – malfunction	**Hose damaged, wiring, EVAP canister purge valve, ECM**
571	Exhaust gas recirculation (EGR) atmospheric solenoid – circuit malfunction during KOEO	**Wiring, EGR atmospheric solenoid, ECM**
572	Exhaust gas recirculation (EGR) vacuum solenoid – circuit malfunction during KOEO	**Wiring, EGR vacuum solenoid, ECM**
578	A/C pressure sensor – short circuit	**Supply wire short ground, A/C pressure sensor, variable control relay module, ECM**
579	A/C pressure sensor – insufficient pressure change	**Wiring, A/C fault, A/C pressure sensor, variable control relay module, wiring to A/C compressor clutch open circuit, ECM**
581	Engine coolant blower motor – excessive current draw	**Wiring short circuit to ground, engine coolant blower motor, variable control relay module, ECM**
582	Engine coolant blower motor – open circuit/short circuit to positive	**Supply wire open circuit/short circuit to positive, ground wire open circuit, engine coolant blower motor, variable control relay module, ECM**
583	Fuel pump – excessive current draw	**Wiring short circuit to ground, fuel pump, ECM**
584	Variable control relay module – ground open circuit	**Wiring open circuit, variable control relay module, ECM**
585	A/C compressor clutch – excessive current draw	**Wiring short circuit to ground, A/C compressor clutch, variable control relay module, ECM**
586	A/C compressor clutch – open circuit/short circuit to positive	**Wiring open circuit/short circuit to positive, A/C compressor clutch, variable control relay module, ECM**
587	Variable control relay module – communication error	**Wiring, variable control relay module, ECM**

Flash code	Fault location	Probable cause
593	Heated oxygen sensor (HO2S) – heater circuit malfunction	**Wiring, HO2S heater relay, HO2S, ECM**
617 6	Gear shift, 1-2 – malfunction	**ATF level/condition, poor connection, wiring, TR sensor, SS, clutch slipping, ECM**
618 6	Gear shift, 2-3 – malfunction	**ATF level/condition, poor connection, wiring, TR sensor, SS, clutch slipping, ECM**
619 6	Gear shift, 3-4 – malfunction	**ATF level/condition, poor connection, wiring, TR sensor, SS, clutch slipping, ECM**
621 6	Shift solenoid (SS) 1 – circuit malfunction during KOEO	**Wiring, SS, ECM**
622 6	Shift solenoid (SS) 2 – circuit malfunction during KOEO	**Wiring, SS, ECM**
623 6	Overdrive indicator lamp – circuit malfunction	**Bulb, wiring, fuse, ECM**
624 6	Transmission fluid pressure (TFP) solenoid – circuit malfunction	**Wiring, poor connection, TFP solenoid, ECM**
625 6	Engine control module (ECM), TFP control – open driver circuit	**ECM**
626 6	Transmission coast clutch solenoid – circuit malfunction during KOEO test	**Wiring, transmission coast clutch solenoid, ECM**
627 6	Torque converter clutch (TCC) solenoid – circuit malfunction	**Wiring, TCC solenoid, ECM**
628 6	Torque converter clutch (TCC) – excessive clutch slip	**ATF level/condition, TCC slipping, ECM**
629 6	Torque converter clutch (TCC) solenoid – circuit malfunction	**Wiring, TCC solenoid, ECM**
631 6	Overdrive indicator lamp – circuit malfunction during KOEO test	**Bulb, wiring, fuse, ECM**
632 6	Overdrive selection switch – no signal change during KOEO test	**Incorrect switch operation, wiring, fuse, overdrive selection switch, ECM**
633	Four wheel drive selector switch – closed during KOEO test	**Wiring short circuit, four wheel drive selector switch, internal damage to transfer case, ECM**
634 6	Transmission range (TR) sensor – voltage high/low	**Incorrect adjustment, wiring, TR sensor, ECM**
636 6	Transmission fluid temperature (TFT) sensor – signal high/low	**ATF level, wiring, TFT sensor, ECM**
637 6	Transmission fluid temperature (TFT) sensor – above maximum voltage	**ATF level, wiring open circuit, TFT sensor, ECM**
638 6	Transmission fluid temperature (TFT) sensor – below minimum voltage	**ATF level, wiring short circuit to ground, TFT sensor, ECM**
639 6	Turbine shaft speed (TSS) sensor/output shaft speed (OSS) sensor – signal low	**Wiring, TSS/OSS sensor, ECM**
641 6	Shift solenoid (SS) 3 – circuit malfunction during KOEO	**Wiring, SS, ECM**
643 6	Torque converter clutch (TCC) solenoid – circuit malfunction during KOEO test	**Wiring, TCC solenoid, ECM**
645 6	Incorrect gear ratio, 1st gear	**ATF level/condition, poor connection, wiring, TR sensor, SS, clutch slipping, ECM**
646 6	Incorrect gear ratio, 2nd gear	**ATF level/condition, poor connection, wiring, TR sensor, SS, clutch slipping, ECM**

Flash code	Fault location	Probable cause
647 6	Incorrect gear ratio, 3rd gear	ATF level/condition, poor connection, wiring, TR sensor, SS, clutch slipping, ECM
648 6	Incorrect gear ratio, 4th gear	ATF level/condition, poor connection, wiring, TR sensor, SS, clutch slipping, ECM
649 6	Transmission fluid pressure (TFP) solenoid – signal high/low	Wiring, poor connection, TFP solenoid, ECM
651 6	Transmission fluid pressure (TFP) solenoid – circuit malfunction	Wiring, poor connection, TFP solenoid, ECM
652 6	Torque converter clutch (TCC) solenoid – circuit malfunction	Wiring, TCC solenoid, ECM
653 6	Overdrive selection switch – no signal change during KOEO test	Incorrect switch operation, wiring, fuse, overdrive selection switch, ECM
654 6	Transmission range (TR) sensor – 'P' signal during KOEO test	Transmission not in 'P' during KOEO test, incorrect adjustment, ECM
655 6	Transmission range (TR) sensor – 'N' signal during KOEO test	Transmission not in 'N' during KOEO test, incorrect adjustment, ECM
656 6	Torque converter clutch (TCC) – continuous clutch slip	ATF level/condition, TCC slipping, ECM
657 6	Automatic transmission fluid (ATF) – temperature high	ATF level, transmission cooling system, towing weight limit exceeded, clutch slipping, wiring, TFT sensor, ECM
659 6	High vehicle speed – transmission in 'P'	TR sensor/PNP switch, ECM
667 6	Transmission range (TR) sensor – below minimum voltage	Wiring open circuit/short circuit to ground, TR sensor, ECM
668 6	Transmission range (TR) sensor – above maximum voltage	Wiring open circuit/short circuit to positive, TR sensor, ECM
675 6	Transmission range (TR) sensor – out of range	Wiring, TR sensor, ECM
691	Four wheel drive selector fault – incorrect engine RPM/vehicle speed	Wiring short circuit, four wheel drive selector switch, internal damage to transfer case, ECM
692 6	Transmission state does not match calculated ratio	Transmission fault, ECM
998	Engine control module (ECM) – hard fault present	ECM

1 Trouble code may be generated if an efficient exhaust extraction system used during KOER test – vehicles with pressure feedback EGR
2 Crown Victoria/Thunderbird 3.8L (California), Thunderbird 3.8L SC
3 Crown Victoria (except California), Taurus
4 Mustang, Thunderbird 5.0L
5 Crown Victoria, Taurus, Thunderbird 3.8L/SC
6 ECM incorporates transmission control function (AT)

FORD

Model:	Engine identification:	Year:
Escort GT 1.8L DOHC	**VIN position 8 = 8**	**1992-95**

System: **MECS**

Self-diagnosis

General information

- Refer to the front of this manual for general test conditions, terminology, detailed descriptions of wiring faults and a general trouble shooter for electrical and mechanical faults.
- Malfunction indicator lamp (MIL) will illuminate continuously if permanent faults are recorded.
- Malfunction indicator lamp (MIL) will illuminate briefly and then switch off if intermittent faults are recorded.
- ECM operates in backup mode if sensors fail, to enable vehicle to be driven to repair shop.
- Trouble codes can be displayed by the MIL or by using an analogue voltmeter or Scan Tool connected to the data link connector (DLC).

Accessing

Using malfunction indicator lamp (MIL)

- Ensure ignition switched OFF.
- Access data link connector (DLC) **1**.
- Jump data link connector (DLC) terminals STI and GND **2**.
- Switch ignition ON.
- Count MIL flashes. Note trouble codes. Compare with trouble code table.
  - 1.2 second flashes indicate the 'tens' of the trouble code **3** [**A**].
  - 0.4 second flashes indicate the 'units' of the trouble code **3** [**C**].
  - A 1.6 second pause (MIL OFF) separates 'tens' and 'units' **3** [**B**].
  - A 0.4 second pause (MIL OFF) separates each flash **3** [**D**].
  - A 4 second pause separates each trouble code **3** [**E**].
  - For example: Trouble code 12 displayed **3**. Throttle position (TP) sensor.
- Switch ignition OFF.
- Trouble codes can also be displayed by using a Scan Tool connected to the data link connector (DLC) **1**.

1

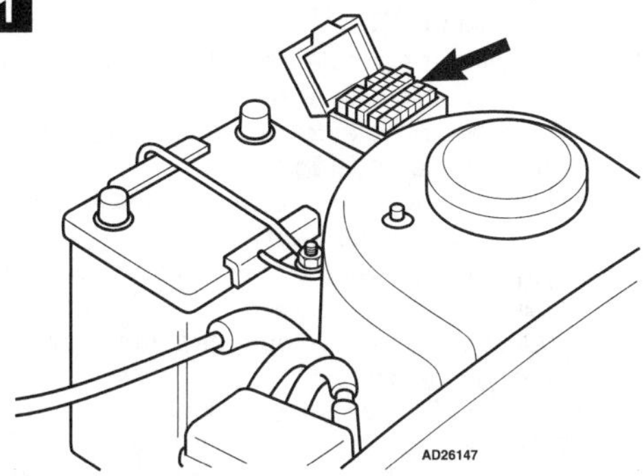

2

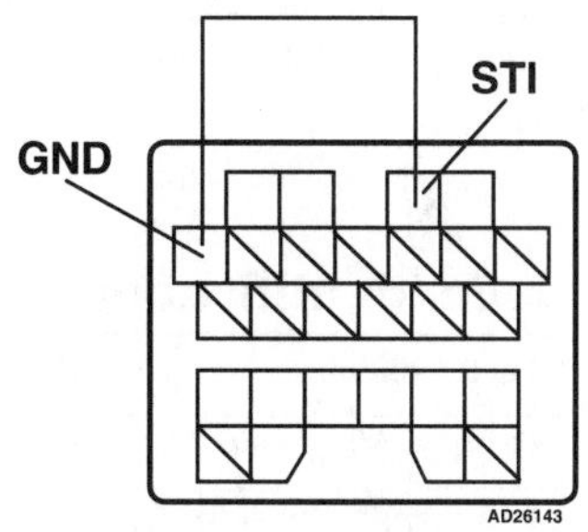

3

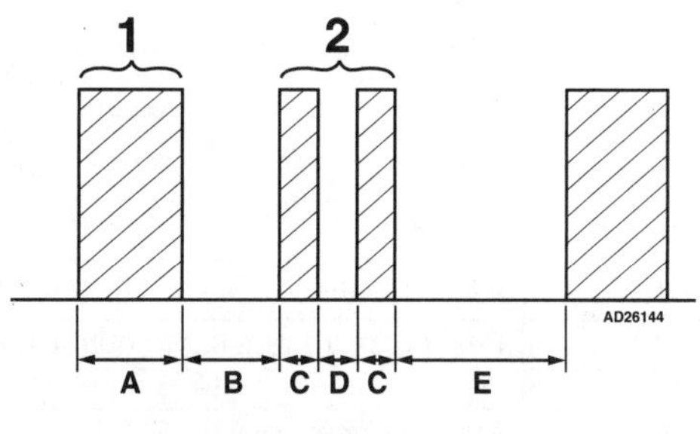

Using analogue voltmeter

- Ensure ignition switched OFF.
- Access data link connector (DLC) **1**.
- Connect an analogue voltmeter between data link connector (DLC) terminal STO and ground **4**.
- Set analogue voltmeter to a DC voltage range of 0-20 V.
- Jump data link connector (DLC) terminals STI and GND **4**.
- Switch ignition ON.
- Count voltmeter needle deflections. Note trouble codes. Compare with trouble code table.
  - 1.2 second needle deflections indicate the 'tens' of the trouble code **3** [**A**].
  - 0.4 second needle deflections indicate the 'units' of the trouble code **3** [**C**].
  - A 1.6 second pause separates 'tens' and 'units' **3** [**B**].
  - A 0.4 second pause separates each needle deflection **3** [**D**].
  - A 4 second pause separates each trouble code **3** [**E**].
  - For example: Trouble code 12 displayed **3**. Throttle position (TP) sensor.
- Switch ignition OFF.
- Trouble codes can also be displayed by using a Scan Tool connected to the data link connector (DLC) **1**.

4

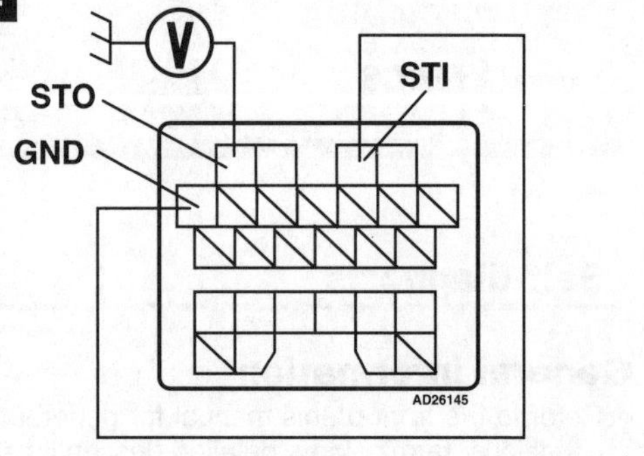

Erasing

- Ensure ignition switched OFF.
- Disconnect battery ground cable.
- Depress brake pedal for 5-10 seconds.
- Reconnect battery ground cable.
- Repeat accessing procedure to ensure no trouble codes remain in ECM fault memory.

NOTE: ***Disconnecting battery ground lead may also erase memory from electronic units, such as radio.***

Trouble code identification

Flash code	Fault location	Probable cause
02	Crankshaft position (CKP) sensor	Wiring, CKP sensor, ECM
03	Crankshaft position (CKP) sensor, cylinder identification (CID) signal – circuit malfunction	Wiring, CKP sensor, ECM
06	Vehicle speed sensor (VSS)	Wiring, VSS, ECM
08	Volume air flow (VAF) sensor	Wiring, VAF sensor, ECM
09	Engine coolant temperature (ECT) sensor	Wiring, ECT sensor, ECM
10	Intake air temperature (IAT) sensor	Wiring, IAT sensor, ECM
12	Throttle position (TP) sensor	Wiring, TP sensor, ECM
14	Barometric pressure (BARO) sensor	ECM
15	Oxygen sensor (O2S) – voltage below 0.55 V	Fuel pressure low, wiring, O2S, ECM
17	Oxygen sensor (O2S) – no voltage change	Wiring, O2S, ECM
25	Fuel pressure regulator control solenoid	Wiring, fuel pressure regulator control solenoid, ECM
26	Evaporative emission (EVAP) canister purge valve	Wiring, EVAP canister purge valve, ECM

Flash code	Fault location	Probable cause
34	Idle air control (IAC) valve	**Wiring, IAC valve, ECM**
41	Intake manifold air control solenoid	**Wiring, intake manifold air control solenoid, ECM**
55 [1]	Transmission speed sensor	**Wiring, transmission speed sensor, ECM**
60 [1]	Transmission shift control valve – 1-2	**Wiring, transmission shift control valve, ECM**
61 [1]	Transmission shift control valve – 2-3	**Wiring, transmission shift control valve, ECM**
62 [1]	Transmission shift control valve – 3-4	**Wiring, transmission shift control valve, ECM**
63 [1]	Torque converter clutch (TCC) solenoid	**Wiring, TCC solenoid, ECM**

[1] ECM incorporates transmission control function

FORD

Model:	Engine identification:	Year:
Aspire 1.3L	**VIN position 8 = H**	**1994-95**
Festiva 1.3L	**VIN position 8 = H**	**1994**

System: **MECS**

Self-diagnosis

General information

- Refer to the front of this manual for general test conditions, terminology, detailed descriptions of wiring faults and a general trouble shooter for electrical and mechanical faults.
- Malfunction indicator lamp (MIL) will illuminate continuously if permanent faults are recorded.
- Malfunction indicator lamp (MIL) will illuminate briefly and then switch off if intermittent faults are recorded.
- ECM operates in backup mode if sensors fail, to enable vehicle to be driven to repair shop.
- Trouble codes can be displayed by the MIL or by using an analogue voltmeter or Scan Tool connected to the data link connector (DLC).

Accessing

Using malfunction indicator lamp (MIL)

- Ensure ignition switched OFF.
- Access data link connector (DLC) **1**.
- Jump data link connector (DLC) terminals STI and GND **2**.
- Switch ignition ON.
- Count MIL flashes. Note trouble codes. Compare with trouble code table.
  - 1.2 second flashes indicate the 'tens' of the trouble code **3** [**A**].
  - 0.4 second flashes indicate the 'units' of the trouble code **3** [**C**].
  - A 1.6 second pause (MIL OFF) separates 'tens' and 'units' **3** [**B**].
  - A 0.4 second pause (MIL OFF) separates each flash **3** [**D**].
  - A 4 second pause separates each trouble code **3** [**E**].
  - For example: Trouble code 12 displayed **3**. Throttle position (TP) sensor.
- Switch ignition OFF.
- Trouble codes can also be displayed by using a Scan Tool connected to the data link connector (DLC) **1**.

1

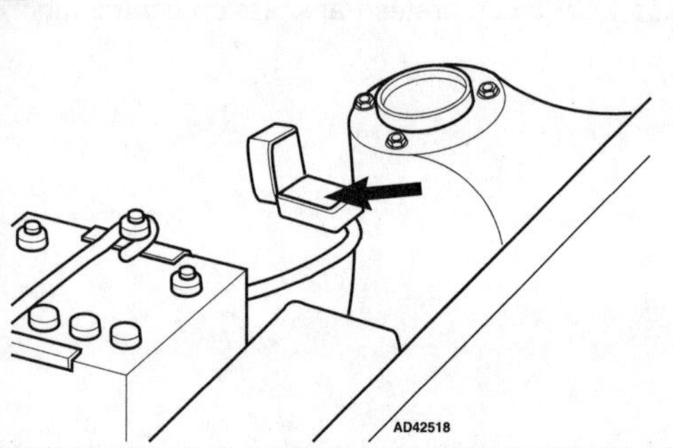

2

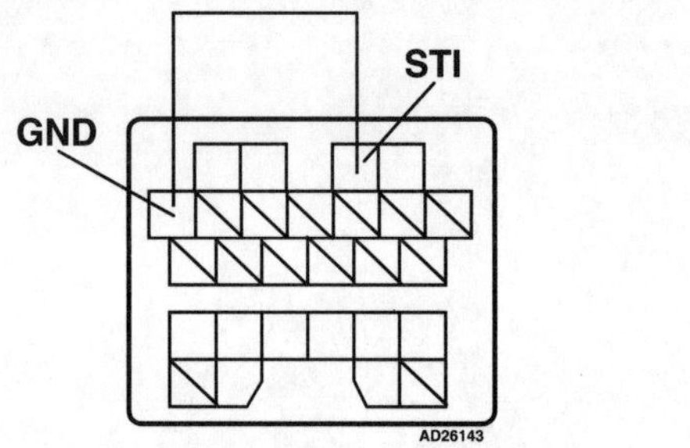

3

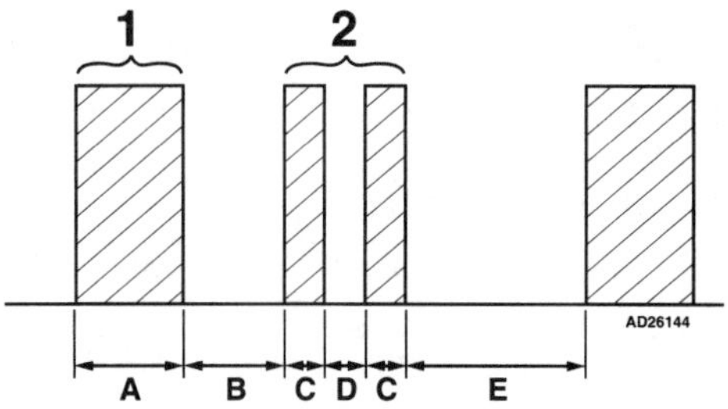

Using analogue voltmeter

- Ensure ignition switched OFF.
- Access data link connector (DLC) 1.
- Connect an analogue voltmeter between data link connector (DLC) terminal STO and ground 4.
- Set analogue voltmeter to a DC voltage range of 0-20 V.
- Jump data link connector (DLC) terminals STI and GND 4.
- Switch ignition ON.
- Count voltmeter needle deflections. Note trouble codes. Compare with trouble code table.
  - 1.2 second needle deflections indicate the 'tens' of the trouble code 3 [**A**].
  - 0.4 second needle deflections indicate the 'units' of the trouble code 3 [**C**].
  - A 1.6 second pause separates 'tens' and 'units' 3 [**B**].
  - A 0.4 second pause separates each needle deflection 3 [**D**].
  - A 4 second pause separates each trouble code 3 [**E**].
  - For example: Trouble code 12 displayed 3. Throttle position (TP) sensor.
- Switch ignition OFF.
- Trouble codes can also be displayed by using a Scan Tool connected to the data link connector (DLC) 1.

4

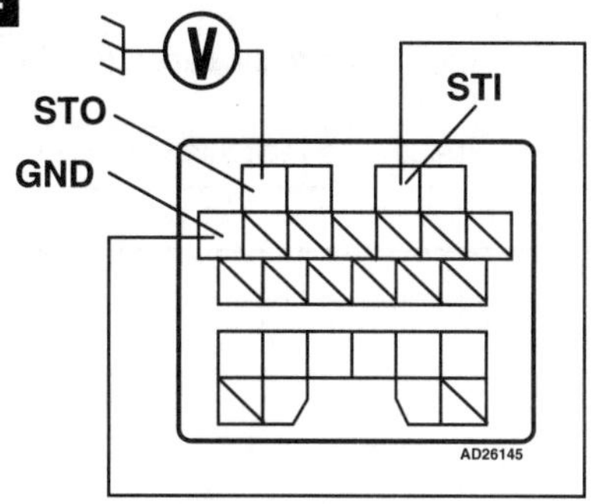

Erasing

- Ensure ignition switched OFF.
- Disconnect battery ground cable.
- Depress brake pedal for 5-10 seconds.
- Reconnect battery ground cable.
- Repeat accessing procedure to ensure no trouble codes remain in ECM fault memory.

NOTE: ***Disconnecting battery ground lead may also erase memory from electronic units, such as radio.***

Trouble code identification

Flash code	Fault location	Probable cause
03	Camshaft position (CMP) sensor	**Wiring, CMP sensor, ECM**
04	Crankshaft position (CKP) sensor	**Wiring, CKP sensor, ECM**
06	Vehicle speed sensor (VSS)	**Speedometer cable broken, wiring, VSS, ECM**
08	Mass air flow (MAF) sensor	**Wiring, MAF sensor, ECM**
09	Engine coolant temperature (ECT) sensor	**Wiring, ECT sensor, ECM**
10	Intake air temperature (IAT) sensor	**Wiring, IAT sensor, ECM**
12	Throttle position (TP) sensor	**Wiring, TP sensor, ECM**

Flash code	Fault location	Probable cause
14	Barometric pressure (BARO) sensor	ECM
15	Oxygen sensor (O2S) – voltage below 0.55 V	Fuel pressure low, wiring, O2S, ECM
16	Exhaust gas recirculation (EGR) valve position sensor	EGR hose(s), wiring, EGR valve position sensor, ECM
17	Oxygen sensor (O2S) – no voltage change	Wiring, O2S, ECM
25	Evaporative emission (EVAP) canister purge valve	Wiring, EVAP canister purge valve, ECM
26	Exhaust gas recirculation (EGR) control solenoid	Wiring, EGR control solenoid, ECM
28	Exhaust gas recirculation (EGR) vent solenoid	Wiring, EGR vent solenoid, ECM
29	Idle air control (IAC) valve	Wiring, IAC valve, ECM

FORD

Model: **Festiva 1.3L**

Engine identification: **VIN position 8 = H**

Year: **1992-93**

System: **MECS**

Self-diagnosis

General information

- Refer to the front of this manual for general test conditions, terminology, detailed descriptions of wiring faults and a general trouble shooter for electrical and mechanical faults.
- Malfunction indicator lamp (MIL) will illuminate continuously if permanent faults are recorded.
- Malfunction indicator lamp (MIL) will illuminate briefly and then switch off if intermittent faults are recorded.
- ECM operates in backup mode if sensors fail, to enable vehicle to be driven to repair shop.
- Trouble codes can be displayed by the MIL or by using an analogue voltmeter or Scan Tool connected to the data link connector (DLC).

Accessing

Using malfunction indicator lamp (MIL)

- Ensure ignition switched OFF.
- Jump data link connector (DLC) 2 terminal STI to ground **1**.
- Switch ignition ON.
- Count MIL flashes. Note trouble codes. Compare with trouble code table.
  - 1.2 second flashes indicate the 'tens' of the trouble code **2** [**A**].
  - 0.4 second flashes indicate the 'units' of the trouble code **2** [**C**].
  - A 1.6 second pause (MIL OFF) separates 'tens' and 'units' **2** [**B**].
  - A 0.4 second pause (MIL OFF) separates each flash **2** [**D**].
  - A 4 second pause separates each trouble code **2** [**E**].
  - For example: Trouble code 14 displayed **2**. Barometric pressure (BARO) sensor.
- Switch ignition OFF.
- Trouble codes can also be displayed by using a Scan Tool connected to the data link connectors (DLC) 1 and 2 **1**.

1

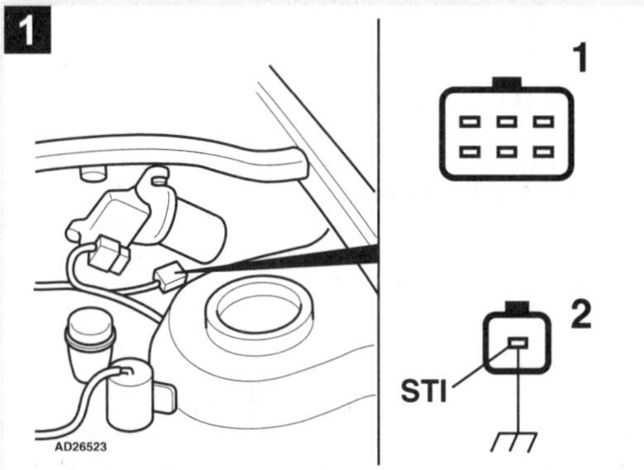

2

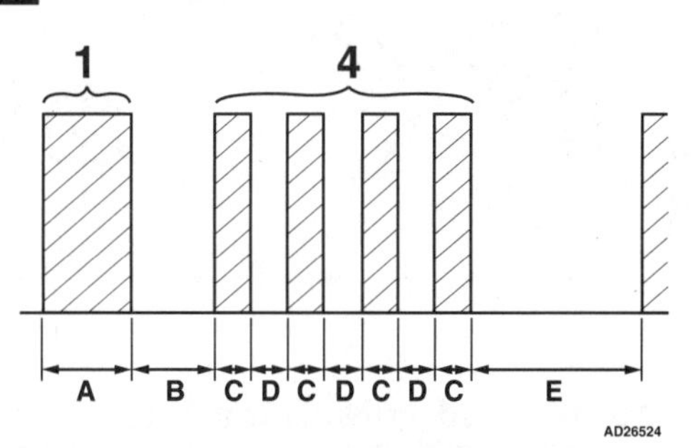

Using analogue voltmeter

- Ensure ignition switched OFF.
- Connect an analogue voltmeter between data link connector (DLC) 1 terminal STO and ground **3**.
- Set analogue voltmeter to a DC voltage range of 0-20 V.
- Jump data link connector (DLC) 2 terminal STI to ground **3**.
- Switch ignition ON.

- Count voltmeter needle deflections. Note trouble codes. Compare with trouble code table.
  - 1.2 second needle deflections indicate the 'tens' of the trouble code **2** [**A**].
  - 0.4 second needle deflections indicate the 'units' of the trouble code **2** [**C**].
  - A 1.6 second pause separates 'tens' and 'units' **2** [**B**].
  - A 0.4 second pause separates each needle deflection **2** [**D**].
  - A 4 second pause separates each trouble code **2** [**E**].
  - For example: Trouble code 14 displayed **2**. Barometric pressure (BARO) sensor.
- Switch ignition OFF.
- Trouble codes can also be displayed by using a Scan Tool connected to the data link connectors (DLC) 1 and 2 **3**.

3

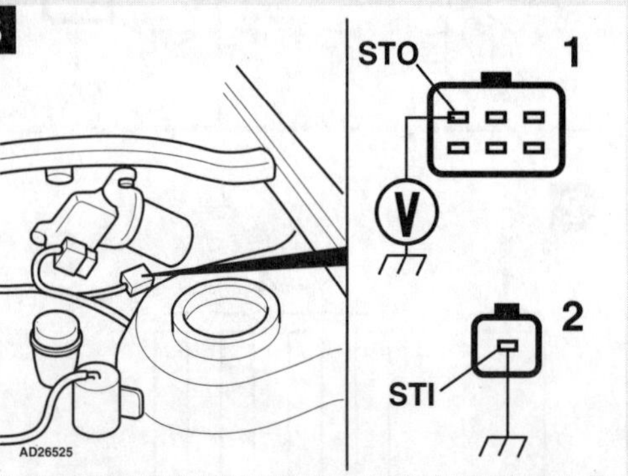

Erasing

- Ensure ignition switched OFF.
- Disconnect battery ground cable.
- Depress brake pedal for 5-10 seconds.
- Reconnect battery ground cable.
- Repeat accessing procedure to ensure no trouble codes remain in ECM fault memory.

NOTE: ***Disconnecting battery ground lead may also erase memory from electronic units, such as radio.***

Trouble code identification

Flash code	Fault location	Probable cause
02	Crankshaft position (CKP) sensor	**Wiring, CKP sensor, ECM**
08	Volume air flow (VAF) sensor	**Wiring, VAF sensor, ECM**
09	Engine coolant temperature (ECT) sensor	**Wiring, ECT sensor, ECM**
10	Intake air temperature (IAT) sensor	**Wiring, IAT sensor, ECM**
14	Barometric pressure (BARO) sensor	**Wiring, BARO sensor, ECM**
15	Oxygen sensor (O2S) – voltage below 0.55 V	**Fuel pressure low, wiring, O2S, ECM**
17	Oxygen sensor (O2S) – no voltage change	**Wiring, O2S, ECM**
26	Evaporative emission (EVAP) canister purge valve	**Wiring, EVAP canister purge valve, ECM**
34	Idle air control (IAC) valve	**Wiring, IAC valve, ECM**

Model: **Probe 2.5L**

Engine identification: **VIN position 8 = B**

Year: **1993-95**

System: **MECS**

Self-diagnosis

General information

- Refer to the front of this manual for general test conditions, terminology, detailed descriptions of wiring faults and a general trouble shooter for electrical and mechanical faults.
- Malfunction indicator lamp (MIL) will illuminate continuously if permanent faults are recorded.
- Malfunction indicator lamp (MIL) will illuminate briefly and then switch off if intermittent faults are recorded.
- ECM operates in backup mode if sensors fail, to enable vehicle to be driven to repair shop.
- Trouble codes can be displayed by the MIL or by using an analogue voltmeter or Scan Tool connected to the data link connector (DLC).

Accessing

Using malfunction indicator lamp (MIL)

- Ensure ignition switched OFF.
- Access data link connector (DLC) **1**.
- Jump data link connector (DLC) terminals STI and GND **2**.
- Switch ignition ON.
- Count MIL flashes. Note trouble codes. Compare with trouble code table.
  - 1.2 second flashes indicate the 'tens' of the trouble code **3** [**A**].
  - 0.4 second flashes indicate the 'units' of the trouble code **3** [**C**].
  - A 1.6 second pause (MIL OFF) separates 'tens' and 'units' **3** [**B**].
  - A 0.4 second pause (MIL OFF) separates each flash **3** [**D**].
  - A 4 second pause separates each trouble code **3** [**E**].
  - For example: Trouble code 12 displayed **3**. Throttle position (TP) sensor.
- Switch ignition OFF.
- Trouble codes can also be displayed by using a Scan Tool connected to the data link connector (DLC) **1**.

1

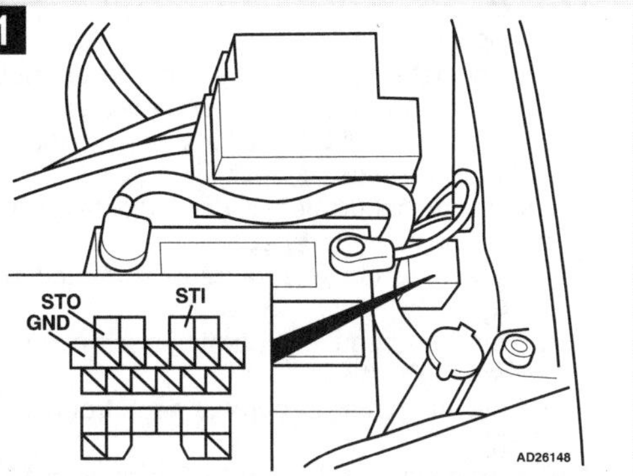

2

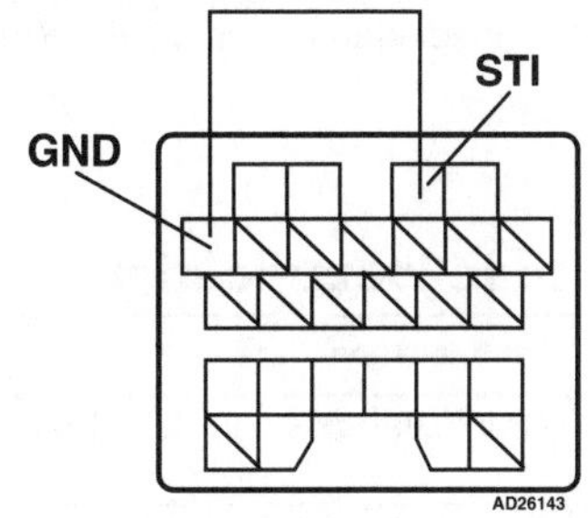

3

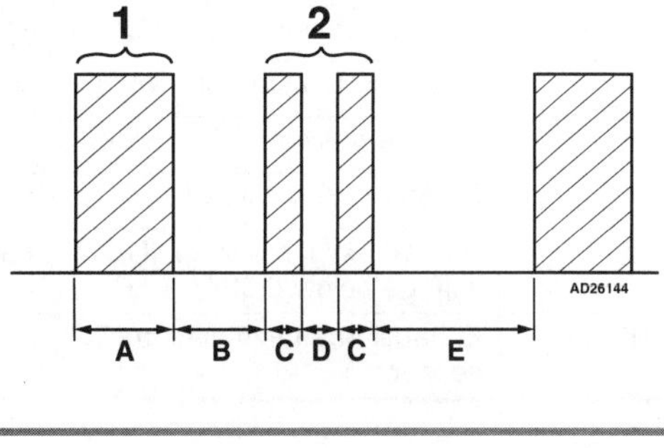

Using analogue voltmeter

- Ensure ignition switched OFF.
- Access data link connector (DLC) **1**.
- Connect an analogue voltmeter between data link connector (DLC) terminal STO and ground **4**.
- Set analogue voltmeter to a DC voltage range of 0-20 V.
- Jump data link connector (DLC) terminals STI and GND **4**.
- Switch ignition ON.
- Count voltmeter needle deflections. Note trouble codes. Compare with trouble code table.
  - 1.2 second needle deflections indicate the 'tens' of the trouble code **3** [**A**].
  - 0.4 second needle deflections indicate the 'units' of the trouble code **3** [**C**].
  - A 1.6 second pause separates 'tens' and 'units' **3** [**B**].
  - A 0.4 second pause separates each needle deflection **3** [**D**].
  - A 4 second pause separates each trouble code **3** [**E**].
  - For example: Trouble code 12 displayed **3**. Throttle position (TP) sensor.
- Switch ignition OFF.
- Trouble codes can also be displayed by using a Scan Tool connected to the data link connector (DLC) **1**.

4

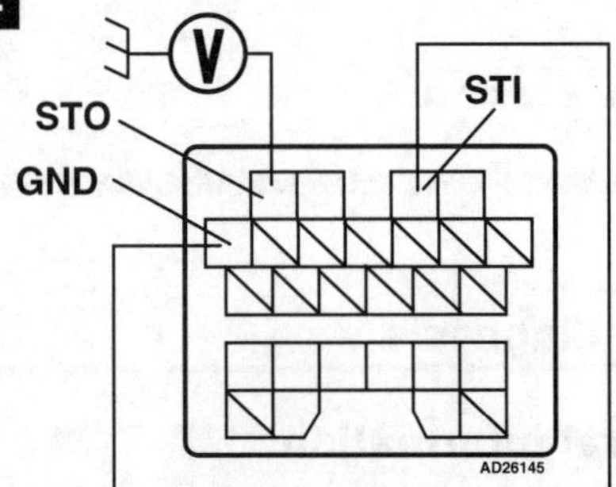

Erasing

- Ensure ignition switched OFF.
- Disconnect battery ground cable.
- Depress brake pedal for 5-10 seconds.
- Reconnect battery ground cable.
- Repeat accessing procedure to ensure no trouble codes remain in ECM fault memory.

NOTE: ***Disconnecting battery ground lead may also erase memory from electronic units, such as radio.***

Trouble code identification

Flash code	Fault location	Probable cause
02	Crankshaft position (CKP) sensor 2	Wiring, CKP sensor (at crankshaft pulley), ECM
03	Camshaft position (CMP) sensor	Wiring, CMP sensor, ECM
04	Crankshaft position (CKP) sensor 1	Wiring, CKP sensor (in distributor), ECM
05	Knock sensor (KS)	Wiring, KS, ECM
08	Volume air flow (VAF) sensor	Wiring, VAF sensor, ECM
09	Engine coolant temperature (ECT) sensor	Wiring, ECT sensor, ECM
10	Intake air temperature (IAT) sensor	Wiring, IAT sensor, ECM
12	Throttle position (TP) sensor	Wiring, TP sensor, ECM
14	Barometric pressure (BARO) sensor	ECM
15	Heated oxygen sensor (HO2S), bank 2 – voltage below 0.55 V	Fuel pressure low, wiring, HO2S, ECM
16	Exhaust gas recirculation (EGR) valve position sensor	EGR hose(s), wiring, EGR valve position sensor, ECM

Flash code	Fault location	Probable cause
17	Heated oxygen sensor (HO2S), bank 2 – no voltage change	**Wiring, HO2S, ECM**
23	Heated oxygen sensor (HO2S), bank 1 – voltage below 0.55 V	**Fuel pressure low, wiring, HO2S, ECM**
24	Heated oxygen sensor (HO2S), bank 1 – no voltage change	**Wiring, HO2S, ECM**
25	Fuel pressure regulator control solenoid	**Wiring, fuel pressure regulator control solenoid, ECM**
26	Evaporative emission (EVAP) canister purge valve	**Wiring, EVAP canister purge valve, ECM**
28	Exhaust gas recirculation (EGR) solenoid, vacuum	**Wiring, EGR solenoid, ECM**
29	Exhaust gas recirculation (EGR) solenoid, vent	**Wiring, EGR solenoid, ECM**
34	Idle air control (IAC) valve	**Wiring, IAC valve, ECM**
41	Intake manifold air control solenoid 1	**Wiring, intake manifold air control solenoid, ECM**
46	Intake manifold air control solenoid 2	**Wiring, intake manifold air control solenoid, ECM**
67	Engine coolant blower motor relay, low speed	**Wiring, engine coolant blower motor relay, ECM**
69	Engine coolant blower motor temperature sensor	**Wiring, engine coolant blower motor temperature sensor, ECM**

GEO

Model:	Engine identification:	Year:
Prizm 1.6L	VIN position 8 = 6	1992-97
Prizm 1.8L	VIN position 8 = 8	1992-97

System: **TCCS**

Self-diagnosis

General information

- Refer to the front of this manual for general test conditions, terminology, detailed descriptions of wiring faults and a general trouble shooter for electrical and mechanical faults.
- Malfunction indicator lamp (MIL) will illuminate if certain faults are recorded.
- ECM operates in backup mode if sensors fail, to enable vehicle to be driven to repair shop.
- 1992-95: Trouble codes can be displayed by the MIL or by using a Scan Tool connected to the data link connector (DLC).
- 1996-97: Trouble codes can only be displayed by using a Scan Tool connected to the data link connector (DLC).

Accessing

1992

- Switch ignition ON.
- Bridge data link connector (DLC) terminals E1 and TE1 **1**.
- Count MIL flashes. Note trouble codes. Compare with trouble code table.
- Each trouble code consists of two groups of one or more flashes **2** [**A**], displayed as follows:
  - The first group of flashes indicate the 'tens' of the trouble code.
  - The second group of flashes indicate the 'units' of the trouble code.
  - A short pause separates 'tens' and 'units' **2** [**B**].
  - A long pause separates each trouble code **2** [**C**].
  - For example: Trouble code 21 displayed **2**.
- Each trouble code will be displayed three times in ascending order, followed by the MIL flashing ON-OFF continuously.
- If no trouble codes are stored the MIL will either:
  - Flash trouble code 51 then flash ON-OFF continuously, indicating correct functioning of diagnostic system.
  - Flash ON-OFF continuously without flashing code 51, indicating a malfunction with either the Idle switch, A/C switch or the park/neutral circuits.
- Trouble codes can also be displayed by using a Scan Tool connected to the data link connector (DLC) **1**.

1

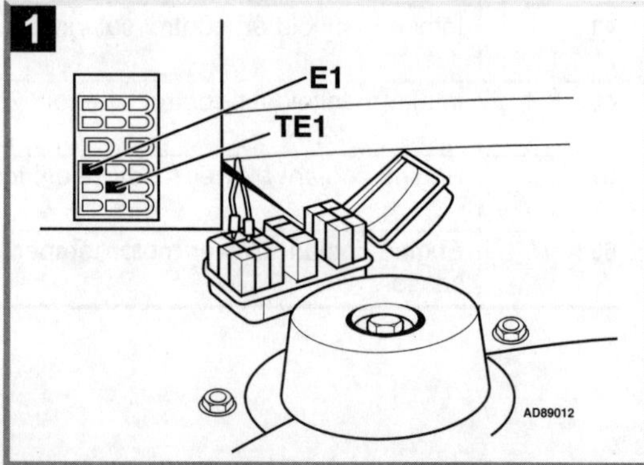

2

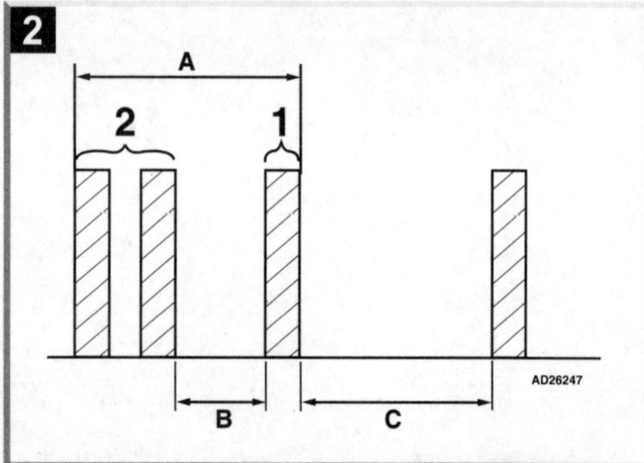

1993-95 – normal mode

- Switch ignition ON.
- Bridge data link connector (DLC) terminals E1 and TE1 **3**.
- Count MIL flashes. Note trouble codes. Compare with trouble code table.
- Each trouble code consists of two groups of one or more flashes **2** [**A**].
  - The first group of flashes indicate the 'tens' of the trouble code.
  - The second group of flashes indicate the 'units' of the trouble code.
  - A short pause separates 'tens' and 'units' **2** [**B**].
  - A long pause separates each trouble code **2** [**C**].
  - For example: Trouble code 21 displayed **2**.
- Trouble codes will be displayed in ascending order.
- If no trouble codes are stored the MIL will Flash ON-OFF continuously.
- Remove bridge wire.
- Switch ignition OFF.
- Rectify faults as necessary.
- Trouble codes can also be displayed by using a Scan Tool connected to the data link connector (DLC) **3**.

3

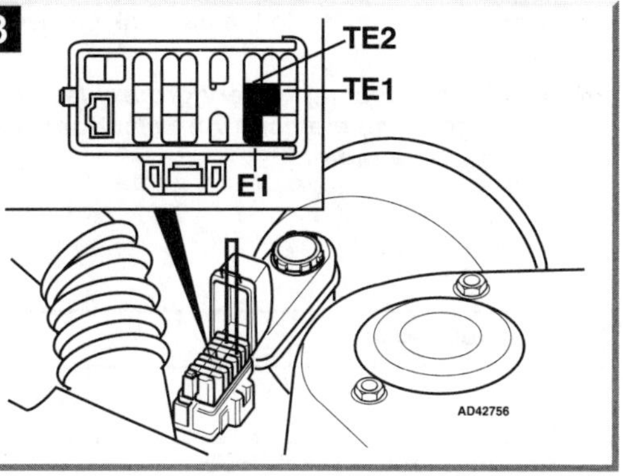

1993-95 – test mode

- Switch ignition OFF.
- Bridge data link connector (DLC) terminals E1 and TE2 **3**.
- Switch ignition ON.
- Ensure malfunction indicator lamp (MIL) flashes ON-OFF continuously.
- Start engine.
- Drive vehicle at over 6 mph to meet conditions of customer complaint.
- Stop vehicle. Do not switch ignition OFF.
- With data link connector (DLC) terminals E1 and TE2 bridged, bridge terminals E1 and TE1 **3**.
- Count MIL flashes. Note trouble codes. Compare with trouble code table.
- Switch ignition OFF.
- Remove bridge wires.
- Rectify faults as necessary.

NOTE: ***Codes 42 and 43 may be produced if this procedure is NOT carried out correctly. Code 51 will be produced during this procedure and can be ignored.***

1996-97

- Trouble codes can only be displayed by using a Scan Tool connected to the data link connector (DLC) **4**.

4

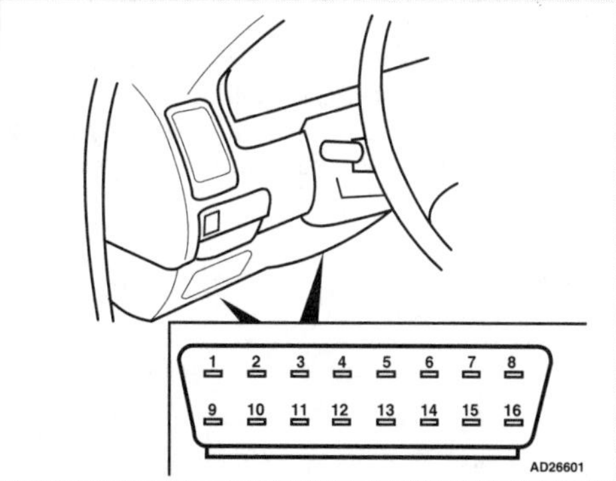

Erasing

Method 1 – 1992

- Ensure ignition switched OFF.
- Remove STOP fuse (15A) from LH kick panel fuse box for at least 30 seconds **5**.
- Reinstall fuse.
- Repeat accessing procedure to ensure no trouble codes remain in ECM fault memory.
- Trouble codes can also be erased by using a Scan Tool connected to the data link connector (DLC).

5
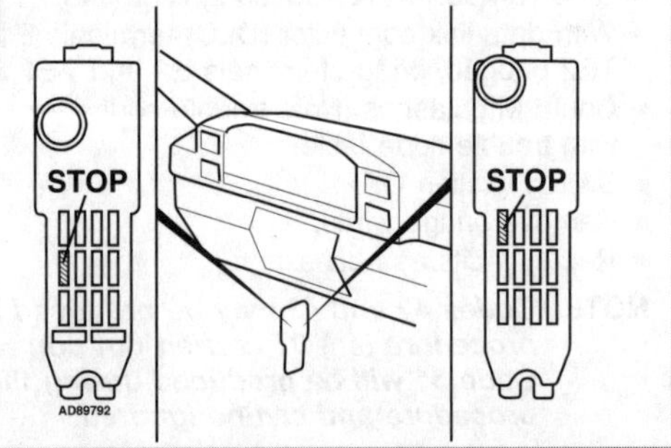

6
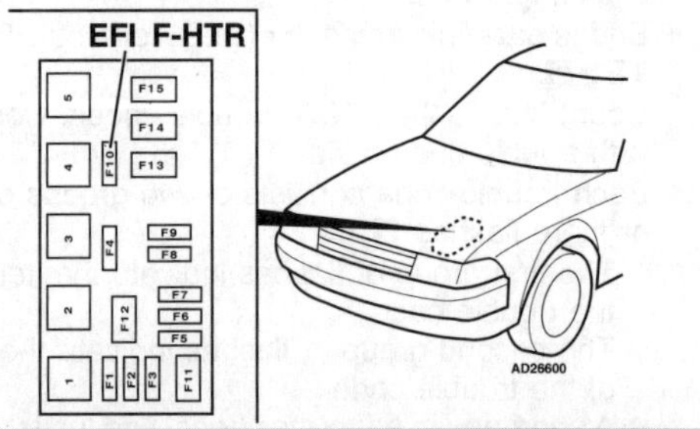

Method 1 - 1993-97

- Ensure ignition switched OFF.
- Remove EFI F-HTR fuse (15A) from engine bay fuse box for at least 30 seconds **6**.
- Reinstall fuse.
- Repeat accessing procedure to ensure no trouble codes remain in ECM fault memory.
- Trouble codes can also be erased by using a Scan Tool connected to the data link connector (DLC).

Method 2

NOTE: ***Disconnecting battery ground lead may cause driveability problems until the engine management system relearns basic values.***

- Ensure ignition switched OFF.
- Disconnect battery ground cable for at least 30 seconds.
- Reconnect battery ground cable.
- Repeat accessing procedure to ensure no trouble codes remain in ECM fault memory.
- Trouble codes can also be erased by using a Scan Tool connected to the data link connector (DLC).

NOTE: ***Disconnecting battery ground lead may also erase memory from electronic units, such as radio.***

Trouble code identification

Scan code	Fault location	Probable cause
P0000-P0999	Refer to OBD-II trouble code table at the front of this manual	–
P1300	Ignition control – circuit malfunction	**Wiring, ignition coil, ECM**
P1335	Crankshaft position (CKP) sensor – circuit malfunction	**Wiring, CKP sensor, ECM**
P1500	Starter signal – circuit malfunction	**Wiring, ignition switch, engine control relay, ECM**
P1600	Engine control module (ECM) – supply voltage	**Wiring, ECM**

Flash type	Fault location	Probable cause
11111	No fault found	–
12	Crankshaft position (CKP) sensor/camshaft position (CMP) sensor – no signal	**Wiring, CKP/CMP sensor, ICM, ECM**
13	Crankshaft position (CKP) sensor – no signal over idle	**Wiring, CKP sensor, ICM, ECM**
14	Ignition signal – no signal	**Wiring, ICM, ECM**
16	Transmission control signal	**Wiring, ECM**
21	Oxygen sensor (O2S) 1/heated oxygen sensor (HO2S) 1 – circuit malfunction	**Wiring open circuit/short circuit, O2S/HO2S, ECM**
22	Engine coolant temperature (ECT) sensor – circuit malfunction	**Wiring open circuit/short circuit, ECT sensor, ECM**
24	Intake air temperature (IAT) sensor – circuit malfunction	**Wiring open circuit/short circuit, IAT sensor, ECM**
25	Weak mixture	**Wiring, air leak, ignition system, fuel pressure, injector(s), MAF sensor, ECT sensor, IAT sensor, O2S/HO2S, ECM**
26	Rich mixture	**Wiring, ignition system, fuel pressure, injector(s), MAF sensor, ECT sensor, IAT sensor, O2S/HO2S, ECM**
27	Oxygen sensor (O2S) 2/heated oxygen sensor (HO2S) 2 – circuit malfunction	**Wiring open circuit/short circuit, O2S/HO2S 2, ECM**
31	Manifold absolute pressure (MAP) sensor – circuit malfunction	**Wiring open circuit/short circuit, MAP sensor, ECM**
41	Throttle position (TP) sensor	**Wiring open circuit/short circuit, TP sensor, ECM**
42	Vehicle speed sensor (VSS)	**Wiring open circuit/short circuit, VSS, ECM**
43	Starter signal	**Wiring, ignition switch, ECM**
51 [1]	Switch signal – A/C switch ON during diagnosis	**Wiring, A/C switch, A/C control module, ECM**
	Switch signal – PNP switch not in P or N during diagnosis	**Wiring, PNP switch, ECM**
	Switch signal – throttle position (TP) sensor, no closed throttle signal during diagnosis	**Wiring, TP sensor, ECM**
52	Knock sensor (KS)	**Wiring, KS, ECM**
71	Exhaust gas recirculation (EGR) system – malfunction	**Wiring open circuit/short circuit, hose leak/ blockage, EGRT sensor, EGR solenoid, ECM**

[1] 1992: Refer to accessing procedure

GEO

Model:	Engine identification:	Year:
Storm 1.6L	VIN position 8 = 6	1992-93
Storm 1.8L	VIN position 8 = 8	1992-93

System: **Isuzu MPi**

Self-diagnosis

General information

- Refer to the front of this manual for general test conditions, terminology, detailed descriptions of wiring faults and a general trouble shooter for electrical and mechanical faults.
- Malfunction indicator lamp (MIL) will illuminate continuously if permanent faults are recorded.
- Malfunction indicator lamp (MIL) will illuminate for 10 seconds and then switch off if intermittent faults are recorded.
- ECM operates in backup mode if sensors fail, to enable vehicle to be driven to repair shop.
- Trouble codes can be displayed by the MIL or by using a Scan Tool connected to the white 3 pin data link connector (DLC).

Accessing

- Access white 3 pin data link connector (DLC).
- Jump data link connector (DLC) terminals 1 and 3 **1**.
- Switch ignition ON. Do NOT start engine.
- Count MIL flashes. Note trouble codes. Compare with trouble code table.
- Each trouble code consists of two groups of one or more flashes **2** [**A**], displayed as follows:
  - The first group of flashes indicate the 'tens' of the trouble code.
  - The second group of flashes indicate the 'units' of the trouble code.
  - A short pause separates 'tens' and 'units' **2** [**B**].
  - A long pause separates each trouble code **2** [**C**].
  - For example: Trouble code 21 displayed **2**.
- First trouble code 12 will be displayed 3 times.
- Each additional trouble code will now be displayed three times, in ascending order.

NOTE: ***If no additional trouble codes are stored, code 12 will continue to be displayed.***

- Trouble codes can also be displayed by using a Scan Tool connected to the data link connector (DLC) **1**.

1

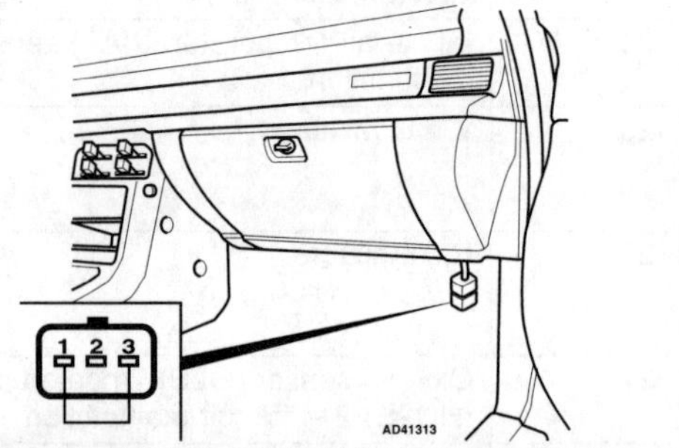

2

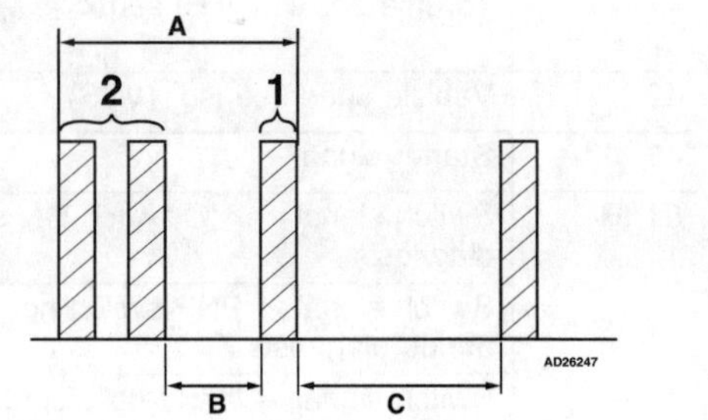

Erasing

Method 1

- Ensure ignition switched OFF.
- Remove ECM fuse (30A) from engine bay fuse box for at least 30 seconds **3**.
- Reinstall fuse.
- Repeat accessing procedure to ensure no trouble codes remain in ECM fault memory.

- Trouble codes can also be erased by using a Scan Tool connected to the data link connector (DLC).

3

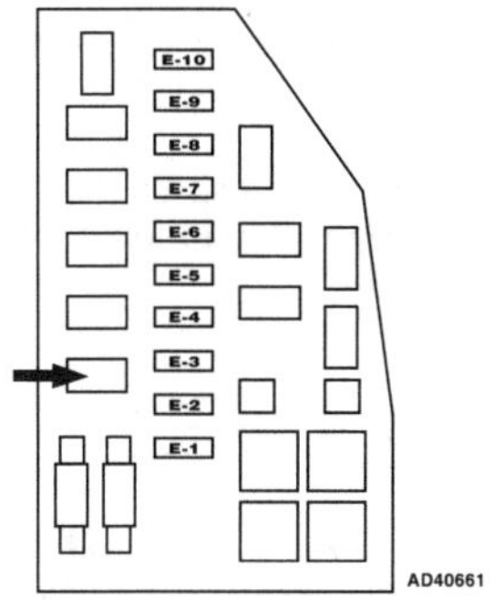

Method 2

NOTE: ***Disconnecting battery ground lead may cause driveability problems until the engine management system relearns basic values.***

- Ensure ignition switched OFF.
- Disconnect battery ground cable for at least 30 seconds.
- Reconnect battery ground cable.
- Repeat accessing procedure to ensure no trouble codes remain in ECM fault memory.
- Trouble codes can also be erased by using a Scan Tool connected to the data link connector (DLC).

NOTE: ***Disconnecting battery ground lead may also erase memory from electronic units, such as radio.***

Trouble code identification

Flash code	Fault location	Probable cause
12	Start of diagnosis	**No engine speed (RPM) signal received**
13	Heated oxygen sensor (HO2S)/oxygen sensor (O2S) – open circuit	**Wiring, HO2S/O2S, ECM**
14	Engine coolant temperature (ECT) sensor	**Wiring open circuit/short circuit, ECT sensor, ECM**
21	Throttle position (TP) sensor – voltage low/high	**Wiring open circuit/short circuit, TP sensor, ECM**
23	Intake air temperature (IAT) sensor	**Wiring, IAT sensor, ECM**
24	Vehicle speed sensor (VSS) – circuit	**Wiring, VSS, ECM**
32	Exhaust gas recirculation (EGR) system	**Wiring, EGR valve, EGRT sensor, ECM**
33	Manifold absolute pressure (MAP) sensor – voltage low/high	**Wiring, MAP sensor, ECM**
42	Ignition control signal – circuit malfunction	**Wiring open circuit/short circuit, ICM, ECM**
44	Heated oxygen sensor (HO2S)/oxygen sensor (O2S) – low voltage	**Wiring short circuit to ground, fuel pressure, intake/exhaust leak, HO2S/O2S**
45	Heated oxygen sensor (HO2S)/oxygen sensor (O2S) – high voltage	**Wiring, fuel pressure, injector control, EVAP canister purge valve, EGR valve, MAP sensor, TP sensor, air filter blocked, HO2S/O2S**
51	Engine control module (ECM) – memory/EEPROM error	**ECM**

GEO

Model:	Engine identification:	Year:
Tracker 1.6L	**G16**	**1992-97**

System: **EPI**

Self-diagnosis

General information

- Refer to the front of this manual for general test conditions, terminology, detailed descriptions of wiring faults and a general trouble shooter for electrical and mechanical faults.
- Malfunction indicator lamp (MIL) will illuminate if certain faults are recorded.
- ECM operates in backup mode if sensors fail, to enable vehicle to be driven to repair shop.
- 1992-95: Trouble codes can be displayed by the MIL or by using a Scan Tool connected to the 12 pin data link connector (DLC).
- 1996-97: Trouble codes can only be displayed by using a Scan Tool connected to the data link connector (DLC).

Accessing

1992-95

- Access data link connector (DLC) **1**.
- Jump data link connector (DLC) terminals 2 and 3 (Federal) **2** [**1**], or terminals 2 and 4 (California) **2** [**2**].
- Switch ignition ON. Do NOT start engine.
- Count MIL flashes. Note trouble codes. Compare with trouble code table.
- Each trouble code consists of two groups of one or more flashes **3** [**A**], displayed as follows:
  - The first group of flashes indicate the 'tens' of the trouble code.
  - The second group of flashes indicate the 'units' of the trouble code.
  - A short pause separates 'tens' and 'units' **3** [**B**].
  - A long pause separates each trouble code **3** [**C**].
  - For example: Trouble code 21 displayed **3**.
- Each trouble code will be displayed three times, in ascending order.

NOTE: ***If no trouble codes are stored, code 12 will be displayed.***

- Trouble codes can also be displayed by using a Scan Tool connected to the 12 pin data link connector (DLC) **4**.

1

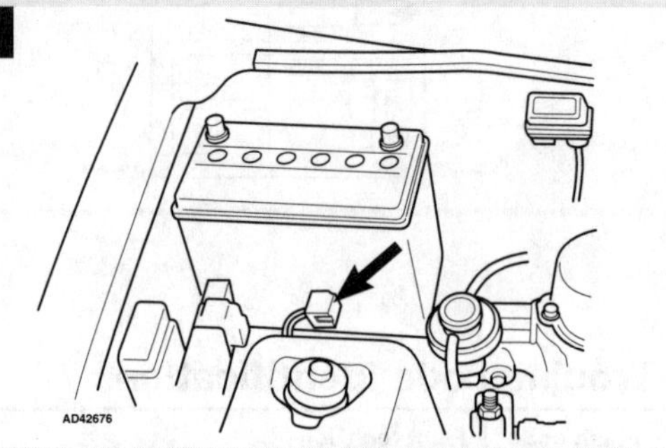

2

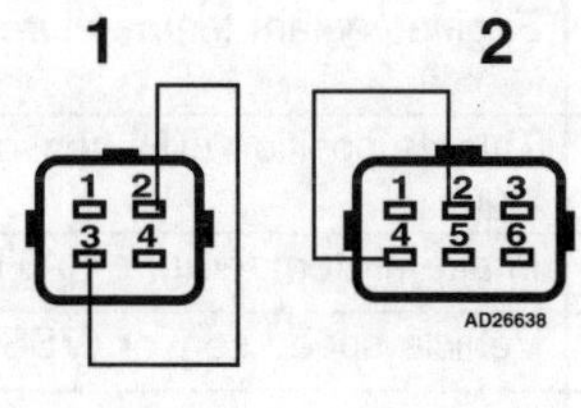

3

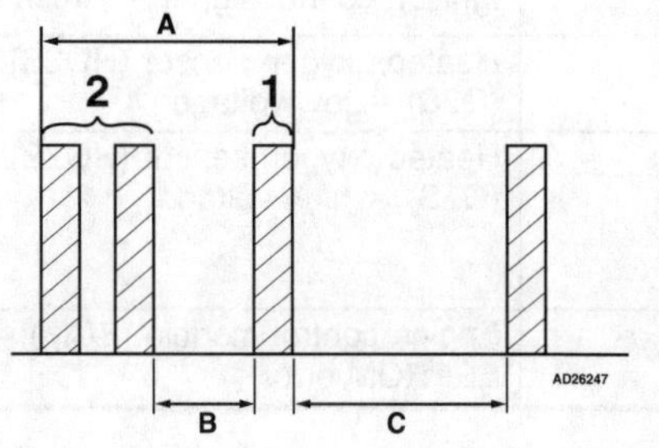

4

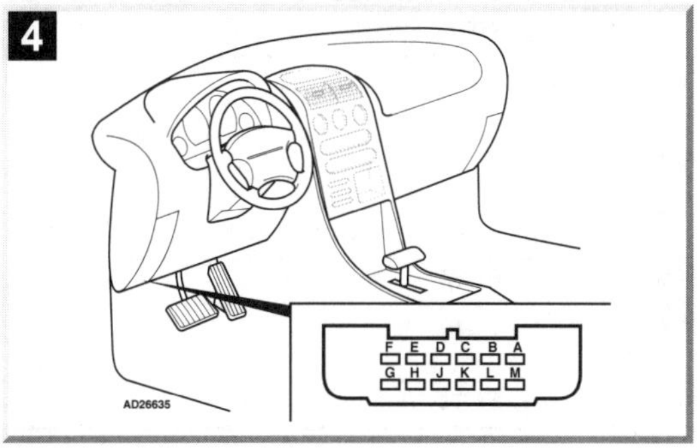

1996-97

- Trouble codes can only be displayed by using a Scan Tool connected to the data link connector (DLC) 5.

5

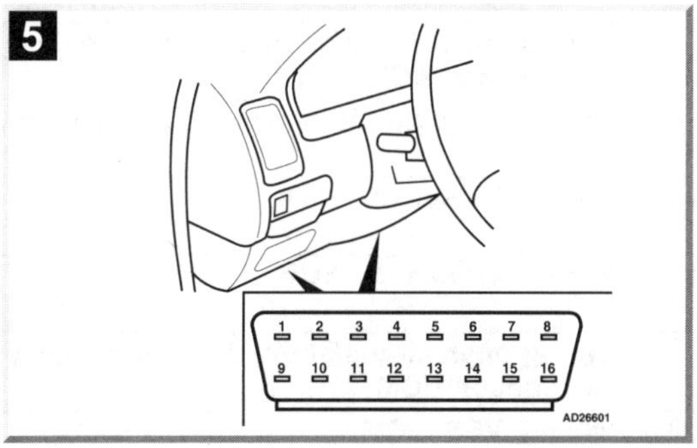

Erasing

Method 1

- Ensure ignition switched OFF.
- Remove TAIL/DOME fuse F10 (15A) from fascia fuse box for at least 30 seconds 6.
- Reinstall fuse.
- Repeat accessing procedure to ensure no trouble codes remain in ECM fault memory.
- Trouble codes can also be erased by using a Scan Tool connected to the data link connector (DLC): 1992-95 – 12 pin, 1996-97 – 16 pin.

6

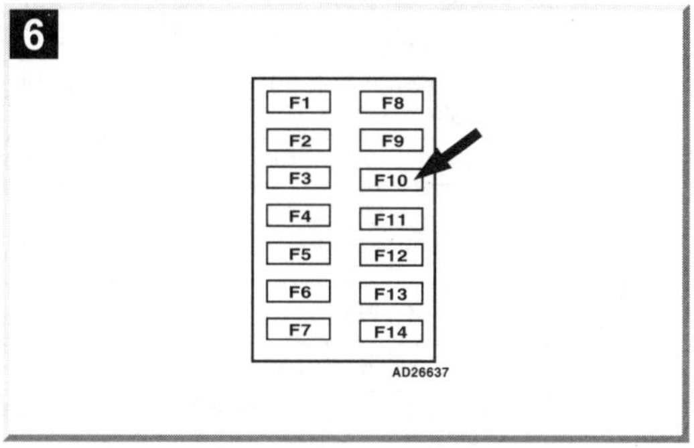

Method 2

NOTE: ***Disconnecting battery ground lead may cause driveability problems until the engine management system relearns basic values.***

- Ensure ignition switched OFF.
- Disconnect battery ground cable for at least 30 seconds.
- Reconnect battery ground cable.
- Repeat accessing procedure to ensure no trouble codes remain in ECM fault memory.
- Trouble codes can also be erased by using a Scan Tool connected to the data link connector (DLC): 1992-95 – 12 pin, 1996-97 – 16 pin.

NOTE: ***Disconnecting battery ground lead may also erase memory from electronic units, such as radio.***

Trouble code identification

Scan code	Fault location	Probable cause
P0000-P0999	Refer to OBD-II trouble code table at the front of this manual	–
P1408	Manifold absolute pressure (MAP) sensor – circuit malfunction	**Wiring, hose leak/blockage, MAP sensor, ECM**
P1410	Fuel tank pressure control system – malfunction	**Wiring, fuel tank pressure control solenoid, ECM**

Scan code	Fault location	Probable cause
P1450	Barometric pressure (BARO) sensor – circuit malfunction	ECM
P1451	Barometric pressure (BARO) sensor – range/ performance problem	ECM
P1500	Starter signal – circuit malfunction	Wiring, ignition switch, ECM
P1510	Engine control module (ECM) – supply voltage	Wiring, ECM
P1530	Ignition timing adjustment connector – circuit malfunction	Wiring, ECM
P1600	Engine control module (ECM) – supply voltage	Wiring, ECM

Trouble code	Fault location	Probable cause
12	Start of diagnosis	No engine speed (RPM) signal received
13	Oxygen sensor (O2S)/heated oxygen sensor (HO2S) – signal not fluctuating	Wiring, O2S/HO2S, ECM
14	Engine coolant temperature (ECT) sensor – high voltage signal	Wiring open circuit/short circuit to positive, ECT sensor, ECM
15	Engine coolant temperature (ECT) sensor – low voltage signal	Wiring short circuit to ground, ECT sensor, ECM
21	Throttle position (TP) sensor – high voltage signal	Wiring, TP sensor, ECM
22	Throttle position (TP) sensor – low voltage signal	Wiring, TP sensor, ECM
23	Intake air temperature (IAT) sensor – high voltage signal	Wiring open circuit/short circuit to positive, IAT sensor, ECM
24	Vehicle speed sensor (VSS) – circuit malfunction	Wiring, VSS, ECM
25	Intake air temperature (IAT) sensor – low voltage signal	Wiring short circuit to ground, IAT sensor, ECM
31	Manifold absolute pressure (MAP) sensor – low voltage signal	Wiring open circuit/short circuit to ground, MAP sensor, ECM
32	Manifold absolute pressure (MAP) sensor – high voltage signal	Wiring short circuit to positive, hose leak, MAP sensor, ECM
41	Ignition fail safe signal	Wiring, noise suppressor, ignition module, ECM
42	Crankshaft position (CKP) sensor – circuit malfunction	Wiring, CKP sensor, ECM
44	Throttle position (TP) sensor – closed throttle position (CTP) switch contacts – circuit malfunction	Incorrectly adjusted TP sensor, wiring open circuit, TP sensor, ECM
45	Throttle position (TP) sensor – closed throttle position (CTP) switch contacts – circuit malfunction	Incorrectly adjusted TP sensor, wiring short circuit, CTP switch, ECM
51	Exhaust gas recirculation (EGR) system – circuit	Hose leak/blockage, wiring, EGR valve, EGRT sensor, ECM
53	Engine control module (ECM) – ground circuit	Wiring open circuit, ECM

GMC

Model:	Engine identification:	Year:
Envoy 4.2/4.3L	VIN position 8 = S/W	1999-02
Jimmy 4.3L – with GM vehicle control module (VCM)	VIN position 8 = W, Z	1995
Jimmy/Safari 4.3L	VIN position 8 = W	1996-02
Rally 5.7/7.4L	VIN position 8 = K/N	1996
Savana 4.3/5.0/5.7/7.4/8.1L	VIN position 8 = G/J/M/R/W	1996-02
Sierra C/K Pickup 4.3/4.8/5.0/5.3/5.7/6.0/7.4/8.1L	VIN position 8 = G/J/M/R/T/U/V/W	1996-02
Sonoma 2.2/4.3L	VIN position 8 = W/X/4	1996-02
Sonoma 4.3L – with GM vehicle control module (VCM)	VIN position 8 = W/Z	1995
Suburban/Yukon 4.8/5.3/5.7/6.0/7.4/8.1L	VIN position 8 = G/J/R/T/U/V	1996-02
Vandura 4.3/5.7/7.4L	VIN position 8 = K/N/Z	1996

System: **GM**

Self-diagnosis

General information

- Refer to the front of this manual for general test conditions, terminology, detailed descriptions of wiring faults and a general trouble shooter for electrical and mechanical faults.
- Malfunction indicator lamp (MIL) will illuminate or flash if certain faults are recorded.
- ECM operates in backup mode if sensors fail, to enable vehicle to be driven to repair shop.

Accessing

- Trouble codes can only be displayed by using a Scan Tool connected to the data link connector (DLC) **1**.

1

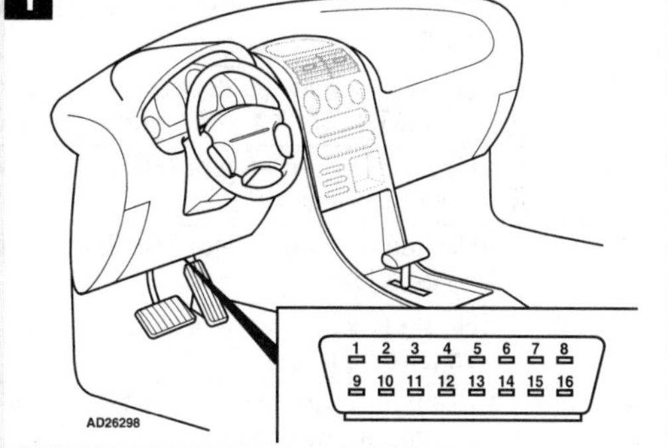

Erasing

- After the faults have been rectified, the manufacturer recommends that trouble codes are erased by using a Scan Tool connected to the data link connector (DLC).

Trouble code identification

Scan code	Fault location	Probable cause
P0000-P0999	Refer to OBD-II trouble code table at the front of this manual	–
P1106	Manifold absolute pressure (MAP) sensor – voltage high, intermittent	**Wiring open circuit/short circuit to positive, poor connections**
P1107	Manifold absolute pressure (MAP) sensor – voltage low, intermittent	**Wiring short circuit to ground, poor connections**
P1111	Intake air temperature (IAT) sensor – voltage high, intermittent	**Wiring open circuit/short to positive, poor connections**
P1112	Intake air temperature (IAT) sensor – voltage low, intermittent	**Wiring short circuit to ground**
P1114	Engine coolant temperature (ECT) sensor – voltage low, intermittent	**Wiring short circuit to ground, poor connections**
P1115	Engine coolant temperature (ECT) sensor, voltage high, intermittent	**Wiring open circuit/short to positive, poor connections**
P1120	Throttle motor position sensor 1 – circuit malfunction	**Wiring, throttle control module**
P1121	Throttle position (TP) sensor – voltage high, intermittent	**Wiring open circuit/short to positive, poor connections**
P1122	Throttle position (TP) sensor – voltage low, intermittent	**Wiring open circuit/short to ground, poor connections**
P1125	Accelerator pedal position (APP) sensor 1/2/3 – signal variation	**Wiring, APP sensor, ECM**
P1133	Oxygen sensor (O2S)/heated oxygen sensor (HO2S) 1, bank 1 – insufficient switching	**Wiring, exhaust system leaking, O2S/HO2S**
P1134	Heated oxygen sensor (HO2S) 1, bank 1 – slow response	**Wiring, oxygen sensor heater, exhaust system leaking, HO2S**
P1137	Heated oxygen sensor (HO2S) 2, bank 1 – voltage low	**Wiring, exhaust system leaking, HO2S**
P1138	Heated oxygen sensor (HO2S) 2, bank 1 – voltage high	**Wiring, HO2S**
P1153	Heated oxygen sensor (HO2S) 1, bank 2 – insufficient switching	**Wiring, oxygen sensor heater, exhaust system leaking, HO2S**
P1154	Heated oxygen sensor (HO2S) 1, bank 2 – slow response	**Wiring, oxygen sensor heater, exhaust system leaking, HO2S**
P1171	Fuel system – lean during acceleration	**Fuel filter, fuel pump, HO2S, injectors**
P1172	Fuel level sensors, primary and secondary – no fuel level change during fuel transfer from secondary to primary tank	**Wiring, secondary fuel pump/relay, fuel level sensors**
P1220	Throttle motor position sensor 1 – circuit malfunction	**Wiring, throttle control module**
P1221 [1]	Fuel pump, run signal – circuit malfunction	**Wiring, engine oil pressure switch, fuel pump relay**
P1221 [2]	Throttle motor position sensor 2 – circuit malfunction	**Wiring, throttle control module**

Scan code	Fault location	Probable cause
P1222	Fuel pump, run signal – circuit malfunction	Wiring, engine oil pressure switch, fuel pump relay
P1258	Engine overheating	Cooling system, ECT sensor
P1271	Accelerator pedal position (APP) sensor 1/2 – signal variation	Wiring, APP sensor, ECM
P1275	Accelerator pedal position (APP) sensor 1 – circuit malfunction	Wiring, APP sensor, ECM
P1275	Accelerator pedal position (APP) sensor 1 – circuit malfunction	Wiring, APP sensor, ECM
P1276	Accelerator pedal position (APP) sensor 1 – range performance	Wiring, APP sensor, ECM
P1280	Accelerator pedal position (APP) sensor 2 – circuit malfunction	Wiring, APP sensor, ECM
P1281	Accelerator pedal position (APP) sensor 2 – range performance	Wiring, APP sensor, ECM
P1285	Accelerator pedal position (APP) sensor 3 – circuit malfunction	Wiring, APP sensor, ECM
P1286	Accelerator pedal position (APP) sensor 3 – range performance	Wiring, APP sensor, ECM
P1336	Crankshaft position sensor (CKP) – variation not learned	Learning procedure not performed
P1345	Crankshaft position sensor (CKP)/camshaft position sensor (CMP) – signal variation	Wiring, CKP sensor, CMP sensor
P1351	Ignition control system, coil – voltage high	Wiring open circuit, ICM, ECM
P1361	Ignition control system, IC circuit – voltage low	Wiring, circuit short circuit to ground, poor connections, ICM, ECM
P1380	Antilock brake system (ABS)/traction control system (TCS), rough road signal to ECM – malfunction	ABS/TCS trouble code(s) stored
P1381	Antilock brake system (ABS)/traction control system (TCS), rough road signal to ECM – no communication	Wiring, poor connections, ABS/TCS
P1404	Exhaust gas recirculation (EGR) valve – not fully closing	Wiring, poor connections, EGR valve position sensor, EGR solenoid
P1406	Exhaust gas recirculation (EGR) valve position sensor – range performance	Wiring, EGR valve tight/sticking, EGR valve position sensor
P1415	Secondary air injection (AIR) system, bank 1	Wiring, hoses blocked/leaking, AIR valve
P1416	Secondary air injection (AIR) system, bank 2	Wiring, hoses blocked/leaking, AIR valve
P1431	Fuel level sensor, secondary – circuit malfunction	Wiring, fuel level sensor
P1432	Fuel level sensor – low voltage	Wiring open circuit, fuel level sensor
P1433	Fuel level sensor – high voltage	Wiring short circuit to positive, fuel level sensor

Scan code	Fault location	Probable cause
P1441	Evaporative emission (EVAP) system – leak detected	**Hose connection(s), EVAP canister, EVAP canister purge valve**
P1442	Evaporative emission (EVAP) system vacuum switch – circuit malfunction	**Hoses blocked/leaking, wiring, EVAP canister, ECM**
P1481	Engine coolant fan speed sensor – signal loss	**Wiring, engine coolant fan speed sensor**
P1482	Engine coolant fan clutch solenoid – circuit malfunction	**Wiring, engine coolant fan clutch solenoid, engine coolant fan clutch solenoid relay**
P1484	Engine coolant fan speed/engine speed – signal variation	**Wiring, engine coolant fan clutch, engine coolant fan speed sensor**
P1504	Engine control module(ECM), vehicle speed sensor (VSS) output signal – circuit malfunction	**Wiring, poor connections, ABS, ECM**
P1508	Idle air control (IAC) valve – low rpm	**Wiring open circuit/short circuit to ground/ short circuit to positive, IAC valve, throttle valve tight/sticking**
P1509	Idle air control (IAC) valve – high rpm	**Wiring open circuit/short circuit to ground/ short circuit to positive, IAC valve, throttle valve tight/sticking**
P1512	Throttle command/actual throttle position – signals out of range	**Wiring, throttle valve sticking, throttle control module**
P1514	Throttle control unit, intake air flow – excessive	**Throttle control unit, MAP sensor, CMP sensor signal, ECM**
P1515	Throttle command/actual throttle position – signal variation	**Wiring, throttle valve sticking, throttle control module**
P1516	Throttle command/actual throttle position – signal variation	**Wiring, throttle valve sticking, throttle control module**
P1517	Throttle control module – malfunction	**Throttle control module**
P1518	Throttle control module/engine control module (ECM) – communication malfunction	**Wiring, throttle control module, ECM**
P1520	Park/neutral position (PNP) switch – circuit malfunction	**Wiring short circuit to ground, PNP switch incorrectly adjusted**
P1523	Throttle control unit, throttle position during battery saving mode	**Throttle valve movement restricted, throttle control unit, ECM**
P1571	Traction control system (TCS), torque request signal – malfunction	**Wiring, poor connections, ABS/TCS**
P1574	Brake pedal position switch (BPP), voltage high	**Wiring, ABS**
P1600	Engine control module (ECM) – malfunction	**ECM**
P1601	CAN data bus, engine control module (ECM)/ antilock brake system (ABS)/traction control system (TCS)/instrumentation control module/ diagnostic control module – some communication lost	**Wiring, poor connections, ECM, ABS/TCS, diagnostic control module, instrumentation control module**
P1602	Engine control module (ECM) to antilock brake system (ABS)/traction control system (TCS), CAN data bus – communication failure	**Wiring, poor connections, ABS control module**

Scan code	Fault location	Probable cause
P1621	Engine control module (ECM) – memory performance	ECM
P1626	Engine control module (ECM) – fuel enable signal lost	Wiring, poor connections, immobilizer control module, ECM
P1627	Engine control module (ECM) – malfunction	ECM
P1630	Engine control module (ECM), immobilizer code – learning mode	ECM immobilizer code programming not completed
P1631	Engine control module (ECM) – incorrect immobilizer code	Incorrect key, immobilizer code not programmed
P1632 1	System voltage – low	Wiring, battery, alternator
P1632 3	Engine control module (ECM) – fuel disable signal received	Fuses, BCM, immobilizer code
P1635	Engine control module (ECM), sensor supply voltage – circuit malfunction	Wiring, poor connections, TP sensor, MAP sensor, EGR valve position sensor, ECM
P1637	Alternator warning lamp, control signal – circuit malfunction	Wiring, instrument panel, ECM
P1638	Alternator 'F' terminal – circuit malfunction	Wiring open circuit/short circuit to ground, alternator, ECM
P1639	Engine control module (ECM), sensor supply voltage – circuit malfunction	Wiring, poor connections, FTP sensor, ECM
P1644	Traction control system (TCS), delivered torque signal – malfunction	Wiring, poor connections, ECM
P1665	Fuel level sensor – signal malfunction	Wiring short circuit to positive, fuel level sensor
P1680	Engine control module (ECM) – malfunction	ECM
P1681	Engine control module (ECM) – malfunction	ECM
P1683	Engine control module (ECM) – malfunction	ECM
P1689	Traction control system (TCS), delivered torque signal – malfunction	Wiring, poor connections, ECM
P1779	Engine control module (ECM), torque reduction signal – out of range	Wiring TCM, ECM
P1781	Transmission control module (TCM), MIL request signal – voltage high/low	Wiring TCM, ECM
P1810	Transmission gear pressure switch, signal – circuit malfunction	Wiring open circuit/short circuit to ground, reverse/D1/D2/D3/D4 gear pressure switch(es), ECM
P1812	Transmission fluid temperature (TFT) – high	Transmission mechanical fault, towing weight limit exceeded
P1860	Torque converter clutch (TCC) solenoid, pulse width modulation (PWM) – circuit malfunction	Wiring open circuit/short circuit to ground, TCC solenoid, ECM
P1864	Torque converter clutch (TCC) solenoid – circuit malfunction	Wiring open circuit/short circuit to ground, TCC solenoid, ECM
P1870	Torque converter clutch (TCC) – slip speed excessive	Transmission fluid level, wiring, TCC solenoid, ECM

Scan code	Fault location	Probable cause
P1875	Transmission, 4WD low selection signal – circuit malfunction	**Wiring, transmission, four wheel drive control module, ECM**
P1886	Transmission 3rd-2nd control solenoid – circuit malfunction	**Wiring, control solenoid, ECM**
U1000	CAN data bus – communication malfunction	**Wiring, BCM, ECM, instrumentation control module, SDM**
U1026	Transmission control module (TCM)/four wheel drive control module, CAN data bus – communication malfunction	**Wiring, BCM**
U1041	Antilock brake system (ABS)/traction control system (TCS), CAN data bus – communication malfunction	**Wiring, ABS/TCS control module**
U1064	Body control module (BCM)/immobilizer control module, CAN data bus – communication malfunction	**Wiring, BCM, immobilizer control module**
U1192	Engine control module (ECM)/immobilizer control module, CAN data bus – signal malfunction	**Wiring, immobilizer control module, ECM**
U1193	Engine control module (ECM)/immobilizer control module, CAN data bus – signal malfunction	**Wiring, immobilizer control module, ECM**

1 1995
2 2001→
3 1998→

GMC

Model:	Engine identification:	Year:
Jimmy 4.3L – without vehicle control module (VCM)	VIN position 8 = W, Z	1992-95
Rally 4.3/5.0/5.7/7.4L	VIN position 8 = H/K/N/Z	1992-95
Safari 4.3L	VIN position 8 = W, Z	1992-95
Sierra C/K Pickup 4.3/5.0L	VIN position 8 = H/Z	1992-95
Sierra C/K Pickup 5.7/7.4L	VIN position 8 = K/N	1992-95
Sonoma 2.2/2.5/2.8L	VIN position 8 = A/R/4	1992-95
Sonoma 4.3L – without vehicle control module (VCM)	VIN position 8 = W, Z	1992-95
Suburban 5.7/7.4L	VIN position 8 = K/N	1992-95
Vandura 4.3/5.0/5.7/7.4L	VIN position 8 = H/K/N/Z	1992-95
Yukon 5.7L	VIN position 8 = K	1992-95

System: **GM**

Self-diagnosis

General information

- Refer to the front of this manual for general test conditions, terminology, detailed descriptions of wiring faults and a general trouble shooter for electrical and mechanical faults.
- Malfunction indicator lamp (MIL) will illuminate if certain faults are recorded.
- ECM operates in backup mode if sensors fail, to enable vehicle to be driven to repair shop.
- Trouble codes can be displayed by the MIL or by using a Scan Tool connected to the data link connector (DLC).

Accessing

- Access data link connector (DLC).
- Models with 12 pin data link connector (DLC).
  - Jump data link connector (DLC) terminals A and B **1**.
- Models with 16 pin data link connector (DLC).
  - Jump data link connector (DLC) terminals 5 and 6 **2**.
- Switch ignition ON. Do NOT start engine.
- Count MIL flashes. Note trouble codes. Compare with trouble code table.
- Firstly trouble code 12 will be displayed 3 times.
- Trouble codes consist of 2 groups, displayed as follows:
  - First group indicates the 'tens' of the trouble code **3** [**A**].
  - Second group indicates the 'units' of the trouble code **3** [**C**].
  - A short pause separates each group **3** [**B**].
  - A long pause separates each trouble code **3** [**D**].
  - For example: Trouble code 12 displayed **3**. Start of diagnosis.
- Each additional trouble code will now be displayed 3 times, in ascending order.

NOTE: ***If no additional trouble codes are stored, code 12 will continue to be displayed.***

- Trouble codes can also be displayed by using a Scan Tool connected to the data link connector (DLC) **1** or **2**.

1

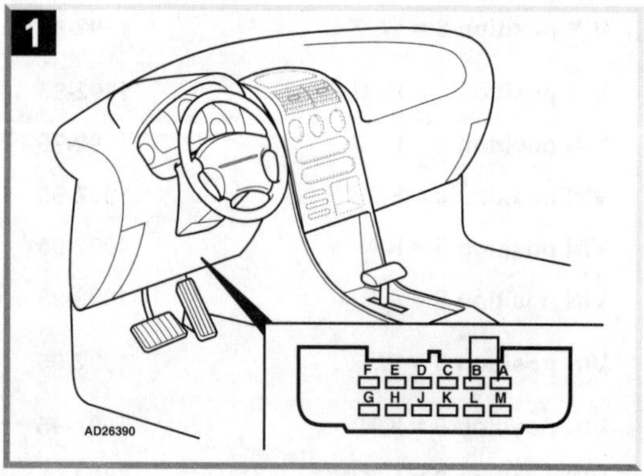

3

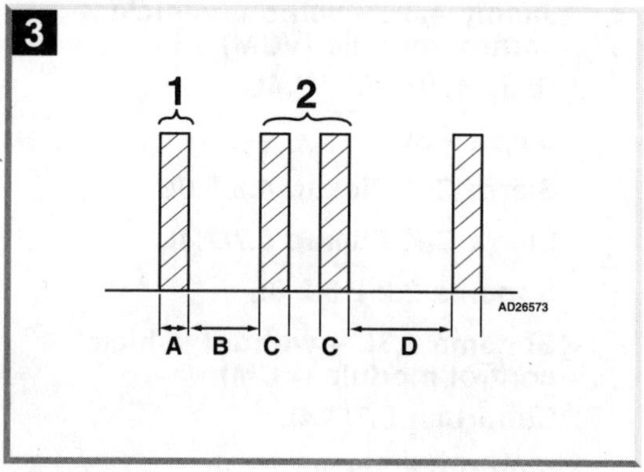

2

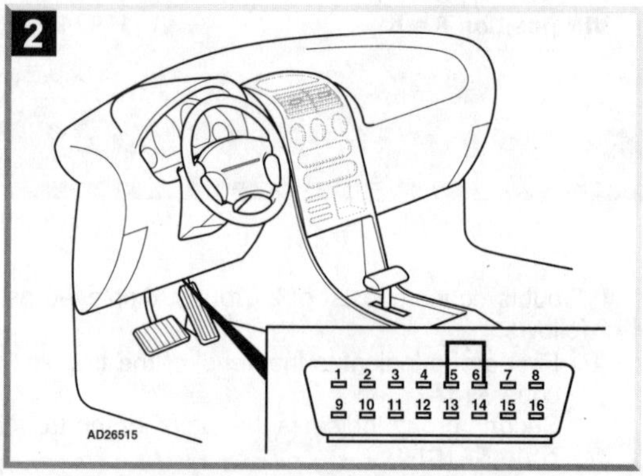

Erasing

NOTE: ***Disconnecting battery ground lead may cause driveability problems until the engine management system relearns basic values.***

- Ensure ignition switched OFF.
- Disconnect battery ground cable for 30 seconds.
- Reconnect battery ground cable.
- Repeat checking procedure to ensure no trouble codes remain in ECM fault memory.

NOTE: ***Disconnecting battery ground lead may also erase memory from electronic units, such as radio.***

- Trouble codes can also be erased by using a Scan Tool connected to the data link connector (DLC).

Trouble code identification

Flash code	Fault location	Probable cause
12	Start of diagnosis	**No engine speed (RPM) signal received**
13	Oxygen sensor (O2S)/heated oxygen sensor (HO2S)	**Wiring open circuit, O2S/HO2S, ECM**
14	Engine coolant temperature (ECT) sensor – voltage high/low	**Wiring, ECT sensor, ECM**
15	Engine coolant temperature (ECT) sensor – voltage high	**Wiring open circuit, ECT sensor, ECM**
16	Vehicle speed sensor (VSS) – no signal	**Wiring, VSS, ECM**
21	Throttle position (TP) sensor – voltage high/low	**Wiring, TP sensor, ECM**

Flash code	Fault location	Probable cause
22	Throttle position (TP) sensor – voltage low	**Wiring open circuit/short circuit to ground, TP sensor, ECM**
23	Intake air temperature (IAT) sensor – voltage high	**Wiring, IAT sensor, ECM**
24	Vehicle speed sensor (VSS) – no signal	**Wiring, VSS, ECM**
25	Intake air temperature (IAT) sensor – voltage low	**Wiring short circuit to ground, IAT sensor, ECM**
26	Engine control module (ECM), quad driver/ output driver – circuit malfunction	**Wiring, ECM controlled components, ECM [4]**
27	Engine control module (ECM), quad driver/ output driver – circuit malfunction	**Wiring, ECM controlled components, ECM [4]**
28	Transmission gear pressure switch, signal – circuit malfunction	**Wiring open circuit/short circuit to ground, reverse/D1/D2/D3/D4 gear pressure switch(es), ECM**
31	Throttle control system – malfunction	**Wiring, throttle control motor, throttle control module, ECM**
32	Exhaust gas recirculation (EGR) solenoid	**Wiring, poor connections, EGR valve, ECM**
33	Manifold absolute pressure (MAP) sensor – voltage high	**Wiring, hose blocked/leaking, MAP sensor, ECM**
34	Manifold absolute pressure (MAP) sensor – voltage low	**Wiring open circuit/short circuit to ground, MAP sensor, ECM**
35	Idle air control (IAC) valve	**Wiring, IAC valve, intake system leaking, fuel system, PCV valve, mechanical fault, ECM**
36	Auxiliary idle air control valve	**Wiring, idle air control valve, vacuum pipes, ECM**
37	Brake pedal position (BPP) switch for torque converter clutch (TCC) – stuck ON	**Wiring, BPP switch, ECM**
38 [2]	Brake pedal position (BPP) switch, torque converter clutch (TCC) – stuck OFF	**Wiring, BPP switch, ECM**
38 [1]	Knock sensor (KS) – circuit malfunction	**Wiring open circuit, KS, ECM**
39 [1]	Knock sensor (KS) – circuit malfunction	**Wiring short circuit, KS, ECM**
39 [3]	Brake pedal position (BPP) switch – stuck OFF	**Wiring, BPP switch, ECM**
41	Crankshaft position (CKP) sensor, 1X reference signal – circuit malfunction	**Wiring, ICM, ECM**
42	Ignition control, bypass/timing signal – circuit malfunction	**Wiring, poor connections, ICM, ECM**
43	Knock sensor (KS)/knock control module – circuit malfunction	**Wiring, KS, knock control module, ECM**

Flash code	Fault location	Probable cause
44	Oxygen sensor (O2S)/heated oxygen sensor (HO2S) – mixture lean	**Wiring open circuit/short circuit to ground, injector(s) blocked, fuel contamination, fuel pressure low, exhaust system leaking, O2S/HO2S, ECM**
45	Oxygen sensor (O2S)/heated oxygen sensor (HO2S) – mixture rich	**Wiring open circuit, fuel pressure high, injector(s) leaking, O2S/HO2S, EVAP purge valve, TP sensor, EGR system, ECM**
51	Engine control module (ECM), PROM – malfunction	**Poor connections, incorrect programming, ECM**
52 **1**	Engine control module (ECM) – malfunction	**ECM, poor connections**
52 **2**	System voltage – high	**Alternator, jump started**
53	System voltage – high	**Alternator, jump started**
54	Fuel pump circuit – voltage low	**Wiring open circuit/short circuit to ground, fuel pump relay, engine oil pressure warning lamp/fuel pump switch**
55	Engine control module (ECM) – malfunction	**ECM**
58	Transmission fluid temperature (TFT) sensor – voltage low	**Wiring short circuit to ground, TFT sensor, ECM**
59	Transmission fluid temperature (TFT) sensor – voltage high	**Wiring open circuit, TFT sensor, ECM**
66	Transmission 3rd-2nd control solenoid – circuit malfunction	**Wiring, control solenoid, ECM**
67	Torque converter clutch (TCC) solenoid – circuit malfunction	**Wiring, TCC solenoid, ECM**
68	Torque converter clutch (TCC) – slipping	**Wiring, TCC solenoid, transmission mechanical fault, ECM**
69	Torque converter clutch (TCC) solenoid – stuck ON	**Wiring open circuit/short circuit to ground, transmission internal wiring, TCC solenoid, transmission mechanical fault, ECM**
72	Vehicle speed sensor (VSS) – loss of signal	**Wiring, VSS, ECM**
73	Transmission fluid pressure (TFP) control solenoid – circuit malfunction	**Wiring, poor connection, TFP solenoid**
74	Transmission input shaft speed sensor – circuit malfunction	**Wiring, input shaft speed sensor, ECM**
75	System voltage, engine control module (ECM) – low	**Wiring, battery, alternator, ECM**
79	Transmission fluid temperature (TFT) – high	**Transmission mechanical fault, towing weight limit exceeded**
81	Transmission shift solenoid (SS), 2nd-3rd – circuit malfunction	**Wiring, SS, ECM**
82	Transmission shift solenoid (SS), 1st-2nd – circuit malfunction	**Wiring, SS, ECM**

Flash code	Fault location	Probable cause
83	Torque converter clutch (TCC) solenoid, pulse width modulated – circuit malfunction	**Wiring, TCC solenoid, ECM**
85	Transmission input/output shaft speed, ratio variation – out of range	**Wiring, input shaft speed sensor, output shaft speed sensor, mechanical fault**
86	Transmission input/output shaft speed, ratio variation – lower than expected	**Wiring, input shaft speed sensor, output shaft speed sensor, mechanical fault**
87	Transmission input/output shaft speed, ratio variation – higher than expected	**Wiring, input shaft speed sensor, output shaft speed sensor, mechanical fault**

1 →1993
2 1994→
3 Automatic transmission 4L80E 1994
4 Depending on model, year and equipment level, the following components may be connected to this circuit. Use of the vehicle wiring diagram is recommended to confirm which components are fitted.

- □ A/C compressor clutch relay
- □ A/C condenser blower motor relay
- □ Engine coolant blower motor relay
- □ Engine coolant 'hot' warning lamp
- □ Evaporative emission (EVAP) canister purge valve
- □ Exhaust gas recirculation (EGR) solenoid
- □ Malfunction indicator lamp (MIL)
- □ Torque converter clutch (TCC) solenoid
- □ Transmission shift solenoids (SS)
- □ Transmission shift warning lamp (MT)

GMC

Model:	Engine identification:	Year:
Rally 6.2L	**VIN position 8 = C**	**1992-93**
Sierra C/K Pickup 6.2L	**VIN position 8 = C**	**1992-93**
Vandura 6.2L	**VIN position 8 = C**	**1992-93**

System: **GM**

Self-diagnosis

General information

- Refer to the front of this manual for general test conditions, terminology, detailed descriptions of wiring faults and a general trouble shooter for electrical and mechanical faults.
- Malfunction indicator lamp (MIL) will illuminate if certain faults are recorded.
- ECM operates in backup mode if sensors fail, to enable vehicle to be driven to repair shop.
- Trouble codes can be displayed by the MIL or by using a Scan Tool connected to the data link connector (DLC).

Accessing

- Access data link connector (DLC).
- Jump data link connector (DLC) terminals A and B **1**.
- Switch ignition ON. Do NOT start engine.
- Count MIL flashes. Note trouble codes. Compare with trouble code table.
- Firstly trouble code 12 will be displayed 3 times.
- Trouble codes consist of 2 groups, displayed as follows:
  - First group indicates the 'tens' of the trouble code **2** [**A**].
  - Second group indicates the 'units' of the trouble code **2** [**C**].
  - A short pause separates each group **2** [**B**].
  - A long pause separates each trouble code **2** [**D**].
  - For example: Trouble code 12 displayed **2**. Start of diagnosis.
- Each additional trouble code will now be displayed 3 times, in ascending order.

NOTE: ***If no additional trouble codes are stored, code 12 will continue to be displayed.***

- Trouble codes can also be displayed by using a Scan Tool connected to the data link connector (DLC) **1**.

1

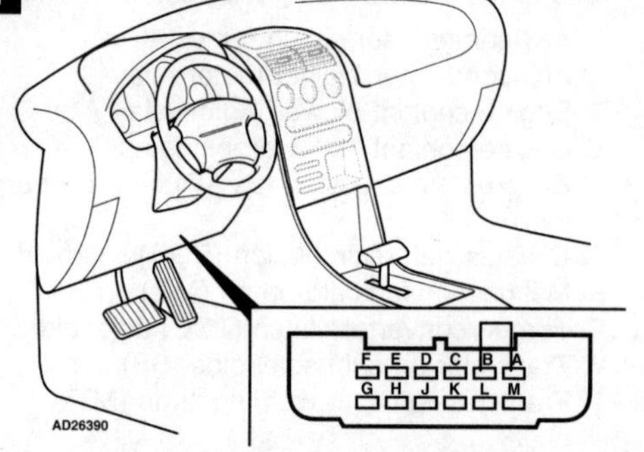

2

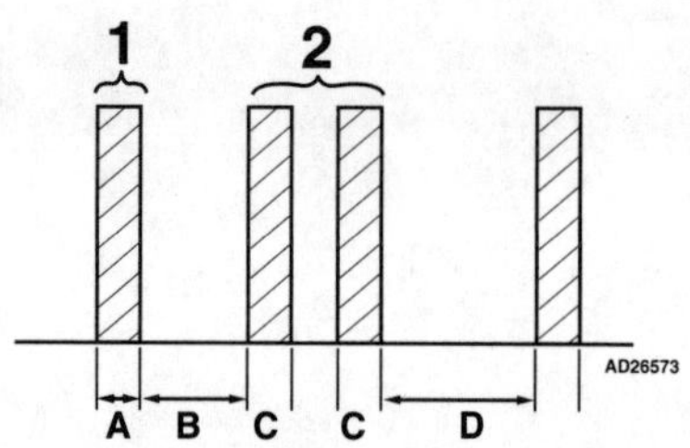

Erasing

NOTE: ***Disconnecting battery ground lead may cause driveability problems until the engine management system relearns basic values.***

- Ensure ignition switched OFF.
- Disconnect battery ground cable for 30 seconds.
- Reconnect battery ground cable.
- Repeat checking procedure to ensure no trouble codes remain in ECM fault memory.

NOTE: ***Disconnecting battery ground lead may also erase memory from electronic units, such as radio.***

- Trouble codes can also be erased by using a Scan Tool connected to the data link connector (DLC).

Trouble code identification

Flash code	Fault location	Probable cause
12	Start of diagnosis	**No engine speed (RPM) signal received**
14	Engine coolant temperature (ECT) sensor – voltage low	**Wiring, ECT sensor, ECM**
15	Engine coolant temperature (ECT) sensor – voltage high	**Wiring open circuit, ECT sensor, ECM**
16	Transmission output speed sensor – signal low	**Wiring, transmission output speed sensor, ECM**
21	Fuel lever position sensor – voltage high	**Wiring open circuit/short circuit to positive, fuel lever position sensor, ECM**
22	Fuel lever position sensor – voltage low	**Wiring open circuit/short circuit to ground, fuel lever position sensor, ECM**
23	Fuel lever position sensor – signal out of range	**Adjustment, wiring, fuel lever position sensor**
24	Vehicle speed sensor (VSS) – no signal	**Wiring, VSS, ECM**
31	Manifold absolute pressure (MAP) sensor – voltage low	**Wiring open circuit/short circuit to ground, MAP sensor, ECM**
32	Exhaust gas recirculation (EGR) solenoid	**Wiring, poor connections, EGR valve, ECM**
33	Manifold absolute pressure (MAP) sensor – voltage high	**Wiring, hose blocked/leaking, MAP sensor, ECM**
51	Engine control module (ECM), PROM – malfunction	**Poor connections, incorrect programming, ECM**
52	Engine control module (ECM) – malfunction	**ECM**
53 [1]	Engine control module (ECM), sensor supply voltage – circuit malfunction	**Wiring, poor connections, MAP sensor**
53 [2]	System voltage – high	**Alternator**
55	Engine control module (ECM) – malfunction	**ECM**

[1] 1992
[2] 1993

GMC

Model:	Engine identification:	Year:
Rally 6.5L	**VIN position 8 = P, Y**	**1994-95**
Sierra C/K Pickup 6.5L	**VIN position 8 = F, P, S**	**1994-95**
Suburban 6.5L	**VIN position 8 = F**	**1994-95**
Vandura 6.5L	**VIN position 8 = P**	**1994-95**
Yukon 6.5L	**VIN position 8 = S**	**1994-95**

System: **GM**

Self-diagnosis

General information

- Refer to the front of this manual for general test conditions, terminology, detailed descriptions of wiring faults and a general trouble shooter for electrical and mechanical faults.
- Malfunction indicator lamp (MIL) will illuminate if certain faults are recorded.
- ECM operates in backup mode if sensors fail, to enable vehicle to be driven to repair shop.
- Trouble codes can be displayed by the MIL or by using a Scan Tool connected to the data link connector (DLC).

Accessing

- Access data link connector (DLC).
- Jump data link connector (DLC) terminals A and B **1**.
- Switch ignition ON. Do NOT start engine.
- Count MIL flashes. Note trouble codes. Compare with trouble code table.
- Firstly trouble code 12 will be displayed 3 times.
- Trouble codes consist of 2 groups, displayed as follows:
  - First group indicates the 'tens' of the trouble code **2** [**A**].
  - Second group indicates the 'units' of the trouble code **2** [**C**].
  - A short pause separates each group **2** [**B**].
  - A long pause separates each trouble code **2** [**D**].
  - For example: Trouble code 12 displayed **2**. Start of diagnosis.
- Each additional trouble code will now be displayed 3 times, in ascending order.

NOTE: ***If no additional trouble codes are stored, code 12 will continue to be displayed.***

- Trouble codes can also be displayed by using a Scan Tool connected to the data link connector (DLC) **1**.

1

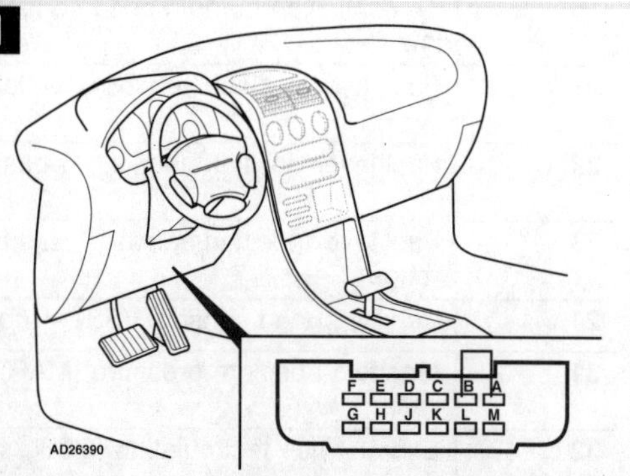

2

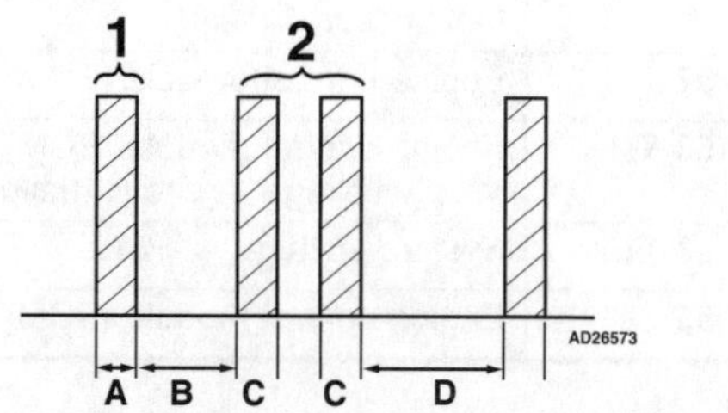

Erasing

NOTE: ***Disconnecting battery ground lead may cause driveability problems until the engine management system relearns basic values.***

- Ensure ignition switched OFF.
- Disconnect battery ground cable for 30 seconds.
- Reconnect battery ground cable.
- Repeat checking procedure to ensure no trouble codes remain in ECM fault memory.

NOTE: ***Disconnecting battery ground lead may also erase memory from electronic units, such as radio.***

- Trouble codes can also be erased by using a Scan Tool connected to the data link connector (DLC).

Trouble code identification

Flash code	Fault location	Probable cause
12	Start of diagnosis	**No engine speed (RPM) signal received**
13	Fuel shut-off solenoid – circuit malfunction	**Wiring open circuit, fuel shut-off solenoid, ECM**
14	Engine coolant temperature (ECT) sensor – voltage low	**Wiring, ECT sensor, ECM**
15	Engine coolant temperature (ECT) sensor – voltage high	**Wiring open circuit, ECT sensor, ECM**
16	Vehicle speed sensor (VSS) – no/low signal	**Wiring, VSS, ECM**
17	Injection pump position sensor, high resolution – circuit malfunction	**Wiring, injection pump position sensor, ECM**
18	Injection pump position sensor, low resolution – circuit malfunction	**Wiring, injection pump position sensor, ECM**
19	Crankshaft position (CKP) sensor – circuit malfunction	**Wiring, CKP sensor, ECM**
21	Accelerator pedal position (APP) sensor 1 – voltage high	**Wiring, APP sensor, ECM**
22	Accelerator pedal position (APP) sensor 1 – voltage low	**Wiring, APP sensor, ECM**
23	Accelerator pedal position (APP) sensor 1 to 2/3 – signal variation	**Wiring, APP sensor, ECM**
24	Vehicle speed sensor (VSS) – no signal	**Wiring, VSS, ECM**
25	Accelerator pedal position (APP) sensor 2 – voltage high	**Wiring, APP sensor, ECM**
26	Accelerator pedal position (APP) sensor 2 – voltage low	**Wiring, APP sensor, ECM**
27	Accelerator pedal position (APP) sensor 2 to 1/3 – signal variation	**Wiring, APP sensor, ECM**
28	Transmission gear pressure switch, signal – circuit malfunction	**Wiring open circuit/short circuit to ground, reverse/D1/D2/D3/D4 gear pressure switch(es), ECM**
29	Glow plug relay – circuit malfunction	**Wiring, glow plug relay, ECM**

Flash code	Fault location	Probable cause
31	Exhaust gas recirculation (EGR) vacuum/ barometric pressure (BARO) sensor – voltage low	**Wiring, EGR vacuum/BARO sensor, ECM**
32	Exhaust gas recirculation (EGR) system – malfunction	**Wiring, vacuum pipes, EGR vacuum sensor, EGR valve**
33	Exhaust gas recirculation (EGR) vacuum/ barometric pressure (BARO) sensor – voltage high	**Wiring, EGR vacuum/BARO sensor, ECM**
34	Fuel injection timing, command/actual – signal variation	**Wiring, fuel injection timing actuator, ECM**
35	Fuel injection duration – pulse width signal shorter than expected	**Wiring, fuel injection duration control solenoid, fuel injection pump control module, ECM**
36	Fuel injection duration – pulse width signal longer than expected	**Wiring, fuel injection duration control solenoid, fuel injection pump control module, ECM**
37	Brake pedal position (BPP) switch – stuck ON	**Wiring, BPP switch, ECM**
38	Brake pedal position (BPP) switch – stuck OFF	**Wiring, BPP switch, ECM**
39	Brake pedal position (BPP) switch – stuck OFF	**Wiring, BPP switch, ECM**
41	Brake pedal position (BPP) switch, torque converter clutch (TCC) – circuit malfunction	**Wiring, BPP switch, ECM**
42	Fuel temperature sensor – voltage low	**Wiring, fuel temperature sensor, ECM**
43	Fuel temperature sensor – voltage high	**Wiring, fuel temperature sensor, ECM**
44	Exhaust gas recirculation (EGR) valve – circuit malfunction	**Wiring, fuses, EGR valve**
45	Exhaust gas recirculation (EGR) vent valve – circuit malfunction	**Wiring, fuses, EGR vent valve**
46	Malfunction indicator lamp (MIL) – circuit malfunction	**Wiring, fuses, MIL**
47	Intake air temperature (IAT) sensor – voltage low	**Wiring short circuit to ground, IAT sensor, ECM**
48	Intake air temperature (IAT) sensor – voltage high	**Wiring, IAT sensor, ECM**
49	Service throttle soon warning lamp – circuit malfunction	**Wiring, fuses, service throttle soon warning lamp**
51	Engine control module (ECM), PROM – malfunction	**Poor connections, incorrect programming, ECM**
52	System voltage – high	**Alternator, jump started**
53	System voltage – high	**Alternator, jump started**
54	Engine control module (ECM) – malfunction	**Check trouble code 17, if not: replace ECM**
56	Fuel injection pump calibration resistor – circuit malfunction	**Wiring, calibration resistor**

Flash code	Fault location	Probable cause
57	Engine control module (ECM), sensor supply voltage – circuit malfunction	**Wiring short to ground, poor connections, MAP sensor**
58	Transmission fluid temperature (TFT) sensor – voltage low	**Wiring short circuit to ground, TFT sensor, ECM**
59	Transmission fluid temperature (TFT) sensor – voltage high	**Wiring open circuit, TFT sensor, ECM**
61	Turbocharger (TC) boost pressure sensor – voltage high	**Wiring, TC boost pressure sensor, ECM**
62	Turbocharger (TC) boost pressure sensor – voltage low	**Wiring short circuit to ground, TC boost pressure sensor, ECM**
63	Accelerator pedal position (APP) sensor 3 – voltage high	**Wiring, APP sensor, ECM**
64	Accelerator pedal position (APP) sensor 3 – voltage low	**Wiring, APP sensor, ECM**
65	Accelerator pedal position (APP) sensor 3 to 1/2 – signal variation	**Wiring, APP sensor, ECM**
66	Transmission 3rd-2nd control solenoid – circuit malfunction	**Wiring, control solenoid, ECM**
67	Torque converter clutch (TCC) solenoid – circuit malfunction	**Wiring, TCC solenoid, ECM**
69	Torque converter clutch (TCC) solenoid – stuck ON	**Wiring open circuit/short circuit to ground, transmission internal wiring, TCC solenoid, transmission mechanical fault, ECM**
71	Cruise control, set/coast switch – circuit malfunction	**Wiring, poor connections, set/coast switch**
72	Vehicle speed sensor (VSS) – loss of signal	**Wiring, VSS, ECM**
73	Transmission fluid pressure (TFP) control solenoid – circuit malfunction	**Wiring, poor connection, TFP solenoid**
74	Transmission input shaft speed sensor – circuit malfunction	**Wiring, input shaft speed sensor, ECM**
75	System voltage, engine control module (ECM) – low	**Wiring, battery, alternator, ECM**
76	Cruise control, resume/acceleration switch – circuit malfunction	**Wiring, poor connections, resume/acceleration switch**
78	Turbocharger (TC) wastegate regulating valve – circuit malfunction	**Wiring, TC wastegate regulating valve, ECM**
79	Transmission fluid temperature (TFT) – high	**Transmission mechanical fault, towing weight limit exceeded**
81	Transmission shift solenoid (SS), 2nd-3rd – circuit malfunction	**Wiring, SS, ECM**
82	Transmission shift solenoid (SS), 1st-2nd – circuit malfunction	**Wiring, SS, ECM**
83	Torque converter clutch (TCC) solenoid, pulse width modulated – circuit malfunction	**Wiring, TCC solenoid, ECM**
84	Accelerator pedal position (APP) sensor 1/2/3 – intermittent circuit malfunction	**Wiring, poor connections, APP sensor**

Flash code	Fault location	Probable cause
85	Transmission input/output shaft speed, ratio variation – out of range	**Wiring, input shaft speed sensor, output shaft speed sensor, mechanical fault**
86	Transmission input/output shaft speed, ratio variation – lower than expected	**Wiring, input shaft speed sensor, output shaft speed sensor, mechanical fault**
87	Transmission input/output shaft speed, ratio variation – higher than expected	**Wiring, input shaft speed sensor, output shaft speed sensor, mechanical fault**
88	Fuel injection pump timing – out of range	**Fuel injection pump timing incorrectly set**
91	Power balance, cylinder 1	**Check other trouble codes, If not: mechanical fault**
92	Power balance, cylinder 2	**Check other trouble codes, If not: mechanical fault**
93	Power balance, cylinder 3	**Check other trouble codes, If not: mechanical fault**
94	Power balance, cylinder 4	**Check other trouble codes, If not: mechanical fault**
95	Power balance, cylinder 5	**Check other trouble codes, If not: mechanical fault**
96	Power balance, cylinder 6	**Check other trouble codes, If not: mechanical fault**
97	Power balance, cylinder 7	**Check other trouble codes, If not: mechanical fault**
98	Power balance, cylinder 8	**Check other trouble codes, If not: mechanical fault**
99	Accelerator pedal position (APP) sensor 2 – supply voltage low	**Wiring, poor connections, APP sensor**

Model:	Engine identification:	Year:
Rally 6.5L	VIN position 8 = Y	1996
Savana 6.5L	VIN position 8 = F, S	1996-02
Sierra C/K Pickup 6.5L	VIN position 8 = F, S	1996-00
Sierra C/K Pickup 6.6L	VIN position 8 = 1	2001-02
Suburban 6.5L	VIN position 8 = F	1996-99
Vandura 6.5L	VIN position 8 = Y	1996
Yukon 6.5L	VIN position 8 = S	1996-99

System: **GM**

Self-diagnosis

General information

- Refer to the front of this manual for general test conditions, terminology, detailed descriptions of wiring faults and a general trouble shooter for electrical and mechanical faults.
- Malfunction indicator lamp (MIL) will illuminate or flash if certain faults are recorded.
- ECM operates in backup mode if sensors fail, to enable vehicle to be driven to repair shop.

Accessing

- Trouble codes can only be displayed by using a Scan Tool connected to the data link connector (DLC) **1**.

1

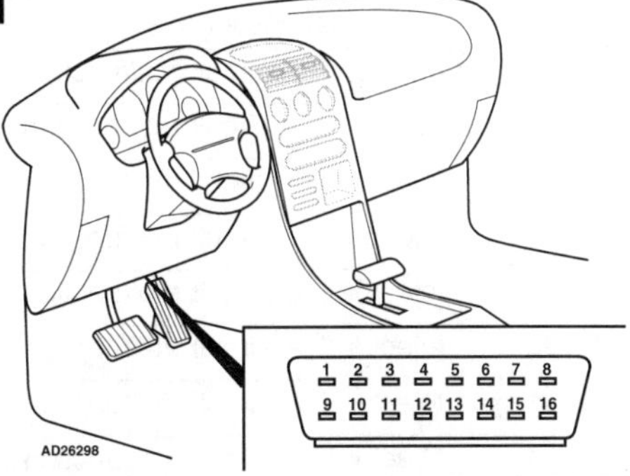

Erasing

- After the faults have been rectified, the manufacturer recommends that trouble codes are erased by using a Scan Tool connected to the data link connector (DLC).

Trouble code identification

Scan code	Fault location	Probable cause
P0000-P0999	Refer to OBD-II trouble code table at the front of this manual	–
P1093	Fuel rail pressure (FRP) sensor, command/ actual fuel pressure – signal variation	Wiring, poor connections, FRP sensor
P1094	Fuel rail pressure (FRP) sensor – signal out of range	Wiring, poor connections, FRP sensor

Scan code	Fault location	Probable cause
P1125	Accelerator pedal position (APP) sensor 1/2/3 – signal variation	Wiring, APP sensor, ECM
P1172	Fuel level sensors, primary and secondary – no fuel level change during fuel transfer from secondary to primary tank	Wiring, secondary fuel pump, fuel level sensors
P1191	Intake manifold system – air leak	Intake hose pipe connections, EGR system, ECM
P1214	Fuel injection pump timing – out of range	Fuel injection pump timing incorrectly set, CKP sensor, ECM
P1216	Fuel injection duration – pulse width signal shorter than expected	Wiring, fuel injection duration control solenoid, fuel injection pump control module, ECM
P1217	Fuel injection duration – pulse width signal longer than expected	Wiring, fuel injection duration control solenoid, fuel injection pump control module, ECM
P1218	Fuel injection pump calibration resistor – circuit malfunction	Wiring, calibration resistor
P1223	Injector 1 – circuit malfunction	Wiring, injector, FICM
P1226	Injector 2 – circuit malfunction	Wiring, injector, FICM
P1229	Injector 3 – circuit malfunction	Wiring, injector, FICM
P1232	Injector 4 – circuit malfunction	Wiring, injector, FICM
P1235	Injector 5 – circuit malfunction	Wiring, injector, FICM
P1238	Injector 6 – circuit malfunction	Wiring, injector, FICM
P1241	Injector 7 – circuit malfunction	Wiring, injector, FICM
P1244	Injector 8 – circuit malfunction	Wiring, injector, FICM
P1261	Injectors 1, 4, 6 & 7 – high voltage	Wiring, injector(s), ECM
P1262	Injectors 2, 3, 5 & 8 – high voltage	Wiring, injector(s), ECM
P1270	Engine control module (ECM), APP sensor signal – analogue/digital process malfunction	Engine control module (ECM)
P1271	Accelerator pedal position (APP) sensor 1/2 – signal variation	Wiring, APP sensor, ECM
P1272	Accelerator pedal position (APP) sensor 2/3 – signal variation	Wiring, APP sensor, ECM
P1273	Accelerator pedal position (APP) sensor 1/3 – signal variation	Wiring, APP sensor, ECM
P1277	Accelerator pedal position (APP) sensor 1 – voltage low	Wiring, APP sensor, ECM
P1278	Accelerator pedal position (APP) sensor 1 – voltage high	Wiring, APP sensor, ECM
P1282	Accelerator pedal position (APP) sensor 2 – voltage low	Wiring, APP sensor, ECM
P1283	Accelerator pedal position (APP) sensor 2 – voltage high	Wiring, APP sensor, ECM

Scan code	Fault location	Probable cause
P1287	Accelerator pedal position (APP) sensor 3 – voltage low	Wiring, APP sensor, ECM
P1288	Accelerator pedal position (APP) sensor 3 – voltage high	Wiring, APP sensor, ECM
P1345	Crankshaft position sensor (CKP)/camshaft position sensor (CMP) – signal variation	Wiring, CKP sensor, CMP sensor
P1404	Exhaust gas recirculation (EGR) vacuum sensor/barometric pressure (BARO) sensor – voltage low	Wiring, EGR vacuum sensor/BARO sensor, ECM
P1406	Exhaust gas recirculation (EGR) valve position sensor – range performance	Wiring, EGR valve tight/sticking, EGR valve position sensor
P1409	Exhaust gas recirculation (EGR) system – malfunction	Wiring, vacuum pipes, EGR vacuum sensor, EGR valve
P1431	Fuel level sensor, secondary – circuit malfunction	Wiring, fuel level sensor
P1432	Fuel level sensor – low voltage	Wiring open circuit, fuel level sensor
P1433	Fuel level sensor – high voltage	Wiring short circuit to positive, fuel level sensor
P1550	System voltage – low	Wiring, battery, alternator
P1600	Engine control module (ECM) – malfunction	ECM
P1621	Engine control module (ECM) – memory performance	ECM
P1626	Engine control module (ECM) – fuel enable signal lost	Wiring, poor connections, immobilizer control module, ECM
P1627	Engine control module (ECM) – analogue/ digital performance	Engine control module (ECM)
P1630	Engine control module (ECM), immobilizer code – learning mode	ECM immobilizer code programming not completed
P1631	Engine control module (ECM) – incorrect immobilizer code	Incorrect key, immobilizer code not programmed
P1635	Engine control module (ECM), sensor supply voltage – circuit malfunction	Wiring short to ground, poor connections, CKP sensor, EGR control/BARO sensor, injection pump position/fuel temperature sensor, TC boost pressure sensor, ECM
P1637	Alternator warning lamp, control signal – circuit malfunction	Wiring, instrument panel, ECM
P1639	Engine control module (ECM), sensor supply voltage – circuit malfunction	Wiring, poor connections, APP sensor, TC boost pressure sensor, BARO sensor, ECM
P1641	Malfunction indicator lamp (MIL) – circuit malfunction	Wiring, fuses, MIL
P1643	Glow plug warning lamp – circuit malfunction	Wiring, glow plug warning lamp, ECM
P1653	Exhaust gas recirculation (EGR) vent valve – circuit malfunction	Wiring, poor connections, EGR vent valve, ECM
P1654	Service throttle soon warning lamp – circuit malfunction	Wiring, fuses, service throttle soon warning lamp

Scan code	Fault location	Probable cause
P1655	Exhaust gas recirculation (EGR) control valve – circuit malfunction	**Wiring, poor connections, EGR control valve, ECM**
P1656	Turbocharger (TC) wastegate regulating valve – circuit malfunction	**Wiring, TC wastegate regulating valve, ECM**
P1658	Fuel injector control module (FICM) – malfunction	**FICM**
P1680	Engine control module (ECM) – malfunction	**ECM**
P1681	Engine control module (ECM) – malfunction	**ECM**
P1683	Engine control module (ECM) – malfunction	**ECM**
P1781	Transmission control module (TCM), malfunction indicator lamp (MIL) request signal – voltage high/low	**Wiring TCM, ECM**
P1810	Transmission gear pressure switch, signal – circuit malfunction	**Wiring open circuit/short circuit to ground, reverse/D1/D2/D3/D4 gear pressure switch(es), ECM**
P1811	Shift time too long – adaptation limit reached	**Fluid level low, transmission mechanical fault**
P1860	Torque converter clutch (TCC) solenoid, pulse width modulation (PWM) – circuit malfunction	**Wiring open circuit/short circuit to ground, TCC solenoid, ECM**
P1870	Torque converter clutch (TCC) – slip speed excessive	**Transmission fluid level, wiring, TCC solenoid, ECM**
P2141	Exhaust gas recirculation (EGR) control valve – voltage low	**Wiring, poor connections, EGR vent valve, ECM**
P2142	Exhaust gas recirculation (EGR) control valve – voltage high	**Wiring, poor connections, EGR vent valve, ECM**
P2144	Exhaust gas recirculation (EGR) vent valve – voltage low	**Wiring, poor connections, EGR vent valve, ECM**
P2145	Exhaust gas recirculation (EGR) vent valve – voltage high	**Wiring, poor connections, EGR vent valve, ECM**
U1026	Transmission control module (TCM)/four wheel drive control module, CAN data bus – communication malfunction	**Wiring, BCM**
U1192	Engine control module (ECM)/immobilizer control module, CAN data bus – signal malfunction	**Wiring, immobilizer control module, ECM**
U1800	Engine control module (ECM)/fuel injector control module (FICM), CAN data bus – communication malfunction	**Wiring, FICM, ECM**
U2104	Engine control module (ECM)/transmission control module (TCM)/Fuel injector control module (FICM), CAN data bus – communication malfunction	**Wiring, TCM, FICM, ECM**
U2106	Engine control module (ECM)/transmission control module (TCM), CAN data bus – communication malfunction	**Wiring, TCM, ECM**

JEEP

Model:	Engine identification:	Year:
Cherokee 2.5L	**VIN position 8 = P**	**1992-95**
Cherokee 4.0L	**VIN position 8 = S**	**1992-95**
Grand Cherokee 4.0L	**VIN position 8 = S**	**1993-95**
Grand Cherokee 5.2L	**VIN position 8 = Y**	**1993-95**
Comanche 2.5L	**VIN position 8 = P**	**1992**
Comanche 4.0L	**VIN position 8 = S**	**1992**
Wrangler 2.5L	**VIN position 8 = P**	**1992-95**
Wrangler 4.0L	**VIN position 8 = S**	**1992-95**

System: **JTEC**

Self-diagnosis

General information

- Refer to the front of this manual for general test conditions, terminology, detailed descriptions of wiring faults and a general trouble shooter for electrical and mechanical faults.
- Malfunction indicator lamp (MIL) will illuminate if certain faults are recorded.
- ECM operates in backup mode if sensors fail, to enable vehicle to be driven to repair shop.
- Trouble codes can be displayed by the MIL or by using a Scan Tool connected to the data link connector (DLC).

Accessing

- Within 5 seconds switch the ignition key ON-OFF-ON-OFF-ON.
- Count MIL flashes. Note trouble codes.
- Each trouble code consists of two groups **1** [**A**].
- The first group of flashes indicate the 'tens' of the trouble code.
- The second group of flashes indicate the 'units' of the trouble code.
- A short pause separates 'tens' and 'units' **1** [**B**].
- A long pause separates each trouble code **1** [**C**].
- For example: Trouble code 21 displayed **1**.
- End of test sequence indicated by trouble code 55.
- Trouble codes can also be displayed by using a Scan Tool connected to the data link connector (DLC):
  - Cherokee/Comanche **2**.
  - Grand Cherokee **3**.
  - Wrangler **4**.

1

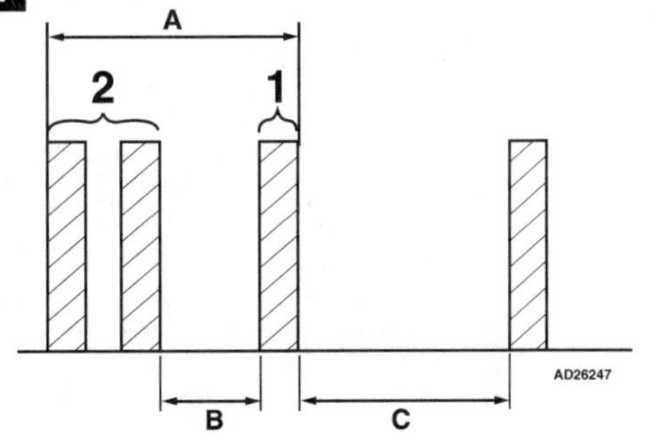

2

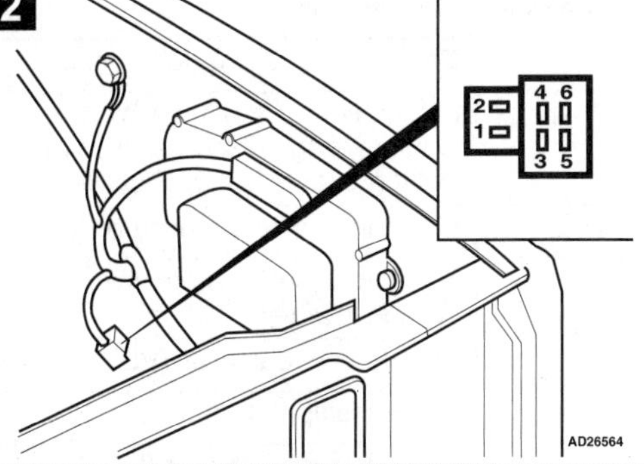

3

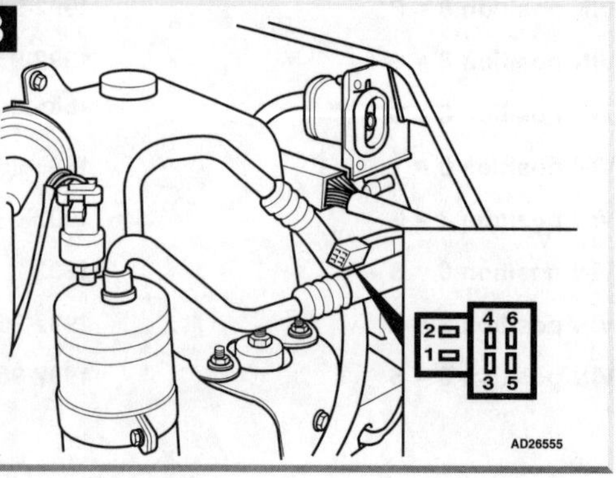

4

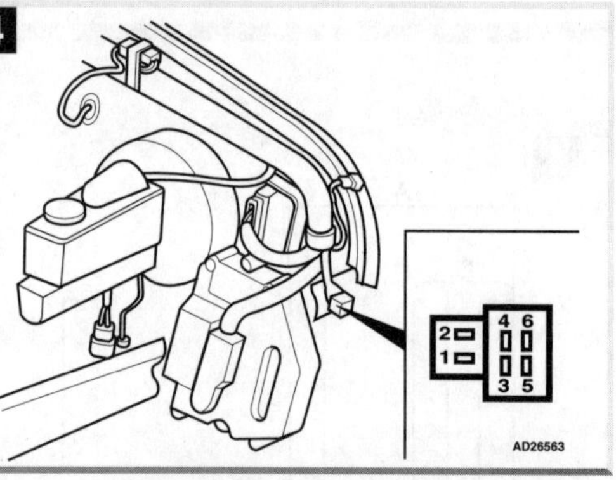

Erasing

NOTE: ***Disconnecting battery ground lead may cause driveability problems until the engine management system relearns basic values.***

- Ensure ignition switched OFF.
- Disconnect battery ground cable for at least 30 seconds.
- Reconnect battery ground cable.
- Repeat accessing procedure to ensure no trouble codes remain in ECM fault memory.

NOTE: ***Disconnecting battery ground lead may also erase memory from electronic units, such as radio.***

- Trouble codes can also be erased by using a Scan Tool connected to the data link connector (DLC).

Trouble code identification

Flash code	Fault location	Probable cause
11	Crankshaft position (CKP) sensor	**Wiring, air gap, CKP sensor**
12	Engine control module (ECM) – supply voltage disconnected within last 50 ignition switch ON-OFF cycles	**Wiring, battery**
13	Manifold absolute pressure (MAP) sensor	**Wiring, MAP sensor**
14	Manifold absolute pressure (MAP) sensor – voltage low/high	**Wiring, MAP sensor**
15	Vehicle speed sensor (VSS) – signal	**Wiring, VSS**
17	Engine temperature – low for long period	**Wiring, coolant thermostat, ECT sensor**
21	Heated oxygen sensor (HO2S) – incorrect signal/voltage high	**Wiring, fuel pressure, injector control, EVAP canister purge valve, compression, air filter blocked, HO2S**
22	Engine coolant temperature (ECT) sensor – voltage low/high	**Wiring, ECT sensor**

Flash code	Fault location	Probable cause
23	Intake air temperature (IAT) sensor – voltage low/high	Wiring, IAT sensor
24	Throttle position (TP) sensor – voltage low/high	Wiring, accelerator cable adjustment, TP sensor
25	Idle air control (IAC) valve	Wiring, IAC valve
27	Injector(s) – circuit malfunction	Wiring, injector
31	Evaporative emission (EVAP) canister purge valve – open/short circuit	Wiring, EVAP canister purge valve
32	Exhaust gas recirculation (EGR) solenoid/system – circuit malfunction	Wiring, hoses, EGR solenoid
33	A/C compressor clutch relay/A/C refrigerant pressure sensor – voltage low/high	Wiring, A/C compressor clutch relay, A/C refrigerant pressure sensor
34	Cruise control	Wiring open circuit/short circuit, hoses, voltage low/high, cruise control master switch, cruise control vacuum switching valve
35	Engine coolant blower motor relay – low/high speed	Wiring, engine coolant blower motor relay – low/high speed
37	Torque converter clutch (TCC) solenoid – circuit malfunction	Wiring, TCC solenoid
41	Alternator	Wiring open circuit/short circuit, alternator
42	Engine control relay	Wiring open circuit/short circuit, engine control relay
44	Battery temperature sensor/engine coolant temperature (ECT) sensor	Wiring open circuit/short circuit, battery temperature sensor, ECT sensor, ECM
45	Overdrive solenoid	Wiring open circuit/short circuit, overdrive solenoid
46	Alternator – voltage high	Wiring, battery, alternator
47	Alternator – voltage low	Wiring, battery, alternator drive belt, alternator
51	System too lean	Wiring, intake/exhaust leak, fuel pressure, injector control, EVAP canister purge valve, HO2S
52	System too rich	Wiring, exhaust blockage, fuel pressure, injector control, EVAP canister purge valve, compression, air filter blocked, ignition system, HO2S
53	Engine control module (ECM) – internal fault	ECM
54	Camshaft position (CMP) sensor – no signal	Wiring, CMP sensor
55	End of MIL code output	–
62	Engine control module (ECM) – EEPROM error	ECM
63	Engine control module (ECM) – EEPROM error	ECM
76	Fuel pump resistor	Wiring open circuit/short circuit, fuel pump resistor

JEEP

Model:	Engine identification:	Year:
Cherokee 2.5L	**VIN position 8 = P**	**1996-97**
Cherokee 4.0L	**VIN position 8 = S**	**1996-97**
Grand Cherokee 4.0L	**VIN position 8 = S**	**1996-97**
Grand Cherokee 5.2L	**VIN position 8 = Y**	**1996-97**
Wrangler 2.5L	**VIN position 8 = P**	**1996-97**
Wrangler 4.0L	**VIN position 8 = S**	**1996-97**

System: **JTEC**

Self-diagnosis

General information

- Refer to the front of this manual for general test conditions, terminology, detailed descriptions of wiring faults and a general trouble shooter for electrical and mechanical faults.
- Malfunction indicator lamp (MIL) will illuminate or flash if certain faults are recorded.
- ECM operates in backup mode if sensors fail, to enable vehicle to be driven to repair shop.
- Trouble codes can be displayed by the MIL, digital odometer or by using a Scan Tool connected to the data link connector (DLC).

Accessing

Flash codes

- Within 5 seconds switch the ignition key ON-OFF-ON-OFF-ON.
- Count MIL flashes. Note trouble codes.
- Each trouble code consists of two groups **1** [**A**].
- The first group of flashes indicate the 'tens' of the trouble code.
- The second group of flashes indicate the 'units' of the trouble code.
- A short pause separates 'tens' and 'units' **1** [**B**].
- A long pause separates each trouble code **1** [**C**].
- For example: Trouble code 21 displayed **1**.
- End of test sequence indicated by trouble code 55.
- Trouble codes can also be displayed by the digital odometer using the same key cycling procedure as above. After code 55 is displayed the odometer will return to its normal mode.

P codes & Hex codes

- Trouble codes can only be displayed by using a Scan Tool connected to the data link connector (DLC):
  - Cherokee/Wrangler **2**.
  - Grand Cherokee **3**.

1

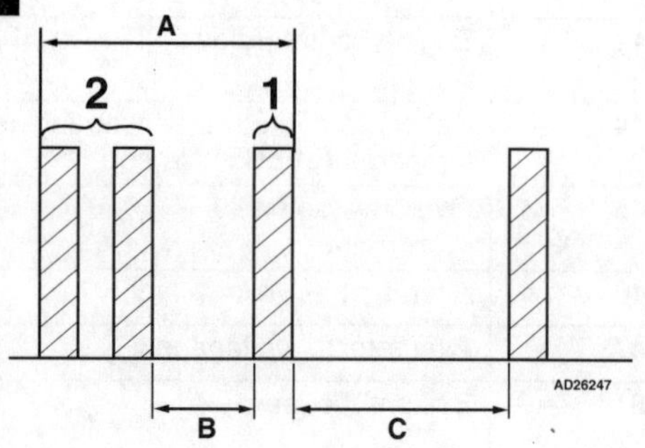

2

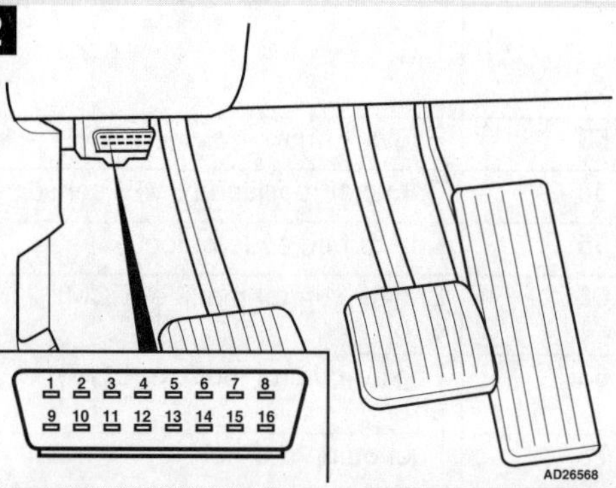

3

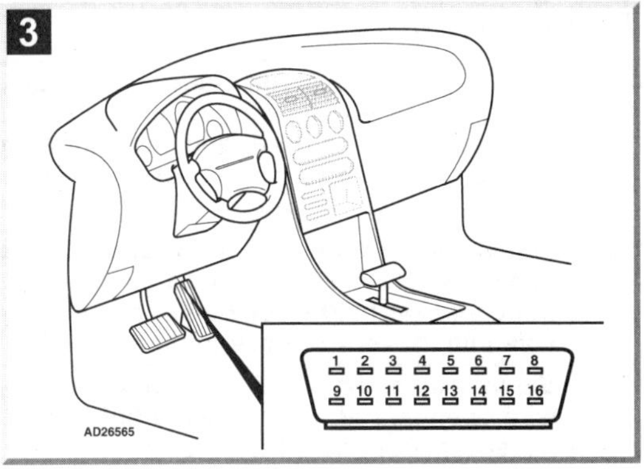

Erasing

- Trouble codes can only be erased by using a Scan Tool connected to the data link connector (DLC).

Trouble code identification

Scan code P type	Fault location	Probable cause
P0000-P0999	Refer to OBD-II trouble code table at the front of this manual	–
P1294	Idle speed control (ISC) system – rpm lower/ higher than expected	**Wiring, air leak, throttle valve tight/sticking, IAC valve**
P1296	Manifold absolute pressure (MAP) sensor – supply voltage	**Wiring, MAP sensor, ECM**
P1297	Manifold absolute pressure (MAP) sensor	**Wiring, hose connection(s), MAP sensor**
P1391	Crankshaft position (CKP) sensor/camshaft position (CMP) sensor – intermittent signal	**Wiring, poor connection, air gap, CKP sensor, CMP sensor**
P1398	Engine control module (ECM) – misfire detection limit reached	**Wiring, poor connection, engine mechanical fault, ignition/fuel system, CKP sensor, ECM**
P1486	Evaporative emission (EVAP) system – blockage detected	**Hose connection(s), hose blocked/trapped, EVAP canister**
P1492	Ambient/battery temperature sensor – voltage high	**Wiring, ambient/battery temperature sensor, ECM**
P1493	Ambient/battery temperature sensor – voltage low	**Wiring, ambient/battery temperature sensor, ECM**
P1494	Evaporative emission (EVAP) leak detection pump – pressure switch	**Wiring, hose leak/blockage, EVAP leak detection pump**
P1495	Evaporative emission (EVAP) leak detection pump solenoid	**Wiring open circuit/short circuit, hose leak/ blockage, EVAP leak detection pump solenoid**
P1698	Engine control module (ECM) – EEPROM error	**ECM**
P1756	Governor pressure not equal to target (15-20 PSI) – mid pressure malfunction	**Signal wire open circuit/short circuit, valve body, sensor supply wire open circuit, transmission fluid level, governor pressure solenoid, transmission mechanical fault, ECM – no 5 volt supply**

Scan code P type	Fault location	Probable cause
P1757	Governor pressure not equal to target – low pressure malfunction	**Signal wire open circuit/short circuit, valve body, sensor supply wire open circuit, transmission fluid level, governor pressure solenoid, transmission mechanical fault, ECM – no 5 volt supply**
P1762	Governor pressure sensor – incorrect signal	**Signal wire short circuit to ground/positive, supply wire open circuit/short circuit, transmission fluid level, governor pressure solenoid, governor pressure sensor, ECM**
P1763	Governor pressure sensor – signal high	**Signal wire short circuit to positive, supply wire open circuit, transmission fluid level, governor pressure sensor, ECM**
P1764	Governor pressure sensor – signal low	**Signal wire short circuit to ground, supply wire open circuit, transmission fluid level, governor pressure sensor, ECM**
P1765	Transmission system relay	**Wiring open circuit/short circuit, transmission system relay, ECM**
P1899	Park/neutral position (PNP) switch – input circuit malfunction	**Wiring, PNP switch, ECM**

Flash code	Scan code Hex type	Fault location	Probable cause
11		Crankshaft position (CKP) sensor/camshaft position (CMP) sensor	**Wiring, poor connection, air gap, CKP sensor, CMP sensor**
	28	Crankshaft position (CKP) sensor – no signal	**Wiring, air gap, CKP sensor**
	9D	Crankshaft position (CKP) sensor/camshaft position (CMP) sensor – intermittent signal	**Wiring, poor connection, air gap, CKP sensor, CMP sensor**
	BA	Engine control module (ECM) – misfire detection limit reached	**Wiring, poor connection, engine mechanical fault, ignition/fuel system, CKP sensor, ECM**
12		Engine control module (ECM) – supply voltage disconnected within last 50 ignition switch ON-OFF cycles	**Wiring, battery**
13	27	Manifold absolute pressure (MAP) sensor	**Wiring, hose connection(s), MAP sensor**
14		Manifold absolute pressure (MAP) sensor – input low/high	**Wiring, MAP sensor**
	24	Manifold absolute pressure (MAP) sensor – low input	**Wiring short circuit to ground, MAP sensor**
	25	Manifold absolute pressure (MAP) sensor – high input	**Wiring short circuit to positive, MAP sensor**
	87	Manifold absolute pressure (MAP) sensor – supply voltage	**Wiring, MAP sensor, ECM**
15		Vehicle speed sensor (VSS) – signal	**Wiring, VSS, ECM**

Flash code	Scan code Hex type	Fault location	Probable cause
	23	Vehicle speed sensor (VSS) – no signal	Wiring, VSS, ECM
	A6	Vehicle speed sensor (VSS) – circuit malfunction	Wiring, VSS, ECM
17		Engine temperature – low for long period	Wiring, cooling system, coolant thermostat, ECT sensor
	21	Engine temperature – low for long period	Wiring, cooling system, coolant thermostat, ECT sensor
	80	Insufficient coolant temperature for closed loop fuel control	Wiring, cooling system, coolant thermostat, ECT sensor
21		Heated oxygen sensor (HO2S) 1/2, bank 1 – circuit/signal malfunction	Wiring, fuel pressure, injector control, EVAP canister purge valve, compression, air filter blocked, catalytic converter, HO2S
	3E	Heated oxygen sensor (HO2S) 1, bank 1 – high voltage	Wiring short circuit to positive, HO2S
	66	Heated oxygen sensor (H02S) 1, bank 1 – slow response	Heating inoperative, wiring, HO2S
	67	Heated oxygen sensor (HO2S) 1, bank 1 – heater control – circuit malfunction	Wiring, HO2S
	69	Heated oxygen sensor (HO2S) 2, bank 1, heater control – circuit malfunction	Wiring, HO2S
	7E	Heated oxygen sensor (HO2S) 2, bank 1 – high voltage	Wiring short circuit to positive, catalytic converter, HO2S
	9B	Heated oxygen sensor (HO2S) 1, bank 1 – low voltage	Exhaust leak, wiring short circuit to ground, HO2S
	9C	Heated oxygen sensor (HO2S) 2, bank 1 – low voltage	Exhaust leak, wiring short circuit to ground, HO2S
	CO	Heated oxygen sensor (HO2S) 1, bank 1 – slow response	Heating inoperative, wiring, HO2S
22		Engine coolant temperature (ECT) sensor – voltage low/high	Wiring, coolant thermostat, ECT sensor
	1E	Engine coolant temperature (ECT) sensor – low input	Coolant thermostat, wiring short circuit to ground, ECT sensor
	1F	Engine coolant temperature (ECT) sensor – high input	Coolant thermostat, wiring open circuit/ short circuit to positive, ground wire defective, ECT sensor
23		Intake air temperature (IAT) sensor – voltage low/high	Wiring, IAT sensor
	39	Intake air temperature (IAT) sensor – low input	Wiring short circuit to ground, IAT sensor
	3A	Intake air temperature (IAT) sensor – high input	Wiring open circuit/short circuit to positive, ground wire defective, IAT sensor
24		Throttle position (TP) sensor – voltage low/high	Wiring, accelerator cable adjustment, TP sensor

Flash code	Scan code Hex type	Fault location	Probable cause
	1A	Throttle position (TP) sensor – low input	**Wiring short circuit to ground, TP sensor**
	1B	Throttle position (TP) sensor – high input	**Wiring short circuit to positive, TP sensor**
	84	Throttle position (TP) sensor – range/ performance problem	**Wiring, accelerator cable adjustment, TP sensor**
25		Idle air control (IAC) valve	**Wiring open circuit/short circuit, IAC valve**
	19	Idle speed control (ISC) system – malfunction	**Wiring open circuit/short circuit, IAC valve**
	8A	Idle speed control (ISC) system – rpm lower/higher than expected	**Wiring, air leak, throttle valve tight/ sticking, IAC valve**
27		Injector(s) – circuit malfunction	**Wiring, injector**
	13	Injector 3 – circuit malfunction	**Wiring, injector**
	14	Injector 2 – circuit malfunction	**Wiring, injector**
	15	Injector 1 – circuit malfunction	**Wiring, injector**
	3D	Injector 4 – circuit malfunction	**Wiring, injector**
	45	Injector 5 – circuit malfunction	**Wiring, injector**
	46	Injector 6 – circuit malfunction	**Wiring, injector**
	4F	Injector 7 – circuit malfunction	**Wiring, injector**
	50	Injector 8 – circuit malfunction	**Wiring, injector**
31		Evaporative emission (EVAP) canister purge valve/system	**Wiring, EVAP canister, EVAP canister purge valve**
	12	Evaporative emission (EVAP) canister purge valve – circuit malfunction	**Wiring open circuit/short circuit, EVAP canister purge valve**
	71	Evaporative emission (EVAP) system – incorrect flow detected	**Hose connection(s), intake leak, EVAP canister purge valve**
	A0	Evaporative emission (EVAP) system – small leak detected	**Hose connection(s), intake leak, EVAP canister, EVAP canister purge valve**
	A1	Evaporative emission (EVAP) system – large leak detected	**Hose connection(s), intake leak, EVAP canister, EVAP canister purge valve**
	BB	Evaporative emission (EVAP) system – blockage detected	**Hose connection(s), hose blocked/ trapped, EVAP canister**
	B8	Evaporative emission (EVAP) leak detection pump – pressure switch	**Wiring, hose leak/blockage, EVAP leak detection pump**
	B7	Evaporative emission (EVAP) leak detection pump solenoid	**Wiring open circuit/short circuit, hose leak/blockage, EVAP leak detection pump solenoid**
33	10	A/C compressor clutch relay	**Wiring open circuit/short circuit, A/C compressor clutch relay**

Flash code	Scan code Hex type	Fault location	Probable cause
34		Cruise control	Wiring open circuit/short circuit, hoses, voltage low/high, cruise control master switch, cruise control vacuum switching valve
	0F	Cruise control	Wiring open circuit/short circuit, hoses, voltage low/high, cruise control vacuum switching valve
	56	Cruise control master switch – high voltage	Wiring, cruise control master switch
	57	Cruise control master switch – low voltage	Wiring, cruise control master switch
35	0E	Engine coolant blower motor relay	Wiring open circuit/short circuit, engine coolant blower motor relay – low/high speed
37		Transmission control system – malfunction	Wiring, transmission fluid level, transmission mechanical fault, PNP switch, TCC solenoid, TFT sensor, TFP solenoid, shift solenoid, governor pressure solenoid, transmission system relay, ECM
	0C	Torque converter clutch (TCC) solenoid – electrical	Wiring open circuit/short circuit, TCC solenoid
	32	Shift solenoid (SS) 3-4 – electrical	Wiring, shift solenoid, ECM
	72	Park/neutral position (PNP) switch – input circuit malfunction	Wiring, PNP switch, ECM
	8D	Governor pressure not equal to target (15-20 PSI) – mid pressure malfunction	Signal wire open circuit/short circuit, valve body, sensor supply wire open circuit, transmission fluid level, governor pressure solenoid, transmission mechanical fault, ECM – no 5 volt supply
	8E	Governor pressure not equal to target – low pressure malfunction	Signal wire open circuit/short circuit, valve body, sensor supply wire open circuit, transmission fluid level, governor pressure solenoid, transmission mechanical fault, ECM – no 5 volt supply
	94	Torque converter clutch (TCC) solenoid – circuit malfunction	Wiring, TCC solenoid
	A4	Transmission fluid temperature (TFT) sensor – range/performance problem	Wiring, TFT sensor
	A5	Gear selection, 3-4 – shift malfunction	Wiring, TR sensor, shift solenoids, transmission mechanical fault
	AB	Transmission fluid pressure (TFP) solenoid – electrical	Wiring, TFP solenoid, ECM
	AD	Transmission system relay	Wiring open circuit/short circuit, transmission system relay, ECM

Flash code	Scan code Hex type	Fault location	Probable cause
	4A	Transmission fluid temperature (TFT) sensor – low input	Wiring short circuit to ground, TFT sensor, ECM
	4B	Transmission fluid temperature (TFT) sensor – high input	Wiring short circuit to positive, TFT sensor, ECM
41	0B	Alternator, field control – circuit malfunction	Wiring open circuit/short circuit, alternator, battery
42		Fuel pump relay/engine control relay/fuel level sensor – circuit malfunction	Wiring open circuit/short circuit, Fuel pump relay, engine control relay, fuel level sensor
	0A	Engine control relay	Wiring open circuit/short circuit, engine control relay
	2C	Engine control relay	Wiring open circuit, engine control relay
	65	Fuel pump relay	Wiring open circuit/short circuit, fuel pump relay
	95	Fuel level sensor – low input	Wiring open circuit, fuel level sensor
	96	Fuel level sensor – high input	Wiring short circuit to positive, fuel level sensor
	97	Fuel level sensor – malfunction	Wiring, fuel level sensor
43		Random/multiple cylinder(s) – misfire detected	Wiring, spark plug(s), HT lead(s), injector(s), ignition coil(s), low compression
	6A	Random/multiple cylinder(s) – misfire detected	Wiring, spark plug(s), HT lead(s), injector(s), ignition coil(s), low compression
	6B	Cylinder 1 – misfire detected	Wiring, engine mechanical fault, ignition/fuel system, injector
	6C	Cylinder 2 – misfire detected	Wiring, engine mechanical fault, ignition/fuel system, injector
	6D	Cylinder 3 – misfire detected	Wiring, engine mechanical fault, ignition/fuel system, injector
	6E	Cylinder 4 – misfire detected	Wiring, engine mechanical fault, ignition/fuel system, injector
	AE	Cylinder 5 – misfire detected	Wiring, engine mechanical fault, ignition/fuel system, injector
	AF	Cylinder 6 – misfire detected	Wiring, engine mechanical fault, ignition/fuel system, injector
	B0	Cylinder 7 – misfire detected	Wiring, engine mechanical fault, ignition/fuel system, injector
	B1	Cylinder 8 – misfire detected	Wiring, engine mechanical fault, ignition/fuel system, injector
	2B	Ignition coil A, primary – circuit malfunction	Wiring, ignition coil
44		Ambient/battery temperature sensor – voltage low/high	Wiring, ambient/battery temperature sensor, ECM

Flash code	Scan code Hex type	Fault location	Probable cause
	9A	Ambient/battery temperature sensor – voltage high	**Wiring, ambient/battery temperature sensor, ECM**
	99	Ambient/battery temperature sensor – voltage low	**Wiring, ambient/battery temperature sensor, ECM**
45		Transmission control system – malfunction	**Wiring, transmission fluid level, transmission mechanical fault, PNP switch, TCC solenoid, TFT sensor, TFP solenoid, shift solenoid, overdrive solenoid/switch, governor pressure solenoid, transmission system relay, ECM**
	32	Shift solenoid (SS) 3-4 – electrical	**Wiring, shift solenoid, ECM**
	4A	Transmission fluid temperature (TFT) sensor – low input	**Wiring short circuit to ground, TFT sensor, ECM**
	4B	Transmission fluid temperature (TFT) sensor – high input	**Wiring short circuit to positive, TFT sensor, ECM**
	8D	Governor pressure not equal to target (15-20 PSI) – mid pressure malfunction	**Signal wire open circuit/short circuit, valve body, sensor supply wire open circuit, transmission fluid level, governor pressure solenoid, transmission mechanical fault, ECM – no 5 volt supply**
	8E	Governor pressure not equal to target – low pressure malfunction	**Signal wire open circuit/short circuit, valve body, sensor supply wire open circuit, transmission fluid level, governor pressure solenoid, transmission mechanical fault, ECM – no 5 volt supply**
	A4	Transmission fluid temperature (TFT) sensor – range/performance problem	**Wiring, TFT sensor**
	A5	Gear selection, 3-4 – shift malfunction	**Wiring, TR sensor, shift solenoids, transmission mechanical fault**
	A7	Governor pressure sensor – signal low	**Signal wire short circuit to ground, supply wire open circuit, transmission fluid level, governor pressure sensor, ECM**
	A8	Governor pressure sensor – signal high	**Signal wire short circuit to positive, supply wire open circuit, transmission fluid level, governor pressure sensor, ECM**
	A9	Governor pressure sensor – incorrect signal	**Signal wire short circuit to ground/ positive, supply wire open circuit/short circuit, transmission fluid level, governor pressure solenoid, governor pressure sensor, ECM**
	AB	Transmission fluid pressure (TFP) solenoid – electrical	**Wiring, TFP solenoid, ECM**

Flash code	Scan code Hex type	Fault location	Probable cause
	AD	Transmission system relay	**Wiring open circuit/short circuit, transmission system relay, ECM**
	BC	Overdrive solenoid/switch – performance problem	**Wiring, overdrive solenoid/switch, transmission mechanical fault, ECM**
46	06	Alternator – voltage high	**Wiring, battery, alternator**
47	05	Alternator – voltage low	**Wiring, battery, alternator drive belt, alternator**
51	77	System too lean	**Wiring, intake/exhaust leak, MAP sensor, fuel pressure/pump, EVAP canister purge valve, injector(s), HO2S**
52	76	System too rich	**Wiring, intake blocked, EVAP canister purge valve, fuel pressure, EGR system, injector(s), air filter blocked, ignition system, HO2S**
53		Engine control module (ECM) – internal fault	**ECM**
	02	Engine control module (ECM) – internal fault	**ECM**
	44	Engine control module (ECM) – internal fault	**ECM**
54	01	Camshaft position (CMP) sensor – no signal	**Wiring, CMP sensor**
55		End of MIL code output	–
63	31	Engine control module (ECM) – EEPROM error	**ECM**
72	70	Catalytic converter system, bank 1 – efficiency below threshold	**Catalytic converter, wiring, HO2S 2**
77	52	Cruise control	**Wiring, cruise control vacuum switching valve**

JEEP

Model:	Engine identification:	Year:
Cherokee 2.5L	VIN position 8 = P	1998
Cherokee 4.0L	VIN position 8 = S	1998
Grande Cherokee 4.0L	VIN position 8 = S	1998
Grande Cherokee 5.2L	VIN position 8 = Y	1998
Grande Cherokee 5.9L	VIN position 8 = Z	1998
Wrangler 2.5L	VIN position 8 = P	1998
Wrangler 4.0L	VIN position 8 = S	1998

System: **JTEC**

Self-diagnosis

General information

- Refer to the front of this manual for general test conditions, terminology, detailed descriptions of wiring faults and a general trouble shooter for electrical and mechanical faults.
- Malfunction indicator lamp (MIL) will illuminate or flash if certain faults are recorded.
- ECM operates in backup mode if sensors fail, to enable vehicle to be driven to repair shop.

Accessing

- Trouble codes can only be displayed by using a Scan Tool connected to the data link connector (DLC):
  - Cherokee/Wrangler **1**.
  - Grand Cherokee **2**.

1

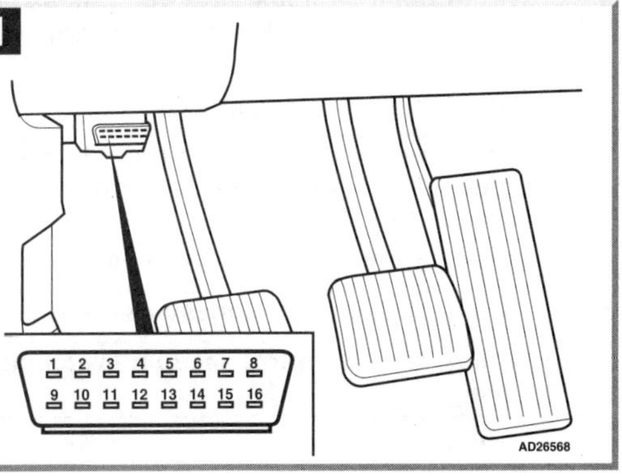

2

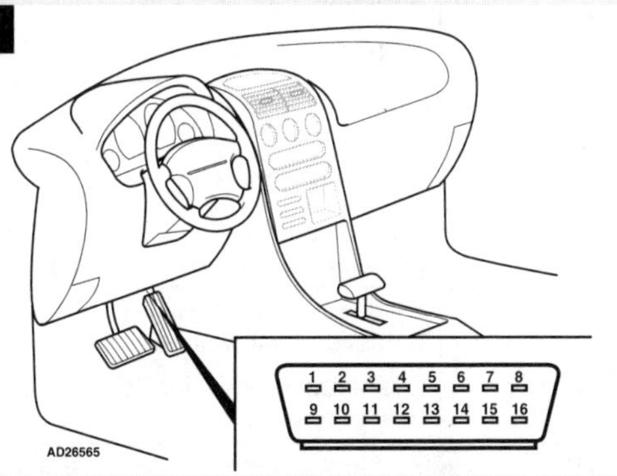

Erasing

- Trouble codes can only be erased by using a Scan Tool connected to the data link connector (DLC).

Trouble code identification

Scan code P type	Fault location	Probable cause
P0000-P0999	Refer to OBD-II trouble code table at the front of this manual	–
P1195	Heated oxygen sensor (HO2S) 1, bank 1 – slow response	**Wiring, heating inoperative, HO2S**
P1197	Heated oxygen sensor (HO2S) 2, bank 1 – slow response	**Wiring, heating inoperative, HO2S**
P1281	Engine temperature – low for long period	**Wiring, cooling system, coolant thermostat, ECT sensor**
P1282	Fuel pump relay	**Wiring open circuit/short circuit, fuel pump relay**
P1294	Idle speed control (ISC) system – rpm lower/higher than expected	**Wiring, air leak, throttle valve tight/sticking, IAC valve**
P1296	Manifold absolute pressure (MAP) sensor – no 5 volt supply	**Wiring, ECM**
P1297	Manifold absolute pressure (MAP) sensor	**Wiring, hose connection(s), MAP sensor**
P1388	Engine control relay	**Wiring open circuit/short circuit, engine control relay**
P1389	Engine control relay	**Wiring open circuit, engine control relay**
P1391	Crankshaft position (CKP) sensor/camshaft position (CMP) sensor – intermittent signal	**Wiring, poor connection, air gap, CKP sensor, CMP sensor**
P1398	Engine control module (ECM) – misfire detection limit reached	**Wiring, poor connection, CKP sensor**
P1486	Evaporative emission (EVAP) system – blockage detected	**Hose connection(s), hose blocked/trapped, EVAP canister**
P1492	Ambient/battery temperature sensor – voltage high	**Wiring, ambient/battery temperature sensor**
P1493	Ambient/battery temperature sensor – voltage low	**Wiring, ambient/battery temperature sensor**
P1494	Evaporative emission (EVAP) leak detection pump – pressure switch	**Wiring, hose leak/blockage, EVAP leak detection pump**
P1495	Evaporative emission (EVAP) leak detection pump solenoid	**Wiring open circuit/short circuit, hose leak/blockage, EVAP leak detection pump solenoid**
P1594	Alternator – voltage high	**Wiring, battery, alternator**
P1595	Cruise control	**Wiring open circuit/short circuit, hoses, cruise control master switch, cruise control vacuum switching valve**
P1596	Cruise control master switch – high voltage	**Wiring, cruise control master switch**
P1597	Cruise control master switch – low voltage	**Wiring, cruise control master switch**
P1683	Cruise control	**Wiring, cruise control vacuum switching valve**
P1696	Engine control module (ECM) – EEPROM error	**ECM**

Scan code P type	Fault location	Probable cause
P1762	Governor pressure sensor – incorrect signal	**Signal wire short circuit to ground/positive, supply wire open circuit/short circuit, transmission fluid level, governor pressure solenoid, governor pressure sensor, ECM**
P1763	Governor pressure sensor – signal high	**Signal wire short circuit to positive, supply wire open circuit, transmission fluid level, governor pressure sensor, ECM**
P1764	Governor pressure sensor – signal low	**Signal wire short circuit to ground, supply wire open circuit, transmission fluid level, governor pressure sensor, ECM**
P1765	Transmission system relay	**Wiring open circuit/short circuit, transmission system relay, ECM**
P1756	Governor pressure not equal to target (15-20 PSI) – mid pressure malfunction	**Signal wire open circuit/short circuit, valve body, sensor supply wire open circuit, transmission fluid level, governor pressure solenoid, transmission mechanical fault, ECM – no 5 volt supply**
P1757	Governor pressure not equal to target – low pressure malfunction	**Signal wire open circuit/short circuit, valve body, sensor supply wire open circuit, transmission fluid level, governor pressure solenoid, transmission mechanical fault, ECM – no 5 volt supply**
P1899	Park/neutral position (PNP) switch – input circuit malfunction	**Wiring, PNP switch, ECM/TCM**

Scan code Hex type	Fault location	Probable cause
00	DTC error	–
01	Camshaft position (CMP) sensor – no signal	**Wiring, CMP sensor**
02	Engine control module (ECM) – internal fault	**ECM**
05	Alternator – voltage low	**Wiring, battery, alternator drive belt, alternator**
06	Alternator – voltage high	**Wiring, battery, alternator**
0A	Engine control relay	**Wiring open circuit/short circuit, engine control relay**
0B	Alternator, field control – circuit malfunction	**Wiring open circuit/short circuit, alternator, battery**
0C	Torque converter clutch (TCC) solenoid – electrical	**Wiring open circuit/short circuit, TCC solenoid**
0F	Cruise control	**Wiring open circuit/short circuit, hoses, cruise control master switch, cruise control vacuum switching valve**
10	A/C compressor clutch relay	**Wiring open circuit/short circuit, A/C compressor clutch relay**

Scan code Hex type	Fault location	Probable cause
12	Evaporative emission (EVAP) canister purge valve – circuit malfunction	**Wiring open circuit/short circuit, EVAP canister purge valve**
13	Injector 3 – circuit malfunction	**Wiring, injector**
14	Injector 2 – circuit malfunction	**Wiring, injector**
15	Injector 1 – circuit malfunction	**Wiring, injector**
19	Idle speed control (ISC) system – malfunction	**Wiring open circuit/short circuit, IAC valve**
1A	Throttle position (TP) sensor – low input	**Wiring short circuit to ground, TP sensor**
1B	Throttle position (TP) sensor – high input	**Wiring short circuit to positive, TP sensor**
1E	Engine coolant temperature (ECT) sensor – low input	**Wiring short circuit to ground, coolant thermostat, ECT sensor**
1F	Engine coolant temperature (ECT) sensor – high input	**Wiring open circuit/short circuit to positive, coolant thermostat, ground wire defective, ECT sensor**
21	Engine temperature – low for long period	**Wiring, cooling system, coolant thermostat, ECT sensor**
23	Vehicle speed sensor (VSS) – signal	**Wiring, VSS**
24	Manifold absolute pressure (MAP) sensor – low input	**Wiring short circuit to ground, MAP sensor**
25	Manifold absolute pressure (MAP) sensor – high input	**Wiring short circuit to positive, MAP sensor**
27	Manifold absolute pressure (MAP) sensor	**Wiring, hose connection(s), MAP sensor**
28	Crankshaft position (CKP) sensor – no signal	**Wiring, air gap, CKP sensor**
2B	Ignition coil A, primary/secondary – circuit malfunction	**Wiring, ignition coil**
2C	Engine control relay	**Wiring open circuit, engine control relay**
31	Engine control module (ECM) – EEPROM error	**ECM**
32	Shift solenoid (SS) A – electrical	**Wiring, shift solenoid, ECM/TCM**
39	Intake air temperature (IAT) sensor – low input	**Wiring short circuit to ground, IAT sensor**
3A	Intake air temperature (IAT) sensor – high input	**Wiring open circuit/short circuit to positive, ground wire defective, IAT sensor**
3D	Injector 4 – circuit malfunction	**Wiring, injector**
3E	Heated oxygen sensor (HO2S) 1, bank 1 – high voltage	**Wiring short circuit to positive, fuel pressure, injector control, EVAP canister purge valve, air filter blocked, HO2S**
44	Engine control module (ECM) – internal fault	**ECM**
45	Injector 5 – circuit malfunction	**Wiring, injector**
46	Injector 6 – circuit malfunction	**Wiring, injector**
4A	Transmission fluid temperature (TFT) sensor – low input	**Wiring short circuit to ground, TFT sensor, ECM**

Scan code Hex type	Fault location	Probable cause
4B	Transmission fluid temperature (TFT) sensor – high input	**Wiring short circuit to positive, TFT sensor, ECM**
4F	Injector 7 – circuit malfunction	**Wiring, injector**
50	Injector 8 – circuit malfunction	**Wiring, injector**
52	Cruise control	**Wiring, cruise control vacuum switching valve**
56	Cruise control master switch – high voltage	**Wiring, cruise control master switch**
57	Cruise control master switch – low voltage	**Wiring, cruise control master switch**
65	Fuel pump relay	**Wiring open circuit/short circuit, fuel pump relay**
66	Heated oxygen sensor (H02S) 1, bank 1 – slow response	**Wiring, heating inoperative, HO2S**
67	Heated oxygen sensor (HO2S) 1, bank 1 – heater control – circuit malfunction	**Wiring, HO2S**
69	Heated oxygen sensor (HO2S) 2, bank 1, heater control – circuit malfunction	**Wiring, HO2S**
6A	Random/multiple cylinder(s) – misfire detected	**Wiring, spark plug(s), HT lead(s), injector(s), ignition coil(s), low compression**
6B	Cylinder 1 – misfire detected	**Wiring, engine mechanical fault, ignition/ fuel system, injector**
6C	Cylinder 2 – misfire detected	**Wiring, engine mechanical fault, ignition/ fuel system, injector**
6D	Cylinder 3 – misfire detected	**Wiring, engine mechanical fault, ignition/ fuel system, injector**
6E	Cylinder 4 – misfire detected	**Wiring, engine mechanical fault, ignition/ fuel system, injector**
70	Catalytic converter system, bank 1 – efficiency below threshold	**Wiring, HO2S 2, catalytic converter**
71	Evaporative emission (EVAP) system – incorrect flow detected	**Hose connection(s), intake leak, EVAP canister purge valve**
72	Park/neutral position (PNP) switch – input circuit malfunction	**Wiring, PNP switch, ECM/TCM**
76	System too rich	**Wiring, intake blocked, EVAP canister purge valve, fuel pressure, EGR system, injector(s), air filter blocked, ignition system, HO2S**
77	System too lean	**Wiring, intake/exhaust leak, MAP sensor, fuel pressure/pump, EVAP canister purge valve, injector(s), HO2S**
7E	Heated oxygen sensor (HO2S) 2, bank 1 – high voltage	**Wiring short circuit to positive, catalytic converter, HO2S**
80	Insufficient coolant temperature for closed loop fuel control	**Wiring, cooling system, coolant thermostat, ECT sensor**
84	Throttle position (TP) sensor – range/ performance problem	**Wiring, accelerator cable adjustment, TP sensor**

Scan code Hex type	Fault location	Probable cause
87	Manifold absolute pressure (MAP) sensor – no 5 volt supply	**Wiring, ECM**
8A	Idle speed control (ISC) system – rpm lower/ higher than expected	**Wiring, air leak, throttle valve tight/sticking, IAC valve**
8D	Governor pressure not equal to target (15-20 PSI) – mid pressure malfunction	**Signal wire open circuit/short circuit, valve body, sensor supply wire open circuit, transmission fluid level, governor pressure solenoid, transmission mechanical fault, ECM – no 5 volt supply**
8E	Governor pressure not equal to target – low pressure malfunction	**Signal wire open circuit/short circuit, valve body, sensor supply wire open circuit, transmission fluid level, governor pressure solenoid, transmission mechanical fault, ECM – no 5 volt supply**
94	Torque converter clutch (TCC) solenoid – circuit malfunction	**Wiring, TCC solenoid**
95	Fuel level sensor – low input	**Wiring open circuit, fuel level sensor**
96	Fuel level sensor – high input	**Wiring short circuit to positive, fuel level sensor**
97	Fuel level sensor – malfunction	**Wiring, fuel level sensor**
99	Ambient/battery temperature sensor – voltage low	**Wiring, ambient/battery temperature sensor**
9A	Ambient/battery temperature sensor – voltage high	**Wiring, ambient/battery temperature sensor**
9B	Heated oxygen sensor (HO2S) 1, bank 1 – low voltage	**Wiring short circuit to ground, exhaust leak, HO2S**
9C	Heated oxygen sensor (HO2S) 2, bank 1 – low voltage	**Wiring short circuit to ground, exhaust leak, HO2S**
9D	Crankshaft position (CKP) sensor/camshaft position (CMP) sensor – intermittent signal	**Wiring, poor connection, air gap, CKP sensor, CMP sensor**
A0	Evaporative emission (EVAP) system – small leak detected	**Hose connection(s), intake leak, EVAP canister, EVAP canister purge valve**
A1	Evaporative emission (EVAP) system – large leak detected	**Hose connection(s), intake leak, EVAP canister, EVAP canister purge valve**
A4	Transmission fluid temperature (TFT) sensor – range/performance problem	**Wiring, TFT sensor**
A5	Gear selection, 3-4 – shift malfunction	**Wiring, TR sensor, shift solenoids, transmission mechanical fault**
A6	Output shaft speed (OSS) sensor – circuit malfunction	**Wiring, OSS sensor, ECM**
A7	Governor pressure sensor – signal low	**Signal wire short circuit to ground, supply wire open circuit, transmission fluid level, governor pressure sensor, ECM**
A8	Governor pressure sensor – signal high	**Signal wire short circuit to positive, supply wire open circuit, transmission fluid level, governor pressure sensor, ECM**

Scan code Hex type	Fault location	Probable cause
A9	Governor pressure sensor – incorrect signal	Signal wire short circuit to ground/positive, supply wire open circuit/short circuit, transmission fluid level, governor pressure solenoid, governor pressure sensor, ECM
AB	Transmission fluid pressure (TFP) solenoid – electrical	Wiring, TFP solenoid, ECM
AD	Transmission system relay	Wiring open circuit/short circuit, transmission system relay, ECM
AE	Cylinder 5 – misfire detected	Wiring, engine mechanical fault, ignition/ fuel system, injector
AF	Cylinder 6 – misfire detected	Wiring, engine mechanical fault, ignition/ fuel system, injector
B0	Cylinder 7 – misfire detected	Wiring, engine mechanical fault, ignition/ fuel system, injector
B1	Cylinder 8 – misfire detected	Wiring, engine mechanical fault, ignition/ fuel system, injector
B7	Evaporative emission (EVAP) leak detection pump solenoid	Wiring open circuit/short circuit, hose leak/ blockage, EVAP leak detection pump solenoid
B8	Evaporative emission (EVAP) leak detection pump – pressure switch	Wiring, hose leak/blockage, EVAP leak detection pump
BA	Engine control module (ECM) – misfire detection limit reached	Wiring, poor connection, CKP sensor
BB	Evaporative emission (EVAP) system – blockage detected	Hose connection(s), hose blocked/trapped, EVAP canister
BC	Overdrive solenoid/switch – performance problem	Wiring, overdrive solenoid/switch, transmission mechanical fault, ECM
C0	Heated oxygen sensor (HO2S) 1, bank 1 – slow response	Wiring, heating inoperative, HO2S
C2	Heated oxygen sensor (HO2S) 2, bank 1 – slow response	Wiring, heating inoperative, exhaust leak, HO2S

JEEP

Model:	Engine identification:	Year:
Cherokee 2.5L	VIN position 8 = P	1999-02
Cherokee 4.0L	VIN position 8 = S	1999-02
Grande Cherokee 4.0L	VIN position 8 = S	1999-02
Grande Cherokee 4.7L	VIN position 8 = N	1999-02
Wrangler 2.5L	VIN position 8 = P	1999-02
Wrangler 4.0L	VIN position 8 = S	1999-02

System: **JTEC**

Self-diagnosis

General information

- Refer to the front of this manual for general test conditions, terminology, detailed descriptions of wiring faults and a general trouble shooter for electrical and mechanical faults.
- Malfunction indicator lamp (MIL) will illuminate or flash if certain faults are recorded.
- ECM operates in backup mode if sensors fail, to enable vehicle to be driven to repair shop.

Accessing

- Trouble codes can only be displayed by using a Scan Tool connected to the data link connector (DLC):
  - Cherokee/Wrangler **1**.
  - Grand Cherokee **2**.

1

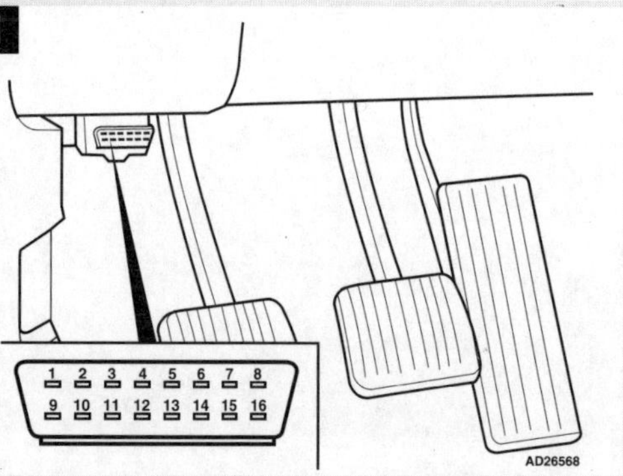

2

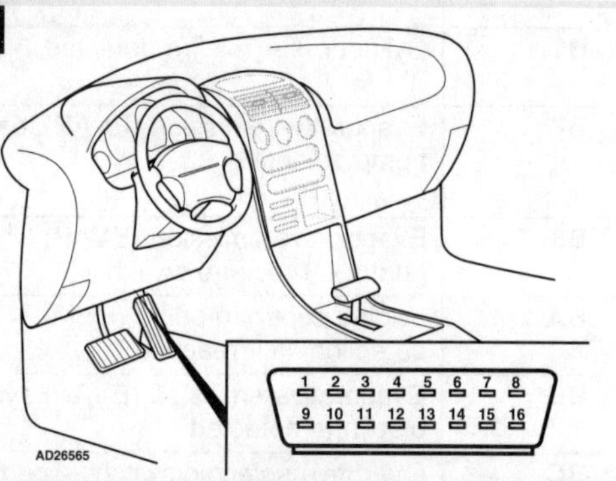

Erasing

- Trouble codes can only be erased by using a Scan Tool connected to the data link connector (DLC).

Trouble code identification

Scan code	Fault location	Probable cause
P0000-P0999	Refer to OBD-II trouble code table at the front of this manual	–
P1110	Decreased engine performance – high intake air temperature	**Wiring open circuit/short to positive, ground wire defective, IAT sensor, ECT sensor, cooling system, high engine bay temperature, ECM**
P1192	Intake air temperature (IAT) sensor – voltage low	**Wiring, IAT sensor, ECM**
P1193	Intake air temperature (IAT) sensor – voltage high	**Wiring, IAT sensor, ECM**
P1194	Heated oxygen sensor (HO2S), heater control – circuit malfunction	**Wiring, HO2S, ECM**
P1195	Heated oxygen sensor (HO2S) 1, bank 1 – slow response	**Wiring, heating inoperative, exhaust leak, HO2S**
P1196	Heated oxygen sensor (HO2S) 1, bank 2 – slow response	**Wiring, heating inoperative, exhaust leak, HO2S**
P1197	Heated oxygen sensor (HO2S) 2, bank 1 – slow response	**Wiring, heating inoperative, exhaust leak, HO2S**
P1280	Fuel pump relay – circuit malfunction	**Wiring, fuel pump relay**
P1281	Engine temperature – low for long period	**Wiring, cooling system, coolant thermostat, ECT sensor**
P1282	Fuel pump relay	**Wiring open circuit/short circuit, fuel pump relay**
P1294	Idle speed control (ISC) system – rpm lower/higher than expected	**Wiring, air leak, throttle valve tight/sticking, IAC valve**
P1295	Throttle position (TP) sensor – no 5 volt supply	**Wiring, TP sensor, ECM**
P1296	Manifold absolute pressure (MAP) sensor – no 5 volt supply	**Wiring, MAP sensor, ECM**
P1297	Manifold absolute pressure (MAP) sensor	**Wiring, hose connection(s), MAP sensor**
P1298	System too lean – at prolonged wide open throttle	**Intake/exhaust leak, MAP sensor, fuel pressure/pump, injector(s), HO2S**
P1299	Manifold absolute pressure (MAP) sensor/throttle position (TP) sensor	**Intake leak, hose connection(s), MAP sensor, TP sensor**
P1388	Engine control relay	**Wiring open circuit/short circuit, engine control relay**
P1389	Engine control relay	**Wiring open circuit, engine control relay**
P1390	Crankshaft position (CKP) sensor/camshaft position (CMP) sensor – signal synchronisation	**Timing belt loose/jumped position, CKP sensor, CMP sensor**
P1391	Crankshaft position (CKP) sensor/camshaft position (CMP) sensor – intermittent signal	**Wiring, poor connection, air gap, CKP sensor, CMP sensor**
P1398	Engine control module (ECM) – misfire detection limit reached	**Wiring, poor connection, CKP sensor**

Scan code	Fault location	Probable cause
P1475	Engine control module (ECM) – auxiliary 5 volt output high	**Wiring, ECM**
P1478	Ambient/battery temperature sensor – range/ performance problem	**Wiring, ambient/battery temperature sensor**
P1486	Evaporative emission (EVAP) system – blockage detected	**Hose connection(s), hose blocked/trapped, EVAP canister**
P1488	Engine control module (ECM) – auxiliary 5 volt output low	**Wiring, ECM**
P1491	Engine coolant blower motor relay – circuit malfunction	**Wiring open circuit/short circuit, engine coolant blower motor**
P1492	Ambient/battery temperature sensor – voltage high	**Wiring, ambient/battery temperature sensor**
P1493	Ambient/battery temperature sensor – voltage low	**Wiring, ambient/battery temperature sensor**
P1494	Evaporative emission (EVAP) leak detection pump – pressure switch	**Wiring, hose leak/blockage, EVAP leak detection pump**
P1495	Evaporative emission (EVAP) leak detection pump solenoid	**Wiring open circuit/short circuit, hose leak/ blockage, EVAP leak detection pump solenoid**
P1496	Engine control module (ECM) – 5 volt output low	**ECM**
P1594	Alternator – voltage high	**Wiring, battery, alternator**
P1595	Cruise control	**Wiring open circuit/short circuit, hoses, cruise control master switch, cruise control vacuum switching valve**
P1596	Cruise control master switch – high voltage	**Wiring, cruise control master switch**
P1597	Cruise control master switch – low voltage	**Wiring, cruise control master switch**
P1598	A/C refrigerant pressure sensor – high input	**Wiring, A/C refrigerant pressure too high (cooling fault/incorrectly charged), A/C refrigerant pressure sensor**
P1599	A/C refrigerant pressure sensor – low input	**Wiring, A/C refrigerant pressure too low (incorrectly charged), A/C refrigerant pressure sensor**
P1602	Engine control module (ECM) – not programmed	**New ECM not programmed, incorrectly programmed, ECM**
P1682	Alternator – voltage low	**Wiring, battery, alternator drive belt, alternator**
P1683	Cruise control	**Wiring, cruise control vacuum switching valve**
P1684	Engine control module (ECM) – supply voltage disconnected within last 50 ignition switch ON-OFF cycles	**Wiring, battery**
P1685	Key not programmed into immobilizer control module	**Wiring, ignition key not matched/damaged, immobilizer control module**
P1686	CAN data bus, Immobilizer control module – no signal	**Wiring, immobilizer control module**

Scan code	Fault location	Probable cause
P1696	Engine control module (ECM) – EEPROM error	ECM
P1756	Governor pressure not equal to target (15-20 PSI) – mid pressure malfunction	Signal wire open circuit/short circuit, valve body, sensor supply wire open circuit, transmission fluid level, governor pressure solenoid, transmission mechanical fault, ECM – no 5 volt supply
P1757	Governor pressure not equal to target (15-20 PSI) – zero pressure malfunction	Signal wire open circuit/short circuit, valve body, sensor supply wire open circuit, transmission fluid level, governor pressure solenoid, transmission mechanical fault, ECM – no 5 volt supply
P1762	Governor pressure sensor – incorrect signal	Signal wire short circuit to ground/positive, supply wire open circuit/short circuit, transmission fluid level, governor pressure solenoid, governor pressure sensor, ECM
P1763	Governor pressure sensor – signal high	Signal wire short circuit to positive, supply wire open circuit, transmission fluid level, governor pressure sensor, ECM
P1764	Governor pressure sensor – signal low	Signal wire short circuit to ground, supply wire open circuit, transmission fluid level, governor pressure sensor, ECM
P1765	Transmission system relay	Wiring open circuit/short circuit, transmission system relay, ECM
P1899	Park/neutral position (PNP) switch – input circuit malfunction	Wiring, PNP switch, ECM

LINCOLN

Model:	Engine identification:	Year:
Continental 3.8L	VIN position 8 = 4	1992-94
Mark VII 5.0L HO	VIN position 8 = E	1992
Mark VIII 4.6L DOHC	VIN position 8 = V/W	1993-95
Town Car 4.6L	VIN position 8 = W	1992-94

System: **Ford EEC-IV**

Self-diagnosis

General information

- Refer to the front of this manual for general test conditions, terminology, detailed descriptions of wiring faults and a general trouble shooter for electrical and mechanical faults.
- Malfunction indicator lamp (MIL) will illuminate if certain faults are recorded.
- ECM operates in backup mode if sensors fail, to enable vehicle to be driven to repair shop.
- Continental:
  - Trouble codes can be displayed by the electronic message center or by using an analogue voltmeter or Scan Tool connected to the data link connectors (DLC) 1 and 2 **1**.
- Mark VII/Mark VIII/Town Car:
  - Trouble codes can be displayed by the MIL or by using an analogue voltmeter or Scan Tool connected to the data link connectors (DLC) 1 and 2:
  - Mark VII **2**.
  - Mark VIII **3**.
  - Town Car **4**.

1

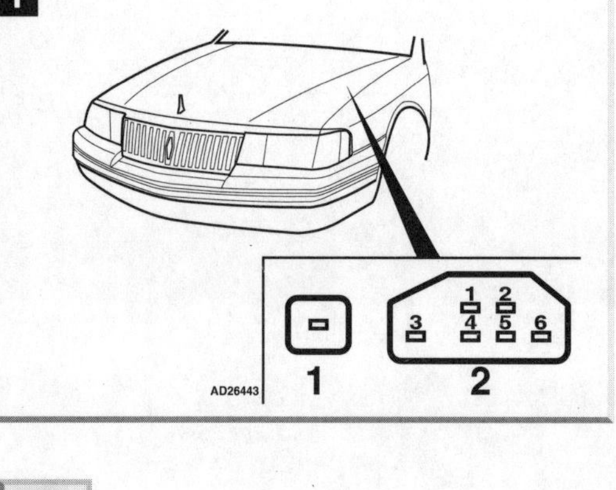

2

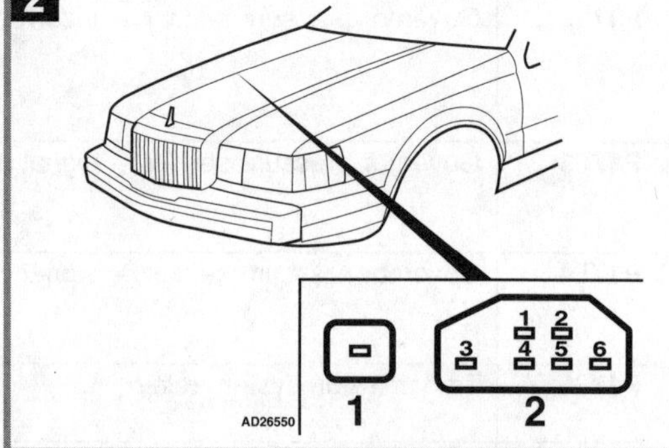

3

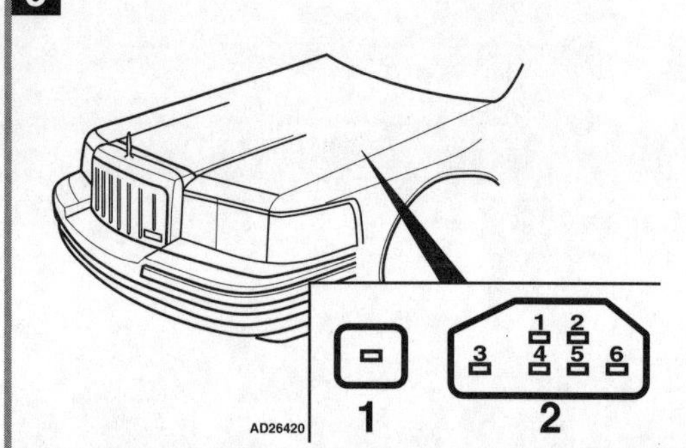

4

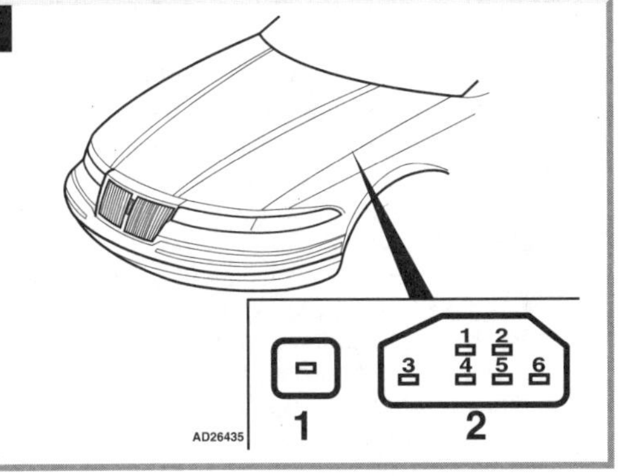

Preparatory conditions

- Automatic transmission in 'PARK'.
- All auxiliary equipment, including air conditioning, switched OFF.

Accessing – Key On Engine Off (KOEO) test

Using malfunction indicator lamp (MIL) – Mark VII/VIII/Town Car

- Start engine.
- Allow to idle until engine at normal operating temperature.
- Switch ignition OFF.
- Wait 10 seconds.
- If engine does not start or stalls: Continue with KOEO test.
- If fitted: Remove 'EEC TEST' cover to access data link connector (DLC).
- Jump DLC 1 terminal (STI) to DLC 2 terminal 2 (SIG RTN) **5**.
- Switch ignition ON.

NOTE: ***MIL may flash briefly while fast DTCs are issued for Scan Tool followed by slow DTC output.***

- Count MIL flashes. Note trouble codes. Compare with trouble code table.
- Each trouble code is displayed twice. Firstly for present faults followed by those for memorized faults as follows:
  - Trouble codes are displayed by illuminated MIL switching OFF.
  - Each trouble code consists of two or three digits.
  - A group of 0.5 second flashes indicate each digit **6** [**A**].
  - A 0.5 second pause separates each flash **6** [**B**].
  - A 2 second pause separates each trouble code digit **6** [**C**].
  - A 4 second pause separates each trouble code **6** [**D**].
  - For example: Three digit trouble code 122 displayed **6**. Throttle position (TP) sensor.

NOTE: ***Two digit trouble codes displayed in a similar pattern.***

  - A 6-9 second pause **7** [**B**] indicates the end of trouble code output for present faults **7** [**A**].
  - A 0.5 second flash (separator code) **7** [**C**], followed by a 6-9 second pause **7** [**D**] indicates the start of trouble code output for memorized faults **7** [**E**].
  - Trouble codes for memorized faults displayed in the same manner as those for present faults.

NOTE: ***Memorized trouble codes refer to faults that have occurred in the last 80 warm-up cycles.***

- Switch ignition OFF.
- Remove jump lead.
- Trouble codes can also be displayed by using a Scan Tool connected to the data link connectors (DLC) 1 and 2.

5

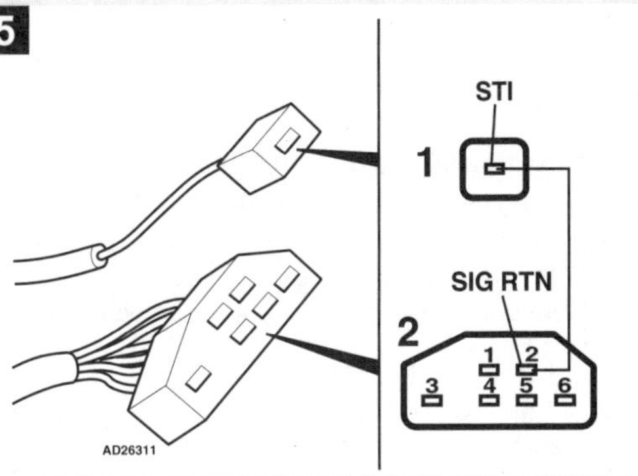

6

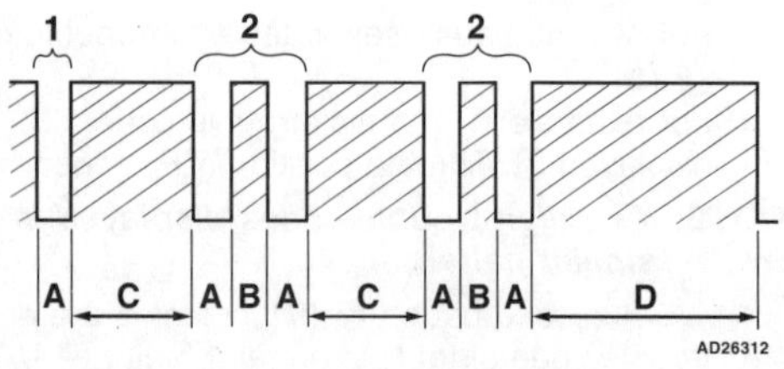

7

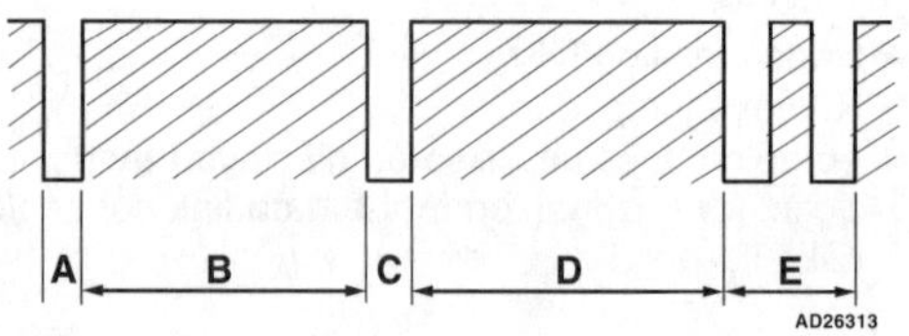

Using analogue voltmeter – all models

- Ensure ignition switched OFF.
- If fitted: Remove 'EEC TEST' cover to access data link connector (DLC).
- Connect an analogue voltmeter between data link connector (DLC) 2 terminal 4 (STO) and battery (12 volt) positive 8.
- Set analogue voltmeter to a DC voltage range of 0-20 V.
- Start engine.
- Allow to idle until engine at normal operating temperature.
- Switch ignition OFF.
- Wait 10 seconds.
- If engine does not start or stalls: Continue with KOEO test.
- Jump DLC 1 terminal (STI) to DLC 2 terminal 2 (SIG RTN) 8.
- Switch ignition ON.

NOTE: ***Voltmeter needle may briefly deflect while fast DTCs are issued for Scan Tool followed by slow DTC output.***

- Count voltmeter needle deflections. Note trouble codes. Compare with trouble code table.
- Each trouble code is displayed twice. Firstly for present faults followed by those for memorized faults as follows:
  - Trouble codes are displayed by voltmeter needle deflecting to zero.
  - Each trouble code consists of two or three digits.
  - A group of 0.5 second needle deflections indicate each digit 6 [**A**].
  - A 0.5 second pause separates each needle deflection 6 [**B**].
  - A 2 second pause separates each trouble code digit 6 [**C**].
  - A 4 second pause separates each trouble code 6 [**D**].
  - For example: Three digit trouble code 122 displayed 6. Throttle position (TP) sensor.

NOTE: ***Two digit trouble codes displayed in a similar pattern.***

  - A 6-9 second pause 7 [**B**] indicates the end of trouble code output for present faults 7 [**A**].
  - A 0.5 second needle deflection (separator code) 7 [**C**], followed by a 6-9 second pause 7 [**D**] indicates the start of trouble code output for memorized faults 7 [**E**].
  - Trouble codes for memorized faults displayed in the same manner as those for present faults.

NOTE: ***Memorized trouble codes refer to faults that have occurred in the last 80 warm-up cycles.***

- Switch ignition OFF.
- Remove jump lead and voltmeter.
- Trouble codes can also be displayed by using a Scan Tool connected to the data link connectors (DLC) 1 and 2.

8

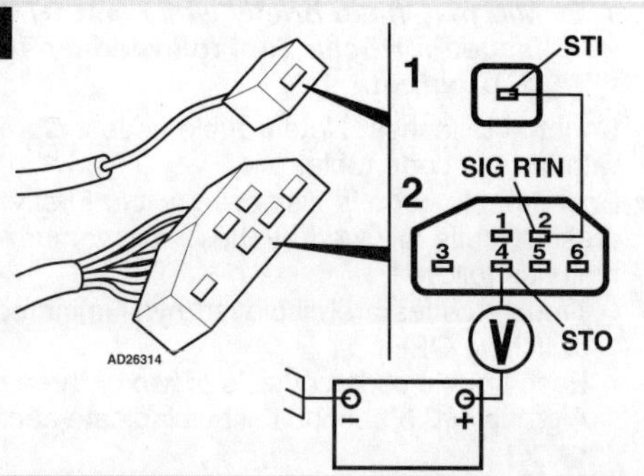

Using electronic message center – Continental

- Start engine.
- Allow to idle until engine at normal operating temperature.
- Switch ignition OFF.
- Wait 10 seconds.
- If engine does not start or stalls: Continue with KOEO test.
- If fitted: Remove 'EEC TEST' cover to access data link connector (DLC).
- Press and hold all 3 message center buttons 'Gauge Select', 'English Metric' and 'Speed Alarm' 9 [**1**].
- Switch ignition ON.
- Release buttons.
- Press 'Gauge Select' button 3 times until 'dEALEr 4' is displayed 9 [**2**].
- Jump DLC 1 terminal (STI) to DLC 2 terminal 2 (SIG RTN) 5.
- Trouble codes displayed on the message center.
- For example: Trouble code 111 displayed 9 [**3**]. No fault found.
- Note trouble codes. Compare with trouble code table.
- Switch ignition OFF.
- Remove jump lead.
- Trouble codes can also be displayed by using a Scan Tool connected to the data link connectors (DLC) 1 and 2.

9

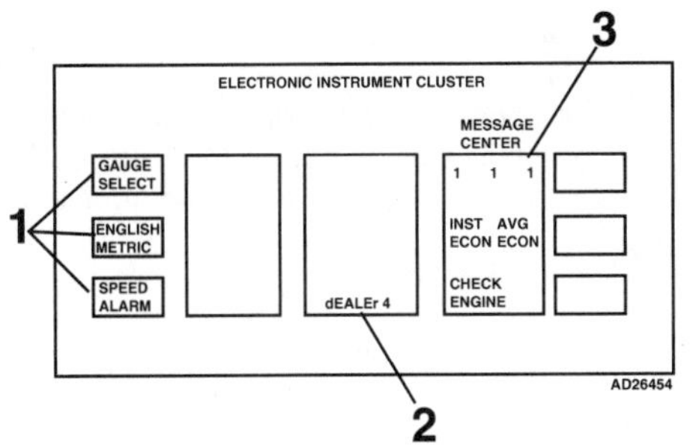

Accessing – Key On Engine Running (KOER) test

Using malfunction indicator lamp (MIL) – Mark VII/VIII/Town Car

- Start engine.
- Allow to idle until engine at normal operating temperature.
- Increase engine speed to over 2000 rpm for 2 minutes.
- Switch ignition OFF.
- Wait 10 seconds.
- If fitted: Remove 'EEC TEST' cover to access data link connector (DLC).
- Jump DLC 1 terminal (STI) to DLC 2 terminal 2 (SIG RTN) 5.
- Start engine.
- Engine identification code displayed as follows:
  - 6 cylinder: Three 0.5 second flashes.
  - 8 cylinder: Four 0.5 second flashes.
  - For example: 6 cylinder engine identified 10 [**A**].
- With PSP switch: Within 1-2 seconds, turn steering wheel at least ½ a turn and release.
- With BPP switch: Depress and release brake pedal.
- If fitted: Operate overdrive selection switch twice.
- If after 6-20 seconds 10 [**B**], a single 0.5 second flash is displayed (dynamic response signal) 10 [**C**]: Briefly fully open throttle.

NOTE: ***DO NOT fully open throttle during KOER test if no dynamic response signal is displayed.***

- Trouble codes displayed for present faults after 4-15 second pause 10 [**D**].
- Fast DTCs are issued for Scan Tool during 4-15 second pause 10 [**E**].
- Count MIL flashes. Note trouble codes. Compare with trouble code table.

NOTE: ***Each trouble code is displayed twice.***

- Trouble codes for KOER test are displayed in the same manner as those for KOEO test 6 [**F**].
- For example: Three digit trouble code 111 is displayed 10. No fault found.
- Switch ignition OFF.
- Remove jump lead.
- Trouble codes can also be displayed by using a Scan Tool connected to the data link connectors (DLC) 1 and 2.

10

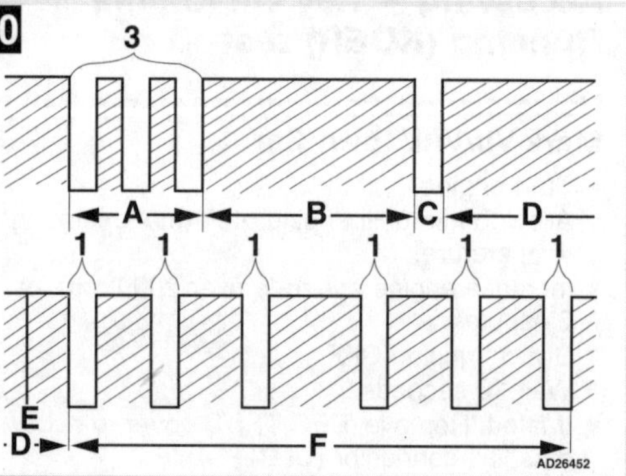

Using analogue voltmeter - all models

- Ensure ignition switched OFF.
- If fitted: Remove 'EEC TEST' cover to access data link connector (DLC).
- Connect an analogue voltmeter between data link connector (DLC) 2 terminal 4 (STO) and battery (12 volt) positive **8**.
- Set analogue voltmeter to a DC voltage range of 0-20 V.
- Start engine.
- Allow to idle until engine at normal operating temperature.
- Increase engine speed to over 2000 rpm for 2 minutes.
- Switch ignition OFF.
- Wait 10 seconds.
- Jump DLC 1 terminal (STI) to DLC 2 terminal 2 (SIG RTN) **8**.
- Start engine.
- Engine identification code displayed as follows:
  - 6 cylinder: Three 0.5 second needle deflections.
  - 8 cylinder: Four 0.5 second needle deflections.
  - For example: 6 cylinder engine identified **10** [**A**].
- With PSP switch: Within 1-2 seconds, turn steering wheel at least ½ a turn and release.
- With BPP switch: Depress and release brake pedal.
- If fitted: Operate overdrive selection switch twice.
- If after 6-20 seconds **10** [**B**], a single 0.5 second needle deflection is displayed (dynamic response signal) **10** [**C**]: Briefly fully open throttle.

NOTE: ***DO NOT fully open throttle during KOER test if no dynamic response signal is displayed.***

- Trouble codes displayed for present faults after 4-15 second pause **10** [**D**].
- Fast DTCs are issued for Scan Tool during 4-15 second pause **10** [**E**].
- Count needle deflections. Note trouble codes. Compare with trouble code table.

NOTE: ***Each trouble code is displayed twice.***

- Trouble codes for KOER test are displayed in the same manner as those for KOEO test **10** [**F**].
- For example: Three digit trouble code 111 is displayed **10**. No fault found.
- Switch ignition OFF.
- Remove jump lead and voltmeter.
- Trouble codes can also be displayed by using a Scan Tool connected to the data link connectors (DLC) 1 and 2.

Using electronic message center - Continental

- Start engine.
- Allow to idle until engine at normal operating temperature.
- Increase engine speed to over 2000 rpm for 2 minutes.
- Switch ignition OFF.
- Wait 10 seconds.
- If fitted: Remove 'EEC TEST' cover to access data link connector (DLC).
- Press and hold all 3 message center buttons 'Gauge Select', 'English Metric' and 'Speed Alarm' **9** [**1**].
- Start engine.
- Release buttons.
- Press 'Gauge Select' button 3 times until 'dEALEr 4' is displayed **9** [**2**].
- Jump DLC 1 terminal (STI) to DLC 2 terminal 2 (SIG RTN) **5**.
- Trouble codes displayed on the message center.
- For example: Trouble code 111 displayed **9** [**3**].
- Note trouble codes. Compare with trouble code table.
- Switch ignition OFF.
- Remove jump lead.
- Trouble codes can also be displayed by using a Scan Tool connected to the data link connectors (DLC) 1 and 2.

Erasing

NOTE: ***Disconnecting battery ground lead may cause driveability problems until the engine management system relearns basic values.***

Using malfunction indicator lamp (MIL) - Mark VII/VIII/Town Car

- Ensure ignition switched OFF.
- If fitted: Remove 'EEC TEST' cover to access data link connector (DLC).
- Jump DLC 1 terminal (STI) to DLC 2 terminal 2 (SIG RTN) **5**.
- Switch ignition ON.
- When trouble codes start to be displayed: Remove jump lead.
- Switch ignition OFF.
- Repeat accessing procedure to ensure no memorized trouble codes remain in ECM fault memory.
- Trouble codes can also be erased by using a Scan Tool connected to the data link connectors (DLC) 1 and 2.
- If emission related component(s) have been replaced:
  - Disconnect battery ground cable for at least 5 minutes to erase adaptive memory from ECM.
  - If required: Drive vehicle for 10 miles or more to allow ECM to relearn basic values.

NOTE: ***Disconnecting battery ground lead may also erase memory from electronic units, such as radio.***

Using analogue voltmeter - all models

- Ensure ignition switched OFF.
- If fitted: Remove 'EEC TEST' cover to access data link connector (DLC).
- Connect an analogue voltmeter between data link connector (DLC) 2 terminal 4 (STO) and battery (12 volt) positive **8**.
- Set analogue voltmeter to a DC voltage range of 0-20 V.
- Jump DLC 1 terminal (STI) to DLC 2 terminal 2 (SIG RTN) **8**.
- Switch ignition ON.
- When trouble codes start to be displayed: Remove jump lead.
- Switch ignition OFF.
- Repeat accessing procedure to ensure no memorized trouble codes remain in ECM fault memory.
- Trouble codes can also be erased by using a Scan Tool connected to the data link connectors (DLC) 1 and 2.
- If emission related component(s) have been replaced:
  - Disconnect battery ground cable for at least 5 minutes to erase adaptive memory from ECM.
  - If required: Drive vehicle for 10 miles or more to allow ECM to relearn basic values.

NOTE: ***Disconnecting battery ground lead may also erase memory from electronic units, such as radio.***

Using electronic message center - Continental

- Ensure ignition switched OFF.
- If fitted: Remove 'EEC TEST' cover to access data link connector (DLC).
- Press and hold all 3 message center buttons 'Gauge Select', 'English Metric' and 'Speed Alarm'.
- Switch ignition ON.
- Release buttons.
- Press 'Gauge Select' button 3 times until 'dEALEr 4' is displayed **9** [**1**].
- Jump DLC 1 terminal (STI) to DLC 2 terminal 2 (SIG RTN) **5**.
- When trouble codes start to be displayed on message center: Remove jump lead.
- Switch ignition OFF.
- Repeat accessing procedure to ensure no memorized trouble codes remain in ECM fault memory.
- Trouble codes can also be erased by using a Scan Tool connected to the data link connectors (DLC) 1 and 2.

Trouble code identification

Two digit

Flash code	Fault location	Probable cause
11	No fault found	–
12	Idle speed, high RPM check – unable to control during KOER test	**Intake system, throttle tight/sticking, IAC valve/ISC actuator, ECM**

Flash code	Fault location	Probable cause
13	Idle speed, low RPM check – unable to control during KOER test	**Intake system, throttle tight/sticking, IAC valve/ISC actuator, ECM**
14	Engine speed (RPM) sensor/profile ignition pickup (PIP) circuit – intermittent signal	**Incorrectly installed RPM sensor, wiring, poor connection, electrical interference, RPM sensor, ECM**
15	Engine control module (ECM) – KAM/ROM fault	**Battery disconnected, wiring, fuses, ECM**
16	Engine RPM too low for heated oxygen sensor (HO2S) test	**Intake system, throttle tight/sticking, IAC valve/ISC actuator, ECM**
	Ignition diagnostic monitor (IDM) circuit – no signal during KOEO test	**Wiring, ICM, ECM**
18	Spark output (SPOUT) circuit open	**Wiring, ICM, ECM**
	Ignition diagnostic monitor (IDM) circuit/spark output (SPOUT) circuit grounded	**Wiring, ICM, ECM**
19	Engine control module (ECM) – internal voltage fault	**ECM**
21	Engine coolant temperature (ECT) sensor – incorrect signal	**Wiring, ECT sensor, ECM**
22	Manifold absolute pressure (MAP) sensor/ barometric pressure (BARO) sensor – incorrect signal	**Vacuum hose damaged, trapped vacuum, high atmospheric pressure, excessive EGR, wiring, MAP/BARO sensor, ECM**
23	Throttle position (TP) sensor – incorrect signal	**Throttle valve tight/sticking, wiring, TP sensor, ECM**
24	Intake air temperature (IAT) sensor – incorrect signal	**Wiring, IAT sensor, ECM**
25	Knock sensor (KS) – no signal during dynamic response test	**Wiring, KS, ECM**
26	Mass air flow (MAF) sensor – incorrect signal	**Intake leak/blockage, wiring, MAF sensor, ECM**
29	Vehicle speed signal – incorrect signal	**Wiring, vehicle speed interface module, VSS, ECM**
31	Exhaust gas recirculation (EGR) valve position sensor – below minimum voltage	**Wiring short circuit to ground, EGR valve/ position sensor, ECM**
32	Exhaust gas recirculation (EGR) valve position sensor – below closed limit	**Wiring short circuit to ground, EGR valve/ position sensor, ECM**
33	Exhaust gas recirculation (EGR) valve – not opening	**Wiring, EGR valve tight/sticking, ECM**
34	Exhaust gas recirculation (EGR) valve position sensor – above closed limit	**Wiring short circuit to positive, EGR valve/ position sensor, ECM**
35	Exhaust gas recirculation (EGR) valve position sensor – above maximum voltage	**Wiring short circuit to positive, EGR valve/ position sensor, ECM**
41	Heated oxygen sensor (HO2S), bank 1 – mixture lean/not switching	**Intake leak, fuel pressure low, fuel filter/ pump, injectors, HO2S, AIR system, MAF/ MAP sensor, ECM**
42	Heated oxygen sensor (HO2S), bank 1 – mixture rich/not switching	**Intake blocked, excessive fuel in engine oil, fuel pressure high, breather system, injectors, HO2S, AIR system, EVAP system, MAF/MAP sensor, ignition system, ECM**

Flash code	Fault location	Probable cause
43	Throttle position (TP) sensor – incorrect signal at idle	**Incorrectly adjusted, TP sensor, ECM**
44	Secondary air injection (AIR) system, bank 1 – inoperative	**Hose connection, wiring, AIR pump motor, AIR solenoid, ECM**
45	Secondary air injection (AIR) – upstream during KOER test	**Hose connection, wiring, AIR solenoid, ECM**
46	Secondary air injection (AIR) – not bypassed during KOER test	**Hose connection, wiring, AIR solenoid, ECM**
47	Four wheel drive selector switch – closed during KOEO test	**Wiring short circuit, four wheel drive selector switch, internal damage to transfer case, ECM**
49 [1]	Gear shift, 1-2 – malfunction	**ATF level/condition, poor connection, wiring, TR sensor, SS, clutch slipping, ECM**
51	Engine coolant temperature (ECT) sensor – open circuit	**Wiring open circuit, ECT sensor, ECM**
52	Power steering pressure (PSP) switch – open circuit/no signal change	**Wiring open circuit, PSP switch, ECM**
53	Throttle position (TP) sensor – above maximum voltage	**TP sensor loose, wiring short circuit to positive, TP sensor, ECM**
54	Intake air temperature (IAT) sensor – open circuit	**Wiring open circuit, IAT sensor, ECM**
56	Mass air flow (MAF) sensor – above maximum voltage	**Wiring short circuit to positive, MAF sensor, ECM**
59 [1]	Gear shift, 2-3 – malfunction	**ATF level/condition, poor connection, wiring, TR sensor, SS, clutch slipping, ECM**
61	Engine coolant temperature (ECT) sensor – short circuit to ground	**Wiring short circuit to ground, ECT sensor, ECM**
62 [1]	Torque converter clutch (TCC) – excessive clutch slip	**ATF level/condition, TCC slipping, ECM**
63	Throttle position (TP) sensor – below minimum voltage	**TP sensor loose, wiring open circuit/short circuit to ground, TP sensor, ECM**
64	Intake air temperature (IAT) sensor – short circuit to ground	**Wiring short circuit to ground, IAT sensor, ECM**
65 [1]	Overdrive selection switch – no signal change during KOEO test	**Incorrect switch operation, wiring, fuse, overdrive selection switch, ECM**
66	Mass air flow (MAF) sensor – below minimum voltage	**Wiring short circuit to ground, MAF sensor, ECM**
67	Neutral position (NP) switch – open circuit	**Incorrect adjustment, wiring open circuit, NP switch, ECM**
	Clutch pedal position (CPP) switch – circuit malfunction	**Incorrect adjustment, wiring, CPP switch, ECM**
	A/C – on during KOEO test	**Wiring, A/C switch, ECM**
67 [1]	Transmission range (TR) sensor – voltage high/low	**Incorrect adjustment, wiring, TR sensor, ECM**
68 [1]	Automatic transmission fluid (ATF) – temperature high	**ATF level, transmission cooling system, towing weight limit exceeded, clutch slipping, wiring, TFT sensor, ECM**

Flash code	Fault location	Probable cause
69 **1**	Gear shift, 3-4 – malfunction	**ATF level/condition, poor connection, wiring, TR sensor, SS, clutch slipping, ECM**
72	Manifold absolute pressure (MAP) sensor/ mass air flow (MAF) sensor – incorrect signal change during dynamic response test	**Intake leak, wiring, MAP sensor, ECM**
73	Throttle position (TP) sensor – incorrect signal change during dynamic response test	**Incorrect throttle operation, wiring, TP sensor, ECM**
74	Brake pedal position (BPP) switch – open circuit/no signal change during KOER test	**Incorrect brake pedal operation, wiring, BPP switch, ECM**
75	Brake pedal position (BPP) switch – closed circuit	**Wiring, BPP switch, ECM**
77	Wide open throttle (WOT) switch – no signal change during dynamic response test	**Incorrect throttle operation, wiring, WOT switch, ECM**
79	A/C – on during KOEO test	**Wiring, A/C switch, ECM**
81	Secondary air injection (AIR) divert solenoid – circuit malfunction during KOEO test	**Wiring, AIR divert solenoid, ECM**
82	Secondary air injection (AIR) bypass solenoid – circuit malfunction during KOEO test	**Wiring, AIR bypass solenoid, ECM**
84	Exhaust gas recirculation (EGR) solenoid – circuit malfunction during KOEO test	**Wiring, EGR solenoid, ECM**
85	Evaporative emission (EVAP) canister purge valve – circuit malfunction during KOEO test	**Wiring, EVAP canister purge valve, ECM**
86	AT – shift solenoid	**Wiring, shift solenoid, ECM**
87	Fuel pump, primary circuit – circuit malfunction	**IFS switch, wiring, fuel pump, ECM**
89	Torque converter clutch (TCC) lock-up solenoid	**Wiring, TCC lock-up solenoid, ECM**
91	Heated oxygen sensor (HO2S), bank 2 – mixture lean/not switching during KOER test	**Intake leak, fuel pressure low, fuel filter/ pump, injectors, HO2S, AIR system, MAF/ MAP sensor, ECM**
91 **1**	Shift solenoid (SS) 1 – circuit malfunction during KOEO test	**Wiring, SS, ECM**
92	Heated oxygen sensor (HO2S), bank 1 – mixture rich/not switching during KOER test	**Intake blocked, excessive fuel in engine oil, fuel pressure high, breather system, injectors, HO2S, AIR system, EVAP system, MAF/MAP sensor, ignition system, ECM**
92 **1**	Shift solenoid (SS) 2 – circuit malfunction during KOEO test	**Wiring, SS, ECM**
93 **1**	Transmission coast clutch solenoid – circuit malfunction during KOEO test	**Wiring, transmission coast clutch solenoid, ECM**
94	Secondary air injection (AIR) system, bank 2 – inoperative during KOER test	**Hose connection, wiring, AIR pump motor, AIR solenoid, ECM**
94 **1**	Torque converter clutch (TCC) solenoid – circuit malfunction during KOEO test	**Wiring, TCC solenoid, ECM**
95	Fuel pump, secondary circuit – circuit malfunction	**IFS switch, wiring, fuel pump, ECM**
96	Fuel pump, secondary circuit – circuit malfunction	**IFS switch, wiring open circuit to battery, fuel pump, ECM**

Flash code	Fault location	Probable cause
97 [1]	Overdrive indicator lamp – circuit malfunction	Bulb, wiring, fuse, ECM
98	Engine control module (ECM) – hard fault detected during KOER test	HO2S bank 2 – indicates lean, ECM
98 [1]	Engine control module (ECM), TFP control – open driver circuit during KOEO test	ECM
99	Engine control module (ECM) – idle speed not learnt during KOER test	Idle speed adjusted, ECM
99 [1]	Transmission fluid pressure (TFP) solenoid – circuit malfunction during KOEO test	Wiring, poor connection, TFP solenoid, ECM

[1] ECM incorporates transmission control function

Three digit

Flash code	Fault location	Probable cause
111	No fault found	–
112	Intake air temperature (IAT) sensor – below minimum voltage	Wiring short circuit to ground, poor connection, IAT sensor, ECM
113	Intake air temperature (IAT) sensor – above maximum voltage	Wiring open circuit, poor connection, IAT sensor, ECM
114	Intake air temperature (IAT) sensor – incorrect signal	Air temperature below 10°C (50°F), poor connection, IAT sensor, ECM
116	Engine coolant temperature (ECT) sensor – incorrect signal	Coolant level low, poor connection, ECT sensor, ECM
117	Engine coolant temperature (ECT) sensor – below minimum voltage	Wiring short circuit to ground, poor connection, ECT sensor, ECM
118	Engine coolant temperature (ECT) sensor – above maximum voltage	Wiring open circuit, poor connection, IAT sensor, ECM
121	Throttle position (TP) sensor – closed throttle voltage high/low	Throttle valve tight/sticking, TP sensor loose/damaged, ECM
	Throttle position (TP) sensor – voltage inconsistent with MAF sensor signal	Throttle valve tight/sticking, TP sensor loose/damaged, ECM
122	Throttle position (TP) sensor – below minimum voltage	TP sensor loose, wiring open circuit/short circuit to ground, TP sensor, ECM
123	Throttle position (TP) sensor – above maximum voltage	TP sensor loose, wiring short circuit to positive, TP sensor, ECM
124	Throttle position (TP) sensor – voltage high	Throttle tight/sticking, intake leak, MAP/BARO sensor, ECM
125	Throttle position (TP) sensor – voltage low	Throttle tight/sticking, intake leak, MAP/BARO sensor, ECM
126	Manifold absolute pressure (MAP) sensor/barometric pressure (BARO) sensor – voltage high/low	Vacuum hose damaged/trapped, high atmospheric pressure, excessive EGR, wiring, MAP/BARO sensor, ECM
128	Manifold absolute pressure (MAP) sensor – loss of vacuum	Vacuum hose damaged/disconnected, MAP sensor, ECM

Flash code	Fault location	Probable cause
129	Manifold absolute pressure (MAP) sensor – insufficient voltage change during KOER dynamic response test	**Incorrect throttle operation, vacuum hose damaged/disconnected, MAP sensor, IAC system, ECM**
	Mass air flow (MAF) sensor – insufficient voltage change during KOER dynamic response test	**Incorrect throttle operation, intake leak, wiring, MAF sensor, IAC system, ECM**
136	Heated oxygen sensor (HO2S), bank 2 – mixture lean during KOER test	**Intake leak, fuel pressure low, fuel filter/pump, injectors, HO2S, AIR system, MAP sensor, ECM**
137	Heated oxygen sensor (HO2S), bank 2 – mixture rich during KOER test	**Intake blocked, excessive fuel in engine oil, fuel pressure high, breather system, injectors, HO2S, AIR system, EVAP system, MAP sensor, ignition system, ECM**
138	Cold start injector – insufficient flow during KOER test	**Fuel pressure, wiring, cold start injector, ECM**
139	Heated oxygen sensor (HO2S), bank 2 – not switching	**Intake leak/blocked, injectors, HO2S, AIR system, breather system, EVAP system, MAP sensor, ignition system, ECM**
141	Fuel system indicates lean	**Intake leak, fuel pressure low, fuel filter/pump, injectors, HO2S, ECM**
144	Heated oxygen sensor (HO2S), bank 1 – not switching	**Intake leak/blocked, injectors, HO2S, AIR system, breather system, EVAP system, MAP sensor, ignition system, ECM**
157	Mass air flow (MAF) sensor – below minimum voltage	**Wiring short circuit to ground, poor connection, MAF sensor, IAC system, ECM**
158	Mass air flow (MAF) sensor – above maximum voltage	**Wiring short circuit to positive, MAF sensor screen blocked, MAF sensor, ECM**
159 1	Mass air flow (MAF) sensor – voltage high/low	**Intake leak, IAC valve, MAF sensor, ECM**
167	Throttle position (TP) sensor – insufficient voltage change during KOER dynamic response test	**Incorrect throttle operation, TP sensor, ECM**
171	Fuel system control, bank 1 – adaptive limits reached	**Intake leak/blocked, breather system, fuel pressure, fuel filter/pump, injectors, HO2S, AIR system, EVAP system, MAP sensor, ignition system, ECM**
172	Heated oxygen sensor (HO2S), bank 1 – mixture lean	**Intake leak, fuel pressure low, fuel filter/pump, injectors, HO2S, AIR system, MAP sensor, ECM**
173	Heated oxygen sensor (HO2S), bank 1 – mixture rich	**Intake blocked, excessive fuel in engine oil, fuel pressure high, breather system, injectors, HO2S, AIR system, EVAP system, MAP sensor, ignition system, ECM**
174	Heated oxygen sensor (HO2S), bank 1 – slow response	**Intake leak, excessive fuel in engine oil, fuel pressure, fuel filter/pump, injectors, HO2S, AIR system, MAF/MAP sensor, ECM**
175	Fuel system control, bank 2 – adaptive limits reached	**Intake leak/blocked, excessive fuel in engine oil, breather system, fuel pressure, fuel filter/pump, injectors, HO2S, AIR system, EVAP system, MAP sensor, ignition system, ECM**

Flash code	Fault location	Probable cause
176	Heated oxygen sensor (HO2S), bank 2 – mixture lean	**Intake leak, fuel pressure low, fuel filter/pump, injectors, HO2S, AIR system, MAP sensor, ECM**
177	Heated oxygen sensor (HO2S), bank 2 – mixture rich	**Intake blocked, fuel pressure high, breather system, injectors, HO2S, AIR system, EVAP system, MAP sensor, ignition system, ECM**
178	Heated oxygen sensor (HO2S), bank 2 – slow response	**Intake leak, excessive fuel in engine oil, fuel pressure, fuel filter/pump, injectors, HO2S, AIR system, MAF/MAP sensor, ECM**
179	Fuel control system, bank 1 – lean adaption limit reached at part throttle, mixture rich	**Intake leak, fuel pressure low, fuel filter blocked, wiring, injector, MAP sensor, ECM**
181	Fuel control system, bank 1 – rich adaption limit reached at part throttle, mixture lean	**Intake blocked, fuel pressure high, wiring, injector, MAP sensor, ECM**
182	Fuel control system, bank 1 – lean adaption limit reached at idle, mixture rich	**Intake leak, fuel pressure low, fuel filter blocked, wiring, injector, MAP sensor, ECM**
183	Fuel control system, bank 1 – rich adaption limit reached at idle, mixture lean	**Intake blocked, fuel pressure high, wiring, injector, MAP sensor, ECM**
184	Mass air flow (MAF) sensor – signal high	**MAF sensor screen blocked, MAF sensor, ECM**
185	Mass air flow (MAF) sensor – signal low	**Intake leak, breather system, MAF sensor, ECM**
186	Injector – pulse width long/mass air flow (MAF) sensor – signal low	**Intake leak, fuel pressure low, MAF sensor, ECM**
187	Injector – pulse width short/mass air flow (MAF) sensor – signal high	**Intake blocked, fuel pressure high, MAF sensor, ECM**
188	Fuel system control, bank 2 – lean adaption limit reached at part throttle, mixture rich	**Intake leak, fuel pressure low, fuel filter blocked, wiring, injector, MAP sensor, ECM**
189	Fuel system control, bank 2 – rich adaption limit reached at part throttle, mixture lean	**Intake blocked, fuel pressure high, wiring, injector, MAP sensor, ECM**
191	Fuel system control, bank 2 – lean adaption limit reached at idle, mixture rich	**Intake leak, fuel pressure low, fuel filter blocked, wiring, injector, MAP sensor, ECM**
192	Fuel system control, bank 2 – rich adaption limit reached at idle, mixture lean	**Intake blocked, fuel pressure high, wiring, injector, MAP sensor, ECM**
193	Fuel quality sensor – circuit malfunction	**Wiring, fuel separation, fuel quality sensor, ECM**
194	Cylinder balance test – malfunction	**Wiring, injector, ECM**
195	Cylinder balance test – malfunction	**Wiring, injector, ECM**
211	Profile ignition pickup (PIP) – circuit malfunction	**Wiring, poor connection, arcing secondary ignition component – coil/cap/rotor/HT leads/plugs, electrical interference, ECM**
212	Ignition control module (ICM) – ignition diagnostic monitor (IDM) signal missing	**Wiring, ICM, ECM**
213	Spark output (SPOUT) – open circuit	**Wiring, ICM, ECM**
214	Camshaft position (CMP) sensor – circuit malfunction	**Wiring, CMP sensor, ICM, ECM**
215	Ignition coil 1 – primary circuit malfunction	**Wiring, ignition coil, ICM, ECM**

LINCOLN

Continental 3.8L • Mark VII 5.0L HO • Mark VIII 4.6L DOHC Town Car 4.6L

Flash code	Fault location	Probable cause
216	Ignition coil 2 – primary circuit malfunction	**Wiring, ignition coil, ICM, ECM**
217	Ignition coil 3 – primary circuit malfunction	**Wiring, ignition coil, ICM, ECM**
218	Ignition diagnostic monitor (IDM), left side – signal missing	**Wiring, ignition coil, ECM**
219	Ignition timing, defaulted to 10° – spark output (SPOUT) circuit open	**Wiring open circuit, ICM, ECM**
221	Ignition timing error	**Wiring, ICM, ECM**
222	Ignition diagnostic monitor (IDM), right side – signal missing	**Wiring, ignition coil, ECM**
223	Loss of dual plug inhibit (DPI) control	**Wiring, ICM, ECM**
224	Ignition coil 1, 2, 3 or 4 – primary circuit malfunction	**Wiring, ignition coil, ICM, ECM**
225	Knock sensor – no signal during KOER dynamic response test	**Wiring, KS, ECM**
226	Ignition control module (ICM) – ignition diagnostic monitor (IDM) signal missing	**Wiring open circuit, ICM, ECM**
232	Ignition coil 1, 2, 3 or 4 – primary circuit malfunction	**Wiring, ignition coil, ICM, ECM**
238	Ignition coil 4 – primary circuit malfunction	**Wiring, ignition coil, ICM, ECM**
241	Ignition control module (ICM) – ignition diagnostic monitor (IDM) signal malfunction	**Wiring, ICM, ECM**
244	Cylinder identification (CID) fault when cylinder balance test requested	**Wiring, CKP sensor, CMP sensor, ICM, ECM**
311	Secondary air injection (AIR) system, bank 1 – inoperative during KOER test	**Hose blocked/leaking, AIR bypass/divert solenoid, AIR pump, ECM**
312	Secondary air injection (AIR) system – misdirected during KOER test	**Hose connections**
313	Secondary air injection (AIR) system – not bypassed during KOER test	**Hose blocked/leaking, AIR bypass solenoid, AIR pump**
314	Secondary air injection (AIR) system, bank 2 – inoperative during KOER test	**Hose blocked/leaking, AIR bypass/divert solenoid, AIR pump, ECM**
326 [2]	Exhaust gas pressure sensor – voltage low	**Hose blocked/leaking, wiring, exhaust gas pressure sensor, ECM**
326 [1] [3]	Exhaust gas pressure sensor – voltage low	**Hose blocked/leaking, EGR solenoid filter, EGR solenoid, wiring, EGR valve, exhaust gas pressure sensor, ECM**
327 [4]	Exhaust gas recirculation (EGR) valve position sensor – minimum voltage exceeded	**Wiring open circuit/short circuit to ground, EGR valve/position sensor, ECM**
327 [5]	Exhaust gas pressure sensor – minimum voltage exceeded	**Hose blocked/leaking, wiring open circuit/ short circuit to ground, exhaust gas pressure sensor, ECM**
328	Exhaust gas recirculation (EGR) valve position sensor, closed position – voltage low	**Wiring, EGR valve/position sensor, ECM**
332 [4]	Exhaust gas recirculation (EGR) system – insufficient flow detected	**Hose blocked/leaking, wiring, EGR solenoid filter, EGR solenoid, EGR valve/position sensor, ECM**

Flash code	Fault location	Probable cause
332 5	Exhaust gas recirculation (EGR) system – insufficient flow detected	**Hose blocked/leaking, EGR solenoid, exhaust gas pressure sensor, EGR valve, ECM**
334	Exhaust gas recirculation (EGR) valve position sensor, closed position – voltage high	**Wiring, EGR valve/position sensor, EGR solenoid, ECM**
335 1	Exhaust gas pressure sensor – voltage high/ low during KOEO test	**Exhaust gas pressure sensor, blocked inlet hose, ECM**
336	Exhaust gas pressure sensor – voltage high	**Exhaust system restricted, wiring, EGR solenoid, exhaust gas pressure sensor, EGR valve, hose routing, ECM**
337 4	Exhaust gas recirculation (EGR) position sensor – above maximum voltage	**Wiring open circuit/short circuit to positive, EGR valve/position sensor, ECM**
337 2	Exhaust gas pressure sensor – above maximum voltage	**Incorrect hose routing, wiring short circuit, signal wire open circuit, exhaust gas pressure sensor, EGR solenoid/valve, ECM**
337 3	Exhaust gas pressure sensor – above maximum voltage	**Exhaust system restricted, wiring short circuit, signal wire open circuit, exhaust gas pressure sensor, ECM**
338	Engine coolant temperature (ECT) – too low	**Thermostat stuck open, coolant leak, ECM**
339	Engine coolant temperature (ECT) – too high	**Coolant level low, thermostat stuck closed, cooling system blocked, coolant blower motor worn/damaged, water pump defective, coolant leak, ECM**
341	Octane adjuster – open circuit	**Wiring, octane adjuster, ECM**
381	A/C compressor clutch – excessive cycling	**A/C system incorrectly charged, cycling switch defective, A/C circuit fault, wiring intermittent open circuit to A/C system, ECM**
411	Idle air control (IAC) valve – no control during KOER low RPM check	**Idle speed adjustment, intake leak, throttle valve tight/sticking, ignition timing incorrect, wiring short circuit to ground, IAC valve, ECM**
412	Idle air control (IAC) valve – no control during KOER high RPM check	**Idle speed adjustment, throttle valve tight/ sticking, wiring open circuit/short circuit to ground, IAC valve, ECM**
415	Idle air control (IAC) valve – lower limit reached	**Intake leak, throttle valve tight/sticking, IAC valve, ECM**
416	Idle air control (IAC) valve – upper limit reached	**Air filter blocked, throttle body dirty/ damaged, IAC valve, ECM**
452	Vehicle speed sensor (VSS) – incorrect signal	**Wiring, VSS, ECM**
453	Servo leaking down – cruise control KOER test	**Servo defective, ECM**
454	Servo leaking up – cruise control KOER test	**Servo defective, ECM**
455	Insufficient RPM increase – cruise control KOER test	**Intake system, ECM**
456	Insufficient RPM decrease – cruise control KOER test	**Intake system, ECM**
457	Cruise control master/selector switch – circuit malfunction during KOEO test	**Wiring, cruise control master/selector switch, ECM**

Flash code	Fault location	Probable cause
458	Cruise control master/selector switch – short circuit to ground during KOEO test	**Wiring short circuit to ground, cruise control master/selector switch, ECM**
459	Cruise control system – ground open circuit during KOEO test	**Wiring open circuit, ECM**
511	Engine control module (ECM) – ROM fault during KOEO test	**ECM**
512	Engine control module (ECM) – KAM fault	**Battery disconnected, wiring, fuses, ECM**
513	Engine control module (ECM) – internal voltage fault	**ECM**
519	Power steering pressure (PSP) switch – circuit malfunction during KOEO test	**PSP switch not fitted – disregard trouble code, wiring, PSP switch, ECM**
521	Power steering pressure (PSP) switch – no signal change during KOEO test	**PSP switch not fitted – disregard trouble code, wiring, PSP switch, ECM**
522 [6]	Park/neutral position (PNP) switch/neutral position (NP) switch – not in park/neutral during KOEO test/open circuit	**Incorrect gear lever position, incorrect adjustment, wiring open circuit, starter motor relay disconnected, PNP/NP switch, ECM**
	Transmission range (TR) sensor – not in park/neutral during KOEO test	**Incorrect shift lever position, incorrect adjustment, wiring, TR sensor, ECM**
524	Fuel pump (FP), high/low speed	**Wiring open circuit/short circuit to ground, left HO2S short circuit to positive, ECM**
525	Vehicle in gear – during KOEO test	**Incorrect shift lever position, incorrect adjustment, PNP switch open, ECM**
	A/C – ON during KOEO test	**A/C ON, A/C wiring short circuit to positive, ECM**
526	Neutral position switch – closed circuit during KOEO test	**Wiring short circuit, NP switch, ECM**
	A/C – ON during KOEO test	**A/C ON, A/C wiring short circuit to positive, A/C switch, ECM**
527 [6]	Park/neutral position (PNP) switch/neutral position switch – open circuit during KOEO test	**Wiring open circuit, starter motor relay disconnected, PNP/NP switch, ECM**
	A/C – ON during KOEO test	**A/C ON, A/C wiring short circuit to positive, A/C switch, ECM**
528	Clutch pedal position (CPP) switch – circuit malfunction	**Wiring open circuit, CPP switch, ECM**
529	Data link/engine control module (ECM) – circuit malfunction	**Wiring, ECM**
532	Instrument panel – circuit malfunction	**Wiring, instrument panel, ECM**
533	Data link/instrument panel – circuit malfunction	**Wiring, instrument panel, ECM**
536	Brake pedal position (BPP) switch – circuit malfunction/no signal change during KOER test	**Incorrect brake pedal operation, wiring, BPP switch, ECM**
538	KOER dynamic response test – insufficient rpm change	**Incorrect throttle operation, ECM**
	KOER cylinder balance test – invalid test	**Throttle operated during test, CMP sensor, ECM**

Flash code	Fault location	Probable cause
539	A/C – ON during KOEO test	Wiring, A/C switch, ECM
542 7	Fuel pump (FP) secondary circuit – engine does not start	IFS switch, supply wire open circuit, ground wire, FP, ECM
	Fuel pump (FP) secondary circuit – engine starts	Supply wire short circuit to positive, high speed FP relay, FP to ECM wire open circuit, left/front HO2S short circuit to positive, variable control relay module, ECM
542 8	Fuel pump (FP) secondary circuit – engine does not start	IFS switch, supply wire open circuit, FP ground wire, FP, ECM
	Fuel pump (FP) secondary circuit – engine starts	Supply wire short circuit to positive, FP relay, FP to ECM wire open circuit, left/front HO2S short circuit to positive, ECM
543 7	Fuel pump (FP) secondary circuit, high/low speed – without DTC 583	Supply wire open circuit, left HO2S short circuit to positive, ECM
	Fuel pump (FP) secondary circuit, high/low speed – with DTC 583	Supply wire short circuit to ground, ECM
543 8	Fuel pump (FP) secondary circuit – engine does not start	Supply wire open circuit, fuel pump relay, ECM
	Fuel pump (FP) secondary circuit – engine starts	Left/front HO2S short circuit to positive, ECM
551	Idle air control (IAC) valve – circuit malfunction during KOEO test	Wiring, IAC valve, ECM
552	Secondary air injection (AIR) bypass solenoid – circuit malfunction during KOEO test	Wiring, AIR bypass solenoid, ECM
553	Secondary air injection (AIR) divert solenoid – circuit malfunction during KOEO test	Wiring, AIR divert solenoid, ECM
554	Fuel pressure control solenoid – circuit malfunction	Hose connection, wiring, fuel pressure control solenoid, ECM
556 7	Fuel pump (FP) primary circuit – engine does not start	Wiring short circuit to positive, variable control relay module, ECM
	Fuel pump (FP) primary circuit – engine starts	Wiring short circuit to ground, variable control relay module, ECM
556 8	Fuel pump (FP) primary circuit	IFS switch, wiring, fuel pump relay, constant control relay module, ECM
557	Fuel pump (FP) primary circuit, low speed	Wiring, FP relay/constant control relay module, ECM
558	Exhaust gas recirculation (EGR) solenoid – circuit malfunction during KOEO test	Hose damaged, wiring, EGR solenoid, ECM
559	A/C relay, primary circuit – circuit malfunction	Wiring, A/C relay, ECM
562	Engine coolant blower motor relay – circuit malfunction	Wiring, engine coolant blower motor relay, ECM
563	Engine coolant blower motor relay, high speed – circuit malfunction	Wiring, engine coolant blower motor relay/ constant control relay module, ECM
564	Engine coolant blower motor relay, low speed – circuit malfunction	Wiring, engine coolant blower motor relay/ constant control relay module, ECM

Flash code	Fault location	Probable cause
565	Evaporative emission (EVAP) canister purge valve 1 – malfunction	**Hose damaged, wiring, EVAP canister purge valve, ECM**
566 6	Shift solenoid (SS) 3-4 – circuit malfunction	**Wiring, SS, ECM**
567	Cruise control vent valve – circuit malfunction	**Wiring, cruise control vent valve, ECM**
568	Cruise control vacuum valve – circuit malfunction	**Wiring, cruise control vacuum valve, ECM**
569	Evaporative emission (EVAP) canister purge valve 2 – malfunction	**Hose damaged, wiring, EVAP canister purge valve, ECM**
571	Exhaust gas recirculation (EGR) atmospheric solenoid – circuit malfunction during KOEO	**Wiring, EGR atmospheric solenoid, ECM**
572	Exhaust gas recirculation (EGR) vacuum solenoid – circuit malfunction during KOEO	**Wiring, EGR vacuum solenoid, ECM**
578	A/C pressure sensor – short circuit	**Supply wire short ground, A/C pressure sensor, variable control relay module, ECM**
579	A/C pressure sensor – insufficient pressure change	**Wiring, A/C fault, A/C pressure sensor, variable control relay module, wiring to A/C compressor clutch open circuit, ECM**
581	Engine coolant blower motor – excessive current draw	**Wiring short circuit to ground, engine coolant blower motor, variable control relay module, ECM**
582	Engine coolant blower motor – open circuit/ short circuit to positive	**Supply wire open circuit/short circuit to positive, ground wire open circuit, engine coolant blower motor, variable control relay module, ECM**
583	Fuel pump – excessive current draw	**Wiring short circuit to ground, fuel pump, ECM**
584	Variable control relay module – ground open circuit	**Wiring open circuit, variable control relay module, ECM**
585	A/C compressor clutch – excessive current draw	**Wiring short circuit to ground, A/C compressor clutch, variable control relay module, ECM**
586	A/C compressor clutch – open circuit/short circuit to positive	**Wiring open circuit/short circuit to positive, A/C compressor clutch, variable control relay module, ECM**
587	Variable control relay module – communication error	**Wiring, variable control relay module, ECM**
593	Heated oxygen sensor (HO2S) – heater circuit malfunction	**Wiring, HO2S heater relay, HO2S, ECM**
617 6	Gear shift, 1-2 – malfunction	**ATF level/condition, poor connection, wiring, TR sensor, SS, clutch slipping, ECM**
618 6	Gear shift, 2-3 – malfunction	**ATF level/condition, poor connection, wiring, TR sensor, SS, clutch slipping, ECM**
619 6	Gear shift, 3-4 – malfunction	**ATF level/condition, poor connection, wiring, TR sensor, SS, clutch slipping, ECM**
621 6	Shift solenoid (SS) 1 – circuit malfunction during KOEO	**Wiring, SS, ECM**

Flash code	Fault location	Probable cause
622 6	Shift solenoid (SS) 2 – circuit malfunction during KOEO	Wiring, SS, ECM
623 6	Overdrive indicator lamp – circuit malfunction	Bulb, wiring, fuse, ECM
624 6	Transmission fluid pressure (TFP) solenoid – circuit malfunction	Wiring, poor connection, TFP solenoid, ECM
625 6	Engine control module (ECM), transmission fluid pressure (TFP) control – open driver circuit	ECM
626 6	Transmission coast clutch solenoid – circuit malfunction during KOEO test	Wiring, transmission coast clutch solenoid, ECM
627 6	Torque converter clutch (TCC) solenoid – circuit malfunction	Wiring, TCC solenoid, ECM
628 6	Torque converter clutch (TCC) – excessive clutch slip	ATF level/condition, TCC slipping, ECM
629 6	Torque converter clutch (TCC) solenoid – circuit malfunction	Wiring, TCC solenoid, ECM
631 6	Overdrive indicator lamp – circuit malfunction during KOEO test	Bulb, wiring, fuse, ECM
632 6	Overdrive selection switch – no signal change during KOEO test	Incorrect switch operation, wiring, fuse, overdrive selection switch, ECM
633 6	Four wheel drive selector switch – closed during KOEO test	Wiring short circuit, four wheel drive selector switch, internal damage to transfer case, ECM
634 6	Transmission range (TR) sensor – voltage high/low	Incorrect adjustment, wiring, TR sensor, ECM
636 6	Transmission fluid temperature (TFT) sensor – signal high/low	ATF level, wiring, TFT sensor, ECM
637 6	Transmission fluid temperature (TFT) sensor – above maximum voltage	ATF level, wiring open circuit, TFT sensor, ECM
638 6	Transmission fluid temperature (TFT) sensor – below minimum voltage	ATF level, wiring short circuit to ground, TFT sensor, ECM
639 6	Turbine shaft speed (TSS) sensor/output shaft speed (OSS) sensor – signal low	Wiring, TSS/OSS sensor, ECM
641 6	Shift solenoid (SS) 3 – circuit malfunction during KOEO	Wiring, SS, ECM
643 6	Torque converter clutch (TCC) solenoid – circuit malfunction during KOEO test	Wiring, TCC solenoid, ECM
645 6	Incorrect gear ratio, 1st gear	ATF level/condition, poor connection, wiring, TR sensor, SS, clutch slipping, ECM
646 6	Incorrect gear ratio, 2nd gear	ATF level/condition, poor connection, wiring, TR sensor, SS, clutch slipping, ECM
647 6	Incorrect gear ratio, 3rd gear	ATF level/condition, poor connection, wiring, TR sensor, SS, clutch slipping, ECM
648 6	Incorrect gear ratio, 4th gear	ATF level/condition, poor connection, wiring, TR sensor, SS, clutch slipping, ECM
649 6	Transmission fluid pressure (TFP) solenoid – signal high/low	Wiring, poor connection, TFP solenoid, ECM

Flash code	Fault location	Probable cause
651 6	Transmission fluid pressure (TFP) solenoid – circuit malfunction	**Wiring, poor connection, TFP solenoid, ECM**
652 6	Torque converter clutch (TCC) solenoid – circuit malfunction	**Wiring, TCC solenoid, ECM**
653 6	Overdrive selection switch – no signal change during KOEO test	**Incorrect switch operation, wiring, fuse, overdrive selection switch, ECM**
654 6	Transmission range (TR) sensor – 'P' signal during KOEO test	**Transmission not in 'P' during KOEO test, incorrect adjustment, ECM**
655 6	Transmission range (TR) sensor – 'N' signal during KOEO test	**Transmission not in 'N' during KOEO test, incorrect adjustment, ECM**
656 6	Torque converter clutch (TCC) – continuous clutch slip	**ATF level/condition, TCC slipping, ECM**
657 6	Automatic transmission fluid (ATF) – temperature high	**ATF level, transmission cooling system, towing weight limit exceeded, clutch slipping, wiring, TFT sensor, ECM**
659 6	High vehicle speed – transmission in 'P'	**TR sensor/PNP switch, ECM**
667 6	Transmission range (TR) sensor – below minimum voltage	**Wiring open circuit/short circuit to ground, TR sensor, ECM**
668 6	Transmission range (TR) sensor – above maximum voltage	**Wiring open circuit/short circuit to positive, TR sensor, ECM**
675 6	Transmission range (TR) sensor – out of range	**Wiring, TR sensor, ECM**
691	Four wheel drive selector fault – incorrect engine RPM/vehicle speed	**Wiring short circuit, four wheel drive selector switch, internal damage to transfer case, ECM**
692 6	Transmission state does not match calculated ratio	**Transmission fault, ECM**
998	Engine control module (ECM) – hard fault present	**ECM**

1 Trouble code may be generated if an efficient exhaust extraction system used during KOER test – vehicle with exhaust gas pressure feedback system
2 Mark VIII, Town Car
3 Continental
4 Mark VII
5 Continental, Mark VIII, Town Car
6 ECM incorporates transmission control function
7 Mark VIII
8 Continental, Mark VII, Town Car

LINCOLN

Model:	Engine identification:	Year:
Continental 4.6L DOHC	**VIN position 8 = V**	**1995-02**
Lincoln LS 3.0L DOHC	**VIN position 8 = S**	**2000-02**
Lincoln LS 3.9L DOHC	**VIN position 8 = A**	**2000-02**
Mark VIII 4.6L DOHC	**VIN position 8 = V**	**1996-98**
Navigator 5.4L SOHC	**VIN position 8 = L/M**	**1998-99**
Navigator 5.4L DOHC	**VIN position 8 = A/R**	**2000-02**
Town Car 4.6L SOHC	**VIN position 8 = W/9**	**1995-02**

System: **Ford EEC-V**

Self-diagnosis

General information

- Refer to the front of this manual for general test conditions, terminology, detailed descriptions of wiring faults and a general trouble shooter for electrical and mechanical faults.
- Malfunction indicator lamp (MIL) will illuminate or flash if certain faults are recorded.
- ECM operates in backup mode if sensors fail, to enable vehicle to be driven to repair shop.

Accessing

- Trouble codes can only be displayed by using a Scan Tool connected to the data link connector (DLC):
  - Continental/Lincoln LS/Mark VIII/Navigator **1**.
  - Town Car **2**.

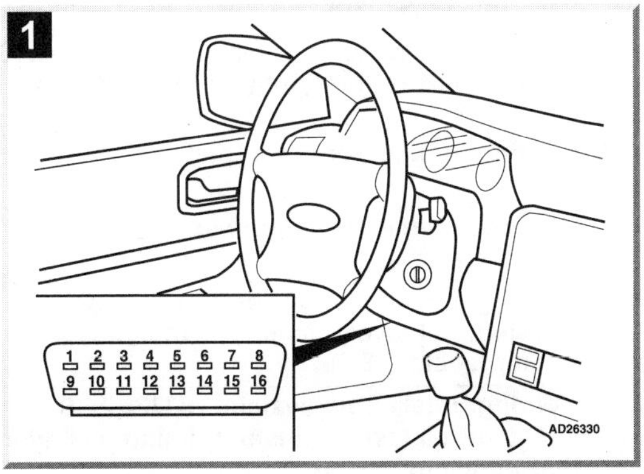

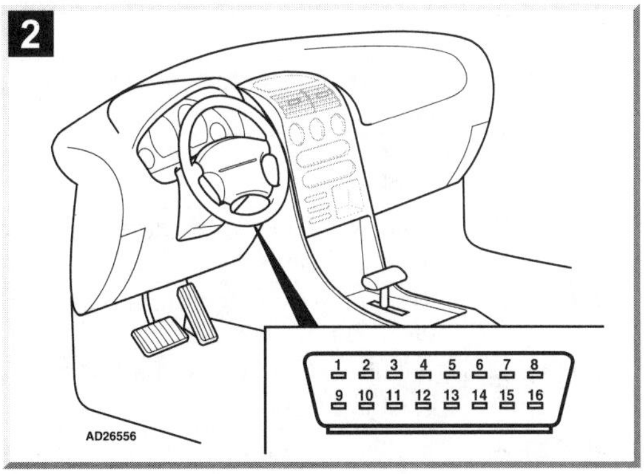

Erasing

- Trouble codes can only be erased by using a Scan Tool connected to the data link connector (DLC).

Trouble code identification

Scan code	Fault location	Probable cause
P0000-P0999	Refer to OBD-II trouble code table at the front of this manual	–
P1000	System readiness test not complete	–
P1001	Self Test	–
P1100	Mass air flow (MAF) sensor – circuit intermittent	**Wiring, MAF sensor**
P1101	Mass air flow (MAF) sensor – signal malfunction	**Wiring, battery, MAF sensor, ECM**
P1109	Intake air temperature (IAT) sensor 2 – circuit intermittent	**Wiring, IAT sensor, ECM**
P1112	Intake air temperature (IAT) – circuit intermittent	**Wiring, IAT sensor, ECM**
P1114	Intake air temperature (IAT) sensor 2 – low input	**Wiring short circuit to ground, IAT sensor, ECM**
P1115	Intake air temperature (IAT) sensor 2 – high input	**Wiring, IAT sensor, ECM**
P1116 1	Cylinder head temperature (CHT) sensor – out of self-test range	**Wiring, cooling system fault, CHT sensor**
P1116 2	Engine coolant temperature (ECT) sensor – signal malfunction	**Wiring, coolant thermostat, cooling system, ECT sensor, ECM**
P1117 1	Cylinder head temperature (CHT) sensor – out of self-test range	**Wiring, cooling system fault, CHT sensor**
P1117 2	Engine coolant temperature (ECT) sensor – circuit intermittent	**Wiring, ECT sensor, cooling system, ECM**
P1120	Throttle position (TP) sensor – signal malfunction, low	**Wiring, TP sensor, ECM**
P1121	Throttle position (TP) sensor – inconsistent with mass air flow (MAF) sensor signal	**Wiring, MAF sensor, TP sensor, air leak**
P1124	Throttle position (TP) sensor – signal malfunction	**Mechanical fault, TP sensor, ECM**
P1125	Throttle position (TP) sensor – circuit intermittent	**Wiring, TP sensor**
P1127	Self test not completed – heated oxygen sensor (HO2S), not tested	**Engine not running long enough prior to self-test**
P1128	Heated oxygen sensor (HO2S) 1, bank 1/ bank 2 – interchanged	**Wiring, connectors swapped**
P1129	Heated oxygen sensor (HO2S) 2, bank 1/ bank 2 – interchanged	**Wiring, connectors swapped**
P1130	Heated oxygen sensor (HO2S) 1, bank 1 – fuel trim (FT), at limit	**Wiring, intake/fuel system, HO2S, EGR system, oil level, camshaft timing, cylinder compression, ECM**
P1131	Heated oxygen sensor (HO2S) 1, bank 1 – not switching, fuel trim (FT) weak mixture	**Wiring, intake/fuel system, HO2S, EGR system, oil level, camshaft timing, cylinder compression, ECM**

Scan code	Fault location	Probable cause
P1132	Heated oxygen sensor (HO2S) 1, bank 1 – not switching, fuel trim (FT) rich mixture	Wiring, intake/fuel system, HO2S, EGR system, oil level, camshaft timing, cylinder compression, ECM
P1137	Heated oxygen sensor (HO2S) 2, bank 1 – not switching, fuel trim (FT) weak mixture	Wiring, exhaust system leaking, HO2S
P1138	Heated oxygen sensor (HO2S) 2, bank 1 – not switching, fuel trim (FT) rich mixture	Wiring, exhaust system leaking, HO2S
P1150	Heated oxygen sensor (HO2S) 1, bank 2 – fuel trim (FT), at limit	Wiring, intake/fuel system, HO2S, EGR system, oil level, camshaft timing, cylinder compression, ECM
P1151	Heated oxygen sensor (HO2S) 1, bank 2 – not switching, fuel trim (FT) weak mixture	Wiring, intake/fuel system, HO2S, EGR system, oil level, camshaft timing, cylinder compression, ECM
P1152	Heated oxygen sensor (HO2S) 1, bank 2 – not switching, fuel trim (FT) rich mixture	Wiring, intake/fuel system, HO2S, EGR system, oil level, camshaft timing, cylinder compression, ECM
P1157	Heated oxygen sensor (HO2S) 2, bank 2 – not switching, fuel trim (FT) weak mixture	Wiring, exhaust system leaking, HO2S
P1158	Heated oxygen sensor (HO2S) 2, bank 2 – not switching, fuel trim (FT) rich mixture	Wiring, exhaust system leaking, HO2S
P1168	Fuel rail pressure sensor – signal low	Wiring, fuel level, low fuel pressure, fuel rail pressure sensor
P1169	Fuel rail pressure sensor – signal high	Wiring, high fuel pressure, fuel rail pressure sensor
P1180	Fuel delivery system – pressure low	Fuel filter blocked, fuel supply blocked
P1181	Fuel delivery system – pressure high	Fuel pressure regulator
P1183	Engine oil temperature sensor – circuit malfunction	Wiring, engine oil temperature sensor, ECM
P1184	Engine oil temperature sensor – out of self-test range	Wiring, engine oil temperature sensor, ECM
P1220	Throttle control system (traction control) – malfunction	Wiring, throttle position motor, mechanical fault, TP sensor, ECM
P1224	Throttle position (TP) sensor (traction control), voltage out of range	Wiring, mechanical fault, TP sensor
P1229	Supercharger (SC) intercooler pump	Wiring, SC intercooler pump, SC intercooler pump relay, ECM
P1230	Fuel pump/variable load control module – circuit intermittent	Wiring
P1231	Fuel pump – circuit malfunction	Wiring, fuel pump relay
P1232	Low speed fuel pump – malfunction	Wiring, low speed fuel pump relay, ECM
P1233	Fuel system – malfunction	Wiring, IFS switch, fuel pump, fuel pump control module, ECM
P1234	Fuel system – malfunction	Wiring, IFS switch, fuel pump, fuel pump control module, ECM

Scan code	Fault location	Probable cause
P1235	Fuel pump control – signal malfunction	**Wiring, fuel pump, fuel pump control module, ECM**
P1236	Fuel pump control – signal malfunction	**Wiring, fuel pump, fuel pump control module, ECM**
P1237	Fuel pump – circuit malfunction	**Wiring, fuel pump, fuel pump relay, fuel pump control module**
P1238	Fuel pump – circuit malfunction	**Wiring, fuel pump, fuel pump relay, fuel pump control module**
P1244	Alternator – low input	**Wiring, alternator, drive belt, ECM**
P1245	Alternator – high input	**Wiring, alternator, ECM**
P1246	Alternator	**Wiring, alternator, drive belt, ECM**
P1260	Theft detected – vehicle immobilized	**–**
P1270	Maximum engine RPM or vehicle speed limit reached	**Engine over revved in neutral, wheel slippage**
P1285	Cylinder head temperature (CHT) sensor – overheat condition	**Wiring, cooling system fault, CHT sensor**
P1288	Cylinder head temperature (CHT) sensor – out of self-test range	**Wiring, cooling system fault, CHT sensor**
P1289	Cylinder head temperature (CHT) sensor – high input	**Wiring, cooling system fault, CHT sensor**
P1290	Cylinder head temperature (CHT) sensor – low input	**Wiring, cooling system fault, CHT sensor**
P1299	Cylinder head temperature (CHT) sensor – over temperature condition	**Wiring, cooling system fault, CHT sensor**
P1309	Misfire monitor disabled	**CMP sensor, CKP sensor, ECT sensor, MAF sensor, ECM**
P1351	Ignition system diagnostic monitor – input circuit malfunction	**Wiring, ECM**
P1352	Ignition coil 1, primary – circuit malfunction	**Wiring, spark plug(s), ignition coil, ICM**
P1353	Ignition coil 2, primary – circuit malfunction	**Wiring, spark plug(s), ignition coil, ICM**
P1354	Ignition coil 3, primary – circuit malfunction	**Wiring, spark plug(s), ignition coil, ICM**
P1355	Ignition coil 4, primary – circuit malfunction	**Wiring, spark plug(s), ignition coil, ICM**
P1358	Ignition system diagnostic monitor – out of self-test range	**Wiring, ignition coil, ECM**
P1359	Ignition system – spark output (SPOUT) signal – circuit malfunction	**Wiring, ECM**
P1364	Ignition coil, primary – circuit malfunction	**Wiring, spark plug(s), ignition coil, ICM**
P1380	Camshaft position (CMP) actuator – circuit malfunction	**Wiring, CMP actuator, ECM**
P1381	Camshaft position (CMP) actuator, bank 1 – timing over advanced	**Wiring, timing belt/chain, CMP actuator, low oil pressure**
P1383	Camshaft position (CMP) actuator, bank 1 – timing over retarded	**Wiring, timing belt/chain, CMP actuator**

Scan code	Fault location	Probable cause
P1390	Octane adjustment	Octane adjustment connector, ECM
P1400	Exhaust gas recirculation (EGR) pressure sensor – circuit malfunction, low input	Wiring, EGR pressure sensor, ECM
P1401	Exhaust gas recirculation (EGR) pressure sensor – circuit malfunction, high input	Wiring, EGR pressure sensor, ECM
P1403	Exhaust gas recirculation (EGR) pressure sensor	Pressure hoses reversed
P1405	Exhaust gas recirculation (EGR) system – upstream hose off or plugged	EGR hoses
P1406	Exhaust gas recirculation (EGR) system – downstream hose off or blocked	EGR hoses
P1407	Exhaust gas recirculation (EGR) solenoid	Wiring, hose leak/blockage, EGR solenoid, EGR valve position sensor, ECM
P1408	Exhaust gas recirculation (EGR) system – out of self-test range	Wiring, EGR hoses, EGR valve, EGR solenoid, ECM
P1409	Exhaust gas recirculation (EGR) solenoid – circuit malfunction	Wiring, EGR solenoid, ECM
P1411	Secondary air injection (AIR) system – malfunction	AIR hoses, AIR pump, AIR solenoid
P1413	Secondary air injection monitor circuit – low input	Wiring, AIR solenoid,AIR pump, ECM
P1414	Secondary air injection monitor circuit – high input	Wiring, AIR solenoid, ECM
P1442	Evaporative emission (EVAP) system – small leak detected	Hose connection(s), intake leak, EVAP canister, EVAP canister purge valve
P1443	Evaporative emission (EVAP) system – malfunction	Wiring, EVAP system hoses, EVAP canister, EVAP canister purge valve
P1444	Evaporative emission (EVAP) system – low input	Wiring, EVAP pressure sensor, ECM
P1445	Evaporative emission (EVAP) system – high input	Wiring, EVAP pressure sensor, ECM
P1450	Evaporative emission (EVAP) system, vent control – circuit malfunction	Wiring, EVAP system hoses, EVAP canister purge valve, fuel filler cap blocked
P1451	Evaporative emission (EVAP) canister purge valve – circuit malfunction	Wiring, EVAP canister purge valve, ECM
P1452	Evaporative emission (EVAP) system – unable to bleed fuel tank vacuum	Wiring, EVAP system hoses, EVAP canister purge valve, fuel filler cap blocked
P1455	Evaporative emission (EVAP) system – malfunction	Fuel filler cap blocked, EVAP system hoses
P1460	Wide open throttle (WOT) relay, A/C – circuit malfunction	Wiring, WOT relay, ECM
P1461	A/C refrigerant pressure sensor – high voltage	Wiring, A/C refrigerant pressure sensor, ECM
P1462	A/C refrigerant pressure sensor – low voltage	Wiring, A/C refrigerant pressure sensor, ECM

Scan code	Fault location	Probable cause
P1463	A/C refrigerant pressure sensor – insufficient pressure change	**Wiring, A/C system – mechanical fault, A/C refrigerant pressure sensor, A/C compressor clutch**
P1464	A/C demand out	**Wiring, A/C ON – during self-test, WOT relay – (A/C cut-out), ECM**
P1469	Frequent A/C compressor clutch cycling	**Wiring, A/C system – mechanical fault**
P1473	Engine coolant blower motor – circuit malfunction	**Wiring open circuit/short circuit to positive, engine coolant blower motor, variable load control module**
P1474	Engine coolant blower motor (low speed) – control circuit malfunction	**Wiring, engine coolant blower motor relay, ECM**
P1477	Engine coolant blower motor (medium speed) – control circuit malfunction	**Wiring, engine coolant blower motor relay, ECM**
P1479	Engine coolant blower motor (high speed) – control circuit malfunction	**Wiring, engine coolant blower motor relay, ECM**
P1480	Engine coolant blower motor (low speed) – with low speed fan ON	**Wiring, engine coolant blower motor relay**
P1481	Engine coolant blower motor (low speed) – with high speed fan ON	**Wiring, engine coolant blower motor relay**
P1483	Engine coolant blower motor – intermittent fault	**Wiring, engine coolant blower motor, variable load control module**
P1484	Variable load control module – malfunction	**Wiring open circuit, variable load control module**
P1500	Vehicle speed sensor (VSS) – circuit intermittent	**Wiring, VSS, ECM**
P1501	Vehicle speed sensor (VSS) – out of self-test range	**Electro-magnetic interference**
P1502	Vehicle speed sensor (VSS) – intermittent signal	**Wiring, VSS/ABS wheel speed sensors**
P1504	Idle air control (IAC) valve – circuit malfunction	**Wiring, IAC valve, intake system leaking, ECM**
P1505	Idle air control (IAC) system – air trim, at limit	**Intake system leaking, air filter blocked, throttle body, IAC valve**
P1506	Idle air control (IAC) valve – over-speed error	**Wiring, IAC valve, vacuum leaks, ECM**
P1507	Idle air control (IAC) valve – under-speed error	**Wiring, IAC valve, hose blocked,ECM**
P1512	Intake manifold runner control (IMRC) module – malfunction, stuck closed –	**Mechanical fault**
P1513	Intake manifold runner control (IMRC) module – malfunction, stuck closed –	**Mechanical fault**
P1516	Intake manifold runner control (IMRC) module – input error (bank 1)	**Wiring, mechanical fault**
P1517	Intake manifold runner control (IMRC) module – input error (bank 2)	**Wiring, mechanical fault**
P1518	Intake manifold runner control (IMRC) module – malfunction, stuck open	**Wiring, IMRC actuator, ECM**

Scan code	Fault location	Probable cause
P1519	Intake manifold runner control (IMRC) module – malfunction, stuck closed	Wiring, IMRC actuator, ECM
P1520	Intake manifold runner control (IMRC) module – circuit malfunction	Mechanical fault, vacuum hoses, IMRC actuator, ECM
P1530	A/C compressor clutch – circuit malfunction	Wiring open circuit/short circuit to positive, A/C compressor clutch, variable load control module
P1537	Intake manifold runner control (IMRC) module – malfunction, stuck closed	Wiring, mechanical fault, ECM
P1538	Intake manifold runner control (IMRC) module – malfunction, stuck closed	Wiring, mechanical fault, ECM
P1539	A/C compressor clutch – circuit malfunction	Wiring, A/C compressor clutch, variable load control module
P1549	Intake manifold air control system – malfunction	Wiring, intake manifold air control solenoid, ECM
P1550	Power steering pressure (PSP) sensor – malfunction	Wiring, PSP sensor, ECM
P1565	Cruise control system	Wiring, CPP switch, BPP switch, cruise control module, cruise control master switch, VSS
P1566	Cruise control system	Wiring, CPP switch, BPP switch, cruise control module, cruise control master switch, VSS
P1567	Cruise control system	Wiring, CPP switch, BPP switch, cruise control module, cruise control master switch, VSS
P1568	Cruise control system	Wiring, CPP switch, BPP switch, cruise control module, cruise control master switch, VSS
P1572	Brake pedal position (BPP) switch – out of self-test range	Wiring, lights control module, BPP switch, ECM
P1605	'Keep alive' memory test failure	Wiring, battery terminal corrosion, loose battery connection, ECM
P1625	Variable load control module, power supply to engine coolant blower motor – circuit malfunction	Wiring, variable load control module
P1626	Variable load control module, power supply to A/C system – circuit malfunction	Wiring, variable load control module
P1633	'Keep alive' power supply – voltage too low	Wiring, ECM
P1635	Tire/axle ratio out of range	Incorrect tire size, incorrect axle ratio, incorrect vehicle ID configuration
P1636	Engine control module (ECM) – internal fault	ECM
P1640	Diagnostic trouble codes stored	P-codes stored in another control module
P1641	Fuel pump – circuit malfunction	Wiring, fuel pump relay, fuel pump, ECM

Scan code	Fault location	Probable cause
P1650	Power steering pressure (PSP) switch – malfunction	Wiring, PSP switch, steering wheel turned during self-test, ECM
P1651	Power steering pressure (PSP) switch – input	Wiring, vehicle towed with engine running, PSP switch, DTC not erased, ECM
P1700	Transmission system – malfunction	Transmission
P1701	Transmission system – reverse engagement error	Transmission
P1702	Transmission range (TR) sensor – circuit intermittent	Wiring, TR sensor
P1703	Brake pedal position (BPP) switch – out of self-test range	Wiring, brake pedal depressed during self-test, rear electronic module/lights control module, BPP switch, ECM
P1704	Transmission range (TR) sensor – malfunction	Wiring, TR sensor
P1705	Park/neutral position (PNP) switch/clutch pedal position (CPP) switch	Wiring, PNP switch, CPP switch, ECM
P1709	Park/neutral position (PNP) switch/clutch pedal position (CPP) switch – out of self-test range	Wiring, PNP switch/CPP switch, ECM
P1710	Transmission fluid temperature (TFT) sensor – mid range failure	Wiring, TFT sensor
P1711	Transmission fluid temperature (TFT) sensor – out of self-test range	Wiring, TFT sensor, fluid level, ECM
P1713	Transmission fluid temperature (TFT) sensor – low range failure	Wiring, TFT sensor
P1714	Shift solenoid (SS) A – inductive malfunction	Wiring, SS, ECM/TCM
P1715	Shift solenoid (SS) B – inductive malfunction	Wiring, SS, ECM/TCM
P1716	Shift solenoid (SS) C – inductive malfunction	Wiring, SS, ECM/TCM
P1717	Shift solenoid (SS) D – inductive malfunction	Wiring, SS, ECM/TCM
P1718	Transmission fluid temperature sensor – high range failure	Wiring, transmission fluid temperature sensor
P1719	Overdrive speed sensor – circuit malfunction	Wiring, overdrive speed sensor, TCM
P1728	Torque converter clutch (TCC) – slip fault	TCC
P1729	4x4L switch – malfunction	Wiring, four wheel drive shift control module, ECM
P1740	Torque converter clutch (TCC) solenoid – circuit malfunction	Wiring, TCC solenoid, ECM/TCM
P1741	Torque converter clutch (TCC) – control error	Wiring, TCC, TCM
P1742	Torque converter clutch (TCC) solenoid	Wiring, TCC solenoid
P1743	Torque converter clutch (TCC) solenoid	Wiring, TCC solenoid
P1744	Torque converter clutch (TCC) – malfunction	Wiring, TCC, TCM
P1745	Transmission fluid pressure (TFP) solenoid – circuit malfunction	Wiring, TFP solenoid, TCM, ECM

Scan code	Fault location	Probable cause
P1746	Transmission fluid pressure (TFP) solenoid – circuit malfunction	Wiring, TFP solenoid, TCM, ECM
P1747	Transmission fluid pressure (TFP) solenoid – circuit malfunction	Wiring short circuit, TFP solenoid, TCM
P1749	Transmission fluid pressure (TFP) solenoid – malfunction	Wiring, TFP solenoid, TCM
P1751	Shift solenoid (SS) A – performance	Wiring, SS
P1754	Coast clutch solenoid – circuit malfunction	Wiring, coast clutch solenoid
P1756	Shift solenoid (SS) B – performance	Wiring, SS
P1760	Transmission fluid pressure (TFP) solenoid – short circuit	Wiring, TFP solenoid, TCM
P1761	Shift solenoid (SS) C – performance	Wiring, SS
P1762	Overdrive band failed/off	Transmission
P1767	Torque converter clutch (TCC) – circuit malfunction	Wiring, TCC, TCM
P1780	Overdrive selection switch	Wiring, transmission control switch, ECM
P1781	4x4L switch – out of self-test range	Wiring, four wheel drive shift control module, 4x4L switch, ECM
P1783	Transmission fluid temperature sensor – overheating condition	Wiring, transmission fluid temperature sensor, TCM, ECM
P1784	Transmission, reverse/first gear – mechanical fault	Transmission
P1785	Transmission, first/second gear – mechanical fault	Transmission
P1786	Gear selection, 3-2 – downshift malfunction	Wiring, TR sensor, shift solenoids, transmission mechanical fault
P1787	Gear selection, 2-1 – downshift malfunction	Wiring, TR sensor, shift solenoids, transmission mechanical fault
P1788	Transmission shift control valve, 3-2 downshift/ coastdown downshift – open circuit	Wiring, transmission shift control valve, TCM
P1789	Transmission shift control valve, 3-2 downshift/ coastdown downshift – short circuit	Wiring, transmission shift control valve, TCM
P1900	Output shaft speed (OSS) sensor – circuit intermittent	Wiring, poor connection, OSS sensor, ECM/ TCM
P1901	Turbine shaft speed (TSS) sensor – circuit intermittent	Wiring, poor connection, TSS sensor, ECM/ TCM
U1020	CAN data bus, A/C – A/C compressor clutch invalid or missing data	Wiring, A/C trouble code(s) stored
U1021	CAN data bus, A/C – communication error	Wiring, A/C trouble code(s) stored, ECM
U1039	CAN data bus, antilock brake system (ABS) – vehicle speed invalid or missing data	Wiring, ABS trouble code(s) stored
U1041	CAN data bus, antilock brake system (ABS) – vehicle speed invalid or missing data	Wiring, ABS trouble code(s) stored

Scan code	Fault location	Probable cause
U1051	CAN data bus, engine control module (ECM) – brake signal invalid or missing data	**Wiring, BPP switch, bulb(s)**
U1073	CAN data bus, cooling system – communication error	**Wiring, variable load control module, ECM (alternative fuel) – if fitted, ECM**
U1089	CAN data bus, active suspension – invalid or missing data	**Wiring, active suspension trouble code(s) stored**
U1098	CAN data bus, antilock brake system (ABS) – vehicle speed invalid or missing data	**Wiring, ABS trouble code(s) stored**
U1130	CAN data bus, fuel pump – communication error	**Wiring, instrument panel, ECM**
U1131	CAN data bus, fuel pump – communication error	**Wiring, instrument panel, ECM**
U1135	CAN data bus, engine control module (ECM) – ignition switch invalid or missing data	**Wiring**
U1147	CAN data bus, alarm	**Wiring, alarm system trouble code(s) stored**
U1243	CAN data bus, A/C – invalid or missing data	**Wiring, A/C system trouble code(s) stored**
U1256	CAN data bus – communication error	**Wiring, trouble code(s) stored in other control modules**
U1262	CAN data bus – communication error	**Wiring, trouble code(s) stored in other control modules**
U1341	CAN data bus, A/C – vehicle speed invalid or missing data	**Wiring, A/C trouble code(s) stored**
U1451	CAN data bus, alarm	**Wiring, alarm system trouble code(s) stored**
U2195	CAN data bus, lights control module – invalid or missing data	**Wiring, lights control module trouble code(s) stored**
U2243	CAN data bus, A/C	**Wiring, A/C system trouble code(s) stored**

1 Navigator/Town Car
2 Except Navigator/Town Car

Model:	Engine identification:	Year:
Capri 1.6L DOHC	**VIN position 8 = Z**	**1992-94**
Capri XR2 1.6L DOHC Turbo	**VIN position 8 = 6**	**1992-94**

System: **MECS**

Self-diagnosis

General information

- Refer to the front of this manual for general test conditions, terminology, detailed descriptions of wiring faults and a general trouble shooter for electrical and mechanical faults.
- Malfunction indicator lamp (MIL) will illuminate continuously if permanent faults are recorded.
- Malfunction indicator lamp (MIL) will illuminate briefly and then switch off if intermittent faults are recorded.
- ECM operates in backup mode if sensors fail, to enable vehicle to be driven to repair shop.
- Trouble codes can be displayed by the MIL or by using an analogue voltmeter or Scan Tool connected to the data link connector (DLC).

Accessing

Using malfunction indicator lamp (MIL)

- Ensure ignition switched OFF.
- Jump data link connector (DLC) 2 terminal STI to ground **1**.
- Switch ignition ON.
- Count MIL flashes. Note trouble codes. Compare with trouble code table.
  - 1.2 second flashes indicate the 'tens' of the trouble code **2** [**A**].
  - 0.4 second flashes indicate the 'units' of the trouble code **2** [**C**].
  - A 1.6 second pause (MIL OFF) separates 'tens' and 'units' **2** [**B**].
  - A 0.4 second pause (MIL OFF) separates each flash **2** [**D**].
  - A 4 second pause separates each trouble code **2** [**E**].
  - For example: Trouble code 12 displayed **2**. Throttle position (TP) sensor.
- Switch ignition OFF.

- Trouble codes can also be displayed by using a Scan Tool connected to the data link connectors (DLC) 1 and 2 **1**.

1

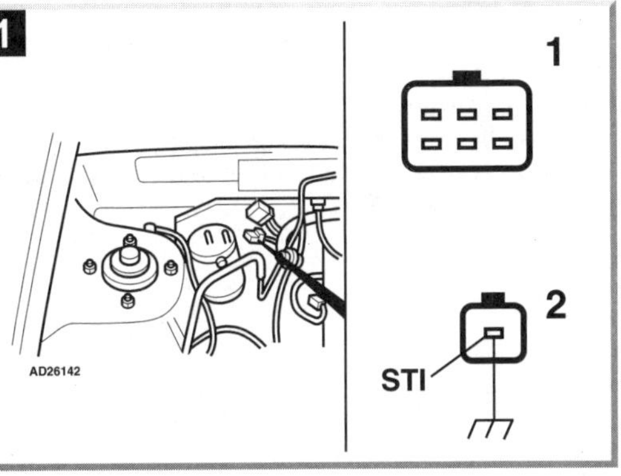

2

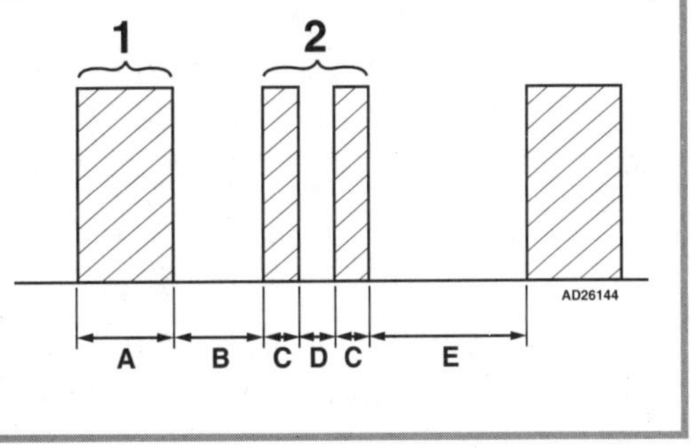

Using analogue voltmeter

- Ensure ignition switched OFF.
- Connect an analogue voltmeter between data link connector (DLC) 1 terminal STO and ground **3**.
- Set analogue voltmeter to a DC voltage range of 0-20 V.

- Jump data link connector (DLC) 2 terminal STI to ground **3**.
- Switch ignition ON.
- Count voltmeter needle deflections. Note trouble codes. Compare with trouble code table.
  - 1.2 second needle deflections indicate the 'tens' of the trouble code **2** [**A**].
  - 0.4 second needle deflections indicate the 'units' of the trouble code **2** [**C**].
  - A 1.6 second pause separates 'tens' and 'units' **2** [**B**].
  - A 0.4 second pause separates each needle deflection **2** [**D**].
  - A 4 second pause separates each trouble code **2** [**E**].
  - For example: Trouble code 12 displayed **2**. Throttle position (TP) sensor.
- Switch ignition OFF.
- Trouble codes can also be displayed by using a Scan Tool connected to the data link connectors (DLC) 1 and 2 **3**.

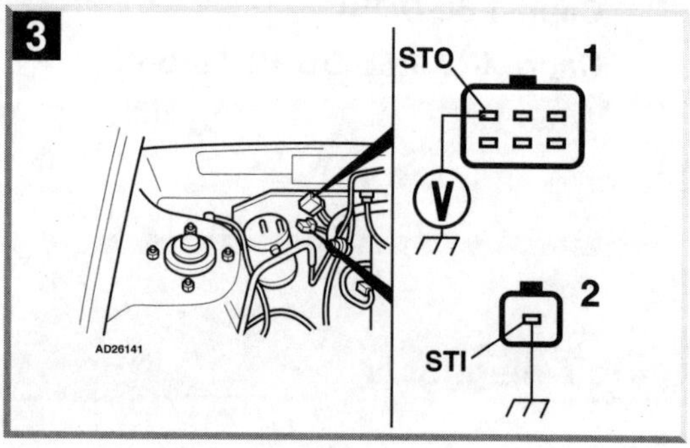

Erasing

- Ensure ignition switched OFF.
- Disconnect battery ground cable.
- Depress brake pedal for 5-10 seconds.
- Reconnect battery ground cable.
- Repeat accessing procedure to ensure no trouble codes remain in ECM fault memory.

NOTE: ***Disconnecting battery ground lead may also erase memory from electronic units, such as radio.***

Trouble code identification

Flash code	Fault location	Probable cause
01	Ignition signal	**Wiring, ignition coil, ECM**
03	Camshaft position (CMP) sensor	**Wiring, CMP sensor, ECM**
08	Volume air flow (VAF) sensor	**Wiring, VAF sensor, ECM**
09	Engine coolant temperature (ECT) sensor	**Wiring, ECT sensor, ECM**
10	Intake air temperature (IAT) sensor	**Wiring, IAT sensor, ECM**
12	Throttle position (TP) sensor	**Wiring, TP sensor, ECM**
14	Barometric pressure (BARO) sensor	**Wiring, BARO sensor, ECM**
15	Oxygen sensor (O2S) – voltage below 0.55 V	**Fuel pressure low, wiring, O2S, ECM**
17	Oxygen sensor (O2S) – no voltage change	**Wiring, O2S, ECM**
25	Fuel pressure regulator control solenoid	**Wiring, fuel pressure regulator control solenoid, ECM**
26	Evaporative emission (EVAP) canister purge valve	**Wiring, EVAP canister purge valve, ECM**
34	Idle air control (IAC) valve	**Wiring, IAC valve, ECM**

MERCURY

Model:	Engine identification:	Year:
Cougar 2.0/2.5L	**VIN position 8 = L/3**	**1998-01**
Cougar/XR7 3.8L	**VIN position 8 = 4**	**1996-97**
Cougar/XR7 4.6L	**VIN position 8 = W**	**1994-97**
Grand Marquis 4.6L SOHC	**VIN position 8 = W/9**	**1995-99**
Mountaineer 4.0/5.0L	**VIN position 8 = E/K/P/X**	**1997-02**
Mystique 2.0/2.5L	**VIN position 8 = L/3**	**1996-00**
Sable 3.0L/24V	**VIN position 8 = N/S/U/1/2**	**1996-02**
Tracer 1.8L DOHC	**VIN position 8 = 8**	**1996**
Tracer 1.9/2.0L	**VIN position 8 = J/P**	**1996-99**

System: **Ford EEC-V**

Self-diagnosis

General information

- Refer to the front of this manual for general test conditions, terminology, detailed descriptions of wiring faults and a general trouble shooter for electrical and mechanical faults.
- Malfunction indicator lamp (MIL) will illuminate or flash if certain faults are recorded.
- ECM operates in backup mode if sensors fail, to enable vehicle to be driven to repair shop.

Accessing

- Trouble codes can only be displayed by using a Scan Tool connected to the data link connector (DLC):
  - Cougar **1**.
  - Sable/Mountaineer **2**.
  - Grand Marquis **3**.
  - Mystique/Tracer **4**.

1

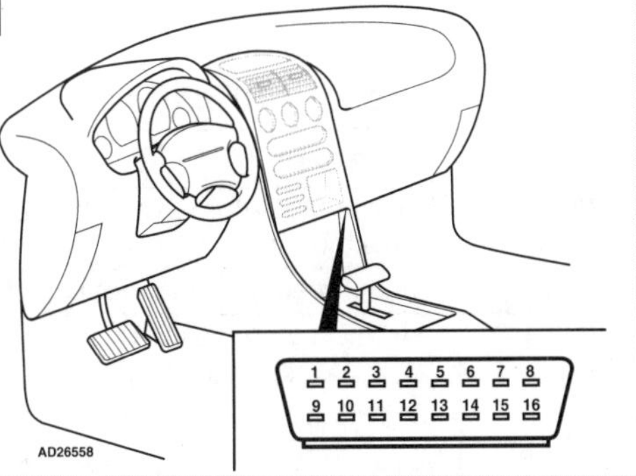

2

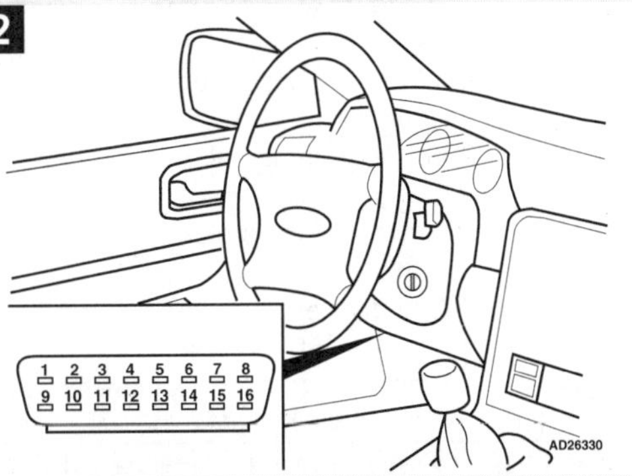

3

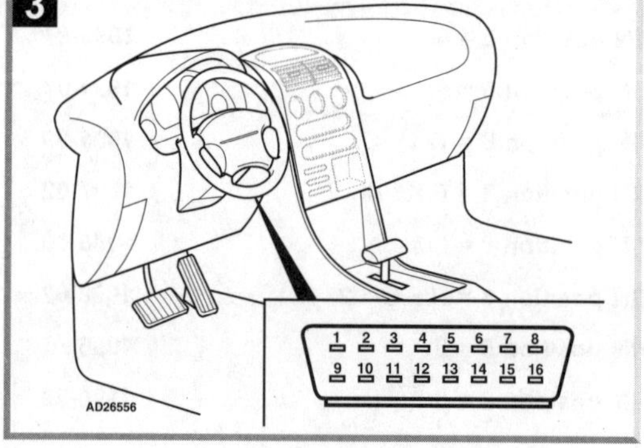

4

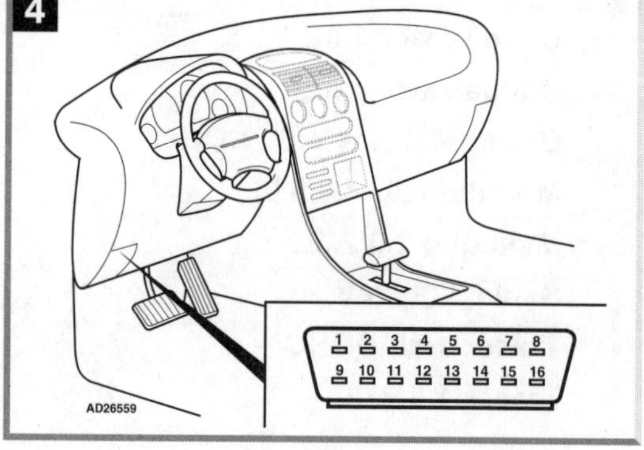

Erasing

- Trouble codes can only be erased by using a Scan Tool connected to the data link connector (DLC).

Trouble code identification

Scan code	Fault location	Probable cause
P0000-P0999	Refer to OBD-II trouble code table at the front of this manual	–
P1000	System readiness test not complete	–
P1001	Self Test	–
P1100	Mass air flow (MAF) sensor – circuit intermittent	**Wiring, MAF sensor**
P1101	Mass air flow (MAF) sensor – signal malfunction	**Wiring, battery, MAF sensor, ECM**
P1109	Intake air temperature (IAT) sensor 2 – circuit intermittent	**Wiring, IAT sensor, ECM**
P1112	Intake air temperature (IAT) – circuit intermittent	**Wiring, IAT sensor, ECM**
P1114	Intake air temperature (IAT) sensor 2 – low input	**Wiring short circuit to ground, IAT sensor, ECM**
P1115	Intake air temperature (IAT) sensor 2 – high input	**Wiring, IAT sensor, ECM**
P1116 1	Cylinder head temperature (CHT) sensor – out of self-test range	**Wiring, cooling system fault, CHT sensor**
P1116 2	Engine coolant temperature (ECT) sensor – signal malfunction	**Wiring, coolant thermostat, cooling system, ECT sensor, ECM**
P1117 1	Cylinder head temperature (CHT) sensor – out of self-test range	**Wiring, cooling system fault, CHT sensor**
P1117 2	Engine coolant temperature (ECT) sensor – circuit intermittent	**Wiring, ECT sensor, cooling system, ECM**

Scan code	Fault location	Probable cause
P1120	Throttle position (TP) sensor – signal malfunction, low	Wiring, TP sensor, ECM
P1121	Throttle position (TP) sensor – inconsistent with mass air flow (MAF) sensor signal	Wiring, MAF sensor, TP sensor, air leak
P1124	Throttle position (TP) sensor – signal malfunction	Mechanical fault, TP sensor, ECM
P1125	Throttle position (TP) sensor – circuit intermittent	Wiring, TP sensor
P1127	Self test not completed – heated oxygen sensor (HO2S), not tested	Engine not running long enough prior to self-test
P1128	Heated oxygen sensor (HO2S) 1, bank 1/bank 2 – interchanged	Wiring, connectors swapped
P1129	Heated oxygen sensor (HO2S) 2, bank 1/bank 2 – interchanged	Wiring, connectors swapped
P1130	Heated oxygen sensor (HO2S) 1, bank 1 – fuel trim (FT), at limit	Wiring, intake/fuel system, HO2S, EGR system, oil level, camshaft timing, cylinder compression, ECM
P1131	Heated oxygen sensor (HO2S) 1, bank 1 – not switching, fuel trim (FT) weak mixture	Wiring, intake/fuel system, HO2S, EGR system, oil level, camshaft timing, cylinder compression, ECM
P1132	Heated oxygen sensor (HO2S) 1, bank 1 – not switching, fuel trim (FT) rich mixture	Wiring, intake/fuel system, HO2S, EGR system, oil level, camshaft timing, cylinder compression, ECM
P1137	Heated oxygen sensor (HO2S) 2, bank 1 – not switching, fuel trim (FT) weak mixture	Wiring, exhaust system leaking, HO2S
P1138	Heated oxygen sensor (HO2S) 2, bank 1 – not switching, fuel trim (FT) rich mixture	Wiring, exhaust system leaking, HO2S
P1150	Heated oxygen sensor (HO2S) 1, bank 2 – fuel trim (FT), at limit	Wiring, intake/fuel system, HO2S, EGR system, oil level, camshaft timing, cylinder compression, ECM
P1151	Heated oxygen sensor (HO2S) 1, bank 2 – not switching, fuel trim (FT) weak mixture	Wiring, intake/fuel system, HO2S, EGR system, oil level, camshaft timing, cylinder compression, ECM
P1152	Heated oxygen sensor (HO2S) 1, bank 2 – not switching, fuel trim (FT) rich mixture	Wiring, intake/fuel system, HO2S, EGR system, oil level, camshaft timing, cylinder compression, ECM
P1157	Heated oxygen sensor (HO2S) 2, bank 2 – not switching, fuel trim (FT) weak mixture	Wiring, exhaust system leaking, HO2S
P1158	Heated oxygen sensor (HO2S) 2, bank 2 – not switching, fuel trim (FT) rich mixture	Wiring, exhaust system leaking, HO2S
P1168	Fuel rail pressure sensor – signal low	Wiring, fuel level, low fuel pressure, fuel rail pressure sensor
P1169	Fuel rail pressure sensor – signal high	Wiring, high fuel pressure, fuel rail pressure sensor
P1180	Fuel delivery system – pressure low	Fuel filter blocked, fuel supply blocked
P1181	Fuel delivery system – pressure high	Fuel pressure regulator

MERCURY

Cougar 2.0/2.5L • Cougar/XR7 3.8L • Cougar/XR7 4.6L
Grand Marquis 4.6L SOHC • Mountaineer 4.0/5.0L • Mystique 2.0/2.5L
Sable 3.0L/24V • Tracer 1.8L DOHC • Tracer 1.9/2.0L

Scan code	Fault location	Probable cause
P1183	Engine oil temperature sensor – circuit malfunction	**Wiring, engine oil temperature sensor, ECM**
P1184	Engine oil temperature sensor – out of self-test range	**Wiring, engine oil temperature sensor, ECM**
P1220	Throttle control system (traction control) – malfunction	**Wiring, throttle position motor, mechanical fault, TP sensor, ECM**
P1224	Throttle position (TP) sensor (traction control), voltage out of range	**Wiring, mechanical fault, TP sensor**
P1229	Supercharger (SC) intercooler pump	**Wiring, SC intercooler pump, SC intercooler pump relay, ECM**
P1230	Fuel pump/variable load control module – circuit intermittent	**Wiring**
P1231	Fuel pump – circuit malfunction	**Wiring, fuel pump relay**
P1232	Low speed fuel pump – malfunction	**Wiring, low speed fuel pump relay, ECM**
P1233	Fuel system – malfunction	**Wiring, IFS switch, fuel pump, fuel pump control module, ECM**
P1234	Fuel system – malfunction	**Wiring, IFS switch, fuel pump, fuel pump control module, ECM**
P1235	Fuel pump control – signal malfunction	**Wiring, fuel pump, fuel pump control module, ECM**
P1236	Fuel pump control – signal malfunction	**Wiring, fuel pump, fuel pump control module, ECM**
P1237	Fuel pump – circuit malfunction	**Wiring, fuel pump, fuel pump relay, fuel pump control module**
P1238	Fuel pump – circuit malfunction	**Wiring, fuel pump, fuel pump relay, fuel pump control module**
P1244	Alternator – low input	**Wiring, alternator, drive belt, ECM**
P1245	Alternator – high input	**Wiring, alternator, ECM**
P1246	Alternator	**Wiring, alternator, drive belt, ECM**
P1260	Theft detected – vehicle immobilized	**–**
P1270	Maximum engine RPM or vehicle speed limit reached	**Engine over revved in neutral, wheel slippage**
P1285	Cylinder head temperature (CHT) sensor – overheat condition	**Wiring, cooling system fault, CHT sensor**
P1288	Cylinder head temperature (CHT) sensor – out of self-test range	**Wiring, cooling system fault, CHT sensor**
P1289	Cylinder head temperature (CHT) sensor – high input	**Wiring, cooling system fault, CHT sensor**
P1290	Cylinder head temperature (CHT) sensor – low input	**Wiring, cooling system fault, CHT sensor**
P1299	Cylinder head temperature (CHT) sensor – over temperature condition	**Wiring, cooling system fault, CHT sensor**
P1309	Misfire monitor disabled	**CMP sensor, CKP sensor, ECT sensor, MAF sensor, ECM**
P1351	Ignition system diagnostic monitor – input circuit malfunction	**Wiring, ECM**

Scan code	Fault location	Probable cause
P1352	Ignition coil 1, primary – circuit malfunction	Wiring, spark plug(s), ignition coil, ICM
P1353	Ignition coil 2, primary – circuit malfunction	Wiring, spark plug(s), ignition coil, ICM
P1354	Ignition coil 3, primary – circuit malfunction	Wiring, spark plug(s), ignition coil, ICM
P1355	Ignition coil 4, primary – circuit malfunction	Wiring, spark plug(s), ignition coil, ICM
P1358	Ignition system diagnostic monitor – out of self-test range	Wiring, ignition coil, ECM
P1359	Ignition system – spark output (SPOUT) signal – circuit malfunction	Wiring, ECM
P1364	Ignition coil, primary – circuit malfunction	Wiring, spark plug(s), ignition coil, ICM
P1380	Camshaft position (CMP) actuator – circuit malfunction	Wiring, CMP actuator, ECM
P1381	Camshaft position (CMP) actuator, bank 1 – timing over advanced	Wiring, timing belt/chain, CMP actuator, low oil pressure
P1383	Camshaft position (CMP) actuator, bank 1 – timing over retarded	Wiring, timing belt/chain, CMP actuator
P1390	Octane adjustment	Octane adjustment connector, ECM
P1400 3	Exhaust gas recirculation (EGR) valve position sensor – low input	Wiring, EGR solenoid, EGR valve position sensor, ECM
P1400 4	Exhaust gas recirculation (EGR) pressure sensor – circuit malfunction, low input	Wiring, EGR pressure sensor, ECM
P1401 3	Exhaust gas recirculation (EGR) valve position sensor – high input	Wiring, EGR solenoid, EGR valve position sensor, ECM
P1401 4	Exhaust gas recirculation (EGR) pressure sensor – circuit malfunction, high input	Wiring, EGR pressure sensor, ECM
P1403	Exhaust gas recirculation (EGR) pressure sensor	Pressure hoses reversed
P1405	Exhaust gas recirculation (EGR) system – upstream hose off or plugged	EGR hoses
P1406	Exhaust gas recirculation (EGR) system – downstream hose off or blocked	EGR hoses
P1407	Exhaust gas recirculation (EGR) solenoid	Wiring, hose leak/blockage, EGR solenoid, EGR valve position sensor, ECM
P1408	Exhaust gas recirculation (EGR) system – out of self-test range	Wiring, EGR hoses, EGR valve, EGR solenoid, ECM
P1409	Exhaust gas recirculation (EGR) solenoid – circuit malfunction	Wiring, EGR solenoid, ECM
P1411	Secondary air injection (AIR) system – malfunction	AIR hoses, AIR pump, AIR solenoid
P1413	Secondary air injection monitor circuit – low input	Wiring, AIR solenoid, AIR pump, ECM
P1414	Secondary air injection monitor circuit – high input	Wiring, AIR solenoid, ECM
P1442	Evaporative emission (EVAP) system – small leak detected	Hose connection(s), intake leak, EVAP canister, EVAP canister purge valve
P1443	Evaporative emission (EVAP) system – malfunction	Wiring, EVAP system hoses, EVAP canister, EVAP canister purge valve

Scan code	Fault location	Probable cause
P1444	Evaporative emission (EVAP) system – low input	**Wiring, EVAP pressure sensor, ECM**
P1445	Evaporative emission (EVAP) system – high input	**Wiring, EVAP pressure sensor, ECM**
P1450	Evaporative emission (EVAP) system, vent control – circuit malfunction	**Wiring, EVAP system hoses, EVAP canister purge valve, fuel filler cap blocked**
P1451	Evaporative emission (EVAP) canister purge valve – circuit malfunction	**Wiring, EVAP canister purge valve, ECM**
P1452	Evaporative emission (EVAP) system – unable to bleed fuel tank vacuum	**Wiring, EVAP system hoses, EVAP canister purge valve, fuel filler cap blocked**
P1455	Evaporative emission (EVAP) system – malfunction	**Fuel filler cap blocked, EVAP system hoses**
P1460	Wide open throttle (WOT) relay, A/C – circuit malfunction	**Wiring, WOT relay, ECM**
P1461	A/C refrigerant pressure sensor – high voltage	**Wiring, A/C refrigerant pressure sensor, ECM**
P1462	A/C refrigerant pressure sensor – low voltage	**Wiring, A/C refrigerant pressure sensor, ECM**
P1463	A/C refrigerant pressure sensor – insufficient pressure change	**Wiring, A/C system – mechanical fault, A/C refrigerant pressure sensor, A/C compressor clutch**
P1464	A/C demand out	**Wiring, A/C ON – during self-test, WOT relay – (AC cut-out), ECM**
P1469	Frequent A/C compressor clutch cycling	**Wiring, A/C system – mechanical fault**
P1473	Engine coolant blower motor – circuit malfunction	**Wiring open circuit/short circuit to positive, engine coolant blower motor, variable load control module**
P1474	Engine coolant blower motor (low speed) – control circuit malfunction	**Wiring, engine coolant blower motor relay, ECM**
P1477	Engine coolant blower motor (medium speed) – control circuit malfunction	**Wiring, engine coolant blower motor relay, ECM**
P1479	Engine coolant blower motor (high speed) – control circuit malfunction	**Wiring, engine coolant blower motor relay, ECM**
P1480	Engine coolant blower motor (low speed) – with low speed fan ON	**Wiring, engine coolant blower motor relay**
P1481	Engine coolant blower motor (low speed) – with high speed fan ON	**Wiring, engine coolant blower motor relay**
P1483	Engine coolant blower motor – intermittent fault	**Wiring, engine coolant blower motor, variable load control module**
P1484	Variable load control module – malfunction	**Wiring open circuit, variable load control module**
P1500	Vehicle speed sensor (VSS) – circuit intermittent	**Wiring, VSS, ECM**
P1501	Vehicle speed sensor (VSS) – out of self-test range	**Electro-magnetic interference**
P1502	Vehicle speed sensor (VSS) – intermittent signal	**Wiring, VSS/ABS wheel speed sensors**

Scan code	Fault location	Probable cause
P1504	Idle air control (IAC) valve – circuit malfunction	Wiring, IAC valve, intake system leaking, ECM
P1505	Idle air control (IAC) system – air trim, at limit	Intake system leaking, air filter blocked, throttle body, IAC valve
P1506	Idle air control (IAC) valve – over-speed error	Wiring, IAC valve, vacuum leaks, ECM
P1507	Idle air control (IAC) valve – under-speed error	Wiring, IAC valve, hose blocked, ECM
P1512	Intake manifold runner control (IMRC) module – malfunction, stuck closed –	Mechanical fault
P1513	Intake manifold runner control (IMRC) module – malfunction, stuck closed –	Mechanical fault
P1516	Intake manifold runner control (IMRC) module – input error (bank 1)	Wiring, mechanical fault
P1517	Intake manifold runner control (IMRC) module – input error (bank 2)	Wiring, mechanical fault
P1518	Intake manifold runner control (IMRC) module – malfunction, stuck open	Wiring, IMRC actuator, ECM
P1519	Intake manifold runner control (IMRC) module – malfunction, stuck closed	Wiring, IMRC actuator, ECM
P1520	Intake manifold runner control (IMRC) module – circuit malfunction	Mechanical fault, vacuum hoses, IMRC actuator, ECM
P1530	A/C compressor clutch – circuit malfunction	Wiring open circuit/short circuit to positive, A/C compressor clutch, variable load control module
P1537	Intake manifold runner control (IMRC) module – malfunction, stuck closed	Wiring, mechanical fault, ECM
P1538	Intake manifold runner control (IMRC) module – malfunction, stuck closed	Wiring, mechanical fault, ECM
P1539	A/C compressor clutch – circuit malfunction	Wiring, A/C compressor clutch, variable load control module
P1549	Intake manifold air control system – malfunction	Wiring, intake manifold air control solenoid, ECM
P1550	Power steering pressure (PSP) sensor – malfunction	Wiring, PSP sensor, ECM
P1565	Cruise control system	Wiring, CPP switch, BPP switch, cruise control module, cruise control master switch, VSS
P1566	Cruise control system	Wiring, CPP switch, BPP switch, cruise control module, cruise control master switch, VSS
P1567	Cruise control system	Wiring, CPP switch, BPP switch, cruise control module, cruise control master switch, VSS
P1568	Cruise control system	Wiring, CPP switch, BPP switch, cruise control module, cruise control master switch, VSS
P1572 5	Brake pedal position (BPP) switch – out of self-test range	Wiring, lights control module, BPP switch, ECM

Scan code	Fault location	Probable cause
P1605	'Keep alive' memory test failure	**Wiring, battery terminal corrosion, loose battery connection, ECM**
P1625	Variable load control module, power supply to engine coolant blower motor – circuit malfunction	**Wiring, variable load control module**
P1626	Variable load control module, power supply to A/C system – circuit malfunction	**Wiring, variable load control module**
P1633	'Keep alive' power supply – voltage too low	**Wiring, ECM**
P1635	Tire/axle ratio out of range	**Incorrect tire size, incorrect axle ratio, incorrect vehicle ID configuration**
P1636	Engine control module (ECM) – internal fault	**ECM**
P1640	Diagnostic trouble codes stored	**P-codes stored in another control module**
P1641	Fuel pump – circuit malfunction	**Wiring, fuel pump relay, fuel pump, ECM**
P1650	Power steering pressure (PSP) switch – malfunction	**Wiring, PSP switch, steering wheel turned during self-test, ECM**
P1651	Power steering pressure (PSP) switch – input	**Wiring, vehicle towed with engine running, PSP switch, DTC not erased, ECM**
P1700	Transmission system – malfunction	**Transmission**
P1701	Transmission system – reverse engagement error	**Transmission**
P1702	Transmission range (TR) sensor – circuit intermittent	**Wiring, TR sensor**
P1703	Brake pedal position (BPP) switch – out of self-test range	**Wiring, brake pedal depressed during self-test, rear electronic module/lights control module, BPP switch, ECM**
P1704	Transmission range (TR) sensor – malfunction	**Wiring, TR sensor**
P1705	Park/neutral position (PNP) switch/clutch pedal position (CPP) switch	**Wiring, PNP switch, CPP switch, ECM**
P1709	Park/neutral position (PNP) switch/clutch pedal position (CPP) switch – out of self-test range	**Wiring, PNP switch/CPP switch, ECM**
P1710	Transmission fluid temperature (TFT) sensor – mid range failure	**Wiring, TFT sensor**
P1711	Transmission fluid temperature (TFT) sensor – out of self-test range	**Wiring, TFT sensor, fluid level, ECM**
P1713	Transmission fluid temperature (TFT) sensor – low range failure	**Wiring, TFT sensor**
P1714	Shift solenoid (SS) A – inductive malfunction	**Wiring, SS, ECM/TCM**
P1715	Shift solenoid (SS) B – inductive malfunction	**Wiring, SS, ECM/TCM**
P1716	Shift solenoid (SS) C – inductive malfunction	**Wiring, SS, ECM/TCM**
P1717	Shift solenoid (SS) D – inductive malfunction	**Wiring, SS, ECM/TCM**
P1718	Transmission fluid temperature (TFT) sensor – high range failure	**Wiring, TFT sensor**
P1719	Overdrive speed sensor – circuit malfunction	**Wiring, overdrive speed sensor, TCM**
P1728	Torque converter clutch (TCC) – slip fault	**TCC**

Scan code	Fault location	Probable cause
P1729	4x4L switch – malfunction	Wiring, four wheel drive shift control module, ECM
P1740	Torque converter clutch (TCC) solenoid – circuit malfunction	Wiring, TCC solenoid, ECM/TCM
P1741	Torque converter clutch (TCC) – control error	Wiring, TCC, TCM
P1742	Torque converter clutch (TCC) solenoid	Wiring, TCC solenoid
P1743	Torque converter clutch (TCC) solenoid	Wiring, TCC solenoid
P1744	Torque converter clutch (TCC) – malfunction	Wiring, TCC, TCM
P1745	Transmission fluid pressure (TFP) solenoid – circuit malfunction	Wiring, TFP solenoid, TCM, ECM
P1746	Transmission fluid pressure (TFP) solenoid – circuit malfunction	Wiring, TFP solenoid, TCM, ECM
P1747	Transmission fluid pressure (TFP) solenoid – circuit malfunction	Wiring short circuit, TFP solenoid, TCM
P1749	Transmission fluid pressure (TFP) solenoid – malfunction	Wiring, TFP solenoid, TCM
P1751	Shift solenoid (SS) A – performance	Wiring, SS
P1754	Coast clutch solenoid – circuit malfunction	Wiring, coast clutch solenoid
P1756	Shift solenoid (SS) B – performance	Wiring, SS
P1760	Transmission fluid pressure (TFP) solenoid – short circuit	Wiring, TFP solenoid, TCM
P1761	Shift solenoid (SS) C – performance	Wiring, SS
P1762	Overdrive band failed/off	Transmission
P1767	Torque converter clutch (TCC) – circuit malfunction	Wiring, TCC, TCM
P1780	Overdrive selection switch	Wiring, transmission control switch, ECM
P1781	4x4L switch – out of self-test range	Wiring, four wheel drive shift control module, 4x4L switch, ECM
P1783	Transmission fluid temperature (TFT) sensor – overheating condition	Wiring, TFT sensor, TCM, ECM
P1784	Transmission, reverse/first gear – mechanical fault	Transmission
P1785	Transmission, first/second gear – mechanical fault	Transmission
P1786	Gear selection, 3-2 – downshift malfunction	Wiring, TR sensor, shift solenoids, transmission mechanical fault
P1787	Gear selection, 2-1 – downshift malfunction	Wiring, TR sensor, shift solenoids, transmission mechanical fault
P1788	Transmission shift control valve, 3-2 downshift/ coastdown downshift – open circuit	Wiring, transmission shift control valve, TCM
P1789	Transmission shift control valve, 3-2 downshift/ coastdown downshift – short circuit	Wiring, transmission shift control valve, TCM
P1900	Output shaft speed (OSS) sensor – circuit intermittent	Wiring, poor connection, OSS sensor, ECM/ TCM

Scan code	Fault location	Probable cause
P1901	Turbine shaft speed (TSS) sensor – circuit intermittent	**Wiring, poor connection, TSS sensor, ECM/ TCM**
U1020	CAN data bus, A/C – A/C compressor clutch invalid or missing data	**Wiring, A/C trouble code(s) stored**
U1021	CAN data bus, A/C – communication error	**Wiring, A/C trouble code(s) stored, ECM**
U1039	CAN data bus, antilock brake system (ABS) – vehicle speed invalid or missing data	**Wiring, ABS trouble code(s) stored**
U1041	CAN data bus, antilock brake system (ABS) – vehicle speed invalid or missing data	**Wiring, ABS trouble code(s) stored**
U1051	CAN data bus, engine control module (ECM) – brake signal invalid or missing data	**Wiring, BPP switch, bulb(s)**
U1073	CAN data bus, cooling system – communication error	**Wiring, variable load control module, ECM (alternative fuel) – if fitted, ECM**
U1089	CAN data bus, active suspension – invalid or missing data	**Wiring, active suspension trouble code(s) stored**
U1098	CAN data bus, antilock brake system (ABS) – vehicle speed invalid or missing data	**Wiring, ABS trouble code(s) stored**
U1130	CAN data bus, fuel pump – communication error	**Wiring, instrument panel, ECM**
U1131	CAN data bus, fuel pump – communication error	**Wiring, instrument panel, ECM**
U1135	CAN data bus, engine control module (ECM) – ignition switch invalid or missing data	**Wiring**
U1147	CAN data bus, alarm	**Wiring, alarm system trouble code(s) stored**
U1243	CAN data bus, A/C – invalid or missing data	**Wiring, A/C system trouble code(s) stored**
U1256	CAN data bus – communication error	**Wiring, trouble code(s) stored in other control modules**
U1262	CAN data bus – communication error	**Wiring, trouble code(s) stored in other control modules**
U1341	CAN data bus, A/C – vehicle speed invalid or missing data	**Wiring, A/C trouble code(s) stored**
U1451	CAN data bus, alarm	**Wiring, alarm system trouble code(s) stored**
U2195	CAN data bus, lights control module – invalid or missing data	**Wiring, lights control module trouble code(s) stored**
U2243	CAN data bus, A/C	**Wiring, A/C system trouble code(s) stored**

1 Grand Marquis
2 Except Grand Marquis
3 Mountaineer
4 Except Mountaineer
5 Mountaineer with 150 pin ECM

MERCURY

Model:	Engine identification:	Year:
Cougar/XR7 3.8L	**VIN position 8 = 4**	**1992-95**
Cougar/XR7 5.0L HO	**VIN position 8 = T**	**1992-93**
Grand Marquis 4.6L	**VIN position 8 = W**	**1992-94**
Mystique 2.0L	**VIN position 8 = 3**	**1995**
Mystique 2.5L	**VIN position 8 = L**	**1995**
Sable 3.0L	**VIN position 8 = U**	**1992-95**
Sable 3.8L	**VIN position 8 = 4**	**1992-95**
Topaz 2.3L	**VIN position 8 = X**	**1992-94**
Topaz 3.0L	**VIN position 8 = U**	**1992-94**
Tracer 1.9L	**VIN position 8 = J**	**1992-95**

System: **Ford EEC-IV**

Self-diagnosis

General information

- Refer to the front of this manual for general test conditions, terminology, detailed descriptions of wiring faults and a general trouble shooter for electrical and mechanical faults.
- Malfunction indicator lamp (MIL) will illuminate if certain faults are recorded.
- ECM operates in backup mode if sensors fail, to enable vehicle to be driven to repair shop.
- Trouble codes can be displayed by the MIL or by using an analogue voltmeter or Scan Tool connected to the data link connectors (DLC) 1 and 2:
  - Cougar/XR7 **1**
  - Grand Marquis **2**.
  - Mystique **3**.
  - Sable **4**.
  - Topaz **5**.
  - Tracer **6**.

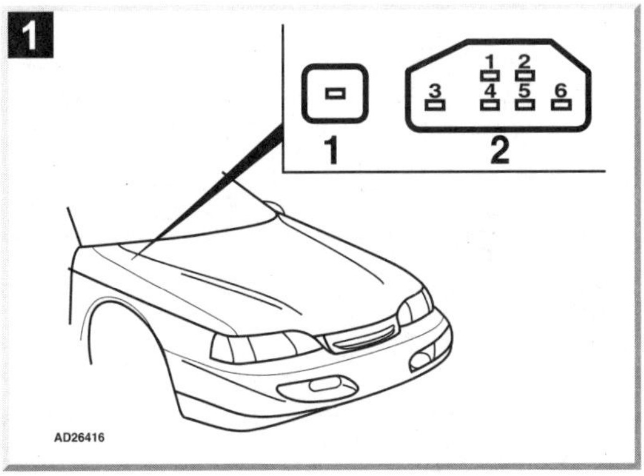

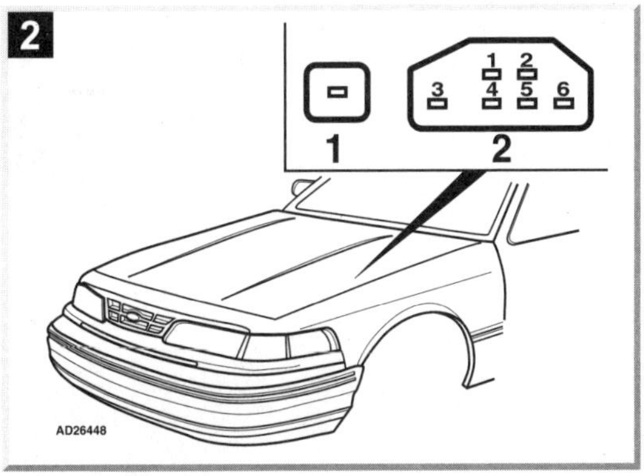

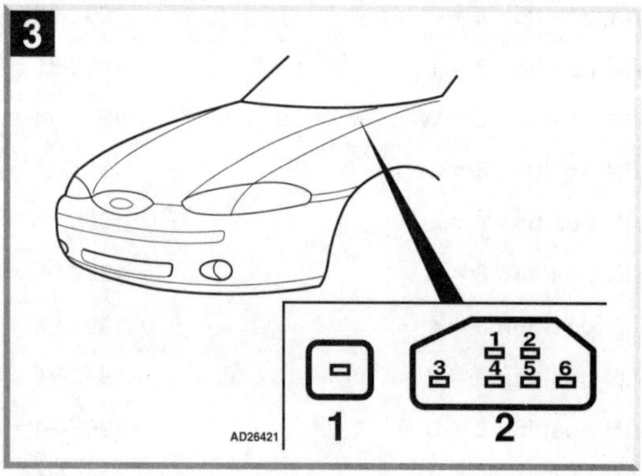

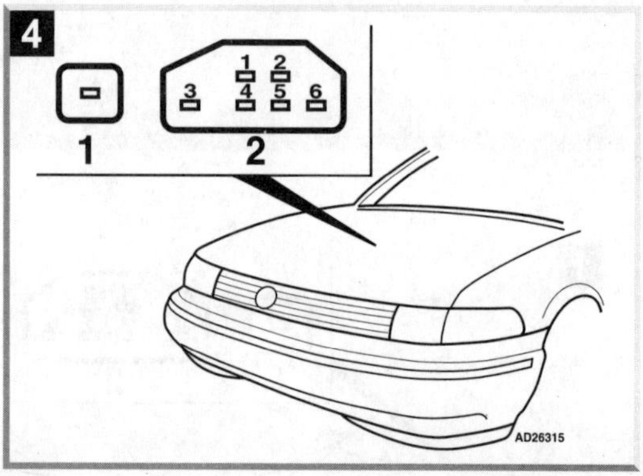

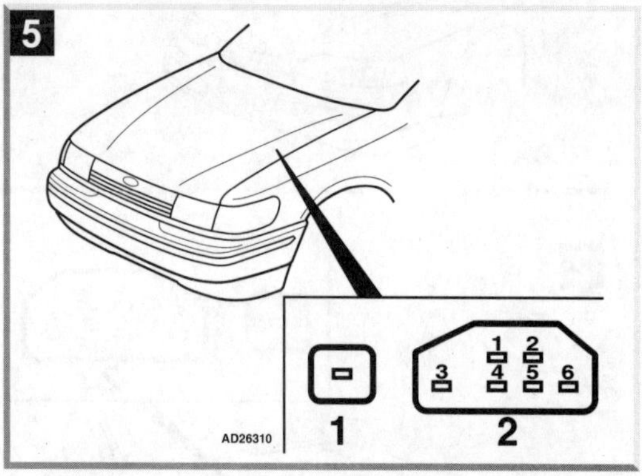

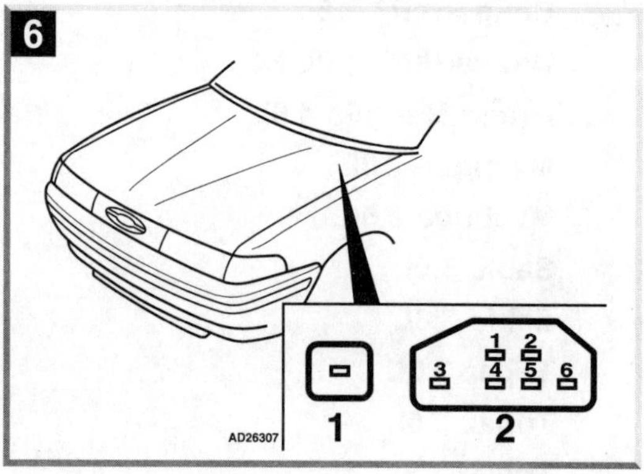

Preparatory conditions

- ○ Automatic transmission in 'PARK'.
- ○ All auxiliary equipment, including air conditioning, switched OFF.

Accessing – Key On Engine Off (KOEO) test

Using malfunction indicator lamp (MIL)

- Start engine.
- Allow to idle until engine at normal operating temperature.
- Switch ignition OFF.
- Wait 10 seconds.
- If engine does not start or stalls: Continue with KOEO test.
- If fitted: Remove 'EEC TEST' cover to access data link connector (DLC).
- Jump DLC 1 terminal (STI) to DLC 2 terminal 2 (SIG RTN) **7**.
- Switch ignition ON.

NOTE: ***MIL may flash briefly while fast DTCs are issued for Scan Tool followed by slow DTC output.***

- Count MIL flashes. Note trouble codes. Compare with trouble code table.
- Each trouble code is displayed twice. Firstly for present faults followed by those for memorized faults as follows:
  - ○ Trouble codes are displayed by illuminated MIL switching OFF.
  - ○ Each trouble code consists of two or three digits.
  - ○ A group of 0.5 second flashes indicate each digit **8** [**A**].
  - ○ A 0.5 second pause separates each flash **8** [**B**].

- A 2 second pause separates each trouble code digit 8 [**C**].
- A 4 second pause separates each trouble code 8 [**D**].
- For example: Three digit trouble code 122 displayed 8. Throttle position (TP) sensor.

NOTE: ***Two digit trouble codes displayed in a similar pattern.***

- A 6-9 second pause 9 [**B**] indicates the end of trouble code output for present faults 9 [**A**].
- A 0.5 second flash (separator code) 9 [**C**], followed by a 6-9 second pause 9 [**D**] indicates the start of trouble code output for memorized faults 9 [**E**].
- Trouble codes for memorized faults displayed in the same manner as those for present faults.

NOTE: ***Memorized trouble codes refer to faults that have occurred in the last 80 warm-up cycles.***

- Switch ignition OFF.
- Remove jump lead.
- Trouble codes can also be displayed by using a Scan Tool connected to the data link connectors (DLC) 1 and 2.

7

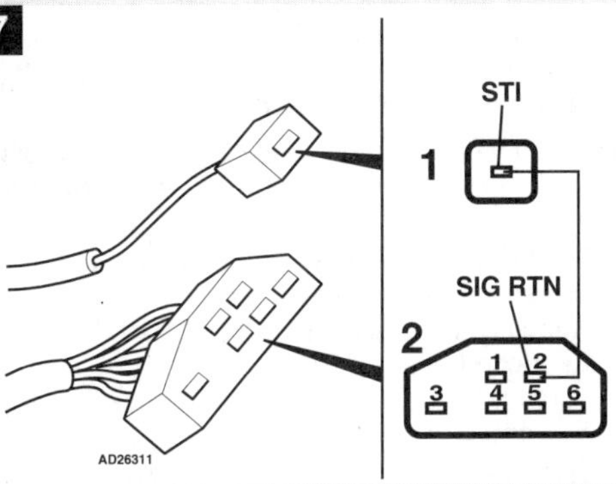

8

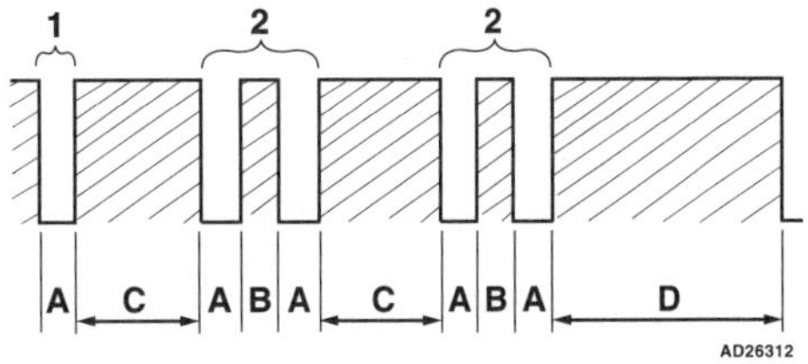

9

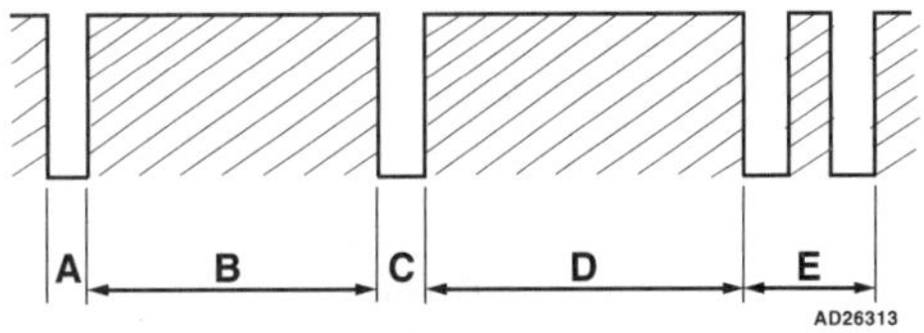

Using analogue voltmeter

- Ensure ignition switched OFF.
- If fitted: Remove 'EEC TEST' cover to access data link connector (DLC).
- Connect an analogue voltmeter between data link connector (DLC) 2 terminal 4 (STO) and battery (12 volt) positive 10.
- Set analogue voltmeter to a DC voltage range of 0-20 V.
- Start engine.
- Allow to idle until engine at normal operating temperature.
- Switch ignition OFF.
- Wait 10 seconds.
- If engine does not start or stalls: Continue with KOEO test.
- Jump DLC 1 terminal (STI) to DLC 2 terminal 2 (SIG RTN) 10.
- Switch ignition ON.

NOTE: ***Voltmeter needle may briefly deflect while fast DTCs are issued for Scan Tool followed by slow DTC output.***

- Count voltmeter needle deflections. Note trouble codes. Compare with trouble code table.
- Each trouble code is displayed twice. Firstly for present faults followed by those for memorized faults as follows:
  - Trouble codes are displayed by voltmeter needle deflecting to zero.
  - Each trouble code consists of two or three digits.
  - A group of 0.5 second needle deflections indicate each digit **8** [**A**].
  - A 0.5 second pause separates each needle deflection **8** [**B**].
  - A 2 second pause separates each trouble code digit **8** [**C**].
  - A 4 second pause separates each trouble code **8** [**D**].
  - For example: Three digit trouble code 122 displayed **8**. Throttle position (TP) sensor.

NOTE: ***Two digit trouble codes displayed in a similar pattern.***

  - A 6-9 second pause **9** [**B**] indicates the end of trouble code output for present faults **9** [**A**].
  - A 0.5 second needle deflection (separator code) **9** [**C**], followed by a 6-9 second pause **9** [**D**] indicates the start of trouble code output for memorized faults **9** [**E**].
  - Trouble codes for memorized faults displayed in the same manner as those for present faults.

NOTE: ***Memorized trouble codes refer to faults that have occurred in the last 80 warm-up cycles.***

- Switch ignition OFF.
- Remove jump lead and voltmeter.
- Trouble codes can also be displayed by using a Scan Tool connected to the data link connectors (DLC) 1 and 2.

10

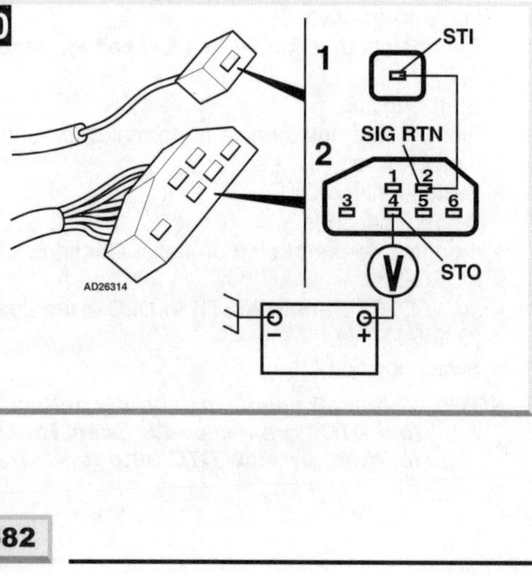

Accessing – Key On Engine Running (KOER) test

Using malfunction indicator lamp (MIL)

- Start engine.
- Allow to idle until engine at normal operating temperature.
- Increase engine speed to over 2000 rpm for 2 minutes.
- Switch ignition OFF.
- Wait 10 seconds.
- If fitted: Remove 'EEC TEST' cover to access data link connector (DLC).
- Jump DLC 1 terminal (STI) to DLC 2 terminal 2 (SIG RTN) **7**.
- Start engine.
- Engine identification code displayed as follows:
  - 4 cylinder: Two 0.5 second flashes.
  - 6 cylinder: Three 0.5 second flashes.
  - 8 cylinder: Four 0.5 second flashes.
  - For example: 6 cylinder engine identified **11** [**A**].
- With PSP switch: Within 1-2 seconds, turn steering wheel at least ½ a turn and release.
- With BPP switch: Depress and release brake pedal.
- If fitted: Operate overdrive selection switch twice.
- If after 6-20 seconds **11** [**B**], a single 0.5 second flash is displayed (dynamic response signal) **11** [**C**]: Briefly fully open throttle.

NOTE: ***DO NOT fully open throttle during KOER test if no dynamic response signal is displayed.***

- Trouble codes displayed for present faults after 4-15 second pause **11** [**D**].
- Fast DTCs are issued for Scan Tool during 4-15 second pause **11** [**E**].
- Count MIL flashes. Note trouble codes. Compare with trouble code table.

NOTE: ***Each trouble code is displayed twice.***

- Trouble codes for KOER test are displayed in the same manner as those for KOEO test **11** [**F**].
- For example: Three digit trouble code 111 is displayed **11**. No fault found.
- Switch ignition OFF.
- Remove jump lead.
- Trouble codes can also be displayed by using a Scan Tool connected to the data link connectors (DLC) 1 and 2.

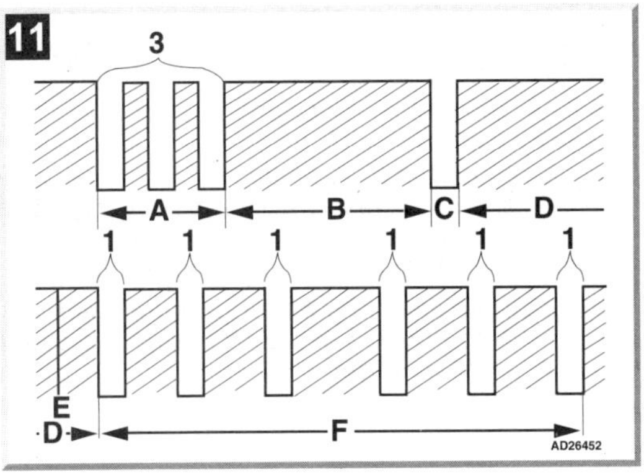

Using analogue voltmeter

- Ensure ignition switched OFF.
- If fitted: Remove 'EEC TEST' cover to access data link connector (DLC).
- Connect an analogue voltmeter between data link connector (DLC) 2 terminal 4 (STO) and battery (12 volt) positive **10**.
- Set analogue voltmeter to a DC voltage range of 0-20 V.
- Start engine.
- Allow to idle until engine at normal operating temperature.
- Increase engine speed to over 2000 rpm for 2 minutes.
- Switch ignition OFF.
- Wait 10 seconds.
- Jump DLC 1 terminal (STI) to DLC 2 terminal 2 (SIG RTN) **10**.
- Start engine.
- Engine identification code displayed as follows:
  - 4 cylinder: Two 0.5 second needle deflections.
  - 6 cylinder: Three 0.5 second needle deflections.
  - 8 cylinder: Four 0.5 second flashes.
  - For example: 6 cylinder engine identified **11** [**A**].
- With PSP switch: Within 1-2 seconds, turn steering wheel at least ½ a turn and release.
- With BPP switch: Depress and release brake pedal.
- If fitted: Operate overdrive selection switch twice.
- If after 6-20 seconds **11** [**B**], a single 0.5 second needle deflection is displayed (dynamic response signal) **11** [**C**]: Briefly fully open throttle.

NOTE: ***DO NOT fully open throttle during KOER test if no dynamic response signal is displayed.***

- Trouble codes displayed for present faults after 4-15 second pause **11** [**D**].
- Fast DTCs are issued for Scan Tool during 4-15 second pause **11** [**E**].
- Count needle deflections. Note trouble codes. Compare with trouble code table.

NOTE: ***Each trouble code is displayed twice.***

- Trouble codes for KOER test are displayed in the same manner as those for KOEO test **11** [**F**].
- For example: Three digit trouble code 111 is displayed **11**. No fault found.
- Switch ignition OFF.
- Remove jump lead and voltmeter.
- Trouble codes can also be displayed by using a Scan Tool connected to the data link connectors (DLC) 1 and 2.

Erasing

NOTE: ***Disconnecting battery ground lead may cause driveability problems until the engine management system relearns basic values.***

Using malfunction indicator lamp (MIL)

- Ensure ignition switched OFF.
- If fitted: Remove 'EEC TEST' cover to access data link connector (DLC).
- Jump DLC 1 terminal (STI) to DLC 2 terminal 2 (SIG RTN) **7**.
- Switch ignition ON.
- When trouble codes start to be displayed: Remove jump lead.
- Switch ignition OFF.
- Repeat accessing procedure to ensure no memorized trouble codes remain in ECM fault memory.
- Trouble codes can also be erased by using a Scan Tool connected to the data link connectors (DLC) 1 and 2.
- If emission related component(s) have been replaced:
  - Disconnect battery ground cable for at least 5 minutes to erase adaptive memory from ECM.
  - If required: Drive vehicle for 10 miles or more to allow ECM to relearn basic values.

NOTE: ***Disconnecting battery ground lead may also erase memory from electronic units, such as radio.***

Using analogue voltmeter

- Ensure ignition switched OFF.
- If fitted: Remove 'EEC TEST' cover to access data link connector (DLC).

- Connect an analogue voltmeter between data link connector (DLC) 2 terminal 4 (STO) and battery (12 volt) positive [10].
- Set analogue voltmeter to a DC voltage range of 0-20 V.
- Jump DLC 1 terminal (STI) to DLC 2 terminal 2 (SIG RTN) [10].
- Switch ignition ON.
- When trouble codes start to be displayed: Remove jump lead.
- Switch ignition OFF.
- Repeat accessing procedure to ensure no memorized trouble codes remain in ECM fault memory.
- Trouble codes can also be erased by using a Scan Tool connected to the data link connectors (DLC) 1 and 2.
- If emission related component(s) have been replaced:
  - Disconnect battery ground cable for at least 5 minutes to erase adaptive memory from ECM.
  - If required: Drive vehicle for 10 miles or more to allow ECM to relearn basic values.

NOTE: ***Disconnecting battery ground lead may also erase memory from electronic units, such as radio.***

Trouble code identification

Two digit

Flash code	Fault location	Probable cause
11	No fault found	–
12	Idle speed, high RPM check – unable to control during KOER test	**Intake system, throttle tight/sticking, IAC valve/ISC actuator, ECM**
13	Idle speed, low RPM check – unable to control during KOER test	**Intake system, throttle tight/sticking, IAC valve/ISC actuator, ECM**
14	Engine speed (RPM) sensor/profile ignition pickup (PIP) circuit – intermittent signal	**Incorrectly installed RPM sensor, wiring, poor connection, electrical interference, RPM sensor, ECM**
15	Engine control module (ECM) – KAM/ROM fault	**Battery disconnected, wiring, fuses, ECM**
16	Engine RPM too low for heated oxygen sensor (HO2S) test	**Intake system, throttle tight/sticking, IAC valve/ISC actuator, ECM**
	Ignition diagnostic monitor (IDM) circuit – no signal during KOEO test	**Wiring, ICM, ECM**
18	Spark output (SPOUT) circuit open	**Wiring, ICM, ECM**
	Ignition diagnostic monitor (IDM) circuit/spark output (SPOUT) circuit grounded	**Wiring, ICM, ECM**
19	Engine control module (ECM) – internal voltage fault	**ECM**
21	Engine coolant temperature (ECT) sensor – incorrect signal	**Wiring, ECT sensor, ECM**
22	Manifold absolute pressure (MAP) sensor/ barometric pressure (BARO) sensor – incorrect signal	**Vacuum hose damaged, trapped vacuum, high atmospheric pressure, excessive EGR, wiring, MAP/BARO sensor, ECM**
23	Throttle position (TP) sensor – incorrect signal	**Throttle valve tight/sticking, wiring, TP sensor, ECM**
24	Intake air temperature (IAT) sensor – incorrect signal	**Wiring, IAT sensor, ECM**

Flash code	Fault location	Probable cause
25	Knock sensor (KS) – no signal during dynamic response test	**Wiring, KS, ECM**
26	Mass air flow (MAF) sensor – incorrect signal	**Intake leak/blockage, wiring, MAF sensor, ECM**
29	Vehicle speed signal – incorrect signal	**Wiring, vehicle speed interface module, VSS, ECM**
31	Exhaust gas recirculation (EGR) valve position sensor – below minimum voltage	**Wiring short circuit to ground, EGR valve/ position sensor, ECM**
32	Exhaust gas recirculation (EGR) valve position sensor – below closed limit	**Wiring short circuit to ground, EGR valve/ position sensor, ECM**
33	Exhaust gas recirculation (EGR) valve – not opening	**Wiring, EGR valve tight/sticking, ECM**
34	Exhaust gas recirculation (EGR) valve position sensor – above closed limit	**Wiring short circuit to positive, EGR valve/ position sensor, ECM**
35	Exhaust gas recirculation (EGR) valve position sensor – above maximum voltage	**Wiring short circuit to positive, EGR valve/ position sensor, ECM**
41	Heated oxygen sensor (HO2S), bank 1 – mixture lean/not switching	**Intake leak, fuel pressure low, fuel filter/ pump, injectors, HO2S, AIR system, MAF/ MAP sensor, ECM**
42	Heated oxygen sensor (HO2S), bank 1 – mixture rich/not switching	**Intake blocked, excessive fuel in engine oil, fuel pressure high, breather system, injectors, HO2S, AIR system, EVAP system, MAF/MAP sensor, ignition system, ECM**
43	Throttle position (TP) sensor – incorrect signal at idle	**Incorrectly adjusted, TP sensor, ECM**
44	Secondary air injection (AIR) system, bank 1 – inoperative	**Hose connection, wiring, AIR pump motor, AIR solenoid, ECM**
45	Secondary air injection (AIR) – upstream during KOER test	**Hose connection, wiring, AIR solenoid, ECM**
46	Secondary air injection (AIR) – not bypassed during KOER test	**Hose connection, wiring, AIR solenoid, ECM**
47	Four wheel drive selector switch – closed during KOEO test	**Wiring short circuit, four wheel drive selector switch, internal damage to transfer case, ECM**
49 **1**	Gear shift, 1-2 – malfunction	**ATF level/condition, poor connection, wiring, TR sensor, SS, clutch slipping, ECM**
51	Engine coolant temperature (ECT) sensor – open circuit	**Wiring open circuit, ECT sensor, ECM**
52	Power steering pressure (PSP) switch – open circuit/no signal change	**Wiring open circuit, PSP switch, ECM**
53	Throttle position (TP) sensor – above maximum voltage	**TP sensor loose, wiring short circuit to positive, TP sensor, ECM**
54	Intake air temperature (IAT) sensor – open circuit	**Wiring open circuit, IAT sensor, ECM**
56	Mass air flow (MAF) sensor – above maximum voltage	**Wiring short circuit to positive, MAF sensor, ECM**
59 **1**	Gear shift, 2-3 – malfunction	**ATF level/condition, poor connection, wiring, TR sensor, SS, clutch slipping, ECM**

Flash code	Fault location	Probable cause
61	Engine coolant temperature (ECT) sensor – short circuit to ground	**Wiring short circuit to ground, ECT sensor, ECM**
62 **1**	Torque converter clutch (TCC) – excessive clutch slip	**ATF level/condition, TCC slipping, ECM**
63	Throttle position (TP) sensor – below minimum voltage	**TP sensor loose, wiring open circuit/short circuit to ground, TP sensor, ECM**
64	Intake air temperature (IAT) sensor – short circuit to ground	**Wiring short circuit to ground, IAT sensor, ECM**
65 **1**	Overdrive selection switch – no signal change during KOEO test	**Incorrect switch operation, wiring, fuse, overdrive selection switch, ECM**
66	Mass air flow (MAF) sensor – below minimum voltage	**Wiring short circuit to ground, MAF sensor, ECM**
67	Neutral position (NP) switch – open circuit	**Incorrect adjustment, wiring open circuit, NP switch, ECM**
	Clutch pedal position (CPP) switch – circuit malfunction	**Incorrect adjustment, wiring, CPP switch, ECM**
	A/C – on during KOEO test	**Wiring, A/C switch, ECM**
67 **1**	Transmission range (TR) sensor – voltage high/low	**Incorrect adjustment, wiring, TR sensor, ECM**
68 **1**	Automatic transmission fluid (ATF) – temperature high	**ATF level, transmission cooling system, towing weight limit exceeded, clutch slipping, wiring, TFT sensor, ECM**
69 **1**	Gear shift, 3-4 – malfunction	**ATF level/condition, poor connection, wiring, TR sensor, SS, clutch slipping, ECM**
72	Manifold absolute pressure (MAP) sensor/ mass air flow (MAF) sensor – incorrect signal change during dynamic response test	**Intake leak, wiring, MAP sensor, ECM**
73	Throttle position (TP) sensor – incorrect signal change during dynamic response test	**Incorrect throttle operation, wiring, TP sensor, ECM**
74	Brake pedal position (BPP) switch – open circuit/no signal change during KOER test	**Incorrect brake pedal operation, wiring, BPP switch, ECM**
75	Brake pedal position (BPP) switch – closed circuit	**Wiring, BPP switch, ECM**
77	Wide open throttle (WOT) switch – no signal change during dynamic response test	**Incorrect throttle operation, wiring, WOT switch, ECM**
79	A/C – on during KOEO test	**Wiring, A/C switch, ECM**
81	Secondary air injection (AIR) divert solenoid – circuit malfunction during KOEO test	**Wiring, AIR divert solenoid, ECM**
82	Secondary air injection (AIR) bypass solenoid – circuit malfunction during KOEO test	**Wiring, AIR bypass solenoid, ECM**
84	Exhaust gas recirculation (EGR) solenoid – circuit malfunction during KOEO test	**Wiring, EGR solenoid, ECM**
85	Evaporative emission (EVAP) canister purge valve – circuit malfunction during KOEO test	**Wiring, EVAP canister purge valve, ECM**
86	AT – shift solenoid	**Wiring, shift solenoid, ECM**

Flash code	Fault location	Probable cause
87	Fuel pump, primary circuit – circuit malfunction	IFS switch, wiring, fuel pump, ECM
89	Torque converter clutch (TCC) lock-up solenoid	Wiring, TCC lock-up solenoid, ECM
91	Heated oxygen sensor (HO2S), bank 2 – mixture lean/not switching during KOER test	Intake leak, fuel pressure low, fuel filter/ pump, injectors, HO2S, AIR system, MAF/ MAP sensor, ECM
91 [1]	Shift solenoid (SS) 1 – circuit malfunction during KOEO test	Wiring, SS, ECM
92	Heated oxygen sensor (HO2S), bank 1 – mixture rich/not switching during KOER test	Intake blocked, excessive fuel in engine oil, fuel pressure high, breather system, injectors, HO2S, AIR system, EVAP system, MAF/MAP sensor, ignition system, ECM
92 [1]	Shift solenoid (SS) 2 – circuit malfunction during KOEO test	Wiring, SS, ECM
93 [1]	Transmission coast clutch solenoid – circuit malfunction during KOEO test	Wiring, transmission coast clutch solenoid, ECM
94	Secondary air injection (AIR) system, bank 2 – inoperative during KOER test	Hose connection, wiring, AIR pump motor, AIR solenoid, ECM
94 [1]	Torque converter clutch (TCC) solenoid – circuit malfunction during KOEO test	Wiring, TCC solenoid, ECM
95	Fuel pump, secondary circuit – circuit malfunction	IFS switch, wiring, fuel pump, ECM
96	Fuel pump, secondary circuit – circuit malfunction	IFS switch, wiring open circuit to battery, fuel pump, ECM
97 [1]	Overdrive indicator lamp – circuit malfunction	Bulb, wiring, fuse, ECM
98	Engine control module (ECM) – hard fault detected during KOER test	HO2S bank 2 – indicates lean, ECM
98 [1]	Engine control module (ECM), transmission fluid pressure (TFP) control – open driver circuit during KOEO test	ECM
99	Engine control module – idle speed not learnt during KOER test	Idle speed adjusted, ECM
99 [1]	Transmission fluid pressure (TFP) solenoid – circuit malfunction during KOEO test	Wiring, poor connection, TFP solenoid, ECM

[1] ECM incorporates transmission control function

Three digit

Flash code	Fault location	Probable cause
111	No fault found	–
112	Intake air temperature (IAT) sensor – below minimum voltage	Wiring short circuit to ground, poor connection, IAT sensor, ECM
113	Intake air temperature (IAT) sensor – above maximum voltage	Wiring open circuit, poor connection, IAT sensor, ECM
114 [1]	Intake air temperature (IAT) sensor – incorrect signal	Air temperature below 10°C (50°F), poor connection, IAT sensor, ECM

Flash code	Fault location	Probable cause
116	Engine coolant temperature (ECT) sensor – incorrect signal	**Coolant level low, poor connection, ECT sensor, ECM**
117	Engine coolant temperature (ECT) sensor – below minimum voltage	**Wiring short circuit to ground, poor connection, ECT sensor, ECM**
118	Engine coolant temperature (ECT) sensor – above maximum voltage	**Wiring open circuit, poor connection, IAT sensor, ECM**
121	Throttle position (TP) sensor – closed throttle voltage high/low	**Throttle valve tight/sticking, TP sensor loose/damaged, ECM**
	Throttle position (TP) sensor – voltage inconsistent with MAF sensor signal	**Throttle valve tight/sticking, TP sensor loose/damaged, ECM**
122	Throttle position (TP) sensor – below minimum voltage	**TP sensor loose, wiring open circuit/short circuit to ground, TP sensor, ECM**
123	Throttle position (TP) sensor – above maximum voltage	**TP sensor loose, wiring short circuit to positive, TP sensor, ECM**
124	Throttle position (TP) sensor – voltage high	**Throttle tight/sticking, intake leak, MAP/BARO sensor, ECM**
125	Throttle position (TP) sensor – voltage low	**Throttle tight/sticking, intake leak, MAP/BARO sensor, ECM**
126	Manifold absolute pressure (MAP) sensor/barometric pressure (BARO) sensor – voltage high/low	**Vacuum hose damaged/trapped, high atmospheric pressure, excessive EGR, wiring, MAP/BARO sensor, ECM**
128	Manifold absolute pressure (MAP) sensor – loss of vacuum	**Vacuum hose damaged/disconnected, MAP sensor, ECM**
129	Manifold absolute pressure (MAP) sensor – insufficient voltage change during KOER dynamic response test	**Incorrect throttle operation, vacuum hose damaged/disconnected, MAP sensor, IAC system, ECM**
	Mass air flow (MAF) sensor – insufficient voltage change during KOER dynamic response test	**Incorrect throttle operation, intake leak, wiring, MAF sensor, IAC system, ECM**
136	Heated oxygen sensor (HO2S), bank 2 – mixture lean during KOER test	**Intake leak, fuel pressure low, fuel filter/pump, injectors, HO2S, AIR system, MAP sensor, ECM**
137	Heated oxygen sensor (HO2S), bank 2 – mixture rich during KOER test	**Intake blocked, excessive fuel in engine oil, fuel pressure high, breather system, injectors, HO2S, AIR system, EVAP system, MAP sensor, ignition system, ECM**
138	Cold start injector – insufficient flow during KOER test	**Fuel pressure, wiring, cold start injector, ECM**
139	Heated oxygen sensor (HO2S), bank 2 – not switching	**Intake leak/blocked, injectors, HO2S, AIR system, breather system, EVAP system, MAP sensor, ignition system, ECM**
141	Fuel system indicates lean	**Intake leak, fuel pressure low, fuel filter/pump, injectors, HO2S, ECM**
144	Heated oxygen sensor (HO2S), bank 1 – not switching	**Intake leak/blocked, injectors, HO2S, AIR system, breather system, EVAP system, MAP sensor, ignition system, ECM**
157	Mass air flow (MAF) sensor – below minimum voltage	**Wiring short circuit to ground, poor connection, MAF sensor, IAC system, ECM**

Flash code	Fault location	Probable cause
158	Mass air flow (MAF) sensor – above maximum voltage	**Wiring short circuit to positive, MAF sensor screen blocked, MAF sensor, ECM**
159 2	Mass air flow (MAF) sensor – voltage high/low	**Intake leak, IAC valve, MAF sensor, ECM**
167	Throttle position (TP) sensor – insufficient voltage change during KOER dynamic response test	**Incorrect throttle operation, TP sensor, ECM**
171	Fuel system control, bank 1 – adaptive limits reached	**Intake leak/blocked, breather system, fuel pressure, fuel filter/pump, injectors, HO2S, AIR system, EVAP system, MAP sensor, ignition system, ECM**
172	Heated oxygen sensor (HO2S), bank 1 – mixture lean	**Intake leak, excessive fuel in engine oil, fuel pressure low, fuel filter/pump, injectors, HO2S, AIR system, MAP sensor, ECM**
173	Heated oxygen sensor (HO2S), bank 1 – mixture rich	**Intake blocked, fuel pressure high, breather system, injectors, HO2S, AIR system, EVAP system, MAP sensor, ignition system, ECM**
174	Heated oxygen sensor (HO2S), bank 1 – slow response	**Intake leak, excessive fuel in engine oil, fuel pressure, fuel filter/pump, injectors, HO2S, AIR system, MAF/MAP sensor, ECM**
175	Fuel system control, bank 2 – adaptive limits reached	**Intake leak/blocked, excessive fuel in engine oil, breather system, fuel pressure, fuel filter/ pump, injectors, HO2S, AIR system, EVAP system, MAP sensor, ignition system, ECM**
176	Heated oxygen sensor (HO2S), bank 2 – mixture lean	**Intake leak, fuel pressure low, fuel filter/ pump, injectors, HO2S, AIR system, MAP sensor, ECM**
177	Heated oxygen sensor (HO2S), bank 2 – mixture rich	**Intake blocked, fuel pressure high, breather system, injectors, HO2S, AIR system, EVAP system, MAP sensor, ignition system, ECM**
178	Heated oxygen sensor (HO2S), bank 2 – slow response	**Intake leak, excessive fuel in engine oil, fuel pressure, fuel filter/pump, injectors, HO2S, AIR system, MAF/MAP sensor, ECM**
179	Fuel control system, bank 1 – lean adaption limit reached at part throttle, mixture rich	**Intake leak, fuel pressure low, fuel filter blocked, wiring, injector, MAP sensor, ECM**
181	Fuel control system, bank 1 – rich adaption limit reached at part throttle, mixture lean	**Intake blocked, fuel pressure high, wiring, injector, MAP sensor, ECM**
182	Fuel control system, bank 1 – lean adaption limit reached at idle, mixture rich	**Intake leak, fuel pressure low, fuel filter blocked, wiring, injector, MAP sensor, ECM**
183	Fuel control system, bank 1 – rich adaption limit reached at idle, mixture lean	**Intake blocked, fuel pressure high, wiring, injector, MAP sensor, ECM**
184	Mass air flow (MAF) sensor – signal high	**MAF sensor screen blocked, MAF sensor, ECM**
185	Mass air flow (MAF) sensor – signal low	**Intake leak, breather system, MAF sensor, ECM**
186	Injector – pulse width long/mass air flow (MAF) sensor – signal low	**Intake leak, fuel pressure low, MAF sensor, ECM**
187	Injector – pulse width short/mass air flow (MAF) sensor – signal high	**Intake blocked, fuel pressure high, MAF sensor, ECM**

Flash code	Fault location	Probable cause
188	Fuel system control, bank 2 – lean adaption limit reached at part throttle, mixture rich	**Intake leak, fuel pressure low, fuel filter blocked, wiring, injector, MAP sensor, ECM**
189	Fuel system control, bank 2 – rich adaption limit reached at part throttle, mixture lean	**Intake blocked, fuel pressure high, wiring, injector, MAP sensor, ECM**
191	Fuel system control, bank 2 – lean adaption limit reached at idle, mixture rich	**Intake leak, fuel pressure low, fuel filter blocked, wiring, injector, MAP sensor, ECM**
192	Fuel system control, bank 2 – rich adaption limit reached at idle, mixture lean	**Intake blocked, fuel pressure high, wiring, injector, MAP sensor, ECM**
193	Fuel quality sensor – circuit malfunction	**Wiring, fuel separation, fuel quality sensor, ECM**
194	Cylinder balance test – malfunction	**Wiring, injector, ECM**
195	Cylinder balance test – malfunction	**Wiring, injector, ECM**
211	Profile ignition pickup (PIP) – circuit malfunction	**Wiring, poor connection, arcing secondary ignition component – coil/cap/rotor/HT leads/plugs, electrical interference, ECM**
212	Ignition control module (ICM) – ignition diagnostic monitor (IDM) signal missing	**Wiring, ICM, ECM**
213	Spark output (SPOUT) – open circuit	**Wiring, ICM, ECM**
214	Camshaft position (CMP) sensor – circuit malfunction	**Wiring, CMP sensor, ICM, ECM**
215	Ignition coil 1 – primary circuit malfunction	**Wiring, ignition coil, ICM, ECM**
216	Ignition coil 2 – primary circuit malfunction	**Wiring, ignition coil, ICM, ECM**
217	Ignition coil 3 – primary circuit malfunction	**Wiring, ignition coil, ICM, ECM**
218	Ignition diagnostic monitor (IDM), left side – signal missing	**Wiring, ignition coil, ECM**
219	Ignition timing, defaulted to 10° – spark output (SPOUT) circuit open	**Wiring open circuit, ICM, ECM**
221	Ignition timing error	**Wiring, ICM, ECM**
222	Ignition diagnostic monitor (IDM), right side – signal missing	**Wiring, ignition coil, ECM**
223	Loss of dual plug inhibit (DPI) control	**Wiring, ICM, ECM**
224	Ignition coil 1, 2, 3 or 4 – primary circuit malfunction	**Wiring, ignition coil, ICM, ECM**
225	Knock sensor (KS) – no signal during KOER dynamic response test	**Wiring, KS, ECM**
226	Ignition control module (ICM) – ignition diagnostic monitor (IDM) signal missing	**Wiring open circuit, ICM, ECM**
232	Ignition coil 1, 2, 3 or 4 – primary circuit malfunction	**Wiring, ignition coil, ICM, ECM**
238	Ignition coil 4 – primary circuit malfunction	**Wiring, ignition coil, ICM, ECM**
241	Ignition control module (ICM) – ignition diagnostic monitor (IDM) signal malfunction	**Wiring, ICM, ECM**
244	Cylinder identification (CID) fault when cylinder balance test requested	**Wiring, CKP sensor, CMP sensor, ICM, ECM**
311	Secondary air injection (AIR) system, bank 1 – inoperative during KOER test	**Hose blocked/leaking, AIR bypass/divert solenoid, AIR pump, ECM**

Flash code	Fault location	Probable cause
312	Secondary air injection (AIR) system – misdirected during KOER test	**Hose connections**
313	Secondary air injection (AIR) system – not bypassed during KOER test	**Hose blocked/leaking, AIR bypass solenoid, AIR pump**
314	Secondary air injection (AIR) system, bank 2 – inoperative during KOER test	**Hose blocked/leaking, AIR bypass/divert solenoid, AIR pump, ECM**
326 [3]	Exhaust gas pressure sensor – voltage low	**Hose blocked/leaking, wiring, exhaust gas pressure sensor, ECM**
326 [2] [4]	Exhaust gas pressure sensor – voltage low	**Hose blocked/leaking, EGR solenoid filter, EGR solenoid, wiring, EGR valve, exhaust gas pressure sensor, ECM**
327 [5]	Exhaust gas recirculation (EGR) valve position sensor – minimum voltage exceeded	**Wiring open circuit/short circuit to ground, EGR valve/position sensor, ECM**
327	Exhaust gas pressure sensor – minimum voltage exceeded	**Hose blocked/leaking, wiring open circuit/ short circuit to ground, exhaust gas pressure sensor, ECM**
328	Exhaust gas recirculation (EGR) valve position sensor, closed position – voltage low	**Wiring, EGR valve/position sensor, ECM**
332 [6]	Exhaust gas recirculation (EGR) system – insufficient flow detected	**Hose blocked/leaking, EGR solenoid, exhaust gas pressure sensor, EGR valve, ECM**
332 [5]	Exhaust gas recirculation (EGR) system – insufficient flow detected	**Hose blocked/leaking, wiring, EGR solenoid filter, EGR solenoid, EGR valve/position sensor, ECM**
334	Exhaust gas recirculation (EGR) valve position sensor, closed position – voltage high	**Wiring, EGR valve/position sensor, EGR solenoid, ECM**
335 [2]	Exhaust gas pressure sensor – voltage high/ low during KOEO test	**Exhaust gas pressure sensor, blocked inlet hose, ECM**
336	Exhaust gas pressure sensor – voltage high	**Exhaust system restricted, wiring, EGR solenoid, exhaust gas pressure sensor, EGR valve, hose routing, ECM**
337 [3]	Exhaust gas pressure sensor – above maximum voltage	**Incorrect hose routing, wiring short circuit, signal wire open circuit, exhaust gas pressure sensor, EGR solenoid/valve, ECM**
337 [4]	Exhaust gas pressure sensor – above maximum voltage	**Exhaust system restricted, wiring short circuit, signal wire open circuit, exhaust gas pressure sensor, ECM**
337 [5]	Exhaust gas recirculation (EGR) position sensor – above maximum voltage	**Wiring open circuit/short circuit to positive, EGR valve/position sensor, ECM**
338	Engine coolant temperature (ECT) – too low	**Thermostat stuck open, coolant leak, ECM**
339	Engine coolant temperature (ECT) – too high	**Coolant level low, thermostat stuck closed, cooling system blocked, coolant blower motor worn/damaged, water pump defective, coolant leak, ECM**
341	Octane adjuster – open circuit	**Wiring, octane adjuster, ECM**
381	A/C compressor clutch – excessive cycling	**A/C system incorrectly charged, cycling switch defective, A/C circuit fault, wiring intermittent open circuit to A/C system, ECM**

Flash code	Fault location	Probable cause
411	Idle air control (IAC) valve – no control during KOER low RPM check	**Idle speed adjustment, intake leak, throttle valve tight/sticking, ignition timing incorrect, wiring short circuit to ground, IAC valve, ECM**
412	Idle air control (IAC) valve – no control during KOER high RPM check	**Idle speed adjustment, throttle valve tight/sticking, wiring open circuit/short circuit to ground, IAC valve, ECM**
415	Idle air control (IAC) valve – lower limit reached	**Intake leak, throttle valve tight/sticking, IAC valve, ECM**
416	Idle air control (IAC) valve – upper limit reached	**Air filter blocked, throttle body dirty/damaged, IAC valve, ECM**
452	Vehicle speed sensor (VSS) – incorrect signal	**Wiring, VSS, ECM**
453	Servo leaking down – cruise control KOER test	**Servo defective, ECM**
454	Servo leaking up – cruise control KOER test	**Servo defective, ECM**
455	Insufficient RPM increase – cruise control KOER test	**Intake system, ECM**
456	Insufficient RPM decrease – cruise control KOER test	**Intake system, ECM**
457	Cruise control master/selector switch – circuit malfunction during KOEO test	**Wiring, cruise control master/selector switch, ECM**
458	Cruise control master/selector switch – short circuit to ground during KOEO test	**Wiring short circuit to ground, cruise control master/selector switch, ECM**
459	Cruise control system – ground open circuit during KOEO test	**Wiring open circuit, ECM**
511	Engine control module (ECM) – ROM fault during KOEO test	**ECM**
512	Engine control module (ECM) – KAM fault	**Battery disconnected, wiring, fuses, ECM**
513	Engine control module (ECM) – internal voltage fault	**ECM**
519	Power steering pressure (PSP) switch – circuit malfunction during KOEO test	**PSP switch not fitted – disregard trouble code, wiring, PSP switch, ECM**
521	Power steering pressure (PSP) switch – no signal change during KOEO test	**PSP switch not fitted – disregard trouble code, wiring, PSP switch, ECM**
522 [7]	Park/neutral position (PNP) switch/neutral position (NP) switch – not in park/neutral during KOEO test/open circuit	**Incorrect gear lever position, incorrect adjustment, wiring open circuit, starter motor relay disconnected, PNP/NP switch, ECM**
	Transmission range (TR) sensor – not in park/neutral during KOEO test	**Incorrect shift lever position, incorrect adjustment, wiring, TR sensor, ECM**
525	Vehicle in gear – during KOEO test	**Incorrect shift lever position, incorrect adjustment, PNP switch open, ECM**
	A/C – ON during KOEO test	**A/C ON, A/C wiring short circuit to positive, ECM**
526	Neutral position switch (NP) – closed circuit during KOEO test	**Wiring short circuit, NP switch, ECM**

Flash code	Fault location	Probable cause
	A/C – ON during KOEO test	**A/C ON, A/C wiring short circuit to positive, A/C switch, ECM**
527 7	Park/neutral position (PNP) switch/neutral position switch – open circuit during KOEO test	**Wiring open circuit, starter motor relay disconnected, PNP/NP switch, ECM**
	A/C – ON during KOEO test	**A/C ON, A/C wiring short circuit to positive, A/C switch, ECM**
528	Clutch pedal position (CPP) switch – circuit malfunction	**Wiring open circuit, CPP switch, ECM**
529	Data link/engine control module (ECM) – circuit malfunction	**Wiring, ECM**
532	Instrument panel – circuit malfunction	**Wiring, instrument panel, ECM**
533	Data link/instrument panel – circuit malfunction	**Wiring, instrument panel, ECM**
536	Brake pedal position (BPP) switch – circuit malfunction/no signal change during KOER test	**Incorrect brake pedal operation, wiring, BPP switch, ECM**
538	KOER dynamic response test – insufficient rpm change	**Incorrect throttle operation, ECM**
	KOER cylinder balance test – invalid test	**Throttle operated during test, CMP sensor, ECM**
539	A/C – ON during KOEO test	**Wiring, A/C switch, ECM**
542	Fuel pump (FP) secondary circuit – engine does not start	**IFS switch, supply wire open circuit, FP ground wire, FP, ECM**
	Fuel pump (FP) secondary circuit – engine starts	**Supply wire short circuit to positive, FP relay, FP to ECM wire open circuit, left/front HO2S short circuit to positive, ECM**
543	Fuel pump (FP) secondary circuit – engine does not start	**Supply wire open circuit, fuel pump relay, ECM**
	Fuel pump (FP) secondary circuit – engine starts	**Left/front HO2S short circuit to positive, ECM**
551	Idle air control (IAC) valve – circuit malfunction during KOEO test	**Wiring, IAC valve, ECM**
552	Secondary air injection (AIR) bypass solenoid – circuit malfunction during KOEO test	**Wiring, AIR bypass solenoid, ECM**
553	Secondary air injection (AIR) divert solenoid – circuit malfunction during KOEO test	**Wiring, AIR divert solenoid, ECM**
554	Fuel pressure control solenoid – circuit malfunction	**Hose connection, wiring, fuel pressure control solenoid, ECM**
556	Fuel pump (FP) primary circuit	**IFS switch, wiring, fuel pump relay, constant control relay module, ECM**
557	Fuel pump (FP) primary circuit, low speed	**Wiring, FP relay/constant control relay module, ECM**
558	Exhaust gas recirculation (EGR) solenoid – circuit malfunction during KOEO test	**Hose damaged, wiring, EGR solenoid, ECM**
559	A/C relay, primary circuit – circuit malfunction	**Wiring, A/C relay, ECM**

Flash code	Fault location	Probable cause
562	Engine coolant blower motor relay – circuit malfunction	**Wiring, engine coolant blower motor relay, ECM**
563	Engine coolant blower motor relay, high speed – circuit malfunction	**Wiring, engine coolant blower motor relay/ constant control relay module, ECM**
564	Engine coolant blower motor relay, low speed – circuit malfunction	**Wiring, engine coolant blower motor relay/ constant control relay module, ECM**
565	Evaporative emission (EVAP) canister purge valve 1 – malfunction	**Hose damaged, wiring, EVAP canister purge valve, ECM**
566 [7]	Shift solenoid (SS) 3-4 – circuit malfunction	**Wiring, SS, ECM**
567	Cruise control vent valve – circuit malfunction	**Wiring, cruise control vent valve, ECM**
568	Cruise control vacuum valve – circuit malfunction	**Wiring, cruise control vacuum valve, ECM**
569	Evaporative emission (EVAP) canister purge valve 2 – malfunction	**Hose damaged, wiring, EVAP canister purge valve, ECM**
571	Exhaust gas recirculation (EGR) atmospheric solenoid – circuit malfunction during KOEO	**Wiring, EGR atmospheric solenoid, ECM**
572	Exhaust gas recirculation (EGR) vacuum solenoid – circuit malfunction during KOEO	**Wiring, EGR vacuum solenoid, ECM**
578	A/C pressure sensor – short circuit	**Supply wire short ground, A/C pressure sensor, variable control relay module, ECM**
579	A/C pressure sensor – insufficient pressure change	**Wiring, A/C fault, A/C pressure sensor, variable control relay module, wiring to A/C compressor clutch open circuit, ECM**
581	Engine coolant blower motor – excessive current draw	**Wiring short circuit to ground, engine coolant blower motor, variable control relay module, ECM**
582	Engine coolant blower motor – open circuit/ short circuit to positive	**Supply wire open circuit/short circuit to positive, ground wire open circuit, engine coolant blower motor, variable control relay module, ECM**
583	Fuel pump – excessive current draw	**Wiring short circuit to ground, fuel pump, ECM**
584	Variable control relay module – ground open circuit	**Wiring open circuit, variable control relay module, ECM**
585	A/C compressor clutch – excessive current draw	**Wiring short circuit to ground, A/C compressor clutch, variable control relay module, ECM**
586	A/C compressor clutch – open circuit/short circuit to positive	**Wiring open circuit/short circuit to positive, A/C compressor clutch, variable control relay module, ECM**
587	Variable control relay module – communication error	**Wiring, variable control relay module, ECM**
593	Heated oxygen sensor (HO2S) – heater circuit malfunction	**Wiring, HO2S heater relay, HO2S, ECM**
617 [7]	Gear shift, 1-2 – malfunction	**ATF level/condition, poor connection, wiring, TR sensor, SS, clutch slipping, ECM**

Flash code	Fault location	Probable cause
618 [7]	Gear shift, 2-3 – malfunction	ATF level/condition, poor connection, wiring, TR sensor, SS, clutch slipping, ECM
619 [7]	Gear shift, 3-4 – malfunction	ATF level/condition, poor connection, wiring, TR sensor, SS, clutch slipping, ECM
621 [7]	Shift solenoid (SS) 1 – circuit malfunction during KOEO	Wiring, SS, ECM
622 [7]	Shift solenoid (SS) 2 – circuit malfunction during KOEO	Wiring, SS, ECM
623 [7]	Overdrive indicator lamp – circuit malfunction	Bulb, wiring, fuse, ECM
624 [7]	Transmission fluid pressure (TFP) solenoid – circuit malfunction	Wiring, poor connection, TFP solenoid, ECM
625 [7]	Engine control module (ECM), TFP control – open driver circuit	ECM
626 [7]	Transmission coast clutch solenoid – circuit malfunction during KOEO test	Wiring, transmission coast clutch solenoid, ECM
627 [7]	Torque converter clutch (TCC) solenoid – circuit malfunction	Wiring, TCC solenoid, ECM
628 [7]	Torque converter clutch (TCC) – excessive clutch slip	ATF level/condition, TCC slipping, ECM
629 [7]	Torque converter clutch (TCC) solenoid – circuit malfunction	Wiring, TCC solenoid, ECM
631 [7]	Overdrive indicator lamp – circuit malfunction during KOEO test	Bulb, wiring, fuse, ECM
632 [7]	Overdrive selection switch – no signal change during KOEO test	Incorrect switch operation, wiring, fuse, overdrive selection switch, ECM
633	Four wheel drive selector switch – closed during KOEO test	Wiring short circuit, four wheel drive selector switch, internal damage to transfer case, ECM
634 [7]	Transmission range (TR) sensor – voltage high/low	Incorrect adjustment, wiring, TR sensor, ECM
636 [7]	Transmission fluid temperature (TFT) sensor – signal high/low	ATF level, wiring, TFT sensor, ECM
637 [7]	Transmission fluid temperature (TFT) sensor – above maximum voltage	ATF level, wiring open circuit, TFT sensor, ECM
638 [7]	Transmission fluid temperature (TFT) sensor – below minimum voltage	ATF level, wiring short circuit to ground, TFT sensor, ECM
639 [7]	Turbine shaft speed (TSS) sensor/output shaft speed (OSS) sensor – signal low	Wiring, TSS/OSS sensor, ECM
641 [7]	Shift solenoid (SS) 3 – circuit malfunction during KOEO	Wiring, SS, ECM
643 [7]	Torque converter clutch (TCC) solenoid – circuit malfunction during KOEO test	Wiring, TCC solenoid, ECM
645 [7]	Incorrect gear ratio, 1st gear	ATF level/condition, poor connection, wiring, TR sensor, SS, clutch slipping, ECM
646 [7]	Incorrect gear ratio, 2nd gear	ATF level/condition, poor connection, wiring, TR sensor, SS, clutch slipping, ECM

Flash code	Fault location	Probable cause
647 7	Incorrect gear ratio, 3rd gear	**ATF level/condition, poor connection, wiring, TR sensor, SS, clutch slipping, ECM**
648 7	Incorrect gear ratio, 4th gear	**ATF level/condition, poor connection, wiring, TR sensor, SS, clutch slipping, ECM**
649 7	Transmission fluid pressure (TFP) solenoid – signal high/low	**Wiring, poor connection, TFP solenoid, ECM**
651 7	Transmission fluid pressure (TFP) solenoid – circuit malfunction	**Wiring, poor connection, TFP solenoid, ECM**
652 7	Torque converter clutch (TCC) solenoid – circuit malfunction	**Wiring, TCC solenoid, ECM**
653 7	Overdrive selection switch – no signal change during KOEO test	**Incorrect switch operation, wiring, fuse, overdrive selection switch, ECM**
654 7	Transmission range (TR) sensor – 'P' signal during KOEO test	**Transmission not in 'P' during KOEO test, incorrect adjustment, ECM**
655 7	Transmission range (TR) sensor – 'N' signal during KOEO test	**Transmission not in 'N' during KOEO test, incorrect adjustment, ECM**
656 7	Torque converter clutch (TCC) – continuous clutch slip	**ATF level/condition, TCC slipping, ECM**
657 7	Automatic transmission fluid (ATF) – temperature high	**ATF level, transmission cooling system, towing weight limit exceeded, clutch slipping, wiring, TFT sensor, ECM**
659 7	High vehicle speed – transmission in 'P'	**TR sensor/PNP switch, ECM**
667 7	Transmission range (TR) sensor – below minimum voltage	**Wiring open circuit/short circuit to ground, TR sensor, ECM**
668 7	Transmission range (TR) sensor – above maximum voltage	**Wiring open circuit/short circuit to positive, TR sensor, ECM**
675 7	Transmission range (TR) sensor – out of range	**Wiring, TR sensor, ECM**
691	Four wheel drive selector fault – incorrect engine RPM/vehicle speed	**Wiring short circuit, four wheel drive selector switch, internal damage to transfer case, ECM**
692 7	Transmission state does not match calculated ratio	**Transmission fault, ECM**
998	Engine control module (ECM) – hard fault present	**ECM**

1 Topaz: Additional warm-up time may be required due to IAT location. Run engine until under hood temperature warm and stabilized

2 Trouble code may be generated if an efficient exhaust extraction system used during KOER test – vehicle with pressure feedback EGR

3 Cougar/XR7 3.8L (California), Grand Marquis (California), Mystique

4 Except Cougar/XR7, Grand Marquis (California), Mystique

5 Cougar-XR7 5.0L

6 Except Cougar-XR7 5.0L

7 ECM incorporates transmission control function

Model: **Tracer LTS 1.8L DOHC**

Engine identification: **VIN position 8 = 8**

Year: **1992-95**

System: **MECS**

Self-diagnosis

General information

- Refer to the front of this manual for general test conditions, terminology, detailed descriptions of wiring faults and a general trouble shooter for electrical and mechanical faults.
- Malfunction indicator lamp (MIL) will illuminate continuously if permanent faults are recorded.
- Malfunction indicator lamp (MIL) will illuminate briefly and then switch off if intermittent faults are recorded.
- ECM operates in backup mode if sensors fail, to enable vehicle to be driven to repair shop.
- Trouble codes can be displayed by the MIL or by using an analogue voltmeter or Scan Tool connected to the data link connector (DLC).

Accessing

Using malfunction indicator lamp (MIL)

- Ensure ignition switched OFF.
- Access data link connector (DLC) **1**.
- Jump data link connector (DLC) terminals STI and GND **2**.
- Switch ignition ON.
- Count MIL flashes. Note trouble codes. Compare with trouble code table.
  - 1.2 second flashes indicate the 'tens' of the trouble code **3** [**A**].
  - 0.4 second flashes indicate the 'units' of the trouble code **3** [**C**].
  - A 1.6 second pause (MIL OFF) separates 'tens' and 'units' **3** [**B**].
  - A 0.4 second pause (MIL OFF) separates each flash **3** [**D**].
  - A 4 second pause separates each trouble code **3** [**E**].
  - For example: Trouble code 12 displayed **3**. Throttle position (TP) sensor.
- Switch ignition OFF.
- Trouble codes can also be displayed by using a Scan Tool connected to the data link connector (DLC) **1**.

1

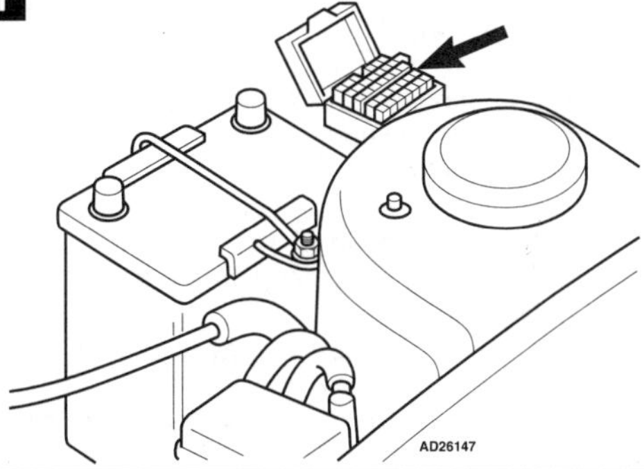

2

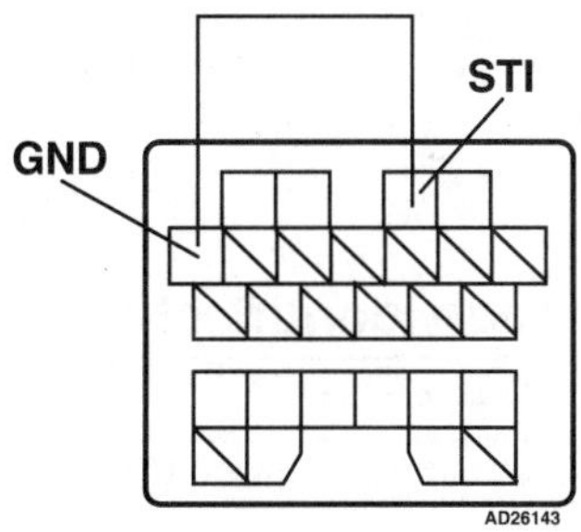

3

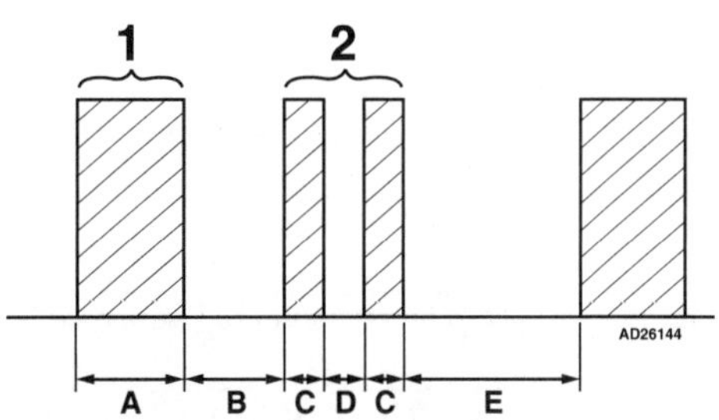

Using analogue voltmeter

- Ensure ignition switched OFF.
- Access data link connector (DLC) **1**.
- Connect an analogue voltmeter between data link connector (DLC) terminal STO and ground **4**.
- Set analogue voltmeter to a DC voltage range of 0-20 V.
- Jump data link connector (DLC) terminals STI and GND **4**.
- Switch ignition ON.
- Count voltmeter needle deflections. Note trouble codes. Compare with trouble code table.
  - 1.2 second needle deflections indicate the 'tens' of the trouble code **3** [**A**].
  - 0.4 second needle deflections indicate the 'units' of the trouble code **3** [**C**].
  - A 1.6 second pause separates 'tens' and 'units' **3** [**B**].
  - A 0.4 second pause separates each needle deflection **3** [**D**].
  - A 4 second pause separates each trouble code **3** [**E**].
  - For example: Trouble code 12 displayed **3**. Throttle position (TP) sensor.
- Switch ignition OFF.
- Trouble codes can also be displayed by using a Scan Tool connected to the data link connector (DLC) **1**.

4

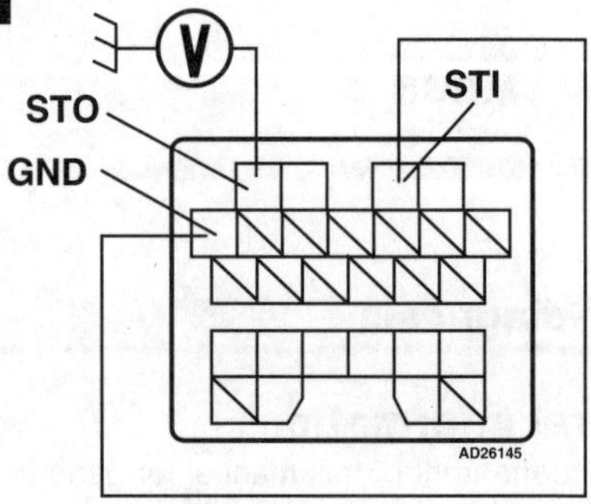

Erasing

- Ensure ignition switched OFF.
- Disconnect battery ground cable.
- Depress brake pedal for 5-10 seconds.
- Reconnect battery ground cable.
- Repeat accessing procedure to ensure no trouble codes remain in ECM fault memory.

NOTE: ***Disconnecting battery ground lead may also erase memory from electronic units, such as radio.***

Trouble code identification

Flash code	Fault location	Probable cause
02	Crankshaft position (CKP) sensor	**Wiring, CKP sensor, ECM**
03	Crankshaft position (CKP) sensor, cylinder identification (CID) signal – circuit malfunction	**Wiring, CKP sensor, ECM**
06	Vehicle speed sensor (VSS)	**Wiring, VSS, ECM**
08	Volume air flow (VAF) sensor	**Wiring, VAF sensor, ECM**
09	Engine coolant temperature (ECT) sensor	**Wiring, ECT sensor, ECM**
10	Intake air temperature (IAT) sensor	**Wiring, IAT sensor, ECM**
12	Throttle position (TP) sensor	**Wiring, TP sensor, ECM**
14	Barometric pressure (BARO) sensor	**ECM**

Flash code	Fault location	Probable cause
15	Oxygen sensor (O2S) – voltage below 0.55 V	Fuel pressure low, wiring, O2S, ECM
17	Oxygen sensor (O2S) – no voltage change	Wiring, O2S, ECM
25	Fuel pressure regulator control solenoid	Wiring, fuel pressure regulator control solenoid, ECM
26	Evaporative emission (EVAP) canister purge valve	Wiring, EVAP canister purge valve, ECM
34	Idle air control (IAC) valve	Wiring, IAC valve, ECM
41	Intake manifold air control solenoid	Wiring, intake manifold air control solenoid, ECM
55 [1]	Transmission speed sensor	Wiring, transmission speed sensor, ECM
60 [1]	Transmission shift control valve – 1-2	Wiring, transmission shift control valve, ECM
61 [1]	Transmission shift control valve – 2-3	Wiring, transmission shift control valve, ECM
62 [1]	Transmission shift control valve – 3-4	Wiring, transmission shift control valve, ECM
63 [1]	Torque converter clutch (TCC) solenoid	Wiring, TCC solenoid, ECM

[1] ECM incorporates transmission control function

MERCURY

Model:	Engine identification:	Year:
Villager 3.0L	**VIN position 8 = W**	**1993-95**

System: **Nissan ECCS**

Self-diagnosis

General information

- Refer to the front of this manual for general test conditions, terminology, detailed descriptions of wiring faults and a general trouble shooter for electrical and mechanical faults.
- Malfunction indicator lamp (MIL) will illuminate if faults are recorded.
- ECM operates in backup mode if sensors fail, to enable vehicle to be driven to repair shop.
- Trouble codes can be displayed by the MIL or by using a Scan Tool connected to the data link connector (DLC).

Accessing

- Start engine.
- Drive vehicle until engine at normal operating temperature.
- Switch ignition OFF.
- If engine does not start or stalls: Continue with test.
- Switch ignition ON.
- Disconnect data link connector (DLC) 1 (in engine compartment) **1** [**1**].
- Jump DLC 1 terminals (blue/white and gray/blue wires) **1** [**2**].
- Wait 2 seconds.
- Remove jump lead.
- Reconnect data link connector (DLC) 1 **1** [**1**].
- Count MIL flashes. Note trouble codes. Compare with trouble code table.
- Trouble codes are displayed as follows:
  - Each trouble code consists of 2 digits.
  - 0.6 second flashes indicate the 'tens' of the trouble code **2** [**A**].
  - A 0.6 second pause separates each flash of the 'tens' digit **2** [**B**].
  - A 0.9 second pause separates each trouble code digit **2** [**C**].
  - 0.3 second flashes indicate the 'units' of the trouble code **2** [**D**].
  - A 0.3 second pause separates each flash of the 'units' digit **2** [**E**].
  - A 2.1 second pause separates each trouble code **2** [**F**].
  - For example: Trouble code 33 displayed **2**. Heated oxygen sensor (HO2S).
- Switch ignition OFF.
- Trouble codes can also be displayed by using a Scan Tool connected to the data link connector (DLC) 2 **3**.

1

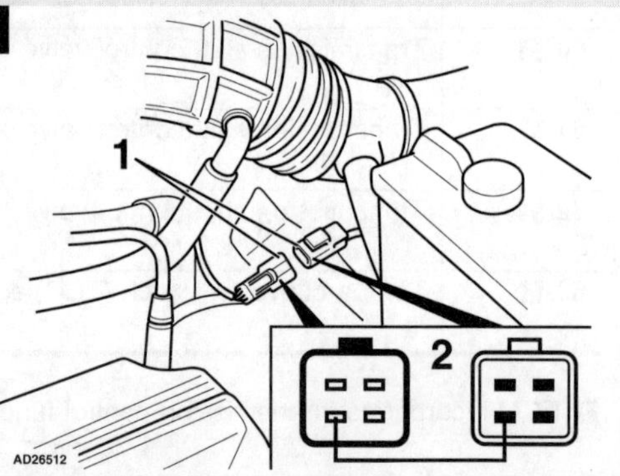

2

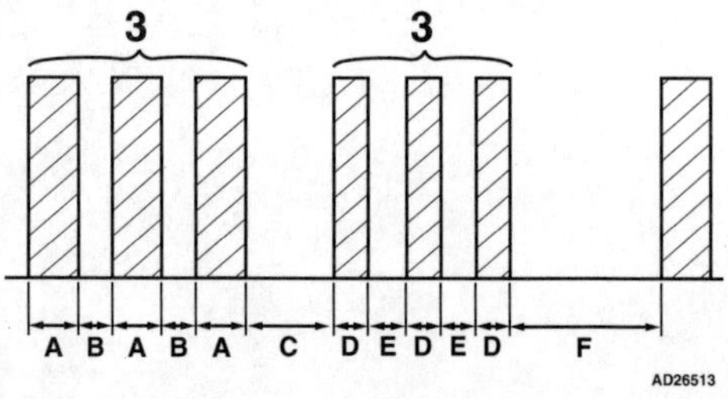

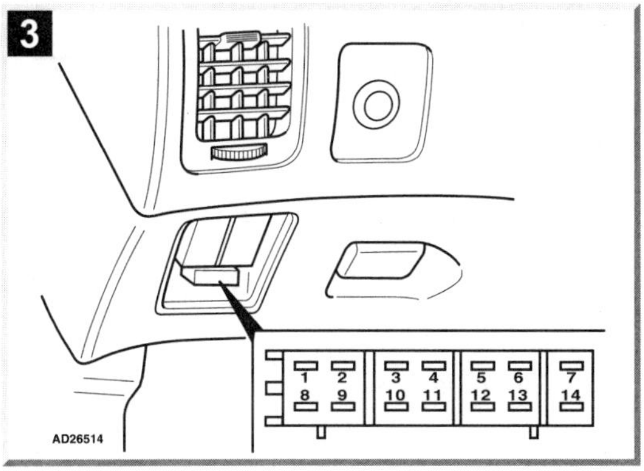

Erasing

Method 1

- Access trouble codes.
- Disconnect data link connector (DLC) 1 (in engine compartment) **1** [**1**].
- Jump DLC 1 terminals (blue/white and gray/blue wires) **1** [**2**].
- Wait 2 seconds.
- Remove jump lead.
- Reconnect data link connector (DLC) 1 **1** [**1**].
- MIL illuminates continuously and trouble codes will be erased.
- Switch ignition OFF.
- Repeat accessing procedure to ensure no trouble codes remain in ECM fault memory.
- Trouble codes can also be erased by using a Scan Tool connected to the data link connector (DLC) 2.

Method 2

NOTE: ***Disconnecting battery ground lead may cause driveability problems until the engine management system relearns basic values.***

- Ensure ignition switched OFF.
- Disconnect battery ground cable for at least 24 hours.
- Repeat accessing procedure to ensure no trouble codes remain in ECM fault memory.
- Trouble codes can also be erased by using a Scan Tool connected to the data link connector (DLC) 2.

NOTE: ***Disconnecting battery ground lead may also erase memory from electronic units, such as radio.***

Trouble code identification

Trouble code	Fault location	Probable cause
11	Crankshaft position (CKP) sensor – circuit malfunction	**Wiring, dirty/damaged rotor, CKP sensor, ECM**
12	Mass air flow (MAF) sensor – range/ performance problem	**Wiring, MAF sensor, ECM**
13	Engine coolant temperature (ECT) sensor – range/performance problem	**Wiring, ECT sensor, ECM**
14	Vehicle speed sensor (VSS) – circuit malfunction	**Wiring to instrument panel, VSS (if speedometer not working), ECM**
21 **1**	Ignition coil – primary circuit malfunction	**Wiring, ignition coil, power transistor, ignition resistor/condenser, ECM**
31	Engine control module (ECM) – malfunction	**ECM**
32	Exhaust gas recirculation (EGR) solenoid – circuit malfunction	**Wiring, EGR solenoid, ECM**
33	Heated oxygen sensor (HO2S) – circuit malfunction	**Wiring, HO2S, ECM**

Trouble code	Fault location	Probable cause
34	Knock sensor (KS) – circuit malfunction	**Wiring, KS, ECM**
35	Exhaust gas recirculation temperature (EGRT) sensor – circuit malfunction	**Wiring, EGRT sensor, ECM**
43	Throttle position (TP) sensor – circuit malfunction	**Wiring, TP sensor, ECM**
45	Injector leak – malfunction	**Injector, ECM**
51	Injector – circuit malfunction	**Wiring, injector, ECM**
55	No fault found	–

1 If trouble code 11 is also displayed, check CKP sensor

Model:	Engine identification:	Year:
Villager 3.0L	VIN position 8 = W	1996-98

System: **Nissan ECCS**

Self-diagnosis

General information

- Refer to the front of this manual for general test conditions, terminology, detailed descriptions of wiring faults and a general trouble shooter for electrical and mechanical faults.
- Malfunction indicator lamp (MIL) will illuminate or flash if certain faults are recorded.
- ECM operates in backup mode if sensors fail, to enable vehicle to be driven to repair shop.
- Trouble codes can be displayed by the MIL or by using a Scan Tool connected to the data link connector (DLC).

Accessing

- Start engine.
- Drive vehicle until engine at normal operating temperature.
- Switch ignition OFF.
- If engine does not start or stalls: Continue with test.
- Switch ignition ON.
- Disconnect data link connector (DLC) 1 (in engine compartment) **1** [**1**].
- Jump DLC 1 terminals 3 & 6 (blue/white and blue wires) **1** [**2**].
- Wait 2 seconds.
- Remove jump lead.
- Reconnect data link connector (DLC) 1 **1** [**1**].
- Count MIL flashes. Note trouble codes. Compare with trouble code table.
- Trouble codes are displayed as follows:
  - Each trouble code consists of 4 digits divided into 2 groups.
  - 0.6 second flashes indicate the 2 digits of the first group **2** [**A**].
  - A 0.6 second pause separates each flash within the first group **2** [**B**].
  - A 0.9 second pause separates the first and second group **2** [**C**].
  - 0.3 second flashes indicate the 2 digits of the second group **2** [**D**].
  - A 0.3 second pause separates each flash within the second group **2** [**E**].
  - A 2.1 second pause separates each trouble code **2** [**F**].

NOTE: ***If total number of flashes within each group is less than 10, a zero must be placed before it to complete the trouble code.***

  - For example: Trouble code 0403 displayed as four 0.6 second flashes, followed by three 0.3 second flashes **2**.
- Switch ignition OFF.
- Trouble codes can also be displayed by using a Scan Tool connected to the data link connector (DLC) 2 or 3 **3**.

1

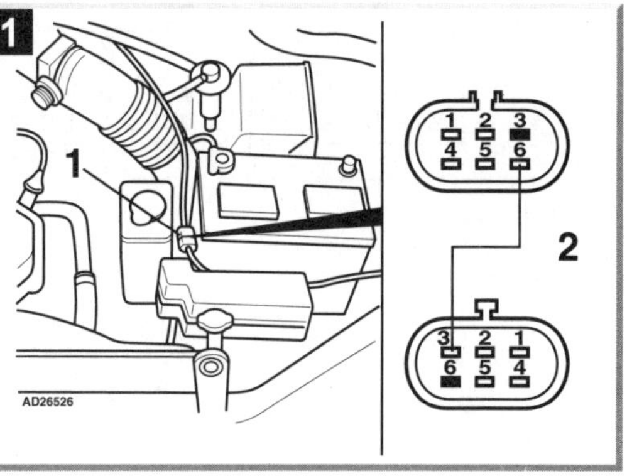

2
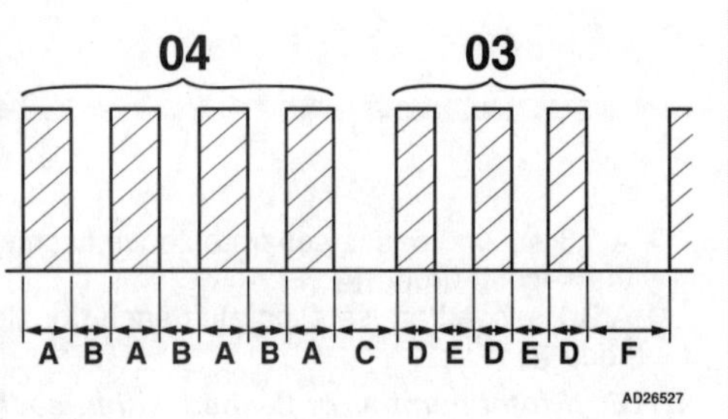

3
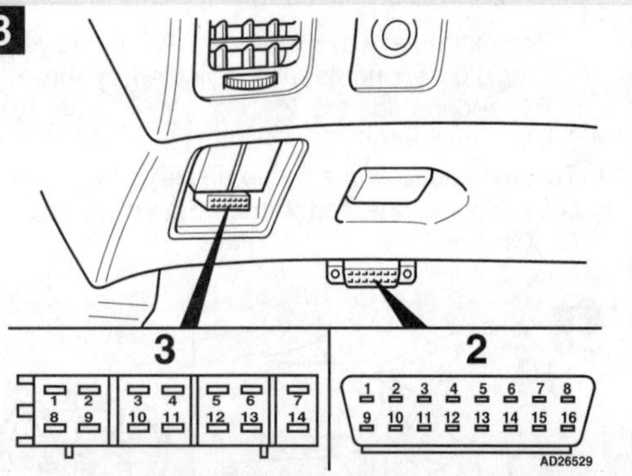

Erasing

Method 1

- Ensure ignition switched OFF for at least 3 seconds.
- Access trouble codes.
- Leave ignition switched ON.
- Disconnect data link connector (DLC) 1 (in engine compartment) **1** [**1**].
- Jump DLC 1 terminals 3 & 6 (blue/white and blue wires) **1** [**2**].
- Wait 2 seconds.
- Remove jump lead.
- Reconnect data link connector (DLC) 1 **1** [**1**].
- Trouble codes will be erased.
- Switch ignition OFF.
- Repeat accessing procedure to ensure no trouble codes remain in ECM fault memory.
- Trouble codes can also be erased by using a Scan Tool connected to the data link connector (DLC) 2 or 3.

Method 2

NOTE: ***Disconnecting battery ground lead may cause driveability problems until the engine management system relearns basic values.***

- Ensure ignition switched OFF.
- Disconnect battery ground cable for at least 24 hours.
- Repeat accessing procedure to ensure no trouble codes remain in ECM fault memory.

NOTE: ***Disconnecting battery ground lead may also erase memory from electronic units, such as radio.***

- Trouble codes can also be erased by using a Scan Tool connected to the data link connector (DLC) 2 or 3.

Trouble code identification

Scan code	Fault location	Probable cause
P0000-P0999	Refer to OBD-II trouble code table at the front of this manual	–
P1105	Manifold absolute pressure (MAP) sensor/ barometric pressure (BARO) sensor	**Wiring, MAP sensor/BARO sensor switch solenoid, hose connection(s), MAP sensor/ BARO sensor**
P1148	Fuel trim (FT) – malfunction	**Wiring, HO2S, fuel system, ECM**
P1217	Engine overheating	**Wiring, ECT sensor, cooling system**

Scan code	Fault location	Probable cause
P1320	Ignition check signal missing	**Wiring, ignition system, CMP sensor**
P1336	Crankshaft position (CKP) sensor	**Wiring, flywheel, CKP sensor**
P1400	Exhaust gas recirculation (EGR) pressure sensor – circuit malfunction, low input	**Wiring, EGR pressure sensor, ECM**
P1401	Exhaust gas recirculation (EGR) temperature sensor	**Wiring, EGR temperature sensor**
P1402	Exhaust gas recirculation (EGR) system – high flow rate	**Wiring, EGR hose(s), EGR valve, ECM**
P1440	Evaporative emission (EVAP) system – small leak detected	**Hose connection(s), fuel filler cap, EVAP canister, EVAP pressure sensor, MAP sensor/BARO sensor, fuel tank vent valve**
P1444	Evaporative emission (EVAP) system – low input	**Wiring, EVAP pressure sensor, ECM**
P1446	Evaporative emission (EVAP) canister purge valve – remains closed	**Wiring, EVAP pressure sensor, EVAP canister purge valve**
P1447	Evaporative emission (EVAP) system – malfunction/leak detected	**Wiring, EVAP canister purge valve, EVAP pressure sensor, EVAP canister, CTP switch, hoses blocked/leaking**
P1448	Evaporative emission (EVAP) canister purge valve – remains open	**Wiring, EVAP pressure sensor, EVAP canister purge valve**
P1490	Evaporative emission (EVAP) canister purge valve bypass solenoid	**Wiring, EVAP canister purge valve bypass solenoid**
P1491	Evaporative emission (EVAP) canister purge valve bypass solenoid – malfunction	**Wiring, EVAP canister, vacuum hoses, EVAP pressure sensor, EVAP canister purge valve bypass solenoid**
P1492	Evaporative emission (EVAP) canister purge valve – circuit malfunction	**Wiring, EVAP canister purge valve**
P1493	Evaporative emission (EVAP) canister purge valve – malfunction, stuck open	**EVAP canister purge valve, vacuum hose(s), EVAP pressure sensor, EVAP canister vent valve**
P1605	Engine control module (ECM)/transmission control module (TCM) – communication error	**Wiring, battery, TCM, ECM**
P1701	Fuel trim (FT) – malfunction	**Intake leak, AIR system, fuel pressure/ pump, injector(s), EGR system, HO2S**
P1705	Throttle position (TP) switch/sensor	**Wiring, TP switch, TCM, ECM**
P1706	Transmission range (TR) switch/park/neutral position (PNP) switch – malfunction	**Wiring, TR switch/PNP switch, TCM**
P1760	Overdrive clutch solenoid valve – circuit malfunction	**Wiring, overdrive clutch solenoid valve**
P1900	Engine coolant blower motor – circuit malfunction	**Wiring, engine coolant blower motor relay, engine coolant blower motor**

Flash code	Fault location	Probable cause
0101	Camshaft position (CMP) sensor – circuit malfunction	**Wiring, CMP sensor, ECM**
0102	Mass air flow (MAF) sensor – circuit malfunction	**Wiring, MAF, ECM**
0103	Engine coolant temperature (ECT) sensor – circuit malfunction	**Wiring, ECT sensor, ECM**
0104	Vehicle speed sensor (VSS) – circuit malfunction	**Wiring, VSS, ECM**
0111	Evaporative emission (EVAP) system – malfunction/leak detected	**Wiring, EVAP canister purge valve, EVAP pressure sensor, EVAP canister, CTP switch, hoses blocked/leaking**
0114	System too rich	**Intake blocked, EVAP canister purge valve, fuel pressure, EGR system, injector(s), HO2S**
0115	Fuel trim (FT) – malfunction	**Wiring, intake/exhaust leak, fuel pressure/ pump, injector(s), EGR system, HO2S, ECM**
0201	Ignition check signal missing	**Wiring, ignition system, CMP sensor**
0203	Closed throttle position (CTP) switch – circuit malfunction	**Wiring, CTP switch, ECM**
0205	Idle speed control (ISC) system – malfunction	**Wiring, ISC actuator/IAC valve, throttle motor, throttle valve tight/sticking, ECM**
0213	Evaporative emission (EVAP) system – small leak detected	**Hose connection(s), fuel filler cap, EVAP canister, EVAP pressure sensor, MAP sensor/BARO sensor, fuel tank vent valve**
0214	Evaporative emission (EVAP) system – low input	**Wiring, EVAP pressure sensor, ECM**
0215	Evaporative emission (EVAP) canister purge valve – remains closed	**Wiring, EVAP pressure sensor, EVAP canister purge valve**
0301	Engine control module (ECM) – ROM error	**ECM**
0302	Exhaust gas recirculation (EGR) system – flow malfunction	**Hose leak/blockage, basic setting not carried out (if applicable), wiring, EGR valve, EGR solenoid, ECM**
0303	Heated oxygen sensor (HO2S) 1, bank 1 – circuit malfunction	**Wiring, heating inoperative, poor connection, HO2S**
0304	Knock sensor (KS) – circuit malfunction	**Wiring, poor connection, KS**
0305	Exhaust gas recirculation (EGR) temperature sensor	**Wiring, EGR temperature sensor**
0306	Exhaust gas recirculation (EGR) system – excessive flow detected	**Hose leak/blockage, basic setting not carried out (if applicable), wiring, EGR valve, EGR solenoid, ECM**
0307 [1]	Heated oxygen sensor (HO2S) 1, bank 1 – circuit malfunction	**Wiring, heating inoperative, poor connection, HO2S**
0307 [2]	Fuel trim (FT) – malfunction	**Wiring, HO2S, fuel system, ECM**
0309	Evaporative emission (EVAP) canister purge valve – remains open	**Wiring, EVAP pressure sensor, EVAP canister purge valve**

Flash code	Fault location	Probable cause
0311	Evaporative emission (EVAP) canister purge valve bypass solenoid – malfunction	**Wiring, EVAP canister, vacuum hoses, EVAP pressure sensor, EVAP canister purge valve bypass solenoid**
0312	Evaporative emission (EVAP) canister purge valve – malfunction, stuck open	**EVAP canister purge valve, vacuum hose(s), EVAP pressure sensor, EVAP canister vent valve**
0401	Intake air temperature (IAT) sensor – circuit malfunction	**Wiring, IAT sensor, ECM**
0402	Fuel temperature sensor A – circuit malfunction	**Wiring, fuel temperature sensor, ECM**
0403	Throttle position (TP) sensor A/accelerator pedal position (APP) sensor A – circuit malfunction	**Wiring, TP/APP sensor, ECM**
0409	Heated oxygen sensor (HO2S) 1, bank 1 – slow response	**Heating inoperative, wiring, HO2S**
0410	Heated oxygen sensor (HO2S) 1, bank 1 – high voltage	**Wiring short circuit to positive, HO2S, ECM**
0411	Heated oxygen sensor (HO2S) 1, bank 1 – low voltage	**Exhaust leak, wiring short circuit to ground, HO2S, ECM**
0412	Heated oxygen sensor (HO2S) 1, bank 1 – no activity detected	**Wiring open circuit, heating inoperative, HO2S**
0504	CAN data bus – malfunction	**Wiring, connected system, ECM**
0505	No fault found	–
0510	Heated oxygen sensor (HO2S) 2, bank 1 – high voltage	**Wiring short circuit to positive, HO2S, ECM**
0511	Heated oxygen sensor (HO2S) 2, bank 1 – low voltage	**Exhaust leak, wiring short circuit to ground, HO2S, ECM**
0512	Heated oxygen sensor (HO2S) 2, bank 1 – no activity detected	**Wiring, heating inoperative, HO2S, ECM**
0514	Exhaust gas recirculation (EGR) system – high flow rate	**Wiring, EGR hose(s), EGR valve, ECM**
0603	Cylinder 6 – misfire detected	**Engine mechanical fault, wiring, ignition/ fuel system, injector, ECT/MAF sensor, ECM**
0604	Cylinder 5 – misfire detected	**Engine mechanical fault, wiring, ignition/ fuel system, injector, ECT/MAF sensor, ECM**
0605	Cylinder 4 – misfire detected	**Engine mechanical fault, wiring, ignition/ fuel system, injector, ECT/MAF sensor, ECM**
0606	Cylinder 3 – misfire detected	**Engine mechanical fault, wiring, ignition/ fuel system, injector, ECT/MAF sensor, ECM**
0607	Cylinder 2 – misfire detected	**Engine mechanical fault, wiring, ignition/ fuel system, injector, ECT/MAF sensor, ECM**
0608	Cylinder 1 – misfire detected	**Engine mechanical fault, wiring, ignition/ fuel system, injector, ECT/MAF sensor, ECM**
0701	Random/multiple cylinder(s) – misfire detected	**Spark plug(s), HT lead(s), injector(s), ignition coil(s), low compression, wiring**

Flash code	Fault location	Probable cause
0702	Catalytic converter system, bank 1 – efficiency below threshold	**Catalytic converter, wiring, HO2S 2**
0704	Evaporative emission (EVAP) pressure sensor – circuit malfunction	**Wiring, EVAP pressure sensor, ECM**
0705	Evaporative emission (EVAP) system – malfunction	**Hose connection(s), intake leak, EVAP canister purge valve**
0707 1	Heated oxygen sensor (HO2S) 2, bank 1 – circuit malfunction	**Heating inoperative, wiring, HO2S, ECM**
0707 2	Heated oxygen sensor (HO2S) 2, bank 1 – slow response	**Heating inoperative, wiring, HO2S**
0714	Evaporative emission (EVAP) system – large leak detected	**Hose connection(s), intake leak, EVAP canister, EVAP canister purge valve**
0801	Evaporative emission (EVAP) canister purge valve bypass solenoid	**Wiring, EVAP canister purge valve bypass solenoid**
0802	Crankshaft position (CKP) sensor – circuit malfunction	**Wiring, CKP sensor, ECM**
0803	Manifold absolute pressure (MAP) sensor/ barometric pressure (BARO) sensor – circuit malfunction	**Wiring, MAP sensor, BARO sensor, ECM**
0804	Engine control module (ECM)/transmission control module (TCM) – communication error	**Wiring, battery, TCM, ECM**
0807	Evaporative emission (EVAP) canister purge valve – circuit malfunction	**Wiring, EVAP canister purge valve**
0901	Heated oxygen sensor (HO2S) 1, bank 1, heater control – circuit malfunction	**Fuse, wiring, HO2S, ECM**
0902	Heated oxygen sensor (HO2S) 2, bank 1, heater control – circuit malfunction	**Wiring, HO2S, ECM**
0903	Evaporative emission (EVAP) system, vent control – circuit malfunction	**Wiring, EVAP canister purge valve, ECM**
0905	Crankshaft position (CKP) sensor	**Wiring, flywheel, CKP sensor**
0908	Insufficient coolant temperature for closed loop fuel control	**Wiring, cooling system, coolant thermostat, ECT sensor**
1003 1	Transmission range (TR) sensor/switch, PRNDL input – circuit malfunction	**Wiring, TR sensor/switch, ECM/TCM**
1003 2	Transmission range (TR) switch/park/neutral position (PNP) switch – malfunction	**Wiring, TR switch/PNP switch, TCM**
1005	Exhaust gas recirculation (EGR) pressure sensor – circuit malfunction, low input	**Wiring, EGR pressure sensor, ECM**
1008	Evaporative emission (EVAP) canister purge valve – circuit malfunction	**Wiring, EVAP canister purge valve, ECM**
1101	Transmission range (TR) sensor/switch, PRNDL input – circuit malfunction	**Wiring, TR sensor/switch, ECM/TCM**
1102	Output shaft speed (OSS) sensor – circuit malfunction	**Wiring, OSS sensor, ECM**
1103	Gear 1 – incorrect ratio	**Wiring, TR sensor/switch, shift solenoids, transmission mechanical fault**

Flash code	Fault location	Probable cause
1104	Gear 2 – incorrect ratio	**Wiring, TR sensor/switch, shift solenoids, transmission mechanical fault**
1105	Gear 3 – incorrect ratio	**Wiring, TR sensor/switch, shift solenoids, transmission mechanical fault**
1106	Gear 4 – incorrect ratio	**Wiring, TR sensor/switch, shift solenoids, transmission mechanical fault**
1107	Torque converter clutch (TCC) solenoid – circuit intermittent	**Wiring, poor connection, TCC solenoid, ECM/TCM**
1108	Shift solenoid (SS) A – circuit malfunction	**Wiring, shift solenoid, ECM/TCM**
1201	Shift solenoid (SS) B – circuit malfunction	**Wiring, shift solenoid, ECM/TCM**
1203	Overdrive clutch solenoid valve – circuit malfunction	**Wiring, overdrive clutch solenoid valve**
1204	Torque converter clutch (TCC) solenoid – circuit malfunction	**Wiring, TCC solenoid, ECM/TCM**
1205	Transmission fluid pressure (TFP) solenoid – circuit malfunction	**Wiring, TFP solenoid, ECM/TCM**
1206	Throttle position (TP) switch/sensor	**Wiring, TP switch, TCM, ECM**
1207	Engine RPM input – circuit malfunction	**Wiring, CKP/RPM sensor, ECM/TCM**
1208	Transmission fluid temperature (TFT) sensor – circuit malfunction	**Wiring, TFT sensor, ECM, ECM/TCM**
1302	Manifold absolute pressure (MAP) sensor/ barometric pressure (BARO) sensor	**Wiring, MAP sensor/BARO sensor switch solenoid, hose connection(s), MAP sensor/ BARO sensor**
1308	Engine coolant blower motor – circuit malfunction	**Wiring, engine coolant blower motor relay, engine coolant blower motor**

1 →1997
2 1998→

MERCURY

Model:	Engine identification:	Year:
Villager 3.3L	**VIN position 8 = T**	**1999-02**

System: **Nissan ECCS**

Self-diagnosis

General information

- Refer to the front of this manual for general test conditions, terminology, detailed descriptions of wiring faults and a general trouble shooter for electrical and mechanical faults.
- Malfunction indicator lamp (MIL) will illuminate or flash if certain faults are recorded.
- ECM operates in backup mode if sensors fail, to enable vehicle to be driven to repair shop.

Accessing

- Trouble codes can only be displayed by using a Scan Tool connected to the data link connector (DLC) **1**.

1

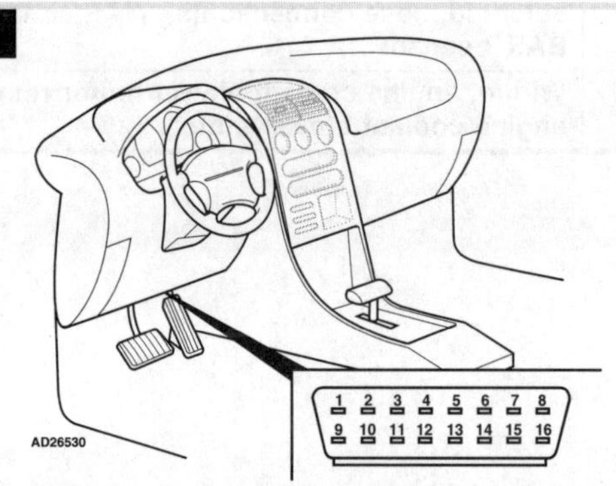

Erasing

Method 1

- Trouble codes can be erased by using a Scan Tool connected to the data link connector (DLC).

Method 2

NOTE: ***Disconnecting battery ground lead may cause driveability problems until the engine management system relearns basic values.***

- Ensure ignition switched OFF.
- Disconnect battery ground cable for at least 24 hours.
- Repeat accessing procedure to ensure no trouble codes remain in ECM fault memory.

NOTE: ***Disconnecting battery ground lead may also erase memory from electronic units, such as radio.***

Trouble code identification

Scan code	Fault location	Probable cause
P0000-P0999	Refer to OBD-II trouble code table at the front of this manual	–
P1105	Manifold absolute pressure (MAP) sensor/ barometric pressure (BARO) sensor	**Wiring, MAP sensor/BARO sensor switch solenoid, hose connection(s), MAP sensor/ BARO sensor**

Scan code	Fault location	Probable cause
P1148	Fuel trim (FT) – malfunction	**Wiring, HO2S, fuel system, ECM**
P1217	Engine overheating	**Wiring, ECT sensor, cooling system**
P1320	Ignition check signal missing	**Wiring, ignition system, CMP sensor**
P1336	Crankshaft position (CKP) sensor	**Wiring, flywheel, CKP sensor**
P1400	Exhaust gas recirculation (EGR) pressure sensor – circuit malfunction, low input	**Wiring, EGR pressure sensor, ECM**
P1401	Exhaust gas recirculation (EGR) temperature sensor	**Wiring, EGR temperature sensor**
P1402	Exhaust gas recirculation (EGR) system – high flow rate	**Wiring, EGR hose(s), EGR valve, ECM**
P1440	Evaporative emission (EVAP) system – small leak detected	**Hose connection(s), fuel filler cap, EVAP canister, EVAP pressure sensor, MAP sensor/BARO sensor, fuel tank vent valve**
P1444	Evaporative emission (EVAP) system – low input	**Wiring, EVAP pressure sensor, ECM**
P1446	Evaporative emission (EVAP) canister purge valve – remains closed	**Wiring, EVAP pressure sensor, EVAP canister purge valve**
P1447	Evaporative emission (EVAP) system – malfunction/leak detected	**Wiring, EVAP canister purge valve, EVAP pressure sensor, EVAP canister, CTP switch, hoses blocked/leaking**
P1448	Evaporative emission (EVAP) canister purge valve – remains open	**Wiring, EVAP pressure sensor, EVAP canister purge valve**
P1490	Evaporative emission (EVAP) canister purge valve bypass solenoid	**Wiring, EVAP canister purge valve bypass solenoid**
P1491	Evaporative emission (EVAP) canister purge valve bypass solenoid – malfunction	**Wiring, EVAP canister, vacuum hoses, EVAP pressure sensor, EVAP canister purge valve bypass solenoid**
P1492	Evaporative emission (EVAP) canister purge valve – circuit malfunction	**Wiring, EVAP canister purge valve**
P1493	Evaporative emission (EVAP) canister purge valve – malfunction, stuck open	**EVAP canister purge valve, vacuum hose(s), EVAP pressure sensor, EVAP canister vent valve**
P1605	Engine control module (ECM)/transmission control module (TCM) – communication error	**Wiring, battery, TCM, ECM**
P1701	Fuel trim (FT) – malfunction	**Intake leak, AIR system, fuel pressure/ pump, injector(s), EGR system, HO2S**
P1705	Throttle position (TP) switch/sensor	**Wiring, TP switch, TCM, ECM**
P1706	Transmission range (TR) switch/park/neutral position (PNP) switch – malfunction	**Wiring, TR switch/PNP switch, TCM**
P1760	Overdrive clutch solenoid valve – circuit malfunction	**Wiring, overdrive clutch solenoid valve**
P1900	Engine coolant blower motor – circuit malfunction	**Wiring, engine coolant blower motor relay, engine coolant blower motor**

OLDSMOBILE

Model:	Engine identification:	Year:
Achieva 2.3L	VIN position 8 = D/A/3	1992-95
Achieva 3.1L	VIN position 8 = M	1994-95
Bravada 4.3L	VIN position 8 = Z	1992
Bravada 4.3L	VIN position 8 = W	1993-95
Cutlass Ciera/Cruiser 2.5L	VIN position 8 = R	1992
Cutlass Ciera/Cruiser 2.2L	VIN position 8 = 4	1994-95
Cutlass Supreme 3.1L	VIN position 8 = M	1993-95
Cutlass Supreme 3.1L	VIN position 8 = T	1992-93
Cutlass Supreme 3.4L	VIN position 8 = X	1992-93
Silhouette 3.1L	VIN position 8 = D	1992-95

System: **GM**

Self-diagnosis

General information

- Refer to the front of this manual for general test conditions, terminology, detailed descriptions of wiring faults and a general trouble shooter for electrical and mechanical faults.
- Malfunction indicator lamp (MIL) will illuminate if certain faults are recorded.
- ECM operates in backup mode if sensors fail, to enable vehicle to be driven to repair shop.
- Trouble codes can be displayed by the MIL (some models), or by using a Scan Tool connected to the data link connector (DLC).

Accessing

Models with data link connector (DLC) terminal B connected

- Access data link connector (DLC).
- Jump data link connector (DLC) terminals A and B **1**.
- Switch ignition ON. Do NOT start engine.
- Count MIL flashes. Note trouble codes. Compare with trouble code table.
- Firstly trouble code 12 will be displayed 3 times.
- Trouble codes consist of 2 groups, displayed as follows:
  - First group indicates the 'tens' of the trouble code **2** [**A**].
  - Second group indicates the 'units' of the trouble code **2** [**C**].
  - A short pause separates each group **2** [**B**].
  - A long pause separates each trouble code **2** [**D**].
  - For example: Trouble code 12 displayed **2**. Start of diagnosis.
- Each additional trouble code will now be displayed 3 times, in ascending order.

NOTE: ***If no additional trouble codes are stored, code 12 will continue to be displayed.***

- Trouble codes can also be displayed by using a Scan Tool connected to the data link connector (DLC) **1**.

1

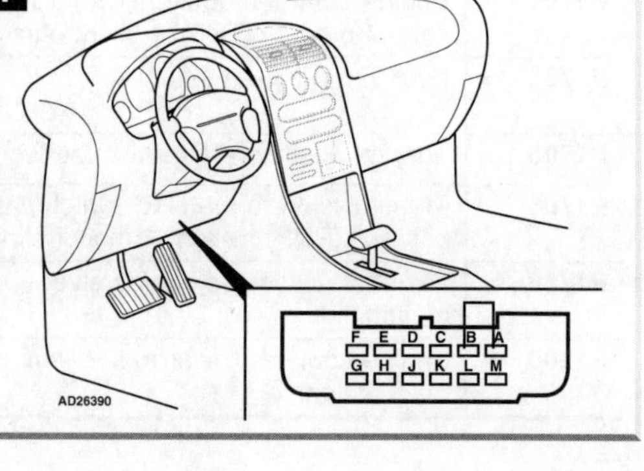

2

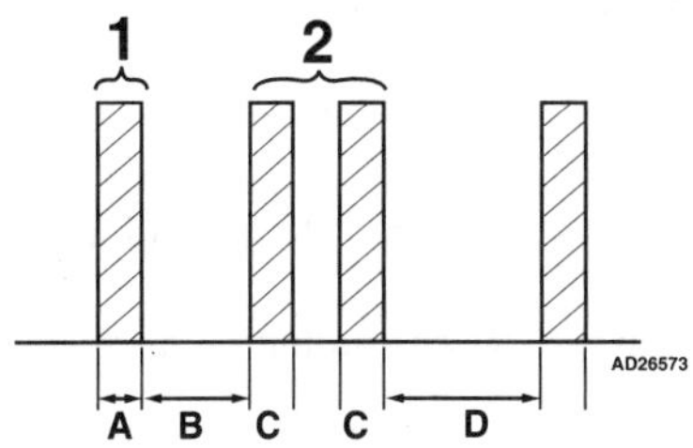

Models without data link connector (DLC) terminal B connected

- Trouble codes can only be displayed by using a Scan Tool connected to the data link connector (DLC) 1.

Erasing

NOTE: ***Disconnecting battery ground lead may cause driveability problems until the engine management system relearns basic values.***

- Ensure ignition switched OFF.
- Disconnect battery ground cable for 30 seconds.
- Reconnect battery ground cable.
- Repeat checking procedure to ensure no trouble codes remain in ECM fault memory.

NOTE: ***Disconnecting battery ground lead may also erase memory from electronic units, such as radio.***

- Trouble codes can also be erased by using a Scan Tool connected to the data link connector (DLC).

Trouble code identification

Flash code	Fault location	Probable cause
12	Start of diagnosis	**No engine speed (RPM) signal received**
13	Oxygen sensor (O2S)/heated oxygen sensor (HO2S)	**Wiring open circuit, oxygen sensor(O2S)/ heated oxygen sensor (HO2S), ECM**
14	Engine coolant temperature (ECT) sensor – high temperature indicated	**Wiring short circuit to ground, ECT sensor, ECM**
15	Engine coolant temperature (ECT) sensor – low temperature indicated	**Wiring open circuit, ECT sensor, ECM**
16 6	Ignition control module (ICM) – 2X reference circuit malfunction	**Wiring open circuit/short circuit to ground/ short circuit to positive, ICM, ECM**
16 7	Battery voltage – low	**Wiring, alternator**
16 9	Vehicle speed sensor (VSS) buffer – signal low	**Wiring, VSS buffer, ECM**
17	Camshaft position (CMP) sensor	**Wiring open circuit/short circuit to ground/ short circuit to positive, CMP sensor, mechanical fault, ECM**
19	Ignition control module (ICM) – 7X reference circuit malfunction	**Wiring, CKP sensor, PROM, ECM**
21	Throttle position (TP) sensor – voltage high	**Wiring open circuit/short circuit to positive, TP sensor, ECM**
22	Throttle position (TP) sensor – voltage low	**Wiring open circuit/short circuit to ground, TP sensor, ECM**

Flash code	Fault location	Probable cause
23	Intake air temperature (IAT) sensor – low temperature indicated	**Wiring open circuit, IAT sensor, ECM**
24	Vehicle speed sensor (VSS)	**Wiring open circuit/short circuit to ground, VSS, ECM**
25	Intake air temperature (IAT) sensor – high temperature indicated	**Wiring short circuit to ground, IAT sensor, ECM**
26 [6]	Engine control module quad driver/output driver	**Wiring, ECM controlled components, ECM [10]**
26 [7]	Engine control module quad driver/output driver	**Wiring, ECM controlled components, ECM [10]**
27	Engine control module quad driver/output driver	**Wiring, ECM controlled components, ECM [10]**
28 [4] [8]	Engine control module quad driver/output driver	**Wiring, ECM controlled components, ECM [10]**
28 [5]	Transmission range (TR) switch	**Wiring open circuit/short circuit to ground/ short circuit to positive, TR switch, ECM**
29	Engine control module quad driver/output driver	**Wiring, ECM controlled components, ECM [10]**
31	Park/neutral position (PNP) switch	**Wiring short circuit to ground, incorrectly adjusted, PNP switch, ECM**
32	Exhaust gas recirculation (EGR) solenoid	**Wiring open circuit/short circuit to ground, supply wire, EGR valve, ECM**
33	Manifold absolute pressure (MAP) sensor – voltage high, low vacuum indicated	**Wiring short circuit to positive, hose blocked/leaking, MAP sensor, ECM**
34	Manifold absolute pressure (MAP) sensor – voltage low, high vacuum indicated	**Wiring open circuit/short circuit to ground, MAP sensor, ECM**
35	Idle air control (IAC) valve	**Wiring open circuit/short circuit to ground/ short circuit to positive, IAC valve, intake system leaking, fuel system, PCV valve, A/C system, mechanical fault, ECM**
36	24X crankshaft position sensor (CKP) – circuit malfunction	**Wiring open circuit/short circuit to ground, 24X CKP sensor, ECM**
37	Brake pedal position (BPP) switch	**Wiring open circuit/short circuit to ground, incorrectly adjusted, BPP switch, ECM**
39	Clutch pedal position (CPP) switch	**Wiring open circuit/short circuit to ground, incorrectly adjusted, CPP switch, ECM**
41 [1]	Engine control module (ECM) – PROM	**insecure PROM, incorrect PROM**
41 [2] [8]	Ignition control module (ICM) – circuit malfunction	**Wiring open circuit, ignition control module (ICM), ECM**
41 [6]	Ignition control module (ICM) – 1X circuit malfunction	**Wiring open circuit/short circuit to ground/ short circuit to positive, ICM, ECM**
42	Ignition control module (ICM) – circuit malfunction	**Wiring open circuit/short circuit to ground, ignition control module (ICM), ECM**
43	Knock sensor (KS)	**Wiring open circuit/short circuit to ground/ short circuit to positive, KS incorrectly fitted, KS, insecure PROM, ECM**

Flash code	Fault location	Probable cause
44	Oxygen sensor (O2S)/heated oxygen sensor (HO2S) – mixture lean	**Wiring open circuit/short circuit to ground, injector(s) blocked, fuel contamination, fuel pressure low, exhaust system leaking, O2S/HO2S, ECM**
45	Oxygen sensor (O2S)/heated oxygen sensor (HO2S) – mixture rich	**Wiring open circuit, fuel pressure high, injector(s) leaking, O2S contamination, evaporative emission (EVAP) canister purge valve, throttle position (TP) sensor, EGR system, O2S/HO2S, ECM**
46	Immobilizer control module – fuel enable signal	**Incorrect ignition key, wiring open circuit/ short circuit to ground/short circuit to positive, immobilizer control module, ECM**
51	Engine control module (ECM) – PROM	**Insecure PROM, incorrect PROM, ECM**
52	Engine control module (ECM)	**ECM**
53	Battery voltage – high/low	**Wiring, alternator**
54	Fuel pump circuit – voltage low	**Wiring open circuit/short circuit to ground, fuel pump relay, engine oil pressure warning lamp switch**
55 4 6 8	Fuel system – lean during acceleration	**Fuel filter, fuel pump, HO2S, injectors**
55 5 7	Engine control module (ECM)	**Ground wiring open circuit, insecure PROM, ECM**
58	Transmission fluid temperature (TFT) sensor – voltage low	**Wiring short circuit to ground, TFT sensor, ECM**
59	Transmission fluid temperature (TFT) sensor – voltage high	**Wiring open circuit, TFT sensor, ECM**
61	Oxygen sensor (O2S)/heated oxygen sensor (HO2S)	**Oxygen sensor (O2S)/heated oxygen sensor (HO2S)**
62	Transmission gear switch signal circuit	**Wiring open circuit/short circuit to ground, 2nd gear switch, 3rd gear switch, ECM**
65	Injectors – circuit malfunction	**Wiring open circuit/short circuit to ground/ short circuit to positive, injectors, ECM**
66	A/C refrigerant pressure sensor – voltage low	**Wiring open circuit/short circuit to ground/ short circuit to positive, A/C refrigerant pressure sensor, ECM**
70	A/C refrigerant pressure sensor – voltage high	**Wiring open circuit/short circuit to positive, A/C refrigerant pressure sensor, ECM**
72	Vehicle speed sensor (VSS)	**Wiring open circuit/short circuit to ground, VSS, ECM**
75	Exhaust gas recirculation (EGR) valve solenoid 1	**Wiring open circuit/short circuit to ground, supply wire, EGR valve, ECM**
76	Exhaust gas recirculation (EGR) valve solenoid 2	**Wiring open circuit/short circuit to ground, supply wire, EGR valve, ECM**
77	Exhaust gas recirculation (EGR) valve solenoid 3	**Wiring open circuit/short circuit to ground, supply wire, EGR valve, ECM**

Flash code	Fault location	Probable cause
79 2 8	Transmission fluid temperature (TFT) – high	**Transmission mechanical fault, towing**
79 3	Vehicle speed sensor (VSS) – high	**Wiring, VSS weight limit exceeded**
80 3	Vehicle speed sensor (VSS) – low	**Wiring, VSS**
80 2 8	Torque converter clutch (TCC) – slipping	**Wiring open circuit/short circuit to ground, supply wire, transmission internal wiring, TCC solenoid, transmission mechanical fault, ECM**
82	Ignition control module (ICM) – 3X reference circuit malfunction	**Wiring open circuit/short circuit to ground, 3X CKP signal sensor, ICM**
85	Engine control module (ECM) – PROM	**insecure PROM, incorrectly coded**
85	Engine control module (ECM) – PROM	**Insecure PROM, incorrectly coded**
86	Engine control module (ECM) – analogue/ digital converter	**A/C refrigerant pressure sensor/knock sensor (KS)/engine coolant temperature (ECT) sensor/fuel pump monitor circuit short circuit to battery voltage**
87	Engine control module (ECM) – EEPROM	**Insecure PROM, incorrectly coded**
90	Torque converter clutch (TCC) – stuck ON	**Wiring open circuit/short circuit to ground, transmission internal wiring, TCC solenoid, transmission mechanical fault, ECM**
96	Engine control module (ECM) – ignition supply voltage low	**Wiring open circuit, alternator, ECM**
98	Engine control module (ECM) – incorrectly coded	**ECM incorrectly coded, ECM**
99	Engine control module (ECM) – incorrectly coded	**ECM incorrectly coded, ECM**

1 Cutlass Supreme – VIN T (except California)/X.
2 Cutlass Supreme – VIN M.
3 Cutlass Supreme – VIN T (California).
4 Cutlass Ciera/Cruiser 2.2L 1994.
5 Except Cutlass Ciera/Cruiser 2.2L 1994.
6 Achieva 2.3L 1992.
7 Except Achieva 2.3L 1992/Bravada 4.3L – VIN W.
8 Achieva 3.1L.
9 Bravada 4.3L – VIN W.
10 Depending on model, year and equipment level, the following components may be connected to this circuit. Use of the vehicle wiring diagram is recommended to confirm which components are fitted

- ☐ A/C compressor clutch relay
- ☐ A/C condenser blower motor relay
- ☐ Engine coolant blower motor relay
- ☐ Engine coolant 'hot' warning lamp
- ☐ Evaporative emission (EVAP) canister purge valve
- ☐ Exhaust gas recirculation (EGR) solenoid
- ☐ Malfunction indicator lamp (MIL)
- ☐ Torque converter clutch (TCC) solenoid
- ☐ Transmission shift solenoids (SS)
- ☐ Transmission shift warning lamp (MT)

OLDSMOBILE

Model:	Engine identification:	Year:
Achieva 2.4L	VIN position 8 = T	1996-98
Alero 2.4L	VIN position 8 = T	1999-00
Alero 2.2L (L61)	VIN position 8 = T	2002
Cutlass Ciera 2.2L	VIN position 8 = 4	1996

System: **GM**

Self-diagnosis

General information

- Refer to the front of this manual for general test conditions, terminology, detailed descriptions of wiring faults and a general trouble shooter for electrical and mechanical faults.
- Malfunction indicator lamp (MIL) will illuminate or flash if certain faults are recorded.
- ECM operates in backup mode if sensors fail, to enable vehicle to be driven to repair shop.

Accessing

- Trouble codes can only be displayed by using a Scan Tool connected to the data link connector (DLC):
  - Achieva **1**.
  - Alero/Cutlass Ciera **2**.

1

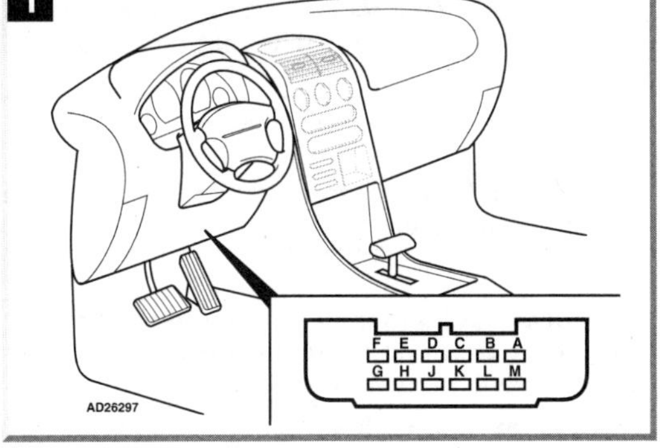

2

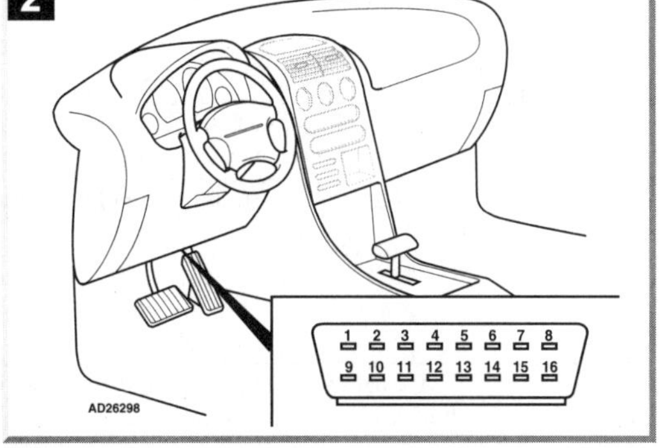

Erasing

- After the faults have been rectified, the manufacturer recommends that trouble codes are erased by using a Scan Tool connected to the data link connector (DLC).

Trouble code identification

Scan code	Fault location	Probable cause
P0000-0999	Refer to OBD-II trouble code table in the front of this manual	–
P1133	Heated oxygen sensor (HO2S)/oxygen sensor (O2S) 1 – insufficient switching	**Wiring, oxygen sensor heater, exhaust system leaking, HO2S**
P1137	Heated oxygen sensor (HO2S)/oxygen sensor (O2S) 2 – insufficient switching	**Wiring,, HO2S/O2S, ECM**
P1138	Heated oxygen sensor (HO2S)/oxygen sensor (O2S) 2 – insufficient switching	**Wiring,, HO2S/O2S, ECM**
P1171	Fuel system – lean during acceleration	**Fuel filter, fuel pump, HO2S/O2S, injectors**
P1336	Crankshaft position sensor (CKP) – variation not learned	**learning procedure not performed**
P1380	Rough road signal – signal error	**ABS control module trouble code(s) stored**
P1381	Rough road signal – no communication from antilock brake system (ABS) CAN data bus	**Wiring open circuit, ABS system**
P1404	Exhaust gas recirculation (EGR) valve – stuck open	**Wiring open circuit/short circuit to positive, EGR valve position sensor, EGR solenoid**
P1406	Exhaust gas recirculation (EGR) valve position sensor – circuit malfunction	**Wiring open circuit/short circuit to ground/ short circuit to positive, EGR valve position sensor, EGR valve**
P1441	Evaporative emission (EVAP) system – leak detected	**Hose connection(s), EVAP canister, EVAP canister purge valve**
P1520	Park/neutral position (PNP) switch	**Wiring short circuit to ground, incorrectly adjusted**
P1601	CAN data bus – circuit malfunction	**Wiring open circuit/short circuit to ground/ short circuit to positive, instrument panel, ECM**
P1600	Engine control module (ECM)	**ECM**
P1601	Data bus – circuit malfunction	**Wiring open circuit/short circuit to ground/ short circuit to positive, instrument panel, ECM**
P1621	Engine control module (ECM) – memory performance	**ECM**
P1626	Data bus, immobilizer	**Wiring, body control module (BCM) trouble code(s) stored**
P1627	Engine control module (ECM)	**ECM**
P1629	Immobilizer control module – fuel enable signal	**Incorrect ignition key, wiring open circuit/ short circuit to ground/short circuit to positive, immobilizer control module, ECM**
P1630	Engine control module (ECM), immobilizer code learning mode	**Immobilizer code not programmed**
P1631	Engine control module (ECM) – incorrect immobilizer code	**Incorrect key, immobilizer code not programmed**
P1632	Engine control module (ECM), fuel disable signal	**Fuses, BCM, immobilizer code**

Scan code	Fault location	Probable cause
P1680	Engine control module (ECM)	**ECM**
P1681	Engine control module (ECM)	**ECM**
P1683	Engine control module (ECM)	**ECM**
U1016	Instrument panel, CAN data bus – communication malfunction	**Wiring open circuit/short circuit to ground/ short circuit to positive, instrument panel, ECM**

OLDSMOBILE

Model:	Engine identification:	Year:
Achieva 3.1L	VIN position 8 = M	1996-97
Aurora 3.8L	VIN position 8 = 1/K	1995
Bravada 4.3L	VIN position 8 = W	1996-02
Cutlass Supreme 3.1L	VIN position 8 = M	1996-97
Cutlass Supreme 3.4L	VIN position 8 = X	1994-97
Cutlass Ciera/Cruiser 3.1L	VIN position 8 = M	1994-96
Eighty Eight 3.8L	VIN position 8 = 1/K/L	1994-97
Ninety Eight/Regency 3.8L	VIN position 8 = 1/L	1994-96
Silhouette 3.4L	VIN position 8 = E	1996-97
Silhouette 3.8L	VIN position 8 = L	1994-95

System: **GM**

Self-diagnosis

General information

- Refer to the front of this manual for general test conditions, terminology, detailed descriptions of wiring faults and a general trouble shooter for electrical and mechanical faults.
- Malfunction indicator lamp (MIL) will illuminate or flash if certain faults are recorded.
- ECM operates in backup mode if sensors fail, to enable vehicle to be driven to repair shop.

Accessing

- Trouble codes can only be displayed by using a Scan Tool connected to the data link connector (DLC).
- 12 pin connector **1** :
  - Cutlass Supreme →95.
  - Cutlass Ciera/Cruiser →95.
  - Silhouette →95.
- 16 pin connector **2** :
  - Cutlass Supreme 96→.
  - Cutlass Ciera/Cruiser 96→.
  - Silhouette 96→.
  - Achieva.
  - Aurora.
  - Bravada.
  - Eighty eight.
  - Ninety eight.
  - Regency.

1

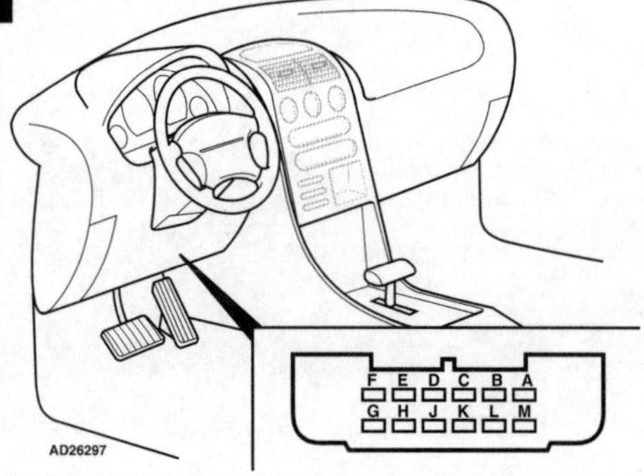

2

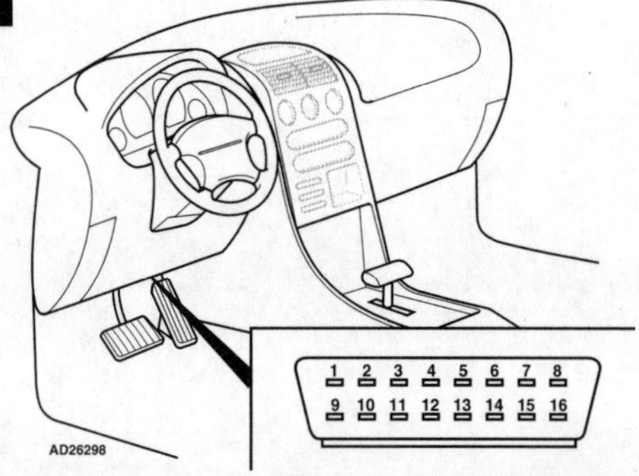

Erasing

12 pin connector

NOTE: ***Disconnecting battery ground lead may cause driveability problems until the engine management system relearns basic values.***

- Ensure ignition switched OFF.
- Disconnect battery ground cable for 30 seconds.
- Reconnect battery ground cable.
- Repeat checking procedure to ensure no trouble codes remain in ECM fault memory.

NOTE: ***Disconnecting battery ground lead may also erase memory from electronic units, such as radio.***

- Trouble codes can also be erased by using a Scan Tool connected to the data link connector (DLC).

16 pin connector

- After the faults have been rectified, the manufacturer recommends that trouble codes are erased by using a Scan Tool connected to the data link connector (DLC).

Trouble code identification

Scan code	Fault location	Probable cause
P0000-0999	Refer to OBD-II trouble code table in the front of this manual	–
P1106	Manifold absolute pressure (MAP) sensor – circuit intermittent – voltage high	**Wiring open circuit/short circuit to ground/ short circuit to positive**
P1107	Manifold absolute pressure (MAP) sensor – circuit intermittent – voltage low	**Wiring short circuit to ground**
P1108	Barometric pressure (BARO) sensor/manifold absolute pressure (MAP) sensor – comparison too high	**Wiring open circuit/short circuit to positive, MAP sensor, ECM**
P1111	Intake air temperature (IAT) sensor – circuit intermittent – voltage high	**Wiring open circuit, ground lead defective**
P1112	Intake air temperature (IAT) sensor – circuit intermittent – voltage low	**Wiring short circuit to ground**
P1114	Engine coolant temperature (ECT) sensor – circuit malfunction – voltage low	**Wiring short circuit to ground**
P1115	Engine coolant temperature (ECT) sensor – circuit malfunction – voltage high	**Wiring open circuit, ground lead defective**
P1121	Throttle position (TP) sensor – circuit malfunction – voltage high	**Wiring short circuit to positive, ground lead defective**
P1120	Throttle motor position sensor 1 – circuit malfunction	**Wiring open circuit/short circuit to ground / short circuit to positive, throttle control unit, ECM**
P1122	Throttle position (TP) sensor – circuit malfunction – voltage low	**Wiring open circuit/short circuit to ground**
P1133	Heated oxygen sensor (HO2S) 1, bank 1 – insufficient switching	**Wiring, oxygen sensor heater, exhaust system leaking, HO2S**
P1134	Heated oxygen sensor (HO2S) 1, bank 1 – slow response	**Wiring, oxygen sensor heater, exhaust system leaking, HO2S**
P1137	Heated oxygen sensor (HO2S) 2 – voltage low	**Wiring, oxygen sensor heater, exhaust system leaking, HO2S**
P1138	Heated oxygen sensor (HO2S) 2 – voltage high	**Wiring short circuit to positive, fuel pressure high, fuel contamination, injector(s) leaking, HO2S, ECM**

Scan code	Fault location	Probable cause
P1153	Heated oxygen sensor (HO2S) 1, bank 2 – insufficient switching	**Wiring, oxygen sensor heater, exhaust system leaking, HO2S**
P1154	Heated oxygen sensor (HO2S) 1, bank 2 – slow response	**Wiring, oxygen sensor heater, exhaust system leaking, HO2S**
P1171	Heated oxygen sensor (HO2S) 1 – voltage low	**Fuel supply blocked, injector(s) blocked, manifold absolute pressure (MAP) sensor wiring**
P1200	Injector(s) – circuit malfunction	**Wiring open circuit/short circuit to ground/ short circuit to positive, injectors, ECM**
P1220	Throttle motor position sensor 2 – circuit malfunction	**Wiring open circuit/short circuit to ground/ short circuit to positive, throttle control unit, ECM**
P1221	Throttle motor position sensor 1/2 – signal variation	**Wiring, throttle control unit, ECM**
P1257	Supercharger (SC) – over-boost	**Hoses blocked/leaking/interchanged/not connected, wiring short circuit to ground, supercharger (SC) bypass valve tight/ sticking, supercharger (SC) bypass valve solenoid, ECM**
P1271	Accelerator pedal position (APP) sensor 1/2 – signal variation	**Wiring, APP sensor, ECM**
P1275	Accelerator pedal position (APP) sensor 1 – circuit malfunction	**Wiring open circuit/short circuit to ground/ short circuit to positive, APP sensor, ECM**
P1280	Accelerator pedal position (APP) sensor 2 – circuit malfunction	**Wiring open circuit/short circuit to ground/ short circuit to positive, APP sensor, ECM**
P1336	Crankshaft position sensor (CKP) – variation not learned	**Learning procedure not performed**
P1345	Crankshaft position sensor (CKP)/camshaft position (CMP) sensor – signal variation	**CKP sensor, distributor/camshaft alignment, mechanical fault**
P1350	Ignition control module (ICM) – bypass line monitor circuit malfunction	**Wiring open circuit/short circuit to ground/ short circuit to positive, ICM, ECM**
P1351	Ignition control module (ICM) – IC circuit, voltage high	**Wiring open circuit, ICM, ECM**
P1361	Ignition control module (ICM) – IC circuit, voltage low	**Wiring short circuit to ground, ICM, ECM**
P1374	Ignition control module (ICM) – 3X reference circuit malfunction	**Wiring open circuit/short circuit to ground/ short circuit to positive, wiring routed near spark plug wires, ignition coils arcing/ tracking, ICM, ECM**
P1380	Rough road signal – signal error	**ABS control module trouble code(s) stored**
P1381	Rough road signal – no communication from ABS CAN data bus	**Wiring open circuit, ABS system**
P1403	Exhaust gas recirculation (EGR) valve solenoid 1	**Wiring open circuit/short circuit to ground, supply wire, EGR valve, ECM**
P1404 5	Exhaust gas recirculation (EGR) valve – stuck open	**Wiring open circuit/short circuit to positive, exhaust gas recirculation (EGR) valve position sensor, EGR solenoid**

Scan code	Fault location	Probable cause
P1404 6	Exhaust gas recirculation (EGR) valve solenoid 2	**Wiring open circuit/short circuit to ground, supply wire, EGR valve, ECM**
P1405	Exhaust gas recirculation (EGR) valve solenoid 3	**Wiring open circuit/short circuit to ground, supply wire, EGR valve, ECM**
P1406	Exhaust gas recirculation (EGR) valve position sensor – circuit malfunction	**Wiring open circuit/short circuit to ground/ short circuit to positive, EGR valve position sensor, EGR valve**
P1415	Secondary air injection (AIR) system	**Wiring, hoses blocked/leaking, AIR valve**
P1416	Secondary air injection (AIR) system	**Wiring, hoses blocked/leaking, AIR valve, AIR pump relay**
P1441	Evaporative emission (EVAP) system – leak detected	**Hose connection(s), EVAP canister, EVAP canister purge valve**
P1442	Evaporative emission (EVAP) system vacuum switch – circuit malfunction	**Hoses blocked/leaking, wiring, EVAP canister, ECM**
P1483	Engine cooling system – performance	**Cooling system, ECT sensor, IAT sensor**
P1508	Idle air control (IAC) valve – low rpm	**Wiring open circuit/short circuit to ground/ short circuit to positive, IAC valve, throttle valve tight/sticking**
P1509	Idle air control (IAC) valve – high rpm	**Wiring open circuit/short circuit to ground/ short circuit to positive, IAC valve, throttle valve tight/sticking**
P1512	Throttle control unit – signal variation	**Wiring, throttle control unit, ECM**
P1514	Intake air flow – excessive flow	**Throttle control unit, MAP sensor, CMP sensor signal, ECM**
P1515	Throttle command/actual throttle position – signal variation	**Wiring, throttle valve sticking, throttle control unit, ECM**
P1516	Throttle command/actual throttle position – signal variation	**Wiring, throttle valve sticking, throttle control unit, ECM**
P1520	Park/neutral position (PNP) switch	**Wiring short circuit to ground, incorrectly adjusted PNP switch**
P1523	Throttle control unit, throttle position during battery saving mode	**Throttle valve movement restricted, throttle control unit, ECM**
P1530 2 7 9	A/C refrigerant pressure switch – engine coolant blower motors	**Wiring open circuit, A/C refrigerant pressure switch, ECM**
P1530 8	A/C refrigerant pressure sensor, circuit malfunction	**A/C refrigerant pressure, wiring open circuit/ short circuit to ground, A/C refrigerant pressure sensor, ECM**
P1531	Low A/C refrigerant charge	**Low A/C refrigerant charge, wiring open circuit/short circuit to ground/short circuit to positive, A/C refrigerant low pressure switch, ECM**
P1550	Cruise control stepper motor, circuit malfunction	**Wiring open circuit/short circuit to ground, cruise control module, ECM**
P1554	Cruise control status circuit – high voltage	**Wiring open circuit/short circuit to ground, cruise control module, ECM**
P1558	Cruise control servo – servo position sensor low	**Hoses leaking, cruise control servo**

Scan code	Fault location	Probable cause
P1561	Cruise control vent solenoid	**Wiring open circuit/short circuit to ground/ short circuit to positive, cruise control vent solenoid, cruise control servo, ECM**
P1562	Cruise control vacuum solenoid	**Wiring open circuit/short circuit to ground/ short circuit to positive, cruise control vacuum solenoid, cruise control servo, ECM**
P1565	Cruise control servo position sensor – circuit malfunction	**Wiring open circuit/short circuit to ground/ short circuit to positive, cruise control servo, ECM**
P1567	Cruise control switch – circuit malfunction	**Wiring open circuit/short circuit to ground/ short circuit to positive, cruise control switch, ECM**
P1568	Cruise control servo – servo position sensor high	**Hoses blocked/leaking, brake pedal vacuum valve, cruise control servo**
P1571	Traction control system (TCS), torque request circuit	**Wiring open circuit/short circuit to ground/ short circuit to positive, ABS control module trouble code(s) stored, ECM**
P1573	CAN data bus – antilock brake system (ABS)/ traction control system (TCS)	**Wiring, ABS control module trouble code(s) stored**
P1599	Cruise control – power management mode	**Cruise control accelerator cable adjustment, hoses blocked, brake pedal vacuum valve, cruise control servo**
P1600	Engine control module (ECM)	**ECM**
P1619	Engine oil life monitor reset switch – circuit malfunction	**Wiring short circuit to ground, engine oil life monitor reset switch, ECM**
P1621	Engine control module (ECM) – memory performance	**ECM**
P1623	Engine control module (ECM) – PROM	**insecure PROM, incorrectly coded**
P1626	Immobilizer control module	**Wiring, body control module (BCM) trouble code(s) stored**
P1627	Engine control module (ECM) – memory performance	**ECM**
P1629	Immobilizer control module – fuel enable signal	**Incorrect ignition key, wiring open circuit/ short circuit to ground/short circuit to positive, immobilizer control module, ECM**
P1630 [2] [7] [9]	Battery voltage – high/low	**Wiring, alternator**
P1630 [8]	Engine control module (ECM), ignition supply voltage	**Fuse, wiring open circuit, ECM**
P1631	Engine control module (ECM) – incorrect immobilizer code	**Incorrect key, Immobilizer code not programmed**
P1633	Engine control module (ECM), ignition supply voltage	**Fuse, wiring open circuit, ECM**
P1635	Sensor supply voltage – circuit A malfunction	**Wiring short circuit to ground/positive, ground wiring open circuit, TP sensor, MAP sensor, EGR valve position sensor, ECM**

Scan code	Fault location	Probable cause
P1639	Sensor supply voltage – circuit B malfunction	Wiring short circuit to ground/positive, ground wiring open circuit, A/C refrigerant pressure sensor, fuel tank pressure (FTP) sensor, ECM
P1640	Engine control module, quad driver/output driver	Wiring, ECM [10]
P1641	Malfunction indicator lamp (MIL) – circuit malfunction	Wiring open circuit/short circuit to ground, MIL, supply wire, ECM
P1642	Secondary air injection (AIR) pump/solenoid relay – circuit malfunction	Wiring open circuit/short circuit to ground/ short circuit to positive, AIR pump/solenoid relay
P1643	Fuel pump control module – circuit malfunction	Wiring open circuit/short circuit to ground/ short circuit to positive, fuel pump control module, ECM
P1645	Supercharger (SC) control valve solenoid – circuit malfunction	Wiring open circuit/short circuit to ground, SC control valve solenoid, ECM
P1650	Engine control module, quad driver/output driver	Wiring, ECM [10]
P1651	Engine coolant blower motor relay 1 – circuit malfunction	Wiring open circuit/short circuit to ground, supply wire, engine coolant blower motor relay, ECM
P1652	Engine coolant blower motor relay 2 – circuit malfunction	Wiring open circuit/short circuit to ground, supply wire, engine coolant blower motor relay(s), ECM
P1653 [1]	Fuel gauge tank sensor – circuit malfunction	Wiring open circuit/short circuit to ground/ short circuit to positive, instrument panel, ECM
P1653 [2]	Traction control system (TCS) – delivered torque circuit	Wiring open circuit/short circuit to ground/ short circuit to positive, ABS control module trouble code(s) stored, ECM
P1654	A/C compressor clutch relay – circuit malfunction	Wiring open circuit/short circuit to ground, A/C compressor clutch relay, ECM
P1655	Evaporative emission (EVAP) canister purge valve – circuit malfunction	Wiring open circuit/short circuit to ground, EVAP canister vent valve, ECM
P1662	Cruise control inhibit circuit – circuit malfunction	Wiring open circuit/short circuit to ground, cruise control motor/actuator, cruise control module, ECM
P1663	Alternator warning lamp – circuit malfunction	Wiring open circuit/short circuit to ground/ short circuit to positive, instrument panel, ECM
P1665	Evaporative emission (EVAP) canister vent valve – circuit malfunction	Wiring open circuit/short circuit to ground, EVAP canister vent valve, ECM
P1670	Engine control module, quad driver/output driver	Wiring, ECM [10]
P1671 [3]	Engine oil change warning lamp – circuit malfunction	Wiring open circuit/short circuit to ground/ short circuit to positive, engine oil change warning lamp, ECM
P1671 [4]	Malfunction indicator lamp (MIL) – circuit malfunction	Wiring open circuit/short circuit to ground/ short circuit to positive, MIL, ECM

Scan code	Fault location	Probable cause
P1672	Engine oil level warning lamp – circuit malfunction	**Wiring open circuit/short circuit to ground/ short circuit to positive, instrument panel, ECM**
P1673	Engine coolant 'hot' warning lamp – circuit malfunction	**Wiring open circuit/short circuit to ground/ short circuit to positive, instrument panel, cooling system, ECM**
P1674	Tachometer – circuit malfunction	**Wiring open circuit/short circuit to ground/ short circuit to positive, instrument panel, ECM**
P1675	Evaporative emission (EVAP) canister vent valve – circuit malfunction	**Wiring open circuit/short circuit to ground, EVAP canister vent valve, ECM**
P1676	Evaporative emission (EVAP) canister purge valve – circuit malfunction	**Wiring open circuit/short circuit to ground, EVAP canister purge valve, ECM**
P1680	Engine control module (ECM)	**ECM**
P1681	Engine control module (ECM)	**ECM**
P1682	Engine control module (ECM), ignition supply voltage	**Fuse, wiring open circuit, ECM**
P1683	Engine control module (ECM)	**ECM**
P1860	Torque converter clutch (TCC) solenoid – circuit malfunction	**Wiring open circuit/short circuit to ground, TCC solenoid, ECM**
U1026	Transmission control module (TCM)/four wheel drive control module, CAN data bus – communication malfunction	**Wiring, BCM**
U1041	Antilock brake system (ABS)/traction control system (TCS), CAN data bus – communication malfunction	**Wiring, ABS/TCS trouble codes stored**
U1064	Body control module (BCM), CAN data bus – communication malfunction	**Wiring, BCM**

1 Silhouette 3.4L
2 Eighty Eight/Ninety Eight/Regency 3.8L
3 Cutlass Supreme 3.1L
4 Except Cutlass Supreme 3.1L
5 Except Cutlass Supreme 3.4L
6 Cutlass Supreme 3.4L
7 Silhouette 3.8L
8 Except Eighty Eight/Ninety Eight/Regency 3.8L and Silhouette 3.8L
9 Aurora 3.8L
10 Use of the vehicle wiring diagram is recommended to confirm which components are fitted. Depending on model, year and equipment level, the following components may be connected to this circuit:

- ☐ A/C compressor clutch relay
- ☐ A/C condenser blower motor relay
- ☐ Engine coolant blower motor relay
- ☐ Engine coolant 'hot' warning lamp
- ☐ Evaporative emission (EVAP) canister purge valve
- ☐ Exhaust gas recirculation (EGR) solenoid
- ☐ Malfunction indicator lamp (MIL)
- ☐ Supercharger (SC) control valve
- ☐ Torque converter clutch (TCC) solenoid
- ☐ Transmission shift solenoids (SS)

OLDSMOBILE

Model:	Engine identification:	Year:
Achieva 3.1L	VIN position 8 = M	1998
Alero 3.4L	VIN position 8 = M	1999-02
Cutlass 3.1L	VIN position 8 = M	1997-99
Eighty Eight 3.8L	VIN position 8 = 1/K	1998-99
Intrigue 3.8L	VIN position 8 = K	1998-00
Silhouette 3.4L	VIN position 8 = E	1998-02

System: **GM**

Self-diagnosis

General information

- Refer to the front of this manual for general test conditions, terminology, detailed descriptions of wiring faults and a general trouble shooter for electrical and mechanical faults.
- Malfunction indicator lamp (MIL) will illuminate or flash if certain faults are recorded.
- ECM operates in backup mode if sensors fail, to enable vehicle to be driven to repair shop.

Accessing

- Trouble codes can only be displayed by using a Scan Tool connected to the data link connector (DLC):
  - Achieva/Alero/Cutlass **1**.
  - Eighty Eight/Intrigue/Silhouette **2**.

1

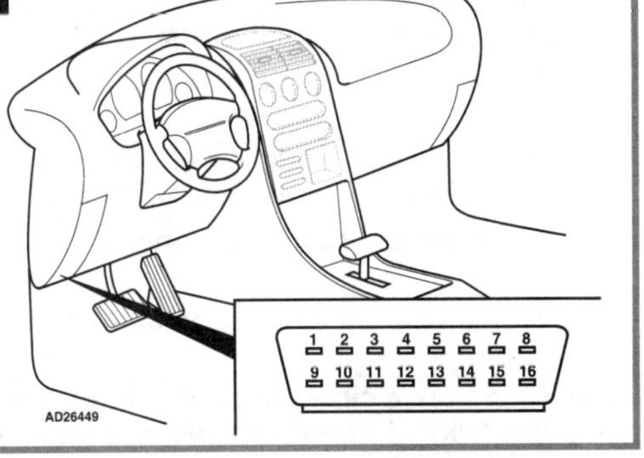

2

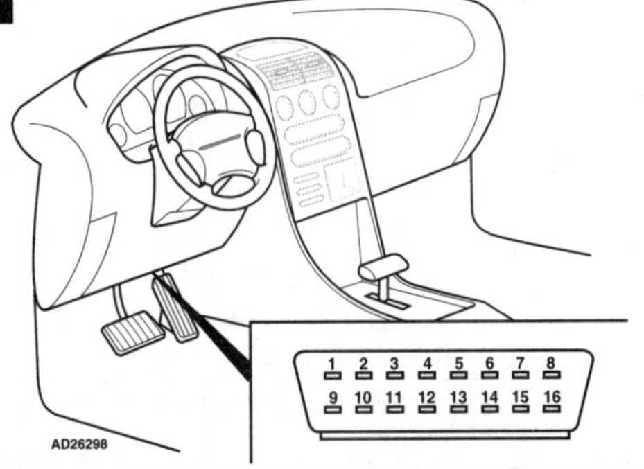

Erasing

- After the faults have been rectified, the manufacturer recommends that trouble codes are erased by using a Scan Tool connected to the data link connector (DLC).

Trouble code identification

Scan code	Fault location	Probable cause
P0000-0999	Refer to OBD-II trouble code table in the front of this manual	–
P1106	Manifold absolute pressure (MAP) sensor – circuit intermittent – voltage high	**Wiring open circuit/short circuit to ground/ short circuit to positive**
P1107	Manifold absolute pressure (MAP) sensor – circuit intermittent – voltage low	**Wiring short circuit to ground**
P1111	Intake air temperature (IAT) sensor – circuit intermittent – voltage high	**Wiring open circuit, ground lead defective**
P1112	Intake air temperature (IAT) sensor – circuit intermittent – voltage low	**Wiring short circuit to ground**
P1114	Engine coolant temperature (ECT) sensor – circuit malfunction – voltage low	**Wiring short circuit to ground**
P1115	Engine coolant temperature (ECT) sensor – circuit malfunction – voltage high	**Wiring open circuit, ground lead defective**
P1121	Throttle position (TP) sensor – circuit malfunction – voltage high	**Wiring short circuit to positive, ground lead defective**
P1122	Throttle position (TP) sensor – circuit malfunction – voltage low	**Wiring open circuit/short circuit to ground**
P1133	Heated oxygen sensor (HO2S) 1, bank 1 – insufficient switching	**Wiring, oxygen sensor heater, exhaust system leaking, HO2S**
P1134	Heated oxygen sensor (HO2S) 1, bank 1 – slow response	**Wiring, oxygen sensor heater, exhaust system leaking, HO2S**
P1189	Engine oil pressure warning switch – circuit malfunction	**Wiring open circuit, engine oil pressure warning switch, ECM**
P1257	Supercharger (SC) – over-boost	**Hoses blocked/leaking/interchanged/not connected, wiring short circuit to ground, supercharger (SC) bypass valve tight/ sticking, supercharger (SC) bypass valve solenoid, ECM**
P1336	Crankshaft position sensor (CKP) – variation not learned	**Learning procedure not performed**
P1350	Ignition control module (ICM) – bypass line monitor, circuit malfunction	**Wiring open circuit/short circuit to ground/ short circuit to positive, ICM, ECM**
P1351	Ignition control module (ICM) – IC circuit, voltage high	**Wiring open circuit, ICM, ECM**
P1352	Ignition control module (ICM) – bypass line, voltage high	**Wiring open circuit/short circuit to ground/ short circuit to positive, ICM, ECM**
P1361	Ignition control module (ICM), IC circuit – voltage low	**Wiring short circuit to ground, ICM, ECM**
P1362	Ignition control module (ICM) – bypass/ IC circuit, voltage low	**Wiring short circuit to ground, ICM, ECM**
P1374	Ignition control module (ICM) – 3X reference circuit malfunction	**Wiring open circuit/short circuit to ground/ short circuit to positive, wiring routed near spark plug wires, ignition coils arcing/ tracking, ICM, ECM**

Scan code	Fault location	Probable cause
P1380	Rough road signal – signal error	**ABS control module trouble code(s) stored**
P1381	Rough road signal – no communication from ABS CAN data bus	**Wiring open circuit, ABS system**
P1404	Exhaust gas recirculation (EGR) valve – stuck open	**Wiring open circuit/short circuit to positive, exhaust gas recirculation (EGR) valve position sensor, EGR solenoid**
P1441	Evaporative emission (EVAP) system – leak detected	**Hose connection(s), EVAP canister, EVAP canister purge valve**
P1483	Engine cooling system – performance	**Cooling system, engine coolant temperature (ECT) sensor, intake air temperature (IAT) sensor**
P1520	Park/neutral position (PNP) switch	**Wiring short circuit to ground, incorrectly adjusted**
P1546	A/C compressor clutch relay – circuit malfunction	**Wiring open circuit/short circuit to ground/ short circuit to positive, A/C compressor clutch relay, ECM**
P1554	Cruise control status circuit – high voltage	**Wiring open circuit/short circuit to ground, cruise control module, ECM**
P1571	Traction control system (TCS), torque request circuit	**Wiring open circuit/short circuit to ground/ short circuit to positive, ABS control module trouble code(s) stored, ECM**
P1573	Data bus – antilock brake system (ABS)/traction control system (TCS)	**Wiring, ABS control module trouble code(s) stored**
P1585	Cruise control system – inhibit output circuit	**Wiring open circuit/short circuit to ground/ short circuit to positive, supply wire, cruise control motor/actuator, ECM**
P1600	Engine control module (ECM)	**ECM**
P1602	CAN data bus – anti-lock brake system (ABS)/traction control system (TCS)	**Wiring, ABS control module trouble code(s) stored**
P1605	CAN data bus – A/C	**Wiring, A/C system trouble code(s) stored**
P1610	CAN data bus – body control module (BCM)	**Wiring, body control module (BCM) trouble code(s) stored**
P1619	Engine oil life monitor reset switch – circuit malfunction	**Wiring short circuit to ground, engine oil life monitor reset switch, ECM**
P1621	Engine control module (ECM) – memory performance	**ECM**
P1626	Data bus, immobilizer	**Wiring, body control module (BCM) trouble code(s) stored**
P1627	Engine control module (ECM) – memory performance	**ECM**
P1629	Immobilizer control module – fuel enable signal	**Incorrect ignition key, wiring open circuit/ short circuit to ground/short circuit to positive, immobilizer control module, ECM**
P1630	Engine control module (ECM), immobilizer code learning mode	**Immobilizer code not programmed**

Scan code	Fault location	Probable cause
P1631	Engine control module (ECM) – incorrect immobilizer code	**Incorrect key, Immobilizer code not programmed**
P1635	Sensor supply voltage – circuit A malfunction	**Wiring short circuit to ground/positive, ground wiring open circuit, TP sensor, MAP sensor, EGR valve position sensor, ECM**
P1637	Alternator 'L' terminal – circuit malfunction	**Wiring open circuit/short circuit to ground, alternator, battery temperature sensor, ECM**
P1639	Sensor supply voltage – circuit B malfunction	**Wiring short circuit to ground/positive, ground wiring open circuit, A/C refrigerant pressure sensor, fuel tank pressure (FTP) sensor, ECM**
P1640	Engine control module quad driver/output driver	**Wiring, ECM controlled components, ECM 3**
P1641	A/C compressor clutch relay – circuit malfunction	**Wiring open circuit/short circuit to ground, supply wire, A/C compressor clutch relay, ECM**
P1646	Supercharger (SC) control valve solenoid – circuit malfunction	**Wiring open circuit/short circuit to ground/ short circuit to positive, SC control valve solenoid, ECM**
P1650	Engine control module quad driver/output driver	**Wiring, ECM controlled components, ECM 3**
P1651	Engine coolant blower motor relay 1 – circuit malfunction	**Wiring open circuit/short circuit to ground, supply wire, engine coolant blower motor relay, ECM**
P1652	Engine coolant blower motor relay 2 – circuit malfunction	**Wiring open circuit/short circuit to ground, supply wire, engine coolant blower motor relay(s), ECM**
P1653 1	Fuel gauge tank sensor – circuit malfunction	**Wiring open circuit/short circuit to ground/ short circuit to positive, instrument panel, ECM**
P1653 2	Engine oil level warning lamp – circuit malfunction	**Wiring open circuit/short circuit to ground/ short circuit to positive, instrument panel, ECM**
P1660	Engine control module quad driver/output driver	**Wiring, ECM controlled components, ECM 3**
P1662	Cruise control inhibit circuit – circuit malfunction	**Wiring open circuit/short circuit to ground, cruise control motor/actuator, cruise control module, ECM**
P1663	Alternator warning lamp – circuit malfunction	**Wiring open circuit/short circuit to ground/ short circuit to positive, instrument panel, ECM**
P1665	Evaporative emission (EVAP) canister vent valve – circuit malfunction	**Wiring open circuit/short circuit to ground, EVAP canister vent valve, ECM**
P1670	Engine control module (ECM) quad driver/ output driver	**Wiring, ECM controlled components, ECM 3**
P1671	Malfunction indicator lamp (MIL) – circuit malfunction	**Wiring open circuit/short circuit to ground/ short circuit to positive, MIL, ECM**
P1672	Engine oil level warning lamp – circuit malfunction	**Wiring open circuit/short circuit to ground/ short circuit to positive, instrument panel, ECM**

Scan code	Fault location	Probable cause
P1673	Engine coolant 'hot' warning lamp – circuit malfunction	**Wiring open circuit/short circuit to ground/ short circuit to positive, instrument panel, cooling system, ECM**
P1676	Evaporative emission (EVAP) canister purge valve – circuit malfunction	**Wiring open circuit/short circuit to ground, EVAP canister purge valve, ECM**
P1680	Engine control module (ECM) – memory performance	**ECM**
P1681	Engine control module (ECM) – memory performance	**ECM**
P1683	Engine control module (ECM) – memory performance	**ECM**
P1689	Traction control system (TCS) – delivered torque output circuit	**Wiring open circuit/short circuit to ground/ short circuit to positive, ABS trouble codes stored, ECM**
P1810	Transmission fluid pressure (TFP) switch – manual valve position	**Fluid level low, wiring open circuit/short circuit to ground, selector lever linkage incorrectly adjusted, ECM**
P1811	Shift time too long – adaptation limit reached	**Fluid level low, transmission mechanical fault**
P1860	Torque converter clutch (TCC) solenoid – circuit malfunction	**Wiring open circuit/short circuit to ground, TCC solenoid, ECM**
P1887	Torque converter clutch (TCC) release switch – circuit malfunction	**Wiring open circuit, TCC solenoid release switch, ECM**

1 Except Silhouette 3.4L 1999

2 Silhouette 3.4L 1999

3 Depending on model, year and equipment level, the following components may be connected to this circuit. Use of the vehicle wiring diagram is recommended to confirm which components are fitted

- □ A/C compressor clutch relay
- □ A/C condenser blower motor relay
- □ Engine coolant blower motor relay
- □ Engine coolant 'hot' warning lamp
- □ Evaporative emission (EVAP) canister purge valve
- □ Exhaust gas recirculation (EGR) solenoid
- □ Malfunction indicator lamp (MIL)
- □ Torque converter clutch (TCC) solenoid
- □ Transmission shift solenoids (SS)
- □ Transmission shift warning lamp (MT)

OLDSMOBILE

Model:	Engine identification:	Year:
Achieva 3.3L	**VIN position 8 = N**	**1992-93**
Cutlass Ciera/Cruiser 2.2L	**VIN position 8 = 4**	**1993**
Cutlass Ciera/Cruiser 3.3L	**VIN position 8 = N**	**1992-93**

System: **GM**

Self-diagnosis

General information

- Refer to the front of this manual for general test conditions, terminology, detailed descriptions of wiring faults and a general trouble shooter for electrical and mechanical faults.
- Malfunction indicator lamp (MIL) will illuminate if certain faults are recorded.
- ECM operates in backup mode if sensors fail, to enable vehicle to be driven to repair shop.
- Trouble codes can be displayed by the MIL (some models), or by using a Scan Tool connected to the data link connector (DLC).

Accessing

- Access data link connector (DLC).
- Jump data link connector (DLC) terminals A and B **1**.
- Switch ignition ON. Do NOT start engine.
- Count MIL flashes. Note trouble codes. Compare with trouble code table.
- Firstly trouble code 12 will be displayed 3 times.
- Trouble codes consist of 2 groups, displayed as follows:
  - First group indicates the 'tens' of the trouble code **2** [**A**].
  - Second group indicates the 'units' of the trouble code **2** [**C**].
  - A short pause separates each group **2** [**B**].
  - A long pause separates each trouble code **2** [**D**].
  - For example: Trouble code 12 displayed **2**. Start of diagnosis.
- Each additional trouble code will now be displayed 3 times, in ascending order.

NOTE: ***If no additional trouble codes are stored, code 12 will continue to be displayed.***

- Trouble codes can also be displayed by using a Scan Tool connected to the data link connector (DLC) **1**.

1

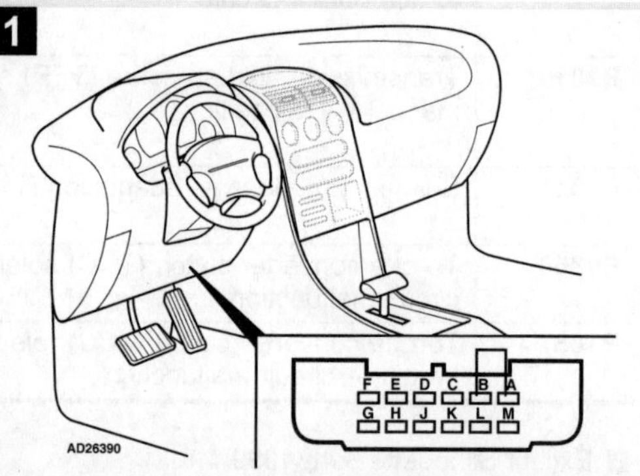

2

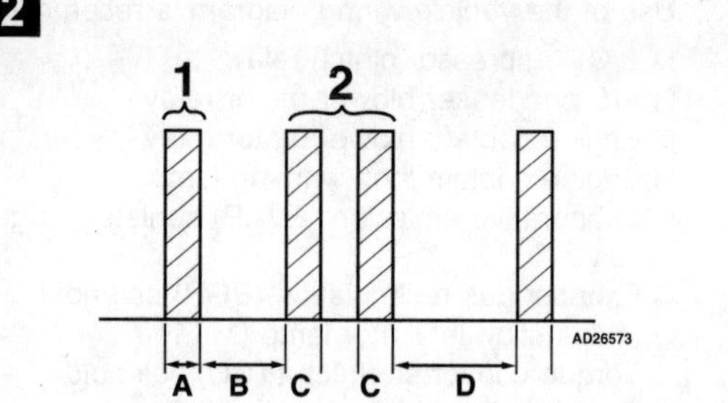

Erasing

NOTE: ***Disconnecting battery ground lead may cause driveability problems until the engine management system relearns basic values.***

- Ensure ignition switched OFF.
- Disconnect battery ground cable for 30 seconds.
- Reconnect battery ground cable.
- Repeat checking procedure to ensure no trouble codes remain in ECM fault memory.

NOTE: ***Disconnecting battery ground lead may also erase memory from electronic units, such as radio.***

- Trouble codes can also be erased by using a Scan Tool connected to the data link connector (DLC) **1**.

Trouble code identification

Flash code	Fault location	Probable cause
12	Start of diagnosis	**No engine speed (RPM) signal received**
13	Oxygen sensor (O2S)	**Wiring open circuit, O2S, ECM**
14 **1**	Engine coolant temperature (ECT) sensor – high temperature indicated	**Wiring short circuit to ground, ECT sensor, ECM**
14 **2**	Engine coolant temperature (ECT) sensor – circuit malfunction	**Wiring open circuit/short circuit to ground/ short circuit to positive, ECT sensor, ECM**
15	Engine coolant temperature (ECT) sensor – low temperature indicated	**Wiring open circuit, ECT sensor, ECM**
16	Battery voltage – low	**Wiring, alternator**
21 **1**	Throttle position (TP) sensor – voltage high	**Wiring open circuit/short circuit to positive, TP sensor, ECM**
21 **2**	Throttle position (TP) sensor – circuit malfunction	**Wiring open circuit/short circuit to ground/ short circuit to positive, TP sensor, ECM**
22	Throttle position (TP) sensor – voltage low	**Wiring open circuit/short circuit to ground, TP sensor, ECM**
23	Intake air temperature (IAT) sensor – circuit malfunction	**Wiring open circuit/short circuit to ground/ short circuit to positive, IAT sensor, ECM**
24	Vehicle speed sensor (VSS)	**Wiring open circuit/short circuit to ground, VSS, ECM**
26	Engine control module quad driver/output driver	**Wiring, ECM controlled components, ECM 3**
27	Transmission gear switch signal circuit 2nd gear	**Wiring open circuit/short circuit to ground, 2nd gear switch, ECM**
28	Transmission gear switch signal circuit 3rd gear	**Wiring open circuit/short circuit to ground, 3rd gear switch, ECM**
29	Transmission gear switch signal circuit 4th gear	**Wiring open circuit/short circuit to ground, 4th gear switch, ECM**
31	Park/neutral position (PNP) switch	**Wiring short circuit to ground, incorrectly adjusted, PNP switch, ECM**
32	Exhaust gas recirculation (EGR) solenoid	**Wiring open circuit/short circuit to ground, supply wire, EGR valve, ECM**
33	Manifold absolute pressure (MAP) sensor – circuit malfunction	**Wiring open circuit/short circuit to ground/ short circuit to positive, hose blocked/ leaking, MAP sensor, ECM**

Flash code	Fault location	Probable cause
34	Mass air flow (MAF) sensor – circuit malfunction	**Wiring open circuit/short circuit to ground, MAF sensor, ECM**
38	Brake pedal position (BPP) switch	**Wiring open circuit/short circuit to ground/ short circuit to positive, incorrectly adjusted, BPP switch, ECM**
39	Torque converter clutch (TCC) – circuit malfunction	**Wiring open circuit, TCC solenoid, 2nd gear switch, ECM**
42	Ignition control module (ICM)	**Wiring open circuit/short circuit to ground, ICM, ECM**
43	Knock sensor (KS)	**Wiring open circuit/short circuit to ground/ short circuit to positive, KS incorrectly fitted, KS, insecure PROM, ECM**
44	Oxygen sensor (O2S) – mixture lean	**Wiring open circuit/short circuit to ground, injector(s) blocked, fuel contamination, fuel pressure low, exhaust system leaking, O2S, ECM**
45	Oxygen sensor (O2S) – mixture rich	**Wiring open circuit, fuel pressure high, injector(s) leaking, O2S contamination, EVAP canister purge valve, TP sensor, EGR system, O2S, ECM**
46	Power steering pressure (PSP) switch – circuit malfunction	**Wiring open circuit/short circuit to ground, PSP switch, ECM**
48	Misfire	**ignition system, fuel system, O2S**
51	Engine control module (ECM) – PROM	**Insecure PROM, incorrect PROM, ECM**
66	A/C refrigerant pressure sensor – circuit malfunction	**Wiring open circuit/short circuit to ground/ short circuit to positive, A/C refrigerant pressure sensor, ECM**

1 Except Cutlass Ciera 2.2L

2 Cutlass Ciera 2.2L

3 Depending on model, year and equipment level, the following components may be connected to this circuit. Use of the vehicle wiring diagram is recommended to confirm which components are fitted

- ☐ A/C compressor clutch relay
- ☐ A/C condenser blower motor relay
- ☐ Engine coolant blower motor relay
- ☐ Engine coolant 'hot' warning lamp
- ☐ Evaporative emission (EVAP) canister purge valve
- ☐ Exhaust gas recirculation (EGR) solenoid
- ☐ Malfunction indicator lamp (MIL)
- ☐ Torque converter clutch (TCC) solenoid
- ☐ Transmission shift solenoids (SS)
- ☐ Transmission shift warning lamp (MT)

OLDSMOBILE

Model:	Engine identification:	Year:
Eighty Eight 3.8L	VIN position 8 = L	1992-93
Ninety Eight/Regency 3.8L	VIN position 8 = L	1992-93
Silhouette 3.8L	VIN position 8 = L	1992-93
Toronado/Troféo 3.8L	VIN position 8 = L	1992

System: **GM**

Self-diagnosis

General information

- Refer to the front of this manual for general test conditions, terminology, detailed descriptions of wiring faults and a general trouble shooter for electrical and mechanical faults.
- Malfunction indicator lamp (MIL) will illuminate if certain faults are recorded.
- ECM operates in backup mode if sensors fail, to enable vehicle to be driven to repair shop.
- Trouble codes can be displayed by the MIL (some models), or by using a Scan Tool connected to the data link connector (DLC).

Accessing

- Access data link connector (DLC):
  - Toronado/Troféo **1**.
  - Except Toronado/Troféo **2**.
- Jump data link connector (DLC) terminals A and B.
- Switch ignition ON. Do NOT start engine.
- Count MIL flashes. Note trouble codes. Compare with trouble code table.
- Firstly trouble code 12 will be displayed 3 times.
- Trouble codes consist of 2 groups, displayed as follows:
  - First group indicates the 'tens' of the trouble code **2** [**A**].
  - Second group indicates the 'units' of the trouble code **2** [**C**].
  - A short pause separates each group **2** [**B**].
  - A long pause separates each trouble code **2** [**D**].
  - For example: Trouble code 12 displayed **2**. Start of diagnosis.
- Each additional trouble code will now be displayed 3 times, in ascending order.

NOTE: ***If no additional trouble codes are stored, code 12 will continue to be displayed.***

- Trouble codes can also be displayed by using a Scan Tool connected to the data link connector (DLC).

1

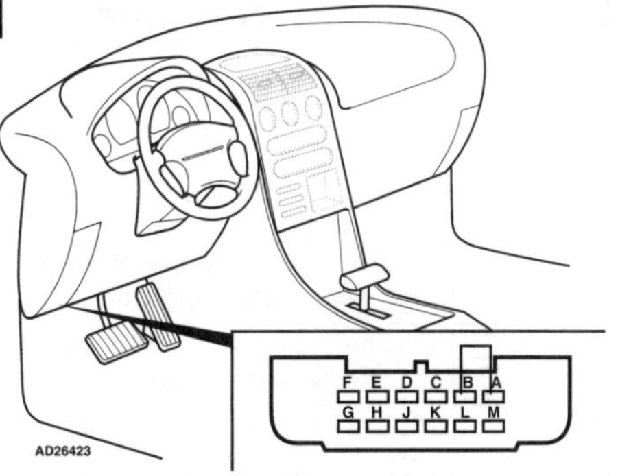

2

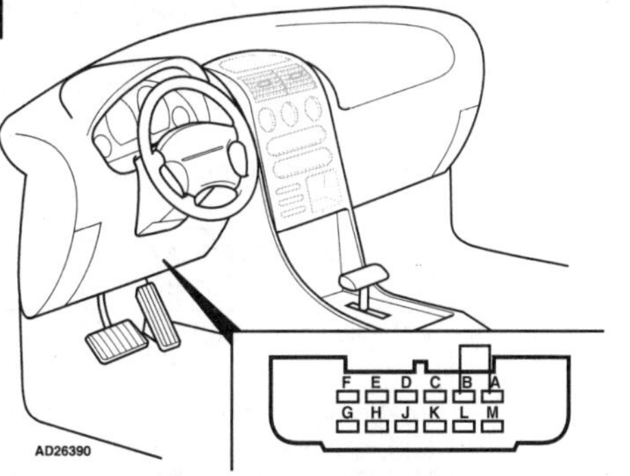

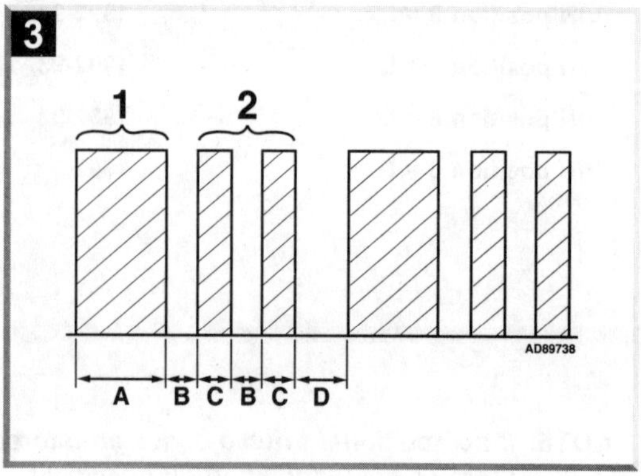

Erasing

NOTE: ***Disconnecting battery ground lead may cause driveability problems until the engine management system relearns basic values.***

- Ensure ignition switched OFF.
- Disconnect battery ground cable for 30 seconds.
- Reconnect battery ground cable.
- Repeat checking procedure to ensure no trouble codes remain in ECM fault memory.

NOTE: ***Disconnecting battery ground lead may also erase memory from electronic units, such as radio.***

- Trouble codes can also be erased by using a Scan Tool connected to the data link connector (DLC).

Trouble code identification

Flash code	Fault location	Probable cause
12	Start of diagnosis	**No engine speed (RPM) signal received**
13	Oxygen sensor (O2S)	**Wiring open circuit, oxygen sensor (O2S), ECM**
14	Engine coolant temperature (ECT) sensor – high temperature indicated	**Wiring short circuit to ground, ECT sensor, ECM**
15	Engine coolant temperature (ECT) sensor – low temperature indicated	**Wiring open circuit, ECT sensor, ECM**
16	Battery voltage – high/low	**Wiring, alternator**
17	Spark reference circuit	**Wiring open circuit/short circuit to ground, ICM, ECM**
18	Camshaft position (CMP) signal/fuel control signal	**Wiring open circuit/short circuit to ground, CKP sensor, CMP sensor, ICM, ECM**
21	Throttle position (TP) sensor – voltage high	**Wiring open circuit/short circuit to positive, TP sensor, ECM**
22	Throttle position (TP) sensor – voltage low	**Wiring open circuit/short circuit to ground, TP sensor, ECM**
23	Intake air temperature (IAT) sensor – low temperature indicated	**Wiring open circuit, IAT sensor, ECM**
24	Vehicle speed sensor (VSS)	**Wiring open circuit/short circuit to ground, VSS, ECM**
25	Intake air temperature (IAT) sensor – high temperature indicated	**Wiring short circuit to ground, IAT sensor, ECM**
26	Engine control module quad driver/output driver	**Wiring, ECM controlled components, ECM 1**
31	Park/neutral position (PNP) switch	**Wiring short circuit to ground, incorrectly adjusted, PNP switch, ECM**

Flash code	Fault location	Probable cause
32	Exhaust gas recirculation (EGR) solenoid	**Wiring open circuit/short circuit to ground, supply wire, EGR valve, ECM**
34	Mass air flow (MAF) sensor	**Wiring open circuit/short circuit to ground/ short circuit to positive, MAF sensor, ECM**
36	Shift solenoids (SS)	**Wiring open circuit/short circuit to ground, SS**
38	Brake pedal position (BPP) switch	**Wiring open circuit/short circuit to ground/ short circuit to positive, BPP switch incorrectly adjusted, BPP switch, ECM**
39	Torque converter clutch (TCC)	**Wiring open circuit/short circuit to ground, TCC solenoid, BPP switch incorrectly adjusted, transmission mechanical fault, ECM**
41	Camshaft position (CMP) sensor	**Wiring, CMP sensor, mechanical fault, ECM**
42	Ignition control module (ICM) – circuit malfunction	**Wiring open circuit/short circuit to ground, ICM, ECM**
43	Knock sensor (KS)	**Wiring open circuit/short circuit to ground/ short circuit to positive, KS incorrectly fitted, KS, insecure PROM, ECM**
44	Oxygen sensor (O2S) – mixture lean	**Wiring open circuit/short circuit to ground, injector(s) blocked, fuel contamination, fuel pressure low, exhaust system leaking, O2S, ECM**
45	Oxygen sensor (O2S) – mixture rich	**Wiring open circuit, fuel pressure high, injector(s) leaking, O2S contamination, EVAP canister purge valve, TP sensor, EGR system, O2S, ECM**
51	Engine control module (ECM) – PROM	**Insecure PROM, incorrect PROM, ECM**
53	Exhaust gas recirculation (EGR) valve solenoid 1	**Wiring open circuit/short circuit to ground, supply wire, EGR valve, ECM**
54	Exhaust gas recirculation (EGR) valve solenoid 2	**Wiring open circuit/short circuit to ground, supply wire, EGR valve, ECM**
55	Exhaust gas recirculation (EGR) valve solenoid 3	**Wiring open circuit/short circuit to ground, supply wire, EGR valve, ECM**
56	Engine control module quad driver/output driver	**Wiring, ECM controlled components, ECM** 1
58	Immobilizer control module – fuel enable signal	**Incorrect ignition key, wiring open circuit/ short circuit to ground/short circuit to positive, immobilizer control module, ECM**
61	Cruise control vent solenoid	**Wiring open circuit/short circuit to ground/ short circuit to positive, cruise control vent solenoid, cruise control servo, ECM**
62	Cruise control vacuum solenoid	**Wiring open circuit/short circuit to ground/ short circuit to positive, cruise control vacuum solenoid, cruise control servo, ECM**
63	Cruise control servo – servo position sensor low	**Hoses leaking, cruise control servo**

Flash code	Fault location	Probable cause
65	Cruise control servo position sensor – circuit malfunction	**Wiring open circuit/short circuit to ground/ short circuit to positive, cruise control servo, ECM**
66	Low A/C refrigerant charge	**Low A/C refrigerant charge, wiring open circuit/short circuit to ground/short circuit to positive, A/C refrigerant low pressure switch, ECM**
67	Cruise control switch – circuit malfunction	**Wiring open circuit/short circuit to ground/ short circuit to positive, cruise control switch, ECM**
68	Cruise control servo – servo position sensor high	**Hoses blocked/leaking, brake pedal vacuum valve, cruise control servo**
69	A/C refrigerant pressure switch – engine coolant blower motors	**Wiring open circuit, A/C refrigerant pressure switch, ECM**
99	Cruise control – power management mode	**Cruise control accelerator cable adjustment, hoses blocked, brake pedal vacuum valve, cruise control servo**

1 Depending on model, year and equipment level, the following components may be connected to this circuit. Use of the vehicle wiring diagram is recommended to confirm which components are fitted

- ☐ A/C compressor clutch relay
- ☐ A/C condenser blower motor relay
- ☐ Engine coolant blower motor relay
- ☐ Engine coolant 'hot' warning lamp
- ☐ Evaporative emission (EVAP) canister purge valve
- ☐ Exhaust gas recirculation (EGR) solenoid
- ☐ Malfunction indicator lamp (MIL)
- ☐ Torque converter clutch (TCC) solenoid
- ☐ Transmission shift solenoids (SS)
- ☐ Transmission shift warning lamp (MT)

OLDSMOBILE

Model:	Engine identification:	Year:
Aurora 3.5L	**VIN position 8 = H**	**2001-02**
Intrigue 3.5L	**VIN position 8 = H**	**1999-00**

System: **GM**

Self-diagnosis

General information

- Refer to the front of this manual for general test conditions, terminology, detailed descriptions of wiring faults and a general trouble shooter for electrical and mechanical faults.
- Malfunction indicator lamp (MIL) will illuminate or flash if certain faults are recorded.
- ECM operates in backup mode if sensors fail, to enable vehicle to be driven to repair shop.

Accessing

- Trouble codes can only be displayed by using a Scan Tool connected to the data link connector (DLC) **1**.

1

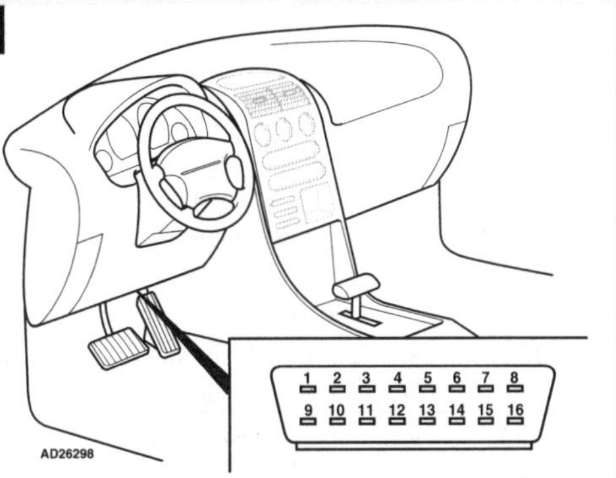

Erasing

- After the faults have been rectified, the manufacturer recommends that trouble codes are erased by using a Scan Tool connected to the data link connector (DLC).

Trouble code identification

Scan code	Fault location	Probable cause
P0000-0999	Refer to OBD-II trouble code table in the front of this manual	–
P1106	Manifold absolute pressure (MAP) sensor – circuit intermittent – voltage high	**Wiring open circuit/short circuit to ground/ short circuit to positive**
P1107	Manifold absolute pressure (MAP) sensor – circuit intermittent – voltage low	**Wiring short circuit to ground**
P1111	Intake air temperature (IAT) sensor – circuit intermittent – voltage high	**Wiring open circuit, ground lead defective**
P1112	Intake air temperature (IAT) sensor – circuit intermittent – voltage low	**Wiring short circuit to ground**
P1114	Engine coolant temperature (ECT) sensor – circuit malfunction – voltage low	**Wiring short circuit to ground**
P1115	Engine coolant temperature (ECT) sensor – circuit malfunction – voltage high	**Wiring open circuit, ground lead defective**

Scan code	Fault location	Probable cause
P1121	Throttle position (TP) sensor – circuit malfunction – voltage high	**Wiring short circuit to positive, ground lead defective**
P1122	Throttle position (TP) sensor – circuit malfunction – voltage low	**Wiring open circuit/short circuit to ground**
P1133	Heated oxygen sensor (HO2S) 1, bank 1 – insufficient switching	**Wiring, oxygen sensor heater, exhaust system leaking, HO2S**
P1134	Heated oxygen sensor (HO2S) 1, bank 1 – slow response	**Wiring, oxygen sensor heater, exhaust system leaking, HO2S**
P1336	Crankshaft position sensor (CKP) – variation not learned	**Learning procedure not performed**
P1351	Ignition control module (ICM) – ignition coil 1 – circuit voltage high	**Wiring open circuit/short circuit to ground/ short circuit to positive, ICM, ECM**
P1352	Ignition control module (ICM) – ignition coil 2 – circuit voltage high	**Wiring open circuit/short circuit to ground/ short circuit to positive, ICM, ECM**
P1353	Ignition control module (ICM) – ignition coil 3 – circuit voltage high	**Wiring open circuit/short circuit to ground/ short circuit to positive, ICM, ECM**
P1354	Ignition control module (ICM) – ignition coil 4 – circuit voltage high	**Wiring open circuit/short circuit to ground/ short circuit to positive, ICM, ECM**
P1355	Ignition control module (ICM) – ignition coil 5 – circuit voltage high	**Wiring open circuit/short circuit to ground/ short circuit to positive, ICM, ECM**
P1356	Ignition control module (ICM) – ignition coil 6 – circuit voltage high	**Wiring open circuit/short circuit to ground/ short circuit to positive, ICM, ECM**
P1361	Ignition control module (ICM) – ignition coil 1 – circuit voltage low	**Wiring short circuit to ground/positive, bypass circuit short circuit to ground, ICM, ECM**
P1362	Ignition control module (ICM) – ignition coil 2 – circuit voltage low	**Wiring short circuit to ground/positive, bypass circuit short circuit to ground, ICM, ECM**
P1363	Ignition control module (ICM) – ignition coil 3 – circuit voltage low	**Wiring short circuit to ground/positive, bypass circuit short circuit to ground, ICM, ECM**
P1364	Ignition control module (ICM) – ignition coil 4 – circuit voltage low	**Wiring short circuit to ground/positive, bypass circuit short circuit to ground, ICM, ECM**
P1365	Ignition control module (ICM) – ignition coil 5 – circuit voltage low	**Wiring short circuit to ground/positive, bypass circuit short circuit to ground, ICM, ECM**
P1366	Ignition control module (ICM) – ignition coil 6 – circuit voltage low	**Wiring short circuit to ground/positive, bypass circuit short circuit to ground, ICM, ECM**
P1372	Crankshaft position sensor (CKP)	**Wiring open circuit/short circuit to ground/ short circuit to positive, CKP, ECM**
P1380	Rough road signal – signal error	**ABS control module trouble code(s) stored**
P1381	Rough road signal – no communication from ABS CAN data bus	**Wiring open circuit, ABS system**

Scan code	Fault location	Probable cause
P1404	Exhaust gas recirculation (EGR) valve – stuck open	Wiring open circuit/short circuit to positive, exhaust gas recirculation (EGR) valve position sensor, EGR solenoid
P1441	Evaporative emission (EVAP) system – leak detected	Hose connection(s), EVAP canister, EVAP canister purge valve
P1546	A/C compressor clutch relay – circuit malfunction	Wiring open circuit/short circuit to ground/ short circuit to positive, A/C compressor clutch relay, ECM
P1554	Cruise control status circuit – high voltage	Wiring open circuit/short circuit to ground, cruise control module, ECM
P1571	Traction control system (TCS) – torque request circuit	Wiring open circuit/short circuit to ground/ short circuit to positive, ABS control module trouble code(s) stored, ECM
P1575	Brake pedal position switch (BPP) – extended travel contacts – voltage high	Wiring, ABS
P1585	Cruise control system – inhibit output circuit	Wiring open circuit/short circuit to ground/ short circuit to positive, supply wire, cruise control motor/actuator, ECM
P1599	Engine stall or near stall detected	ECM trouble code(s) stored, mechanical fault, transmission mechanical fault
P1602	CAN data bus – ABS/TCS	Wiring, ABS control module trouble code(s) stored
P1605	Data bus, A/C	Wiring – A/C system trouble code(s) stored
P1610	CAN data bus – body control module (BCM)	Wiring, body control module (BCM) trouble code(s) stored
P1626	Data bus, immobilizer	Wiring, body control module (BCM) trouble code(s) stored
P1630	Engine control module (ECM),immobilizer code learning mode	Immobilizer code not programmed
P1631	Engine control module (ECM) – incorrect immobilizer code	Incorrect key, immobilizer code not programmed
P1635	Sensor supply voltage – circuit A malfunction	Wiring short circuit to ground/positive, ground wiring open circuit, TP sensor, MAP sensor, EGR valve position sensor, ECM
P1639	Sensor supply voltage – circuit B malfunction	Wiring short circuit to ground/positive, ground wiring open circuit, A/C refrigerant pressure sensor, fuel tank pressure (FTP) sensor, ECM
P1640	Engine control module quad driver/output driver	Wiring, ECM controlled components, ECM 1
P1650	Engine control module quad driver/output driver	Wiring, ECM controlled components, ECM 1
P1660	Engine control module quad driver/output driver	Wiring, ECM controlled components, ECM 1
P1670	Engine control module quad driver/output driver	Wiring, ECM controlled components, ECM 1
P1683	Engine control module	ECM

Scan code	Fault location	Probable cause
P1689	Traction control system (TCS) – delivered torque output circuit	**Wiring open circuit/short circuit to ground/ short circuit to positive, ABS trouble codes stored, ECM**
P1810	Transmission fluid pressure (TFP) switch – manual valve position	**fluid level low, wiring open circuit/short circuit to ground, selector lever linkage incorrectly adjusted, ECM**
P1811	Shift time too long – adaptation limit reached	**fluid level low, transmission mechanical fault**
P1814	Torque converter clutch (TCC) – over-stressed	**Fluid level low, fluid condition, transmission mechanical fault**
P1860	Torque converter clutch (TCC) solenoid – circuit malfunction	**Wiring open circuit/short circuit to ground, TCC solenoid, ECM**
P1887	Torque converter clutch (TCC) release switch – circuit malfunction	**Wiring open circuit, TCC solenoid release switch, ECM**

1 Depending on model, year and equipment level, the following components may be connected to this circuit. Use of the vehicle wiring diagram is recommended to confirm which components are fitted.

- ☐ A/C compressor clutch relay
- ☐ A/C condenser blower motor relay
- ☐ Engine coolant blower motor relay
- ☐ Engine coolant 'hot' warning lamp
- ☐ Evaporative emission (EVAP) canister purge valve
- ☐ Exhaust gas recirculation (EGR) solenoid
- ☐ Malfunction indicator lamp (MIL)
- ☐ Torque converter clutch (TCC) solenoid
- ☐ Transmission shift solenoids (SS)
- ☐ Transmission shift warning lamp (MT)

OLDSMOBILE

Model:	Engine identification:	Year:
Aurora 4.0L	**VIN position 8 = C**	**1995**

System: **GM**

Self-diagnosis

General information

- Refer to the front of this manual for general test conditions, terminology, detailed descriptions of wiring faults and a general trouble shooter for electrical and mechanical faults.
- Malfunction indicator lamp (MIL) will illuminate if certain faults are recorded.
- ECM operates in backup mode if sensors fail, to enable vehicle to be driven to repair shop.

Accessing

- Trouble codes can only be displayed by using a Scan Tool connected to the data link connector (DLC) **1**.

1

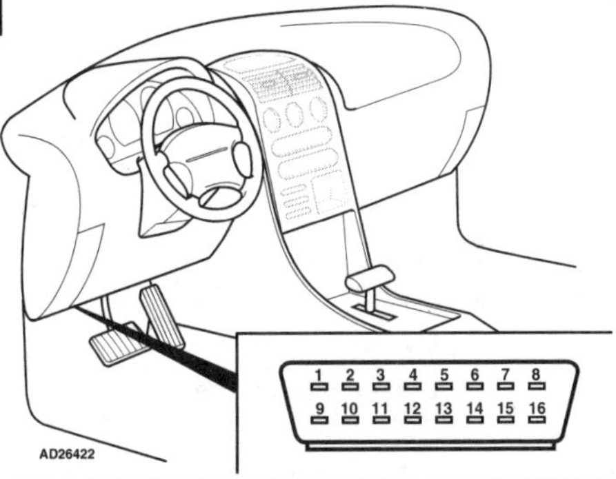

Erasing

- After the faults have been rectified, the manufacturer recommends that trouble codes are erased by using a Scan Tool connected to the data link connector (DLC).

Trouble code identification

Flash code	Fault location	Probable cause
12	Ignition control module (ICM) – 4X reference circuit malfunction	**Wiring open circuit/short circuit to ground/ short circuit to positive, ICM, ECM**
13	Heated oxygen sensor (HO2S) 1, bank 1 – open circuit	**Wiring open circuit/short circuit to ground/ short circuit to positive, HO2S, ECM**
14	Engine coolant temperature (ECT) sensor – high temperature indicated	**Wiring short circuit to ground, ECT sensor, ECM**
15	Engine coolant temperature (ECT) sensor – low temperature indicated	**Wiring open circuit, ECT sensor, ECM**
16	Battery voltage – high/low	**Wiring, alternator, ECM**
17	Heated oxygen sensor (HO2S) 1, bank 2 – open circuit	**Wiring open circuit/short circuit to ground/ short circuit to positive, HO2S, ECM**
19	Fuel pump circuit – short circuit	**Wiring short circuit to positive, fuel pump relay, engine oil pressure warning lamp switch**

Flash code	Fault location	Probable cause
20	Fuel pump circuit – open circuit	Wiring open circuit/short circuit to ground, fuel pump relay, fuel pump, engine oil pressure warning lamp switch, engine oil pressure too low
21	Throttle position (TP) sensor – voltage high	Wiring open circuit/short circuit to positive, TP sensor, ECM
22	Throttle position (TP) sensor – voltage low	Wiring open circuit/short circuit to ground, TP sensor, ECM
23	Ignition control module (ICM) – IC circuit malfunction	Wiring open circuit/short circuit to positive, ICM, ECM
24	Vehicle speed sensor (VSS)	Wiring open circuit/short circuit to ground, VSS, ECM
25	Ignition control module (ICM) – 24X signal circuit malfunction	Wiring open circuit/short circuit to ground, ICM, ECM
26	Closed throttle position (CTP) switch – voltage low	Wiring short circuit to ground, ISC actuator, ECM
27	Closed throttle position (CTP) switch – voltage high	Wiring open circuit, ISC actuator, ECM
28	Transmission range (TR) switch	Wiring open circuit/short circuit to ground, TR switch, ECM
29	Shift solenoid (SS) B	Wiring open circuit/short circuit to ground, transmission mechanical fault, SS, ECM
30	Idle speed (ISC) actuator	Wiring open circuit/short circuit to ground, throttle valve tight/sticking, ISC actuator, ECM
31	Manifold absolute pressure (MAP) sensor – voltage high, low vacuum indicated	Wiring short circuit to positive, hose blocked/leaking, MAP sensor, ECM
32	Manifold absolute pressure (MAP) sensor – voltage low, high vacuum indicated	Wiring open circuit/short circuit to ground, MAP sensor, ECM
33	Brake pedal position switch (BPP) – extended travel contacts – circuit malfunction	Wiring, ABS
34	Manifold absolute pressure (MAP) sensor, low vacuum indicated	Wiring open circuit, vacuum leak, ECM
35	Ignition control module (ICM) – reference low circuit malfunction	Wiring open circuit/short circuit to positive, ICM, ECM
36	Exhaust gas recirculation (EGR) solenoid position sensor	Wiring open circuit/short circuit to ground/ short circuit to positive, EGR valve, ECM
37	Intake air temperature (IAT) sensor – high temperature indicated	Wiring short circuit to ground, IAT sensor, ECM
38	Intake air temperature (IAT) sensor – low temperature indicated	Wiring open circuit, IAT sensor, ECM
39	Torque converter clutch (TCC) – not engaging	Wiring open circuit/short circuit to ground, BPP switch (TCC), supply wire, TCC solenoid, transmission mechanical fault, ECM

Flash code	Fault location	Probable cause
40	Power steering pressure (PSP) switch – open circuit	**Wiring open circuit/short circuit to ground, PSP switch, ECM**
41	Ignition control module (ICM) – camshaft position (CMP) signal	**Wiring open circuit/short circuit to ground, CMP sensor, ICM, ECM**
42	Heated oxygen sensor (HO2S) 1, bank 2 – mixture lean	**Wiring short circuit to ground, injector(s) blocked, fuel contamination, fuel pressure low, exhaust system leaking, HO2S**
43	Heated oxygen sensor (HO2S) 1, bank 2 – mixture rich	**Wiring open circuit, fuel pressure high, injector(s) leaking, HO2S contamination, EVAP canister purge valve, TP sensor, EGR system, HO2S**
44	Heated oxygen sensor (HO2S) 1, bank 1 – mixture lean	**Wiring short circuit to ground, injector(s) blocked, fuel contamination, fuel pressure low, exhaust system leaking, HO2S**
45	Heated oxygen sensor (HO2S) 1, bank 1 – mixture rich	**Wiring open circuit, fuel pressure high, injector(s) leaking, HO2S contamination, EVAP canister purge valve, TP sensor, EGR system, HO2S**
46	Long term fuel trim – imbalance bank 1/ bank 2	**Injector(s) leaking, HO2S, intake/exhaust system leaking**
48	Exhaust gas recirculation (EGR) valve – flow	**Wiring open circuit/short circuit to ground, EGR system blocked, EGR valve, ECM**
51	Engine control module (ECM) – PROM	**Insecure PROM, incorrect PROM, ECM**
52	Engine control module (ECM) – memory reset	**Battery disconnected, wiring**
53	Ignition control module (ICM) – 4X reference circuit malfunction	**Wiring open circuit/short circuit to ground, ICM, ECM**
55	Closed throttle position (CPT) – range	**Throttle valve tight/sticking, ISC actuator, TP sensor, cruise control cable**
56	Turbine shaft speed (TSS) sensor – circuit malfunction	**Wiring, TSS sensor, ECM**
57	Transmission fluid temperature (TFT) sensor – high temperature indicated	**Wiring short circuit to ground, TFT sensor, ECM**
58	Immobilizer control module – fuel enable signal	**Wiring open circuit/short circuit to ground, immobilizer control module, ECM**
59	Transmission fluid temperature (TFT) sensor – low temperature indicated	**Wiring open circuit, TFT sensor, ECM**
60	Park/neutral input with cruise control engaged	**Driver error, wiring open circuit/short circuit to ground, TFP switch**
61	Cruise control vent solenoid	**Wiring open circuit/short circuit to ground/ short circuit to positive, cruise control vent solenoid, cruise control servo, ECM**
62	Cruise control vacuum solenoid	**Wiring open circuit/short circuit to ground/ short circuit to positive, cruise control vacuum solenoid, cruise control servo, ECM**

Flash code	Fault location	Probable cause
63	Cruise control – set speed/vehicle speed mismatch	**Wiring open circuit/short circuit to ground/ short circuit to positive, hoses leaking, cruise control servo, ECM**
64	Cruise control – acceleration too high	**Slippery road conditions**
65	Cruise control servo position sensor – circuit malfunction	**Wiring open circuit/short circuit to ground/ short circuit to positive, cruise control servo, ECM**
66	Cruise control – engine speed too high	**Slippery road conditions, transmission mechanical fault**
67	Cruise control switch – circuit malfunction	**Wiring open circuit/short circuit to ground/ short circuit to positive, cruise control switch, ECM**
68	Cruise control servo position sensor – range	**Wiring, hoses blocked/leaking, brake pedal vacuum valve, cruise control servo, ECM**
69	Cruise control disabled – TCS active	**Slippery road conditions**
70	Throttle position (TP) sensor – circuit malfunction	**Wiring open circuit/short circuit to ground/ short circuit to positive, TP sensor**
71	Manifold absolute pressure (MAP) sensor – circuit malfunction	**Wiring open circuit/short circuit to ground/ short circuit to positive, MAP sensor**
73	Engine coolant temperature (ECT) sensor – circuit malfunction	**Wiring open circuit/short circuit to ground/ short circuit to positive, ECT sensor**
74	Intake air temperature (IAT) sensor – circuit malfunction	**Wiring open circuit/short circuit to ground/ short circuit to positive, IAT sensor**
75	Vehicle speed sensor (VSS) – circuit malfunction	**Wiring open circuit/short circuit, VSS**
76	Transmission pressure control (PC) solenoid – circuit malfunction	**Wiring, PC solenoid, ECM**
80	Throttle position (TP) sensor – idle learned incomplete	**TP sensor/idle learn procedure not completed**
83	Ignition control module (ICM) – 4X/24X signal circuit malfunction	**Electromagnetic interference, wiring, CKP sensor(s), ICM, ECM**
85	Idle learn throttle position – maximum position	**Throttle valve tight/sticking, TP sensor, ISC actuator**
86	Transmission gear ratio	**Turbine shaft speed (TSS) sensor, VSS, transmission mechanical fault, ECM**
88	Torque converter clutch (TCC) – stuck ON	**Wiring short circuit to ground, transmission internal wiring, TCC solenoid, ECM**
89	Extended shift – adaptation limit reached	**Transmission mechanical fault**
90	Brake pedal position (BPP) switch TCC	**Wiring open circuit/short circuit to ground, incorrectly adjusted, BPP switch, ECM**
91	Park/neutral position (PNP) switch – circuit malfunction	**Wiring open circuit/short circuit to ground, incorrectly adjusted, PNP switch, ECM**
93	Traction control system (TCS) – torque request circuit	**Wiring open circuit/short circuit to ground/ short circuit to positive, ABS control module trouble code(s) stored, ECM**

Flash code	Fault location	Probable cause
94	Shift solenoid (SS) A – ratio problem	**Wiring open circuit/short circuit to ground, SS, transmission mechanical fault, ECM**
95	Engine stall or near stall detected	**ECM trouble code(s) stored, wiring, mechanical fault, transmission mechanical fault**
96	Torque converter clutch (TCC) – over stressed	**ECM trouble codes stored, driver error**
99	Cruise control servo – position high after disengagement	**Wiring, throttle valve tight/sticking, cruise control servo**
106	Brake pedal position (BPP) switch, brake lamps – circuit malfunction	**Wiring, BPP switch**
108	Engine control module (ECM) – PROM	**Insecure PROM, incorrect PROM**
109	Engine control module (ECM) – memory reset	**Battery disconnected, wiring**
110	Alternator 'L' terminal – circuit malfunction	**Wiring open circuit/short circuit to ground, alternator, battery temperature sensor, ECM**
112	Engine control module (ECM) – EEPROM	**ECM**
117	Shift solenoid (SS) A/B – circuit malfunction	**Wiring open circuit/short circuit to ground, transmission mechanical fault, ECM**
119	Injector – circuit malfunction	**Wiring open circuit/short circuit to ground, injectors, ECM**
120	A/C refrigerant temperature sensor low pressure side – circuit malfunction	**Wiring open circuit/short circuit to ground, A/C refrigerant temperature sensor, ECM**
121	A/C refrigerant temperature sensor high pressure side – circuit malfunction	**Wiring open circuit/short circuit to ground, A/C refrigerant temperature sensor, ECM**
122	Engine coolant temperature – high with A/C request	**Cooling system**
123	Low A/C refrigerant charge	**Low A/C refrigerant charge, wiring open circuit/short circuit to ground/short circuit to positive, A/C refrigerant low pressure switch, A/C amplifier, ECM**
124	A/C request circuit – voltage high	**Wiring short circuit to positive, A/C refrigerant low pressure switch, A/C amplifier, ECM**
125	A/C refrigerant high pressure switch – circuit malfunction	**Wiring short circuit to ground, A/C refrigerant high pressure switch, ECM**
126	A/C refrigerant pressure – high	**Wiring short circuit to ground, A/C refrigerant temperature sensor (high side), A/C system**
128	A/C high discharge temperature/pressure	**Wiring short circuit to ground, A/C refrigerant high pressure switch, A/C compressor clutch**
131	Knock sensor (KS) – test failure	**Wiring, KS incorrectly fitted, KS, ECM**
132	Knock sensor (KS) – circuit malfunction	**Wiring open circuit/short circuit to ground/ short circuit to positive, KS incorrectly fitted, KS, insecure PROM, ECM**

Flash code	Fault location	Probable cause
133	CAN data bus – ABS/TCS	**Wiring, ABS control module trouble code(s) stored**
134	CAN data bus – instrument panel	**Wiring open circuit, instrument panel**
135	CAN data bus – A/C system	**Wiring open circuit, A/C amplifier**
136	CAN data bus – diagnostic energy reserve module (DERM)	**Wiring open circuit, DERM**

Model: **Aurora 4.0L**

Engine identification: **VIN position 8 = C**

Year: **1996-02**

System: **GM**

Self-diagnosis

General information

- Refer to the front of this manual for general test conditions, terminology, detailed descriptions of wiring faults and a general trouble shooter for electrical and mechanical faults.
- Malfunction indicator lamp (MIL) will illuminate or flash if certain faults are recorded.
- ECM operates in backup mode if sensors fail, to enable vehicle to be driven to repair shop.

Accessing

- Trouble codes can only be displayed by using a Scan Tool connected to the data link connector (DLC) **1**.

1

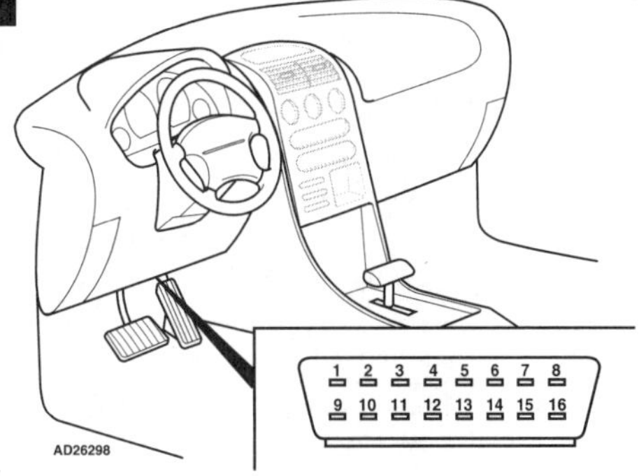

Erasing

- After the faults have been rectified, the manufacturer recommends that trouble codes are erased by using a Scan Tool connected to the data link connector (DLC).

Trouble code identification

Scan code	Fault location	Probable cause
P0000-P0999	Refer to OBD-II trouble code table at the front of this manual	–
P1106	Manifold absolute pressure (MAP) sensor – circuit intermittent – voltage high	**Wiring open circuit/short circuit to ground/ short circuit to positive**
P1107	Manifold absolute pressure (MAP) sensor – circuit intermittent – voltage low	**Wiring short circuit to ground**
P1108	Barometric pressure (BARO) sensor/manifold absolute pressure (MAP) sensor – comparison too high	**Wiring open circuit/short circuit to positive, MAP sensor, ECM**
P1111	Intake air temperature (IAT) sensor – circuit intermittent – voltage high	**Wiring open circuit, ground lead defective**
P1112	Intake air temperature (IAT) sensor – circuit intermittent – voltage low	**Wiring short circuit to ground**
P1114	Engine coolant temperature (ECT) sensor – circuit malfunction – voltage low	**Wiring short circuit to ground**

Scan code	Fault location	Probable cause
P1115	Engine coolant temperature (ECT) sensor – circuit malfunction – voltage high	**Wiring open circuit, ground lead defective**
P1121	Throttle position (TP) sensor – circuit malfunction – voltage high	**Wiring short circuit to positive, ground lead defective**
P1122	Throttle position (TP) sensor – circuit malfunction – voltage low	**Wiring open circuit/short circuit to ground**
P1133	Heated oxygen sensor (HO2S) 1, bank 1 – insufficient switching	**Wiring, oxygen sensor heater, exhaust system leaking, HO2S**
P1134	Heated oxygen sensor (HO2S) 1, bank 1 – slow response	**Wiring, oxygen sensor heater, exhaust system leaking, HO2S**
P1139	Heated oxygen sensor (HO2S) 2, bank 1 – insufficient switching	**Wiring, oxygen sensor heater, exhaust system leaking, HO2S**
P1140	Heated oxygen sensor (HO2S) 2, bank 1 – slow response	**Wiring, oxygen sensor heater, exhaust system leaking, HO2S**
P1153	Heated oxygen sensor (HO2S) 1, bank 2 – insufficient switching	**Wiring, oxygen sensor heater, exhaust system leaking, HO2S**
P1154	Heated oxygen sensor (HO2S) 1, bank 2 – slow response	**Wiring, oxygen sensor heater, exhaust system leaking, HO2S**
P1258	Engine overheating	**Cooling system, engine coolant temperature (ECT) sensor**
P1320	Ignition control module (ICM) – 4X reference circuit, intermittent	**Wiring open circuit/short circuit to ground, ICM, ECM**
P1323	Ignition control module (ICM) – 24X reference circuit, low frequency	**Wiring open circuit/short circuit to ground, ICM, ECM**
P1336	Crankshaft position sensor (CKP) – variation not learned	**Learning procedure not performed**
P1350	Ignition control module (ICM) – bypass line monitor circuit malfunction	**Wiring open circuit/short circuit to ground/ short circuit to positive, ICM, ECM**
P1359	Ignition control module (ICM) bank 1 – circuit malfunction	**Wiring open circuit/short circuit to positive, ICM, ECM**
P1360	Ignition control module (ICM) bank 2 – circuit malfunction	**Wiring open circuit/short circuit to positive, ICM, ECM**
P1370	Ignition control module (ICM) – 4X reference circuit, too many pulses	**Wiring, electromagnetic interference, ICM**
P1371	Ignition control module (ICM) – 4X reference circuit, too few pulses	**Wiring, electromagnetic interference, ICM**
P1372	Crankshaft position sensor (CKP)	**Wiring open circuit/short circuit to ground/ short circuit to positive, CKP, ECM**
P1375	Ignition control module (ICM) – 24X reference circuit, voltage high	**Wiring open circuit/short circuit to ground/ short circuit to positive, CKP sensor(s), ICM**
P1376	Ignition control module (ICM) – ground circuit	**Wiring open circuit/short circuit to positive, ground lead defective, ICM**
P1377	Ignition control module (ICM) – cam pulse/ 4X reference pulse	**Wiring open circuit/short circuit to ground, electromagnetic interference, valve timing**
P1380	Rough road signal – signal error	**ABS control module trouble code(s) stored**

Scan code	Fault location	Probable cause
P1381	Rough road signal – no communication from ABS CAN data bus	**Wiring open circuit, ABS system**
P1404	Exhaust gas recirculation (EGR) valve – stuck open	**Wiring open circuit/short circuit to positive, exhaust gas recirculation (EGR) valve position sensor, EGR solenoid**
P1406	Exhaust gas recirculation (EGR) valve position sensor – circuit malfunction	**Wiring open circuit/short circuit to ground/ short circuit to positive, exhaust gas recirculation (EGR) valve position sensor, EGR valve**
P1415	Secondary air injection (AIR) system	**Wiring, hoses blocked/leaking, AIR valve**
P1416	Secondary air injection (AIR) system	**Wiring, hoses blocked/leaking, AIR valve, AIR pump relay**
P1441	Evaporative emission (EVAP) system – leak detected	**Hose connection(s), EVAP canister, EVAP canister purge valve**
P1483	Engine cooling system – performance	**Cooling system, engine coolant temperature (ECT) sensor, intake air temperature (IAT) sensor**
P1508	Idle air control (IAC) valve – rpm low	**Wiring open circuit/short circuit to ground, intake system blocked, throttle valve, IAC valve, ECM**
P1509	Idle air control (IAC) valve – rpm high	**Wiring open circuit/short circuit to ground, intake system blocked, throttle valve, IAC valve, ECM**
P1520	Park/neutral position (PNP) switch	**Wiring short circuit to ground, incorrectly adjusted**
P1524	Throttle position (TP) sensor – learned closed throttle angle	**Accelerator cable adjustment, throttle valve tight/sticking, throttle valve incorrectly adjusted**
P1526	Minimum throttle position not learned	**Initial learn procedure not carried out**
P1533	A/C refrigerant temperature sensor – circuit malfunction low pressure side	**Wiring open circuit/short circuit to ground, A/C refrigerant temperature sensor, ECM**
P1535	A/C refrigerant temperature sensor – circuit malfunction high pressure side	**Wiring open circuit/short circuit to ground, A/C refrigerant temperature sensor, ECM**
P1536	A/C system – ECT over temperature	**Cooling system**
P1537	A/C request circuit – voltage low	**Wiring open circuit/short circuit to ground, A/C refrigerant low pressure switch, A/C control module, ECM**
P1538	A/C request circuit – voltage high	**Wiring open circuit/short circuit to positive, A/C control module, ECM**
P1540	A/C system – high pressure	**Wiring open circuit, A/C refrigerant high pressure switch, A/C system, ECM**
P1542	A/C system – high pressure/high temperature	**High ambient temperature, A/C cooling system**
P1554	Cruise control status circuit – high voltage	**Wiring open circuit/short circuit to ground, cruise control module, ECM**

Scan code	Fault location	Probable cause
P1560	Cruise control – transmission not in drive	Driver error, transmission fluid pressure (TFP) switch circuit
P1564	Cruise control – acceleration too high	Slippery road conditions
P1566	Cruise control – engine speed too high	Slippery road conditions, transmission mechanical fault
P1567	Cruise control – active brake control system active	Active brake control system engaged with cruise control on
P1570	Cruise control – traction control system (TCS) active	Slippery road surface
P1571	Traction control system (TCS), torque request circuit	Wiring open circuit/short circuit to ground/ short circuit to positive, ABS control module trouble code(s) stored, ECM
P1574	Brake pedal position switch (BPP) – voltage high	Wiring, ABS
P1575	Brake pedal position switch (BPP) – extended travel contacts – voltage high	Wiring, ABS
P1579	Transmission shifted from P/N to D/R at high throttle angle	Driver error, wiring, transmission range (TR) switch
P1599	Engine stall or near stall detected	ECM trouble code(s) stored, mechanical fault, transmission mechanical fault
P1602	CAN data bus – antilock brake system (ABS)/ traction control system (TCS)	Wiring, ABS control module trouble code(s) stored
P1603	CAN data bus – sensing/diagnostic module (SDM)	Wiring
P1604	CAN data bus – instrument panel	Wiring, body control module (BCM) trouble code(s) stored
P1605	CAN data bus – A/C	Wiring, A/C system trouble code(s) stored
P1617	Engine oil level switch – circuit malfunction	Wiring short circuit to ground, engine oil level, ECM
P1621	Engine control module (ECM) – memory performance	ECM
P1624	Customer snapshot function operated	ECM trouble codes stored
P1626	CAN data bus,immobilizer	Wiring, body control module (BCM) trouble code(s) stored
P1633	Engine control module (ECM) ignition supply voltage, voltage low	Wiring open circuit, ECM
P1634	Engine control module (ECM) – ignition supply voltage – voltage low	Wiring open circuit, ECM
P1635	Sensor supply voltage – circuit A malfunction	Wiring short circuit to ground/positive, ground wiring open circuit, TP sensor, MAP sensor, EGR valve position sensor, ECM
P1637	Alternator 'L' terminal – circuit malfunction	Wiring open circuit/short circuit to ground, alternator, battery temperature sensor, ECM
P1638	Alternator 'F' terminal – circuit malfunction	Wiring open circuit/short circuit to ground, alternator, ECM

Scan code	Fault location	Probable cause
P1639	Sensor supply voltage – circuit B malfunction	**Wiring short circuit to ground/positive, ground wiring open circuit, A/C refrigerant pressure sensor, fuel tank pressure (FTP) sensor, ECM**
P1640	Engine control module quad driver/output driver	**Wiring, ECM controlled components, ECM 1**
P1641	Malfunction indicator lamp (MIL) – circuit malfunction	**Wiring open circuit/short circuit to ground, MIL, supply wire, ECM**
P1642	Vehicle speed sensor (VSS) – circuit malfunction	**Wiring open circuit/short circuit to ground/ short circuit to positive, VSS, ECM**
P1644	Delivered torque – circuit malfunction	**Wiring open circuit/short circuit to ground/ short circuit to positive, ABS trouble codes stored, ECM**
P1645	Evaporative emission (EVAP) solenoid – circuit malfunction	**Wiring open circuit/short circuit to ground, EVAP solenoid, ECM**
P1646	Evaporative emission (EVAP) canister purge valve – circuit malfunction	**Wiring open circuit/short circuit to ground, EVAP canister purge valve, ECM**
P1650	Engine control module quad driver/output driver	**Wiring, ECM controlled components, ECM 1**
P1653	A/C compressor clutch relay – circuit malfunction	**Wiring open circuit/short circuit to ground, supply wire, A/C compressor clutch relay, ECM**
P1654	Cruise control disable circuit – circuit malfunction	**Wiring open circuit/short circuit to ground/ short circuit to positive, cruise control module, ECM**
P1660	Engine coolant blower motor relay(s) – circuit malfunction	**Wiring open circuit/short circuit to ground, engine coolant blower motor relay(s), ECM**

1 Depending on model, year and equipment level, the following components may be connected to this circuit. Use of the vehicle wiring diagram is recommended to confirm which components are fitted.

- □ A/C compressor clutch relay
- □ A/C condenser blower motor relay
- □ Engine coolant blower motor relay
- □ Engine coolant 'hot' warning lamp
- □ Evaporative emission (EVAP) canister purge valve
- □ Exhaust gas recirculation (EGR) solenoid
- □ Malfunction indicator lamp (MIL)
- □ Torque converter clutch (TCC) solenoid
- □ Transmission shift solenoids (SS)
- □ Transmission shift warning lamp (MT)

PLYMOUTH

Model:	Engine identification:	Year:
Acclaim 2.5L	**EDB/EDM/EDN**	**1992-95**
Acclaim 3.0L	**EFA**	**1992-95**
Sundance 2.2L	**EDF/EDS**	**1992-94**
Sundance 2.5L	**EDB/EDM/EDT/EFA**	**1992-94**
Sundance 3.0L	**EFA**	**1992-94**

System: **SBEC II**

Self-diagnosis

General information

- Malfunction indicator lamp (MIL) will illuminate if certain faults are recorded.
- ECM operates in backup mode if sensors fail, to enable vehicle to be driven to repair shop.
- Trouble codes can be displayed by the MIL or by using a Scan Tool connected to the data link connector (DLC).

Accessing

- Within 5 seconds switch the ignition key ON-OFF-ON-OFF-ON.
- Count MIL flashes. Note trouble codes.
- Each trouble code consists of two groups **1** [**A**].
- The first group of flashes indicate the 'tens' of the trouble code.
- The second group of flashes indicate the 'units' of the trouble code.
- A short pause separates 'tens' and 'units' **1** [**B**].
- A long pause separates each trouble code **1** [**C**].
- For example: Trouble code 21 displayed **1**.
- End of test sequence indicated by trouble code 55.
- Trouble codes can also be displayed by using a Scan Tool connected to the data link connector (DLC) **2**.

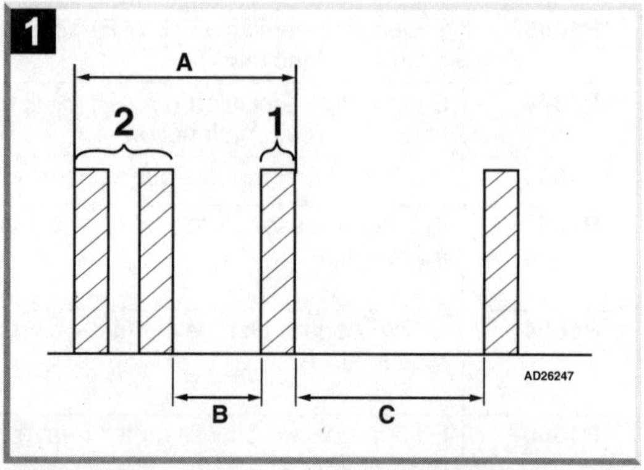

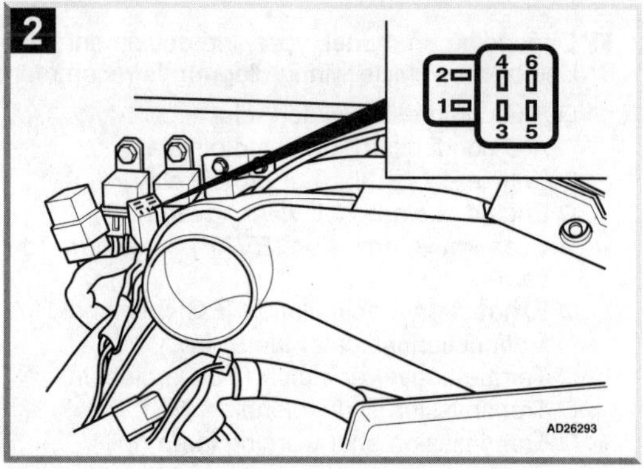

Erasing

NOTE: ***Disconnecting battery ground lead may cause driveability problems until the engine management system relearns basic values.***

- Ensure ignition switched OFF.
- Disconnect battery ground cable for at least 30 seconds.
- Reconnect battery ground cable.
- Repeat accessing procedure to ensure no trouble codes remain in ECM fault memory.

NOTE: ***Disconnecting battery ground lead may also erase memory from electronic units, such as radio.***

- Trouble codes can also be erased by using a Scan Tool connected to the data link connector (DLC).

Trouble code identification

Flash code	Fault location	Probable cause
11	Crankshaft position (CKP) sensor	**Wiring, air gap, CKP sensor**
12	Engine control module (ECM) – supply voltage disconnected within last 50 ignition switch ON-OFF cycles	**Wiring, battery**
13	Manifold absolute pressure (MAP) sensor	**Wiring, MAP sensor**
14	Manifold absolute pressure (MAP) sensor – voltage low/high	**Wiring, MAP sensor**
15	Vehicle speed sensor (VSS) – signal	**Wiring, VSS**
16	Knock sensor (KS) – wiring	**Wiring open circuit/short circuit, KS**
17	Engine temperature – low for long period	**Wiring, coolant thermostat, ECT sensor**
21	Heated oxygen sensor (HO2S) – incorrect signal/voltage high	**Wiring, fuel pressure, injector control, EVAP canister purge valve, compression, air filter blocked, heated oxygen sensor (HO2S)**
22	Engine coolant temperature (ECT) sensor – voltage low/high	**Wiring, ECT sensor**
23	Intake air temperature (IAT) sensor	**Wiring, IAT sensor**
24	Throttle position (TP) sensor	**Wiring, accelerator cable adjustment, TP sensor**
25	Idle air control (IAC) valve	**Wiring, IAC valve**
27	Injector(s) – circuit malfunction	**Wiring, injector**
31	Evaporative emission (EVAP) canister purge valve – open/short circuit	**Wiring, EVAP canister purge valve**
32	Exhaust gas recirculation (EGR) solenoid/ system – circuit malfunction	**Wiring, hoses, EGR solenoid**
33	A/C compressor clutch relay/A/C refrigerant pressure sensor – voltage low/high	**Wiring, A/C compressor clutch relay, A/C refrigerant pressure sensor**
34	Cruise control	**Wiring open circuit/short circuit, hoses, voltage low/high, cruise control master switch, cruise control vacuum switching valve**
35	Engine coolant blower motor relay – low/high speed	**Wiring, engine coolant blower motor relay – low/high speed**

Flash code	Fault location	Probable cause
36	Turbocharger (TC) wastegate regulating valve – circuit malfunction	**Wiring, TC wastegate regulating valve, ECM**
37	Torque converter clutch (TCC) solenoid – circuit malfunction	**Wiring, TCC solenoid**
41	Alternator	**Wiring open circuit/short circuit, alternator**
42	Engine control relay	**Wiring open circuit/short circuit, engine control relay**
43	Ignition coil – primary circuit	**Wiring, ignition coil**
44	Battery temperature sensor/engine coolant temperature (ECT) sensor	**Wiring open circuit/short circuit, battery temperature sensor, ECT sensor**
45	Turbo boost limit exceeded	**Wiring, hoses, TC wastegate, TC wastegate regulating valve, MAP sensor, ECM**
46	Alternator – voltage high	**Wiring, battery, alternator**
47	Alternator – voltage low	**Wiring, battery, alternator drive belt, alternator**
51	System too lean	**Wiring, intake/exhaust leak, fuel pressure, injector control, EVAP canister purge valve, heated oxygen sensor (HO2S)**
52	System too rich	**Wiring, exhaust blockage, fuel pressure, injector control, EVAP canister purge valve, compression, air filter blocked, ignition system, heated oxygen sensor (HO2S)**
53	Engine control module (ECM) – internal fault	**ECM**
54	Camshaft position (CMP) sensor – no signal	**Wiring, CMP sensor**
55	End of MIL code output	**–**
61	Barometric pressure (BARO) read solenoid – circuit malfunction	**Wiring, BARO solenoid**
62	Engine control module (ECM) – EEPROM error	**ECM**
63	Engine control module (ECM) – EEPROM error	**ECM**
64	Compressed natural gas pressure sensor – pressure high/low	**Compressed natural gas system, wiring, compressed natural gas pressure sensor**
65	Intake manifold air control solenoid	**Wiring open circuit/short circuit, intake manifold air control solenoid**
66	CAN data bus, transmission control module (TCM)/ body control module (BCM) – no signal	**Wiring, TCM, BCM**
71	Engine control module (ECM) – 5 volt output low	**ECM**
77	Cruise control system relay	**Wiring open circuit/short circuit, cruise control system relay**

PLYMOUTH

Model:	Engine identification:	Year:
Breeze 2.0L	**ECB**	**1996-97**
Neon 2.0L	**ECB/ECC**	**1996-97**

System: **SBEC III**

Self-diagnosis

General information

- Refer to the front of this manual for general test conditions, terminology, detailed descriptions of wiring faults and a general trouble shooter for electrical and mechanical faults.
- Malfunction indicator lamp (MIL) will illuminate or flash if certain faults are recorded.
- ECM operates in backup mode if sensors fail, to enable vehicle to be driven to repair shop.
- Trouble codes can be displayed by the MIL or by using a Scan Tool connected to the data link connector (DLC).

Accessing

Flash codes

- Within 5 seconds switch the ignition key ON-OFF-ON-OFF-ON.
- Count MIL flashes. Note trouble codes.
- Each trouble code consists of two groups **1** [**A**].
- The first group of flashes indicate the 'tens' of the trouble code.
- The second group of flashes indicate the 'units' of the trouble code.
- A short pause separates 'tens' and 'units' **1** [**B**].
- A long pause separates each trouble code **1** [**C**].
- For example: Trouble code 21 displayed **1**.
- End of test sequence indicated by trouble code 55.

P codes & Hex codes

- Trouble codes can be displayed by using a Scan Tool connected to the data link connector (DLC):
  - Breeze **2**.
  - Neon **3**.

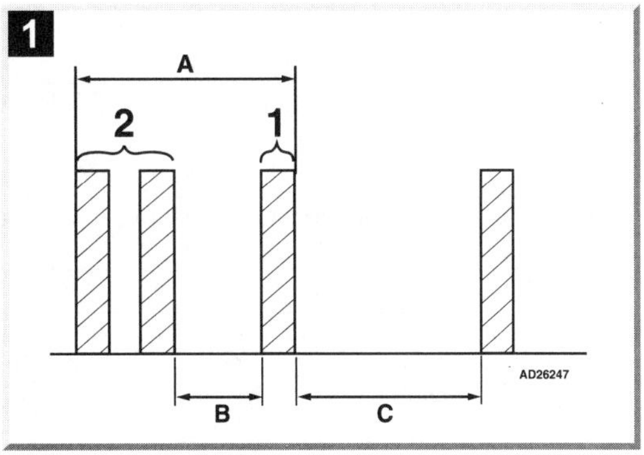

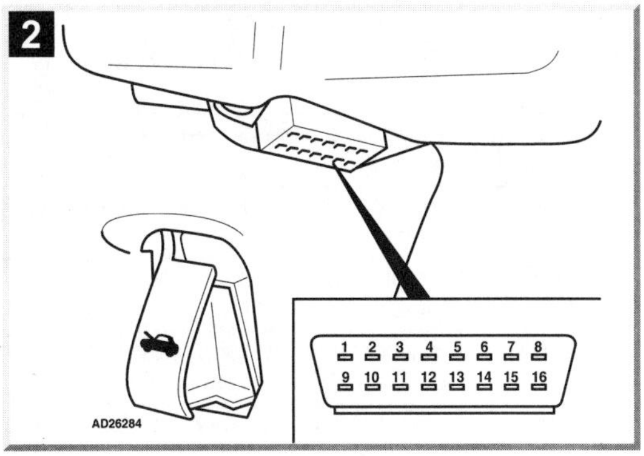

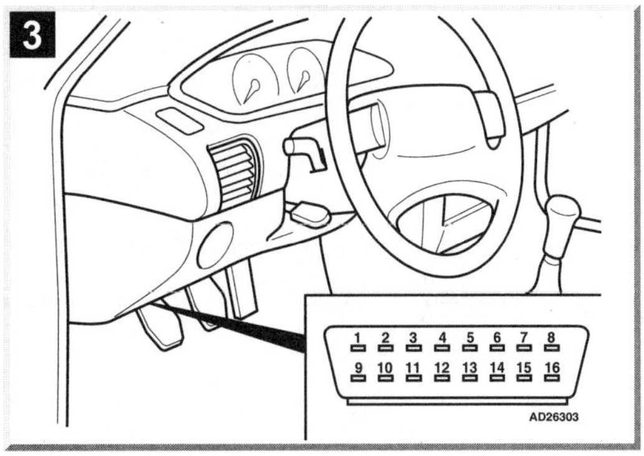

Erasing

- Trouble codes can only be erased by using a Scan Tool connected to the data link connector (DLC).

Trouble code identification

Scan code P type	Fault location	Probable cause
P0000-P0999	Refer to OBD-II trouble code table at the front of this manual	–
P1294	Idle speed control (ISC) system – rpm lower/higher than expected	**Wiring, air leak, throttle valve tight/sticking, IAC valve**
P1296	Manifold absolute pressure (MAP) sensor – no 5 volt supply	**Wiring, ECM**
P1297	Manifold absolute pressure (MAP) sensor	**Wiring, hose connection(s), MAP sensor**
P1299	Manifold absolute pressure (MAP) sensor/throttle position (TP) sensor	**Intake leak, hose connection(s), MAP sensor, TP sensor**
P1390	Timing belt jumped 1 tooth or more	**Timing belt loose/jumped position**
P1391	Crankshaft position (CKP) sensor/camshaft position (CMP) sensor – intermittent signal	**Wiring, poor connection, air gap, CKP sensor, CMP sensor**
P1398	Engine control module (ECM) – misfire detection limit reached	**Wiring, poor connection, CKP sensor**
P1486	Evaporative emission (EVAP) system – blockage detected	**Hose connection(s), hose blocked/trapped, EVAP canister**
P1489	Engine coolant blower motor relay – high speed	**Wiring open/short circuit, engine coolant blower motor relay – high speed**
P1490	Engine coolant blower motor relay – low speed	**Wiring open/short circuit, engine coolant blower motor relay – low speed**
P1492	Ambient/battery temperature sensor – voltage high	**Wiring, ambient/battery temperature sensor**
P1493	Ambient/battery temperature sensor – voltage low	**Wiring, ambient/battery temperature sensor**
P1494	Evaporative emission (EVAP) leak detection pump – pressure switch	**Wiring, hose leak/blockage, EVAP leak detection pump**
P1495	Evaporative emission (EVAP) leak detection pump solenoid	**Wiring open/short circuit, hose leak/blockage, EVAP leak detection pump solenoid**
P1496	Manifold absolute pressure (MAP) sensor – 5 volt output low	**Wiring, MAP sensor**
P1496	Engine control module (ECM) – 5 volt output low	**ECM**
P1696	Engine control module (ECM) – EEPROM error	**ECM**
P1697	Engine control module (ECM) – EEPROM error	**ECM**
P1698	CAN data bus, TCM – no signal	**Wiring, TCM**
P1899	Park/neutral position (PNP) switch – input circuit malfunction	**Wiring, PNP switch, ECM/TCM**

Flash code	Scan code Hex type	Fault location	Probable cause
11		Crankshaft position (CKP) sensor/camshaft position (CMP) sensor	**Wiring, poor connection, air gap, CKP sensor, CMP sensor**
	85	Timing belt jumped 1 tooth or more	**Timing belt loose/jumped position**
	9D	Crankshaft position (CKP) sensor/camshaft position (CMP) sensor – intermittent signal	**Wiring, poor connection, air gap, CKP sensor, CMP sensor**
	28	Crankshaft position (CKP) sensor – no signal	**Wiring, air gap, CKP sensor**
	BA	Engine control module (ECM) – misfire detection limit reached	**Wiring, poor connection, CKP sensor**
12		Engine control module (ECM) – supply voltage disconnected within last 50 ignition switch ON-OFF cycles	**Wiring, battery**
13	27	Manifold absolute pressure (MAP) sensor	**Wiring, hose connection(s), MAP sensor**
14		Manifold absolute pressure (MAP) sensor – input low/high	**Wiring, MAP sensor**
	24	Manifold absolute pressure (MAP) sensor – low input	**Wiring short circuit to ground, MAP sensor**
	25	Manifold absolute pressure (MAP) sensor – high input	**Wiring short circuit to positive, MAP sensor**
	87	Manifold absolute pressure (MAP) sensor – no 5 volt supply	**Wiring, ECM**
	92	Manifold absolute pressure (MAP) sensor – 5 volt output low	**Wiring, MAP sensor**
15	23	Vehicle speed sensor (VSS) – signal	**Wiring, VSS**
16	3B	Knock sensor (KS) 1 – no signal	**Wiring, KS incorrectly tightened,KS**
17		Engine temperature – low for long period	**Wiring, cooling system, coolant thermostat, ECT sensor**
	21	Engine temperature – low for long period	**Wiring, cooling system, coolant thermostat, ECT sensor**
	80	Insufficient coolant temperature for closed loop fuel control	**Wiring, cooling system, coolant thermostat, ECT sensor**
21		Heated oxygen sensor (HO2S) 1/2, bank 1 – circuit malfunction/incorrect signal/ voltage low/high	**Wiring, fuel pressure, injector control, EVAP canister purge valve, compression, air filter blocked, HO2S**
	9B	Heated oxygen sensor (HO2S) 1, bank 1 – low voltage	**Wiring short circuit to ground, exhaust leak, HO2S**
	3E	Heated oxygen sensor (HO2S) 1, bank 1 – high voltage	**Wiring short circuit to positive, fuel pressure, injector control, EVAP canister purge valve, air filter blocked, HO2S**
	66	Heated oxygen sensor (H02S) 1, bank 1 – slow response	**Wiring, heating inoperative, HO2S**
	20	Heated oxygen sensor (H02S) 1, bank 1 – no activity detected	**Wiring open circuit, heating inoperative, HO2S**
	67	Heated oxygen sensor (HO2S) 1, bank 1 – heater control – circuit malfunction	**Wiring, HO2S**

Flash code	Scan code Hex type	Fault location	Probable cause
	9C	Heated oxygen sensor (HO2S) 2, bank 1 – low voltage	**Wiring short circuit to ground, exhaust leak, HO2S**
	7E	Heated oxygen sensor (HO2S) 2, bank 1 – high voltage	**Wiring short circuit to positive, catalytic converter, HO2S**
	81	Heated oxygen sensor (HO2S) 2, bank 1 – no activity detected	**Wiring, heating inoperative, HO2S**
	69	Heated oxygen sensor (HO2S) 2, bank 1, heater control – circuit malfunction	**Wiring, HO2S**
22		Engine coolant temperature (ECT) sensor – voltage low/high	**Wiring, coolant thermostat, ECT sensor**
	1E	Engine coolant temperature (ECT) sensor – low input	**Wiring short circuit to ground, coolant thermostat, ECT sensor**
	1F	Engine coolant temperature (ECT) sensor – high input	**Wiring open circuit/short circuit to positive, coolant thermostat, ground wire defective, ECT sensor**
23		Intake air temperature (IAT) sensor – voltage low/high	**Wiring, IAT sensor**
	39	Intake air temperature (IAT) sensor – low input	**Wiring short circuit to ground, IAT sensor**
	3A	Intake air temperature (IAT) sensor – high input	**Wiring open circuit/short circuit to positive, ground wire defective, IAT sensor**
24		Throttle position (TP) sensor – voltage low/high	**Wiring, accelerator cable adjustment, TP sensor**
	84	Throttle position (TP) sensor – range/performance problem	**Wiring, accelerator cable adjustment, TP sensor**
	1A	Throttle position (TP) sensor – low input	**Wiring short circuit to ground, TP sensor**
	1B	Throttle position (TP) sensor – high input	**Wiring short circuit to positive, TP sensor**
25		Idle air control (IAC) valve	**Wiring open/short circuit, air leak, IAC valve**
	19	Idle speed control (ISC) system – malfunction	**Wiring open/short circuit, IAC valve**
	8A	Idle speed control (ISC) system – rpm lower/higher than expected	**Wiring, air leak, throttle valve tight/sticking, IAC valve**
	91	Manifold absolute pressure (MAP) sensor/throttle position (TP) sensor	**Intake leak, hose connection(s), MAP sensor, TP sensor**
27		Injector – circuit malfunction	**Wiring, injector**
	15	Injector 1 – circuit malfunction	**Wiring, injector**
	14	Injector 2 – circuit malfunction	**Wiring, injector**
	13	Injector 3 – circuit malfunction	**Wiring, injector**
	3D	Injector 4 – circuit malfunction	**Wiring, injector**
31		Evaporative emission (EVAP) canister purge valve – open/short circuit	**Wiring, EVAP canister purge valve**
	71	Evaporative emission (EVAP) system – incorrect flow detected	**Hose connection(s), intake leak, EVAP canister purge valve**

Flash code	Scan code Hex type	Fault location	Probable cause
	A0	Evaporative emission (EVAP) system – small leak detected	**Hose connection(s), intake leak, EVAP canister, EVAP canister purge valve**
	12	Evaporative emission (EVAP) canister purge valve – circuit malfunction	**Wiring open/short circuit, EVAP canister purge valve**
	A1	Evaporative emission (EVAP) system – large leak detected	**Hose connection(s), intake leak, EVAP canister, EVAP canister purge valve**
	BB	Evaporative emission (EVAP) system – blockage detected	**Hose connection(s), hose blocked/ trapped, EVAP canister**
	B8	Evaporative emission (EVAP) leak detection pump – pressure switch	**Wiring, hose leak/blockage, EVAP leak detection pump**
	B7	Evaporative emission (EVAP) leak detection pump solenoid	**Wiring open/short circuit, hose leak/ blockage, EVAP leak detection pump solenoid**
32		Exhaust gas recirculation (EGR) solenoid/ system – circuit malfunction	**Wiring, hoses, EGR solenoid**
	2E	Exhaust gas recirculation (EGR) system – insufficient flow detected	**Hose leak/blockage, wiring, EGR valve, EGR solenoid**
	11	Exhaust gas recirculation (EGR) – circuit malfunction	**Wiring open/short circuit, EGR solenoid**
33		A/C compressor clutch relay/A/C refrigerant pressure sensor – voltage low/high	**Wiring open/short circuit, A/C refrigerant pressure too low/high (cooling fault/incorrectly charged), A/C compressor clutch relay, A/C refrigerant pressure sensor**
	10	A/C compressor clutch relay	**Wiring open/short circuit, A/C compressor clutch relay**
	5A	A/C refrigerant pressure sensor – high input	**Wiring, A/C refrigerant pressure too high (cooling fault/incorrectly charged), A/C refrigerant pressure sensor**
	5B	A/C refrigerant pressure sensor – low input	**Wiring, A/C refrigerant pressure too low (incorrectly charged), A/C refrigerant pressure sensor**
34		Cruise control	**Wiring open/short circuit, voltage low/ high, cruise control master switch, cruise control vacuum switching valve**
	0F	Cruise control	**Wiring open/short circuit, voltage low/ high, cruise control vacuum switching valve**
	56	Cruise control master switch – high voltage	**Wiring, cruise control master switch**
	57	Cruise control master switch – low voltage	**Wiring, cruise control master switch**
35		Engine coolant blower motor relay – low/high speed	**Wiring open/short circuit, engine coolant blower motor relay – low/high speed**
	5D	Engine coolant blower motor relay – high speed	**Wiring open/short circuit, engine coolant blower motor relay – high speed**
	5C	Engine coolant blower motor relay – low speed	**Wiring open/short circuit, engine coolant blower motor relay – low speed**

Flash code	Scan code Hex type	Fault location	Probable cause
37		Torque converter clutch (TCC) solenoid/ park/neutral position (PNP) switch – circuit malfunction	**Wiring, TCC solenoid**
	94	Torque converter clutch (TCC) solenoid – circuit malfunction	**Wiring, TCC solenoid**
	0C	Torque converter clutch (TCC) solenoid – electrical	**Wiring open/short circuit, TCC solenoid**
	72	Park/neutral position (PNP) switch – input circuit malfunction	**Wiring, PNP switch, ECM/TCM**
41	0B	Alternator, field control – circuit malfunction	**Wiring open/short circuit, alternator, battery**
42		Fuel pump relay/engine control relay/fuel level sensor – circuit malfunction	**Wiring open/short circuit, fuel pump relay, engine control relay, fuel level sensor**
	65	Fuel pump relay	**Wiring open/short circuit, fuel pump relay**
	0A	Engine control relay	**Wiring open/short circuit, engine control relay**
	2C	Engine control relay	**Wiring open circuit, engine control relay**
	95	Fuel level sensor	**Wiring open circuit, fuel level sensor**
	96	Fuel level sensor	**Wiring short circuit to positive, fuel level sensor**
	97	Fuel level sensor	**Wiring, fuel level sensor**
43		Random/multiple cylinder(s) – misfire detected	**Wiring, spark plug(s), HT lead(s), injector(s), ignition coil(s), low compression**
	6A	Random/multiple cylinder(s) – misfire detected	**Wiring, spark plug(s), HT lead(s), injector(s), ignition coil(s), low compression**
	6B	Cylinder 1 – misfire detected	**Wiring, engine mechanical fault, ignition/fuel system, injector**
	6C	Cylinder 2 – misfire detected	**Wiring, engine mechanical fault, ignition/fuel system, injector**
	6D	Cylinder 3 – misfire detected	**Wiring, engine mechanical fault, ignition/fuel system, injector**
	6E	Cylinder 4 – misfire detected	**Wiring, engine mechanical fault, ignition/fuel system, injector**
	2B	Ignition coil A, primary/secondary – circuit malfunction	**Wiring, ignition coil**
	2A	Ignition coil B, primary/secondary – circuit malfunction	**Wiring, ignition coil**
44		Ambient/battery temperature sensor – voltage low/high	**Wiring, ambient/battery temperature sensor**
	9A	Ambient/battery temperature sensor – voltage high	**Wiring, ambient/battery temperature sensor**
	99	Ambient/battery temperature sensor – voltage low	**Wiring, ambient/battery temperature sensor**

Flash code	Scan code Hex type	Fault location	Probable cause
45	89	Transmission control system – malfunction	**Wiring, transmission control module (TCM)**
46	06	Alternator – voltage high	**Wiring, battery, alternator**
47	05	Alternator – voltage low	**Wiring, battery, alternator drive belt, alternator**
51	77	System too lean	**Wiring, intake/exhaust leak, fuel pressure, injector control, EVAP canister purge valve, heated oxygen sensor (HO2S)**
52	76	System too rich	**Wiring, exhaust blockage, fuel pressure, injector control, EVAP canister purge valve, compression, air filter blocked, ignition system, heated oxygen sensor (HO2S)**
53		Engine control module (ECM) – internal fault	**ECM**
	02	Engine control module (ECM) – internal fault	**ECM**
	44	Engine control module (ECM) – internal fault	**ECM**
54	01	Camshaft position (CMP) sensor – no signal	**Wiring, CMP sensor**
55		End of MIL code output	–
62	30	Engine control module (ECM) – EEPROM error	**ECM**
63	31	Engine control module (ECM) – EEPROM error	**ECM**
65		Power steering pressure (PSP) switch/ brake pedal position (BPP) switch/Intake manifold air control solenoid	**Wiring, PAS system, PSP switch, BPP switch, intake manifold air control solenoid**
	73	Power steering pressure (PSP) switch – range/performance problem	**Wiring, PAS system, PSP switch**
	98	Brake pedal position (BPP) switch B – circuit malfunction	**Wiring, BPP switch**
66		CAN data bus, transmission control module (TCM)/body control module (BCM) – no signal	**Wiring, TCM, BCM**
	60	CAN data bus, transmission control module (TCM) – no signal	**Wiring, TCM**
	61	CAN data bus, body control module (BCM) – no signal	**Wiring, BCM**
71	92	Engine control module (ECM) – 5 volt output low	**ECM**
72	70	Catalytic converter system – efficiency below threshold	**Wiring, catalytic converter, heated oxygen sensor (HO2S) 2**
77	52	Cruise control	**Wiring, cruise control vacuum switching valve**

PLYMOUTH

Model:	Engine identification:	Year:
Breeze 2.0L	**ECB**	**1998**
Breeze 2.4L	**EDZ**	**1998**
Neon 2.0L	**ECB/ECC**	**1998**
Prowler 3.5L	**VIN position 8 = G**	**1997-00**

System: **SBEC III**

Self-diagnosis

General information

- Refer to the front of this manual for general test conditions, terminology, detailed descriptions of wiring faults and a general trouble shooter for electrical and mechanical faults.
- Malfunction indicator lamp (MIL) will illuminate or flash if certain faults are recorded.
- ECM operates in backup mode if sensors fail, to enable vehicle to be driven to repair shop.

Accessing

- Trouble codes can only be displayed by using a Scan Tool connected to the data link connector (DLC):
  - Breeze **1**.
  - Neon **2**.
  - Prowler **3**.

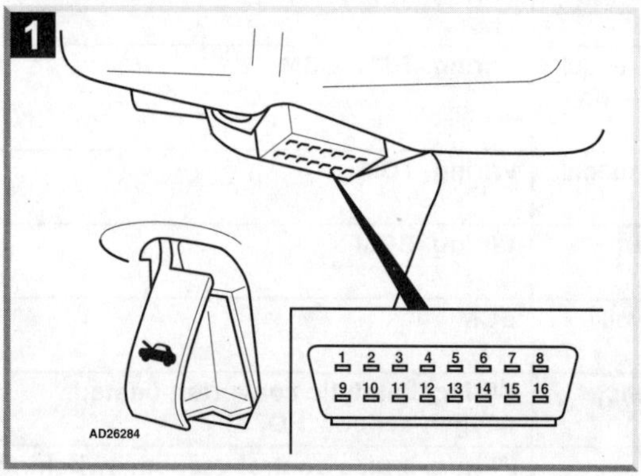

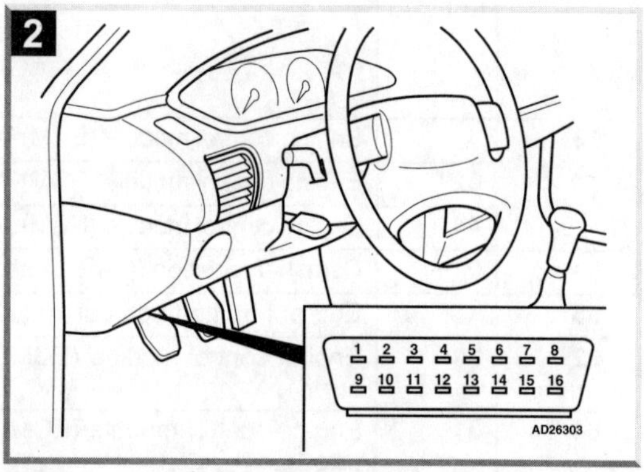

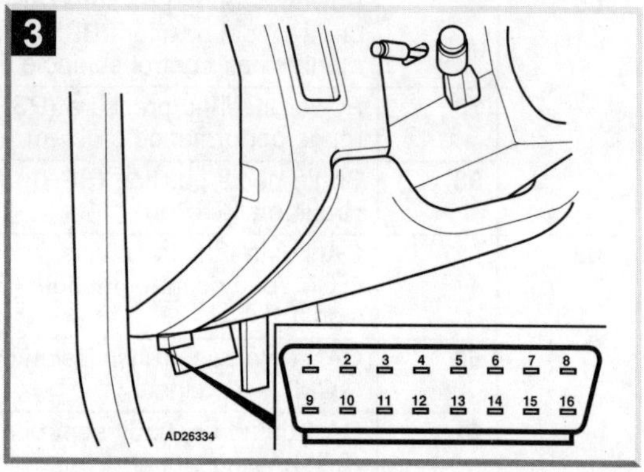

Erasing

- Trouble codes can only be erased by using a Scan Tool connected to the data link connector (DLC).

Trouble code identification

Scan code P type	Fault location	Probable cause
P0000-P0999	Refer to OBD-II trouble code table at the front of this manual	–
P1195	Heated oxygen sensor (HO2S) 1, bank 1 – slow response	Wiring, heating inoperative, HO2S
P1196	Heated oxygen sensor (HO2S) 1, bank 2 – slow response	Wiring, heating inoperative, HO2S
P1281	Engine temperature – low for long period	Wiring, cooling system, coolant thermostat, ECT sensor
P1282	Fuel pump relay	Wiring open circuit/short circuit, fuel pump relay
P1289	Intake manifold air control solenoid	Wiring open circuit/short circuit, intake manifold air control solenoid
P1294	Idle speed control (ISC) system – rpm lower/higher than expected	Wiring, air leak, throttle valve tight/sticking, IAC valve
P1297	Manifold absolute pressure (MAP) sensor	Wiring, hose connection(s), MAP sensor
P1299	Manifold absolute pressure (MAP) sensor/throttle position (TP) sensor	Intake leak, hose connection(s), MAP sensor, TP sensor
P1388	Engine control relay	Wiring open circuit/short circuit, engine control relay
P1389	Engine control relay	Wiring open circuit, engine control relay
P1390	Timing belt jumped 1 tooth or more	Timing belt loose/jumped position
P1391	Crankshaft position (CKP) sensor/camshaft position (CMP) sensor – intermittent signal	Wiring, poor connection, air gap, CKP sensor, CMP sensor
P1398	Engine control module (ECM) – misfire detection limit reached	Wiring, poor connection, CKP sensor
P1478	Ambient/battery temperature sensor	Wiring, ambient/battery temperature sensor
P1486	Evaporative emission (EVAP) system – blockage detected	Hose connection(s), hose blocked/trapped, EVAP canister
P1489	Engine coolant blower motor relay – high speed	Wiring open circuit/short circuit, engine coolant blower motor relay – high speed
P1490	Engine coolant blower motor relay – low speed	Wiring open circuit/short circuit, engine coolant blower motor relay – low speed
P1491	Engine coolant blower motor relay – low/high speed	Wiring open circuit/short circuit, engine coolant blower motor relay – low/high speed
P1492	Ambient/battery temperature sensor – voltage high	Wiring, ambient/battery temperature sensor
P1493	Ambient/battery temperature sensor – voltage low	Wiring, ambient/battery temperature sensor
P1494	Evaporative emission (EVAP) leak detection pump – pressure switch	Wiring, hose leak/blockage, EVAP leak detection pump
P1495	Evaporative emission (EVAP) leak detection pump solenoid	Wiring open circuit/short circuit, hose leak/blockage, EVAP leak detection pump solenoid

Scan code P type	Fault location	Probable cause
P1496	Engine control module (ECM) – 5 volt output low	ECM
P1594	Alternator – voltage high	Wiring, battery, alternator
P1595	Cruise control	Wiring open circuit/short circuit, hoses, cruise control master switch, cruise control vacuum switching valve
P1596	Cruise control master switch – high voltage	Wiring, cruise control master switch
P1597	Cruise control master switch – low voltage	Wiring, cruise control master switch
P1598	A/C refrigerant pressure sensor – high input	Wiring, A/C refrigerant pressure too high (cooling fault/incorrectly charged), A/C refrigerant pressure sensor
P1599	A/C refrigerant pressure sensor – low input	Wiring, A/C refrigerant pressure too low (incorrectly charged), A/C refrigerant pressure sensor
P1682	Alternator – voltage low	Wiring, battery, alternator drive belt, alternator
P1683	Cruise control	Wiring, cruise control vacuum switching valve
P1685	Key not programmed into immobilizer control module	Wiring, ignition key not matched/damaged, immobilizer control module
P1686	CAN data bus, Immobilizer control module – no signal	Wiring, immobilizer control module
P1695	CAN data bus, body control module (BCM) – no signal	Wiring, BCM
P1696	Engine control module (ECM) – EEPROM error	ECM
P1697	Engine control module (ECM) – EEPROM error	ECM
P1698	CAN data bus, transmission control module (TCM) – no signal	Wiring, TCM
P1899	Park/neutral position (PNP) switch – input circuit malfunction	Wiring, PNP switch, ECM/TCM

Scan code Hex type	Fault location	Probable cause
01	Camshaft position (CMP) sensor – no signal	Wiring, CMP sensor
02	Engine control module (ECM) – internal fault	ECM
05	Alternator – voltage low	Wiring, battery, alternator drive belt, alternator
06	Alternator – voltage high	Wiring, battery, alternator
0A	Engine control relay	Wiring open circuit/short circuit, engine control relay
0B	Alternator, field control – circuit malfunction	Wiring open circuit/short circuit, alternator, battery
0C	Torque converter clutch (TCC) solenoid – electrical	Wiring open circuit/short circuit, TCC solenoid

Scan code Hex type	Fault location	Probable cause
0E	Engine coolant blower motor relay – low/high speed	Wiring open circuit/short circuit, engine coolant blower motor relay – low/high speed
0F	Cruise control	Wiring open circuit/short circuit, hoses, cruise control master switch, cruise control vacuum switching valve
10	A/C compressor clutch relay	Wiring open circuit/short circuit, A/C compressor clutch relay
11	Exhaust gas recirculation (EGR) – circuit malfunction	Wiring open circuit/short circuit, EGR solenoid
12	Evaporative emission (EVAP) canister purge valve – circuit malfunction	Wiring open circuit/short circuit, EVAP canister purge valve
13	Injector 3 – circuit malfunction	Wiring, injector
14	Injector 2 – circuit malfunction	Wiring, injector
15	Injector 1 – circuit malfunction	Wiring, injector
19	Idle speed control (ISC) system – malfunction	Wiring open circuit/short circuit, IAC valve
1A	Throttle position (TP) sensor – low input	Wiring short circuit to ground, TP sensor
1B	Throttle position (TP) sensor – high input	Wiring short circuit to positive, TP sensor
1E	Engine coolant temperature (ECT) sensor – low input	Wiring short circuit to ground, coolant thermostat, ECT sensor
1F	Engine coolant temperature (ECT) sensor – high input	Wiring open circuit/short circuit to positive, coolant thermostat, ground wire defective, ECT sensor
20	Heated oxygen sensor (H02S) 1, bank 1 – no activity detected	Wiring open circuit, heating inoperative, HO2S
21	Engine temperature – low for long period	Wiring, cooling system, coolant thermostat, ECT sensor
23	Vehicle speed sensor (VSS) – signal	Wiring, VSS
24	Manifold absolute pressure (MAP) sensor – low input	Wiring short circuit to ground, MAP sensor
25	Manifold absolute pressure (MAP) sensor – high input	Wiring short circuit to positive, MAP sensor
27	Manifold absolute pressure (MAP) sensor	Wiring, hose connection(s), MAP sensor
28	Crankshaft position (CKP) sensor – no signal	Wiring, air gap, CKP sensor
29	Ignition coil C, primary/secondary – circuit malfunction	Wiring, ignition coil
2A	Ignition coil B, primary/secondary – circuit malfunction	Wiring, ignition coil
2B	Ignition coil A, primary/secondary – circuit malfunction	Wiring, ignition coil
2C	Engine control relay	Wiring open circuit, engine control relay
2E	Exhaust gas recirculation (EGR) system – insufficient flow detected	Wiring, hose leak/blockage, EGR valve, EGR solenoid

Scan code Hex type	Fault location	Probable cause
30	Engine control module (ECM) – EEPROM error	**ECM**
31	Engine control module (ECM) – EEPROM error	**ECM**
39	Intake air temperature (IAT) sensor – low input	**Wiring short circuit to ground, IAT sensor**
3A	Intake air temperature (IAT) sensor – high input	**Wiring open circuit/short circuit to positive, ground wire defective, IAT sensor**
3B	Knock sensor (KS) 1 – no signal	**Wiring, KS incorrectly tightened, KS**
3C	Manifold absolute pressure (MAP) sensor – barometric pressure range/performance problem	**Intake/exhaust leak, wiring, MAP sensor**
3D	Injector 4 – circuit malfunction	**Wiring, injector**
3E	Heated oxygen sensor (HO2S) 1, bank 1 – high voltage	**Wiring short circuit to positive, fuel pressure, injector control, EVAP canister purge valve, air filter blocked, HO2S**
41	Heated oxygen sensor (HO2S) 1, bank 2 – no activity detected	**Wiring, HO2S**
42	Heated oxygen sensor (HO2S) 1, bank 2 – high voltage	**Wiring short circuit to positive, fuel pressure, injector control, EVAP canister purge valve, air filter blocked, HO2S**
44	Engine control module (ECM) – internal fault	**ECM**
45	Injector 5 – circuit malfunction	**Wiring, injector**
46	Injector 6 – circuit malfunction	**Wiring, injector**
47	Ambient/battery temperature sensor	**Wiring, ambient/battery temperature sensor**
4C	Ignition coil D, primary/secondary – circuit malfunction	**Wiring, ignition coil**
4D	Ignition coil E, primary/secondary – circuit malfunction	**Wiring, ignition coil**
52	Cruise control	**Wiring, cruise control vacuum switching valve**
56	Cruise control master switch – high voltage	**Wiring, cruise control master switch**
57	Cruise control master switch – low voltage	**Wiring, cruise control master switch**
58	Intake manifold air control solenoid	**Wiring open circuit/short circuit, intake manifold air control solenoid**
5A	A/C refrigerant pressure sensor – high input	**Wiring, A/C refrigerant pressure too high (cooling fault/incorrectly charged), A/C refrigerant pressure sensor**
5B	A/C refrigerant pressure sensor – low input	**Wiring, A/C refrigerant pressure too low (incorrectly charged), A/C refrigerant pressure sensor**
5C	Engine coolant blower motor relay – low speed	**Wiring open circuit/short circuit, engine coolant blower motor relay – low speed**
5D	Engine coolant blower motor relay – high speed	**Wiring open circuit/short circuit, engine coolant blower motor relay – high speed**
60	CAN data bus, transmission control module (TCM) – no signal	**Wiring, TCM**

Scan code Hex type	Fault location	Probable cause
61	CAN data bus, body control module (BCM) – no signal	Wiring, BCM
65	Fuel pump relay	Wiring open circuit/short circuit, fuel pump relay
66	Heated oxygen sensor (H02S) 1, bank 1 – slow response	Wiring, heating inoperative, HO2S
67	Heated oxygen sensor (HO2S) 1, bank 1 – heater control – circuit malfunction	Wiring, HO2S
69	Heated oxygen sensor (HO2S) 2, bank 1, heater control – circuit malfunction	Wiring, HO2S
6A	Random/multiple cylinder(s) – misfire detected	Wiring, spark plug(s), HT lead(s), injector(s), ignition coil(s), low compression
6B	Cylinder 1 – misfire detected	Wiring, engine mechanical fault, ignition/ fuel system, injector
6C	Cylinder 2 – misfire detected	Wiring, engine mechanical fault, ignition/ fuel system, injector
6D	Cylinder 3 – misfire detected	Wiring, engine mechanical fault, ignition/ fuel system, injector
6E	Cylinder 4 – misfire detected	Wiring, engine mechanical fault, ignition/ fuel system, injector
70	Catalytic converter system, bank 1 – efficiency below threshold	Wiring, heated oxygen sensor (HO2S) 2, catalytic converter
71	Evaporative emission (EVAP) system – incorrect flow detected	Hose connection(s), intake leak, EVAP canister purge valve
72	Park/neutral position (PNP) switch – input circuit malfunction	Wiring, PNP switch, ECM/TCM
73	Power steering pressure (PSP) switch – range/performance problem	Wiring, PAS system, PSP switch
76	System too rich	Wiring, intake blocked, EVAP canister purge valve, fuel pressure, EGR system, injector(s), air filter blocked, ignition system, heated oxygen sensor (HO2S)
77	System too lean	Wiring, intake/exhaust leak, MAP sensor, fuel pressure/pump, EVAP canister purge valve, injector(s), heated oxygen sensor (HO2S)
78	System too rich, bank 2	Intake blocked, EVAP canister purge valve, fuel pressure, EGR system, injector(s), heated oxygen sensor (HO2S)
79	System too lean, bank 2	Intake/exhaust leak, fuel pressure/pump, injector(s), AIR system, hose connection(s)
7A	Heated oxygen sensor (HO2S) 1, bank 2 – slow response	Wiring, heating inoperative, HO2S
7C	Heated oxygen sensor (HO2S) 1, bank 2, heater control – circuit malfunction	Wiring, HO2S

PLYMOUTH

Breeze 2.0L • Breeze 2.4L • Neon 2.0L • Prowler 3.5L

Scan code Hex type	Fault location	Probable cause
7D	Heated oxygen sensor (HO2S) 2, bank 2, heater control – circuit malfunction	**Wiring, HO2S**
7E	Heated oxygen sensor (HO2S) 2, bank 1 – high voltage	**Wiring short circuit to positive, catalytic converter, HO2S**
7F	Heated oxygen sensor (HO2S) 2, bank 2 – high voltage	**Wiring short circuit to positive, catalytic converter, HO2S**
80	Insufficient coolant temperature for closed loop fuel control	**Wiring, cooling system, coolant thermostat, ECT sensor**
81	Heated oxygen sensor (HO2S) 2, bank 1 – no activity detected	**Wiring, heating inoperative, HO2S**
82	Heated oxygen sensor (HO2S) 2, bank 2 – no activity detected	**Wiring, HO2S**
84	Throttle position (TP) sensor – range/ performance problem	**Wiring, accelerator cable adjustment, TP sensor**
85	Timing belt jumped 1 tooth or more	**Timing belt loose/jumped position**
89	Transmission control system – malfunction	**Wiring, ECM/TCM**
8A	Idle speed control (ISC) system – rpm lower/ higher than expected	**Wiring, air leak, throttle valve tight/sticking, IAC valve**
91	Manifold absolute pressure (MAP) sensor/ throttle position (TP) sensor	**Intake leak, hose connection(s), MAP sensor, TP sensor**
92	Engine control module (ECM) – 5 volt output low	**ECM**
94	Torque converter clutch (TCC) solenoid – circuit malfunction	**Wiring, TCC solenoid**
95	Fuel level sensor – low input	**Wiring open circuit, fuel level sensor**
96	Fuel level sensor – high input	**Wiring short circuit to positive, fuel level sensor**
97	Fuel level sensor – malfunction	**Wiring, fuel level sensor**
98	Brake pedal position (BPP) switch – circuit malfunction	**Wiring, BPP switch**
99	Ambient/battery temperature sensor – voltage low	**Wiring, ambient/battery temperature sensor**
9A	Ambient/battery temperature sensor – voltage high	**Wiring, ambient/battery temperature sensor**
9B	Heated oxygen sensor (HO2S) 1, bank 1 – low voltage	**Wiring short circuit to ground, exhaust leak, HO2S**
9C	Heated oxygen sensor (HO2S) 2, bank 1 – low voltage	**Wiring short circuit to ground, exhaust leak, HO2S**
9D	Crankshaft position (CKP) sensor/camshaft position (CMP) sensor – intermittent signal	**Wiring, poor connection, air gap, CKP sensor, CMP sensor**
A0	Evaporative emission (EVAP) system – small leak detected	**Hose connection(s), intake leak, EVAP canister, EVAP canister purge valve**
A1	Evaporative emission (EVAP) system – large leak detected	**Hose connection(s), intake leak, EVAP canister, EVAP canister purge valve**

Scan code Hex type	Fault location	Probable cause
AE	Cylinder 5 – misfire detected	Wiring, engine mechanical fault, ignition/fuel system, injector
AF	Cylinder 6 – misfire detected	Wiring, engine mechanical fault, ignition/fuel system, injector
B4	Main catalytic converter, bank 2 – efficiency below threshold	Catalytic converter, wiring, HO2S 2
B5	Heated oxygen sensor (HO2S) 1, bank 2 – low voltage	Wiring short circuit to ground, exhaust leak, HO2S
B6	Heated oxygen sensor (HO2S) 2, bank 2 – low voltage	Wiring short circuit to ground, exhaust leak, HO2S
B7	Evaporative emission (EVAP) leak detection pump solenoid	Wiring open circuit/short circuit, hose leak/blockage, EVAP leak detection pump solenoid
B8	Evaporative emission (EVAP) leak detection pump – pressure switch	Wiring, hose leak/blockage, EVAP leak detection pump
BA	Engine control module (ECM) – misfire detection limit reached	Wiring, poor connection, CKP sensor
BB	Evaporative emission (EVAP) system – blockage detected	Hose connection(s), hose blocked/trapped, EVAP canister
C0	Heated oxygen sensor (HO2S) 1, bank 1 – slow response	Wiring, heating inoperative, HO2S
C1	Heated oxygen sensor (HO2S) 1, bank 2 – slow response	Wiring, heating inoperative, HO2S
D4	Exhaust gas recirculation (EGR) valve position sensor A – low input	Wiring short circuit to ground, EGR valve position sensor
D5	Exhaust gas recirculation (EGR) valve position sensor A – high input	Wiring short circuit to positive, EGR valve position sensor
D7	Exhaust gas recirculation (EGR) system – range/performance problem	Wiring, hose leak/blockage,EGR valve/solenoid
D8	Ignition coil F, primary/secondary – circuit malfunction	Wiring, ignition coil
D9	Intake manifold air control solenoid (short runner)	Wiring open circuit/short circuit, intake manifold air control solenoid (short runner)
E1	CAN data bus, instrumentation – no signal	Wiring, instrumentation control module (mechanical)
E2	CAN data bus, Immobilizer control module – no signal	Wiring, immobilizer control module
E8	Key not programmed into immobilizer control module	Wiring, ignition key not matched/damaged, immobilizer control module

PLYMOUTH

Model:	Engine identification:	Year:
Breeze 2.0L	ECB	1999-00
Breeze 2.4L	EDZ	1999-00
Neon 2.0L	ECB/ECC	1999-00

System: **SBEC III**

Self-diagnosis

General information

- Refer to the front of this manual for general test conditions, terminology, detailed descriptions of wiring faults and a general trouble shooter for electrical and mechanical faults.
- Malfunction indicator lamp (MIL) will illuminate or flash if certain faults are recorded.
- ECM operates in backup mode if sensors fail, to enable vehicle to be driven to repair shop.

Accessing

- Trouble codes can only be displayed by using a Scan Tool connected to the data link connector (DLC):
  - Breeze **1**.
  - Neon **2**.

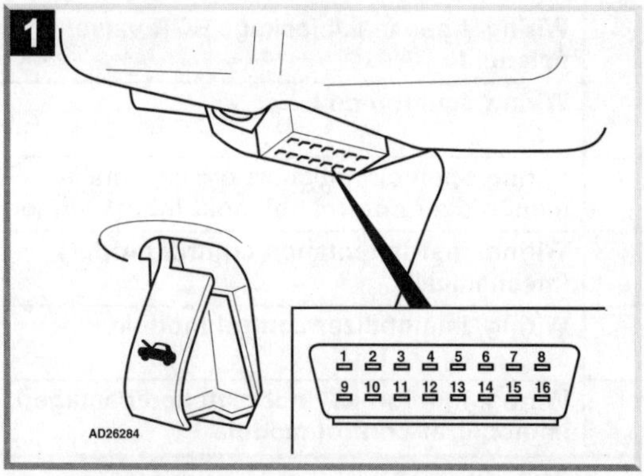

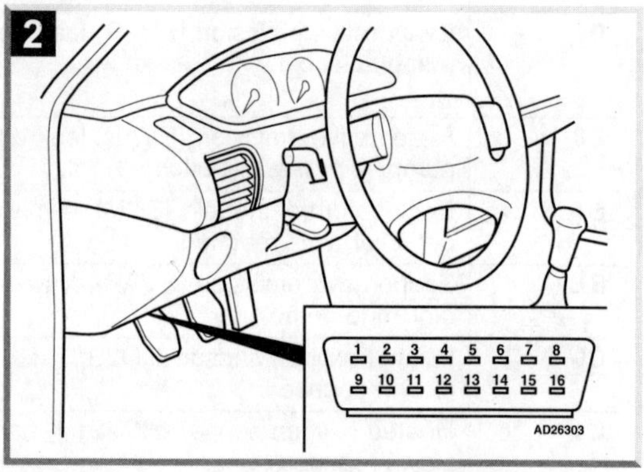

Erasing

- Trouble codes can only be erased by using a Scan Tool connected to the data link connector (DLC).

Trouble code identification

Scan code	Fault location	Probable cause
P0000-P0999	Refer to OBD-II trouble code table at the front of this manual	–
P1192	Intake air temperature (IAT) sensor – voltage low	Wiring, IAT sensor
P1193	Intake air temperature (IAT) sensor – voltage high	Wiring, IAT sensor
P1195	Heated oxygen sensor (H02S) 1, bank 1 – slow response	Wiring, heating inoperative, HO2S
P1197	Heated oxygen sensor (HO2S) 2, bank 1 – slow response	Wiring, heating inoperative, HO2S
P1281	Engine temperature – low for long period	Wiring, cooling system, coolant thermostat, ECT sensor
P1282	Fuel pump relay	Wiring open circuit/short circuit, fuel pump relay
P1294	Idle speed control (ISC) system – rpm lower/ higher than expected	Wiring, air leak, throttle valve tight/sticking, IAC valve
P1295	Throttle position (TP) sensor – no 5 volt supply	Wiring, TP sensor, ECM
P1296	Manifold absolute pressure (MAP) sensor – no 5 volt supply	Wiring, MAP sensor, ECM
P1297	Manifold absolute pressure (MAP) sensor	Wiring, hose connection(s), MAP sensor
P1298	System too lean – at prolonged wide open throttle	Intake/exhaust leak, MAP sensor, fuel pressure/pump, injector(s), HO2S
P1299	Manifold absolute pressure (MAP) sensor/ throttle position (TP) sensor	Intake leak, hose connection(s), MAP sensor, TP sensor
P1388	Engine control relay	Wiring open circuit/short circuit, engine control relay
P1389	Engine control relay	Wiring open circuit, engine control relay
P1390	Crankshaft position (CKP) sensor/camshaft position (CMP) sensor – signal synchronisation	Timing belt loose/jumped position, CKP sensor, CMP sensor
P1391	Crankshaft position (CKP) sensor/camshaft position (CMP) sensor – intermittent signal	Wiring, poor connection, air gap, CKP sensor, CMP sensor
P1398	Engine control module (ECM) – misfire detection limit reached	Wiring, poor connection, CKP sensor
P1403	Exhaust gas recirculation (EGR) valve position sensor – no 5 volt supply	Wiring, ECM
P1478	Ambient/battery temperature sensor	Wiring, ambient/battery temperature sensor
P1486	Evaporative emission (EVAP) system – blockage detected	Hose connection(s), hose blocked/trapped, EVAP canister
P1488	Engine control module (ECM) – auxiliary 5 volt output low	Wiring, ECM
P1489	Engine coolant blower motor relay – high speed	Wiring open circuit/short circuit, engine coolant blower motor relay – high speed

Scan code	Fault location	Probable cause
P1490	Engine coolant blower motor relay – low speed	**Wiring open circuit/short circuit, engine coolant blower motor relay – low speed**
P1491	Engine coolant blower motor relay – circuit malfunction	**Wiring open circuit/short circuit, engine coolant blower motor relay**
P1492	Ambient/battery temperature sensor – voltage high	**Wiring, ambient/battery temperature sensor**
P1493	Ambient/battery temperature sensor – voltage low	**Wiring, ambient/battery temperature sensor**
P1494	Evaporative emission (EVAP) leak detection pump – pressure switch	**Wiring, hose leak/blockage, EVAP leak detection pump**
P1495	Evaporative emission (EVAP) leak detection pump solenoid	**Wiring open circuit/short circuit, hose leak/ blockage, EVAP leak detection pump solenoid**
P1496	Engine control module (ECM) – 5 volt output low	**ECM**
P1594	Alternator – voltage high	**Wiring, battery, alternator**
P1595	Cruise control	**Wiring open circuit/short circuit, hoses, cruise control master switch, cruise control vacuum switching valve**
P1596	Cruise control master switch – high voltage	**Wiring, cruise control master switch**
P1597	Cruise control master switch – low voltage	**Wiring, cruise control master switch**
P1598	A/C refrigerant pressure sensor – high input	**Wiring, A/C refrigerant pressure too high (cooling fault/incorrectly charged), A/C refrigerant pressure sensor**
P1599	A/C refrigerant pressure sensor – low input	**Wiring, A/C refrigerant pressure too low (incorrectly charged), A/C refrigerant pressure sensor**
P1602	Engine control module (ECM) – not programmed	**New ECM not programmed, incorrectly programmed, ECM**
P1603	Engine control module (ECM) – RAM link error	**ECM**
P1604	Engine control module (ECM) – RAM read/ write error	**ECM**
P1607	Engine control module (ECM) – shut down timer error	**ECM**
P1681	CAN data bus, instrumentation – no signal	**Wiring, instrumentation control module**
P1682	Alternator – voltage low	**Wiring, battery, alternator drive belt, alternator**
P1683	Cruise control	**Wiring, cruise control vacuum switching valve**
P1684	Engine control module (ECM) – supply voltage disconnected within last 50 ignition switch ON-OFF cycles	**Wiring, battery**
P1685	Key not programmed into immobilizer control module	**Wiring, ignition key not matched/damaged, immobilizer control module**

Scan code	Fault location	Probable cause
P1686	CAN data bus, Immobilizer control module – no signal	Wiring, immobilizer control module
P1687	CAN data bus, instrumentation – no signal	Wiring, instrumentation control module (mechanical)
P1695	CAN data bus, body control module (BCM) – no signal	Wiring, BCM
P1696	Engine control module (ECM) – EEPROM error	ECM
P1697	Engine control module (ECM) – EEPROM error	ECM
P1698	CAN data bus, transmission control module (TCM) – no signal	Wiring, TCM
P1765	Transmission system relay	Wiring open circuit/short circuit, transmission system relay, ECM/TCM
P1899	Park/neutral position (PNP) switch – input circuit malfunction	Wiring, PNP switch, ECM/TCM
P2302	Ignition coil 1, secondary – circuit malfunction	Wiring, ignition coil
P2305	Ignition coil 2, secondary – circuit malfunction	Wiring, ignition coil
P2308	Ignition coil 3, secondary – circuit malfunction	Wiring, ignition coil
P2311	Ignition coil 4, secondary – circuit malfunction	Wiring, ignition coil
P2503	Charging system voltage – low	Wiring, poor connection, battery, alternator

PLYMOUTH

Model:	Engine identification:	Year:
Colt 1.5L	4G15	1992-93
Colt 1.8L	4G93	1992-93
Colt Vista 1.8L	4G93	1992-93
Colt Vista 2.4L	4G64	1992-93
Laser 1.8L	4G37	1992-94
Laser 2.0L	4G63	1992-94
Laser 2.0L Turbo	4G63	1992-94

System: **Mitsubishi MPI**

Self-diagnosis

General information

- Refer to the front of this manual for general test conditions, terminology, detailed descriptions of wiring faults and a general trouble shooter for electrical and mechanical faults.
- Malfunction indicator lamp (MIL) will illuminate if certain faults are recorded.
- ECM operates in backup mode if sensors fail, to enable vehicle to be driven to repair shop.
- 1992: Trouble codes can be displayed by using an analogue voltmeter connected to the data link connector (DLC).
- 1993 →: Trouble codes can be displayed by using an analogue voltmeter or Scan Tool connected to the data link connector (DLC).

Accessing

- Ensure ignition switched OFF.
- Access data link connector (DLC) **1**.
- Connect an analogue voltmeter between data link connector (DLC) terminals 1 and 12 **2**.
- Switch ignition ON.
- Count voltmeter needle deflections. Note trouble codes. Compare with trouble code table.
- Each trouble code consists of two groups of one or more needle deflections **3** :
  - Long needle deflections indicate the 'tens' of the trouble code **3** [**A**].
  - Short needle deflections indicate the 'units' of the trouble code **3** [**D**].
  - A short pause separates each trouble code group **3** [**C**].
  - A long pause separates each trouble code **3** [**E**].
  - For example: Trouble code 22 displayed **3**.
- If a constant deflection is displayed: Suspect engine control module (ECM).
- If constant short deflections are displayed: No trouble code is stored.
- 1993 →: Trouble codes can also be displayed by using a Scan Tool connected to the data link connector (DLC).

1

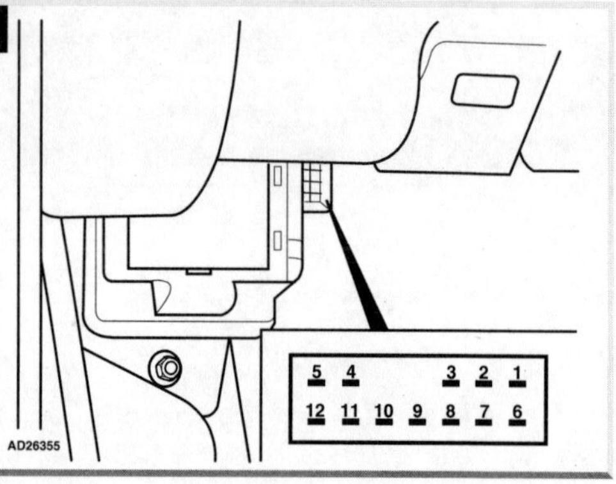

2

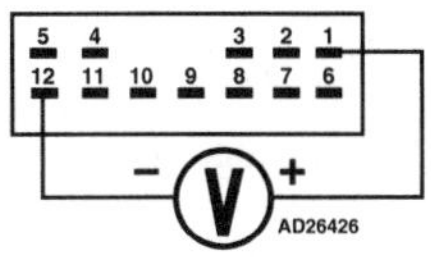

3

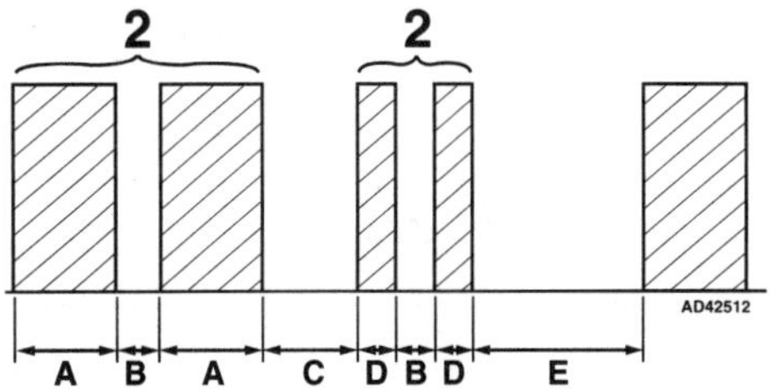

Erasing

NOTE: ***Disconnecting battery ground lead may cause driveability problems until the engine management system relearns basic values.***

- Ensure ignition switched OFF.
- Disconnect battery ground cable for at least 15 seconds.
- Reconnect battery ground lead.
- Repeat checking procedure to ensure no data remains in ECM fault memory.
- 1993 →: Trouble codes can also be erased by using a Scan Tool connected to the data link connector (DLC).

NOTE: ***Disconnecting battery ground lead may erase memory from electronic units such as radio.***

Trouble code identification

Flash code	Fault location	Probable cause
11	Heated oxygen sensor (HO2S)/oxygen sensor (O2S) 1, bank 1 – circuit malfunction	**Wiring, fuel pressure, injector(s), intake leak, HO2S/O2S, ECM**
12	Volume air flow (VAF) sensor – malfunction	**Wiring, VAF sensor, ECM**
13	Intake air temperature (IAT) sensor – circuit malfunction	**Wiring, IAT sensor, ECM**
14	Throttle position (TP) sensor – circuit malfunction	**Wiring, TP sensor, CTP switch, ECM**
15	Idle speed control (ISC) actuator position sensor – circuit malfunction	**Wiring, ISC actuator, ISC actuator position sensor, ECM**
21	Engine coolant temperature (ECT) sensor – circuit malfunction	**Wiring, ECT sensor, ECM**
22	Crankshaft position (CKP) sensor – circuit malfunction	**Wiring, CKP sensor, ECM**
23	Camshaft position (CMP) sensor – circuit malfunction	**Wiring, CMP sensor, ECM**

Flash code	Fault location	Probable cause
24	Vehicle speed sensor (VSS) – circuit malfunction	**Wiring, VSS, ECM**
25	Barometric pressure (BARO) sensor – circuit malfunction	**Wiring, BARO sensor, ECM**
31	Knock sensor (KS) – circuit malfunction	**Wiring, KS, ECM**
32	Manifold absolute pressure (MAP) sensor – circuit malfunction	**Wiring, MAP sensor, ECM**
36	Ignition timing adjustment	**Wiring, ignition timing connector short to ground, ECM**
41	Injectors – circuit malfunction	**Wiring, injectors, ECM**
42	Fuel pump – circuit malfunction	**Wiring, engine control relay, fuel pump, ECM**
43	Exhaust gas recirculation (EGR) system – circuit malfunction	**Wiring, hose leak/blockage, EGR valve, EGR temperature sensor, EGR solenoid, ECM**
44	Ignition coil/ignition control module (ICM) – circuit malfunction	**Wiring, ignition coil/ ICM, ECM**
55	Idle air control (IAC) valve position sensor – circuit malfunction	**Wiring, IAC valve, IAC valve position sensor, ECM**
59	Heated oxygen sensor (HO2S) 2, bank 1 – circuit malfunction	**Wiring, HO2S, ECM**

PLYMOUTH

Model:	Engine identification:	Year:
Colt 1.5L	**4G15**	**1994**

System: **Mitsubishi MPI**

Self-diagnosis

General information

- Refer to the front of this manual for general test conditions, terminology, detailed descriptions of wiring faults and a general trouble shooter for electrical and mechanical faults.
- Malfunction indicator lamp (MIL) will illuminate if certain faults are recorded.
- ECM operates in backup mode if sensors fail, to enable vehicle to be driven to repair shop.
- Trouble codes can be displayed by using an analogue voltmeter or Scan Tool connected to the data link connector(s) (DLC).

Accessing

- Ensure ignition switched OFF.
- Access 12 pin and 16 pin data link connectors (DLC) **1**.
- Connect an analogue voltmeter between 12 pin data link connector (DLC) terminal 5 and 16 pin data link connector (DLC) terminal 4 or 5 **2**
- Switch ignition ON.
- Count voltmeter needle deflections. Note trouble codes. Compare with trouble code table.
- Each trouble code consists of two groups of one or more needle deflections **3** :
  - Long needle deflections indicate the 'tens' of the trouble code **3** [**A**].
  - Short needle deflections indicate the 'units' of the trouble code **3** [**D**].
  - A short pause separates each trouble code group **3** [**C**].
  - A long pause separates each trouble code **3** [**E**].
  - For example: Trouble code 22 displayed **3**.
- If a constant deflection is displayed: Suspect engine control module (ECM).
- If constant short deflections are displayed: No trouble code is stored.
- Trouble codes can also be displayed by using a Scan Tool connected to the data link connector(s) (DLC).

1

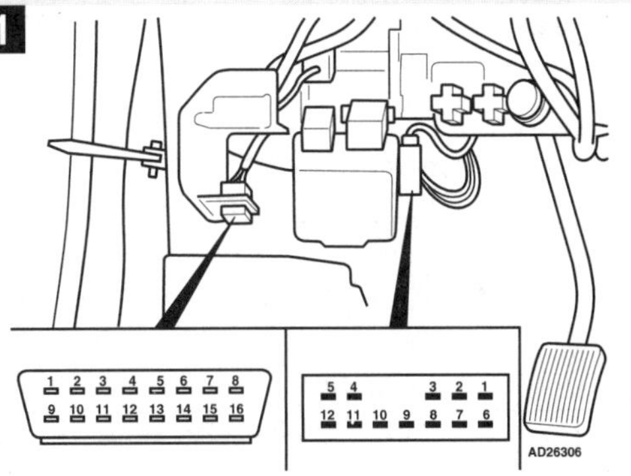

2

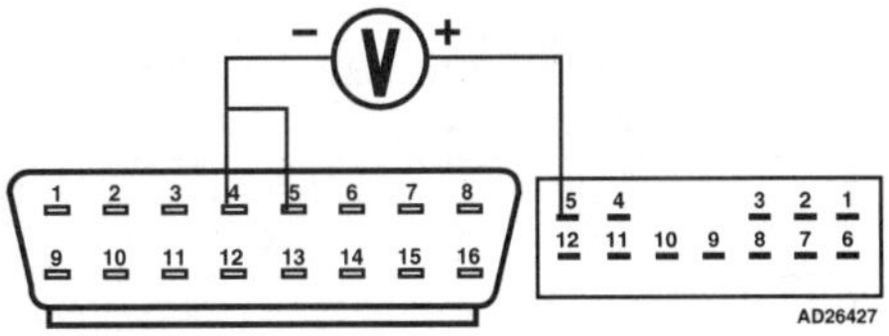

3

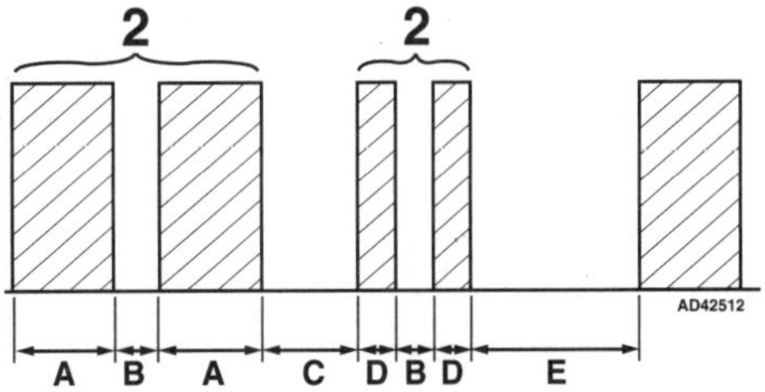

Erasing

NOTE: ***Disconnecting battery ground lead may cause driveability problems until the engine management system relearns basic values.***

- Ensure ignition switched OFF.
- Disconnect battery ground cable for at least 15 seconds.
- Reconnect battery ground lead.
- Repeat checking procedure to ensure no data remains in ECM fault memory.
- Trouble codes can also be erased by using a Scan Tool connected to the data link connector(s) (DLC).

NOTE: ***Disconnecting battery ground lead may erase memory from electronic units such as radio.***

Trouble code identification

Flash code	Fault location	Probable cause
11	Heated oxygen sensor (HO2S) 1, bank 1 – circuit malfunction	**Wiring, fuel pressure, injector(s), intake leak, HO2S, ECM**
13	Intake air temperature (IAT) sensor – circuit malfunction	**Wiring, IAT sensor, ECM**
14	Throttle position (TP) sensor – circuit malfunction	**Wiring, TP sensor, CTP switch, ECM**
21	Engine coolant temperature (ECT) sensor – circuit malfunction	**Wiring, ECT sensor, ECM**
22	Crankshaft position (CKP) sensor – circuit malfunction	**Wiring, CKP sensor, ECM**
23	Camshaft position (CMP) sensor – circuit malfunction	**Wiring, CMP sensor, ECM**
24	Vehicle speed sensor (VSS) – circuit malfunction	**Wiring, VSS, ECM**
32	Manifold absolute pressure (MAP) sensor – circuit malfunction	**Wiring, MAP sensor, ECM**
36	Ignition timing adjustment	**Wiring, ignition timing connector short to ground, ECM**
41	Injectors – circuit malfunction	**Wiring, injectors, ECM**
43	Exhaust gas recirculation (EGR) system – circuit malfunction	**Wiring, hose leak/blockage, EGR valve, EGR temperature sensor, EGR solenoid, ECM**
55	Idle air control (IAC) valve position sensor – circuit malfunction	**Wiring, IAC valve, IAC valve position sensor, ECM**
59	Heated oxygen sensor (HO2S) 2, bank 1 – circuit malfunction	**Wiring, HO2S, ECM**

PLYMOUTH

Model:	Engine identification:	Year:
Colt 1.8L	4G93	1994
Colt Vista 1.8L	4G93	1994
Colt Vista 2.4L	4G64	1994

System: **Mitsubishi MPI**

Self-diagnosis

General information

- Refer to the front of this manual for general test conditions, terminology, detailed descriptions of wiring faults and a general trouble shooter for electrical and mechanical faults.
- Malfunction indicator lamp (MIL) will illuminate if certain faults are recorded.
- ECM operates in backup mode if sensors fail, to enable vehicle to be driven to repair shop.
- Trouble codes can be displayed by the MIL or by using a Scan Tool connected to the data link connector(s) (DLC).

Accessing

- Ensure ignition switched OFF.
- Access 16 pin data link connector (DLC) **1**.
- Jump data link connector (DLC) terminal 1 and ground **2**.
- Switch ignition ON.
- Count MIL flashes. Note trouble codes. Compare with trouble code table.
- Each trouble code consists of two groups **3** :
  - Long flashes indicate the 'tens' of the trouble code **3** [**A**].
  - Short flashes indicate the 'units' of the trouble code **3** [**D**].
  - A short pause separates each trouble code group **3** [**C**].
  - A long pause separates each trouble code **3** [**E**].
  - For example: Trouble code 22 displayed **3**.
- If MIL remains ON: Suspect faulty engine control module (ECM).
- If constant short flashes are displayed, no trouble code is stored.
- Trouble codes can also be displayed by using a Scan Tool connected to the data link connector(s) (DLC).

1

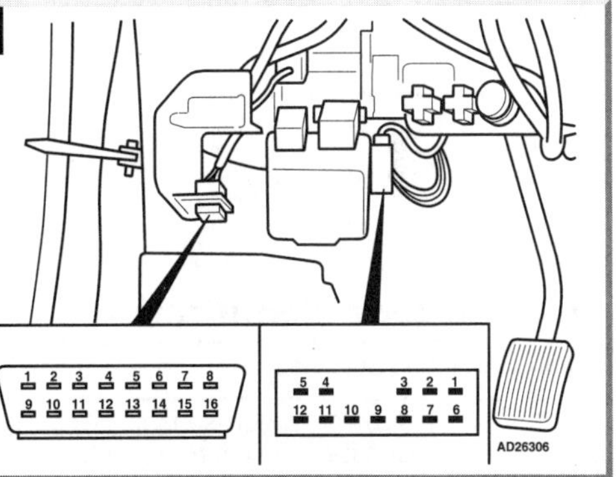

2

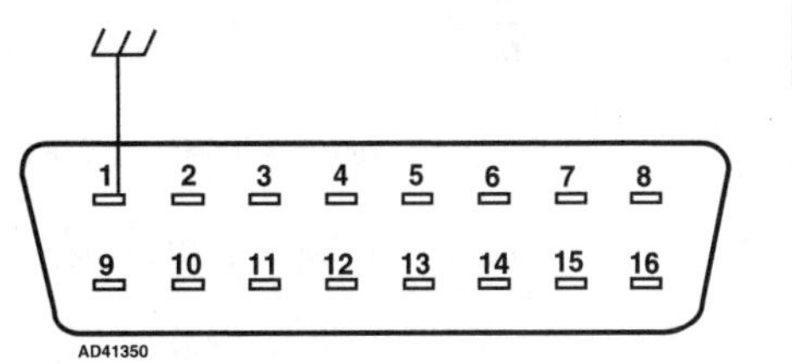

3

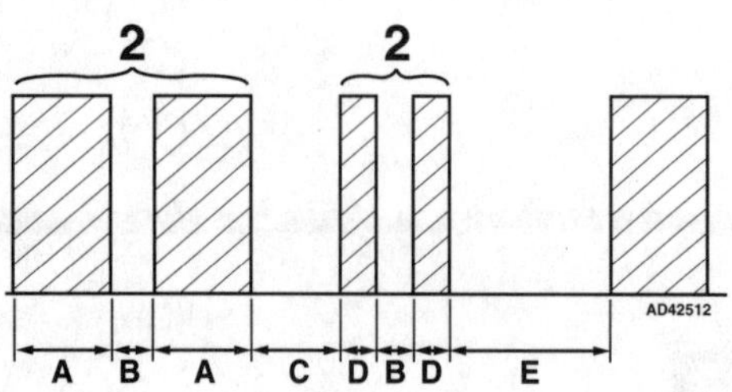

Erasing

NOTE: ***Disconnecting battery ground lead may cause driveability problems until the engine management system relearns basic values.***

- Ensure ignition switched OFF.
- Disconnect battery ground cable for at least 15 seconds.
- Reconnect battery ground lead.
- Repeat checking procedure to ensure no data remains in ECM fault memory.
- Trouble codes can also be erased by using a Scan Tool connected to the data link connector(s) (DLC).

NOTE: ***Disconnecting battery ground lead may erase memory from electronic units such as radio.***

Trouble code identification

Flash code	Fault location	Probable cause
11	Heated oxygen sensor (HO2S) 1, bank 1 – circuit malfunction	**Wiring, fuel pressure, injector(s), intake leak, HO2S, ECM**
12	Volume air flow (VAF) sensor – malfunction	**Wiring, VAF sensor, ECM**
13	Intake air temperature (IAT) sensor – circuit malfunction	**Wiring, IAT sensor, ECM**
14	Throttle position (TP) sensor – circuit malfunction	**Wiring, TP sensor, CTP switch, ECM**
21	Engine coolant temperature (ECT) sensor – circuit malfunction	**Wiring, ECT sensor, ECM**
22	Crankshaft position (CKP) sensor – circuit malfunction	**Wiring, CKP sensor, ECM**
23	Camshaft position (CMP) sensor – circuit malfunction	**Wiring, CMP sensor, ECM**
24	Vehicle speed sensor (VSS) – circuit malfunction	**Wiring, VSS, ECM**
25	Barometric pressure (BARO) sensor – circuit malfunction	**Wiring, BARO sensor, ECM**
36	Ignition timing adjustment	**Wiring, ignition timing connector short to ground, ECM**
41	Injectors – circuit malfunction	**Wiring, injectors, ECM**
43	Exhaust gas recirculation (EGR) system – circuit malfunction	**Wiring, hose leak/blockage, EGR valve, EGR temperature sensor, EGR solenoid, ECM**
55	Idle air control (IAC) valve position sensor – circuit malfunction	**Wiring, IAC valve, IAC valve position sensor, ECM**
59	Heated oxygen sensor (HO2S) 2, bank 1 – circuit malfunction	**Wiring, HO2S, ECM**

PLYMOUTH

Model: **Neon 2.0L** | Engine identification: **ECB/ECC** | Year: **1995**

System: **SBEC II**

Self-diagnosis

General information

- Malfunction indicator lamp (MIL) will illuminate or flash if certain faults are recorded.
- ECM operates in backup mode if sensors fail, to enable vehicle to be driven to repair shop.
- Trouble codes can be displayed by the MIL or by using a Scan Tool connected to the data link connector (DLC).

Accessing

Flash codes

- Within 5 seconds switch the ignition key ON-OFF-ON-OFF-ON.
- Count MIL flashes. Note trouble codes.
- Each trouble code consists of two groups **1** [**A**].
- The first group of flashes indicate the 'tens' of the trouble code.
- The second group of flashes indicate the 'units' of the trouble code.
- A short pause separates 'tens' and 'units' **1** [**B**].
- A long pause separates each trouble code **1** [**C**].
- For example: Trouble code 21 displayed **1**.
- End of test sequence indicated by trouble code 55.

P codes

- Trouble codes can be displayed by using a Scan Tool connected to the data link connector (DLC) **2**.

1

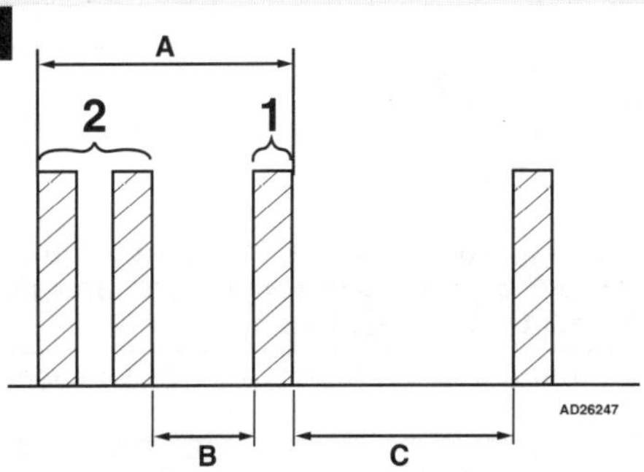

2

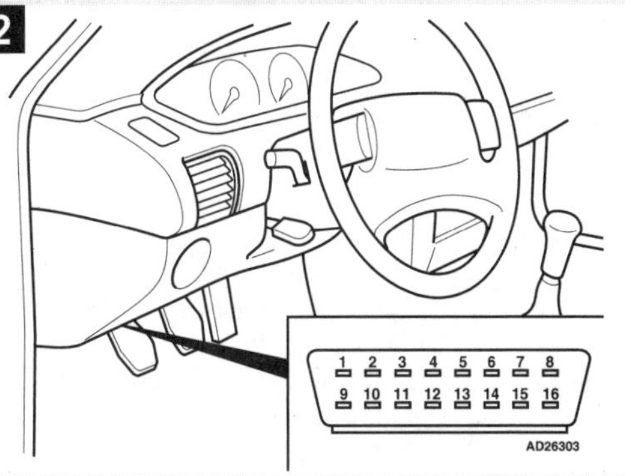

Erasing

NOTE: ***Disconnecting battery ground lead may cause driveability problems until the engine management system relearns basic values.***

- Ensure ignition switched OFF.
- Disconnect battery ground cable for at least 30 seconds.
- Reconnect battery ground cable.
- Repeat accessing procedure to ensure no trouble codes remain in ECM fault memory.

NOTE: ***Disconnecting battery ground lead may also erase memory from electronic units, such as radio.***

- Trouble codes can also be erased by using a Scan Tool connected to the data link connector (DLC).

Trouble code identification

Scan code P type	Fault location	Probable cause
P0000-P0999	Refer to OBD-II trouble code table at the front of this manual	–
P1294	Idle speed control (ISC) system – rpm lower/higher than expected	**Wiring, air leak, throttle valve tight/sticking, IAC valve**
P1297	Manifold absolute pressure (MAP) sensor	**Wiring, hose connection(s), MAP sensor**
P1299	Manifold absolute pressure (MAP) sensor/throttle position (TP) sensor	**Intake leak, hose connection(s), MAP sensor, TP sensor**
P1390	Crankshaft position (CKP) sensor/camshaft position (CMP) sensor – signal synchronisation	**Timing belt loose/jumped position, CKP sensor, CMP sensor**
P1391	Crankshaft position (CKP) sensor/camshaft position (CMP) sensor – intermittent signal	**Wiring, poor connection, air gap, CKP sensor, CMP sensor**
P1491	Engine coolant blower motor relay – circuit malfunction	**Wiring open circuit/short circuit, engine coolant blower motor relay**
P1899	Park/neutral position (PNP) switch – input circuit malfunction	**Wiring, PNP switch, ECM/TCM**

Flash code	Fault location	Probable cause
11	Crankshaft position (CKP) sensor/camshaft position (CMP) sensor	**Wiring, poor connection, air gap, CKP sensor, CMP sensor**
	Crankshaft position (CKP) sensor/camshaft position (CMP) sensor – signal synchronisation	**Timing belt loose/jumped position, CKP sensor, CMP sensor**
	Crankshaft position (CKP) sensor/camshaft position (CMP) sensor – intermittent signal	**Wiring, poor connection, air gap, CKP sensor, CMP sensor**
	Crankshaft position (CKP) sensor – no signal	**Wiring, air gap, CKP sensor**
12	Engine control module (ECM) – supply voltage disconnected within last 50 ignition switch ON-OFF cycles	**Wiring, battery**
13	Manifold absolute pressure (MAP) sensor	**Wiring, hose connection(s), MAP sensor**
14	Manifold absolute pressure (MAP) sensor – input low/high	**Wiring, MAP sensor**
15	Vehicle speed sensor (VSS) – signal	**Wiring, VSS**
17	Engine temperature – low for long period/insufficient coolant temperature for closed loop fuel control	**Wiring, cooling system, coolant thermostat, ECT sensor**
21	Heated oxygen sensor (HO2S) 1/2, bank 1 – circuit malfunction/incorrect signal/voltage low/high	**Wiring, fuel pressure, injector control, EVAP canister purge valve, compression, air filter blocked, HO2S**
22	Engine coolant temperature (ECT) sensor – voltage low/high	**Wiring, coolant thermostat, ECT sensor**

Flash code	Fault location	Probable cause
23	Intake air temperature (IAT) sensor – voltage low/high	**Wiring, IAT sensor**
24	Throttle position (TP) sensor – voltage low/high	**Wiring, accelerator cable adjustment, TP sensor**
25	Idle air control (IAC) valve	**Wiring open circuit/short circuit, IAC valve**
	Idle speed control (ISC) system – rpm lower/higher than expected	**Wiring, air leak, throttle valve tight/sticking, IAC valve**
	Manifold absolute pressure (MAP) sensor/throttle position (TP) sensor	**Intake leak, hose connection(s), MAP sensor, TP sensor**
27	Injector(s) – circuit malfunction	**Wiring, injector**
31	Evaporative emission (EVAP) solenoid/system	**Wiring, hose connection(s), intake leak, EVAP canister purge valve**
	Evaporative emission (EVAP) system – insufficient/excessive flow detected	**Hose leak/blockage, wiring, intake leak, EVAP canister purge valve**
	Evaporative emission (EVAP) solenoid – circuit malfunction	**Wiring open circuit/short circuit, EVAP canister purge valve**
32	Exhaust gas recirculation (EGR) solenoid/system – circuit malfunction	**Wiring, hoses, EGR valve, EGR solenoid**
	Exhaust gas recirculation (EGR) system – insufficient flow detected	**Hose leak/blockage, wiring, EGR valve, EGR solenoid**
	Exhaust gas recirculation (EGR) solenoid – circuit malfunction	**Wiring open circuit/short circuit, EGR solenoid**
33	A/C compressor clutch relay – voltage low/high	**Wiring open circuit/short circuit, A/C compressor clutch relay**
34	Cruise control	**Wiring open circuit/short circuit, hoses, voltage low/high, cruise control master switch, cruise control vacuum switching valve**
35	Engine coolant blower motor relay – circuit malfunction	**Wiring open circuit/short circuit, engine coolant blower motor relay**
37	Torque converter clutch (TCC) solenoid/park/neutral position (PNP) switch – circuit malfunction	**Wiring, TCC solenoid**
	Park/neutral position (PNP) switch – input circuit malfunction	**Wiring, PNP switch, ECM/TCM**
	Torque converter clutch (TCC) solenoid – circuit malfunction	**Wiring, TCC solenoid**
41	Alternator, field control – circuit malfunction	**Wiring open circuit/short circuit, alternator, battery**
42	Fuel pump relay/engine control relay/fuel level sensor – circuit malfunction	**Wiring open circuit/short circuit, Fuel pump relay, engine control relay, fuel level sensor**
43	Random/multiple cylinder(s) – misfire detected	**Wiring, spark plug(s), HT lead(s), injector(s), ignition coil(s), low compression**
44	Ambient/battery temperature sensor – voltage low/high	**Wiring, ambient/battery temperature sensor**

Flash code	Fault location	Probable cause
46	Alternator – voltage high	**Wiring, battery, alternator**
47	Alternator – voltage low	**Wiring, battery, alternator drive belt, alternator**
51	System too lean	**Wiring, intake/exhaust leak, MAP sensor, fuel pressure/pump, EVAP canister purge valve, injector(s), heated oxygen sensor (HO2S)**
52	System too rich	**Wiring, intake blocked, EVAP canister purge valve, fuel pressure, EGR system, injector(s), air filter blocked, ignition system, heated oxygen sensor (HO2S)**
53	Engine control module (ECM) – internal fault	**ECM**
54	Camshaft position (CMP) sensor – no signal	**Wiring, CMP sensor**
55	End of MIL code output	**–**
62	Engine control module (ECM) – EEPROM error	**ECM**
63	Engine control module (ECM) – EEPROM error	**ECM**
64	Catalytic converter system – efficiency below threshold	**Catalytic converter, wiring, HO2S**
65	Power steering pressure (PSP) switch – circuit malfunction	**Wiring, PAS system, PSP switch, ECM**

PLYMOUTH

Model:	Engine identification:	Year:
Voyager 2.4L	**EDZ**	**1996-97**
Voyager 3.0L	**EFA**	**1996-97**
Voyager 3.3L	**EGA/EGM**	**1996-97**
Voyager 3.8L	**EGH**	**1996-97**

System: **SBEC III**

Self-diagnosis

General information

- Refer to the front of this manual for general test conditions, terminology, detailed descriptions of wiring faults and a general trouble shooter for electrical and mechanical faults.
- Malfunction indicator lamp (MIL) will illuminate or flash if certain faults are recorded.
- ECM operates in backup mode if sensors fail, to enable vehicle to be driven to repair shop.
- Trouble codes can be displayed by the MIL or by using a Scan Tool connected to the data link connector (DLC).

Accessing

Flash codes

- Within 5 seconds switch the ignition key ON-OFF-ON-OFF-ON.
- Count MIL flashes. Note trouble codes.
- Each trouble code consists of two groups **1** [**A**].
- The first group of flashes indicate the 'tens' of the trouble code.
- The second group of flashes indicate the 'units' of the trouble code.
- A short pause separates 'tens' and 'units' **1** [**B**].
- A long pause separates each trouble code **1** [**C**].
- For example: Trouble code 21 displayed **1**.
- End of test sequence indicated by trouble code 55.

P codes & Hex codes

- Trouble codes can be displayed by using a Scan Tool connected to the data link connector (DLC) **2**.

1

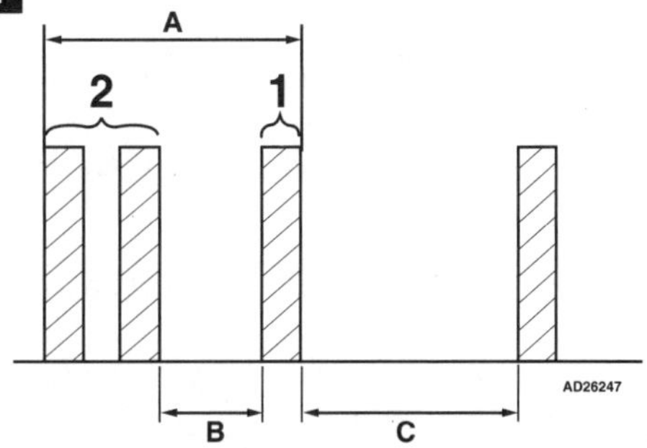

2

Erasing

- Trouble codes can only be erased by using a Scan Tool connected to the data link connector (DLC).

Trouble code identification

Scan code P type	Fault location	Probable cause
P0000-P0999	Refer to OBD-II trouble code table at the front of this manual	–
P1290	Compressed natural gas pressure sensor – pressure high	**Compressed natural gas system, wiring, compressed natural gas pressure sensor**
P1294	Idle speed control (ISC) system – rpm lower/higher than expected	**Wiring, air leak, throttle valve tight/sticking, IAC valve**
P1297	Manifold absolute pressure (MAP) sensor	**Wiring, hose connection(s), MAP sensor**
P1391	Crankshaft position (CKP) sensor/camshaft position (CMP) sensor – intermittent signal	**Wiring, poor connection, air gap, CKP sensor, CMP sensor**
P1398	Engine control module (ECM) – misfire detection limit reached	**Wiring, poor connection, CKP sensor**
P1486	Evaporative emission (EVAP) system – blockage detected	**Hose connection(s), hose blocked/trapped, EVAP canister**
P1491	Engine coolant blower motor relay – low/high speed	**Wiring open circuit/short circuit, engine coolant blower motor relay – low/high speed**
P1492	Ambient/battery temperature sensor – voltage high	**Wiring, ambient/battery temperature sensor**
P1493	Ambient/battery temperature sensor – voltage low	**Wiring, ambient/battery temperature sensor**
P1494	Evaporative emission (EVAP) leak detection pump – pressure switch	**Wiring, hose leak/blockage, EVAP leak detection pump**
P1495	Evaporative emission (EVAP) leak detection pump solenoid	**Wiring open circuit/short circuit, hose leak/blockage, EVAP leak detection pump solenoid**
P1496	Engine control module (ECM) – 5 volt output low	**ECM**
P1696	Engine control module (ECM) – EEPROM error	**ECM**
P1697	Engine control module (ECM) – EEPROM error	**ECM**
P1698	CAN data bus, TCM – no signal	**Wiring, TCM**
P1899	Park/neutral position (PNP) switch – input circuit malfunction	**Wiring, PNP switch, ECM/TCM**

Flash code	Scan code Hex type	Fault location	Probable cause
11		Crankshaft position (CKP) sensor/camshaft position (CMP) sensor	**Wiring, poor connection, air gap, CKP sensor, CMP sensor**
	9D	Crankshaft position (CKP) sensor/camshaft position (CMP) sensor – intermittent signal	**Wiring, poor connection, air gap, CKP sensor, CMP sensor**
	28	Crankshaft position (CKP) sensor – no signal	**Wiring, air gap, CKP sensor**

Flash code	Scan code Hex type	Fault location	Probable cause
	BA	Engine control module (ECM) – misfire detection limit reached	Wiring, poor connection, CKP sensor
12		Engine control module (ECM) – supply voltage disconnected within last 50 ignition switch ON-OFF cycles	Wiring, battery
13	27	Manifold absolute pressure (MAP) sensor	Wiring, hose connection(s), MAP sensor
14		Manifold absolute pressure (MAP) sensor – input low/high	Wiring, MAP sensor
	24	Manifold absolute pressure (MAP) sensor – low input	Wiring short circuit to ground, MAP sensor
	25	Manifold absolute pressure (MAP) sensor – high input	Wiring short circuit to positive, MAP sensor
15	23	Vehicle speed sensor (VSS) – signal	Wiring, VSS
17		Engine temperature – low for long period	Wiring, cooling system, coolant thermostat, ECT sensor
	21	Engine temperature – low for long period	Wiring, cooling system, coolant thermostat, ECT sensor
	80	Insufficient coolant temperature for closed loop fuel control	Wiring, cooling system, coolant thermostat, ECT sensor
21		Heated oxygen sensor (HO2S) 1/2, bank 1 – circuit malfunction/incorrect signal/voltage low/high	Wiring, fuel pressure, injector control, EVAP canister purge valve, compression, air filter blocked, HO2S
	9B	Heated oxygen sensor (HO2S) 1, bank 1 – low voltage	Exhaust leak, wiring short circuit to ground, HO2S
	3E	Heated oxygen sensor (HO2S) 1, bank 1 – high voltage	Wiring short circuit to positive, fuel pressure, injector control, EVAP canister purge valve, air filter blocked, HO2S
	66	Heated oxygen sensor (H02S) 1, bank 1 – slow response	Wiring, heating inoperative, HO2S
	20	Heated oxygen sensor (H02S) 1, bank 1 – no activity detected	Wiring open circuit, heating inoperative, HO2S
	67	Heated oxygen sensor (HO2S) 1, bank 1 – heater control – circuit malfunction	Wiring, HO2S
	9C	Heated oxygen sensor (HO2S) 2, bank 1 – low voltage	Exhaust leak, wiring short circuit to ground, HO2S
	7E	Heated oxygen sensor (HO2S) 2, bank 1 – high voltage	Wiring short circuit to positive, catalytic converter, HO2S
	81	Heated oxygen sensor (HO2S) 2, bank 1 – no activity detected	Wiring, heating inoperative, HO2S
	69	Heated oxygen sensor (HO2S) 2, bank 1, heater control – circuit malfunction	Wiring, HO2S
	C0	Heated oxygen sensor (HO2S) 1, bank 1 – slow response	Wiring, heating inoperative, HO2S

Flash code	Scan code Hex type	Fault location	Probable cause
22		Engine coolant temperature (ECT) sensor – voltage low/high	Wiring, coolant thermostat, ECT sensor
	1E	Engine coolant temperature (ECT) sensor – low input	Coolant thermostat, wiring short circuit to ground, ECT sensor
	1F	Engine coolant temperature (ECT) sensor – high input	Coolant thermostat, wiring open circuit/ short circuit to positive, ground wire defective, ECT sensor
23		Intake air temperature (IAT) sensor – voltage low/high	Wiring, IAT sensor
	39	Intake air temperature (IAT) sensor – low input	Wiring short circuit to ground, IAT sensor
	3A	Intake air temperature (IAT) sensor – high input	Wiring open circuit/short circuit to positive, ground wire defective, IAT sensor
24		Throttle position (TP) sensor – voltage low/ high	Wiring, accelerator cable adjustment, TP sensor
	84	Throttle position (TP) sensor – range/ performance problem	Wiring, accelerator cable adjustment, TP sensor
	1A	Throttle position (TP) sensor – low input	Wiring short circuit to ground, TP sensor
	1B	Throttle position (TP) sensor – high input	Wiring short circuit to positive, TP sensor
25		Idle air control (IAC) valve	Wiring open circuit/short circuit, IAC valve
	19	Idle speed control (ISC) system – malfunction	Wiring open circuit/short circuit, IAC valve
	8A	Idle speed control (ISC) system – rpm lower/higher than expected	Wiring, air leak, throttle valve tight/ sticking, IAC valve
27		Injector(s) – circuit malfunction	Wiring, injector
	15	Injector 1 – circuit malfunction	Wiring, injector
	14	Injector 2 – circuit malfunction	Wiring, injector
	13	Injector 3 – circuit malfunction	Wiring, injector
	3D	Injector 4 – circuit malfunction	Wiring, injector
	45	Injector 5 – circuit malfunction	Wiring, injector
	46	Injector 6 – circuit malfunction	Wiring, injector
31		Evaporative emission (EVAP) canister purge valve – open/short circuit	Wiring, EVAP canister purge valve
	71	Evaporative emission (EVAP) system – incorrect flow detected	Hose connection(s), intake leak, EVAP canister purge valve
	A0	Evaporative emission (EVAP) system – small leak detected	Hose connection(s), intake leak, EVAP canister, EVAP canister purge valve
	12	Evaporative emission (EVAP) canister purge valve – circuit malfunction	Wiring open circuit/short circuit, EVAP canister purge valve

Flash code	Scan code Hex type	Fault location	Probable cause
	A1	Evaporative emission (EVAP) system – large leak detected	**Hose connection(s), intake leak, EVAP canister, EVAP canister purge valve**
	BB	Evaporative emission (EVAP) system – blockage detected	**Hose connection(s), hose blocked/trapped, EVAP canister**
	B8	Evaporative emission (EVAP) leak detection pump – pressure switch	**Wiring, hose leak/blockage, EVAP leak detection pump**
	B7	Evaporative emission (EVAP) leak detection pump solenoid	**Wiring open circuit/short circuit, hose leak/blockage, EVAP leak detection pump solenoid**
32		Exhaust gas recirculation (EGR) solenoid/system – circuit malfunction	**Wiring, hoses, EGR solenoid**
	2E	Exhaust gas recirculation (EGR) system – insufficient flow detected	**Hose leak/blockage, wiring, EGR valve, EGR solenoid**
	11	Exhaust gas recirculation (EGR) – circuit malfunction	**Wiring open circuit/short circuit, EGR solenoid**
33		A/C compressor clutch relay/A/C refrigerant pressure sensor – voltage low/high	**A/C refrigerant pressure too low/high (cooling fault/incorrectly charged), wiring open circuit/short circuit, A/C compressor clutch relay, A/C refrigerant pressure sensor**
	10	A/C compressor clutch relay	**Wiring open circuit/short circuit, A/C compressor clutch relay**
	5A	A/C refrigerant pressure sensor – high input	**A/C refrigerant pressure too high (cooling fault/incorrectly charged), wiring, A/C refrigerant pressure sensor**
	5B	A/C refrigerant pressure sensor – low input	**A/C refrigerant pressure too low (incorrectly charged), wiring, A/C refrigerant pressure sensor**
34	0F	Cruise control	**Wiring open circuit/short circuit, hoses, voltage low/high, cruise control master switch, cruise control vacuum switching valve**
35	0E	Engine coolant blower motor relay – low/high speed	**Wiring open circuit/short circuit, engine coolant blower motor relay – low/high speed**
37		Torque converter clutch (TCC) solenoid/park/neutral position (PNP) switch – circuit malfunction	**Wiring, TCC solenoid**
	72	Park/neutral position (PNP) switch – input circuit malfunction	**Wiring, PNP switch, ECM/TCM**
	94	Torque converter clutch (TCC) solenoid – circuit malfunction	**Wiring, TCC solenoid**
	0C	Torque converter clutch (TCC) solenoid – electrical	**Wiring open circuit/short circuit, TCC solenoid**
41	0B	Alternator, field control – circuit malfunction	**Wiring open circuit/short circuit, alternator, battery**

Flash code	Scan code Hex type	Fault location	Probable cause
42		Fuel pump relay/engine control relay/fuel level sensor – circuit malfunction	**Wiring open circuit/short circuit, fuel pump relay, engine control relay, fuel level sensor**
	65	Fuel pump relay	**Wiring open circuit/short circuit, fuel pump relay**
	0A	Engine control relay	**Wiring open circuit/short circuit, engine control relay**
	2C	Engine control relay	**Wiring open circuit, engine control relay**
	95	Fuel level sensor – low input	**Wiring open circuit, fuel level sensor**
	96	Fuel level sensor – high input	**Wiring short circuit to positive, fuel level sensor**
	97	Fuel level sensor – malfunction	**Wiring, fuel level sensor**
43		Random/multiple cylinder(s) – misfire detected	**Wiring, spark plug(s), HT lead(s), injector(s), ignition coil(s), low compression**
	6A	Random/multiple cylinder(s) – misfire detected	**Wiring, spark plug(s), HT lead(s), injector(s), ignition coil(s), low compression**
	6B	Cylinder 1 – misfire detected	**Wiring, engine mechanical fault, ignition/fuel system, injector**
	6C	Cylinder 2 – misfire detected	**Wiring, engine mechanical fault, ignition/fuel system, injector**
	6D	Cylinder 3 – misfire detected	**Wiring, engine mechanical fault, ignition/fuel system, injector**
	6E	Cylinder 4 – misfire detected	**Wiring, engine mechanical fault, ignition/fuel system, injector**
	AE	Cylinder 5 – misfire detected	**Wiring, engine mechanical fault, ignition/fuel system, injector**
	AF	Cylinder 6 – misfire detected	**Wiring, engine mechanical fault, ignition/fuel system, injector**
	2B	Ignition coil A, primary/secondary – circuit malfunction	**Wiring, ignition coil**
	2A	Ignition coil B, primary/secondary – circuit malfunction	**Wiring, ignition coil**
	29	Ignition coil C, primary/secondary – circuit malfunction	**Wiring, ignition coil**
44		Battery temperature sensor – voltage low/high	**Wiring, battery temperature sensor**
	9A	Ambient/battery temperature sensor – voltage high	**Wiring, ambient/battery temperature sensor**
	99	Ambient/battery temperature sensor – voltage low	**Wiring, ambient/battery temperature sensor**
45	89	Transmission control system – malfunction	**Wiring, ECM/TCM**
46	06	Alternator – voltage high	**Wiring, battery, alternator**

Flash code	Scan code Hex type	Fault location	Probable cause
47	05	Alternator – voltage low	**Wiring, battery, alternator drive belt, alternator**
51	77	System too lean	**Wiring, intake/exhaust leak, MAP sensor, fuel pressure/pump, EVAP canister purge valve, injector(s), heated oxygen sensor (HO2S)**
52	76	System too rich	**Wiring, intake blocked, EVAP canister purge valve, fuel pressure, EGR system, injector(s), air filter blocked, ignition system, heated oxygen sensor (HO2S)**
53		Engine control module (ECM) – internal fault	**ECM**
	02	Engine control module (ECM) – internal fault	**ECM**
	44	Engine control module (ECM) – internal fault	**ECM**
54	01	Camshaft position (CMP) sensor – no signal	**Wiring, CMP sensor**
55		End of MIL code output	–
61	3C	Manifold absolute pressure (MAP) sensor – barometric pressure range/ performance problem	**Intake/exhaust leak, wiring, MAP sensor**
62	30	Engine control module (ECM) – EEPROM error	**ECM**
63	31	Engine control module (ECM) – EEPROM error	**ECM**
64	BE	Compressed natural gas pressure sensor – pressure high	**Compressed natural gas system, wiring, compressed natural gas pressure sensor**
65	98	Brake pedal position (BPP) switch – circuit malfunction	**Wiring, BPP switch**
66		CAN data bus, transmission control module (TCM)/body control module (BCM) – no signal	**Wiring, TCM, BCM**
	60	CAN data bus, transmission control module (TCM) – no signal	**Wiring, TCM**
	61	CAN data bus, body control module (BCM) – no signal	**Wiring, BCM**
71	92	Engine control module (ECM) – 5 volt output low	**ECM**
72	70	Catalytic converter system, bank 1 – efficiency below threshold	**Catalytic converter, wiring, heated oxygen sensor (HO2S) 2**
77	52	Cruise control	**Wiring, cruise control vacuum switching valve**

PLYMOUTH

Model:	Engine identification:	Year:
Voyager 2.4L	EDZ	1998
Voyager 3.0L	EFA	1998
Voyager 3.3L	EGA/EGM	1998
Voyager 3.8L	EGH	1998

System: **SBEC III**

Self-diagnosis

General information

- Refer to the front of this manual for general test conditions, terminology, detailed descriptions of wiring faults and a general trouble shooter for electrical and mechanical faults.
- Malfunction indicator lamp (MIL) will illuminate or flash if certain faults are recorded.
- ECM operates in backup mode if sensors fail, to enable vehicle to be driven to repair shop.

Accessing

- Trouble codes can only be displayed by using a Scan Tool connected to the data link connector (DLC) **1**.

1

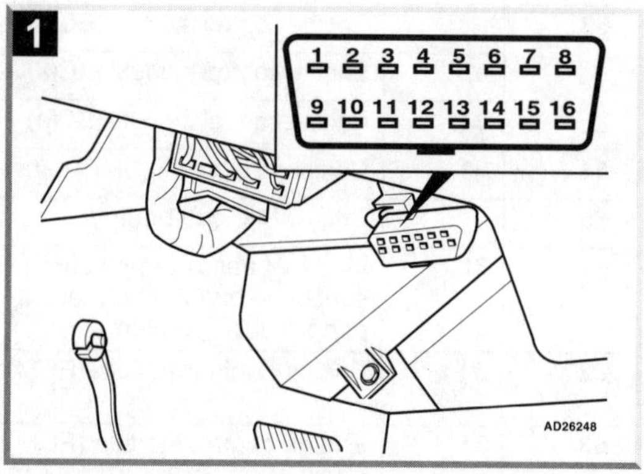

Erasing

- Trouble codes can only be erased by using a Scan Tool connected to the data link connector (DLC).

Trouble code identification

Scan code P type	Fault location	Probable cause
P0000-P0999	Refer to OBD-II trouble code table at the front of this manual	–
P1290	Compressed natural gas pressure sensor – pressure high	**Compressed natural gas system, wiring, compressed natural gas pressure sensor**
P1294	Idle speed control (ISC) system – rpm lower/higher than expected	**Wiring, air leak, throttle valve tight/sticking, IAC valve**
P1391	Crankshaft position (CKP) sensor/camshaft position (CMP) sensor – intermittent signal	**Wiring, poor connection, air gap, CKP sensor, CMP sensor**
P1398	Engine control module (ECM) – misfire detection limit reached	**Wiring, poor connection, CKP sensor**
P1486	Evaporative emission (EVAP) system – blockage detected	**Hose connection(s), hose blocked/trapped, EVAP canister**

Scan code P type	Fault location	Probable cause
P1491	Engine coolant blower motor relay – low/high speed	Wiring open circuit/short circuit, engine coolant blower motor relay – low/high speed
P1492	Ambient/battery temperature sensor – voltage high	Wiring, ambient/battery temperature sensor
P1493	Ambient/battery temperature sensor – voltage low	Wiring, ambient/battery temperature sensor
P1494	Evaporative emission (EVAP) leak detection pump – pressure switch	Wiring, hose leak/blockage, EVAP leak detection pump
P1495	Evaporative emission (EVAP) leak detection pump solenoid	Wiring open circuit/short circuit, hose leak/ blockage, EVAP leak detection pump solenoid
P1496	Engine control module (ECM) – 5 volt output low	ECM
P1595	Cruise control	Wiring open circuit/short circuit, hoses, cruise control master switch, cruise control vacuum switching valve
P1696	Engine control module (ECM) – EEPROM error	ECM
P1698	CAN data bus, transmission control module (TCM) – no signal	Wiring, TCM
P1899	Park/neutral position (PNP) switch – input circuit malfunction	Wiring, PNP switch, ECM/TCM

Scan code Hex type	Fault location	Probable cause
01	Camshaft position (CMP) sensor – no signal	Wiring, CMP sensor
02	Engine control module (ECM) – internal fault	ECM
05	Alternator – voltage low	Wiring, battery, alternator drive belt, alternator
06	Alternator – voltage high	Wiring, battery, alternator
0A	Engine control relay	Wiring open circuit/short circuit, engine control relay
0B	Alternator, field control – circuit malfunction	Wiring open circuit/short circuit, alternator, battery
0C	Torque converter clutch (TCC) solenoid – electrical	Wiring open circuit/short circuit, TCC solenoid
0E	Engine coolant blower motor relay – low/high speed	Wiring open circuit/short circuit, engine coolant blower motor relay – low/high speed
0F	Cruise control	Wiring open circuit/short circuit, hoses, cruise control master switch, cruise control vacuum switching valve
10	A/C compressor clutch relay	Wiring open circuit/short circuit, A/C compressor clutch relay
11	Exhaust gas recirculation (EGR) – circuit malfunction	Wiring open circuit/short circuit, EGR solenoid
12	Evaporative emission (EVAP) canister purge valve – circuit malfunction	Wiring open circuit/short circuit, EVAP canister purge valve
13	Injector 3 – circuit malfunction	Wiring, injector

PLYMOUTH

Voyager 2.4L • Voyager 3.0L • Voyager 3.3L • Voyager 3.8L

Scan code Hex type	Fault location	Probable cause
14	Injector 2 – circuit malfunction	**Wiring, injector**
15	Injector 1 – circuit malfunction	**Wiring, injector**
19	Idle speed control (ISC) system – malfunction	**Wiring open circuit/short circuit, IAC valve**
1A	Throttle position (TP) sensor – low input	**Wiring short circuit to ground, TP sensor**
1B	Throttle position (TP) sensor – high input	**Wiring short circuit to positive, TP sensor**
1E	Engine coolant temperature (ECT) sensor – low input	**Coolant thermostat, wiring short circuit to ground, ECT sensor**
1F	Engine coolant temperature (ECT) sensor – high input	**Coolant thermostat, wiring open circuit/short circuit to positive, ground wire defective, ECT sensor**
20	Heated oxygen sensor (H02S) 1, bank 1 – no activity detected	**Wiring open circuit, heating inoperative, HO2S**
21	Engine temperature – low for long period	**Wiring, cooling system, coolant thermostat, ECT sensor**
23	Vehicle speed sensor (VSS) – signal	**Wiring, VSS**
24	Manifold absolute pressure (MAP) sensor – low input	**Wiring short circuit to ground, MAP sensor**
25	Manifold absolute pressure (MAP) sensor – high input	**Wiring short circuit to positive, MAP sensor**
27	Manifold absolute pressure (MAP) sensor	**Wiring, hose connection(s), MAP sensor**
28	Crankshaft position (CKP) sensor – no signal	**Wiring, air gap, CKP sensor**
29	Ignition coil C, primary/secondary – circuit malfunction	**Wiring, ignition coil**
2A	Ignition coil B, primary/secondary – circuit malfunction	**Wiring, ignition coil**
2B	Ignition coil A, primary/secondary – circuit malfunction	**Wiring, ignition coil**
2C	Engine control relay	**Wiring open circuit, engine control relay**
2E	Exhaust gas recirculation (EGR) system – insufficient flow detected	**Hose leak/blockage, wiring, EGR valve, EGR solenoid**
30	Engine control module (ECM) – EEPROM error	**ECM**
31	Engine control module (ECM) – EEPROM error	**ECM**
39	Intake air temperature (IAT) sensor – low input	**Wiring short circuit to ground, IAT sensor**
3A	Intake air temperature (IAT) sensor – high input	**Wiring open circuit/short circuit to positive, ground wire defective, IAT sensor**
3C	Manifold absolute pressure (MAP) sensor – barometric pressure range/performance problem	**Intake/exhaust leak, wiring, MAP sensor**
3D	Injector 4 – circuit malfunction	**Wiring, injector**
3E	Heated oxygen sensor (HO2S) 1, bank 1 – high voltage	**Wiring short circuit to positive, fuel pressure, injector control, EVAP canister purge valve, air filter blocked, HO2S**
44	Engine control module (ECM) – internal fault	**ECM**
45	Injector 5 – circuit malfunction	**Wiring, injector**
46	Injector 6 – circuit malfunction	**Wiring, injector**

Scan code Hex type	Fault location	Probable cause
52	Cruise control	Wiring, cruise control vacuum switching valve
5A	A/C refrigerant pressure sensor – high input	A/C refrigerant pressure too high (cooling fault/incorrectly charged), wiring, A/C refrigerant pressure sensor
5B	A/C refrigerant pressure sensor – low input	A/C refrigerant pressure too low (incorrectly charged), wiring, A/C refrigerant pressure sensor
60	CAN data bus, transmission control module (TCM) – no signal	Wiring, TCM
61	CAN data bus, body control module (BCM) – no signal	Wiring, BCM
65	Fuel pump relay	Wiring open circuit/short circuit, fuel pump relay
66	Heated oxygen sensor (H02S) 1, bank 1 – slow response	Wiring, heating inoperative, HO2S
67	Heated oxygen sensor (HO2S) 1, bank 1 – heater control – circuit malfunction	Wiring, HO2S
69	Heated oxygen sensor (HO2S) 2, bank 1, heater control – circuit malfunction	Wiring, HO2S
6A	Random/multiple cylinder(s) – misfire detected	Wiring, spark plug(s), HT lead(s), injector(s), ignition coil(s), low compression
6B	Cylinder 1 – misfire detected	Wiring, engine mechanical fault, ignition/fuel system, injector
6C	Cylinder 2 – misfire detected	Wiring, engine mechanical fault, ignition/fuel system, injector
6D	Cylinder 3 – misfire detected	Wiring, engine mechanical fault, ignition/fuel system, injector
6E	Cylinder 4 – misfire detected	Wiring, engine mechanical fault, ignition/fuel system, injector
70	Catalytic converter system, bank 1 – efficiency below threshold	Catalytic converter, wiring, heated oxygen sensor (HO2S) 2
71	Evaporative emission (EVAP) system – incorrect flow detected	Hose connection(s), intake leak, EVAP canister purge valve
72	Park/neutral position (PNP) switch – input circuit malfunction	Wiring, PNP switch, ECM/TCM
76	System too rich	Wiring, intake blocked, EVAP canister purge valve, fuel pressure, EGR system, injector(s), air filter blocked, ignition system, heated oxygen sensor (HO2S)
77	System too lean	Wiring, intake/exhaust leak, MAP sensor, fuel pressure/pump, EVAP canister purge valve, injector(s), heated oxygen sensor (HO2S)
7E	Heated oxygen sensor (HO2S) 2, bank 1 – high voltage	Wiring short circuit to positive, catalytic converter, HO2S
80	Insufficient coolant temperature for closed loop fuel control	Wiring, cooling system, coolant thermostat, ECT sensor
81	Heated oxygen sensor (HO2S) 2, bank 1 – no activity detected	Wiring, heating inoperative, HO2S

Scan code Hex type	Fault location	Probable cause
84	Throttle position (TP) sensor – range/ performance problem	Wiring, accelerator cable adjustment, TP sensor
89	Transmission control system – malfunction	Wiring, ECM/TCM
8A	Idle speed control (ISC) system – rpm lower/ higher than expected	Wiring, air leak, throttle valve tight/sticking, IAC valve
92	Engine control module (ECM) – 5 volt output low	ECM
94	Torque converter clutch (TCC) solenoid – circuit malfunction	Wiring, TCC solenoid
95	Fuel level sensor – low input	Wiring open circuit, fuel level sensor
96	Fuel level sensor – high input	Wiring short circuit to positive, fuel level sensor
97	Fuel level sensor – malfunction	Wiring, fuel level sensor
98	Brake pedal position (BPP) switch – circuit malfunction	Wiring, BPP switch
99	Ambient/battery temperature sensor – voltage low	Wiring, ambient/battery temperature sensor
9A	Ambient/battery temperature sensor – voltage high	Wiring, ambient/battery temperature sensor
9B	Heated oxygen sensor (HO2S) 1, bank 1 – low voltage	Exhaust leak, wiring short circuit to ground, HO2S
9C	Heated oxygen sensor (HO2S) 2, bank 1 – low voltage	Exhaust leak, wiring short circuit to ground, HO2S
9D	Crankshaft position (CKP) sensor/camshaft position (CMP) sensor – intermittent signal	Wiring, poor connection, air gap, CKP sensor, CMP sensor
A0	Evaporative emission (EVAP) system – small leak detected	Hose connection(s), intake leak, EVAP canister, EVAP canister purge valve
A1	Evaporative emission (EVAP) system – large leak detected	Hose connection(s), intake leak, EVAP canister, EVAP canister purge valve
AE	Cylinder 5 – misfire detected	Wiring, engine mechanical fault, ignition/fuel system, injector
AF	Cylinder 6 – misfire detected	Wiring, engine mechanical fault, ignition/fuel system, injector
B7	Evaporative emission (EVAP) leak detection pump solenoid	Wiring open circuit/short circuit, hose leak/ blockage, EVAP leak detection pump solenoid
B8	Evaporative emission (EVAP) leak detection pump – pressure switch	Wiring, hose leak/blockage, EVAP leak detection pump
BA	Engine control module (ECM) – misfire detection limit reached	Wiring, poor connection, CKP sensor
BB	Evaporative emission (EVAP) system – blockage detected	Hose connection(s), hose blocked/trapped, EVAP canister
BE	Compressed natural gas pressure sensor – pressure high	Compressed natural gas system, wiring, compressed natural gas pressure sensor
C0	Heated oxygen sensor (HO2S) 1, bank 1 – slow response	Wiring, heating inoperative, HO2S

PLYMOUTH

Model:	Engine identification:	Year:
Voyager 2.4L	EDZ	1999-00
Voyager 3.0L	EFA	1999-00
Voyager 3.3L	EGA/EGM	1999-00
Voyager 3.8L	EGH	1999-00

System: **SBEC III**

Self-diagnosis

General information

- Refer to the front of this manual for general test conditions, terminology, detailed descriptions of wiring faults and a general trouble shooter for electrical and mechanical faults.
- Malfunction indicator lamp (MIL) will illuminate or flash if certain faults are recorded.
- ECM operates in backup mode if sensors fail, to enable vehicle to be driven to repair shop.

Accessing

1999-01

- Trouble codes can only be displayed by using a Scan Tool connected to the data link connector (DLC) 1999-00 **1** & 2000→ **2**.

1

2

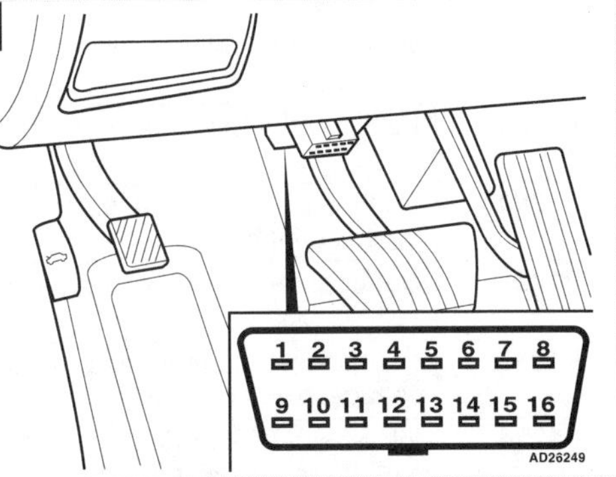

2001→

- Within 5 seconds switch the ignition key ON-OFF-ON-OFF-ON.
- Odometer will display trouble codes.
- If no trouble codes are stored in ECM, odometer will display 'P1684' and 'done' or only 'done'.
- Trouble codes can also be displayed by using a Scan Tool connected to the data link connector (DLC) **2**.

Erasing

- Trouble codes can only be erased by using a Scan Tool connected to the data link connector (DLC).

Trouble code identification

Scan code	Fault location	Probable cause
P0000-P0999	Refer to OBD-II trouble code table at the front of this manual	–
P1192	Intake air temperature (IAT) sensor – voltage low	**Wiring, IAT sensor**
P1193	Intake air temperature (IAT) sensor – voltage high	**Wiring, IAT sensor**
P1195	Heated oxygen sensor (H02S) 1, bank 1 – slow response	**Heating inoperative, wiring, HO2S**
P1196	Heated oxygen sensor (H02S) 2, bank 1 – slow response	**Heating inoperative, wiring, HO2S**
P1281	Engine temperature – low for long period	**Wiring, cooling system, coolant thermostat, ECT sensor**
P1282	Fuel pump relay	**Wiring open circuit/short circuit, fuel pump relay**
P1290	Compressed natural gas pressure sensor – pressure high	**Compressed natural gas system, wiring, compressed natural gas pressure sensor**
P1292	Compressed natural gas pressure sensor – voltage high	**Wiring, compressed natural gas pressure sensor**
P1293	Compressed natural gas pressure sensor – voltage low	**Wiring, compressed natural gas pressure sensor**
P1294	Idle speed control (ISC) system – rpm lower/higher than expected	**Wiring, air leak, throttle valve tight/sticking, IAC valve**
P1295	Throttle position (TP) sensor – supply voltage	**Wiring, TP sensor, ECM**
P1296	Manifold absolute pressure (MAP) sensor – supply voltage	**Wiring, MAP sensor, ECM**
P1297	Manifold absolute pressure (MAP) sensor	**Wiring, hose connection(s), MAP sensor**
P1298	System too lean – at prolonged wide open throttle	**Intake/exhaust leak, MAP sensor, fuel pressure/pump, injector(s), HO2S**
P1299	Manifold absolute pressure (MAP) sensor/throttle position (TP) sensor	**Intake leak, hose connection(s), MAP sensor, TP sensor**
P1388	Engine control relay	**Wiring open circuit/short circuit, engine control relay**
P1389	Engine control relay	**Wiring open circuit, engine control relay**
P1390	Crankshaft position (CKP) sensor/camshaft position (CMP) sensor – signal synchronisation	**Timing belt loose/jumped position, CKP sensor, CMP sensor**
P1391	Crankshaft position (CKP) sensor/camshaft position (CMP) sensor – intermittent signal	**Wiring, poor connection, air gap, CKP sensor, CMP sensor**
P1398	Engine control module (ECM) – misfire detection limit reached	**Wiring, poor connection, CKP sensor**
P1486	Evaporative emission (EVAP) system – blockage detected	**Hose connection(s), hose blocked/trapped, EVAP canister**
P1488	Engine control module (ECM) – auxiliary 5 volt output low	**Wiring, ECM**

Scan code	Fault location	Probable cause
P1489	Engine coolant blower motor relay	Wiring open circuit/short circuit, engine coolant blower motor relay – high speed
P1490	Engine coolant blower motor relay	Wiring open circuit/short circuit, engine coolant blower motor relay – low speed
P1491	Engine coolant blower motor relay – circuit malfunction	Wiring open circuit/short circuit, engine coolant blower motor relay
P1492	Ambient/battery temperature sensor – high input	Wiring short circuit to positive, ambient/ battery temperature sensor, ECM
P1493	Ambient/battery temperature sensor – voltage low	Wiring, ambient/battery temperature sensor
P1494	Evaporative emission (EVAP) leak detection pump – pressure switch	Wiring, hose leak/blockage, EVAP leak detection pump
P1495	Evaporative emission (EVAP) leak detection pump solenoid	Wiring open circuit/short circuit, hose leak/ blockage, EVAP leak detection pump solenoid
P1496	Engine control module (ECM) – 5 volt output low	ECM
P1594	Alternator – voltage high	Wiring, battery, alternator
P1595	Cruise control	Wiring open circuit/short circuit, hoses, cruise control master switch, cruise control vacuum switching valve
P1596	Cruise control master switch – high voltage	Wiring, cruise control master switch
P1597	Cruise control master switch – low voltage	Wiring, cruise control master switch
P1598	A/C refrigerant pressure sensor – high input	Wiring, A/C refrigerant pressure too high (cooling fault/incorrectly charged), A/C refrigerant pressure sensor
P1599	A/C refrigerant pressure sensor – low input	Wiring, A/C refrigerant pressure too low (incorrectly charged), A/C refrigerant pressure sensor
P1602	Engine control module (ECM) – not programmed	New ECM not programmed, incorrectly programmed, ECM
P1603	Engine control module (ECM) – RAM link error	ECM
P1604	Engine control module (ECM) – RAM read/ write error	ECM
P1607	Engine control module (ECM) – shut down timer error	ECM
P1681	CAN data bus, instrumentation – no signal	Wiring, instrumentation control module
P1682	Alternator – voltage low	Wiring, battery, alternator drive belt, alternator
P1683	Cruise control	Wiring, cruise control vacuum switching valve
P1684	Engine control module (ECM) – supply voltage disconnected within last 50 ignition switch ON-OFF cycles	Wiring, battery

Scan code	Fault location	Probable cause
P1685	Key not programmed into immobilizer control module	**Wiring, ignition key not matched/damaged, immobilizer control module**
P1686	CAN data bus, Immobilizer control module – no signal	**Wiring, immobilizer control module**
P1687	CAN data bus, instrumentation – no signal	**Wiring, instrumentation control module mechanical**
P1695	CAN data bus, body control module (BCM) – no signal	**Wiring, BCM**
P1696	Engine control module (ECM) – EEPROM error	**ECM**
P1697	Engine control module (ECM) – EEPROM error	**ECM**
P1698	CAN data bus, transmission control module (TCM) – no signal	**Wiring, TCM**
P1765	Transmission system relay	**Wiring open circuit/short circuit, transmission system relay, ECM/TCM**
P1899	Park/neutral position (PNP) switch – input circuit malfunction	**Wiring, PNP switch, ECM/TCM**
P2503	Charging system voltage – low	**Wiring, poor connection, battery, alternator**

PLYMOUTH

Model:	Engine identification:	Year:
Voyager 2.5L	EDM	1992-95
Voyager 3.0L	EFA	1992-95
Voyager 3.3L	EGA/EGM	1992-95
Voyager 3.8L	EGH	1992-95

System: **SBEC II**

Self-diagnosis

General information

- Refer to the front of this manual for general test conditions, terminology, detailed descriptions of wiring faults and a general trouble shooter for electrical and mechanical faults.
- Malfunction indicator lamp (MIL) will illuminate if certain faults are recorded.
- ECM operates in backup mode if sensors fail, to enable vehicle to be driven to repair shop.
- Trouble codes can be displayed by the MIL or by using a Scan Tool connected to the data link connector (DLC).

Accessing

- Within 5 seconds switch the ignition key ON-OFF-ON-OFF-ON.
- Count MIL flashes. Note trouble codes.
- Each trouble code consists of two groups **1** [**A**].
- The first group of flashes indicate the 'tens' of the trouble code.
- The second group of flashes indicate the 'units' of the trouble code.
- A short pause separates 'tens' and 'units' **1** [**B**].
- A long pause separates each trouble code **1** [**C**].
- For example: Trouble code 21 displayed **1**.
- End of test sequence indicated by trouble code 55.
- Trouble codes can also be displayed by using a Scan Tool connected to the data link connector (DLC) 1992-93 **2** & 1994-95 **3**.

1

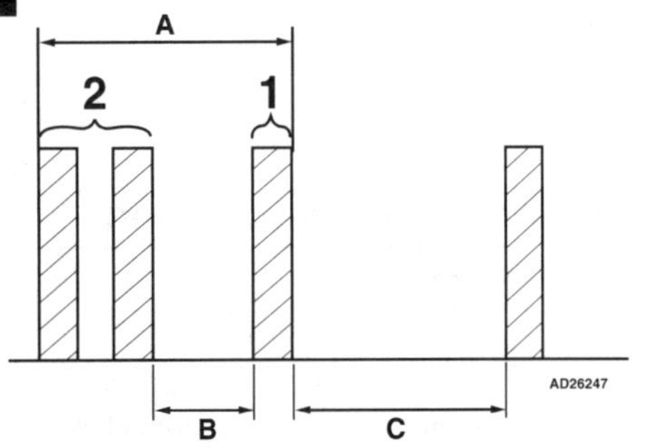

2

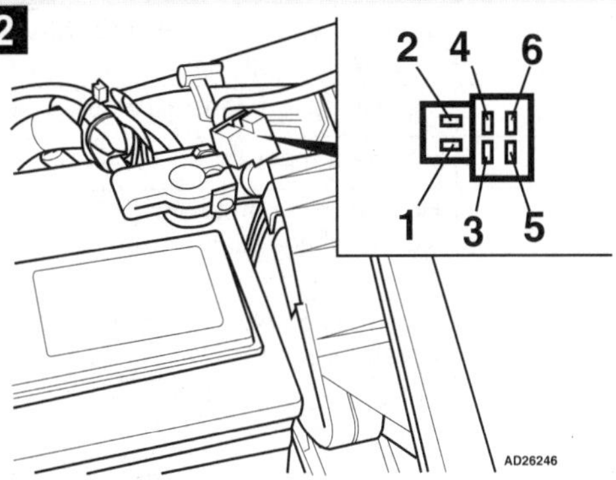

3

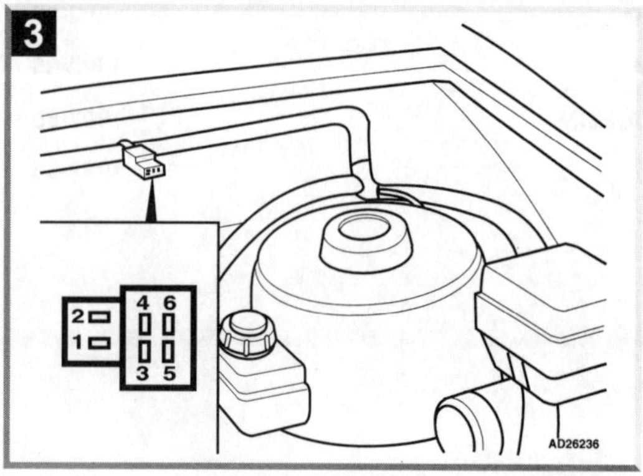

Erasing

NOTE: ***Disconnecting battery ground lead may cause driveability problems until the engine management system relearns basic values.***

- Ensure ignition switched OFF.
- Disconnect battery ground cable for at least 30 seconds.
- Reconnect battery ground cable.
- Repeat accessing procedure to ensure no trouble codes remain in ECM fault memory.

NOTE: ***Disconnecting battery ground lead may also erase memory from electronic units, such as radio.***

- Trouble codes can also be erased by using a Scan Tool connected to the data link connector (DLC).

Trouble code identification

Flash code	Fault location	Probable cause
11	Crankshaft position (CKP) sensor	**Wiring, air gap, CKP sensor**
12	Engine control module (ECM) – supply voltage disconnected within last 50 ignition switch ON-OFF cycles	**Wiring, battery**
13	Manifold absolute pressure (MAP) sensor	**Wiring, MAP sensor**
14	Manifold absolute pressure (MAP) sensor – voltage low/high	**Wiring, MAP sensor**
15	Vehicle speed sensor (VSS) – signal	**Wiring, VSS**
17	Engine temperature – low for long period	**Wiring, coolant thermostat, ECT sensor**
21	Heated oxygen sensor (HO2S) – incorrect signal/voltage high	**Wiring, fuel pressure, injector control, EVAP canister purge valve, compression, air filter blocked, heated oxygen sensor (HO2S)**
22	Engine coolant temperature (ECT) sensor – voltage low/high	**Wiring, ECT sensor**
24	Throttle position (TP) sensor	**Wiring, accelerator cable adjustment, TP sensor**
25	Idle air control (IAC) valve	**Wiring, IAC valve**
27	Injector(s) – circuit malfunction	**Wiring, injector**
31	Evaporative emission (EVAP) canister purge valve – open/short circuit	**Wiring, EVAP canister purge valve**
32	Exhaust gas recirculation (EGR) solenoid/ system – circuit malfunction	**Wiring, hoses, EGR solenoid**
33	A/C compressor clutch relay	**Wiring, A/C compressor clutch relay**

Flash code	Fault location	Probable cause
34	Cruise control	**Wiring open circuit/short circuit, hoses, voltage low/high, cruise control master switch, cruise control vacuum switching valve**
35	Engine coolant blower motor relay – low/high speed	**Wiring, engine coolant blower motor relay – low/high speed**
37	Torque converter clutch (TCC) solenoid	**Wiring, torque converter clutch (TCC) solenoid**
41	Alternator	**Wiring open circuit/short circuit, alternator**
42	Engine control relay	**Wiring open circuit/short circuit, engine control relay**
43	Ignition coil – primary circuit	**Wiring, ignition coil**
44	Battery temperature sensor/engine coolant temperature (ECT) sensor	**Wiring open circuit/short circuit, battery temperature sensor, ECT sensor, ECM**
46	Alternator – voltage high	**Wiring, battery, alternator**
47	Alternator – voltage low	**Wiring, battery, alternator drive belt, alternator**
51	System too lean	**Wiring, intake/exhaust leak, fuel pressure, injector control, EVAP canister purge valve, heated oxygen sensor (HO2S)**
52	System too rich	**Wiring, exhaust blockage, fuel pressure, injector control, EVAP canister purge valve, compression, air filter blocked, ignition system, heated oxygen sensor (HO2S)**
53	Engine control module (ECM) – internal fault – defective	**ECM**
54	Camshaft position (CMP) sensor – no signal	**Wiring, CMP sensor**
55	End of MIL code output	–
62	Engine control module (ECM) – EEPROM error	**ECM**
63	Engine control module (ECM) – EEPROM error	**ECM**

PONTIAC

Model:	Engine identification:	Year:
Aztek 3.4L	**VIN position 8 = E**	**2001-02**
Bonneville 3.8L	**VIN position 8 = L, K, 1**	**1994-02**
Grand AM 2.2L	**VIN position 8 = T**	**2002-**
Grand AM 2.4L	**VIN position 8 = T**	**1996-01**
Grand AM 3.1L	**VIN position 8 = M**	**1996-98**
Grand AM 3.4L	**VIN position 8 = E**	**1999-02**
Grand Prix 3.4L	**VIN position 8 = X**	**1994-95**
Trans Sport 3.8L	**VIN position 8 = L**	**1994-95**

System: **GM**

Self-diagnosis

General information

- Refer to the front of this manual for general test conditions, terminology, detailed descriptions of wiring faults and a general trouble shooter for electrical and mechanical faults.
- Malfunction indicator lamp (MIL) will illuminate or flash if certain faults are recorded.
- ECM operates in backup mode if sensors fail, to enable vehicle to be driven to repair shop.

Accessing

- Trouble codes can only be displayed by using a Scan Tool connected to the data link connector (DLC) **1** or **2**.

1

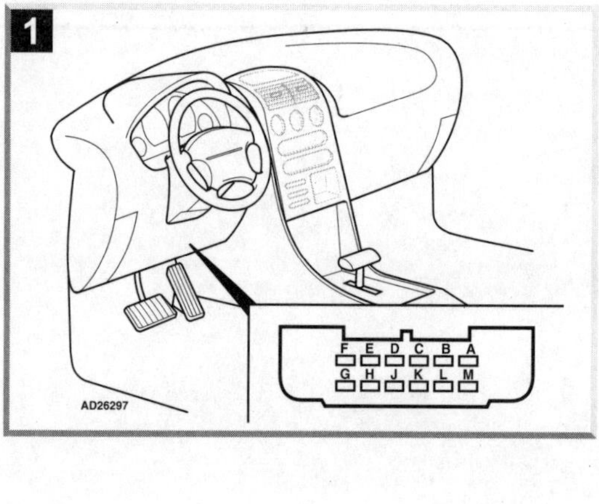

2

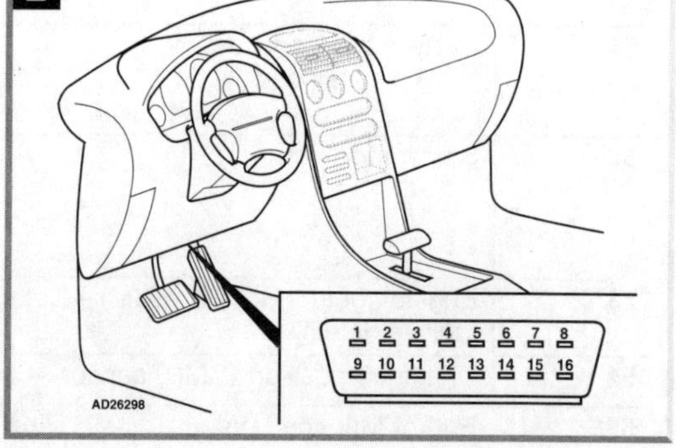

Erasing

- After the faults have been rectified, the manufacturer recommends that trouble codes are erased by using a Scan Tool connected to the data link connector (DLC).

Trouble code identification

Scan code	Fault location	Probable cause
P0000-P0999	Refer to OBD-II trouble code table at the front of this manual	–
P1106	Manifold absolute pressure (MAP) sensor – voltage high, intermittent	**Wiring open circuit/short circuit to positive, poor connections**
P1107	Manifold absolute pressure (MAP) sensor – voltage low, intermittent	**Wiring short circuit to ground, poor connections**
P1111	Intake air temperature (IAT) sensor – voltage high, intermittent	**Wiring open circuit/short to positive, poor connections**
P1112	Intake air temperature (IAT) sensor – voltage low, intermittent	**Wiring short circuit to ground**
P1114	Engine coolant temperature (ECT) sensor – voltage low, intermittent	**Wiring short circuit to ground, poor connections**
P1115	Engine coolant temperature (ECT) sensor, voltage high, intermittent	**Wiring open circuit/short to positive, poor connections**
P1121	Throttle position (TP) sensor – voltage high, intermittent	**Wiring open circuit/short to positive, poor connections**
P1122	Throttle position (TP) sensor – voltage low, intermittent	**Wiring open circuit/short to ground, poor connections**
P1133	Oxygen sensor (O2S)/heated oxygen sensor (HO2S) 1, bank 1 – insufficient switching	**Wiring, exhaust system leaking, O2S/HO2S**
P1134	Heated oxygen sensor (HO2S) 1, bank 1 – slow response	**Wiring, oxygen sensor heater, exhaust system leaking, HO2S**
P1137	Heated oxygen sensor (HO2S) 2, bank 1 – voltage low	**Wiring, exhaust system leaking, HO2S**
P1138	Heated oxygen sensor (HO2S) 2, bank 1 – voltage high	**Wiring, HO2S**
P1171	Fuel system – lean during acceleration	**Fuel filter, fuel pump, HO2S, injectors**
P1189	Engine oil pressure warning lamp switch – circuit malfunction	**Wiring, engine oil pressure warning lamp switch**
P1200	Injectors – circuit malfunction	**Wiring open circuit/short circuit to ground/ short circuit to positive, injectors, ECM**
P1257	Supercharger (SC), boost – too high	**Supercharger (SC) bypass valve hoses blocked/leaking/interchanged/not connected, wiring short circuit to ground, supercharger (SC) bypass valve solenoid, ECM**
P1336	Crankshaft position sensor (CKP) – variation not learned	**Learning procedure not performed**
P1350	Ignition control system, bypass monitor circuit – malfunction	**Wiring, poor connections, ICM, ECM**
P1351	Ignition control system, coil – voltage high	**Wiring open circuit, ICM, ECM**
P1352	Ignition control system, bypass circuit – voltage high	**Wiring open circuit, ICM, ECM**

Scan code	Fault location	Probable cause
P1361	Ignition control system, IC circuit – voltage low	**Wiring, circuit short circuit to ground, poor connections, ICM, ECM**
P1362	Ignition control, bypass circuit – voltage low	**Wiring short circuit to ground, ICM, ECM**
P1374	Crankshaft position sensor (CKP), high to low resolution frequency – signal variation	**Wiring, wiring routed near spark plug wires, ignition coils arcing/tracking, ICM, ECM**
P1380	Antilock brake system (ABS)/traction control system (TCS), rough road signal to ECM – malfunction	**ABS/TCS trouble code(s) stored**
P1381	Antilock brake system (ABS)/traction control system (TCS), rough road signal to ECM – no communication	**Wiring, poor connections, ABS/TCS**
P1403	Exhaust gas recirculation (EGR) solenoid 1 – circuit malfunction	**Wiring, EGR valve, ECM**
P1404 [3]	Exhaust gas recirculation (EGR) valve – not fully closing	**Wiring, poor connections, EGR valve position sensor, EGR solenoid**
P1404 [4]	Exhaust gas recirculation (EGR) solenoid 2 – circuit malfunction	**Wiring, EGR valve, ECM**
P1405	Exhaust gas recirculation (EGR) solenoid 3 – circuit malfunction	**Wiring, EGR valve, ECM**
P1406	Exhaust gas recirculation (EGR) valve position sensor – range performance	**Wiring, EGR valve tight/sticking, EGR valve position sensor**
P1441	Evaporative emission (EVAP) system – leak detected	**Hose connection(s), EVAP canister, EVAP canister purge valve**
P1442	Evaporative emission (EVAP) system vacuum switch – circuit malfunction	**Hoses blocked/leaking, wiring, EVAP canister, ECM**
P1483	Engine cooling system – performance	**Cooling system, engine coolant temperature (ECT) sensor, intake air temperature (IAT) sensor**
P1520	Park/neutral position (PNP) switch – circuit malfunction	**Wiring short circuit to ground, PNP switch incorrectly adjusted**
P1530	A/C refrigerant pressure sensor/switch – circuit malfunction	**Wiring, A/C refrigerant pressure sensor/ switch, ECM**
P1531	A/C refrigerant pressure sensor – voltage low	**Incorrect refrigerant charge, wiring, A/C refrigerant pressure sensor, ECM**
P1546	A/C compressor clutch relay – circuit malfunction	**Wiring, A/C compressor clutch relay, ECM**
P1550	Cruise control system, motor actuator – circuit malfunction	**Wiring, poor connections, cruise control motor/actuator, ECM**
P1554	Cruise control status circuit – high voltage	**Wiring open circuit/short circuit to ground, cruise control module, ECM**
P1558	Cruise control system – performance low	**Wiring, poor connections, accelerator cable sticking**
P1561	Cruise control, vent solenoid – circuit malfunction	**Wiring, solenoid valve**
P1562	Cruise control, vacuum solenoid – circuit malfunction	**Wiring, solenoid valve**

Scan code	Fault location	Probable cause
P1565	Cruise control servo position sensor – circuit malfunction	**Wiring, poor connections, servo position sensor**
P1567	Cruise control master switch – circuit malfunction	**Wiring, poor connections, cruise control master switch**
P1568	Cruise control system – performance high	**Wiring, poor connections, accelerator cable sticking**
P1571	Traction control system (TCS), torque request signal – malfunction	**Wiring, poor connections, ABS/TCS**
P1573	Engine control module (ECM), ABS/TCS communication – malfunction	**Wiring, poor connections, ABS control module, ECM**
P1575	Brake pedal position switch (BPP), extended travel contacts – voltage high	**Wiring, ABS**
P1585	Cruise control system, inhibit output circuit	**Wiring, poor connections, cruise control motor/actuator, ECM**
P1599	Cruise control system – engine management has operated in limited power mode	**Accelerator cable sticking, brake vacuum release valve, cruise control actuator**
P1600	Engine control module (ECM) – malfunction	**ECM**
P1601	CAN data bus, engine control module (ECM)/ antilock brake system (ABS)/traction control system (TCS)/instrumentation control module/ diagnostic control module – some communication lost	**Wiring, poor connections, ECM, ABS/TCS, diagnostic control module, instrumentation control module**
P1602	CAN data bus, engine control module (ECM) to antilock brake system (ABS)/traction control system (TCS) – signal malfunction	**Wiring, ABS control module**
P1610	CAN data bus, engine control module (ECM) to body control module (BCM) – communication failure	**Wiring, poor connections**
P1619	Engine oil life monitor reset switch – circuit malfunction	**Wiring short circuit to ground, engine oil life monitor reset switch, ECM**
P1621	Engine control module (ECM) – memory performance	**ECM**
P1623	Engine control module (ECM), PROM – malfunction	**Poor connections, incorrect programming, ECM**
P1626	Engine control module (ECM) – fuel enable signal lost	**Wiring, poor connections, immobilizer control module, ECM**
P1627	Engine control module (ECM) – malfunction	**Engine control module (ECM)**
P1629	Engine control module (ECM) – fuel enable signal not received	**Incorrect ignition key, wiring, poor connections, immobilizer control module, ECM**
P1630 5	System voltage – high/low	**Wiring, battery, alternator**
P1630 6	Engine control module (ECM), immobilizer code – learning mode	**ECM immobilizer code programming not completed**
P1631	Engine control module (ECM) – incorrect immobilizer code	**Incorrect key, immobilizer code not programmed**

Scan code	Fault location	Probable cause
P1632	Engine control module (ECM) – fuel disable signal received	**Fuses, BCM, immobilizer code**
P1635	Engine control module (ECM), sensor supply voltage – circuit malfunction	**Wiring, poor connections, TP sensor, MAP sensor, EGR valve position sensor, fuel tank pressure (FTP) sensor, ECM**
P1639	Engine control module (ECM), sensor supply voltage – circuit malfunction	**Wiring, poor connections, A/C refrigerant pressure sensor, ECM**
P1640	Engine control module (ECM), quad driver/ output driver – circuit malfunction	**Wiring, ECM controlled components, ECM 7**
P1641 2	A/C compressor clutch relay – circuit malfunction	**Wiring open circuit/short circuit to ground, A/C compressor clutch relay, ECM**
P1641 1	Malfunction indicator lamp (MIL) – circuit malfunction	**Wiring, poor connections, MIL, ECM**
P1643	Fuel pump, control signal – circuit malfunction	**Wiring, fuel pump control module, ECM**
P1644	Traction control system (TCS), delivered torque signal – malfunction	**Wiring, poor connections, ECM**
P1645 1	Supercharger (SC) bypass valve, control signal – circuit malfunction	**Wiring, SC bypass valve, ECM**
P1646 2	Supercharger (SC) bypass valve, control signal – circuit malfunction	**Wiring, SC bypass valve, ECM**
P1650	Engine control module (ECM), quad driver/ output driver – circuit malfunction	**Wiring, ECM controlled components, ECM 7**
P1651	Engine coolant blower motor relay 1 – circuit malfunction	**Wiring, engine coolant blower motor relay, ECM**
P1652	Engine coolant blower motor relay 2 & 3 – circuit malfunction	**Wiring, engine coolant blower motor relay(s), ECM**
P1653	Traction control system (TCS), delivered torque signal – malfunction	**Wiring, poor connections, ECM**
P1654	A/C compressor clutch relay – circuit malfunction	**Wiring open circuit/short circuit to ground, A/C compressor clutch relay, ECM**
P1655	Evaporative emission (EVAP) canister purge valve – circuit malfunction	**Wiring open circuit/short circuit to ground, EVAP canister vent valve, ECM**
P1660	Engine control module (ECM), quad driver/ output driver – circuit malfunction	**Wiring, ECM controlled components, ECM 7**
P1662	Cruise control, inhibit output circuit – malfunction	**Wiring open circuit/short circuit to ground, cruise control motor/actuator, cruise control module, ECM**
P1663	Alternator warning lamp, control signal – circuit malfunction	**Wiring, instrument panel, ECM**
P1665	Evaporative emission (EVAP) canister vent valve – circuit malfunction	**Wiring open circuit/short circuit to ground, EVAP canister vent valve, ECM**
P1667	Fuel pump, control signal – circuit malfunction	**Wiring, fuel pump control module, ECM**
P1670	Engine control module (ECM), quad driver/ output driver – circuit malfunction	**Wiring, ECM controlled components, ECM 7**

Scan code	Fault location	Probable cause
P1671	Malfunction indicator lamp (MIL) – circuit malfunction	**Wiring open circuit/short circuit to ground/ short circuit to positive, MIL, ECM**
P1672	Engine oil level 'low' warning lamp, control signal – circuit malfunction	**Wiring short circuit to ground, engine oil level, ECM**
P1673	Engine coolant 'hot' warning lamp, control signal – circuit malfunction	**Wiring, instrument panel, cooling system, ECM**
P1675	Evaporative emission (EVAP) canister vent valve – circuit malfunction	**Wiring, poor connection, EVAP canister vent valve, ECM**
P1676	Evaporative emission (EVAP) canister purge valve – circuit malfunction	**Wiring open circuit/short circuit to ground, EVAP canister purge valve, ECM**
P1680	Engine control module (ECM) – malfunction	**ECM**
P1681	Engine control module (ECM) – malfunction	**ECM**
P1683	Engine control module (ECM) – malfunction	**ECM**
P1689	Traction control system (TCS), delivered torque signal – malfunction	**Wiring, poor connections, ECM**
P1810	Transmission fluid pressure (TFP) switch – manual valve position	**Fluid level low, wiring, selector lever linkage incorrectly adjusted, ECM**
P1811	Shift time too long – adaptation limit reached	**Fluid level low, transmission mechanical fault**
P1814	Torque converter clutch (TCC) – over-stressed	**Fluid level low, fluid condition, transmission mechanical fault**
P1819	Transmission range (TR) switch, signals 'A/B/C & P' – signal variation	**Wiring, TR switch**
P1820	Transmission range (TR) switch, signal 'A' – voltage low	**Wiring short to ground, TR switch**
P1822	Transmission range (TR) switch, signal 'B' – voltage high	**Wiring, TR switch**
P1823	Transmission range (TR) switch, signal 'P' – voltage low	**Wiring short to ground, TR switch**
P1825	Transmission range (TR) switch, signals 'A/B/C & P' – incorrect signal	**Wiring, TR switch**
P1826	Transmission range (TR) switch, signal 'C' – voltage high	**Wiring, TR switch**
P1860	Torque converter clutch (TCC) solenoid – circuit malfunction	**Wiring open circuit/short circuit to ground, TCC solenoid, ECM**
P1887	Torque converter clutch (TCC) release switch – circuit malfunction	**Wiring open circuit, TCC solenoid release switch, ECM**
U1000	CAN data bus – communication malfunction	**Wiring, BCM, ECM, instrumentation control module, SDM**
U1016	CAN data bus, engine control module (ECM)/ antilock brake system (ABS)/traction control system (TCS)/instrumentation control module/ diagnostic control module – loss of serial communication	**Wiring, poor connections, ECM, ABS/TCS, diagnostic control module, instrumentation control module**

Scan code	Fault location	Probable cause
U1254	CAN data bus – loss of serial communication	**Wiring, BCM, ECM, instrumentation control module, SDM**
U1255	CAN data bus – communication malfunction	**Wiring, BCM, ECM, instrumentation control module, SDM**
U1300	CAN data bus – signal low	**Wiring**
U1301	CAN data bus – signal high	**Wiring**
U1305	CAN data bus – signal high or low	**Wiring**

1 →1997
2 1998→
3 Except Grand Prix 3.4L
4 Grand Prix 3.4L
5 →1995
6 1996→
7 Depending on model, year and equipment level, the following components may be connected to this circuit. Use of the vehicle wiring diagram is recommended to confirm which components are fitted

- ☐ A/C compressor clutch relay
- ☐ A/C condenser blower motor relay
- ☐ Engine coolant blower motor relay
- ☐ Engine coolant 'hot' warning lamp
- ☐ Evaporative emission (EVAP) canister purge valve
- ☐ Exhaust gas recirculation (EGR) solenoid
- ☐ Malfunction indicator lamp (MIL)
- ☐ Torque converter clutch (TCC) solenoid
- ☐ Transmission shift solenoids (SS)
- ☐ Transmission shift warning lamp (MT)

PONTIAC

Model:	Engine identification:	Year:
Bonneville 3.8L	**VIN position 8 = L**	**1992-93**
Grand AM 3.3L	**VIN position 8 = N**	**1992-93**
Trans Sport 3.8L	**VIN position 8 = L**	**1992-93**

System: **GM**

Self-diagnosis

General information

- Refer to the front of this manual for general test conditions, terminology, detailed descriptions of wiring faults and a general trouble shooter for electrical and mechanical faults.
- Malfunction indicator lamp (MIL) will illuminate if certain faults are recorded.
- ECM operates in backup mode if sensors fail, to enable vehicle to be driven to repair shop.
- Trouble codes can be displayed by the MIL or by using a Scan Tool connected to the data link connector (DLC).

Accessing

- Access data link connector (DLC).
- Jump data link connector (DLC) terminals A and B **1**.
- Switch ignition ON. Do NOT start engine.
- Count MIL flashes. Note trouble codes. Compare with trouble code table.
- Firstly trouble code 12 will be displayed 3 times.
- Trouble codes consist of 2 groups, displayed as follows:
  - First group indicates the 'tens' of the trouble code **2** [**A**].
  - Second group indicates the 'units' of the trouble code **2** [**C**].
  - A short pause separates each group **2** [**B**].
  - A long pause separates each trouble code **2** [**D**].
  - For example: Trouble code 12 displayed **2**. Start of diagnosis.
- Each additional trouble code will now be displayed 3 times, in ascending order.

NOTE: ***If no additional trouble codes are stored, code 12 will continue to be displayed.***

- Trouble codes can also be displayed by using a Scan Tool connected to the data link connector (DLC) **1**.

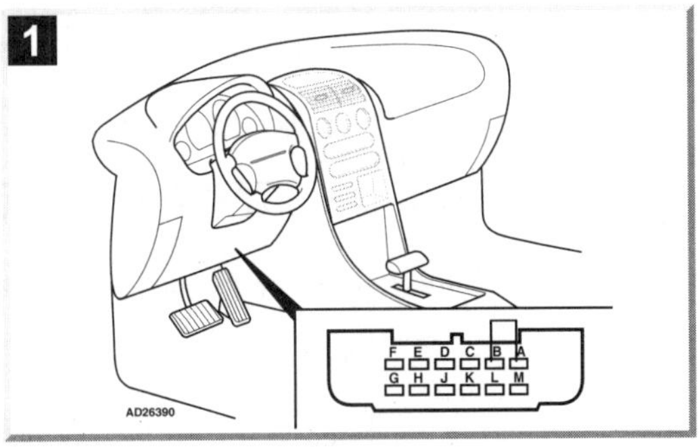

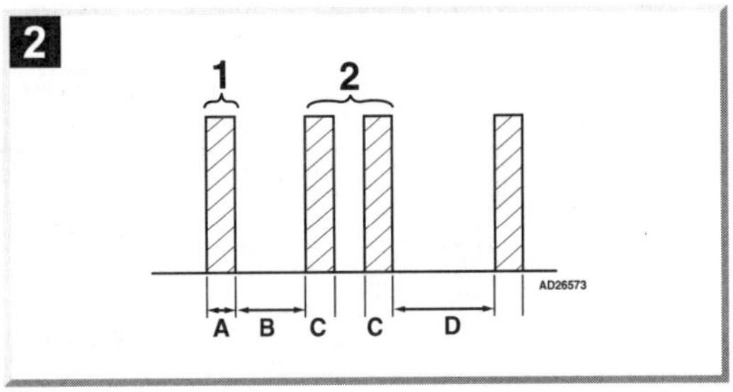

Erasing

NOTE: ***Disconnecting battery ground lead may cause driveability problems until the engine management system relearns basic values.***

- Ensure ignition switched OFF.
- Disconnect battery ground cable for 30 seconds.
- Reconnect battery ground cable.
- Repeat checking procedure to ensure no trouble codes remain in ECM fault memory.

NOTE: ***Disconnecting battery ground lead may also erase memory from electronic units, such as radio.***

- Trouble codes can also be erased by using a Scan Tool connected to the data link connector (DLC) **1**.

Trouble code identification

Flash code	Fault location	Probable cause
12	Start of diagnosis	**No engine speed (RPM) signal received**
13	Oxygen sensor (O2S)/heated oxygen sensor (HO2S)	**Wiring open circuit, O2S/HO2S, ECM**
14	Engine coolant temperature (ECT) sensor – voltage low	**Wiring short circuit to ground, ECT sensor, ECM**
15	Engine coolant temperature (ECT) sensor – voltage high	**Wiring open circuit, ECT sensor, ECM**
16	System voltage – high/low	**Wiring, battery, alternator**
17	Ignition control system, spark reference signal – circuit malfunction	**Wiring, ICM, ECM**
18	Camshaft position (CMP) sensor – circuit malfunction	**Wiring, CMP sensor, mechanical fault, ECM**
21	Throttle position (TP) sensor – voltage high	**Wiring open circuit/short circuit to positive, TP sensor, ECM**
22	Throttle position (TP) sensor – voltage low	**Wiring open circuit/short circuit to ground, TP sensor, ECM**
23	Intake air temperature (IAT) sensor – voltage high	**Wiring open circuit, IAT sensor, ECM**
24	Vehicle speed sensor (VSS) – no signal	**Wiring, VSS, ECM**
25	Intake air temperature (IAT) sensor – voltage low	**Wiring short circuit to ground, IAT sensor, ECM**
26	Engine control module (ECM), quad driver/ output driver – circuit malfunction	**Wiring, ECM controlled components, ECM** 1
27	Transmission gear switch, signal – circuit malfunction	**Wiring open circuit/short circuit to ground, 2nd gear switch, 3rd gear switch, ECM**
28	Transmission gear switch, signal – circuit malfunction	**Wiring open circuit/short circuit to ground, 2nd gear switch, 3rd gear switch, ECM**
31	Park/neutral position (PNP) switch – malfunction	**Wiring, PNP switch, ECM**
34	Mass airflow (MAF) sensor – voltage low	**Wiring, poor connections, MAF sensor, ECM**
36	Transmission shift solenoid (SS) – circuit malfunction	**Wiring, SS, ECM**
38	Brake pedal position (BPP) switch, torque converter clutch (TCC)	**Wiring, incorrectly adjusted, BPP switch, ECM**
39	Torque converter clutch (TCC) – slipping	**Wiring, TCC solenoid, transmission mechanical fault, ECM**
41	Camshaft position (CMP) sensor – circuit malfunction	**Wiring, CMP sensor, mechanical fault, ECM**
42	Ignition control, timing signal – circuit malfunction	**Wiring, poor connections, ICM, ECM**
43	Knock sensor (KS) – circuit malfunction	**Wiring, KS, ECM**
44	Oxygen sensor (O2S)/heated oxygen sensor (HO2S) – mixture lean	**Wiring open circuit/short circuit to ground, injector(s) blocked, fuel contamination, fuel pressure low, exhaust system leaking, O2S/ HO2S, ECM**

Flash code	Fault location	Probable cause
45	Oxygen sensor (O2S)/heated oxygen sensor (HO2S) – mixture rich	**Wiring open circuit, fuel pressure high, injector(s) leaking, O2S/HO2S, EVAP purge valve, throttle position (TP) sensor, EGR system, ECM**
48	Misfire detected	**Ignition system, fuel system, O2S, engine mechanical fault**
51	Engine control module (ECM), PROM – malfunction	**Poor connections, incorrect programming, ECM**
53	Exhaust gas recirculation (EGR) solenoid 1 – circuit malfunction	**Wiring, EGR valve, ECM**
54	Exhaust gas recirculation (EGR) solenoid 2 – circuit malfunction	**Wiring, EGR valve, ECM**
55	Exhaust gas recirculation (EGR) solenoid 3 – circuit malfunction	**Wiring, EGR valve, ECM**
56	Engine control module (ECM), quad driver/ output driver – circuit malfunction	**Wiring, ECM controlled components, ECM** [1]
61	Cruise control, vent solenoid – circuit malfunction	**Wiring, solenoid valve**
62	Cruise control, vacuum solenoid – circuit malfunction	**Wiring, solenoid valve**
63	Cruise control system – performance low	**Wiring, poor connections, accelerator cable sticking**
65	Cruise control servo position sensor – circuit malfunction	**Wiring, poor connections, servo position sensor**
66	A/C refrigerant pressure sensor – voltage low	**Incorrect refrigerant charge, wiring, A/C refrigerant pressure sensor, ECM**
67	Cruise control master switch – circuit malfunction	**Wiring, poor connections, cruise control master switch**
68	Cruise control system – performance high	**Wiring, poor connections, accelerator cable sticking**
69	A/C refrigerant pressure switch, engine coolant blower motors – circuit malfunction	**Wiring open circuit, A/C refrigerant pressure switch, ECM**
99	Cruise control system – engine management has operated in limited power mode	**Accelerator cable sticking, brake vacuum release valve, cruise control actuator**

[1] Depending on model, year and equipment level, the following components may be connected to this circuit. Use of the vehicle wiring diagram is recommended to confirm which components are fitted

- ☐ A/C compressor clutch relay
- ☐ A/C condenser blower motor relay
- ☐ Engine coolant blower motor relay
- ☐ Engine coolant 'hot' warning lamp
- ☐ Evaporative emission (EVAP) canister purge valve
- ☐ Exhaust gas recirculation (EGR) solenoid
- ☐ Malfunction indicator lamp (MIL)
- ☐ Torque converter clutch (TCC) solenoid
- ☐ Transmission shift solenoids (SS)
- ☐ Transmission shift warning lamp (MT)

PONTIAC

Model:	Engine identification:	Year:
Firebird 3.1/5.0/5.7L	VIN position 8 = E/F/T/8	1992
Grand AM 2.3L	VIN position 8 = D, A, 3	1992-95
Grand Prix 3.1/3.4L	VIN position 8 = T/M/X	1992-93
LeMans 1.6L	VIN position 8 = 6	1992-93
Sunbird 2.0/3.1L	VIN position 8 = H/T	1992-94
Sunfire 2.2L	VIN position 8 = 4	1995
Sunfire 2.3L	VIN position 8 = D	1992-95
Trans Sport 3.1L	VIN position 8 = D	1992-95

System: **GM**

Self-diagnosis

General information

- Refer to the front of this manual for general test conditions, terminology, detailed descriptions of wiring faults and a general trouble shooter for electrical and mechanical faults.
- Malfunction indicator lamp (MIL) will illuminate if certain faults are recorded.
- ECM operates in backup mode if sensors fail, to enable vehicle to be driven to repair shop.
- Trouble codes can be displayed by the MIL or by using a Scan Tool connected to the data link connector (DLC).

Accessing

Models with data link connector (DLC) terminal B connected

- Access data link connector (DLC).
- Jump data link connector (DLC) terminals A and B **1**.
- Switch ignition ON. Do NOT start engine.
- Count MIL flashes. Note trouble codes. Compare with trouble code table.
- Firstly trouble code 12 will be displayed 3 times.
- Trouble codes consist of 2 groups, displayed as follows:
  - First group indicates the 'tens' of the trouble code **2** [**A**].
  - Second group indicates the 'units' of the trouble code **2** [**C**].
  - A short pause separates each group **2** [**B**].
  - A long pause separates each trouble code **2** [**D**].
  - For example: Trouble code 12 displayed **2**. Start of diagnosis.
- Each additional trouble code will now be displayed 3 times, in ascending order.

NOTE: ***If no additional trouble codes are stored, code 12 will continue to be displayed.***

- Trouble codes can also be displayed by using a Scan Tool connected to the data link connector (DLC) **1**.

1

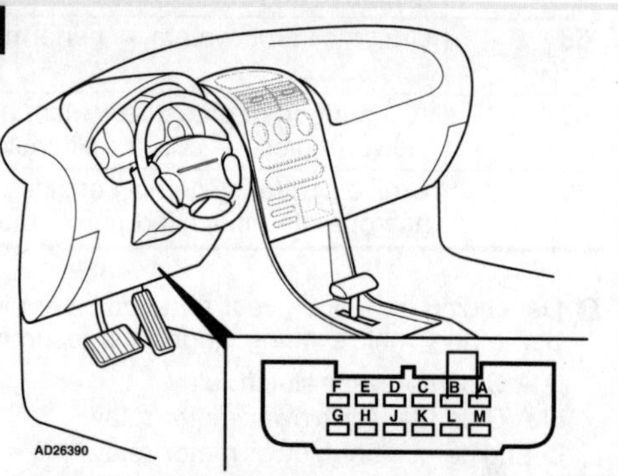

2

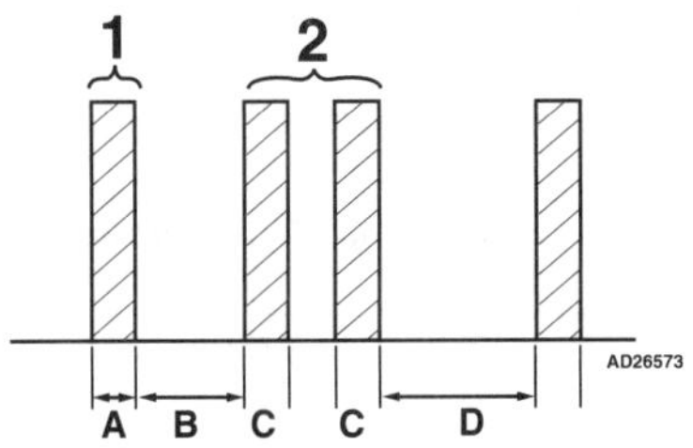

Models without data link connector (DLC) terminal B connected

- Trouble codes can only be displayed by using a Scan Tool connected to the data link connector (DLC) **1**.

Erasing

NOTE: ***Disconnecting battery ground lead may cause driveability problems until the engine management system relearns basic values.***

- Ensure ignition switched OFF.
- Disconnect battery ground cable for 30 seconds.
- Reconnect battery ground cable.
- Repeat checking procedure to ensure no trouble codes remain in ECM fault memory.

NOTE: ***Disconnecting battery ground lead may also erase memory from electronic units, such as radio.***

- Trouble codes can also be erased by using a Scan Tool connected to the data link connector (DLC).

Trouble code identification

Flash code	Fault location	Probable cause
12	Start of diagnosis	**No engine speed (RPM) signal received**
13	Oxygen sensor (O2S)/heated oxygen sensor (HO2S)	**Wiring open circuit, O2S/HO2S, ECM**
14	Engine coolant temperature (ECT) sensor – voltage high/low	**Wiring, ECT sensor, ECM**
15	Engine coolant temperature (ECT) sensor – voltage high	**Wiring open circuit, ECT sensor, ECM**
16 **3**	System voltage – high/low	**Wiring, battery, alternator**
16 **4**	Crankshaft position (CKP) sensor, 2X reference signal – circuit malfunction	**Wiring, ICM, ECM**
17	Camshaft position (CMP) sensor – circuit malfunction	**Wiring, CMP sensor, mechanical fault, ECM**
19	Crankshaft position (CKP) sensor, 7X/58X signal – circuit malfunction	**Wiring, CKP sensor, ECM**
21	Throttle position (TP) sensor – voltage high/low	**Wiring, TP sensor, ECM**
22	Throttle position (TP) sensor – voltage low	**Wiring open circuit/short circuit to ground, TP sensor, ECM**
23	Intake air temperature (IAT) sensor – voltage high/low	**Wiring, IAT sensor, ECM**
24	Vehicle speed sensor (VSS) – no signal	**Wiring, VSS, ECM**
25	Intake air temperature (IAT) sensor – voltage low	**Wiring short circuit to ground, IAT sensor, ECM**

Flash code	Fault location	Probable cause
26	Engine control module (ECM), quad driver/ output driver – circuit malfunction	Wiring, ECM controlled components, ECM 5
27	Engine control module (ECM), quad driver/ output driver – circuit malfunction	Wiring, ECM controlled components, ECM 5
28	Engine control module (ECM), quad driver/ output driver – circuit malfunction	Wiring, ECM controlled components, ECM 5
29	Engine control module (ECM), quad driver/ output driver – circuit malfunction	Wiring, ECM controlled components, ECM 5
31	Transmission range (TR) switch – malfunction	Wiring, TR switch, ECM
32	Exhaust gas recirculation (EGR) solenoid	Wiring, poor connections, EGR valve, ECM
33	Manifold absolute pressure (MAP) sensor – voltage high/low	Wiring, hose blocked/leaking, MAP sensor, ECM
34	Manifold absolute pressure (MAP) sensor – voltage low	Wiring open circuit/short circuit to ground, MAP sensor, ECM
35	Idle air control (IAC) valve	Wiring, IAC valve, intake system leaking, fuel system, PCV valve, A/C system, mechanical fault, ECM
36	Crankshaft position (CKP) sensor, at crankshaft pulley – 24X signal circuit malfunction	Wiring, CKP sensor, ECM
39	Clutch pedal position (CPP) switch	Wiring, incorrectly adjusted, CPP switch, ECM
41 3	Engine control module (ECM), programming – malfunction	ECM incorrect programming, ECM failure
41 4	Crankshaft position (CKP) sensor, 1X reference signal – circuit malfunction	Wiring, ICM, ECM
42	Ignition control, timing signal – circuit malfunction	Wiring, poor connections, ICM, ECM
43	Knock sensor (KS) – circuit malfunction	Wiring, KS, ECM
44	Oxygen sensor (O2S)/heated oxygen sensor (HO2S) – mixture lean	Wiring open circuit/short circuit to ground, injector(s) blocked, fuel contamination, fuel pressure low, exhaust system leaking, O2S/ HO2S, ECM
45	Oxygen sensor (O2S)/heated oxygen sensor (HO2S) – mixture rich	Wiring open circuit, fuel pressure high, injector(s) leaking, O2S/HO2S, EVAP purge valve, TP sensor, EGR system, ECM
46	Immobilizer system, code – malfunction	Incorrect key, ECM
51	Engine control module (ECM), PROM – malfunction	Poor connections, incorrect programming, ECM
52	Engine control module (ECM) – malfunction	ECM
53	System voltage – high	Alternator
54	Fuel pump circuit – voltage low	Wiring open circuit/short circuit to ground, fuel pump relay, engine oil pressure warning lamp/fuel pump switch
55 1	Engine control module (ECM) – malfunction	ECM

Flash code	Fault location	Probable cause
55 **2**	Engine control module (ECM), fuel lean monitor	**Fuel filter/lines blocked, fuel contamination**
61	Oxygen sensor (O2S) – malfunction	**O2S**
62	Transmission gear switch, signal – circuit malfunction	**Wiring open circuit/short circuit to ground, 2nd gear switch, 3rd gear switch, ECM**
65	Injector(s) – circuit malfunction	**Wiring, injector(s)**
66	A/C refrigerant pressure sensor – voltage low	**Wiring open circuit/short circuit to ground, A/C refrigerant pressure sensor, ECM**
70	A/C refrigerant pressure sensor – voltage high	**Wiring open circuit/short circuit to positive, A/C refrigerant pressure sensor, ECM**
72	CAN data bus, ECM to instrumentation control module – lost communication	**Wiring, poor connections, instrumentation control module, ECM**
75	Exhaust gas recirculation (EGR) solenoid 1 – circuit malfunction	**Wiring, EGR valve, ECM**
76	Exhaust gas recirculation (EGR) solenoid 2 – circuit malfunction	**Wiring, EGR valve, ECM**
77	Exhaust gas recirculation (EGR) solenoid 3 – circuit malfunction	**Wiring, EGR valve, ECM**
79	Vehicle speed sensor (VSS) – voltage high	**Wiring, VSS, ECM**
80	Vehicle speed sensor (VSS) – voltage low	**Wiring, VSS, ECM**
81	Brake pedal position (BPP) switch, TCC – circuit malfunction	**Wiring, BPP switch, ECM**
82	Crankshaft position (CKP) sensor, in cylinder block – 3X signal circuit malfunction	**Wiring open circuit/short circuit to ground, CKP sensor, ICM**
85	Engine control module (ECM), PROM – malfunction	**Poor connections, incorrect programming, ECM**
86	Engine control module (ECM) – analogue/ digital converter	**A/C refrigerant pressure sensor/knock sensor (KS)/engine coolant temperature (ECT) sensor/fuel pump monitor circuit short circuit to battery voltage**
87	Engine control module (ECM), EEPROM – malfunction	**Poor connections, incorrect programming, ECM**

1 Except 1995 Sunfire/1994-95 Grand AM 2.3L
2 1995 Sunfire/1994-95 Grand AM 2.3L
3 Except 1992 Grand AM 2.3L
4 1992 Grand AM 2.3L
5 Depending on model, year and equipment level, the following components may be connected to this circuit. Use of the vehicle wiring diagram is recommended to confirm which components are fitted

- ☐ A/C compressor clutch relay
- ☐ A/C condenser blower motor relay
- ☐ Engine coolant blower motor relay
- ☐ Engine coolant 'hot' warning lamp
- ☐ Evaporative emission (EVAP) canister purge valve
- ☐ Exhaust gas recirculation (EGR) solenoid
- ☐ Malfunction indicator lamp (MIL)
- ☐ Torque converter clutch (TCC) solenoid
- ☐ Transmission shift solenoids (SS)
- ☐ Transmission shift warning lamp (MT)

PONTIAC

Model:	Engine identification:	Year:
Firebird 3.4L	VIN position 8 = S	1993-95
Grand AM 3.1L	VIN position 8 = M	1994-95
Grand Prix 3.1L	VIN position 8 = T, M	1994-95

System: **GM**

Self-diagnosis

General information

- Refer to the front of this manual for general test conditions, terminology, detailed descriptions of wiring faults and a general trouble shooter for electrical and mechanical faults.
- Malfunction indicator lamp (MIL) will illuminate if certain faults are recorded.
- ECM operates in backup mode if sensors fail, to enable vehicle to be driven to repair shop.
- Trouble codes can be displayed by the MIL or by using a Scan Tool connected to the data link connector (DLC).

Accessing

Models with data link connector (DLC) terminal B connected

- Access data link connector (DLC).
- Jump data link connector (DLC) terminals A and B **1**.
- Switch ignition ON. Do NOT start engine.
- Count MIL flashes. Note trouble codes. Compare with trouble code table.
- Firstly trouble code 12 will be displayed 3 times.
- Trouble codes consist of 2 groups, displayed as follows:
  - First group indicates the 'tens' of the trouble code **2** [**A**].
  - Second group indicates the 'units' of the trouble code **2** [**C**].
  - A short pause separates each group **2** [**B**].
  - A long pause separates each trouble code **2** [**D**].
  - For example: Trouble code 12 displayed **2**. Start of diagnosis.
- Each additional trouble code will now be displayed 3 times, in ascending order.

NOTE: ***If no additional trouble codes are stored, code 12 will continue to be displayed.***

- Trouble codes can also be displayed by using a Scan Tool connected to the data link connector (DLC) **1**.

1

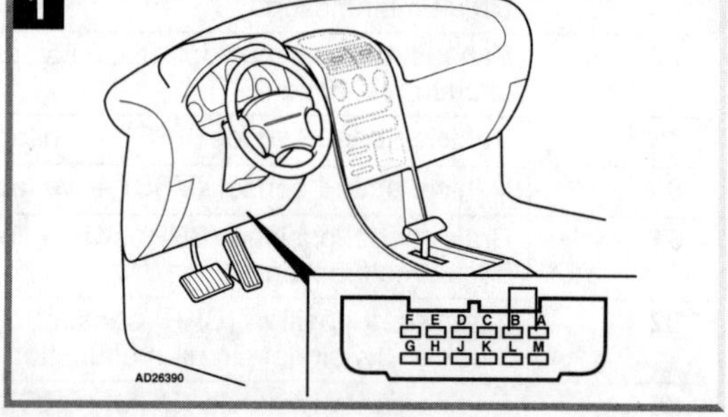

2

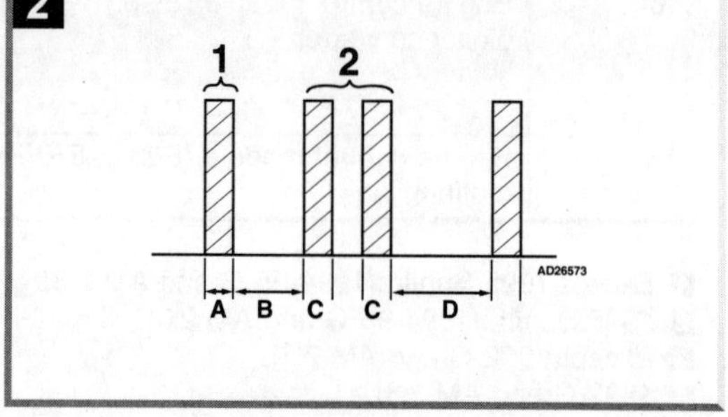

Models without data link connector (DLC) terminal B connected

- Trouble codes can only be displayed by using a Scan Tool connected to the data link connector (DLC) **1**.

Erasing

NOTE: ***Disconnecting battery ground lead may cause driveability problems until the engine management system relearns basic values.***

- Ensure ignition switched OFF.
- Disconnect battery ground cable for 30 seconds.
- Reconnect battery ground cable.
- Repeat checking procedure to ensure no trouble codes remain in ECM fault memory.

NOTE: ***Disconnecting battery ground lead may also erase memory from electronic units, such as radio.***

- Trouble codes can also be erased by using a Scan Tool connected to the data link connector (DLC) **1**.

Trouble code identification

Flash code	Fault location	Probable cause
12	Start of diagnosis	**No engine speed (RPM) signal received**
13 **1**	Heated oxygen sensor (HO2S)	**Wiring open circuit, HO2S, ECM**
13 **2**	Oxygen sensor (O2S), bank 2	**Wiring open circuit, O2S, ECM**
14	Engine coolant temperature (ECT) sensor – voltage low	**Wiring short circuit to ground, ECT sensor, ECM**
15	Engine coolant temperature (ECT) sensor – voltage high	**Wiring open circuit, ECT sensor, ECM**
16	System voltage – low	**Wiring, alternator**
17	Camshaft position (CMP) sensor – circuit malfunction	**Wiring, CMP sensor, mechanical fault, ECM**
21	Throttle position (TP) sensor – voltage high	**Wiring open circuit/short circuit to positive, TP sensor, ECM**
22	Throttle position (TP) sensor – voltage low	**Wiring open circuit/short circuit to ground, TP sensor, ECM**
23	Intake air temperature (IAT) sensor – voltage high	**Wiring open circuit, IAT sensor, ECM**
24	Vehicle speed sensor (VSS) – no signal	**Wiring, VSS, ECM**
25	Intake air temperature (IAT) sensor – voltage low	**Wiring short circuit to ground, IAT sensor, ECM**
28	Transmission range (TR) switch – malfunction	**Wiring, TR switch, ECM**
33	Manifold absolute pressure (MAP) sensor – voltage high	**Wiring short circuit to positive, hose blocked/leaking, MAP sensor, ECM**
34	Manifold absolute pressure (MAP) sensor – voltage low	**Wiring open circuit/short circuit to ground, MAP sensor, ECM**
35	Idle air control (IAC) valve	**Wiring open circuit/short circuit to ground/ short circuit to positive, IAC valve, intake system leaking, fuel system, PCV valve, A/C system, mechanical fault, ECM**
36	Crankshaft position (CKP) sensor, at crankshaft pulley – 24X signal circuit malfunction	**Wiring, CKP sensor, ECM**
37	Brake pedal position (BPP) switch for torque converter clutch (TCC)	**Wiring, incorrectly adjusted, BPP switch, ECM**
39	Clutch pedal position (CPP) switch	**Wiring, incorrectly adjusted, CPP switch, ECM**
41	Ignition control, timing signal – circuit malfunction	**Wiring, poor connections, ICM, ECM**

Flash code	Fault location	Probable cause
42	Ignition control, bypass signal – circuit malfunction	**Wiring, poor connections, ICM, ECM**
43	Knock sensor (KS) – circuit malfunction	**Wiring, KS, ECM**
44 1	Heated oxygen sensor (HO2S) – mixture lean	**Wiring open circuit/short circuit to ground, injector(s) blocked, fuel contamination, fuel pressure low, exhaust system leaking, HO2S, ECM**
44 2	Oxygen sensor (O2S), bank 2 – mixture lean	**Wiring open circuit/short circuit to ground, injector(s) blocked, fuel contamination, fuel pressure low, exhaust system leaking, O2S, ECM**
45 1	Heated oxygen sensor (HO2S) – mixture rich	**Wiring open circuit, fuel pressure high, injector(s) leaking, HO2S, EVAP purge valve, throttle position (TP) sensor, EGR system, ECM**
45 2	Oxygen sensor (O2S), bank 2 – mixture rich	**Wiring open circuit, fuel pressure high, injector(s) leaking, O2S, EVAP purge valve, throttle position (TP) sensor, EGR system, ECM**
46	Immobilizer system, code – malfunction	**Incorrect key, immobilizer control module, ECM**
51	Engine control module (ECM), PROM – malfunction	**Poor connections, incorrect programming, ECM**
53	System voltage – high	**Alternator**
54	Fuel pump circuit – voltage low	**Wiring open circuit/short circuit to ground, fuel pump relay, engine oil pressure warning lamp/fuel pump switch**
58	Transmission fluid temperature (TFT) sensor – voltage low	**Wiring short circuit to ground, TFT sensor, ECM**
59	Transmission fluid temperature (TFT) sensor – voltage high	**Wiring open circuit, TFT sensor, ECM**
61	A/C system – performance	**Refrigerant low, wiring, A/C refrigerant pressure sensor, A/C evaporator temperature sensor, ECM**
63	Oxygen sensor (O2S), bank 1	**Wiring open circuit, O2S, ECM**
64	Oxygen sensor (O2S), bank 1 – mixture lean	**Wiring open circuit/short circuit to ground, injector(s) blocked, fuel contamination, fuel pressure low, exhaust system leaking, O2S, ECM**
65	Oxygen sensor (O2S), bank 1 – mixture rich	**Wiring open circuit, fuel pressure high, injector(s) leaking, O2S, EVAP purge valve, throttle position (TP) sensor, EGR system, ECM**
66	A/C refrigerant pressure sensor – voltage low	**Wiring open circuit/short circuit to ground, A/C refrigerant pressure sensor, ECM**
67	A/C refrigerant pressure sensor – circuit malfunction	**Wiring open circuit, A/C refrigerant pressure sensor, ECM**

Flash code	Fault location	Probable cause
68	A/C compressor clutch relay – circuit malfunction	**Wiring short circuit, A/C compressor clutch relay, ECM**
69	A/C compressor clutch relay – circuit malfunction	**Wiring open circuit, A/C compressor clutch relay, ECM**
70	A/C refrigerant pressure sensor – voltage high	**Wiring open circuit/short circuit to positive, A/C refrigerant pressure sensor, ECM**
71	A/C evaporator temperature sensor – voltage high	**Wiring, poor connections, A/C evaporator temperature sensor**
72	Vehicle speed sensor (VSS) – circuit malfunction	**Wiring, VSS, ECM**
73	A/C evaporator temperature sensor – voltage low	**Wiring, poor connections, A/C evaporator temperature sensor**
75	Exhaust gas recirculation (EGR) solenoid 1 – circuit malfunction	**Wiring, EGR valve, ECM**
76	Exhaust gas recirculation (EGR) solenoid 2 – circuit malfunction	**Wiring, EGR valve, ECM**
77	Exhaust gas recirculation (EGR) solenoid 3 – circuit malfunction	**Wiring, EGR valve, ECM**
79	Transmission fluid temperature (TFT) – high	**Transmission mechanical fault, towing weight limit exceeded**
80	Torque converter clutch (TCC) – slipping	**Wiring, TCC solenoid, transmission mechanical fault, ECM**
81	Brake pedal position (BPP) switch, torque converter clutch (TCC) – circuit malfunction	**Wiring, BPP switch, ECM**
82	Crankshaft position (CKP) sensor, in cylinder block – 3X signal circuit malfunction	**Wiring open circuit/short circuit to ground, CKP sensor, ICM**
85	Engine control module (ECM), PROM – malfunction	**Poor connections, incorrect programming, ECM**
86	Engine control module (ECM) – analogue/ digital converter	**A/C refrigerant pressure sensor/knock sensor (KS)/engine coolant temperature (ECT) sensor/fuel pump monitor circuit short circuit to battery voltage**
87	Engine control module (ECM), EEPROM – malfunction	**Poor connections, incorrect programming, ECM**
90	Torque converter clutch (TCC) – stuck ON	**Wiring open circuit/short circuit to ground, transmission internal wiring, TCC solenoid, transmission mechanical fault, ECM**
93	Transmission fluid pressure (TFP) control solenoid – circuit malfunction	**Wiring, poor connection, TFP solenoid**
96	Engine control module (ECM) – ignition supply voltage low	**Wiring open circuit, alternator, ECM**
98	Engine control module (ECM) – incorrectly coded	**ECM incorrectly coded, ECM**
99	Engine control module (ECM) – incorrectly coded	**ECM incorrectly coded, ECM**

1 Except Firebird 3.4L 1993-95
2 Firebird 3.4L 1993-95

PONTIAC

Model:	Engine identification:	Year:
Firebird 3.8L	VIN position 8 = K	1995-02
Firebird 5.7L	VIN position 8 = G, P	1996-02
Grand Prix 3.1L	VIN position 8 = M	1996-02
Grand Prix 3.4L	VIN position 8 = X	1996
Grand Prix 3.8L	VIN position 8 = K, 1	1997-02
Montana/Extended Montana 3.4L	VIN position 8 = E	1999-02
Sunfire 2.2L	VIN position 8 = 4, 5, F	1995-02
Sunfire 2.3L	VIN position 8 = D	1995
Sunfire 2.4L	VIN position 8 = T	1996-02
Trans Sport 3.4L	VIN position 8 = E	1996-98

System: **GM**

Self-diagnosis

General information

- Refer to the front of this manual for general test conditions, terminology, detailed descriptions of wiring faults and a general trouble shooter for electrical and mechanical faults.
- Malfunction indicator lamp (MIL) will illuminate or flash if certain faults are recorded.
- ECM operates in backup mode if sensors fail, to enable vehicle to be driven to repair shop.

Accessing

- Trouble codes can only be displayed by using a Scan Tool connected to the data link connector (DLC) **1** or **2**.

1

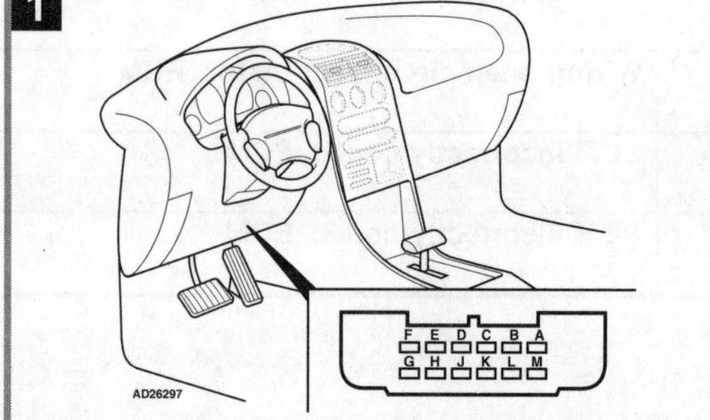

2

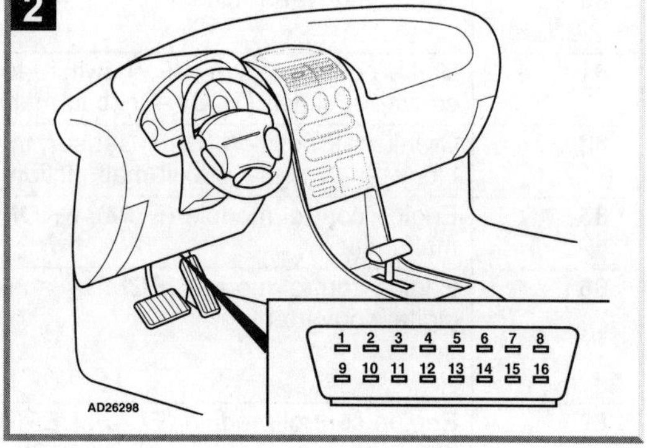

Erasing

- After the faults have been rectified, the manufacturer recommends that trouble codes are erased by using a Scan Tool connected to the data link connector (DLC).

Trouble code identification

Scan code	Fault location	Probable cause
P0000-P0999	Refer to OBD-II trouble code table at the front of this manual	–
P1106	Manifold absolute pressure (MAP) sensor – voltage high, intermittent	**Wiring open circuit/short circuit to positive, poor connections**
P1107	Manifold absolute pressure (MAP) sensor – voltage low, intermittent	**Wiring short circuit to ground, poor connections**
P1111	Intake air temperature (IAT) sensor – voltage high, intermittent	**Wiring open circuit/short to positive, poor connections**
P1112	Intake air temperature (IAT) sensor – voltage low, intermittent	**Wiring short circuit to ground**
P1114	Engine coolant temperature (ECT) sensor – voltage low, intermittent	**Wiring short circuit to ground, poor connections**
P1115	Engine coolant temperature (ECT) sensor, voltage high, intermittent	**Wiring open circuit/short to positive, poor connections**
P1120	Throttle motor position sensor 1 – circuit malfunction	**Wiring, throttle control module**
P1121	Throttle position (TP) sensor – voltage high, intermittent	**Wiring open circuit/short to positive, poor connections**
P1122	Throttle position (TP) sensor – voltage low, intermittent	**Wiring open circuit/short to ground, poor connections**
P1125	Accelerator pedal position (APP) sensor 1/2/3 – signal malfunction	**Wiring, APP sensor, ECM**
P1133	Oxygen sensor (O2S)/heated oxygen sensor (HO2S) 1, bank 1 – insufficient switching	**Wiring, exhaust system leaking, O2S/HO2S**
P1134	Heated oxygen sensor (HO2S) 1, bank 1 – slow response	**Wiring, oxygen sensor heater, exhaust system leaking, HO2S**
P1137	Heated oxygen sensor (HO2S) 2, bank 1 – voltage low	**Wiring, exhaust system leaking, HO2S**
P1138	Heated oxygen sensor (HO2S) 2, bank 1 – voltage high	**Wiring, HO2S**
P1153	Heated oxygen sensor (HO2S) 1, bank 2 – insufficient switching	**Wiring, oxygen sensor heater, exhaust system leaking, HO2S**
P1154	Heated oxygen sensor (HO2S) 1, bank 2 – slow response	**Wiring, oxygen sensor heater, exhaust system leaking, HO2S**
P1171	Fuel system – lean during acceleration	**Fuel filter, fuel pump, HO2S, injectors**
P1200	Injectors – circuit malfunction	**Wiring open circuit/short circuit to ground/ short circuit to positive, injectors, ECM**
P1220	Throttle motor position sensor 2 – circuit malfunction	**Wiring, throttle control module**
P1221	Throttle motor position sensor 1/2 – signal variation	**Wiring, throttle control module**
P1222	Injector control, circuit monitor – malfunction	**Injector, wiring**
P1258	Engine overheating	**Cooling system, engine coolant temperature (ECT) sensor**

Scan code	Fault location	Probable cause
P1260	Fuel pump, control signal – circuit malfunction	**Wiring, fuel pump control module, ECM**
P1271	Accelerator pedal position (APP) sensor 1/2 – signal variation	**Wiring, APP sensor, ECM**
P1272	Accelerator pedal position (APP) sensor 2/3 – signal variation	**Wiring, APP sensor, ECM**
P1273	Accelerator pedal position (APP) sensor 1/3 – signal variation	**Wiring, APP sensor, ECM**
P1275	Accelerator pedal position (APP) sensor 1 – circuit malfunction	**Wiring, APP sensor, ECM**
P1276	Accelerator pedal position (APP) sensor 1 – range performance	**Wiring, APP sensor, ECM**
P1280	Accelerator pedal position (APP) sensor 2 – circuit malfunction	**Wiring, APP sensor, ECM**
P1281	Accelerator pedal position (APP) sensor 2 – range performance	**Wiring, APP sensor, ECM**
P1285	Accelerator pedal position (APP) sensor 3 – circuit malfunction	**Wiring, APP sensor, ECM**
P1286	Accelerator pedal position (APP) sensor 3 – range performance	**Wiring, APP sensor, ECM**
P1336	Crankshaft position sensor (CKP) – variation not learned	**Learning procedure not performed**
P1350	Ignition control system, bypass circuit – circuit malfunction	**Wiring, poor connections, ICM, ECM**
P1351	Ignition control system, coil – voltage high	**Wiring open circuit, ICM, ECM**
P1352	Ignition control system, bypass circuit – voltage high	**Wiring open circuit, ICM, ECM**
P1361	Ignition control system, IC circuit – voltage low	**Wiring, circuit short circuit to ground, poor connections, ICM, ECM**
P1362	Ignition control, bypass circuit – voltage low	**Wiring short circuit to ground, ICM, ECM**
P1371	Ignition control reference signal, low/high resolution – low signal lost	**Wiring, ICM**
P1374	Crankshaft position sensor (CKP), high to low resolution frequency – signal variation	**Wiring, wiring routed near spark plug wires, ignition coils arcing/tracking, ICM, ECM**
P1380	Antilock brake system (ABS)/traction control system (TCS), rough road signal to ECM – malfunction	**ABS trouble code(s) stored**
P1381	Antilock brake system (ABS)/traction control system (TCS), rough road signal to ECM – no communication	**Wiring, poor connections, ABS**
P1404	Exhaust gas recirculation (EGR) valve – not fully closing	**Wiring, poor connections, EGR valve position sensor, EGR solenoid**
P1406	Exhaust gas recirculation (EGR) valve position sensor – range performance	**Wiring, EGR valve tight/sticking, EGR valve position sensor**
P1415	Secondary air injection (AIR) system, bank 1	**Wiring, hoses blocked/leaking, AIR valve**
P1416	Secondary air injection (AIR) system, bank 2	**Wiring, hoses blocked/leaking, AIR valve**

Scan code	Fault location	Probable cause
P1441	Evaporative emission (EVAP) system – leak detected	Hose connection(s), EVAP canister, EVAP canister purge valve
P1442	Evaporative emission (EVAP) system vacuum switch – circuit malfunction	Hoses blocked/leaking, wiring, EVAP canister, ECM
P1483	Engine cooling system – performance	Cooling system, engine coolant temperature (ECT) sensor, intake air temperature (IAT) sensor
P1508	Idle air control(IAC) valve – low rpm	Wiring open circuit/short circuit to ground/ short circuit to positive, IAC valve, throttle valve tight/sticking
P1509	Idle air control(IAC) valve – high rpm	Wiring open circuit/short circuit to ground/ short circuit to positive, IAC valve, throttle valve tight/sticking
P1514	Throttle control unit, intake air flow – excessive	Throttle control unit, MAP sensor, CMP sensor signal, ECM
P1515	Throttle command/actual throttle position – signal variation	Wiring, throttle valve sticking, throttle control module
P1516	Throttle command/actual throttle position – signal variation	Wiring, throttle valve sticking, throttle control module
P1517	Throttle control module – malfunction	Throttle control module
P1518	Throttle control module/ECM – communication malfunction	Wiring, throttle control module, ECM
P1519	Throttle motor position sensor 1 – signal malfunction	Throttle control module
P1523	Throttle control unit, closed throttle position – performance	Throttle valve movement restricted, throttle control unit
P1532	A/C refrigerant temperature sensor – voltage low	Wiring open circuit/short circuit to ground, A/C refrigerant temperature sensor, ECM
P1533	A/C refrigerant temperature sensor – voltage high	Wiring open circuit/short circuit to ground, A/C refrigerant temperature sensor, ECM
P1539	A/C compressor clutch relay – voltage high	Wiring, A/C compressor clutch relay, ECM
P1543	A/C system – performance	Refrigerant low, wiring, A/C refrigerant pressure sensor, A/C evaporator temperature sensor, ECM
P1545	A/C compressor clutch relay – voltage low	Wiring, A/C compressor clutch relay, ECM
P1546	A/C compressor clutch relay – circuit malfunction	Wiring, A/C compressor clutch relay, ECM
P1554	Cruise control status circuit – high voltage	Wiring open circuit/short circuit to ground, cruise control module, ECM
P1571	Traction control system (TCS), torque request signal – malfunction	Wiring, poor connections, ABS/TCS
P1572	Traction control system (TCS), active signal – malfunction	Wiring short to ground, ABS/TCS, ECM
P1573	Engine control module (ECM), antilock brake system (ABS)/traction control system (TCS) communication – malfunction	Wiring, poor connections, ABS control module, ECM

Scan code	Fault location	Probable cause
P1574	Brake pedal position switch (BPP), voltage high	**Wiring, ABS**
P1585	Cruise control system, inhibit output circuit	**Wiring, poor connections, cruise control motor/actuator, ECM**
P1586	Cruise control system, brake switch 2 – circuit malfunction	**Wiring, poor connections, BPP switch**
P1600	Engine control module (ECM) – malfunction	**Engine control module (ECM)**
P1601	CAN data bus, engine control module (ECM)/ antilock brake system (ABS)/traction control system (TCS)/instrumentation control module/ diagnostic control module – not all modules communicating	**Wiring, poor connections, ECM, ABS/TCS, diagnostic control module, instrumentation control module**
P1621	Engine control module (ECM) – memory performance	**ECM**
P1626	Engine control module (ECM) – fuel enable signal lost	**Wiring, poor connections, immobilizer control module, ECM**
P1627	Engine control module (ECM) – malfunction	**Engine control module (ECM)**
P1629	Engine control module (ECM) – fuel enable signal not received	**Incorrect ignition key, wiring, poor connections, immobilizer control module, ECM**
P1630	Engine control module (ECM), immobilizer code – learning mode	**ECM immobilizer code programming not completed**
P1631	Engine control module (ECM) – incorrect immobilizer code	**Incorrect key, immobilizer code not programmed**
P1632	Engine control module (ECM) – fuel disable signal received	**Fuses, BCM, immobilizer code**
P1635	Engine control module (ECM), sensor supply voltage – circuit malfunction	**Wiring, poor connections, TP sensor, MAP sensor, EGR valve position sensor, fuel tank pressure (FTP) sensor, ECM**
P1637	Alternator warning lamp, control signal – circuit malfunction	**Wiring, instrument panel, ECM**
P1639	Engine control module (ECM), sensor supply voltage – circuit malfunction	**Wiring, poor connections, A/C refrigerant pressure sensor, ECM**
P1640	Engine control module (ECM), quad driver/ output driver – circuit malfunction	**Wiring, ECM controlled components, ECM 9**
P1641 1	Malfunction indicator lamp (MIL) – circuit malfunction	**Wiring, poor connections, MIL, ECM**
P1641 2	A/C compressor clutch relay – circuit malfunction	**Wiring open circuit/short circuit to ground, A/C compressor clutch relay, ECM**
P1641 6	Engine coolant blower motor relay 1 – circuit malfunction	**Wiring, engine coolant blower motor relay, ECM**
P1642 6	Engine coolant blower motor relay 2 & 3 – circuit malfunction	**Wiring, engine coolant blower motor relay(s), ECM**
P1642 7	Engine oil change warning lamp – circuit malfunction	**Wiring, ECM**
P1642 8	Secondary air injection (AIR) pump relay – circuit malfunction	**Wiring, AIR pump relay, ECM**

Scan code	Fault location	Probable cause
P1643 6	Engine speed (RPM), engine control module (ECM) output signal – circuit malfunction	Wiring, tachometer, ECM
P1643 7	Fuel pump, control signal – circuit malfunction	Wiring, fuel pump control module, ECM
P1644	Traction control system (TCS), delivered torque signal – malfunction	Wiring, poor connections, ECM
P1645	Supercharger (SC) bypass valve, control signal – circuit malfunction	Wiring, SC bypass valve, ECM
P1650	Engine control module (ECM), quad driver/ output driver – circuit malfunction	Wiring, ECM controlled components, ECM 9
P1651	Engine coolant blower motor relay 1 – circuit malfunction	Wiring, engine coolant blower motor relay, ECM
P1652 6	Engine control module(ECM), vehicle speed sensor (VSS) output signal – circuit malfunction	Wiring, poor connections, ECM
P1652 7	Engine coolant blower motor relay 2 & 3 – circuit malfunction	Wiring, engine coolant blower motor relay(s), ECM
P1653 3	Fuel gauge, control signal – circuit malfunction	Wiring, ECM
P1653 4	Engine oil level 'low' warning lamp, control signal – circuit malfunction	Wiring short circuit to ground, ECM
P1653 5	Traction control system (TCS), delivered torque signal – malfunction	Wiring, poor connections, ECM
P1654	A/C compressor clutch relay – circuit malfunction	Wiring open circuit/short circuit to ground, A/C compressor clutch relay, ECM
P1655	Evaporative emission (EVAP) canister purge valve – circuit malfunction	Wiring open circuit/short circuit to ground, EVAP canister vent valve, ECM
P1657	Skip shift solenoid – circuit malfunction	Wiring, skip shift solenoid, ECM
P1660	Engine control module (ECM), quad driver/ output driver – circuit malfunction	Wiring, ECM controlled components, ECM 9
P1661	Malfunction indicator lamp (MIL) – circuit malfunction	Wiring, MIL, ECM
P1662	Cruise control, inhibit output circuit – malfunction	Wiring open circuit/short circuit to ground, cruise control motor/actuator, cruise control module, ECM
P1663	Alternator warning lamp, control signal – circuit malfunction	Wiring, instrument panel, ECM
P1664	Skip shift warning lamp – circuit malfunction	Wiring, skip shift warning lamp, ECM
P1665	Evaporative emission (EVAP) canister vent valve – circuit malfunction	Wiring open circuit/short circuit to ground, EVAP canister vent valve, ECM
P1667 6	Reverse inhibit solenoid – circuit malfunction	Wiring, reverse inhibit solenoid, ECM
P1667 7	Fuel pump, control signal – circuit malfunction	Wiring, fuel pump control module, ECM
P1670	Engine control module (ECM), quad driver/ output driver – circuit malfunction	Wiring, ECM controlled components, ECM 9
P1671	Malfunction indicator lamp (MIL) – circuit malfunction	Wiring open circuit/short circuit to ground/ short circuit to positive, MIL, ECM
P1672	Engine oil level 'low' warning lamp, control signal – circuit malfunction	Wiring short circuit to ground, engine oil level, ECM

Scan code	Fault location	Probable cause
P1673	Engine coolant 'hot' warning lamp, control signal – circuit malfunction	**Wiring, instrument panel, cooling system, ECM**
P1676	Evaporative emission (EVAP) canister purge valve – circuit malfunction	**Wiring open circuit/short circuit to ground, EVAP canister purge valve, ECM**
P1680	Engine control module (ECM) – malfunction	**Engine control module (ECM)**
P1681	Engine control module (ECM) – malfunction	**Engine control module (ECM)**
P1683	Engine control module (ECM) – malfunction	**Engine control module (ECM)**
P1689	Traction control system (TCS), delivered torque signal – malfunction	**Wiring, poor connections, ECM**
P1810	Transmission fluid pressure (TFP) switch – manual valve position	**Fluid level low, wiring, selector lever linkage incorrectly adjusted, ECM**
P1811	Shift time too long – adaptation limit reached	**Fluid level low, transmission mechanical fault**
P1814	Torque converter clutch (TCC) – over-stressed	**Fluid level low, fluid condition, transmission mechanical fault**
P1860	Torque converter clutch (TCC) solenoid – circuit malfunction	**Wiring open circuit/short circuit to ground, TCC solenoid, ECM**
P1870	Torque converter clutch (TCC) – slip speed excessive	**Transmission fluid level, wiring, TCC solenoid, ECM**
P1887	Torque converter clutch (TCC) release switch – circuit malfunction	**Wiring open circuit, TCC solenoid release switch, ECM**
U1016	CAN data bus, engine control module (ECM)/ antilock brake system (ABS)/traction control system (TCS)/instrumentation control module/ diagnostic control module – loss of serial communication	**Wiring, poor connections, ECM, ABS/TCS, diagnostic control module, instrumentation control module**

1 →1997
2 1998→
3 1998
4 1999→
5 1997 Grand Prix 3.8L
6 Firebird 5.7L
7 Except Firebird 5.7L
8 1996 Grand Prix 3.4L
9 Depending on model, year and equipment level, the following components may be connected to this circuit. Use of the vehicle wiring diagram is recommended to confirm which components are fitted

- A/C compressor clutch relay
- A/C condenser blower motor relay
- Engine coolant blower motor relay
- Engine coolant 'hot' warning lamp
- Evaporative emission (EVAP) canister purge valve
- Exhaust gas recirculation (EGR) solenoid
- Malfunction indicator lamp (MIL)
- Torque converter clutch (TCC) solenoid
- Transmission shift solenoids (SS)
- Transmission shift warning lamp (MT)

PONTIAC

Model:	Engine identification:	Year:
Firebird 5.7L	**VIN position 8 = P**	**1993-95**

System: **GM**

Self-diagnosis

General information

- Refer to the front of this manual for general test conditions, terminology, detailed descriptions of wiring faults and a general trouble shooter for electrical and mechanical faults.
- Malfunction indicator lamp (MIL) will illuminate if certain faults are recorded.
- ECM operates in backup mode if sensors fail, to enable vehicle to be driven to repair shop.
- Trouble codes can be displayed by the MIL or by using a Scan Tool connected to the data link connector (DLC).

Accessing

- Access data link connector (DLC).
- Jump data link connector (DLC) terminals A and B **1**.
- Switch ignition ON. Do NOT start engine.
- Count MIL flashes. Note trouble codes. Compare with trouble code table.
- Firstly trouble code 12 will be displayed 3 times.
- Trouble codes consist of 2 groups, displayed as follows:
  - First group indicates the 'tens' of the trouble code **2** [**A**].
  - Second group indicates the 'units' of the trouble code **2** [**C**].
  - A short pause separates each group **2** [**B**].
  - A long pause separates each trouble code **2** [**D**].
  - For example: Trouble code 12 displayed **2**. Start of diagnosis.
- Each additional trouble code will now be displayed 3 times, in ascending order.

NOTE: ***If no additional trouble codes are stored, code 12 will continue to be displayed.***

- Trouble codes can also be displayed by using a Scan Tool connected to the data link connector (DLC) **1**.

1

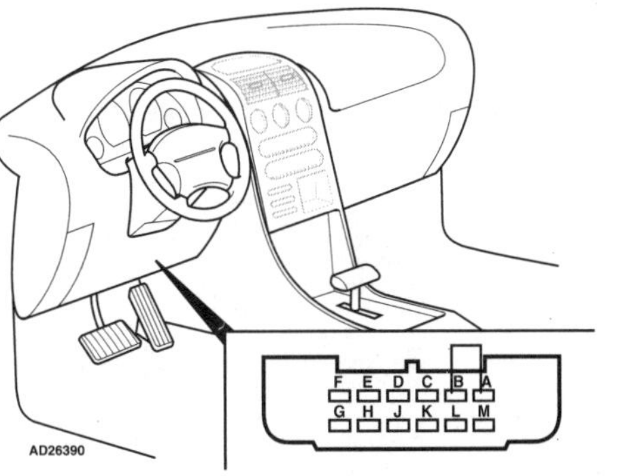

2

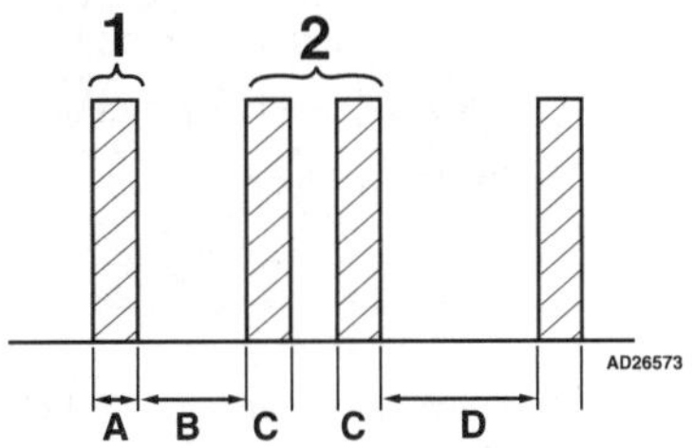

Erasing

NOTE: ***Disconnecting battery ground lead may cause driveability problems until the engine management system relearns basic values.***

- Ensure ignition switched OFF.
- Disconnect battery ground cable for 30 seconds.
- Reconnect battery ground cable.
- Repeat checking procedure to ensure no trouble codes remain in ECM fault memory.

NOTE: ***Disconnecting battery ground lead may also erase memory from electronic units, such as radio.***

- Trouble codes can also be erased by using a Scan Tool connected to the data link connector (DLC).

Trouble code identification

Flash code	Fault location	Probable cause
11	Malfunction indicator lamp (MIL) – circuit malfunction	**Wiring open circuit/short circuit to ground, MIL, ECM**
12	Start of diagnosis	**No engine speed (RPM) signal received**
13	Oxygen sensor (O2S)/heated oxygen sensor (HO2S), bank 1	**Wiring open circuit, O2S/HO2S, ECM**
14	Engine coolant temperature (ECT) sensor – voltage high/low	**Wiring, ECT sensor, ECM**
15	Engine coolant temperature (ECT) sensor – voltage high	**Wiring open circuit, ECT sensor, ECM**
16	Distributor ignition system, low resolution signal – circuit malfunction	**Wiring, distributor, ECM**
18	Injector(s) – circuit malfunction	**Wiring, injector(s)**
21	Throttle position (TP) sensor – voltage high/low	**Wiring, TP sensor, ECM**
22	Throttle position (TP) sensor – voltage low	**Wiring open circuit/short circuit to ground, TP sensor, ECM**
23	Intake air temperature (IAT) sensor – voltage high/low	**Wiring, IAT sensor, ECM**
24	Vehicle speed sensor (VSS) – no signal	**Wiring, VSS, ECM**
25	Intake air temperature (IAT) sensor – voltage low	**Wiring short circuit to ground, IAT sensor, ECM**
26	Engine control module (ECM), quad driver/output driver – circuit malfunction	**Wiring, ECM controlled components, ECM** 5
27	Engine control module (ECM), quad driver/output driver – circuit malfunction	**Wiring, ECM controlled components, ECM** 5
28 1	Engine control module (ECM), quad driver/output driver – circuit malfunction	**Wiring, ECM controlled components, ECM** 5
28 2	Transmission gear pressure switch, signal – circuit malfunction	**Wiring open circuit/short circuit to ground, reverse/D1/D2/D3/D4 gear pressure switch(es), ECM**
29	Secondary air injection (AIR) pump relay – circuit malfunction	**Wiring, AIR pump relay, ECM**
32	Exhaust gas recirculation (EGR) solenoid	**Wiring, poor connections, EGR valve, ECM**

Flash code	Fault location	Probable cause
33	Manifold absolute pressure (MAP) sensor – voltage high/low	**Wiring, hose blocked/leaking, MAP sensor, ECM**
34	Manifold absolute pressure (MAP) sensor – voltage low	**Wiring open circuit/short circuit to ground, MAP sensor, ECM**
36	Distributor ignition system, high resolution signal – circuit malfunction	**Wiring, distributor, ECM**
37	Brake pedal position (BPP) switch for torque converter clutch (TCC) – high voltage	**Wiring short to positive, incorrectly adjusted, BPP switch, ECM**
38	Brake pedal position (BPP) switch, torque converter clutch (TCC) – low voltage	**Wiring open circuit, incorrectly adjusted, BPP switch, ECM**
41	Ignition control, reference signal – circuit malfunction	**Wiring open circuit, poor connections, ICM, ECM**
42	Ignition control, reference signal – circuit malfunction	**Wiring short to ground, ICM, ECM**
43	Knock sensor (KS) – circuit malfunction	**Wiring, KS, ECM**
44	Oxygen sensor (O2S)/heated oxygen sensor (HO2S), bank 1 – mixture lean	**Wiring open circuit/short circuit to ground, injector(s) blocked, fuel contamination, fuel pressure low, exhaust system leaking, O2S/ HO2S, ECM**
45	Oxygen sensor (O2S)/heated oxygen sensor (HO2S), bank 1 – mixture rich	**Wiring open circuit, fuel pressure high, injector(s) leaking, O2S/HO2S, EVAP purge valve, throttle position (TP) sensor, EGR system, ECM**
46	Immobilizer system, code – malfunction	**Incorrect key, ECM**
47	Knock control module, removable part in ECM – circuit malfunction	**Wiring, poor connection, missing knock control module, ECM**
48	Mass airflow (MAF) sensor – no signal	**Wiring, poor connections, MAF sensor, ECM**
50	System voltage – low	**Alternator**
51	Engine control module (ECM), PROM – malfunction	**Poor connections, incorrect programming, ECM**
53	System voltage – high	**Alternator**
55	Engine control module (ECM), fuel lean monitor	**Fuel filter/lines blocked, fuel contamination**
58	Transmission fluid temperature (TFT) sensor – voltage low	**Wiring short circuit to ground, TFT sensor, ECM**
59	Transmission fluid temperature (TFT) sensor – voltage high	**Wiring open circuit, TFT sensor, ECM**
61	A/C system – performance	**Refrigerant low, wiring, A/C refrigerant pressure sensor, A/C evaporator temperature sensor, ECM**
63	Oxygen sensor (O2S)/heated oxygen sensor (HO2S), bank 2	**Wiring open circuit, O2S/HO2S, ECM**
64	Oxygen sensor (O2S)/heated oxygen sensor (HO2S), bank 2 – mixture lean	**Wiring open circuit/short circuit to ground, injector(s) blocked, fuel contamination, fuel pressure low, exhaust system leaking, O2S/HO2S, ECM**

Flash code	Fault location	Probable cause
65	Oxygen sensor (O2S)/heated oxygen sensor (HO2S), bank 2 – mixture rich	**Wiring open circuit, fuel pressure high, injector(s) leaking, O2S/HO2S, EVAP purge valve, throttle position (TP) sensor, EGR system, ECM**
66	A/C refrigerant pressure sensor – voltage low	**Wiring open circuit/short circuit to ground, A/C refrigerant pressure sensor, ECM**
67	A/C refrigerant pressure sensor – circuit malfunction	**Wiring open circuit, A/C refrigerant pressure sensor, ECM**
68	A/C compressor clutch relay – circuit malfunction	**Wiring short circuit, A/C compressor clutch relay, ECM**
69	A/C compressor clutch relay – circuit malfunction	**Wiring open circuit, A/C compressor clutch relay, ECM**
70	A/C compressor clutch relay, control signal – circuit malfunction	**Wiring, A/C compressor clutch relay, ECM**
71	A/C evaporator temperature sensor – voltage high	**Wiring, poor connections, A/C evaporator temperature sensor**
72	Vehicle speed sensor (VSS) – loss of signal	**Wiring, VSS, ECM**
73	Transmission fluid pressure (TFP) control solenoid – circuit malfunction	**Wiring, poor connection, TFP solenoid**
74	Traction control system (TCS), active signal – malfunction	**Wiring short to ground, ABS/TCS, ECM**
75	System voltage, engine control module (ECM) – low	**Wiring, battery, alternator, ECM**
77	Engine coolant blower motor relay – circuit malfunction	**Wiring, engine coolant blower motor relay, ECM**
79	Transmission fluid temperature (TFT) – high	**Transmission mechanical fault, towing weight limit exceeded**
80	Torque converter clutch (TCC) – slipping	**Wiring, TCC solenoid, transmission mechanical fault, ECM**
81	Transmission shift solenoid (SS), 2nd-3rd – circuit malfunction	**Wiring, SS, ECM**
82	Transmission shift solenoid (SS), 1st-2nd – circuit malfunction	**Wiring, SS, ECM**
83	Reverse gear inhibit solenoid – circuit malfunction	**Wiring, reverse gear inhibit solenoid, ECM**
84 [3]	Transmission 3rd-2nd control solenoid – circuit malfunction	**Wiring, control solenoid, ECM**
84 [4]	Skip shift solenoid – circuit malfunction	**Wiring, skip shift solenoid, ECM**
85	Torque converter clutch (TCC) solenoid – stuck ON	**Wiring open circuit/short circuit to ground, transmission internal wiring, TCC solenoid, transmission mechanical fault, ECM**
90	Torque converter clutch (TCC) solenoid – circuit malfunction	**Wiring, TCC solenoid, ECM**
91	Skip shift warning lamp – circuit malfunction	**Wiring, skip shift warning lamp, ECM**

Flash code	Fault location	Probable cause
97	Engine control module (ECM), vehicle speed sensor (VSS) output signal – circuit malfunction	**Wiring, poor connections, ECM**
99	Engine control module (ECM), tachometer output signal – circuit malfunction	**Wiring, poor connections, ECM**

1 1993 Firebird 5.7L
2 1994 Firebird 5.7L
3 Automatic transmission
4 Manual transmission
5 Depending on model, year and equipment level, the following components may be connected to this circuit. Use of the vehicle wiring diagram is recommended to confirm which components are fitted

- □ A/C compressor clutch relay
- □ A/C condenser blower motor relay
- □ Engine coolant blower motor relay
- □ Engine coolant 'hot' warning lamp
- □ Evaporative emission (EVAP) canister purge valve
- □ Exhaust gas recirculation (EGR) solenoid
- □ Malfunction indicator lamp (MIL)
- □ Torque converter clutch (TCC) solenoid
- □ Transmission shift solenoids (SS)
- □ Transmission shift warning lamp (MT)

SATURN

Model:	Engine identification:	Year:
L-Series 2.2L	**VIN position 8 = F**	**2000-02**
L-Series 3.0L	**VIN position 8 = R**	**2000-02**
S-Series 1.9L	**VIN position 8 = 7, 8**	**1996-02**

System: **GM**

Self-diagnosis

General information

- Refer to the front of this manual for general test conditions, terminology, detailed descriptions of wiring faults and a general trouble shooter for electrical and mechanical faults.
- Malfunction indicator lamp (MIL) will illuminate or flash if certain faults are recorded.
- ECM operates in backup mode if sensors fail, to enable vehicle to be driven to repair shop.

Accessing

- Trouble codes can only be displayed by using a Scan Tool connected to the data link connector (DLC) **1**.

1

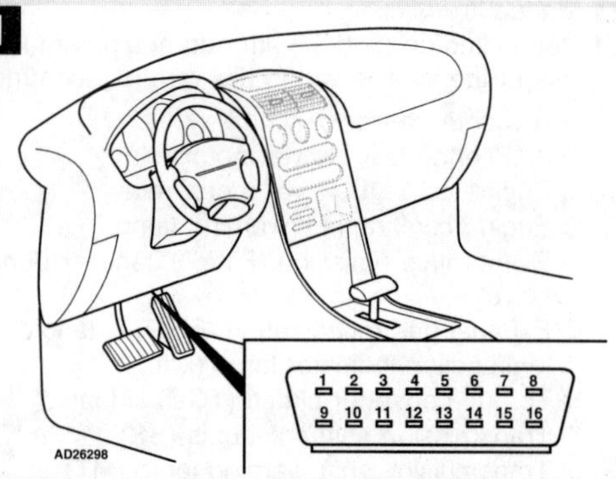

Erasing

- Trouble codes can only be erased by using a Scan Tool connected to the data link connector (DLC).

Trouble code identification

Scan code	Fault location	Probable cause
P1106	Manifold absolute pressure (MAP) sensor, circuit intermittent, voltage high	**Wiring open circuit/short circuit to positive**
P1107	Manifold absolute pressure (MAP) sensor, circuit intermittent, voltage low	**Wiring short circuit to ground**
P1111	Intake air temperature (IAT) sensor, circuit intermittent, voltage high	**Wiring open circuit, ground lead defective**
P1112	Intake air temperature (IAT) sensor, circuit intermittent, voltage low	**Wiring short circuit to ground**
P1114	Engine coolant temperature (ECT) sensor, circuit malfunction, voltage low	**Wiring short circuit to ground**
P1115	Engine coolant temperature (ECT) sensor, circuit malfunction, voltage high	**Wiring open circuit, ground lead defective**
P1121	Throttle position (TP) sensor, circuit malfunction, voltage high	**Wiring open circuit/short circuit to positive, TP sensor**

Scan code	Fault location	Probable cause
P1122	Throttle position (TP) sensor, circuit malfunction, voltage low	**Wiring open circuit/short circuit to ground, TP sensor**
P1133	Oxygen sensor (O2S)/Heated oxygen sensor (HO2S) 1, bank 1 – insufficient switching	**Wiring, oxygen sensor heater, exhaust system leaking, O2S/HO2S**
P1134	Oxygen sensor (O2S)/heated oxygen sensor (HO2S) 1, bank 1, slow response	**Wiring, oxygen sensor heater, exhaust system leaking, O2S/HO2S**
P1137	Heated oxygen sensor (HO2S) 2, bank 1 – voltage low	**Wiring, exhaust system leaking, HO2S**
P1138	Heated oxygen sensor (HO2S) 2, bank 1 – voltage high	**Wiring, HO2S**
P1171	Fuel system – lean during acceleration	**Fuel filter, fuel pump, HO2S, injectors**
P1215	Engine control module (ECM), ground detection – malfunction	**Wiring, ECM**
P1245	Intake manifold air control solenoid – circuit malfunction	**Wiring, intake manifold air control solenoid, ECM**
P1271	Accelerator pedal position (APP) sensor 1/2 – signal variation	**APP sensor**
P1277	Accelerator pedal position (APP) sensor 1 – voltage low	**Wiring, APP sensor**
P1278	Accelerator pedal position (APP) sensor 1 – voltage high	**Wiring, APP sensor**
P1282	Accelerator pedal position (APP) sensor 2 – voltage low	**Wiring, APP sensor**
P1283	Accelerator pedal position (APP) sensor 2 – voltage high	**Wiring, APP sensor**
P1336	Crankshaft position sensor (CKP) – variation not learned	**Learning procedure not performed**
P1351	Ignition control module (ICM),control – cylinder 1 & 4	**Wiring open circuit/short circuit to ground/ short circuit to positive, ICM, ECM**
P1352	Ignition control module (ICM),control – cylinder 2 & 3	**Wiring open circuit/short circuit to ground/ short circuit to positive, ICM, ECM**
P1380	Rough road signal error	**ABS control module trouble code(s) stored**
P1403	Exhaust gas recirculation (EGR) solenoid – voltage low	**Wiring short circuit to ground, EGR solenoid**
P1404	Exhaust gas recirculation (EGR) valve, stuck open	**Wiring open circuit/short circuit to positive, EGR valve position sensor, EGR solenoid**
P1405	Exhaust gas recirculation (EGR) solenoid – voltage high	**Wiring short circuit to positive, EGR solenoid**
P1406	Exhaust gas recirculation (EGR) valve position sensor – circuit malfunction	**Wiring, EGR valve position sensor, EGR valve**
P1441	Evaporative emission (EVAP) system – leak detected	**Hose connection(s), EVAP canister, EVAP canister purge valve**
P1508	Idle air control (IAC) valve, rpm low	**Wiring open circuit/short circuit to ground, intake system blocked, throttle valve, IAC valve, ECM**

Scan code	Fault location	Probable cause
P1509	Idle air control (IAC) valve, rpm high	**Wiring open circuit/short circuit to ground, intake system blocked, throttle valve, IAC valve, ECM**
P1510	Throttle control unit – performance	**Wiring, throttle motor, throttle spring, TP sensor**
P1511	Throttle control unit – back up performance	**Wiring, throttle motor, throttle spring, TP sensor**
P1516	Throttle command/actual throttle position – signal variation	**Wiring, throttle control unit**
P1519	Engine control module (ECM), throttle actuator control – output driver malfunction	**Wiring, throttle control unit, ECM**
P1523	Throttle control unit, closed throttle position – performance	**Throttle valve movement restricted, throttle control unit**
P1526	Throttle control unit – replaced or flat battery	**Initial learn procedure not carried out**
P1530	Throttle control system – malfunction	**TP sensor, ECM**
P1555	Electronic variable orifice – circuit malfunction	**Wiring, electronic variable orifice, ECM**
P1580	Cruise control system, move circuit – voltage low	**Wiring, poor connections**
P1581	Cruise control system, move circuit – voltage high	**Wiring, poor connections**
P1582	Cruise control system, direction circuit – voltage low	**Wiring, poor connections**
P1583	Cruise control system, direction circuit – voltage high	**Wiring, poor connections**
P1584	Cruise control system, cancel circuit – malfunction	**Wiring, poor connections, master switch**
P1599	Engine stall or near stall detected	**ECM trouble code(s) stored, mechanical fault, transmission mechanical fault**
P1601	CAN data bus, engine control module (ECM)/ antilock braking system (ABS)/traction control system (TCS)/instrumentation control module/ diagnostic control module – some communication lost	**Wiring, poor connections, ECM, ABS/TCS, diagnostic control module, instrumentation control module**
P1602 1	CAN data bus, antilock braking system (ABS)/ traction control system (TCS) to engine control module (ECM) – lost communication	**Wiring, poor connections, ABS control module trouble code(s) stored**
P1602 2	Engine control module (ECM), knock sensor – circuit malfunction	**ECM**
P1620	Engine coolant 'low' warning lamp switch	**Coolant level low, wiring, engine coolant 'low' warning lamp switch**
P1621	Engine control module (ECM) – memory performance	**ECM**
P1623	Engine control module (ECM), transaxle temperature – malfunction	**ECM**
P1625	Engine control module (ECM) – memory malfunction	**ECM**
P1627	Engine control module (ECM) – malfunction	**ECM**

Scan code	Fault location	Probable cause
P1628	Engine control module (ECM), engine coolant temperature – malfunction	Engine control module (ECM)
P1630	Engine control module (ECM) – immobilizer code learning mode	Immobilizer code not programmed
P1631	Engine control module (ECM) – incorrect immobilizer code	Incorrect key, immobilizer code not programmed
P1632	Engine control module (ECM) – fuel disable signal lost	Fuses, BCM, immobilizer code
P1635	Sensor supply voltage, circuit malfunction	Wiring short circuit to ground/positive, ground wiring open circuit, TP sensor, MAP sensor, EGR valve position sensor, ECM
P1637	Alternator 'L' terminal, circuit malfunction	Wiring open circuit/short circuit to ground, alternator, battery temperature sensor, ECM
P1639	Engine control module (ECM) – malfunction	ECM
P1640	Engine control module (ECM), quad driver/ output driver – circuit malfunction	Wiring, ECM controlled components, ECM [3]
P1641	Engine control module (ECM), quad driver/ output driver – circuit malfunction	Wiring, ECM controlled components, ECM [3]
P1650	Engine control module (ECM), quad driver/ output driver – circuit malfunction	Wiring, ECM controlled components, ECM [3]
P1651	Engine control module (ECM), quad driver/ output driver – circuit malfunction	Wiring, ECM controlled components, ECM [3]
P1660	Transmission control module (TCM), ground detection – malfunction	Wiring, TCM
P1669	Engine control module (ECM), programmed for ABS – incorrectly programmed or ABS signal not received	Without ABS reprogram ECM, with ABS check wiring
P1691	Engine control module (ECM), coolant temperature gauge output signal – circuit malfunction	Wiring, poor connections, ECM
P1693	Engine control module (ECM), tachometer output signal – circuit malfunction	Wiring, poor connections, ECM
P1694	Engine control module (ECM), speedometer output signal – circuit malfunction	Wiring, poor connections, ECM
P1695	Remote central locking module, entry signal – voltage low	Wiring, poor connections, central locking module, ECM
P1696	Remote central locking module, entry signal – voltage low	Wiring, poor connections, central locking module, ECM
P1780	Transmission control module (TCM), MIL ON request – transmission malfunction	Transmission trouble codes stored
P1810	Transmission fluid pressure (TFP) switch – manual valve position	Fluid level low, wiring open circuit/short circuit to ground, selector lever linkage incorrectly adjusted, ECM
P1811	Shift time too long – adaptation limit reached	Fluid level low, transmission mechanical fault
P1845	Engine control module (ECM), torque management – torque higher than expected	MAF sensor, ECT sensor, fuel injectors

Scan code	Fault location	Probable cause
P1860	Torque converter clutch (TCC) solenoid – circuit malfunction	**Wiring open circuit/short circuit to ground, TCC solenoid, ECM**
P1887	Torque converter clutch (TCC) release switch – circuit malfunction	**Wiring open circuit, TCC solenoid release switch, ECM**
U1000	CAN data bus – communication malfunction	**Wiring, BCM, ECM, instrumentation control module, SDM**
U1016	Engine control module (ECM), CAN data bus – communication malfunction	**Wiring, ECM**
U1064	Body control module (BCM), CAN data bus – communication malfunction	**Wiring, BCM**
U1096	Instrumentation control module, CAN data bus – communication malfunction	**Wiring, instrumentation control module**
U1300	CAN data bus – signal low	**Wiring**
U1301	CAN data bus – signal high	**Wiring**
U2100	CAN data bus – communication malfunction	**Wiring**
U2103	CAN data bus – not all programmed modules detected	**Wiring, ECM, BCM, TCM**
U2104	CAN data bus – any one control module not communicating	**Wiring, ECM, BCM, TCM**
U2105	Engine control module (ECM), CAN data bus – lost communication with body control module (BCM)/transmission control module (TCM)	**Wiring**
U2106	Transmission control module (TCM), CAN data bus – lost communication with body control module (BCM)/engine control module (ECM)	**Wiring**
U2107	Body control module (BCM), CAN data bus – lost communication with engine control module (ECM)/transmission control module (TCM)	**Wiring**

1 Saturn 1.9L
2 Saturn 3.0L
3 Depending on model, year and equipment level, the following components may be connected to this circuit. Use of the vehicle wiring diagram is recommended to confirm which components are fitted
- A/C compressor clutch relay
- A/C condenser blower motor relay
- Engine coolant blower motor relay
- Engine coolant 'hot' warning lamp
- Evaporative emission (EVAP) canister purge valve
- Exhaust gas recirculation (EGR) solenoid
- Malfunction indicator lamp (MIL)
- Torque converter clutch (TCC) solenoid
- Transmission shift solenoids (SS)
- Transmission shift warning lamp (MT)

SATURN

Model:	Engine identification:	Year:
S-Series 1.9L	VIN position 8 = 7, 9	1992-94
S-Series 1.9L	VIN position 8 = 7, 8	1995

System: **GM**

Self-diagnosis

General information

- Refer to the front of this manual for general test conditions, terminology, detailed descriptions of wiring faults and a general trouble shooter for electrical and mechanical faults.
- Malfunction indicator lamp (MIL) will illuminate if certain faults are recorded.
- ECM operates in backup mode if sensors fail, to enable vehicle to be driven to repair shop.
- Trouble codes can be displayed by the MIL or by using a Scan Tool connected to the data link connector (DLC).

Accessing

Models with data link connector (DLC) terminal B connected

- Access data link connector (DLC).
- Jump data link connector (DLC) terminals A and B **1**.
- Switch ignition ON. Do NOT start engine.
- Count MIL flashes. Note trouble codes. Compare with trouble code table.
- Firstly trouble code 12 will be displayed 3 times.
- Trouble codes consist of 2 groups, displayed as follows:
  - First group indicates the 'tens' of the trouble code **2** [**A**].
  - Second group indicates the 'units' of the trouble code **2** [**C**].
  - A short pause separates each group **2** [**B**].
  - A long pause separates each trouble code **2** [**D**].
  - For example: Trouble code 12 displayed **2**. Start of diagnosis.
- Each additional trouble code will now be displayed 3 times, in ascending order.

NOTE: ***If no additional trouble codes are stored, code 12 will continue to be displayed.***

- Trouble codes can also be displayed by using a Scan Tool connected to the data link connector (DLC) **1**.

1

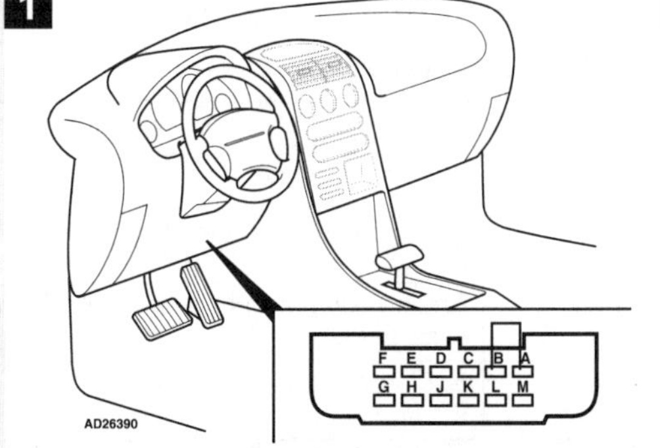

2

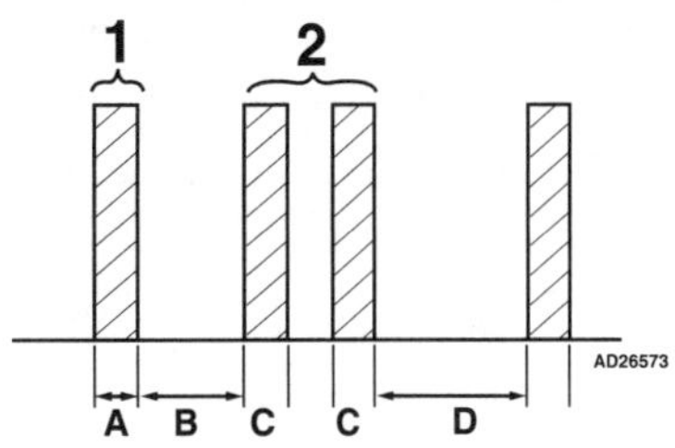

Models without data link connector (DLC) terminal B connected

- Trouble codes can only be displayed by using a Scan Tool connected to the data link connector (DLC).

Erasing

NOTE: ***Disconnecting battery ground lead may cause driveability problems until the engine management system relearns basic values.***

- Ensure ignition switched OFF.
- Disconnect battery ground cable for 30 seconds.
- Reconnect battery ground cable.
- Repeat checking procedure to ensure no trouble codes remain in ECM fault memory.

NOTE: ***Disconnecting battery ground lead may also erase memory from electronic units, such as radio.***

- Trouble codes can also be erased by using a Scan Tool connected to the data link connector (DLC) **1**.

Trouble code identification

Flash code	Fault location	Probable cause
11	Transaxle trouble code logged	**Check transaxle**
12	Start of diagnosis	**No engine speed (RPM) signal received**
13	Oxygen sensor (O2S)/heated oxygen sensor (HO2S)	**Wiring open circuit, O2S/HO2S, ECM**
14	Engine coolant temperature (ECT) sensor – voltage low	**Wiring, ECT sensor, ECM**
15	Engine coolant temperature (ECT) sensor – voltage high	**Wiring open circuit, ECT sensor, ECM**
16	System voltage – high/low	**Wiring, battery, alternator**
19	Crankshaft position (CKP) sensor, 6X signal – circuit malfunction	**Wiring, CKP sensor, ECM**
21	Throttle position (TP) sensor – voltage high	**Wiring, TP sensor, ECM**
22	Throttle position (TP) sensor – voltage low	**Wiring open circuit/short circuit to ground, TP sensor, ECM**
23	Intake air temperature (IAT) sensor – voltage high	**Wiring, IAT sensor, ECM**
24	Vehicle speed sensor (VSS) – no signal	**Wiring, VSS, ECM**
25	Intake air temperature (IAT) sensor – voltage low	**Wiring short circuit to ground, IAT sensor, ECM**
26	Engine control module (ECM), quad driver/ output driver – circuit malfunction	**Wiring, ECM controlled components, ECM 1**
27	Engine control module (ECM), quad driver/ output driver – circuit malfunction	**Wiring, ECM controlled components, ECM 1**
32	Exhaust gas recirculation (EGR) solenoid	**Wiring, poor connections, EGR valve, ECM**
33	Manifold absolute pressure (MAP) sensor – voltage high/low	**Wiring, hose blocked/leaking, MAP sensor, ECM**

Flash code	Fault location	Probable cause
34	Manifold absolute pressure (MAP) sensor – voltage low	Wiring open circuit/short circuit to ground, MAP sensor, ECM
35	Idle air control (IAC) valve	Wiring, IAC valve, intake system leaking, fuel system, PCV valve, A/C system, mechanical fault, ECM
41	Ignition control, bypass signal – circuit malfunction	Wiring, poor connections, ICM, ECM
42	Ignition control, timing signal – circuit malfunction	Wiring, poor connections, ICM, ECM
43	Knock sensor (KS) – circuit malfunction	Wiring, KS, ECM
44	Oxygen sensor (O2S)/heated oxygen sensor (HO2S) – mixture lean	Wiring open circuit/short circuit to ground, injector(s) blocked, fuel contamination, fuel pressure low, exhaust system leaking, O2S/HO2S, ECM
48	Ignition control system, reference signal – circuit malfunction	Wiring, ICM, ECM
49	Idle air control (IAC) valve – malfunction	Wiring, IAC valve, intake system leaking, fuel system, PCV valve, A/C system, ECM
51	Engine control module (ECM), PROM – malfunction	Poor connections, incorrect programming, ECM
52	System voltage – high/low	Wiring, alternator, battery
53	Knock sensor (KS), knock present – unable to rectify	Wiring, mechanical fault, ECM
54	Engine control module (ECM), sensor supply voltage – circuit malfunction	Wiring, poor connections, ECM
55	Engine control module (ECM) – malfunction	Engine control module (ECM)
58	System voltage – sharp voltage changes	Wiring, alternator
61	Ignition control system, 6X signal – circuit malfunction	Wiring, ICM, ECM
63	Cruise control system – performance low	Wiring, poor connections, accelerator cable sticking
67	Steering wheel position sensor – voltage high/low	Wiring, poor connections, steering wheel position sensor
71	Engine coolant temperature (ECT) sensor – voltage low	Wiring, ECT sensor, ECM
72	Engine coolant temperature (ECT) sensor – voltage high	Wiring open circuit, ECT sensor, ECM
73	Engine coolant temperature (ECT) sensor – sharp voltage changes	Wiring, ECT sensor, ECM
74	Engine coolant temperature to transmission fluid temperature – signal variation	Wiring, ECT sensor, TFT sensor, ECM
75	Intake air temperature (IAT) sensor – sharp voltage changes	Wiring, IAT sensor, ECM

Flash code	Fault location	Probable cause
76	Throttle position to manifold absolute pressure – signal variation	**Wiring, TP sensor, MAP sensor, ECM**
83	Engine coolant 'low' warning lamp switch	**Coolant level low, wiring, engine coolant 'low' warning lamp switch**
84	Engine control module (ECM) to transmission control module (TCM) – communication malfunction	**Poor connections, incorrect programming, ECM**

1 Depending on model, year and equipment level, the following components may be connected to this circuit. Use of the vehicle wiring diagram is recommended to confirm which components are fitted

- ☐ A/C compressor clutch relay
- ☐ A/C condenser blower motor relay
- ☐ Engine coolant blower motor relay
- ☐ Engine coolant 'hot' warning lamp
- ☐ Evaporative emission (EVAP) canister purge valve
- ☐ Exhaust gas recirculation (EGR) solenoid
- ☐ Malfunction indicator lamp (MIL)
- ☐ Torque converter clutch (TCC) solenoid
- ☐ Transmission shift solenoids (SS)
- ☐ Transmission shift warning lamp (MT)